# ENCYCLOPEDIA OF EMULSION TECHNOLOGY

VOLUME 1

Basic Theory

# ENCYCLOPEDIA OF EMULSION TECHNOLOGY

## VOLUME 1

## Basic Theory

*Edited by* PAUL BECHER
*Paul Becher Associates Ltd.*
*Wilmington, Delaware*



MARCEL DEKKER, INC. NEW YORK and BASEL

Library of Congress Cataloging in Publication Data
Main entry under title:

Encyclopedia of emulsion technology.

Includes index.
Contents: v. 1. Basic theory.
1. Emulsions. I. Becher, Paul.
TP156.E6E62 1983 660.2'94514 82-18257
ISBN 0-8247-1876-3



MARCEL DEKKER, INC.
270 Madison Avenue, New York, New York 10016

Current printing (last digit):
10 9 8 7 6 5 4 3 2 1

PRINTED IN THE UNITED STATES OF AMERICA

FOR JANE

Proverbs 31: 10-31

## Preface

Twenty-four years have passed since the first edition of *Emulsions: Theory and Practice* appeared; 16 years since the second edition. The second edition, which is still in print (and selling nicely, thank you), continues, I think, to fulfill its avowed purpose as a unified discussion of—as its title indicates—the theory and practice of emulsions.

Nonetheless, this work has suffered somewhat from the passage of time. One seeks in the index in vain for any mention of steric stabilization, although to be sure it is discussed in a qualitative way, but not under that designation. The discussion of microemulsions barely fills two pages. Many of the examples of emulsion formulations are out of date, and some have suffered from regulatory changes.

But as I considered, some years ago, the question of a revision versus an extensive rewriting—in effect, a totally new book—I concluded that while the proper course was a new book, such a book would contain not less than one million words. Immediately, I ran into the traditional conflict between the flesh and the spirit.

However, at the crucial moment, I had the good fortune to have a long conversation with Dr. Maurits Dekker, who cut the Gordian knot by suggesting a collective treatise. While this idea had been floating around somewhere at the subliminal level, Dr. Dekker's obvious willingness to consider the publication of such a large undertaking triggered my thinking, and the plan of this work emerged.

As a result, we have a new work in which the various topics are discussed by the leading authorities in the field, with a depth of expertise that no single author could have hoped to achieve. On the other hand, the *Encyclopedia of Emulsion Technology* can hardly serve as an introduction for the tyro emulsion chemist. Fortunately, there is still *Emulsions: Theory and Practice* to fall back on.

I have taken the role of editor seriously. Every page of the manuscript has been closely scanned. I have tried to keep the style, terminology, and formalism reasonably consistent, but I have, I trust, avoided Emerson's hobgoblin. I have also strived to overcome the greatest difficulty in collective works, unnecessary duplication. Note, however, the word *unnecessary*; some duplication, if a different point of view is being expressed, is no doubt useful. Accordingly, I have not hesitated to wield the editorial red pencil when I felt it appropriate. I hope that my contributors who have been thus assaulted will forgive me when they see the final result. *Finis coronat opus.*

Finally, I should like to express my gratitude to my contributors, who were incredibly cooperative and prompt (any delays being entirely attributable to the dilatory working habits of the editor). My thanks to Dr. Maurits Dekker, whose far-from-dilatory working habits are an eternal inspiration, to Marcel Dekker, and the entire staff at Marcel Dekker, Inc. And, finally, to my wife, to whom I express my gratitude more formally on another page.

Paul Becher

## Contents of Volume 1

## Contributors

MARC CLAUSSE*
Université de Pau et des Pays de l'Adour, Pau, France

RAYMOND S. FARINATO†
University of Massachusetts, Amherst, Massachusetts

STIG E. FRIBERG
University of Missouri-Rolla, Rolla, Missouri

HIRONOBU KUNIEDA
Yokohama National University, Yokohama, Japan

CLYDE ORR‡
Georgia Institute of Technology, Atlanta, Georgia

ROBERT L. ROWELL
University of Massachusetts, Amherst, Massachusetts

PHILIP SHERMAN
Queen Elizabeth College, University of London, London, England

KŌZŌ SHINODA
Yokohama National University, Yokohama, Japan

THARWAT F. TADROS
Imperial Chemical Industries, PLC Ltd., Bracknell, Berkshire, England

RAYMOND L. VENABLE
University of Missouri-Rolla, Rolla, Missouri

BRIAN VINCENT
University of Bristol, Bristol, England

PIETER WALSTRA
Agricultural University, Wageningen, The Netherlands

*Current affiliation: Université de Technologie de Compiègne, Compiègne, France
†Current affiliation: American Cyanamid Company, Stamford, Connecticut
‡Current affiliation: Micromeritics Instrument Corporation, Norcross, Georgia

# Contents of Volume 2: Applications of Emulsions

# Introduction

In an earlier work [1], the author opened with the statement that "the thoery of emulsions has grown in a rather haphazard way." In the intervening years (and, I hope, partly as a result of the systematic account given in that work), the study of emulsions has become deeper and more fundamental, without losing touch with the manifold applications, practical and theoretical, which make emulsions such an interesting area of study.

This has been accompanied by an incredible growth in the literature of the field. For example, Clayton's treatise of 1923 [2] contains a bibliography of 204 items. The second edition of *Emulsions: Theory and Practice* lists more than 2000 citations. In contrast, the first volume of the present work, which barely covers the basic theory, contains more than 1500 citations to the literature.

One of the fascinations of emulsions as a research activity (and not the less for their practical applications) is the interconnectedness of the factors governing their stability and other properties. Walstra [3] has communicated to the writer a useful table in which he attempts to show these interactions. I have attempted to modify this table into a synoptic key to this encyclopedia, but have regretfully concluded that nothing less than a multidimensional plot would surface.

However, it is this very complexity which makes emulsions such a fruitful field for the study of dispersions. Our ability to modify, for example, the liquid state and nature of the dispersed phase (including, as a special but nonetheless important case, the behavior of emulsions at temperatures between the melting point of the disperse phase and that of the continuous phase), as well as our ability to modify the nature of surface charges and concentration by the appropriate choice among the literally thousands of surface-active agents and stabilizers available to us, are but two ways in which advantage of this complexity may be taken.

In the first volume attention is devoted to the background of surface-chemical theory and physical behavior. Chapter 1, by Tadros and Vincent, sketches the requisite background of theory for an understanding of events occurring at the liquid/liquid interface. Chapter 2, by Walstra, considers next the mechanical requirements for the creation of such interfaces, the large interfacial area of emulsions being of course one of their most significant properties.

Chapter 3 (Tadros and Vincent) next considers the significance of this background in defining and explaining the stability, or lack thereof, of emulsion systems, while Chap. 4, contributed by Friberg and Venable, discusses the special case of microemulsions (a subject which, for all practical purposes, did not exist 15 years ago, when our earlier work appeared).

Chapter 5, by Shinoda and Kunieda, is devoted primarily to the phase behavior of emulsion systems, as affected by the nature of the disperse phase and surface-active agents employed. In Chap. 6, by Orr, we turn to the description of emulsions systems in terms of the particle size distribution and its mathematical form. Chapter 7, by Sherman, completes this section of Volume 1 by a discussion of the rheological properties of emulsions, which depend on factors discussed in many of the preceeding chapters.

The final portion of Volume 1 consists of two long chapters which may be broadly described as discussions of the interaction of electromagnetic radiation or currents with emulsions. The first of these (Chap. 8, by Farinato and Rowell) is concerned with the optical properties of emulsions, while Chap. 9, by Clausse, is an extensive treatment of the dielectric properties of emulsions.

Further volumes will devote more space to the applications of emulsions, in industry as in life, and as a research tool.

## References

1. P. Becher, *Emulsions: Theory and Practice*, 1st ed. (ACS Monograph No. 135), Reinhold, New York, 1957; 2d ed. (ACS Monograph No. 162), Reinhold, New York, 1965 (2d ed. reprinted, Robert E. Krieger, Melbourne, Florida, 1977).
2. W. Clayton, *The Theory of Emulsions and Emulsification*, Blakiston's, Philadelphia, Pa., 1923.
3. P. Walstra, private communication, 1979.

# ENCYCLOPEDIA OF EMULSION TECHNOLOGY

VOLUME 1

Basic Theory

# 1
# Liquid/Liquid Interfaces

THARWAT F. TADROS / Imperial Chemical Industries, PLC Ltd., Bracknell, Berkshire, England

and

BRIAN VINCENT / University of Bristol, Bristol, England



## I. The Interface Between Pure Fluids: Interfacial Tension

### A. Introduction

The most fundamental thermodynamic property of any interface is the interfacial tension, yet its origin has been the cause of much confusion and controversy.

Many authors and teachers present a diagram showing the imbalance of forces at a liquid-vapor interface, together with a statement implying that the net inward attraction on molecules in the surface layers of the liquid causes the surface to contract. There are two serious objections to this simplistic view.

First, the surface cannot be in mechanical equilbirium if there is a net force normal to the interface. Second, an inward attraction on surface molecules does not explain the tension parallel to the interface which is invoked to explain such phenomena as the rise of liquids in capillary tubes and the attachment of soap films to wire frames. Indeed, Adam [1], one of the leading figures in the early days of surface "science" went as far as to say that "surface tension does not exist as a physical reality; it is only the mathematical equivalent of free surface energy."

One concludes that it is not possible simply to interpret interfacial tension purely in terms of some net force on molecules perpendicular to the surface. Rather one has to build a detailed statistical mechanical model of the interfacial region, based on some knowledge of the intermolecular forces operating, and hence establish the necessary partition function. The interfacial tension may then be derived from the interfacial free energy. Alternatively, one can use a computer simulation approach to derive, for example, the density and pressure profiles across the interface [2]. Once the pressure profile has been established then the interfacial tension is readily obtained, as we shall see. The question of how macroscopic (thermodynamic) properties vary at an interface is considered in the following section.

## B. Macroscopic Properties at an Interface

In the last 10 years or so our understanding of the liquid state has progressed significantly, at least with regard to simple, nonpolar molecules having spherical symmetry.† Appreciation of the structure and molecular dynamics at liquid interfaces is following in the wake of this, but it is still at a primitive level. Unfortunately, the theories that are emerging are difficult to test because there are very few experimental techniques available for probing interfacial structure and dynamics. In this section we briefly review some of the progress that has been made in understanding the way in which thermodynamic properties, such as the density, pressure, chemical potential, and temperature behave in the region of an interface.

---

†See, for example, reviews by Hansen and McDonald [3] and by Barker and Henderson [4].

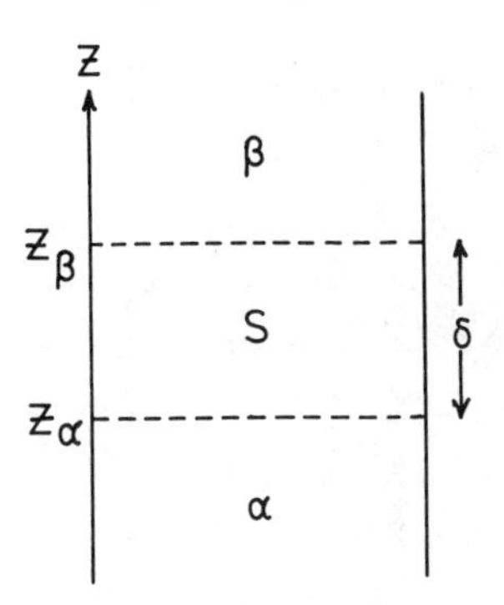


Figure 1. Schematic representation of an interface between two bulk phases.

The basic model of an interface is depicted in Fig. 1. It shows two bulk phases $\alpha$ and $\beta$ having uniform thermodynamic properties, separated by an interfacial region s of thickness $\delta$ and over which the thermodynamic variables may possibly vary. For a liquid in equilibrium with its own vapor, $\delta$ is known to increase on traversing the liquid temperature range from the triple point to the critical point. At the triple-point temperature, when the density of the vapor is generally less than $\sim 10^{-3}$ of its value for the corresponding liquid, $\delta$ is of the order of a few molecular diameters (i.e., effectively the range of the intermolecular forces). On approaching the critical temperature, the density difference between the vapor and liquid phases disappears, and $\delta$ may be of the order of several tens or even several hundreds of molecular diameters.

Both statistical mechanical and computer simulation models have been used to determine $\rho(z)$, the density profile across s, for liquid-vapor interfaces. With regard to the former, the penetrable sphere model [5] has proved particularly suitable since it ensures that the molecules can move freely through space without restriction to cells on lattice sites. Also, although it assumes a simple characteristic interaction energy between neighboring molecules, it does, like the Ising lattice model, predict phase separation and also leads to tractable equations for $\rho(z)$. In the simplest case this function turns out to be a hyperbolic tangent [6]. For a general review of recent statistical mechanical theories of interfaces the reader is referred to an article by Toxvaerd [7].

In computer simulation studies of interfaces [2] a small portion of the system is modeled by establishing within the computer store a set of coordinates defining the positions of some $10^2$ or $10^3$ molecules (depending on the capacity of the computer available). The intermolecular pair potential energy function $\phi(r)$ must also be defined. The one most frequently used in this context is the Lennard-Jones function

$$\phi(r) = \varepsilon \left[ \left(\frac{\sigma}{r}\right)^{12} - 2\left(\frac{\sigma}{r}\right)^{6} \right] \tag{1}$$

where $\varepsilon$ is the interaction energy at the equilibrium separation distance $\sigma$ between two molecules. Two computer simulation techniques have been used: the so-called Monte Carlo and molecular dynamic methods.

In the Monte Carlo method the temperature of the system is specified, and the molecules are then moved in such a way that each configuration appears with a probability proportional to its Boltzmann factor exp $(-U_{config}/kT)$ where $U_{config}$ is the sum of the intermolecular pairwise potential energies for that configuration. By averaging over many such configurations the equilibrium properties of the system may be established.

In a molecular dynamic simulation each molecule is allotted a randomly chosen velocity, and Newton's equations of motion are solved to establish how the system evolves in time. The initial total internal energy is fixed, and when the system has come to equilibrium, the average temperature reaches a steady value (although subject to statistical fluctuations). The equilibrium properties of the system are thus found from an average over time. Moreover, the *dyanmic* properties of the system may also be established since the successive states of the system are generated in their correct time sequence.

Over the last few years several workers [2] have applied these two types of computer simulation techniques to study liquid-vapor interfaces. There are several problems to overcome, most of these being associated with the very limited size of the computed experimental system compared to real systems. For example, in the real world, the earth's gravitational field, although weak, is sufficient to maintain a reasonably flat interface since it is acting on a system whose dimensions are large compared with the range of the intermolecular forces. This is not so for a computed system of even 1000 molecules, since over a distance of 10 (i.e., $1000^{1/3}$) molecular diameters or so the change in the earth's gravitational potential is only $\sim 10^{-6}$ of the intermolecular potential energy. Several sophisticated techniques have been devised in attempts to overcome this problem, but it is still not certain just how large a sample one needs before the thermodynamic parameters, such as density, "settle" to their equilibrium profiles across the interface. Typical profiles, evaluated by computer simulation [2], are shown in Fig. 2 for three different reduced temperatures $\tau$:

$$\rho = \frac{N\sigma^3}{V} \tag{2}$$

$$\tau = \frac{kT}{\varepsilon} \tag{3}$$

where $\varepsilon$ and $\sigma$ are the Lennard-Jones interaction energy and separation, respectively, referred to in Eq. (1), N is the number of molecules in the sample, and V the total volume of the system. The values of $\tau$ selected lie between the triple

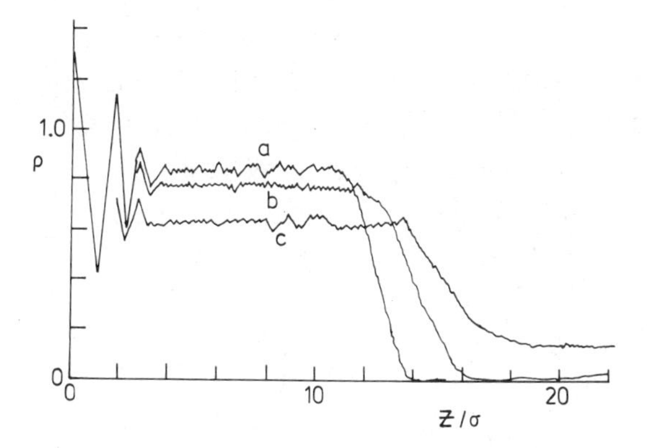


**Figure 2.** Density profile $\rho(z)$ for a liquid/vapor interface, obtained by computer simulation, for three different reduced temperatures $\tau(= kT/\varepsilon)$: (a) 0.701, (b) 0.918, (c) 1.127. $\varepsilon$ is the minimum value of the energy and $\sigma$ the corresponding distance in the Lennard-Jones equation. (From Ref. 2.)

point at $\tau = 0.61$ and the critical point at $\tau = 1.34$. The oscillations in $\rho$ at $z < 4\sigma$ are an artifact. Once the density profile is established, one may then proceed to speculate about, and in some cases calculate, the equivalent profiles for other thermodynamic properties.

A local temperature may be defined in terms of the local density of kinetic energy (at least for a classical system of monatomic molecules). Computer simulation studies have confirmed [2] that the distribution of molecular velocities, and hence also of kinetic energies, is the same for molecules in the interfacial region as in either of the neighboring bulk phases. This implies that the temperature is uniform in the interface. Moreover, it implies that the *thermal* part of the entropy is also uniform and equal to that in the neighboring phases.

Whether the chemical potential $\mu(z)$ is uniform or not has been the subject of debate [8] ever since Gibbs [9] first assumed that $\mu^s = \mu^\alpha = \mu^\beta$ (without clearly defining what he meant by $\mu^s$, however). This debate was settled in 1963 using statistical mechanical arguments [10] in Gibbs' favor, and recent computer simulation work also implies, without rigorously confirming, that indeed the chemical potential across a liquid-vapor interface appears to be uniform.

Pressure, unlike temperature and chemical potential, is not a scalar quantity, and in the interfacial region there is no a priori reason why we should assume that the tangential component of the pressure tansor $P_t$ is necessarily the same as the normal component $P_n$ (in a *bulk* fluid they are, of course, identical). However, in order to maintain mechanical equilibrium, $P_n$ must be equal to the uniform pressure P in each neighboring bulk phase (assuming a flat interface and neglecting hydrostatic, i.e., gravitational, effects in the immediate neighborhood of the interface). There is no such restriction on $P_t$, however; it is, therefore, necessary to establish $P_t(z)$.

The pressure at a "point" is a somewhat nebulous quantity [11], and it is better to think of that associated with a small volume element. Such a pressure will have two contributions: a kinetic part $P^k$ and a potential part $P^p$. $P^k$ is associated with the motion and hence the kinetic energy of the molecules, while $P^p$ is associated with the position and hence the potential energy of the molecules in that volume element. For an *ideal* fluid, only the kinetic contribution is present, and this is given by

$$P^k = \rho kT \tag{4}$$

where $\rho$ is the local number density. One might then suppose that $P(z)$ would have the same form as $\rho(z)$. However, for an ideal fluid, phase separation and, therefore, interface formation does not occur! Thus, for a *real* fluid one must consider the role of $P^p$ and hence of the intermolecular forces. Clearly, $P^p$ is going to be anisotropic in the interfacial region, since the intermolecular forces themselves are anisotropic.

It is of interest, at this point, to examine the signs of $P^k$ and $P^p$. Clearly, the former must be positive, and in general, the latter will be negative (net attraction between molecules: $r > \sigma$ on average), although at very high densities and pressures it is possible for $P^p$ also to be positive (net repulsion between molecules: $r < \sigma$, on average). In a bulk liquid phase both $P^p$ and $P^k$ are large with respect to their corresponding values in the coexisting vapor phase, but the sum $P^p + P^k$ is the same in both, and this will also equal $P_n^k(z) + P_n^p(z)$ at any plane z in the interface.

In order to establish $P_t(z) [= P_t^k(z) + P_t^p(z)]$, it is necessary to know not only the singlet distribution function [i.e., the density distribution function $\rho(z)$] but also the pair or radial distribution function $g(r,z)$. For bulk fluids $g(r)$ [$= \rho(r)/\bar{\rho}$, where $\bar{\rho}$ is the mean, macroscopic number density] may be obtained experimentally from x-ray scattering or, more recently, from neutron scattering data. It is also necessary to know $\phi(r)$.

It turns out that the $P_t(z)$ profile for a liquid vapor interface has the form shown in Fig. 3. $P^p$ changes more rapidly than $P_t^k$ [$= kT\rho(z)$], since the greatest variation in intermolecular forces occurs in the layer(s) of molecules immediately adjacent to the vapor phase. This gives rise to an important feature of the $P_t(z)$ profile, namely, that it may assume negative values over certain values of z. In that case there exists a tangential *tensile* stress which, even for rare gas liquids where the interactions are relatively weak, may amount to values exceeding $10^3$ atm [12]. The surface layers of a liquid, in contrast with the bulk, must possess rigidity in order to resist the shear stress that arises because $P_t$ is different from $P_n$. This is the basis of the statements appearing in certain older texts that liquids behave as if their surfaces are covered by an "elastic skin."

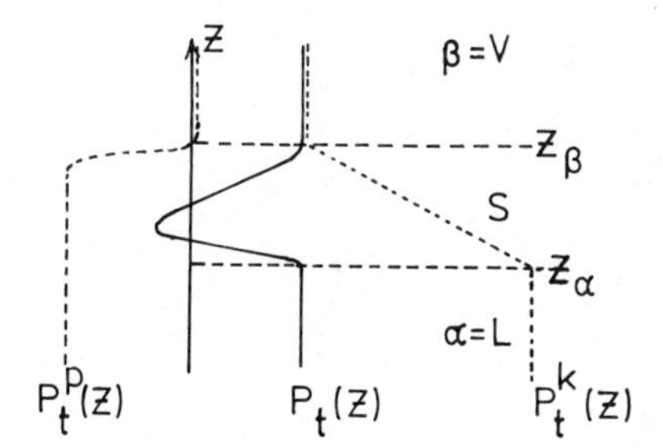


**Figure 3.** Contributions to the total tangential pressure profile $P_t(z)$ for a liquid/vapor interface; $P_t^k(z)$ is the kinetic part, and $P_E^p(z)$ is the potential part.

## C. Theories of Interfacial Tension

There have been three main approaches for evaluating interfacial tension:

1. Through distribution functions, in particular $P_t(z)$
2. From partition functions, via some model of the liquid surface
3. Directly from summation (or integration) of the interaction forces in the whole interfacial region, i.e., in terms of the cohesive free energy (for a free surface) or the adhesive free energy (for a liquid/liquid interface)

### 1. *Distribution Function Approach*

The *interfacial tension* is defined in terms of $P_t(z)$ as follows:

$$\gamma = \int_{z_\alpha}^{z_\beta} [P - P_t(z)]\, dz \tag{5}$$

Consideration of Eq. (5) and Fig. 3 leads one to conclude that $\gamma$ is always positive. This is the case even if $P_t(z)$ does not actually go negative at any point, and this is usually the case for interfaces between two condensed phases, i.e., between two nominally immiscible liquids.

Kirkwood and Buff [13], in 1949, were the first to consider Eq. (5) as the basis of a statistical mechanical route to $\gamma$. They used the Lennard-Jones form of $\phi(r)$ [see Eq. (1)]. They assumed that $\rho^V = 0$ and, since its form was not then established, that $\rho(z)$ was *uniform* over the whole of s and equal to $\rho^L$. On this basis they were effectively able to evaluate $g(r,z)$, and hence $P_t(z)$, assuming the form of $g(r)$ for the bulk liquid obtained experimentally from x-ray scattering. Thus, for liquid Ar at 90 K they obtained a value for $\gamma$ of 14.9 mN m$^{-1}$, as compared with the experiment value of 11.9 mN m$^{-1}$. Berry et al. [14] have more recently extended the Kirkwood-Buff theory by assuming an *experimental* form for $\rho(z)$ across the interfacial zone s. The thickness $\delta$ of

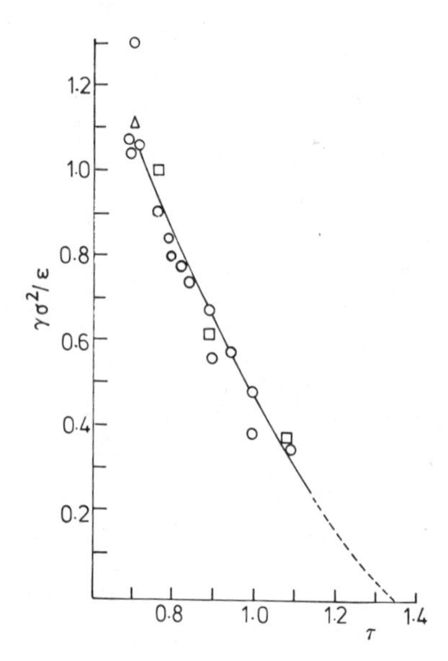


**Figure 4.** Computer simulation results for the reduced interfacial tension $\gamma\sigma^2/\varepsilon$ as a function of reduced temperature $\tau$: o, Ref. 17; △, Ref. 16; □, Ref. 15— results from perturbation theory. (From Ref. 2.)

this zone was chosen as an adjustable variable to give a best fit with experimental values of $\gamma$. The number of molecular layers in the transition zone was found to be of the order of 1 to 3, which, as we have seen, is reasonable for a liquid near its triple-point temperature.

In Fig. 4 are shown some recent results for $\gamma$ calculated using computer simulation predictions for $\rho(z)$ [15–17]. Also shown (continuous line) are the results of a perturbation theory calculation of $\gamma$ developed by Toxvaerd [7]. This theory is not an a priori one since, like the Berry et al. theory [14] referred to above, it assumes a particular functional form for $\rho(z)$. The results are all presented in the form of reduced surface tension versus reduced temperature. One conclusion is that since the agreement between the various approaches is so good, $\gamma$ is not very sensitive to the precise form of $p(z)$. All the theories predict that $\gamma \to 0$ as $\tau \to \tau_{crit}$ ($\tau_{crit} = 1.34$), although computed simulation studies cannot, as yet, be used to establish $\gamma$ values in the region of $\tau_{crit}$ since the size of the sample required is much too large to handle on the present generation of computers.

### 2. *Partition Function Approach*

This approach is based on extensions of the cell model of a liquid, in which the forces exerted by nearest neighbors are thought of as effectively confining any given molecule within a potential well or cell.† This gives rise to a quasi-lattice structure for the liquid. A total partition function Z for the system, containing thermal and configurational contributions, is then set up, and the total Helmholtz free energy A of the liquid is then calculated from the standard expression

$$A = kT \ln Z \tag{6}$$

In order to take account of surface effects, it is necessary to define a plane boundary to the lattice and to recognize that molecules in this surface layer (a monolayer in the simplest case) have a different partition function $Z^s$. The major difficulty is in evaluating the configurational contribution to $Z^s$. However, by making various simplifying assumptions, it is possible to evaluate $Z^s$ and hence $A^s$. As will be shown in Sec. I.D, Eq. (36), $\gamma$ is then simply obtained from

$$\gamma = \left(\frac{\partial A^s}{\partial \mathcal{A}}\right)_{T,n^s} \tag{7}$$

i.e., the rate of change of $A^s$ with area $\mathcal{A}$ at constant temperature and number of molecules in the surface.

Despite the gross features of this approach, reasonable agreement between calculated and experimental values of $\gamma$ has, in general, been obtained, at least at temperatures close to the triple point. Close to the critical point, of course, the concept of a cell model and the associated restricted diffusion for a liquid become questionable. As an example, the simple approach outlined above yields a value of $\gamma = 9$ mN m$^{-1}$ for liquid Ar at 85 K compared to an experimental value of 13.2 mN m$^{-1}$. (Note that for Ar, $T_{trip} = 83.9$ K and $T_{crit} = 150.7$ K.) Prigogine and Scrage [18] obtained much closer agreement ($\gamma = 13$ mN m$^{-1}$) by introducing the concept of "holes" (or vacant cells) into the quasi-lattice structure of the liquid surface. They used a square-well form for $\phi(r)$ rather than the Lennard-Jones form.

Ono and Kondo [19] extended the Prigogine theory to allow for the possibility of holes in more layers than the one immediately adjacent to the vapor phase. This theory indicated that two such layers were involved (i.e., had effective densities less than the bulk liquid) at $T = 0.982\ T_{crit}$. Good agreement between theory and experiment was obtained for carbon tetrachloride over the temperature range 330 to 550 K. A further refinement to the cell theory has been considered by Eyring and Jhon [20]. When a solid melts there is a large

†Typically a molecule will oscillate ~$10^4$ to $10^5$ times about its mean position in such a cell before it "escapes" by diffusion to a new cell.

increase in molar volume (~12% for Ar) but little or no change in the nearest neighbor distance [i.e., the first peak in g(r); 0.38 nm for Ar]. Eyring suggested the retention of small regions of solidlike structure interdispersed with "gaslike" regions. These ideas are incorporated in Eyring's so-called significant structure theory of liquids, and calculations of interfacial tension have been made using this theory [21].

### 3. *From Free Energies of Cohesion and Adhesion*

The extension of the theories outlined in Secs. I.A.1 and I.A.2 to polar molecules (i.e., having dipole and/or quadrupole moments) or to chainlike molecules is complex. Their extension to calculate the interfacial tension between two *condensed* phases is also difficult. More success has been achieved in these areas by consideration of the free energies (works) of cohesion (for LV interfaces) and adhesion (for LL interfaces). These, in turn, are evaluated from a consideration of the net interactions between two macroscopic bulk phases.

The general situation is illustrated schematically in Fig. 5, where two macroscopic bulk phases $\alpha$ and $\beta$ (containing molecular species 1 and 2, respectivelv) are separated by a gap b. The free energy of adhesion, $G_A^{\alpha\beta}$ is defined by,

$$G_A^{\alpha\beta} = \gamma^{\alpha} + \gamma^{\beta} - \gamma^{\alpha\beta} \tag{8}$$

and therefore

$$\gamma^{\alpha\beta} = \gamma^{\alpha} + \gamma^{\beta} - G_A^{\alpha\beta} \tag{9}$$

where $\gamma^{\alpha\beta}$ is the interfacial tension between phase $\alpha$ and $\beta$ in contact; $\gamma^{\alpha}$ and $\gamma^{\beta}$ are the "free" surface tensions of $\alpha$ and $\beta$ in isolation. If $\alpha$ and $\beta$ are identical then the the corresponding free energies of cohesion, $G_c^{\alpha}$ and $G_c^{\beta}$ are defined by:

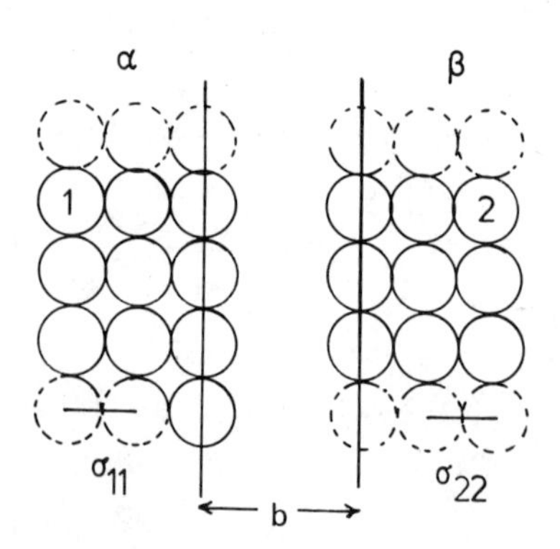


Figure 5. Interaction of two parallel semi-infinite bulk phases across a gap with thickness b.

$$G_c^{\alpha} = 2\gamma^{\alpha} \tag{10}$$

and

$$G_c^{\beta} = 2\gamma^{\beta} \tag{11}$$

Thus, if $G_c^{\alpha}$, $G_c^{\beta}$ and $G_A^{\alpha\beta}$ can be calculated, then $\gamma^{\alpha}$ and $\gamma^{\beta}$ may be estimated from Eqs. (10) and (11), and hence $\gamma^{\alpha\beta}$ from Eq. (9).

Estimation of $G_A^{\alpha\beta}$ (and similarly $G_c^{\alpha}$ or $G_c^{\beta}$) is based on evaluating G(b) in the limit of $b = \sigma_{12}$ (Fig. 5). The calculation of G(b) is considered in much greater detail in Chap. 4 and only the essential details are repeated here. In general, the forces operating in liquids are the London dispersion forces and, in the case of polar liquids, the dipole-dipole (Keesom) forces, dipole-induced dipole (Debye) forces, etc. In the case of liquid metals and associating liquids (e.g., water), other types of forces also have to be considered (i.e., for water, H bonds). For the case of nonpolar liquids, a simple pairwise integration technique, following deBoer [22] and Hamaker [23], for evaluating $G_c$ and, hence, $\gamma$ for LV interfaces gives

$$\gamma = \frac{A}{24\pi\sigma^2} \tag{12}$$

Note that this follows from Eq. (27) of Chap. 3, and Eq. (10) or (11) above, Here $\sigma$ is the intermolecular separation in the liquid phase, and A is the Hamaker constant for the liquid, defined by

$$A = \pi^2\rho^2\lambda \tag{13}$$

where $\rho$ is the number density in the liquid phase and $\lambda$ is the London intermolecular interaction constant

$$\lambda = \frac{3}{4}h\nu\alpha^2 \tag{14}$$

where h is Planck's constant, $\alpha$ the static polarizability of the molecules, and $\nu$ some characteristic absorption frequency, usually in the ultraviolet. Taking for Ar at 90 K values of $\lambda = 5.2 \times 10^{-78}$ J m$^6$ [24], $\rho$ (mass density) = $1.374 \times 10^3$ kg m$^{-3}$ [25], and $\sigma$ = 0.369 nm (from neutron scattering [26]) gives a value for $\gamma$ of 2.1 mN m$^{-1}$ compared to the experimental value of 11.9 mN m$^{-1}$, which seems very poor agreement. Padday and Uffindell [27], on the other hand, have calculated the surface tensions of series of *n*-alkanes using Eqs. (12) and (13), and the Moelwyn-Hughes [28] expression for $\lambda$, rather than the one due to London [Eq. (14)]. They obtained quite good agreement with experimental values by assuming the fundamental volume elements of the alkanes to be $CH_2$ and $CH_3$ groups.

Consider now the interfacial tension $\gamma^{\alpha\beta}$ bewteen phase $\alpha$, a nonpolar liquid, and phase $\beta$, which may be polar or nonpolar (Fig. 5). The only significant forces that operate across the "gap" (b) will be the dispersion forces. $G_A^{\alpha\beta}$ is then given by

$$G_A^{\alpha\beta} = \frac{A_{12}}{12\pi\sigma_{12}^2} \tag{15}$$

where $\sigma_{12}$ is the intermolecular separation between a 1 molecule and a 2 molecule in "contact," and $A_{12}$ is the cross-interaction Hamaker constant between species 1 and 2, which may be expressed in terms of the Hamaker constants for the interactions between like molecules, as follows [29]:

$$A_{12} = (A_{11}A_{22})^{1/2} \tag{16}$$

Assuming that we can make the approximation $\sigma_{11} = \sigma_{22} = \sigma_{12}$, then Eq. (15) may be expressed in the form

$$G_A^{\alpha\beta} = 2\left(\frac{A_{11}}{24\pi\sigma_{11}^2}\right)^{1/2}\left(\frac{A_{22}}{24\pi\sigma_{22}^2}\right)^{1/2} \tag{17}$$

Combining Eqs. (12) and (17) leads to

$$G_A^{\alpha\beta} = 2(\gamma^{\alpha}\gamma_d^{\beta})^{1/2} \tag{18}$$

(Note that the subscript d in $\gamma_d^{\beta}$ refers to the fact that this represents only the *dispersion* force contribution to $\gamma^{\beta}$. Remember that phase $\beta$ may be a polar liquid, whereas phase $\alpha$ is nonpolar, i.e., $\gamma^{\alpha} \sim \gamma_d^{\alpha}$).

Combining Eq. (18) with Eq. (8) leads to the following expression for $\gamma^{\alpha\beta}$:

$$\gamma^{\alpha\beta} = \gamma^{\alpha} + \gamma^{\beta} - 2(\gamma^{\alpha}\gamma_d^{\beta})^{1/2} \tag{19}$$

Equation (19), first introduced by Fowkes [30], therefore, gives a method for estimating interfacial tensions between immiscible liquids, provided one of the pair is nonpolar. Girafalco and Good [31] had previously suggested a more general form of Eq. (19), i.e.,

$$\gamma^{\alpha\beta} = \gamma^{\alpha} + \gamma^{\beta} - 2\Phi(\gamma^{\alpha}\gamma^{\beta})^{1/2} \tag{20}$$

where $\Phi$ is a parameter related to the molar volumes of the two liquids.

Equations (19) and (20) have been widely used to estimate interfacial tensions. For example, Padday and Uffindel [27] calculated $\gamma^{\alpha\beta}$ values for various *n*-alkane/water interfaces using Eq. (19). For the $C_5$ to $C_{16}$ series they calculated theoretical values in the range of 53.7 to 54.4 mN m$^{-1}$, compared to corresponding experimental values which vary from 50.2 to 53.8 mN $m^{-1}$. Fowkes [30] derived $\gamma_d^\alpha$ values for water and mercury using Eq. (19) and the experimental $\gamma^{\alpha\beta}$ values for both liquids against a series of saturated hydrocarbons. A reasonably constant value of $\gamma_d = 21.8 \pm 0.7$ mN $m^{-1}$ seemed to emerge for water, but when more precise experimental determinations of $\gamma^{\alpha\beta}$ were established, it was found that $\gamma_d$ for water varies smoothly with alkane chain length [32–35].

There are many inherent weaknesses of the deBoer-Hamaker approach (despite its apparent success). The chief of these in the current context concerns the validity of the G(b) function [Eq. (15)] for values of b of the order of molecular diameters. Strictly speaking, the integration procedure used is invalid. A summation procedure is better [36–38], yet to do this properly the pair distribution function g(r,z) near the surface is really required. Of course, the presence of the second liquid phase will modify this compared to the surface of the free liquid!

The so-called *macroscopic* approach for calculating the G(b) function (see Chap. 3, Sec. IV.B.1) initially developed by Lifshitz [39], suffers from a similar weakness. It is necessary to know the dielectric permittivities $\varepsilon(z)$ of the liquid concerned over the whole experimental frequency range, and it is an inherent assumption that $\varepsilon(z)$ is uniform in the surface region. However the Lifshitz approach, and developments thereof,† do overcome the pairwise additivity limitation of the deBoer-Hamaker approach (also implicit in the theories discussed under Secs. I.C.1 and I.C.2).

In addition, since zero frequency contributions are also included, then, in principle, Keesom and Debye forces are automatically accounted for. Furthermore, G(b) is now a true *free* energy (i.e., contains entropy contributions) and not just a potential energy, as is the case with the de Boer-Hamaker approach [Eq. (15)].

Israelachvili [29] has suggested a simplified macroscopic approach, based on the method of imaging in electrostatics, for calculating $\gamma$. Again, as in the de Boer-Hamaker approach, it is assumed that each material involved exhibits a single absorption peak centered at some frequency $\omega$ (rad $sec^{-1}$), in the ultraviolet. Thus the dielectric permittivity $\varepsilon$ at this frequency may be equated with $n^2$, where n is the corresponding refractive index. This leads to the following expression for $G_A^{\alpha\beta}$:

---

†Several reviews on this subject have now appeared. The reader is referred to very readable reviews by Winterton [40] and by Israelachvili [41]; more advanced accounts have been given by Israelachvili and Tabor [42] and by Parsegian [43].

$$G_A^{\alpha\beta} = \frac{h}{64\sqrt{2}\,\pi^2 D^2}$$

$$\times \frac{[(n^\alpha)^2 - 1][(n^\beta)^2 - 1]\omega^\alpha\omega^\beta}{[(n^\alpha)^2 + 1]^{1/2}[n^\beta + 1]^{1/2}\{[(n^\alpha)^2 + 1]^{1/2}\omega^\alpha + [(n^\beta)^2 + 1]^{1/2}\omega^\beta\}} \qquad (21)$$

where D is the distance between the planes through the outermost polarizable electrons of the two surfaces (thus $D^2$ replaces $\sigma_{12}^2$ in the de Boer-Hamaker theory). Aveyard and Saleem [44] have shown that Eq. (21) correctly reproduces the chain-length dependence of $G_A^{\alpha\beta}$ for the *n*-alkane/water series if D is assumed to be independent of chain length, and its value is obtained by equating the theoretical and experimental values of $G_A^{\alpha\beta}$ for the dodecane/water interface. For the case of a free surface tension Eq. (21) reduces to

$$\gamma = \frac{(n^2 - 1)h\omega}{256\sqrt{2}\,\pi^2 D^2 (n^2 + 1)^{3/2}} \qquad (22)$$

Israelachvili has used Eq. (22) to calculate values for the surface tension of various hydrocarbons, obtaining good agreement with experimental values.

Finally, in this section on calculating surface and interfacial tensions, it is worth pointing out one feature which has largely been neglected by all the theories to date; this concerns the question of molecular orientation at interfaces. This can be an important effect for polar molecules. For example, the orientation of water molecules at the water/vapor or water/second-liquid interface is still largely an unanswered question. Considerable surface orientation may occur in the case of chain molecules (e.g., the *n*-alkanes), and this may extend over tens, if not hundreds, of molecular layers. Such layers are known, for example, to exhibit optical birefringence [45]. Theoretical work in this area awaits development.

### D. Interfacial Thermodynamics

This subject often appears to be complex, mainly because different authors have adopted different conventions. However, a concise and clear summary of the subject is given by Aveyard and Haydon [46]; they also point out which of the principle authors in the subject have used which conventions. Here we shall only outline the essential features of the subject necessary for an understanding of later chapters.

The first convention concerns the model to be used to describe the interface. Two such models have been commonly used: one associated with the name of Gibbs [9], the other with that of Guggenheim [47]. The latter model is essentially that depicted in Fig. 1 and involves the concept of surface phase(s). The major weakness of this model is the assignment of a value to $\delta$, its *thickness*,

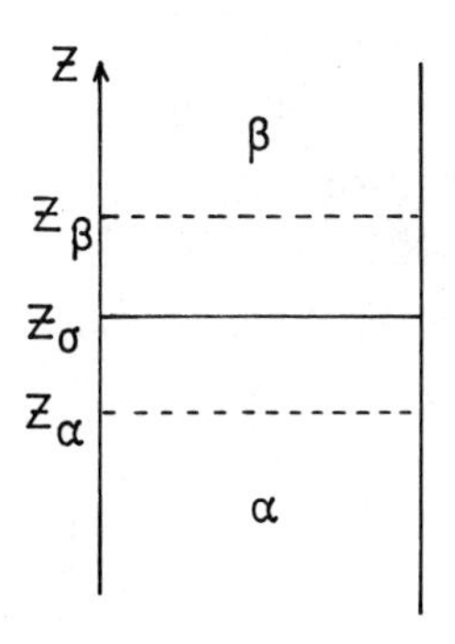


Figure 6. The Gibbs convention for an interface.

and also $V^s$, its volume. The extensive thermodynamic functions which appear in thermodynamic expressions relating to this surface phase are *total* quantities and, where appropriate, will be designated here by the superscript s.

In the Gibbs convention the interface is regarded as a mathematical dividing plane at $z_\sigma$ (Fig. 6), often referred to as the *Gibbs surface*. The two bulk phases $\alpha$ and $\beta$ are assumed to have uniform thermodynamic properties up to $z_\sigma$, and any variation in these properties (e.g., density or tangential pressure tensor) is assumed to be a delta function at $z_\sigma$. Where these parameters exceed their value in either of the adjacent bulk phases, they are then *excess* quantities with respect to that bulk phase (rather than total quantities as in the Guggenheim model). Parameters defined in this way will be designated here by the superscript $\sigma$.

The choice of location of $z_\sigma$ is arbitrary (it does not even have to lie between $z_\alpha$ and $z_\beta$). However, one very convenient location for a liquid/vapor interface is shown in Fig. 7 in terms of the density profile $\rho(z)$. If $z_\sigma$ is chosen such

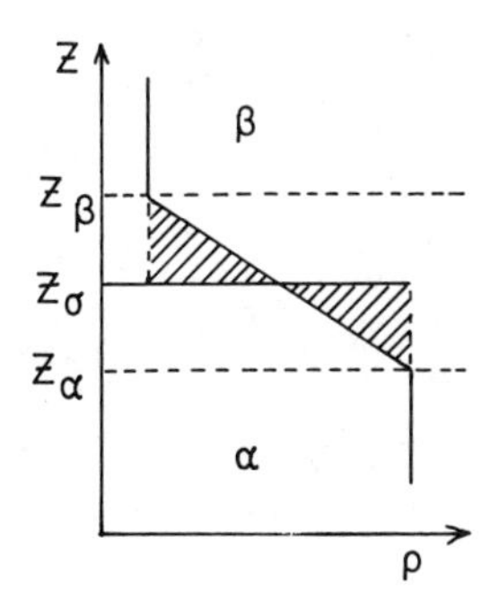


Figure 7. Location of the Gibbs dividing plane $2\sigma$ in terms of the density profile $\rho(z)$.

that the two shaded areas in Fig. 7 are equal, then the excess number of molecules allotted to phase $\alpha$ just balances the deficit in phase $\beta$. Thus, the excess density in the plane $z_\sigma$ is zero, i.e., $\rho^\sigma = 0$.[†] As we shall see (Sec. II), a similar location of the Gibbs surface is generally chosen when discussing adsorption from solution, i.e., $z_\sigma$ is chosen such that the excess number concentration of solvent molecules is zero.

The Gibbs model overcomes the problem of deciding on a value for $\delta$. It is not, however, as conceptually simple as the Guggenheim model. For example, it would be wrong to deduce that the plane in which the surface tension "acts" in the Gibbs convention is necessarily the same plane where $\rho^\sigma = 0$. The choice of which convention to use in a given situation is purely a matter of convenience.

The second convention which causes some confusion concerns the choice of the state function to describe interfacial free energy. There are four possibilities: two forms of the Helmholtz free energy and two forms of the Gibbs free energy [46]. This in turn has led to some confusion in the past over the definition of the chemical potential of molecules in the interface and, indeed, over the question of whether chemical potentials of molecules in an interface are the same as those in the bulk [46]. Although this point will not be developed here, it should be realized that this can be a point of misunderstanding. For our purposes we shall simply designate the chemical potential of molecules, species i, in the interface as $\mu_i^\sigma$, and assume that

$$\mu_i^\sigma = \mu_i^\alpha = \mu_i^\beta \tag{23}$$

With regard to the interfacial tension, its thermodynamic definition is probably best approached through a consideration of the first law. For a closed system, neglecting for the moment the presence of any interfaces,

$$dU = dQ + dW \tag{24}$$

where dU is the change in internal energy brought about by an exchange of energy dQ with the surroundings ($dQ = T\,dS$ for a reversible change) and/or by work done on the system $dW$ ($dW = -P\,dV$). All the changes are assumed to be infinitesimally small.

For an open system, Eq. (24) becomes

$$dU = T\,dS - P\,dV + \sum_i \mu_i\,dn_i \tag{25}$$

---

[†]For a liquid/liquid interface, it is not usually possible to define a single plane where $\rho^\sigma = 0$ for *both* liquids (see Sec. II).

where $\mu_i$ and $n_i$ are the chemical potential and number of moles of species i in the system, respectively. If an interface is present in the system, then additional work (= $\gamma$ dA) will be done on expanding the interface. Thus Eq. (25) becomes

$$dU = T\,dS - P\,dV + \gamma\,dA + \sum_i \mu_i\,dn_i \tag{26}$$

or for a one-component system,

$$dU = T\,dS - p\,dV + \gamma\,d\mathcal{A} + \mu\,dn \tag{27}$$

This leads to one definition of the surface tension of a pure liquid, i.e.,

$$\gamma = \left(\frac{dU}{d\mathcal{A}}\right)_{s,v,n} \tag{28}$$

U is made up of contributions from the two bulk phases and the interface. Therefore, adopting the Gibbs convention,

$$U = U^{\alpha} + U^{\beta} + U^{\sigma} \tag{29}$$

and from Eq. (27)

$$dU^{\sigma} = T\,ds^{\sigma} + \gamma\,d\mathcal{A} + \mu\,dn^{\sigma} \tag{30}$$

Note that in Eq. (30), $U^{\sigma}$, $S^{\sigma}$ and $n^{\sigma}$ are *excess* quantities. Also the term $PV^{\sigma}$ does not appear since $V^{\sigma} = 0$ in the Gibbs convention. An alternative definition of $\gamma$ is thus

$$\gamma = \left(\frac{dU^{\sigma}}{d\mathcal{A}}\right)_{S^{\sigma},n^{\sigma}} \tag{31}$$

Similar definitions for $\gamma$ may be given in terms of any of the four *free*-energy functions referred to above. By way of example, we select only the normal Helmholtz free-energy function A. The analogous equations to Eqs. (27) and (30) above are then

$$dA = -S\,dT - P\,dV + \gamma\,d\mathcal{A} + \mu\,dn \tag{32}$$

$$dA^{\sigma} = -S^{\sigma}\,dT + \gamma\,d\mathcal{A} + \mu\,dn \tag{33}$$

leading to the following definition of $\gamma$:

$$\gamma = \left(\frac{dA}{d\mathcal{A}}\right)_{T,V,n} \tag{34}$$

$$\gamma = \left(\frac{dA^{\sigma}}{d\mathcal{A}}\right)_{T.n^{\sigma}} \tag{35}$$

or in terms of the Guggenheim convention

$$\gamma = \left(\frac{dA^s}{d\mathcal{A}}\right)_{T,n^s} \tag{35a}$$

Equation (36), as we have seen (Sec. I.C), is the equation used in statistical mechanical calculations of $\gamma$ via the partition function approach.

Integration of Eq. (33), holding the intensive variables T, $\gamma$, and $\mu$ constant, yields for a one-component system

$$A^\sigma = \gamma\mathcal{A} + \mu n^\sigma \tag{36}$$

or for a multicomponent system

$$A^\sigma = \gamma\mathcal{A} + \sum_i \mu_i n_i^\sigma \tag{37}$$

If, in the case of a one-component system, the Gibbs surface is chosen such that $n^\sigma = 0$, as discussed above, then

$$\gamma = \frac{A^\sigma}{\mathcal{A}} = \overline{A}^\sigma \tag{38}$$

One may thus identify the interfacial tension with the excess Helmholtz free energy per unit area of interface. This is not the case for multicomponent systems, however, since it is not generally possible to choose a single Gibbs surface such that $n_i^\sigma = 0$ for all i. Complete differentiation of Eq. (36) leads to

$$dA^\sigma = \gamma\, d\mathcal{A} + \mathcal{A}\, d\gamma + n^\sigma\, d\mu + \mu\, dn^\sigma \tag{39}$$

Comparison of Eqs. (33) and (39) gives a relationship of the Gibbs-Duhem type

$$-S^\sigma\, dT + \mathcal{A}\, d\gamma + n^\sigma\, d\mu = 0 \tag{40}$$

for a one-component system. Again choosing the Gibbs surface such that $n^\sigma = 0$, then

$$\frac{d\gamma}{dT} = -\frac{S^\sigma}{\mathcal{A}} = -\overline{S}^\sigma \tag{41}$$

For most liquids $d\gamma/dT$ is negative (typically ~0.1 mN $m^{-1}$ $K^{-1}$) and $\overline{S}^\sigma$, the excess entropy per unit area of interface, is positive. This implies that the surface entropy is greater than the bulk entropy. Since, as discussed earlier (Sec. I.B), the *thermal* part of the entropy (for simple liquids) is the same in the interface as in the bulk, this increase in entropy must reflect a greater *configurational* entropy in the surface than in bulk liquid. This is not a direct result of a lower density of molecules in the interfacial region, however, since we have chosen the

Gibbs surface where $n^\sigma = 0$ (i.e., the excess of interfacial molecules is zero). It has rather to do with the more subtle effect that configuration entropy is related to potential energy; there is a difference in potential energy between surface and bulk molecules because of intermolecular forces. This excess energy† is given by

$$\bar{U}^\sigma = \bar{A}^\sigma + TS^\sigma \tag{42}$$

or

$$\bar{U}^\sigma = \gamma - T\frac{d\gamma}{dT} \tag{43}$$

Thus $\bar{U}^\sigma$ may be evaluated if $\gamma$ and $d\gamma/dT$ are known. Moreover, from Eq. (43),

$$\frac{d\bar{U}^\sigma}{dT} = T\frac{d^2\gamma}{dT^2} \tag{44}$$

Experimental data and theoretical plots (e.g., Fig. 4) indicate that $\gamma$ approximates a linear function of T, at least in the region of the triple point, and thus $\bar{U}^\sigma$ is independent of temperature (as one might expect for a potential energy term). In the region of the critical point, however, $\gamma$ versus T plots show a distinct curvature as $\gamma \to 0$, but here all the excess thermodynamic functions also tend to zero.

### E. Curved Surfaces

The discussion so far has concentrated on planar interfaces. In many cases, however, one is concerned with *curved* interfaces, and, of course, this is the case with emulsion systems. Consider (Fig. 8) a spherical droplet of radius r of phase $\alpha$ immersed in phase $\beta$, where r is assumed to be large with respect to $\delta$, the interfacial thickness, such that $\gamma$ may be assumed to be independent of r. When $r \sim \delta$, then this assumption is no longer valid. For example, for a water droplet of radius 1 nm (approximately seven water molecules across the diameter), $\gamma$ is estimated to be 0.755 of its value for a plane surface [48], whereas for a droplet of radius 10 nm, $\gamma$ is the same within experimental error. There is, however, a *pressure* difference across the interface (see below) which suggests that the density profile $\rho(z)$ may be somewhat perturbed. This fact lends support to the argument (Sec. I.C) that $\gamma$ is seemingly not too sensitive to the exact form of $\rho(z)$. It is readily shown (see virtually any

---

†Again this is a *potential* energy, since there is no excess kinetic energy for the Gibbs surface at $n^\sigma = 0$.

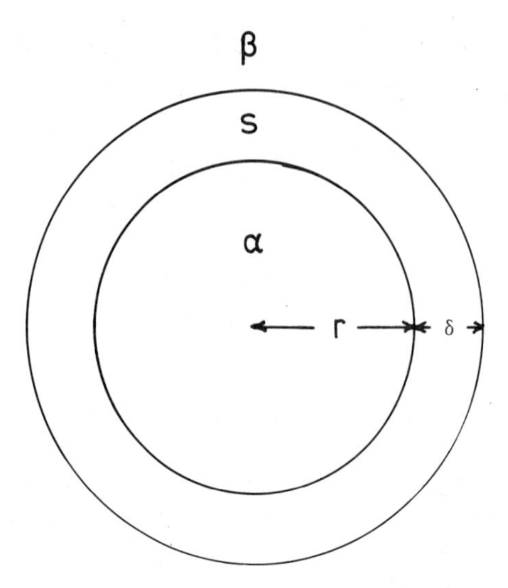


**Figure 8.** Model for a spherical droplet.

standard physics text) that the pressure difference $\Delta P$ across a spherical droplet, of the type shown in Fig. 8, is given by

$$\Delta P = P^{\alpha} - P^{\beta} = 2\,\frac{\gamma}{r} \tag{45}$$

For a general surface, whose curvature at any point is defined by two radii $r_1$ and $r_2$, we have the general form of the Laplace equation

$$\Delta P = \gamma\left(\frac{1}{r_1} + \frac{1}{r_2}\right) \tag{46}$$

Note that the pressure is always greater on the concave side of an interface. For a hydrocarbon droplet of radius 100 nm in water, assuming $\gamma$ = 50 mN $m^{-1}$, $\Delta P$ is ~$10^6$ Pa (i.e., ~10 atm).

A second standard relationship for curved interfaces is the Kelvin equation. For a liquid-vapor interface this relates the chemical potential of the molecules and the vapor pressure to the radius of curvature of the droplet:

$$\mu^{\alpha} - \mu^{\beta} = kT \ln \frac{P^{r}}{P^{\infty}} = \frac{2v\gamma}{r} \tag{47}$$

where the superscripts r and $\infty$ represent the value at radius r and $\infty$ (plane interface), respectively, and v is the molar volume of the molecules in the liquid phase. Thus, for a water droplet, radius 10 nm, taking $\gamma$ to be 72.0 mN $m^{-1}$ at 298 K, $P^r$ is ~10% higher than $P^{\infty}$.

An equivalent expression, for the case of a liquid droplet of material 1 immersed in a second liquid in which it has a finite solubility, is given by

$$\mu_1^r - \mu_1^\infty = kT \ln \frac{a_1^r}{a_1^\infty} - 2 v_1^0 \frac{\gamma}{r} \tag{48}$$

Here, a is the activity of species 1 in the continuous phase and $v_1^0$ is the molar volume of species 1 in the pure liquid state. By way of example, for a hydrocarbon droplet (species 1) in water, taking $\gamma = 50$ mN $m^{-1}$ and $v_1^0 = 10^{-4}$ $m^3$ $mol^{-1}$, then for a droplet of 400 nm radius there is a 1 per cent increase in $a_1$ and for a droplet of 40 nm radius (i.e., in the microemulsion range) there is a 10 per cent increase in $a_1$.

## F. Experimental Techniques for Interfacial Tension

Although we are primarily concerned here with interfaces between pure fluids, the techniques described below also apply, in general, for interfaces between multicomponent fluids. The purpose of this section is to summarize the techniques available, without repeating too much of the experimental details which are very adequately reviewed elsewhere [49–51].

In theory all methods used for measuring the surface tension can be applied to measure the interfacial tension. In practice, however, the choice is limited owing to the relatively high viscosity of liquids (compared with air) and other factors. Existing methods fall roughly into two categories: those in which the properties of the meniscus is measured at equilibrium, e.g., pendant drop shape, sessile drop shape, and Wilhelmy plate methods, and nonequilibrium methods such as the drop volume (weight) and du Noüy ring methods. The latter methods are faster, although they suffer from the disadvantage of premature rupture and expansion of the surface, causing adsorption depletion. They are also inconvenient for measuring interfacial tensions in the presence of macromolecular species, where adsorption is slow. This problem is overcome in the equilibrium (static) methods.

### 1. *Du Noüy's Method* [52]

This is probably the most commonly used method in practice in view of its simplicity and the fact that a fairly simple (and cheap) instrument is sufficient for carrying out the measurements. Basically one measures the force required to detach a ring or loop of wire from the liquid/liquid interface. As a first approximation the detachment force is taken to be equal to the interfacial tension $\gamma$ multiplied by the periphery of the ring. Thus, if the ring is attached to a sensitive balance (e.g., tension balance or microbalance), the total force F in detaching the ring from the interface is the sum of its weight W and the interfacial force

$$F = W + 4\pi R\gamma \tag{49}$$

where R is the radius of the ring.

However Harkins and Jordan [53] found that Eq. (49) leads to significant errors in calculating $\gamma$; they introduced an empirical correction factor f which is a function of the meniscus volume V and r the radius of the wire. Thus

$$f = \frac{\gamma}{\gamma_{ideal}} = f\left(\frac{R^3}{V}, \frac{R}{r}\right) \tag{50}$$

where $\gamma_{ideal}$ is the surface tension calculated by Eq. (49). Values of the correction factor f have been tabulated by Harkins and Jordan [53]. Extensive tables that cover higher density and lower $\gamma$ values are also available [54]. Detailed theories to account for f have been put forward by several authors, e.g., by Freud and Freud [55].

One of the difficulties in using the du Noüy ring method is the disturbance of the interface as the point of detachment is approached. Moreover, the ring should be kept horizontal, and it should be free from any contaminants. This is usually achieved by flaming the ring before use. Moreover, a zero (or nearly zero) contact angle is required; otherwise, the measured $\gamma$ will be too low. This is particularly a problem in measuring $\gamma$ in the presence of surfactants, where adsorption on the ring can alter its wetting characteristics. In these cases the use of the du Noüy ring method for measuring $\gamma$ or the surface pressure of spread monolayers at the liquid/liquid interface is not recommended.

### 2. *Drop-volume or Drop-weight Method*

This is again a versatile method and is as widely used as the du Noüy method. It can be carried out with very simple apparatus. Basically, one determines the volume V (or weight W) of a drop of liquid [immersed in the second (less dense) liquid] which becomes detached from a vertically mounted capillary tip having circular cross section (radius r). The capillary is usually attached to a micrometer syringe for accurate determination of the drop volume. The ideal drop weight $W_{ideal}$ can be equated to the product of the interfacial tension $\gamma$ and the perimeter of the tube tip (Tate's law)

$$W_{ideal} = 2\pi r\gamma \tag{51}$$

In practice, a weight W is obtained which is less than $W_{ideal}$ owing to the fact that a portion of the drop remains attached to the tube tip. This arises from the fact the drops are formed as a result of the mechanical instability of the cylindrical neck which develops at the capillary tip. Thus, only a portion of the drop which has reached the point of instability actually becomes detached; as much as 40 per cent of the liquid may remain attached to the tip. Therefore, Eq. (51) should include a correction factor $\phi$ that is a function of the tube radius r and some linear dimension of the drop, i.e., $V^{1/3}$. Thus,

$$W = 2\pi r \gamma \phi\left(\frac{r}{V^{1/3}}\right) \tag{52}$$

Values of $\phi(r/V^{1/3})$ have been tabulated by Harkins and Brown [56], who obtained them by measuring the surface tension of various liquids using the absolute capillary rise method and then measuring the drop volume of these liquids falling from tips of brass and glass capillaries of various radii. Lando and Oakley [57] used an empirical quadratic equation to fit the correction function to $r/V^{1/3}$. However, the fit with experimental data is not very good, and more recently Wilkinson and Kidwell [58] produced an equation which fits the experimental data better.

One of the major problems with the drop volume method is the preparation of a flat tip, which has to be ground smooth at the end and should be free from any nicks. Moreover, care should be taken that the liquid wets the surface of the tip, in which case the external diameter of the tip is used in Eq. (52). For liquids which do not wet the tip, the internal diameter of the tip is used in calculating $\gamma$. The drop should be formed slowly at the tip, and vibration should be avoided.

### 3. *The Wilhelmy Plate Method* [59]

In this method a thin plate made from glass (e.g., a microscope cover slide) or platinum foil is either detached from the interface or its weight measured statically using an accurate microbalance. The total force F is then given by the sum of the weight of the plate W and the interfacial tension force

$$F = W + \gamma p \tag{53}$$

where p is the "contact length" of the plates with the liquid, i.e., the plate perimeter. Providing the contact angle of the liquid with the plate is zero, no correction is required for Eq. (53). Thus, the Wilhelmy plate can be used in the same manner as the du Noüy detachment method.

Alternatively, in the so-called static method, the plate is suspended from one arm of a microbalance and allowed to penetrate the top liquid until it touches the interface, or the whole vessel containing the liquid is raised until the interface touches the plate. The increase in weight $\Delta W$ is given by

$$\Delta W = \gamma p \cos \theta \tag{54}$$

where $\theta$ is the contact angle. Thus, if the plate is completely wetted by the lower liquid as it penetrates ($\theta = 0$), $\gamma$ may be simply calculated from $\Delta W$. In those cases when one is measuring $\gamma$ for the oil/water interface, where the density of the oil is lower than water, care should always be taken that the plate is completely wetted by the aqueous solution. For that purpose a roughened platinum or glass plate, which normally has a zero contact angle with water, is used. However, if the oil is more dense than water, a hydrophobic plate is used, so that when the plate penetrates through the upper aqueous layer and touches the interface, it is completely wetted by the oil phase.

The static Wilhelmy plate method is very useful for following the variation of interfacial tension with time and in particular in the study of monolayers spread at an oil/water interface.

#### 4. *Pendant Drop Method*

If a drop is allowed to hang from the end of a vertically mounted capillary tube, it will adopt an equilibrium profile that is a unique function of the tube radius, the surface tension of the liquid, its density, and the gravitational field. The shape of bounded menisci† can be described by a single parameter $\beta$, derived by Bashforth and Adams [60] according to the equation

$$\beta = \frac{\Delta\rho\, g b^2}{\gamma} \tag{55}$$

where b is the radius of a drop at its apex. (See Fig. 9.) Padday [61] suggested the more general equation

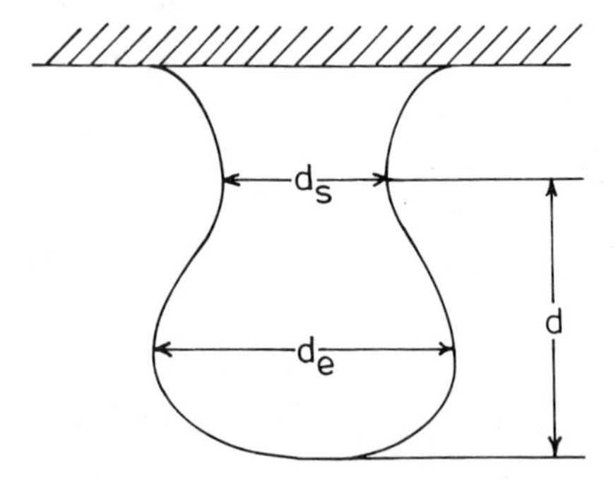


**Figure 9.** Schematic representation of a pendant drop.

---

†Bounded menisci are those with only one solid surface supporting the liquid forming the meniscus, as in the present case of the pendant drop.

$$\beta' = \frac{\Delta\rho\, gR_v^2(90°)}{\gamma} \tag{56}$$

where $R_v(90°)$ is the principal vertical radius of curvature of the meniscus at its neck or narrow point. The quantities $\beta$ and $R_v$ bear a fixed relationship to each other for a given shape.

Andreas et al. [62] suggested that the most conveniently measurable shape-dependent quantity is $S = d_s/d_e$ (cf. Fig. 9), where $d_e$ is the equatorial diameter and $d_s$ is the diameter measured at a distance $d_e$ from the bottom of the drop. The difficulty in measuring b is overcome by combining it with $\beta$ and defining a new quantity H, given by

$$H = -\beta\left(\frac{d_e}{b}\right)^2 \tag{57}$$

Thus

$$\gamma = \frac{-\Delta\rho\, gh^2}{\beta} = \frac{-\Delta\rho\, gd_e^2}{\beta(d_e/b)^2} = \frac{\Delta\rho\, gd_e^2}{H} \tag{58}$$

The relationship between H and the experimentally accessible parameter S has been obtained empirically using pendant drops of water. Accurate values of 1/H versus S have been obtained by Niederhauser and Bartell [63] and tables of 1/H as a function of S are given by Adamson [49]. The pendant drop method is widely used for accurate ($\pm$ 0.1%) measurements of interfacial tension. All that is needed is good optical equipment for taking photographs of the drop.

Recently Levin et al. [64] suggested a simple method which is particularly suitable for measuring interfacial tensions, based on the experimental determination of the maximum height attained by a drop hanging from a vertical tube. It was shown many years ago by Lehnstein [65] that as the drop grows in volume, its height increases, until eventually the volume reaches a maximum. At greater volumes, an equilibrium profile is not possible, and so the drop must break.

Lehnstein assumed that at all volumes less than this maximum, the drop had its stable equilibrium profile, so that the drop detached from the tip when its theoretical maximum volume was attained. Thus, the maximum volume of a drop hanging from a tube of known radius determines a unique height, from which, by comparison to the equivalent theoretical value, the interfacial tension may be calculated. It is convenient to introduce the characteristic length $\ell$ given by

$$\ell = \left(\frac{\gamma}{g\,\Delta\rho}\right)^{1/2} \tag{59}$$

where $\Delta\rho$ is the difference in density between the drop and the surrounding medium. If h is the height of the drop, V its volume, and r the radius of the tube, then the following dimensionless quantities are defined:

$$\lambda = \frac{r}{\ell} \qquad K = \frac{h}{\ell} \qquad v = \frac{V}{\pi \ell^3}$$

By a numerical procedure described by Levin et al. [64], it is possible to obtain the value of K corresponding to the maximum value of v for a given $\lambda$. The procedure for determining the interfacial tension is then straightforward. The height h of the drop when it is just about to become detached is measured (e.g., by using a traveling microscope) and from the radius of the tube, r/h is calculated; using the numerical values given by Levin et al., $\lambda$ may be interpolated and hence $\gamma$ calculated.

This method was assessed by the authors by measuring the surface tension of *n*-hexane and the interfacial tension of hexane/water. The results obtained were in excellent agreement with recent determinations by other methods. Thus the method is simple (only requiring measurement of length) and offers the advantage of being absolute, and thus does not depend on empirical correction factors.

5. *Sessile Drop Method*

This is similar to the pendant drop method except in this case a drop of the liquid with the higher density is placed on a flat plate immersed in the second liquid (Fig. 10). Basically, Eq. (55) is used for calculation of the interfacial tension $\gamma$,

$$\gamma = \frac{\Delta\rho\, gb^2}{\beta} \tag{55}$$

While b is difficult to determine, the Bashforth-Adam tables [60] give $x_e/b$ as a function of $\beta$, where $x_e$ is the equatorial radius, so that Eq. (55) may be written in the form

$$\gamma = \Delta\rho \frac{gx_e^2}{[f(\beta)]^2} \tag{60}$$

with

$$f(\beta) = \frac{x_e}{b}$$

Since $x_e$ can be determined accurately, the experimental problem reduces to the determination of $\beta$. This is achieved by comparing the drop profile to a theoretical set of profiles, for various values of $\beta$ as given by tables.

From a knowledge of $\beta$, $f(\beta)$ is read from the same tables, allowing the calculation of $\gamma$ from Eq. (60). This method, like the pendant drop, is very accurate ($\pm$ 0.1%). It also allows measurement of the variation of interfacial ten-

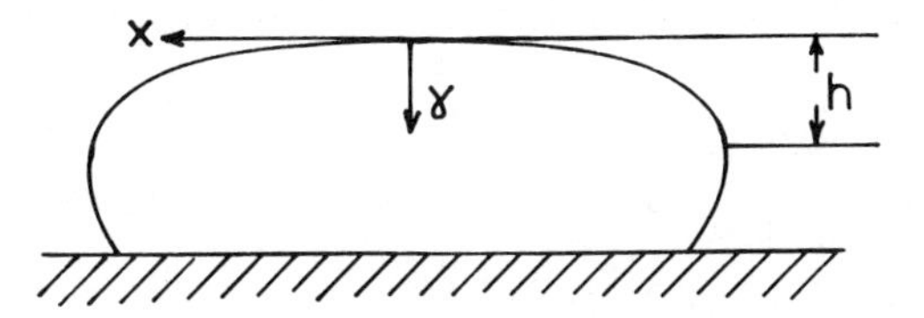


Figure 10. Schematic representation of a sessile drop.

sion with time, and thus allows one to follow the kinetics of adsorption of surfactants or macromolecules at the liquid/liquid interface.

*6. Spinning Drop Method*

This method is particularly useful for the measurement of very low interfacial tensions (~$10^{-4}$ to $10^{-2}$ mN $m^{-1}$), which are particularly important in many applications such as microemulsion formation (see Chap. 4), enhanced oil recovery, etc. Basically a drop of the less dense liquid $\alpha$ is suspended in a tube containing the second liquid $\beta$. On rotating the whole mass (Fig. 11), the drop of the liquid $\alpha$ moves to the center. With increasing speed of revolution, the drop $\alpha$ elongates as the centrifugal force opposes the interfacial tension force that tends to maintain the spherical shape (i.e., that having minimum surface area), until an equilibrium shape is reached at a given speed of rotation. At moderate speeds of rotation, the drop approximates to a prolate ellipsoid, whereas at very high speeds of revolution, the drop approximates to an elongated cylinder. This latter limiting case makes it easier to analyze for the calculation of $\gamma$.

Consider a section of the elongated cylinder of volume V (Fig. 11b). The centrifugal force on a volume element is $\omega^2 r \Delta\rho$, where $\omega$ is the speed of rotation, $\Delta\rho$ is the density difference between the two liquids $\alpha$ and $\beta$, and r is the distance from the axis of revolution. The potential energy at r is $\omega^2 r^2 \Delta\rho/2$

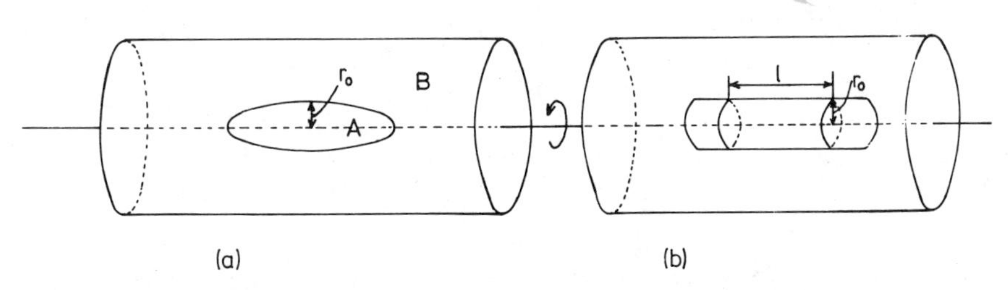


Figure 11. Schematic representation of a spinning drop: (a) prolate ellipsoid and (b) elongated cylinder.

and the total potential energy U for the cylinder of length $\ell$ is then given by

$$U = \ell \int_0^{r_0} \omega^2 r^2 \frac{\Delta\rho}{2} 2\pi r \, dr = \pi\omega^2 \Delta\rho \frac{r_0^4}{4} \qquad (61)$$

The interfacial free energy for the cylinder is

$$G^\sigma = 2\pi r_0 \ell \gamma \qquad (62)$$

Thus the total free energy of the system is given by

$$G = \pi\omega^2 \Delta\rho \frac{r_0^4}{4} + 2\pi r_0 \ell \gamma = \omega^2 \Delta\rho \, r_0^2 \frac{V}{4} + 2V \frac{\gamma}{r_0} \qquad (63)$$

where V is the volume ($= \pi r_0^2 \ell$). At equilibrium

$$\frac{dG}{dr_0} = 0$$

leading to the following expression for $\gamma$:

$$\gamma = \frac{\omega^2 \Delta\rho r_0^3}{4} \qquad (64)$$

Equation (63) was first derived by Vonnegut [66].

The spinning drop method has been used by many authors for measuring low interfacial tensions, e.g., Wade and co-workers [67].

## II. The Interface Between Multicomponent Fluids: Adsorption

### A. Interfacial Concentration and Interfacial Tension

Figure 12 represents, schematically, two immiscible bulk liquid phases $\alpha$ and $\beta$, containing molecules of types 1 and 2, respectively, to which has been added a third component 3 which distributes itself between $\alpha$ and $\beta$. The interfacial region S contains 1, 2, and 3. The total adsorption of species 3 in this interfacial region $\Gamma_3^s$ is defined by

$$\Gamma_3^s = \frac{n_3^s}{A} = n_3 - \frac{c_3^\alpha V^\alpha + c_3^\beta V^\beta}{A} \qquad (65)$$

where $n_3$ is the total number of molecules of 3 added to the system, and $c_3^\alpha$ and $c_3^\beta$ are the equilibrium concentrations (mol vol$^{-1}$) of 3 in $\alpha$ and $\beta$. The reasons that $\Gamma_3^s$ is preferred to $c_3^s$ are that, first, $V^s$ is, as we have seen, difficult to define, and, second, $c_1^s$, $c_2^s$, and $c_3^s$ are not uniform in the interfacial region.

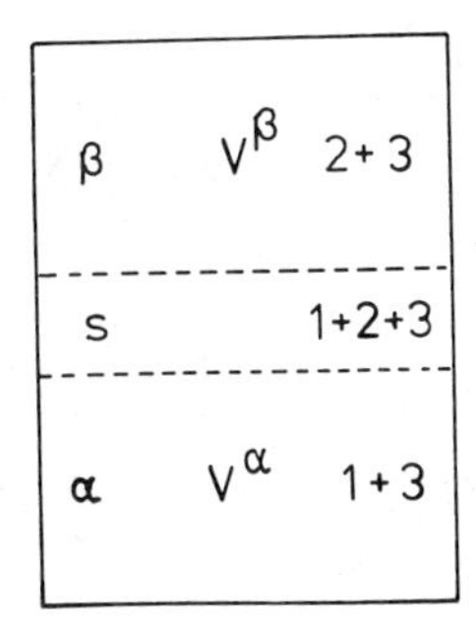


**Figure 12.** Model for adsorption at a liquid/liquid interface. Species 3 is present in both bulk phases as well as the interface.

These variations in concentration are illustrated in Fig. 13. $\Gamma_3^s$ is related to $c_3^s$ through the relationship,

$$\Gamma_3^s = \int_{z_\alpha}^{z_\beta} c_3^s(z)\, dz \tag{66}$$

($\Gamma_1^s$ and $\Gamma_2^s$ are similarly defined.) A more useful definition of the adsorption of species 3 is in terms of the *excess* number of molecules per unit area of species 3 in the interfacial region, defined with respect to some chosen Gibbs surface, $z_\sigma$. $\Gamma_3^\sigma$ is then defined as follows:

$$\Gamma_3^\sigma = \int_{z_\alpha}^{z_\sigma} [c_3(z) - c_3^\alpha]\, dz + \int_{z_\sigma}^{z_\beta} [c_3(z) - c_3^\beta]\, dz \tag{67}$$

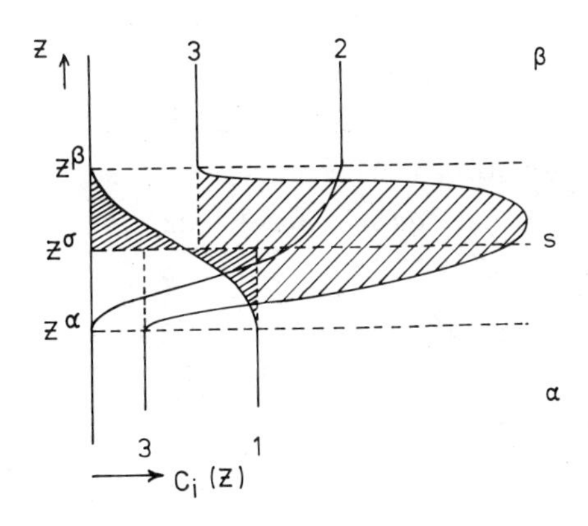


**Figure 13.** Concentration profiles $c_1(z)$, $c_2(z)$, and $c_3(z)$ at a liquid/liquid interface.

A convenient Gibbs surface to choose is either the one where $\Gamma_1^\sigma = 0$ or where $\Gamma_2^\sigma = 0$ [$\Gamma_1^\sigma$ and $\Gamma_2^\sigma$ are defined similarly to $\Gamma_3^\sigma$ in Eq. (51)]. In Fig. 13 the plane $\Gamma_1^\sigma = 0$ is chosen. $\Gamma_1^\sigma$ is zero when $z_\sigma$ is chosen such that the two heavily shaded areas are of equal area. One obvious problem with three (or more) -component systems of this kind is that we cannot usually choose a plane where both $\Gamma_1^\sigma$ *and* $\Gamma_2^\sigma = 0$, simultaneously. This leads to complex sets of thermodynamic equations [48], but as will be seen, certain simplifying assumptions can often be made. In terms of Fig. 13 when $z_\sigma$ is chosen such that $\Gamma_1^\sigma = 0$, then $\Gamma_3^\sigma$ is designated $\Gamma_{3,1}^\sigma$.[†]

We saw in Sec. I.B that for *pure* interfaces if the singlet density (or concentration) profile, pair profile, and intermolecular energies are known, then the tangential pressure tensor profile $P_t(z)$ may be evaluated. For a multicomponent interface, the evaluation of $P_t(z)$ by a statistical mechanical or computer simulation approach is obviously an extremely difficult problem [although a start has been made in obtaining $\rho_1(z)$ and $\rho_2(z)$ for the two-component liquid/vapor (LV) interface, e.g., argon and krypton, by computer simulation techniques [17]]. However, $\gamma$ would still be defined in terms of $P_t(z)$ through Eq. (5).

Statistical mechanical approaches for evaluating $\gamma$ for mixed LV interfaces in terms of simple models of the interface have been devised [68,69] (cf. the cell model discussed in Sec. I.C.). For example, for an ideal mixture of species 1 and 2, Guggenheim [70] has derived the following equation for the LV interfacial tension $\gamma$ in terms of the LV interfacial tensions of the pure components $\gamma_1^0$ and $\gamma_2^0$,

$$\exp\left(-\frac{\gamma a}{kT}\right) = x_1^L \exp\left(-\frac{\gamma_1^0 a}{kT}\right) + x_2^L \exp\left(-\frac{\gamma_2^0 a}{kT}\right) \tag{68}$$

where it is assumed that the molar surface areas of both species are equal ($a_1 = a_2 = a$).

Belton and Evans [71] have examined Eq. (68) and found good agreement with experiment for the LV interfacial tensions of mixtures of molecules which are structurally very similar (e.g., $D_2O$ and $H_2O$, chlorobenzene and bromobenzene). Prigogine and Marechal [72] have given a more complex expression for molecules of different size.

The *thermodynamic* relationship between interfacial tension and adsorption was first derived by Gibbs [9]. One starting point for the derivation of this relationship is Eq. (40):

---

†In many texts the superscript $\sigma$ is dropped since this designation implies we are dealing with *excess* adsorption and not total adsorption.

$$-S^{\sigma}\, dT + A\, d\gamma + \sum_i n_i\, d\mu_i = 0 \tag{69}$$

Thus, at constant T,

$$d\gamma = -\frac{\Sigma_i n_i^{\sigma}\, d\mu_i}{A} = -\sum_i \Gamma_i^{\sigma}\, d\mu_i \tag{70}$$

Equation (54) is known as the Gibbs adsorption equation or isotherm. For a two-component system (e.g., the LV interface for a mixture of 1 and 2), then

$$d\gamma = -\Gamma_{2,1}^{\sigma}\, d\mu_2 = -\Gamma_{2,1}^{\sigma} RT\, d(\ln a_2^L) \tag{71}$$

where $a_2^L$ is the activity of species 2 in the liquid phase and $\mu_2 = \mu_2^{\sigma} = \mu_2^{\alpha} = \mu_2^{\beta}$ [cf. Eq. (23)], or

$$\Gamma_{2,1}^{\sigma} = -\frac{1}{RT}\, \frac{d\gamma}{d(\ln a_2^L)} \tag{72}$$

Equation (72) then gives an *exact* relationship for $\Gamma_{2,1}^{\sigma}$ in terms of the change of $\gamma$ with $\ln a_2^L$. For three-component LL interfaces such a simple exact relationship is not attainable because of the problems associated with defining the Gibbs surface, as discussed above.

For example, if we choose $z_\sigma$ such that $\Gamma_1^{\sigma} = 0$, then Eq. (70) becomes

$$d\gamma = -\Gamma_{2,1}^{\sigma}\, d\mu_2 - \Gamma_{3,1}^{\sigma}\, d\mu_3 \tag{73}$$

If one can assume that $\Gamma_{3,1}^{\sigma} \gg \Gamma_{2,1}^{\sigma}$ we can approximate Eq. (71) to give

$$d\gamma \simeq -\Gamma_{3,1}^{\sigma}\, d\mu_3 = -\Gamma_{3,1}^{\sigma} RT\, d(\ln a_3^{\alpha})$$

or

$$\Gamma_{3,1}^{\sigma}\, RT\, d(\ln a_3^{\beta}) \tag{74}$$

or if $a_3 \approx c_3$, then

$$\Gamma_{3,1}^{\sigma} \simeq -\frac{1}{RT}\, \frac{d\gamma}{d(\ln c_2^{\alpha})} = -\frac{1}{RT}\, \frac{d\gamma}{d(\ln c_2^{\beta})} \tag{75}$$

The adsorption of strongly surface-active nonelectrolytes from dilute solution fulfils these conditions (see Sec. II.B).

The question now arises as to how $\Gamma_{i,1}^{\sigma}$ relates to $\Gamma_i^{s}$. For two-component LV interfaces, it may be shown fairly readily [73] that

$$\Gamma_{2,1}^{\sigma} = \Gamma_2^{s} - \frac{a_2^L}{a_1^L}\, \Gamma_1^{s} \tag{76}$$

although to evaluate $\Gamma_1^s$ and $\Gamma_2^s$ a nonthermodynamic assumption has to be made. For example, if it is known that the mixture is a monomeric layer with respect to both components, one may then write

$$\Gamma_1^s \bar{a}_1 + \Gamma_2^s \bar{a}_2 = 1 \tag{77}$$

where $\bar{a}_1$ and $\bar{a}_2$ are the partial molar surface areas of 1 and 2 (assumed constant).

Again for a three-component LL interface, one can readily handle only the case of strongly surface-active nonelectrolytes. Then

$$\Gamma_{3,1}^{\sigma} \simeq \Gamma_3^s \tag{78}$$

moreover, $\Gamma_3^s$ may be evaluated from interfacial tension-concentration data [Eq. (75)], but it must be emphasized that neither Eq. (75) nor (78) is an exact relation.

Alternatively, $\Gamma_3^s$ may of course be determined directly from Eq. (65) if the equilibrium concentrations $c_3^{\alpha}$ and $c_3^{\beta}$ can be analytically determined. The real problem with this method is the accurate estimation of $A$, particularly when $\alpha$ is dispersed in $\beta$ (or vice versa), as in an emulsion system.

### B. Adsorption of Nonionic Molecules from Dilute Solutions

There are basically two theoretical approaches for considering interfaces where adsorption is occurring: (1) by means of a derived adsorption isotherm or (2) by means of a proposed equation of state for the interface. For a three-component LL interface the adsorption isotherm relates $\Gamma_3^s$ to $c_3^{\alpha}$ or $c_3^{\beta}$, whereas the equation of state relates $\pi$, the surface pressure, to $\Gamma_3^s$, where

$$\pi = \gamma_0 - \gamma \tag{79}$$

$\gamma_0$ is the (pure) LL interfacial tension in the absence of component 3.

Once the adsorption isotherm *or* equation of state has been derived, then the other may be obtained by use of the Gibbs adsorption equation. An elementary example would be to assume an *ideal* nonlocalized monolayer for species 3 adsorbed at a liquid/liquid interface. The approximate two-dimensional equation of state is then

$$\pi A = n_3^s RT \tag{80}$$

or

$$\pi = \Gamma_3^s RT \tag{81}$$

[Equation (80) is the direct analog of the three-dimensional equation of state for an ideal gas: PV = nRT.] Differentiation of Eq. (81) at constant T gives

$$d\pi = RT\ d\Gamma_3^s \tag{82}$$

But the Gibbs adsorption equation gives

$$d\pi = -d\gamma \simeq \Gamma_3^s RT\ d(\ln c_3^\alpha) \tag{83}$$

Equating Eqs. (82) and (83) and integrating leads to

$$d(\ln \Gamma_3^s) = d(\ln c_3^\alpha) \tag{84}$$

or

$$\Gamma_3^s = Kc_3^\alpha \quad \text{or} \quad K'c_3^\beta \tag{85}$$

Equation (85) represents the adsorption isotherm corresponding to the equation of state given by Eq. (80). Equation (85) is known as the Henry's law isotherm, which predicts that $\Gamma_3^s$ rises linearly with $c_3^\alpha$. Clearly, this is unreasonable and a more realistic model for the LL interface is required. Possible choices are obviously a two-dimensional van der Waals analog or a virial expansion of $\pi$, again by analogy with the gas phase. Both of these are nonlocalized models, i.e., the time between diffusional jumps of a type 3 molecule in the interface is short compared to its residence time in the interface. Localized models (e.g., those that lead to the Langmuir isotherm and its extensions) are normally not appropriate to solid interfaces where there are specific binding sites.

The two-dimensional van der Waals equation of state is usually written in the form

$$\left[\pi + \frac{(n_3^s)^2 \alpha}{A^2}\right](A - n_3^s a_3^0) = n_3^s RT \tag{86}$$

where $a_3^0$ is the excluded area or co-area of a type 3 molecule in the interface and $\alpha$ is a parameter which allows for lateral interactions in the interface. Equation (86) leads to the following theoretical adsorption isotherms, via the Gibbs adsorption equation [74]:

$$\Gamma_3^s = K_1\left(\frac{\theta}{1-\theta}\right)\exp\left(\frac{\theta}{1-\theta} - \frac{2\alpha\theta}{a_3^0 RT}\right) \tag{87}$$

where $\theta$ is the coverage and is given by

$$\theta = \frac{a_3^0}{a_3} = \frac{\Gamma_3^s}{\Gamma_{3,max}^s} \tag{88}$$

where $a_3 = A/n_3^s$ and $\Gamma_{3,max}^s$ are the values at maximum coverage (i.e., a complete monolayer of species 3 in the interface). $K_1$ is a constant related to the net free energy of adsorption of a type 3 molecule at the interface. A simplified form of Eq. (86) which has been quite widely used for adsorption at LL interfaces is the so-called Volmer equation of state [75]. This neglects lateral interactions and is equivalent to Eq. (86) with $\alpha = 0$. The equivalent derived adsorption isotherm is similarly given by Eq. (87) with $\alpha = 0$.

Any proposed theoretical model may obviously be tested against experimental data, either in terms of the adsorption isotherm or the equivalent equation of state. The latter is normally better if interfacial tension data are being used since $\pi$ is tested directly rather than $d\pi/d(\ln c_3)$. Thus, for example, the Volmer equation of state has been tested for the *n*-alkanols adsorbed at the *n*-alkane/water interface [76]. Intuitively one would expect the *n*-alkanol molecule to be oriented at the interface so that the OH group is "in the water" and the hydrocarbon chain is "in the *n*-alkane." This is a somewhat crude description, but nevertheless leads one to suspect that lateral interactions will be negligibly small (i.e., $\alpha \sim 0$), thus justifying the choice of the Volmer equation of state. Aveyard and Briscoe [76] found excellent agreement with experimental data for the *n*-alkanols at the *n*-alkane/water interface. For example, they found that for *n*-octanol, a plot of the Volmer equation in the linear form ($1/\pi$ versus $a_3$)

$$\frac{1}{\pi} = \frac{a_3}{RT} - \frac{a_3^0}{RT} \tag{89}$$

indeed gave a slope of $1/RT$ and an $a_3^0$ value of 0.24 $nm^2$ per molecule.

Ross and Chen [77] have reviewed the work done up to 1965 on the adsorption of simple nonionic materials at the LL interface. The more recent work has been reviewed by Aveyard and Vincent [78]. The latter authors also discuss the evaluation of standard free energies of adsorption (from $K_1$) and the standard enthalpies and entropies of adsorption from the temperature dependence of interfacial tensions in these types of systems. For example, referring again to the work of Aveyard and Briscoe [76], these authors found a value of $-31.4$ kJ $mol^{-1}$ for the mean standard enthalpy of adsorption of *n*-octanol (over the T range 15 to 35°C) from dodecane at the dodecane/water interface. This value is in close agreement with the value of $-30$ kJ $mol^{-1}$ for the standard enthalpy of transfer of alkanols from dilute solution in alkanes to water. The latter value is thought to result mainly from the transfer of the OH group. Thus, the results for the transfer and adsorption processes seem to be consistent. The agreement is much poorer at the air/water interface presumably because of the effect of lateral interactions of the hydrocarbon tails. These are shielded in a hydrocarbon solvent.

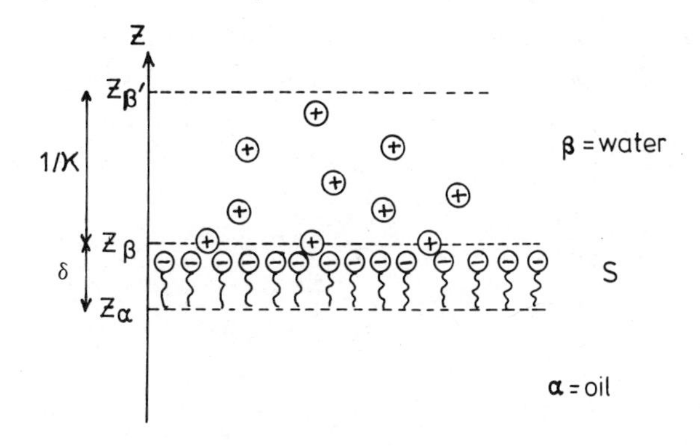


**Figure 14.** Model for the adsorption of anionic surfactant molecules at the oil/water interface.

C. Adsorption of Ionic Molecules from Dilute Solutions

When a surface-active molecule with an ionizable group adsorbs at, say, a hydrocarbon/water interface, an electrical double layer is set up. The situation is depicted schematically in Fig. 14. It is convenient to choose a kind of pseudo-Guggenheim model for the interface in this case, in which the interfacial region as such (S) comprises the adsorbed hydrocarbon anions, specifically adsorbed (possibly partially dehydrated) cations, and any bound solvent (water or hydrocarbon) molecules. The remaining counter-ions are distributed in the diffuse part of the double layer. It turns out, in general, to be more convenient to ascribe the diffuse layer to the bulk water phase (β) rather than the interfacial region itself. Hence, the term pseudo-Guggenheim model because this implies that the physical properties of the bulk phase β are *not* uniform up to the plane at $z = z_\beta$. One could, of course, define the upper boundary of S (Fig. 14) at $z_\beta$, but the location of this boundary is not easily defined, although an arbitrary choice might be at $z_{\beta'}$, where $z_{\beta'} - z_\beta = 1/\kappa$ [see Eq. (91)].

An additional important thermodynamic parameter for charged interfaces is σ the net charge per unit area of interface.† Associated with this charge are the electrical potentials at the planes $z_\alpha$ and $z_\beta$, i.e., $\Psi_\alpha$ and $\Psi_\beta$. $\Psi_\beta$ is commonly referred to as the Stern potential. Two immediate questions are: what is the form of $\Psi(z)$ in the water (β) phase and the hydrocarbon (α) phase, and, second, what is the relationship between σ and $\Psi_\alpha$ and σ and $\Psi_\beta$?

†Note that σ is the total charge ($\sigma_0$) due to adsorbed hydrocarbon anions, less that due to specifically adsorbed counter-ions.

By making certain simplifying assumptions (among which are that the ions of valency Z in the aqueous part of the diffuse double layer behave as point charges and that the interfacial charge is evenly smeared out), it may be shown that for $Z\Psi_\beta \ll kT$,

$$\Psi(z) = \Psi_\beta \exp(-\kappa z) \tag{90}$$

where

$$\kappa = \left( \frac{2Z^2e^2c}{\varepsilon_r\varepsilon_0 kT} \right)^{1/2} \tag{91}$$

and where e is the electronic charge, c is the number concentration of counter-ions in the *bulk* aqueous solution, $\varepsilon_r$ is the relative permittivity (dielectric constant) of the aqueous solution, and $\varepsilon_0$ is the permittivity of free space. $\Psi(z)$ thus falls exponentially with z in the aqueous phase, and $1/\kappa$ is commonly taken as a measure of the extent of the diffuse layer. From Eq. (91) it is seen that $\kappa$ is proportional to the square root of electrolyte concentration c in the aqueous phase. For 1:1 electrolytes at 25°C, $\kappa^{-1}$ (nm) $\simeq 3 \times 10^{-1}/ZM^{1/2}$, where M is the molar concentration of the electrolyte. Thus, $\kappa^{-1}$ is about 1 nm for $10^{-1}$ molar electrolyte, about 10 nm for $10^{-3}$ molar electrolyte and about 100 nm for $10^{-5}$ molar electrolyte. For $Z\Psi_\beta \ll kT$ it may also be shown that

$$\sigma = \varepsilon_r\varepsilon_0\kappa\Psi_\beta \tag{92}$$

This is the required relationship between $\Psi_\beta$ and $\sigma$.

The question arises as to whether there is a diffuse double layer on the *hydrocarbon* side of the interface. Because the concentration of free ions is so negligibly small, the use of equations such as Eqs. (90) to (92) is questionable [79]. In the absence of ions in the oil phase, $\Psi_\alpha$ is an *unscreened* potential. The general form of the potential profile across the oil/water interface is shown in Fig. 15. It is to be noted that the total potential drop $\phi$ from the bulk aqueous phase to the bulk oil phase is given by

$$\phi = \Psi_\alpha + \chi + \Psi_\beta \tag{93}$$

where $\phi$ is the work done in taking a unit charge from $z = +\infty$ to $z = -\infty$, $\chi$ is the potential drop across the interfacial region owing to oriented permanent dipoles (e.g., water molecules) and to induced dipoles, associated with the polarization of ions and molecules in the interfacial region, both resulting from the electric field across it. Note that if $\sigma = 0$, then $\phi = \chi$.

Assuming that the ionic concentration in the oil phase is finite (if small), then it may be shown that the ratio of $\Psi_\alpha$ to $\Psi_\beta$ is given by

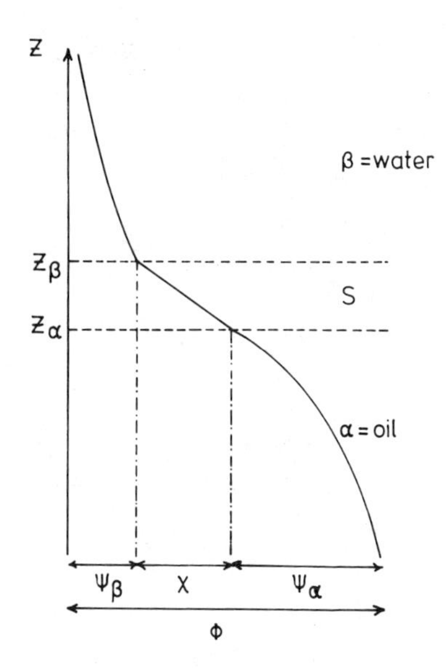


**Figure 15.** Potential profile $\phi(z)$ at an oil/water interface.

$$\frac{\Psi_\alpha}{\Psi_\beta} = \left(\frac{c^\beta \varepsilon_r^\beta}{c^\alpha \varepsilon_r^\alpha}\right)^{1/2} \tag{94}$$

where $c^\alpha$ and $c^\beta$ are the ionic concentrations in the bulk oil and water phases, respectively, and $\varepsilon_r^\alpha$ and $\varepsilon_r^\beta$ the relative permittivities in these two phases. Since $c^\beta$ and $\varepsilon_r^\beta$ are normally much smaller than $c^\alpha$ and $\varepsilon_r^\alpha$, it can be seen that $\Psi_\alpha << \Psi_\beta$, i.e., most of the potential drop is across the *oil* phase. It is worth noting in passing that the potentials $\Psi_\beta$ and $\Psi_\alpha$ will be approximately equal to the zeta potentials determined for water/oil (W/O) and oil/water (O/W) emulsion droplets, respectively. This approximation rests on the assumption that the plane of shear around a droplet is located at (or close to) $z_\beta$ or $z_\alpha$, respectively. $\sigma_0$ may be determined from $\Gamma_3^s$ through the relationship

$$\sigma_0 = ZF\,\Gamma_3^s \tag{95}$$

where F is the Faraday constant, $\Gamma_3^s$ is expressed in moles per unit area, and Z is the charge on the head groups. To obtain $\sigma$ as such, it is necessary to know the number p (generally <1) of specifically bound counter-ions per head group. If $p = 0$, then $\sigma = \sigma_0$.

The presence of charged groups in an interface may be expected to increase the surface pressure $\pi$. This effect has been demonstrated by Pal et al. [80], who compared $\pi$-A curves for sodium sebacate and neutral sebacic acid, both adsorbed at the heptane/water interface. At all values of A, $\pi$ was found to be greater for the adsorbed sebacate anion.

The Gibbs adsorption Eq. (75) has to be modified to take account of ionization. For the adsorption of fully ionized (p = 0) univalent surfactant molecules from water at, say, a water/hydrocarbon interface, Eq. (75) becomes

$$\Gamma_3^s \simeq \Gamma_{3,1}^\sigma = \frac{1}{2\,RT} \frac{d\gamma}{d(\ln a_3^\beta)} \tag{96}$$

where $a_3^\beta$ is the activity of the surfactant in aqueous solution.

However, if there is a relatively high concentration of surface-*inactive* electrolyte present in the aqueous phase (i.e., $c_e \gg c_3$), which is a common situation in practive, then the Gibbs adsorption equation reverts, for all practical purposes, to

$$\Gamma_3^s \simeq \Gamma_{3,1}^\sigma = \frac{1}{RT} \frac{d\gamma}{d(\ln c_3^\beta)} \tag{97}$$

i.e., Eq. (75).

With regard to the interfacial equation of state it can be shown, making certain simplifying assumptions, that for an ionized, nonlocalized monolayer (p = 0),

$$\pi = \frac{RT}{a - a^0} + \int_0^{\Gamma_3^s} \sigma_0 \, d\Psi_0 \tag{98}$$

$\Psi_0$ is the *surface* potential, i.e., the potential in the plane containing the charged head groups. Equation (98) is thus a modified Volmer equation of state where the integral term represents the increase in $\pi$ caused by the lateral coulombic repulsion between head groups. The integral term can be evaluated if a suitable relationship between $\sigma_0$ and $\Psi_0$ can be established. Normally this is complex, but it turns out that in the limiting case for *high* surface potentials (i.e., $\Psi_0 > 100$ mV) Eq. (98) reduces simply to

$$\pi = \frac{RT}{a - a^0} + \frac{2RT}{a} \tag{99}$$

Davies [81] has derived the following expression for $\pi$ for charged interfaces, based on the Gouy-Chapman relationship between $\sigma_0$ and $\Psi_0$,

$$\pi = \frac{RT}{a - a^0} + \frac{2RT}{F}(8c_e \varepsilon_r \varepsilon_0 \, kT)^{1/2} \left[\left(\frac{\sigma_0}{8c_e \varepsilon_r \varepsilon_0 \, kT} + 1\right)^{1/2} - 1\right] \tag{100}$$

The validity of this equation has been tested for spread monolayers (see Sec. II.F).

One problem with charged interfaces is that $a^0$ cannot be obtained directly from linearized plots of the equation of state as can be done with uncharged interfaces [Eq. (89)]; $a^0$ has to be obtained independently, e.g., by the use of molecular models. The adsorption isotherm analogous to Eq. (87) is

$$\Gamma_3^s = K_1\left(\frac{\theta}{1-\theta}\right)\exp\left(\frac{\theta}{1-\theta}\right)\exp\left(\frac{e\Psi_0}{kT}\right) \tag{100a}$$

Eq. (100a) indicates the way in which the electrical potential energy ($e\Psi_0$) of the adsorbed surfactant ions influences adsorption. Assuming that the bulk electrolyte concentration remains constant, then $\Psi_0$ increases as $\theta$ increases. Thus, $[\theta/(1-\theta)]\exp[\theta/(1-\theta)]$ increases less rapidly with $c_3$, i.e., adsorption is inhibited as a result of ionization.

## D. The Adsorption of Macromolecules

With regard to the adsorption of macromolecules at interfaces, more attention, both from the theoretical and experimental viewpoints, has been paid to adsorption at the solid/liquid interface than at the liquid/liquid interface. For a discussion of the former the reader is referred to the book by Lipatov and Seergreva [82] and the recent review by Vincent and Whittington [83].

At a solid/liquid interface a polymer is generally thought to adsorb in a loop-and-train type of conformation, with tails at each end of the molecule. Segments in trains are in contact with the solid surface, while those in loops or tails are immersed in the liquid phase. At a liquid/liquid interface segments of a given homopolymer chain will partition themselves between the two liquid phases, in a manner which reflects the two segment-solvent interactions parameters $\varepsilon_{13}$ and $\varepsilon_{23}$ (where 1 and 2 refer to the two liquids and 3 to the polymer segments). For a copolymer the number of such interaction energies to be considered is obviously greater. The time-average conformation adopted will be governed by the net energy-entropy balance for the polymer at the interface. A typical situation for an adsorbed homopolymer is illustrated in Fig. 16.

The interfacial region, thickness $\delta$, is the region depicted as s in Fig. 16. There are two major distinctions to be made in comparing the adsorption of macromolecules with that of small molecules (Sec. II.B). First, we have to distinguish between the adsorption of the *whole* molecule $\Gamma_p^s$ and that of segments $\Gamma_3^s$, etc.† Second, because in many cases the adsorption of macromolecules is

†We shall designate the whole polymer molecule as p. Homopolymer segments are designated as 3; for copolymers containing different segments 3, 4, etc., will be used.

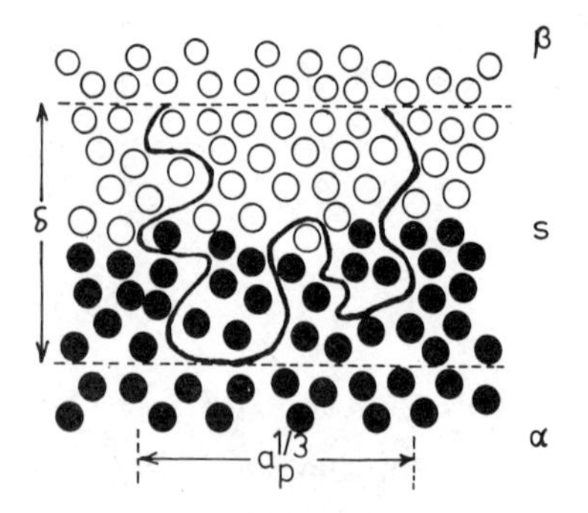


**Figure 16.** Typical conformation of a homopolymer chain adsorbed at a liquid (α)/liquid (β) interface.

irreversible (on a measurable time scale), we have to question the use of thermodynamic analysis, e.g., application of the Gibbs equation, Eq. (70).

$\Gamma_p^s$ is related to $c_p^\alpha$ and $c_p^\beta$ by an equation similar to Eq. (65), i.e.,

$$\Gamma_p^s = \frac{n_p^s}{A} = \frac{n_p - (c_p^\alpha V^\alpha + c_p^\beta V^\beta)}{A} \tag{101}$$

The effective area per polymer molecule $a_p$ is simply given by $a_p = 1/\Gamma_p^s$. Determination of $\Gamma_p^s$ as a function of $c_p^\alpha$ or $c_p^\beta$ is one of the major considerations in any study of polymer adsorption. Establishment of polymer adsorption isotherms is more complex for the liquid/liquid interface than the solid/liquid interface, however, because, in the first place, both the equilibrium values $c_p^\alpha$ and $c_p^\beta$ are required. In the second place, $A$ has to be known. In general, if a dispersed system (i.e., an emulsion) is being used in order to achieve a high value for $A$ (thereby increasing sensitivity), $A$ is determined after the emulsion has been formed in the presence of the polymer. This will mean that one can effectively only determine $\Gamma_p^s$ at the maximum coverage concentration. Establishment of the rising (low coverage) part of the isotherm has to be carried out using a macroscopic, planar liquid/liquid interface. Ellipsometry [84] is probably the only technique suitable. It also has the advantage that being an optical method, it is nonperturbing and, second, that the thickness of the interfacial region δ may be measured simultaneously. If has the disadvantage that a uniform average segment concentration $\bar{\rho}_3^s$ has to be assumed for the interfacial region, or some form for $c_3^s(z)$ has to be assumed. There are no experimental techniques available as yet for measuring $c_s(z)$, although neutron scattering from emulsion droplets, using the contrast matching technique [85,86] may offer one possibility.

$\Gamma_3^s$ is related to $c_3^s$ through Eq. (66). Similarly, one could define a value for the *excess* adsorption of segments, $\Gamma_3^\sigma$, i.e., Eq. (67), by choosing some arbitrary Gibbs dividing surface $z = \sigma$ (see Fig. 13), but this is of little practical use. $\Gamma_3^s$ and $\Gamma_p^s$ are, of course, simply related through

$$\Gamma_p^s = \frac{\Gamma_3^s}{x} \tag{102}$$

where x is the number of segments per chain. Implicit in Eq. (102), however, is the assumption that only *monolayer* adsorption of the polymer is occurring.

Knowledge of $\delta$ is perhaps the most important interfacial parameter to be established, particularly in regard to emulsion stability considerations (see Chap. 3). As we have seen, ellipsometry is a method for obtaining this value. A further general class of techniques are hydrodynamic methods, based on measuring some property (e.g., the sedimentation coefficient, the electrophoretic mobility, the specific viscosity, or the diffusion coefficient) of the emulsion droplets. These techniques have mainly been used with solid particles, where the hydrodynamic thickness of the adsorbed polymer layer is obtained from the difference in the property measured, with and without the polymer layer present [83]. For emulsion droplets, of course, it is not possible to measure the droplet size without the polymer layer being present. Some in situ method is required. Packer and Rees [87] have suggested a novel way of measuring the internal diameter of droplets using a magnetic field gradient, spin-echo technique. In essence, the internal diameter of the droplet is measured in terms of the self-diffusion coefficient of the molecules constituting the internal phase. Probably the most satisfactory technique for measuring the *external* diameter of the droplets is photon correlation spectroscopy [88,89]. This method is nonperturbing but does require reasonably monodisperse droplets. Essentially one measures the diffusion coefficient D of the droplets (including the adsorbed polymer layer), and this is related to the dimension of the droplet through the Stokes equation

$$D = \frac{kT}{6\pi\eta(a + \delta_h)} \tag{103}$$

where $\eta$ is the viscosity of the continuous phase, a is the internal radius of the droplet, $\delta_h$ is the hydrodynamic thickness of the adsorbed layer, and kT has its usual significance. As far as the authors are aware, no measurements, as yet, have been reported in the literature for the thickness of adsorbed polymer layers on emulsion droplets. An alternative way to proceed is to form a thin liquid film between two large droplets, each covered with the adsorbed polymer and immersed in the second fluid. Interferometic techniques may then be used to assess the thickness of the liquid film. As with ellipsometry, however, some model for the film (e.g., a three-layer model) would be required. Sonntag [90] has made measurements on this type of system.

One of the first systematic studies of the adsorption of a nonionic polymer at the liquid/liquid interface was that by Lankveld and Lyklema [91], who studied the adsorption of poly(vinyl alcohol) (PVA) at the liquid-paraffin/water interface. PVA is in fact a copolymer of vinyl alcohol and vinyl acetate and changes in the acetate content can lead to changes in conformation, since the water-acetate and water-alcohol interactions are different. Interfacial tensions were measured as a function of PVA concentration ($c_p$) in the aqueous phase and as a function of time, using a Wilhelmy plate technique [91]. Prolonged time dependencies of the interfacial tension were observed. At low polymer concentrations, the adsorption process seems to be diffusion controlled, i.e., by the rate of supply of polymer to the interface. At high polymer concentrations, the rate-determining step seems to be the rearrangement of polymer molecules at the interface to their steady-state conformation. There also seems to be a difference between PVA samples containing only 2% acetate groups and these containing more than 10% acetate groups. In the latter case the curves of $\gamma$ versus $c_p$ were convex with respect to the $c_p$ axis; in the former case, plots of $\gamma$ versus log $c_p$ consisted of two linear regions, in the lower $c_p$ region $\gamma$ being virtually independent of $c_p$. An explanation of these effects was offered in terms of the probable conformations adopted by the polymers at different coverage. No thermodynamic analysis was attempted for the reasons stated earlier concerning the irreversibility of polymer adsorption. In a second paper [92] Lankveld and Lyklema studied the properties of a PVA layer *spread* at the liquid-paraffin/water interface. Bohm and Lyklema [93] extended the studies to investigate the behavior of polyelectrolytes at the liquid-paraffin/water interface. Interfacial tensions were again measured as a function of polyelectrolyte type and concentration, pH, and time. Measurements of interfacial tension, unfortunately, do not seem to be very informative with regard to the structure of the interfacial layer. The adsorption of polyelectrolyte at the petroleum-ether/water interface has been studied by Gabrielli and Puggelli [94].

The role of adsorbed proteins at liquid/liquid interface is well established in biological systems, e.g., milk and blood. Phillips [95] has recently reviewed the conformation and properties of proteins at liquid interfaces.

### E. Adsorption Kinetics

According to Ward and Tordai [96] the rate of adsorption of molecules (species 3) from a bulk phase, for a purely diffusion-controlled process (i.e., no stirring and no energy barriers to surmount), is given by

$$\frac{d\Gamma_3^s}{dt} = \left(\frac{D_3}{\pi}\right)^{1/2} \frac{N_A}{1000} c_3 t^{-1/2} \tag{104}$$

where $D_3$ is the diffusion coefficient of species 3 in the bulk solution being considered, $c_3$ is the concentration in that solution (mol $dm^{-3}$), and $N_A$ is Avogadro's constant. Integration of Eq. (104) leads to

$$\Gamma_3^s = 2\left(\frac{D_3}{\pi}\right)^{1/2} \frac{N_A}{1000} c_3 t^{1/2} \tag{105}$$

No account is taken of desorption in Eqs. (104) and (105), and hence they only apply in the early stages of adsorption. Ward and Tordai [96] obtained the following equation when desorption was included in the analysis:

$$\Gamma_3^s = 2\left(\frac{D_3}{\pi}\right)^{1/2} \frac{N_A}{1000} \left[c_3 t^{1/2} - \int_0^{t^{1/2}} \phi(Z)\, d(t-Z)^{1/2}\right] \tag{106}$$

where $\phi(Z)$ represents the time variable concentration of species 3 in the solution immediately adjacent to the interfacial film, t is the time elapsed since adsorption began, and Z varies from 0 to t. Equation (106) predicts lower values of $\Gamma_3^s$ than does Eq. (105) and also predicts a limiting value of $\Gamma_3^s$ at infinite time. Considering the simpler Eq. (105) this suggests that a surfactant molecule having $D_3 \sim 6 \times 10^{-12}$ $cm^2$ $s^{-1}$, at a concentration $c_3$ of $1.5 \times 10^{-3}$ mol $dm^{-3}$, would reach a $\Gamma_3^s$ value of $2 \times 10^{14}$ mol $cm^{-2}$ in about 7 ms.

When concentration gradients are set up in the system, or when the system is stirred, then diffusion to the interface may be expressed in terms of Fick's law. In that case the following relations were derived [97,98]:

$$\frac{d\Gamma_3^s}{dt} = \frac{D_3}{\delta} \frac{N_A}{1000} c_3 \tag{107}$$

or

$$\frac{d\Gamma_3^s}{dt} = B_1 c_3 \tag{108}$$

where $B_1 = (D_3/\delta)(N_A/1000)$ and $\delta$ is the thickness of the diffusion layer. A correction term should be included in these equations to allow for the fraction $\theta$ of the surface already covered with adsorbed molecules, i.e., Eq. (108) should read

$$\frac{d\Gamma_3^s}{dt} = B_1 c_3 (1 - \theta) \tag{109}$$

Equation (108) is probably a reasonable approximation for $\theta < 0.3$.

Experiments on the rate of adsorption of surfactant molecules or macromolecules at the liquid/liquid interface are scarce. In the case of surfactant molecules, this is due to the experimental difficulties encountered in measuring the interfacial tension at short time intervals. Although, in principle, the oscillating-jet method (commonly used to study rate of adsorption at the air/water interface [97]) may be adapted to the oil/water interface; the technique is laborious. For that reason most of the information on the rate of adsorption at the liquid/liquid interface has been obtained using the drop-weight method. In principle, one may study the liquid/liquid interface for times as short of 0.1 s after their formation. Basically, the drop-weight method is modified such that drops are formed at various rates, instead of infinitely slowly. At appreciable drop formation rates, however, the drop weight is no longer a function only of interfacial tension, but depends on the circulation currents caused by the stream of falling drops. This leads to easier detachment of the drops, i.e., they are smaller than when formed slowly. It is possible to take account of this effect by using pure liquids (where the interfacial tension is independent of time) to establish calibration curves. One may then use the variation of drop weight in the presence of surfactants with time for measuring the rate of change of interfacial tension.

Using the above technique Davies et al. [99] measured the rate of adsorption of CTAB at the carbon tetrachloride/water interface. The results showed, as expected, that $d\Gamma_3^s/dt$ is proportional to $c_3$ in the early stages of adsorption, i.e., before $\theta$ and the desorption rate became appreciable. Thus, Eq. (108) is applicable, even at small times.

F. Spread Monolayers

All interfacial film between the bulk phases may be formed by "spreading" a surfactant or polymer solution directly at the interface. Curves of $\pi$ vs. A may then be established. Studies on the liquid/liquid interface are not as easy as with the liquid/air interface. There are a number of reasons for this. For example, difficulties are often encountered in the spreading of the film of ensuring uniform coverage of the available interface and with retention of the spreading solvent at the interface. Brooks and Pethica [100] designed a Langmuir trough for studying the oil/water interface and this was later improved by Taylor and Mingins [101]. In the latter case the trough barriers were carefully designed and a null-bouyancy Wilhelmy plate method was used to monitor the interfacial tension. This was linked to an electronic microbalance and recorder. Using this apparatus Taylor and Mingins [101] were able to make careful measurements of the surface tensions of water and heptane, and the water/heptane interfacial tension. Excellent agreement with current literature values was obtained.

An Agla micrometer syringe was used to deliver the spreading solution at the interface. For spreading solvents immiscible with the oil phase, the tip of the needle is placed in the oil phase, close to the interface, and droplets (~1 μℓ in volume) are touched into various parts of interface. With oil-miscible solvents the needle tip is held at the interface, raised to form a meniscus, and the solution very slowly expelled while preserving the meniscus. Using this technique Taylor and Mingins [101] studied monolayers of $C_{18}OSO_3^-$ ions spread from ethanol-water solutions. They found that the $\pi$-A curve obtained depends on the proportion of ethanol in the spreading solution. For example, the authors found that using 90:10 or 80:20 ethanol-water mixtures leads to larger $\pi$ values than those obtained using 70:30 mixtures. This may be attributed to the reluctance of the spreading solvent to leave the interface. However, that this is not the only problem has been demonstrated by Taylor and Mingins [101] using aqueous CTAB solutions. In that case equilibrium $\pi$ values were reached only after long times depending on the area per molecule i.e., 2 to 3 min at areas of 1000 to 4000 $Å^2$ $mol^{-1}$ and 10 to 15 mins at 10,000 to 20,000 $Å^2$ $mol^{-1}$. These long equilibration times were attributed to the slow diffusion of the molecules to the interface or aggregation of the surfactant species in solution. The $\pi$ values obtained were also lower than those previously obtained by Brooks and Pethica [100], who spread $C_{18}$TAB from 25:75 ethanol-water mixtures. This was attributed to the retantion of ethanol in the monolayers.

Taylor and Mingins [101] pointed out that when a layer spreads it must displace other species residing in the interface. These could be, for example, oriented solvent molecules in the interfacial region. For this reason, monolayers spread more easily at high A (low $\pi$), than at low A, but as shown by the authors, the spreading solvents are not easily removed from the dilute film. Some films can be spread quantitatively over a large range of A, with the spreading solvent readily leaving the surface. However, there are certain films which spread well at high A, yet do not spread quantitatively at low A, even though the spreading solvent disappears rapidly to give stable $\pi$ values. This was illustrated for films of L-$\alpha$-di-$C_{22}$ lecithin spread from heptane-ethanol mixtures. The monolayers spread satisfactorily at high A, giving stable and reproducible values of $\pi$. However, below a certain value of A, depending on the temperature, the concentration of the spreading solutions, the bore of the Agla needle used, and the rate of spreading, discordant results were obtained in that some of the film disappeared. Once the spreading solvent was removed, $\pi$ at these low A values became stable with time and successive compression-expansion cycles reproduced the same $\pi$-A curve.

The points raised above have been mentioned in some detail to illustrate the difficulties involved in establishing $\pi$-$A$ curves for spread monolayers. For this reason work [102–108] prior to that of Taylor and Mingins must be questioned.

Mingins et al. [109] have studied dilute charged monolayers and investigated the validity of the equation for $\pi$ given in Sec. II.C, Eq. (100). The presence of the charged groups leads to an increase in $\pi$ and many spread films which are solid or liquid condensed monolayers in the natural state become gaseous or vapor expanded when ionized. The surfactants studied by Mingins et al. [109] were $C_{18}TAB$ and $C_{18}OSO_3Na$ spread at the *n*-heptane/0.1 mol dm$^{-3}$ aqueous NaCl solution. The data of Brooks and Pethica [100] for $C_{18}TAB$ and of Robb and Alexander [110] for $C_{22}TAB$ were included for comparison purposes. The $\pi$-$A$ curves were of the expanded form and plots of $\pi A$ versus $A$ gave initial slopes which yield a mean value for $a^0$ in the range 0.2 to 0.35 nm$^2$. The $\pi$ values were also compared to theoretical values predicted by the Davies equation [Eq. (100)]. With the exception of $C_{18}OSO_3Na$ at 20°C and 0.1 mol dm$^{-3}$ NaCl and at $A$ values in excess of 15 nm$^2$ mol$^{-1}$, in all cases the experimental $\pi$ values were lower than those predicted by the Davies equation. The dependence of $\pi$ on electrolyte concentration is also less than that predicted by the Davies equation. These differences were attributed to an overestimate of the lateral electrical repulsion term in $\pi$ and possible neglect of a small attractive contribution to $\pi$ associated with the cohesion of the surfactant tails.

Mingins et al. [111] have estimated compression entropies for charged monolayers by measuring $\pi$-$A$ curves at two different temperatures (5 and 10°C). The entropy values obtained were negative and found to vary linearly with log $A$. The $C_{18}TAB$ and $C_{18}OSO_3Na$ systems behaved similarly in this respect. There was no detectable dependence on ionic strength over the electrolyte concentration range 0.001 to 0.1 mol dm$^{-3}$.

If the expanded monolayers are considered as sets of molecules having two-dimensional translational motion, then compression from $A_1$ to $A_2$ gives a change in translational entropy:

$$\Delta S_{trans} = R \ln \frac{A_2}{A_1} \tag{110}$$

However, entropies calculated from Eq. (110) are too small when compared with the experimental values. For that reason, Mingins et al. [111] introduced an excluded area term $a^0$ to allow for the hard sphere repulsions of the surfactant molecules. This gives an increase in the change in the translateral entropy described by

$$\Delta S_{trans}^{HS} = R \ln \left( \frac{A_2 - A^0}{A_1 - A^0} \right) \tag{111}$$

Using a reasonable value for $A^0$, namely, 0.3 $nm^2$, the divergence between theory and experiment decreased only marginally. Higher values of $A^0$ (1 and 1.8 $nm^2$) gave moderate agreement, but only over a limited range. Other contributions to the entropy change are required. One of these contributions considered by the authors was the entropy of compression of a smeared-out head-group charge and its associated diffuse double-layer counter-ion charge ($\Delta S_{DL}$). This term was calculated from established expressions for the Helmholtz free energy of the double layer. $\Delta S_{DL}$ was found to vary in nonlinear way with log A and leveled off at high A. In contrast to the experimental results, $\Delta S_{DL}$ depends on electrolyte concentration. Incorporating the $\Delta S_{DL}$ term into the translational entropy, and choosing $A^0 = 0.3\ nm^2$, led to values that were closer, but still lower than the experimentally derived values for $\Delta S$. The discrepancy could be due to change in chain conformation as the monolayers were compressed at the O/W interface.

## G. Interfacial Rheology

Two types of basic rheological measurements may be made with interfacial films: shear measurements and dilational (or compressional) measurements.

The interfacial shear viscosity is an important parameter in discussing the properties of a spread monolayer. A monolayer is resistant to shear stress in the plane of the surface (cf. retardation of liquid flow by viscous forces), and hence the interfacial viscosity of a monolayer can be measured (in two dimensions) from, for example, its drag on a ring or a needle located at the interface. The surface viscosity $\eta_s$ can be calculated from a knowledge of the tangential force per unit length and rate of shear, i.e.,

$$\eta_s = \frac{\text{tangential force per unit length}}{\text{rate of shear}} \tag{112}$$

$\eta_s$ is therefore expressed in units of $MT^{-1}$ (surface poise), whereas the bulk viscosity $\eta$ is expressed in units of $ML^{-1}T^{-1}$ (poise); thus, the surface viscosity is in effect the bulk viscosity multiplied by the thickness of the surface phase.

Most of the earlier measurements of interfacial viscosity were carried out using an oscillating needle located at the interface. However, this method suffers from the disadvantage of disturbing the interfacial film and also affects its behavior by creating interfacial gradients as a result of the movement of the needle. Davies and Meyers [112] described a sensitive interfacial viscometer in which rings of stainless steel wires (0.064 cm in diameter), forming concentric circles of diameters 12.5 cm and 11.6 cm, were placed at the interface. When the vessel containing the liquid phases is rotated, the canal formed by the space between these rings retards the motion of the interface; the retardation is greater the higher the viscosity of the monolayer. By carefully placing talc particles at the interface, it is possible to measure the movement of the film in the canal by accurately timing the rotation of talc particles moving in the center of the canal.

Many interfacial films, particularly those of the macromolecular type, show viscoelastic properties, in the sense that the film has a viscosity and an elastic modulus. The viscoelastic properties of surfactant films, particularly those of the mixed type, as well as those of macromolecules, play an important role in stabilizing emulsions. As pointed out by Biswas and Haydon [113], there are two principal ways in which monomolecular films may be deformed. The first involves shearing by a ring or a disk in the plane of the interface while the area of the film remains constant. The second may occur under the influence of forces causing dilation or compression of the film. The latter effect is likely to be of particular importance in controlling the stability of thin liquid films in emulsions.

Rheological measurements with condensed viscoelastic films are rendered difficult by the highly nonideal character of the film. However, Biswas and Haydon [113] described a rheometer, fitted with an electromagnetic driving technique for measuring the shear modulus G and viscosity $\eta$ of viscoelastic films of macromolecules. The apparatus is easy to operate and allows the application of a wide range of shearing stress and is operated without disturbing the film. Using this apparatus both creep and stress relaxation experiments were performed. In the former a constant stress $\tau$ (mN $m^{-1}$) is applied and the resulting deformation $\gamma$ (radians) is recorded as a function of time. In the latter experiments, a certain deformation $\gamma$ is produced in the film by applying an initial stress $\tau_0$, and the deformation is maintained constant by decreasing the stress. In this manner both G and $\eta$ can be estimated.

The two-dimensional compressibility of a monolayer for any area A, at the liquid/liquid interface is simply defined by

$$C_s = \frac{d(\ln A)}{d\pi} \tag{113}$$

$C_s$ can be directly calculated from the slope of the $\pi$-A curve. It is more convenient to use the reciprocal of $C_s$, namely, the interfacial dilational modulus $\varepsilon$ (Mn $m^{-1}$), in discussing the properties of interfacial films, i.e.,

$$\varepsilon = \frac{d\pi}{d(\ln A)} \tag{114}$$

In a sufficiently thick film ($b > 10^{-5}$ cm), where the interaction between film surfaces can be regarded as being negligible, the elasticity of the film is entirely due to its two surfaces. The Gibbs elasticity E or the modulus of elasticity of the film is simply twice the surface dilational modulus of each of the surfaces

$$E = 2\varepsilon = \frac{d\pi}{d(\ln A)} \tag{115}$$

Gibbs [114] and Prins and van den Tempel [115] suggested that the elasticity of a thin film is related to its stability. In a series of experiments using the air/water interface van den Tempel and co-workers [116,117] and Mysels and collaborators [118,119] have shown, both from theory and experiment, that the film elasticity strongly depends on the composition of the film. Clearly the dilational modulus, which is a measure of the compressional elasticity of the film [120,121] is of vital importance in determining the kinetic properties of monolayers. Moreover, as discussed in Chap. 3, the film elasticity of adsorbed macromolecular layers plays a vital role in the stabilization of emulsions against coalescence. Unfortunately, results of elastic moduli for surfactant films at the O/W interface are unavailable at present and measurements have only been made at the air/water interface using a Langmuir trough with two movable barriers [122]. Obviously measurements at the oil/water interface will suffer from the same difficulties described in Sec. II.F for the measurement of the $\pi$-A curves. Clearly, further studies of interfacial rheological properties, in the case of liquid/liquid films, are required.

### H. Interfacial Potentials

The interfacial potential which results from the spreading of a charged monolayer can be determined either by measuring a change in compensation potential [123,124] or by measuring the small current drawn through a very high resistance electrometer. When the oil phase is nonpolar, e.g., saturated hydrocarbon, no compensating double layer can be built up in the oil phase, owing to the very low solubility of ionic species in hydrocarbon solvents. The differences in the Galvani potential $\Delta\phi$ between the two phases can be identified with the interfacial potential $\Delta V$. However, since nonpolar oils do not conduct electricity, the interfacial potential cannot be measured directly. Thus, a capacity method [125,126] is generally used. In this method, alternating current is generated by small amplitude vibrations of a gold plate immersed in the oil; these are amplified and balanced to zero with a potentiometer connected to the electrode in water. The potentials of the gold plate and the aqueous surface are then equal. The amplitude of vibration of the gold plate is kept to about 0.01 cm, the frequency to few hundred hertz and the plate is kept at a distance greater than 0.5 mm from the interface in order to avoid disturbing the oil close to the monolayer at the interface. A description of the apparatus used has been given by Davies and Rideal [127]. In order to increase the sensitivity, the signal produced is amplified and then analyzed using an oscilloscope. Using the above apparatus, surface potentials of spread monolayers of $C_{18}TA^+$ ions were measured by Davies [128,129] at the petroleum-ether (high-boiling fraction)/1 mol $dm^{-3}$ aqueous NaCl solution interface. The value of $\Delta V$ obtained was in the region of 200 mV at an area of 92 $\AA^2$ per chain. These potentials were stable with time.

If the oil phase is more polar, e.g., benzene, such that the ionic species has only a small solubility in the oil phase, a spread monolayer tends to give a measured potential difference which decreases with time to zero. In that case, $\Delta\phi$ may be determined either by the vibrating plate method or by using a radioactive source attached to a stationary electrode suspended in the air above a thin oil layer [130]. For very polar oils, e.g., nitrobenzene, an ordinary dc circuit may be used. In this case only a transient potential change would be expected when a monolayer is spread. Using a carefully designed apparatus, Mingins et al. [131] obtained results for $\Delta V$ using a number of polar oils. These authors confirmed the previous findings that the potentials are established rapidly after spreading, and disappear when the film is swept off. The decay of these potentials, given complete spreading, is caused either by desorption of the monolayers (as confirmed by surface pressure measurements) or by electrical leakages. Thus, Mingins et al. [131] concluded that adsorption potentials may be measured at the polar-oil/water interface and are equivalent to those measured on films at the nonpolar-oil/water interface. They also found that the decay of $\Delta V$ for "stable monolayers" is extremely slow. They confirmed the thermodynamic argument of Dean et al. [132] to the effect that when a monolayer, insoluble in both bulk phases, is spread at a plane interface, the potential difference between the two phases, at equilibrium, is not altered if the passage of any ionic species across the interface is not restricted by the monolayer. Thus, $\Delta V$ would eventually decay to zero as a result of the diffusive redistribution of ions in the bulk phases adjacent to the interface. Moreover, Mingins et al. [131] found that $\Delta V$ depends on the nature of polar oil phase, suggesting that the double layers in the oil phase make significant contributions to the surface dipole strength.

It is important to consider the various contributions to $\Delta V$. For charged monolayers $\Delta V$ comprises the following contributions:

$$\Delta V = 4\pi n'\mu_1 + 4\pi n''\mu_2 + 4\pi n'''\mu_3 + \Psi_0 \tag{116}$$

where n refer to the number of dipoles per square centimeter, $\mu_1$ is the contribution of the dipole moment resulting from the reorientation of water dipoles as a result of the introduction of the film, $\mu_2$ is the dipole moment of the film-forming molecules, e.g., $=CH-NH_2$ in a long-chain amine, and $\mu_3$ is the moment of the bond, e.g., $\equiv C-H$ at the upper limit of the monolayer. $\Psi_0$ is the electrostatic contribution arising from the unequal distribution of ions in the vicinity of the adsorbed monolayer; it represents the electrostatic potential at the interface relative to the adjacent aqueous phase. Clearly for uncharged films, $\Psi_0 = 0$, but there is still a contribution to $\Delta V$ from oriented dipoles.

## Symbols

| | |
|---|---|
| a | activity |
| b | radius of a drop at its apex |
| c | concentration |
| d | diameter |
| e | electronic charge |
| g | acceleration due to gravity |
| h | Planck's constant; height of a drop |
| k | Boltzmann constant |
| l | length |
| n | refractive index |
| $n_i$ | number of moles of species i |
| r | radius |
| s | interfacial region |
| t | time |
| v | molar volume |
| w | weight |
| x | mole fraction |
| $x_e$ | equatorial radius |
| A | Helmholtz free energy |
| $A_{12}$ | Hamaker constant between a 1 molecule and a 2 molecule in contact |
| $\mathcal{A}$ | area |
| D | distance between planes; diffusion coefficient |
| E | Gibbs elasticity |
| F | force; Faraday constant |
| G | Gibbs free energy |
| M | molecular weight |
| N | number of molecules |
| P | pressure |
| R | radius; gas constant |
| S | entropy |
| T | absolute temperature (K) |
| U | internal energy |
| $U_{config}$ | sum of intermolecular pairwise potential energies |
| V | volume of the system |

| | |
|---|---|
| $\Delta V$ | interfacial potential |
| W | work; weight |
| X | number of segments per chain |
| Z | partition function; charge on head groups |
| $\alpha$ | bulk phase |
| $\beta$ | bulk phase |
| $\gamma$ | interfacial tension |
| $\Gamma$ | surface excess |
| $\delta$ | thickness of interfacial region; hydrodynamic thickness of adsorbed layer |
| $\varepsilon$ | interaction energy; dielectric constant; surface dilational modulus |
| $\eta$ | viscosity |
| $\theta$ | contact angle; coverage |
| $\kappa$ | Debye-Hückel parameter |
| $\mu$ | chemical potential; dipole moment |
| $\pi$ | surface pressure |
| $\rho(z)$ | density profile across the interfacial region |
| $\sigma$ | equilibrium separation distance between two molecules; net charge per unit area of interface |
| $\tau$ | reduced temperature |
| $\chi$ | potential drop across the interfacial region due to oriented permanent dipoles |
| $\psi$ | electric potential |
| $\omega$ | frequency; speed of rotation |
| $\phi(r)$ | intermolecular pair potential-energy function |

## References


1. N. K. Adam, *The Physics and Chemistry of Surfaces*, 3rd ed., University Press, Oxford, 1941, p. 5.
2. J. S. Rowlinson, *Chem. Soc. Rev.* *7*:329 (1978).
3. J. P. Hansen and I. R. McDonald, *Theory of Simple Liquids*, Academic Press, London, 1976.
4. J. A. Barker and D. Henderson, *Rev. Mod. Phys.* *48*:587 (1976).
5. B. Widom and J. S. Rowlinson, *J. Chem. Phys.* *52*:1670 (1970).
6. C. A. Leng, J. S. Rowlinson, and S. M. Thompson, *Proc. Roy Soc.* *A352*:1 (1976).
7. S. Toxvaerd, in *Chem. Soc. Spec. Periodical Rep.* (K. Singer, ed.) *2*:256 (1975).
8. J. C. Eriksson, *Adv. Chem. Phys.* *6*:145 (1964).
9. J. W. Gibbs, *Collected Works*, Longmans, New York, 1928, Vol. 1, p. 219.
10. B. Widom, *J. Chem. Phys.* *39*:2808 (1963).

11. A. Harasima, *Adv. Chem. Phys. 1*:203 (1958).
12. M. V. Berry, *Phys. Educ. 6*:79 (1971).
13. J. E. Kirkwood and F. P. Buff, *J. Chem. Phys. 17*:338 (1949).
14. M. V. Berry, R. F. Durrans, and R. Evans, *J. Phys. A5*:166 (1972).
15. A. C. Optiz, *Phys. Lett. A47*:439 (1974).
16. F. F. Abraham, D. E. Schreiber, and J. A. Barker, *J. Chem. Phys. 62*:1958 (1975).
17. G. A. Chapel, E. Saville, S. M. Thompson, and J. S. Rowlinson, *J. Chem. Soc. Faraday Trans. II 73*:1133 (1977).
18. I. Prigogine and L. Scrage, *J. Chem. Phys. 49*:399 (1952).
19. S. Ono and S. Kondo, in *Handbuch der Physik* (S. Flugge, ed.) Springer-Verlag, Berlin, 1960, Vol. 10, p. 134.
20. H. Eyring and M. S. Jhon, *Significant Liquid Structures*, Wiley, New York, 1969, p. 28.
21. F. C. Goodrich, in *Surface and Colloid Science* (E. Matijević, ed.), Wiley-Interscience, New York, *3*, p. 1.
22. J. H. de Boer, *Trans. Faraday Soc. 32*:17 (1936).
23. H. C. Hamaker, *Physika 4*:1058 (1937).
24. *International Critical Tables*, McGraw-Hill, New York, 1928, Vol. 3, p. 20.
25. Chemical Rubber Company, *Handbook of Chemistry and Physics 53*:134, 1973.
26. A. Eisenstein and N. S. Gingrich, *Phys. Rev. 58*:307 (1940); *62*:261 (1940).
27. J. F. Paddy and U. B. Uffindell, *J. Phys. Chem. 72*:1407 (1968).
28. E. A. Moelwyn-Hughes, *Physical Chemistry*, 2d ed., Pergamon, London, 1962, p. 392.
29. J. N. Israelachvili, *Proc. Roy. Soc. A331*:39 (1972).
30. F. M. Fowkes, *J. Phys. Chem. 67*:2538 (1963).
31. L. A. Girafalco and R. J. Good, *J. Phys. Chem. 61*:904 (1957).
32. R. E. Johnson and R. H. Detre, *J. Colloid Interface Sci. 21*:610 (1966).
33. W. R. Gillap, N. D. Weiner, and M. Gibaldi, *J. Amer. Oil Chem. Soc. 35*:114 (1967).
34. D. K. Owens, *J. Phys. Chem. 74*:3305 (1970).
35. R. Aveyard, *J. Colloid Interface Sci. 52*:631 (1975).
36. F. M. Fowkes, *J. Phys. Chem. 72*:3700 (1968).
37. J. F. Padday and N. D. Uffindell, *J. Phys. Chem. 72*:1407 (1968).
38. F. M. Fowkes, *J. Colloid Interface Sci. 28*:493 (1968).
39. E. M. Lifshitz, *Zhur. Eksp. Teor. Fiz. 29*:94 (1955).
40. R. S. Winterton, *Contemp. Phys. 11*:559 (1970).
41. J. N. Israelachvili, *Contemp. Phys. 15*:159 (1974).
42. J. N. Israelachvili and D. Tabor, *Prog. Surface Membrane Sci. 7*:1 (1973).
43. V. A. Parsegian, in *Physical Chemistry: Enriching Topics from Surface and Colloid Science* (H. van Olphen and K. J. Mysels, eds.), Theorex, La Jolla, Calif., 1975, Chap. 4.
44. R. Aveyard and S. M. Saleem, *J. Chem. Soc. Faraday Trans. I 72*:1609 (1976).
45. R. C. Brown, *Contemp. Phys. 15*:301 (1924).
46. R. Aveyard and D. Haydon, *An Introduction to the Principles of Surface Chemistry*, University Press, Cambridge, 1973, p. 7.
47. E. A. Guggenheim, *Thermodynamics*, North-Holland, Amsterdam, 5th ed., 1967, p. 45.

48. R. Defay, I. Prigogine, A. Bellamans, and D. H. Everett, *Surface Tension and Adsorption*, Longmans, London, 1966, p. 256.
49. A. W. Adamson, *Physical Chemistry of Surfaces,* 3d ed., Wiley-Interscience, New York, 1976.
50. J. F. Padday, in *Surface and Colloid Science* (E. Matijević, ed.), Wiley-Interscience, New York, 1969, Vol. 1, p. 39.
51. A. E. Alexander and J. B. Hayes, in *Techniques of Chemistry* (A. Weissberger and B. W. Rossitor, eds.), Wiley-Interscience, New York, 1971, Vol. 1, Part V, Chap. IX.
52. P. L. du Noüy, *J. Gen. Physiol. 1*:521 (1919).
53. W. D. Harkins and H. F. Jordan, *J. Amer. Chem. Soc. 52*:1715 (1930).
54. H. W. Fox and C. H. Chrisman, Jr., *J. Phys. Chem. 56*:784 (1952).
55. B. B. Freud and H. Z. Freud, *J. Amer. Chem. Soc. 52*:1772 (1930).
56. W. D. Harkins and F. E. Brown, *J. Amer. Chem. Soc. 41*:499 (1919).
57. J. L. Lando and H. T. Oakley, *J. Colloid Interface Sci. 25*:526 (1967).
58. M. C. Wilkinson and R. L. Kidwell, *J. Colloid Interface Sci. 35*:114 (1971).
59. L. Wilhelmy, *Ann. Phys. 119*:177 (1863).
60. F. Bashforth and J. C. Adams, *An Attempt to Test the Theories of Capillary Action*, University Press, Cambridge, 1883.
61. J. F. Padday, *Phil Trans. Roy. Sci. (London) A269*:265 (1971).
62. J. M. Andreas, E. A. Hauser, and W. B. Tucker, *J. Phys. Chem. 42*:1001 (1938).
63. D. O. Neiderhauser and F. E. Bartell, *Report of Progress—Fundamental Research on the Occurrence of Petroleum,* Publication of the American Petroleum Institute, Lord Baltimore Press, Balitmore, Md., 1950, p. 114.
64. P. F. Levin, E. Pitts, and G. C. Terry, *J. Chem. Soc. Faraday Trans. I 72*:1519 (1976).
65. T. Lohnstein, *Ann. Physik 20*:237 (1906).
66. B. Vonnegut, *Rev. Sci. Instrum. 13*:6 (1942).
67. J. L. Cayias, R. S. Schecter, and W. H. Wade, *Adsorption at Interfaces* (K. L. Mittal, ed.), ACS Symposium Series No. 8, American Chemical Society, Washington, 1975, p. 234; L. Cash, J. L. Cayias, G. Fournier, D. MacAllister, T. Schares, R. S. Schechter, and W. H. Wade, *J. Colloid Interface Sci. 59*:39 (1977).
68. Reference 46, p. 78.
69. Reference 48, p. 166.
70. E. A. Guggenheim, *Trans. Faraday Soc. 41*:50 (1945).
71. J. W. Belton and M. E. Evans, *Trans. Faraday Soc. 41*:1 (1945).
72. I. Prigogine and J. Marechal, *J. Colloid Sci. 7*:122 (1952).
73. Reference 46, p. 17.
74. T. L. Giu, *Adv. Catalysis 4*:211 (1952).
75. M. Volmer, *Z. Phys. Chem. 115*:253 (1925).
76. R. Aveyard and B. J. Briscoe, *Trans. Faraday Soc. 66*:2911 (1970).
77. S. Ross and E. S. Chen, *Ind. Eng. Chem. 57(7)*:40 (1965).
78. R. Aveyard and B. Vincent, *Progr. Surface Sci. 8*:59 (1977).
79. D. W. J. Osmond, *Disc. Faraday Soc. 42*:247 (1966).
80. R. P. Pal, A. K. Chaterjee, and D. K. Cattoraj, *J. Colloid Interface Sci. 52*:46 (1975).
81. Reference 46, p. 99.
82. Yu. S. Lipatov and L. M. Seergeeva, *Adsorption of Polymers*, Wiley, New York, 1974.

83. B. Vincent and S. Whittington, *Surface and Colloid Science* (E. Matijević, ed.), Plenum, New York *12*:1 (1982).
84. Reference 82, p. 6.
85. D. Cebula, R. K. Thomas, N. M. Harris, J. Tabony, and J. W. White, *Faraday Disc. Chem. Soc. 65*:76 (1978).
86. K. Barnett, T. Cosgrove, B. Vincent, T. L. Crowley, and T. F. Tadros, *The Effect of Polymers on Dispersion Properties* (Th. F. Tadros, ed.), Academic Press, London, 183 (1982).
87. K. J. Packer and C. Rees, *J. Colloid Interface Sci. 40*:206 (1972).
88. H. C. Cummins and E. R. Pike (eds.), *Photon Correlation and Light Beating Spectroscopy*, Plenum, New York, 1974.
89. M. J. Garvey, Th. F. Tadros, and B. Vincent, *J. Colloid Interface Sci. 55*:440 (1976).
90. H. Sonntag, *Colloid Polymer Sci. 257*:286 (1979).
91. J. M. G. Lankveld and J. Lyklema, *J. Colloid Interface Sci. 41*:454 (1972).
92. J. M. G. Lankveld and J. Lyklema, *J. Colloid Interface Sci. 41*:466 (1972).
93. J. Th. C. Bohm and J. Lyklema, *J. Colloid Interface Sci. 50*:559 (1974).
94. G. Gabrielli and M. Puggelli, *J. Colloid Interface Sci. 35*:460 (1971); *37*:503 (1971); *45*:217 (1973).
95. M. C. Phillips, *Chem. Ind. 1977*:170.
96. A. F. H. Ward and L. Tordai, *J. Chem. Phys. 14*:453 (1946).
97. Lord Rayleigh, *Proc. Roy. Soc. 29*:71 (1879).
98. F. H. Garner, P. Mina, and V. G. Jenson, *Trans. Faraday Soc. 55*:1607, 1616, 1627 (1959).
99. J. T. Davies, J. A. Collis Smith, and D. G. Humphreys, *Proc. Int. Conf. Surface Activity*, Butterworths, London, 1957, Vol. 1, p. 281.
100. J. H. Brooks and B. A. Pethica, *Trans. Faraday Soc. 60*:208 (1964).
101. J. A. G. Taylor and J. Mingins, *J. Chem. Soc. Faraday Trans. I 71*:1161 (1975).
102. A. E. Alexander and T. Teorell, *Trans. Faraday Soc. 35*:729 (1939).
103. F. A. Askew and J. F. Danielli, *Trans. Faraday Soc. 36*:785 (1940).
104. J. T. Davies, *Proc. Roy. Soc. A208*:224 (1951).
105. J. N. Philips and E. K. Rideal, *Proc. Roy. Soc. A232*:149 (1955).
106. J. H. Brooks and F. MacRitchie, *J. Colloid Sci. 16*:442 (1961).
107. T. G. Jones, D. A. Walker and B. A. Pethica, *J. Colloid Sci. 18*:485 (1963).
108. L. Blight, C. W. N. Cumper and V. Kyte, *J. Colloid Sci. 20*:393 (1961).
109. J. Mingins, J. A. G. Taylor, N. F. Owens, and J. H. Brooks, *Adv. Chem. Sci. 144*:28 (1975).
110. I. D. Robb and A. E. Alexander, *J. Colloid Interface Sci. 1*:28 (1968).
111. J. Mingins, N. F. Owens, J. A. G. Taylor, J. H. Brooks, and B. A. Pethica, *Adv. Chem. Ser. 144*:14 (1975).
112. J. T. Davies and G. R. A. Mayers, *Trans. Farad. Soc. 54*:69 (1960).
113. B. Biswas and D. A. Haydon, *Proc. Roy. Soc. A271*:296 (1963).
114. J. W. Gibbs, *Collected Works*, Vol. I., Dover, New York, 1961, p. 301.
115. A. Prins and M. van den Tempel, *Proc. 4th Int. Congr. Surface Active Substances, II*:1119 (1964).
116. M. van den Tempel, J. Lucassen, and E. H. Lucassen-Reynders, *J. Phys. Chem. 69*:1798 (1965).
117. A. Prins, C. Arcuri, and M. van den Tempel, *J. Colloid Interface Sci. 24*:84 (1967).

118. K. J. Mysels, M. C. Cox, and J. D. Skewis, *J. Phys. Chem.*, *65*:1107 (1961).
119. R. I. Razouk and K. J. Mysels, *J. Amer. Oil Chem. Soc.* *45*:381 (1968).
120. N. W. Tschoegl, *J. Colloid Sci.* *13*:500 (1958).
121. A. Cheesman and A. Sten Kundsen, *Biochem. Biophys. Acta* *33*:158 (1959).
122. J. Lucassen and D. Giles, *J. Chem. Soc. Faraday I* *71*:217 (1975).
123. E. Lange, *Z. Elektrochem.* *55*:76 (1951).
124. R. Parsons, *Modern Aspects of Electrochemistry* (J. O'M. Bockris and B. E. Conway, eds.), Butterworths, London, 1954.
125. J. T. Davies, *J. Electrochem.* *55*:559 (1951); *Nature* *167*:193 (1951).
126. C. D. Kinloch and A. I. McMullen, *J. Sci. Instrum.* *36*:347 (1959).
127. J. T. Davies and E. K. Rideal, *Interfacial Phenomena*, Academic, London, 1963.
128. J. T. Davies, *Nature L* *167*:193 (1951); *Biochem Biophys. Acta* *11*:165 (1953); *Biochem. J.* *56*:509 (1954).
129. J. T. Davies, *Proc. Roy. Soc. A* *208*:224 (1951).
130. J. T. Davies, *Trans. Faraday Soc.* *49*:683, 949 (1953).
131. J. Mingins, F. G. R. Zobel, B. A. Pethica, and C. Smart, *Proc. Roy. Soc. A* *324*:99 (1971).
132. R. B. Dean, O. Gatty, and E. K. Rideal, *Trans. Faraday Soc.* *36*:161 (1940).

# 2

# Formation of Emulsions

PIETER WALSTRA / Agricultural University, Wageningen, the Netherlands

## I. Introduction

### A. Processes during Emulsification

There are many ways to produce an emulsion from two liquid phases that are mutually not or only slightly soluble. It is usually achieved by applying mechanical energy. First, the interface between the two phases is deformed to such an extent that droplets form. These droplets are mostly far too large, and they are subsequently broken up or disrupted into smaller ones. Hence, the disruption of droplets is a critical step in emulsification.

Particularly if the volume fraction $\phi$ of the future disperse phase is high, the formation and disruption of droplets may be considered to be at least initiated by making films of the continuous phase, sandwiched between the disperse phase. But very little study has been made of emulsification from this point of view.

In any case, droplets must be deformed to achieve their disruption. This deformation is opposed by the Laplace pressure. The pressure at the concave side of a curved interface with interfacial tension $\gamma$ is higher than that at the convex side by an amount

$$\Delta p = \gamma \left( \frac{1}{R_1} + \frac{1}{R_2} \right) = \frac{2\gamma}{R} \tag{1}$$

where $R_1$ and $R_2$ are the principal radii of curvature (cf. Chap 1, Sec. I.E). For a spherical droplet of radius r this becomes $2\gamma/r$, and any deformation of the droplet leads to an increase in $\Delta p$. To disrupt the droplet, such a pressure must be externally applied over a distance r, which means a pressure gradient of the order of $2\gamma/r^2$. The droplets can also be deformed by viscous forces exerted by the surrounding liquid, and the viscous stress $G\eta_C$ should overcome, i.e., should be of the same magnitude as, the Laplace pressure; G is the velocity gradient (or rate of strain), and $\eta$ is the viscosity. The velocity gradients or pressure gradients needed are mostly supplied by agitation. The smaller the droplets, the more intense the agitation should be to disrupt them. Hence, liquid motion during emulsification is usually turbulent, unless $\eta_C$ is high.

A suitable surfactant is needed to produce an emulsion, hence the name emulsifier. The surfactant lowers $\gamma$ (e.g., from 40 to 5 mN $m^{-1}$), and thereby the Laplace pressure, which facilitates deformation, hence disruption. Presence of a surfactant is also prerequisite for making a film of continuous phase. The role of the surfactant is manifold and complicated, but essential. It depends on the nature and the concentration of the surfactant. To be active, it must be transported to the interface where it is adsorbed and forms a surface layer. This transport is greatly speeded up by the intense agitation. Because a large interface is formed, the bulk phase containing the surfactant becomes depleted during emulsification.

Droplets may coalesce again after being formed, and this may be another critical step. Unfortunately, coalescence mechanisms are only partly understood, and certainly in the complicated situation during emulsification, we cannot predict whether and how fast coalescence will happen. The surfactant is of paramount importance. It largely determines which phase is going to be the continuous one: it is commonly the one in which the surfactant is soluble (Bancroft's rule). The explanation is presumably that droplets of the phase in which the surfactant is soluble are very unstable to coalescence or that a film devoid of surfactant cannot be made. The coalescence rate of droplets of the disperse phase is very variable and depends on many factors, but mainly on the nature and the concentration of the surfactant.

Droplets may also flocculate during emulsification, but the floccules are mostly disrupted again in a very short time.

The processes mentioned occur simultaneously, and their rates depend on several factors, each in a different way. Moreover, they influence each other. See Fig. 1.

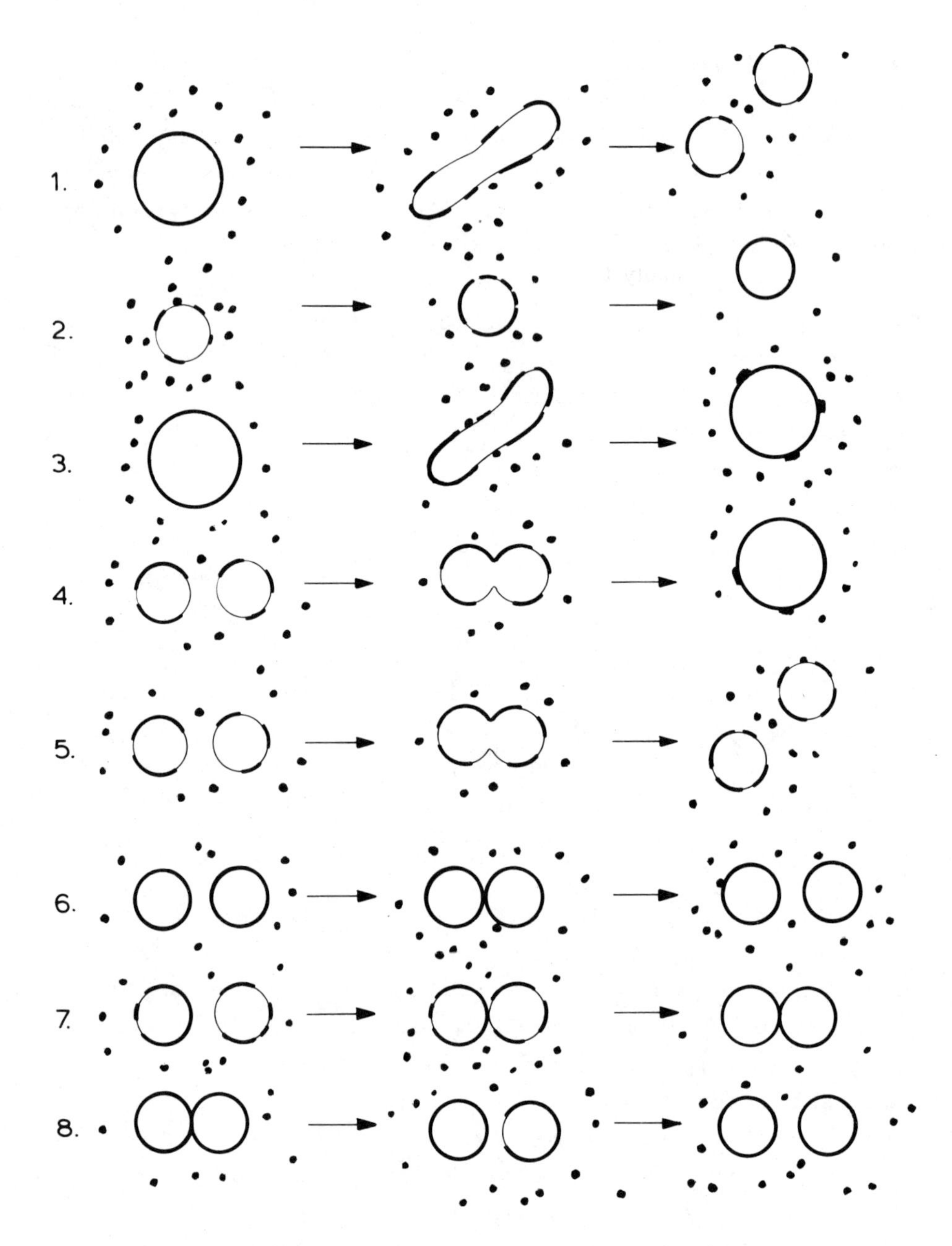


Figure 1. Processes taking place during the later stages of emulsification. Surfactant material is depicted by heavy lines and dots. Not to scale and highly schematic!

B. Energy Relations

During emulsification, the interfacial area A becomes much larger, and the surface free energy of the system is increased by an amount $\gamma \, \Delta A$. Assuming that oil droplets with a radius $r = 1\ \mu m$ are formed in water, that $\phi = 0.1$ and that $\gamma = 0.01\ N\ m^{-1}$, the surface free energy amounts to $3\ kJ\ m^{-3}$. This is a very small quantity of energy; it would be insufficient to heat the emulsion by $10^{-3}$ K. But the energy actually needed to produce the emulsion would be at least $3\ MJ\ m^{-3}$, hence 1000 times as much.

The large excess of energy needed is predominantly due to the high Laplace pressure gradient: some $2 \times 10^{10}\ Pa\ m^{-1}$ would be needed in the given case; or a velocity gradient of some $10^7\ s^{-1}$, assuming $\eta_C = 1$ mPa s. Such conditions can only be produced by very intense agitation, which, in turn, costs much energy. Except for the minute fraction that is needed for the surface free energy, this energy is dissipated into heat.

The intensity of agitation is largely governed by the energy density $\varepsilon$ in the liquid, i.e., the amount of mechanical energy dissipated per unit volume and unit time (or the net power input per unit volume). Hence, the effectiveness of emulsification (for instance, measured by the specific surface area of the disperse phase $A_D$ obtained) strongly depends on $\varepsilon$. In our example, $\varepsilon$ should be of the order of $10^{10}\ W\ m^{-3}$, that is, sufficient to heat the emulsion to the boiling point in 0.03 s. It will be clear that such energy densities can occur only locally and for very short times. To achieve efficient emulsification it is therefore advantageous to dissipate the available energy in a very short time, since this gives a high $\varepsilon$.

Adding a suitable surfactant may considerably reduce the agitation energy needed to obtain a certain droplet size. A reduction by a factor 10, or even more, can in some cases be obtained. Very roughly one may state that (1) with a higher surfactant concentration, less energy is needed, and (2) at low $\varepsilon$, type and concentration of surfactant very much affect droplet size, but for high $\varepsilon$ these effects are smaller. There are, however, exceptions to these rules.

C. Time Scales

The various processes taking place during emulsification each have their own rates or time scales. Mostly, they are difficult to predict since conditions are not precisely known or the theory is not fully developed. However, we can estimate the order of magnitude, and the examples given below are just that.

The relaxation time for a small deformation of a droplet, assuming constant $\gamma$ and no inertial effects is given by [1,2]

$$\tau_{def} \approx \frac{\eta_D r}{\gamma} \tag{2}$$

In addition to this, $\tau_{def}$ can never be smaller than $G^{-1}$ for deformation in laminar flow. A more rigorous treatment is given in Refs. 3 and 4. The time needed for disruption is presumably a few times $\tau_{def}$.

The transport of surfactant molecules to the newly created interface where they are adsorbed is largely determined by convection, rather than diffusion. Assuming a (rather high) surface excess $\Gamma$ of $10^{-5}$ mol m$^{-2}$, we obtain for convection in laminar flow toward a small droplet of radius r [1]:

$$\tau_{ads} \approx \frac{3\pi\Gamma}{rm_C G} \approx (10^4 rm_C G)^{-1} \text{ (s)} \tag{3a}$$

and in isotropic turbulence [1,5]:

$$\tau_{ads} \approx \frac{5\Gamma \eta_C^{1/2}}{rm_C \varepsilon^{1/2}} \approx (10^4 rm_C)^{-1} \left(\frac{\varepsilon}{\eta_C}\right)^{-1/2} \text{ (s)} \tag{3b}$$

Here $m_C$ is the concentration of the surfactant solution, usually between 1 and 100 mol m$^{-3}$. For monolayer coverage $m_C = 3\Gamma\phi/r(1-\phi)$ would be needed, assuming complete adsorption.

Results for these characteristic times are given in Table 1 for a fairly viscous emulsion in laminar flow and for an emulsion of low viscosity under isotropic turbulence. It is seen that in the beginning of the process (large droplets) adsorption of surfactant to a new interface can keep up with its creation, but that this is not necessarily true when droplets become smaller. Moreover, the bulk phase becomes depleted of surfactant, further slowing down the adsorption. Spreading of surfactant, i.e., lateral transport in the interface, will occur during the enlargement of the interface caused by deformation. Assuming the spreading rate to be 0.1 m s$^{-1}$, this would give a $\tau_{spreading} \approx 10r$, which is often comparable in magnitude to $\tau_{def}$. Anyhow, during the final stages of droplet breakup, surface concentration of surfactant will usually be below, and sometimes far below, equilibrium.

Coalescence rates cannot be predicted. Collision rate theory [1,5] leads for low $\phi$ to an average time between collisions of a droplet; in the case of laminar flow,

$$\tau_{col} = \frac{\pi}{8\phi G} \tag{4a}$$

and in the case of isotropic turbulence,

$$\tau_{col} \approx \frac{r^{2/3} \rho_C^{1/3}}{10\,\phi\,\varepsilon^{1/3}} \tag{4b}$$

**Table 1.** Approximate Time Scales of Processes during Emulsification

| Condition[a] | Laminar flow | | Turbulence | |
|---|---|---|---|---|
| $G$ ($s^{-1}$) | $10^4$ | $10^5$ | — | — |
| $\varepsilon$ ($W\ m^{-3}$) | — | — | $3 \times 10^7$ | $10^{10}$ |
| $\eta_C$ (mPa s) | 100 | 100 | 1 | 1 |
| $\eta_D$ (mPa s) | 1 | 1 | 100 | 100 |
| $r_{cr}$ (μm) | 10 | 1 | 10 | 1 |
| $\tau_{def}$ (ms) for r = 1 mm | (0.1) | 0.1 | 10 | 10 |
| $\tau_{def}$ (ms) for $r = r_{cr}$ | (0.1) | (0.01) | 0.1 | 0.01 |
| $\tau_{ads}$ (ms) for r = 1 mm | $10^{-3}$ | $10^{-4}$ | 0.5[b] | 0.05[b] |
| $\tau_{ads}$ (ms) for $r = r_{cr}$ | 0.1 | 0.1 | 6 | 3 |
| $\tau_{col}$ (ms) | 0.4 | 0.04 | 0.01[c] | 0.0005[c] |

[a]Other conditions: $\gamma = 10\ mN\ m^{-1}$, $\rho_D = \rho_C = 10^3\ kg\ m^{-3}$, $\phi = 0.1$, $m = 10\ mol\ m^{-3}$.

[b]Rough estimate.

[c]For $r = r_{cr}$.

Examples are given in Table 1, and it is seen that the collision frequency is very high. But *collision* is in fact a wrong term: the droplets only meet each other. Normally, the viscosity of the liquid between them and mutual repulsion prevent actual contact. Other mechanisms will often prevent rupture of the film of continuous phase between droplets, hence coalescence. Even if coalescence sets in, it takes a time comparable to $\tau_{def}$ for completion, and before that time the "aggregate" may have been disrupted again. All these factors depend on many conditions and mostly in un unknown way.

## D. Aspects Covered

It follows from the aspects mentioned that emulsion formation involves several process steps. Some occur simultaneously, others must take place in succession; moreover, they influence each other. Hence, emulsification is a complicated process, and explanation of all the effects in one all-embracing theory is clearly impossible. We will therefore try to derive a few general rules from theories covering certain aspects or branches.

The formation of droplets will be discussed, though it is poorly understood. Disruption of droplets under different conditions has been studied extensively, and we can give some semiquantitative relations. The role of the surfactant is essential, but difficult to explain, partly because the coalescence mechanism remains largely obscure. Still we will try to give at least a few rules of thumb for practical operation. The same holds for emulsification rate.

If the "surfactants" used adsorb irreversibly (e.g., polymers, solid particles), the method of emulsification may affect the composition of the surface layers. This will be treated briefly since it may, in turn, affect many properties of the emulsion.

Not treated are technical details of apparatus or preparation methods, unusual ways of producing emulsions, properties of emulsifiers, spontaneous emulsification, microemulsions, multiple emulsions, and specific problems when making "high internal-phase emulsions."

An earlier treatment of roughly the same aspects as covered here is by Gopal [6], to whom we will often refer. Many fundamental points are treated by Levich [1]. Mason and co-workers have added important contributions to our knowledge of droplet break up, e.g., in Refs. 2, 35, and 43. Van den Tempel and his school have particularly contributed to our understanding of dynamic surface properties, e.g., see Refs. 23, 26, 89, and 96. Much practical information is in Becher's book [7].

Some of the author's results that will be used have not been published before, or only in a condensed form [8]; the latter reference also treats some aspects on a more applied level.

## II. Ways to Produce Emulsions

### A. Apparatus

Many devices have been designed to produce emulsions [7]. A summary is given in Table 2; it is not complete and several intermediate types and variants exist (see also Fig. 2). We will briefly discuss the methods mentioned in the table.

1. Is self-evident.
2. This may include constrictions of the flow and baffles on which the liquid can impinge, to increase velocity gradient or turbulence.
3. The disperse phase is injected into the continuous one as a cylindrical jet, where it is broken up into fairly large droplets. It can be combined with a variant of method 2 to give a "static mixer."
4. Many variants exist, for instance, pumps; revolution rates are up to 300 $s^{-1}$. In particular, rotor-stator machines exist in great variety

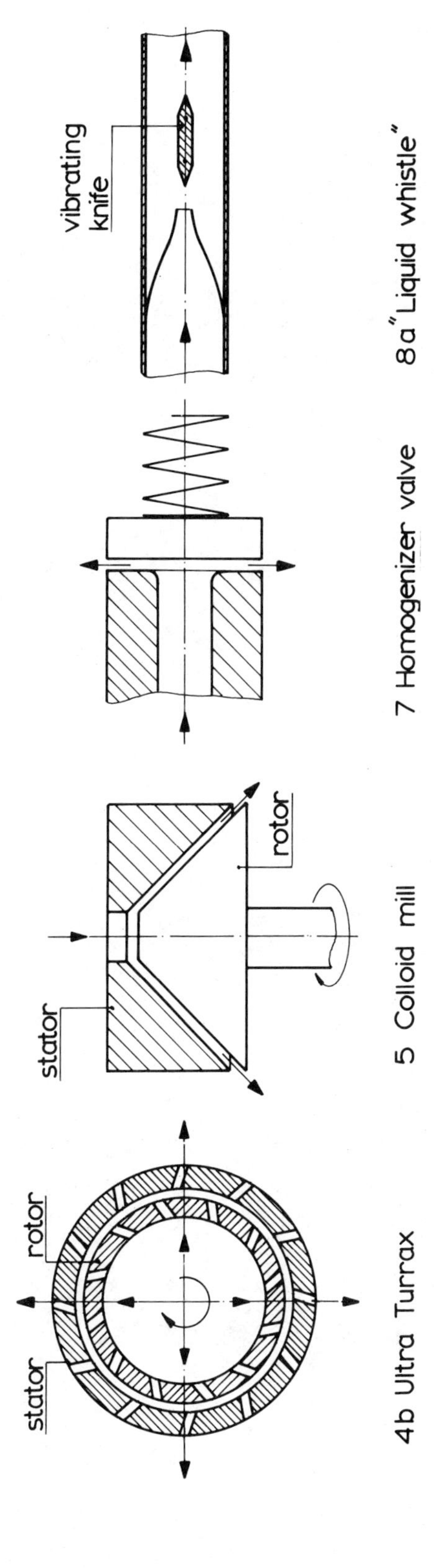


Figure 2. Principle of some emulsifying machines. The numbers refer to Table 2. The slit width in the homogenizer valve is very much exaggerated. Arrows indicate the directions of flow and rotation.

Table 2. Methods and Apparatus to Produce Emulsions

| Method | Related to method | Drop formation | Drops mainly disrupted by[a] | Energy density[b] | Mode of operation[c] | Restrictions[d] |
|---|---|---|---|---|---|---|
| 1. Shaking | 4a | + | (T) | L | B | N |
| 2. Pipe flow | | | | | | |
| a. Laminar | 5 | (+) | V | L – M | C | V |
| b. Turbulent | 4a | + | T | L – M | C | N |
| 3. Injection | 10a | + | — | L | C | |
| 4. Stirring | | | | | | |
| a. Simple stirrer | 1, 2b | + | T, V | L | B, C | |
| b. Rotor-stator | (5) | + | T, V | M – H | B, C | |
| c. Scraper | 5 | + | V | L – M | B, C | V |
| d. Vibrator | 8a | + | ? | L | B, C | N |
| 5. Colloid mill | 2a, 4c, 6 | (+) | V | M – H | C | V |
| 6. Ball and roller mills | 5 | + | V | M | B (C) | V |
| 7. High-press. homogenizer | (2b) | — | T, C, V | H | C | N |
| 8. Ultrasonic | | | | | | |
| a. Vibrating knife | 4d | + | C, T | M – H | C | W |
| b. Magneto-striction | | + | C | M – H | B, C | W |
| 9. Electrical | 10b | — | Electric charge | M | B (C) | Several |
| 10. Aerosol to liquid | | | | | | |
| a. Mechanical | 3 | + | — | L – M | B, C | |
| b. Electrical | 9 | + | — | M | B, C | Several |
| 11. Foaming or boiling | | (+) | Spreading | L – M | | (W) |
| 12. Condensation | | + | — | | | Several |

[a] V = viscous forces in laminar flow, T = turbulence, C = cavitation.
[b] L = low, M = moderate, and H = high.
[c] B = batch and C = continuous.
[d] The continuous phase should be V = viscous, N = not too viscous, W = aqueous.

and are described in numerous patents, often under the name *homogenizer*. In the laboratory the "Ultra Turrax" is often used; a detailed description is in Ref. 9.

5. A colloid mill is also a rotor-stator device, but with a very narrow slit (e.g., 0.1 mm) and designed to achieve very high (simple) shear (up to $10^7$ $s^{-1}$). Variants with corrugated rotor and stator surfaces exist, but these are more like 4b, since turbulence occurs.
6. More suited for disruption of solid particles or very viscous emulsion droplets.
7. In the homogenizer the liquid is brought under a high pressure $p_h$ (e.g., 10 to 40 MPa) by a positive pump and is forced through a narrow (e.g., 0.1 mm) valve slit; owing to the pressure, the valve opens against a spring. The potential energy is converted into kinetic energy as the liquid obtains a high velocity u ($p_h \approx \rho u^2/2$; u is, e.g., 200 m $s^{-1}$). The kinetic energy is dissipated into heat during passage through the valve. This takes a very short time (<0.1 ms); so energy density is very high (up to $10^{12}$ W $m^{-3}$). A general description is in Ref. 8, and a detailed analysis of the flow conditions is in Ref. 10. Widely different types of valves are used.
8. Ultrasonic waves can be generated in many ways. A common one is the "liquid whistle" or Pholmann generator. The liquid stream impinges on a knife or blade, which is then brought to vibrate at high frequency (6 to 40 kHz); the liquid is forced through a narrow slit at speeds over 50 m $s^{-1}$, which requires a pressure of some 1 MPa or more. Magnetostriction devices often work at a frequency of 20 kHz; very high $\varepsilon$ can be obtained.
9. See Ref. 11.
10. One may atomize the disperse phase in air and let the droplets be taken up by the continuous phase. See Refs. 12 and 13 for mechanical atomization (particularly from nozzles) and Ref. 6 for electrical dispersion.
11. Some oils will spread over a water/gas interface. If air is beaten in or the water boiled, the thinly spread oil layer is disrupted, and very small droplets may result [8]. This may happen as an additional mechanism during various stirring operations. Also by steam injection, droplets can be disrupted to very small ones, though at a high cost of steam [14].
12. See, e.g., Ref. 6.

Some methods are almost exclusively used in the laboratory: 1, 4d, 8b, 9, 10, and 12. For large-scale production of emulsions methods 4b, 5, and 7 are mostly used, and more recently 8a. Several specific devices are in use for specific products. Often two methods are combined, e.g., 3 and 7 or 4 and 7.

B. Mechanisms Involved

Droplets can be deformed and disrupted by viscous or inertial forces. Viscous forces generate tangential and normal stresses at the drop surface. Inertial forces generate pressure differences. In practice, it is useful to distinguish three situations: laminar flow, turbulent flow, and cavitation.

In laminar flow viscous forces are predominant. The flow can cause shear or elongation, and elongational flow is generally more effective (see Sec. V.A). Elongation always occurs if the liquid is (locally) accelerated. The existing theories can be applied if a fairly constant flow pattern exists, i.e., constant during times much larger than $\tau_{def}$ and over distances much larger than drop size.

In turbulent flow inertial forces are usually predominant, but viscous forces may be involved too (see Sec. VI.B). Hence, there is often only a gradual difference with conditions in laminar flow. However, flow conditions vary over small distances and short times. For some conditions, theory predicts the results rather well.

During cavitation, small vapor bubbles are formed which subsequently collapse extremely fast, causing heavy shock waves in the liquid (continuous phase). These may disrupt droplets. The liquid is intensely agitated and flow is turbulent. Hence, the situation is somehow comparable to that during disruption by "turbulence." A prerequisite is that the pressure falls, at least locally, below the vapor pressure.

The occurrence or preponderance of any of these mechanisms does not only depend on the type of apparatus. It may, of course, depend on $\eta_C$ and on constructional details. The scale of the apparatus may considerably affect the operation, a larger machine often giving more turbulence (Re higher). Furthermore, the intensity of agitation can be varied, for instance, by altering flow rate (in methods 2 and 3), revolution rate (4 and 5) or slit width (5 and 7). In the colloid mill, decreasing slit width increases shear rate; in the homogenizer it increases pressure. In a homogenizer, for instance, the main mechanism may be elongation in laminar flow, turbulence, or cavitation, depending on size, pressure, and valve design. Hence, it is often difficult to predict from laboratory experiments what will happen in a production machine.

Finally, it should be mentioned that an additional mechanism for droplet disruption may occur, namely, interfacial instability caused by surface-tension gradients (Sec. VIII.A, point 5). This may happen in any machine, and it depends largely on the surfactant.

### C. Other Variables

For batchwise operation, the duration of treatment is an obvious variable.

Composition of the mixture to be emulsified (e.g., volume fraction of the disperse phase, nature and concentration of surfactant, and viscosity of both phases) naturally affects the result.

In addition, the way the substances are added is important. Usually the surfactant, which should be most soluble in the continuous phase, is indeed dissolved in that phase, but if it is possible to dissolve it in the disperse phase, smaller droplets often result. In the so-called nascent-soap method, a fatty acid is dissolved in the oil phase and cations (alkali) in the aqueous phase, so that the soap stabilizing the emulsion is formed in situ; this often gives small droplets (see also Sec. VIII.A, point 5).

Adding the disperse phase slowly to the continuous one during emulsification is often advantageous; it may enable one to add the surfactant to the disperse phase. If both phases are mixed in bulk, the one containing the surfactant usually becomes the continuous phase, unless it amounts to less than, say 20% of the total mixture.

In producing an emulsion by shaking, it is often advantageous to wait a few seconds between shakes: a few (e.g., 10) shakes by hand may then suffice to produce an emulsion, while uninterrupted shaking costs much more effort. The explanation may be that newly formed drops or threads of disperse phase need some time to acquire an equilibrium surface layer, and that they are more stable to coalescence once this has happened. But this asks, in turn, for the coalescence mechanism, which is unknown.

All the rules given here have exceptions. They apply for particular compositions and particular apparatus. Variables may have different effects under different conditions or in different machines. Remember that temperature affects many variables and that it often rises during emulsification. Unfortunately, much knowledge about emulsification is still of a cookbook nature.

## III. Evaluation of Emulsification

### A. Different Methods

Three groups of methods have mostly been used to study the results of emulsification procedures.

i. The *emulsion capacity*, or the maximum amount of disperse phase that can be emulsified under specified conditions, was originally developed by Swift et al. [19] and exists in many modifications [20,21]. It tells very little about droplet size, and the results depend on the combination of apparatus and composition in ways not understood. A variant is observing the time needed to emulsify a certain amount of disperse phase under specified conditions.

ii. The *emulsion stability* is, for instance, the amount of phase separation taking place. Mostly the sedimentation rate of the emulsion is somehow estimated, either under gravity or by centrifugation [20,21]. The idea is that sedimentation rate depends on droplet size. This may be true under ideal conditions (dilute emulsions; suitable combination of time, sedimentation depth, and centrifugal force), but even then the relation may not be simple [22]. More important is that sedimentation rate depends on flocculation of the droplets, and on the apparent viscosity at very low velocity gradient (or the existence of a yield stress) of the continuous phase. These conditions may also be affected by the variables studied, leading to misinterpretation.

iii. *Droplet size*, which may be some kind of average or a full distribution. Sometimes the number of drops is determined, or some characteristic depending on drop size, such as turbidity under specified conditions. It should be realized that several methods tend to underestimate the number of very small droplets; consequently, some literature results about the effect of, for instance, agitation intensity on droplet size may have a serious bias. Droplet size distribution may, of course, alter after emulsification. This can be due to coalescence, but it also may be due to isothermal distillation (Ostwald ripening), unless the disperse phase is completely insoluble in the continuous one.

Determining droplet size distribution as a function of process and product variables is by far the best method to study emulsification. But there are other aspects of the emulsions, such as type oil in water (O/W) or water in oil (W/O), and various kinds of stability which may be affected by the way of making them, and these should be studied separately.

## B. Aspects of Size Distributions

Mainly to help interpretation of the data presented in this chapter, we give some details about size distributions and their characterization (see also Chap. 6, where this is treated in more detail).

We consider spherical drops with diameter $x$. If the number of drops per unit volume with diameter $< x$ is given by $F(x)$, then the number frequency distribution $f(x) = dF(x)/dx$. Volume frequency is given by $(1/6)\pi x^3 f(x)$.

As an auxiliary parameter, we define the nth moment of the distribution as

$$S_n = \int_0^\infty x^n f(x)\, dx \tag{5}$$

Hence, $S_0$ = total number of droplets per unit volume. Any type average diameter is thus given by

$$x_{nm} = \left(\frac{S_n}{S_m}\right)^{1/(n-m)} \tag{6}$$

so $x_{10}$ = number average, $x_{32}$ = volume/surface or Sauter average, etc. $x_{nm}$ increases with n + m, the more so as the emulsion is more polydisperse. We will use $\bar{x}$ to denote an unspecified type of average x.

Polydispersity is best expressed as the relative width, i.e., the variation coefficient $\sigma$, of the distribution given by

$$\sigma_n = \left(\frac{S_n S_{n+2}}{S_{n+1}^2} - 1\right)^{1/2} \tag{7}$$

Hence $\sigma_0$ is number standard deviation divided by $x_{10}$, $\sigma_2$ corresponds to $x_{32}$, etc.

We will also use the specific surface area, either of the whole emulsion or of the disperse phase:

$$A_E = \pi S_2 = \frac{6\phi}{x_{32}} \tag{8a}$$

$$A_D = \frac{6S_2}{S_3} = \frac{6}{x_{32}} \tag{8b}$$

Several authors have derived expressions for the size distribution from theory or from experimental results, but it should be realized that a suitable three-parameter equation will fit almost any distribution found. Quite often, the distributions are reasonably log-normal, which implies positive skewness and $\sigma_n$ independent of n. Mostly, $\sigma_2$ is between 0.5 and 1, which implies that the emulsions are markedly polydisperse. Other types of size distribution may also occur, e.g., bimodal ones; in such a case, $\sigma$ may be very high, say, $\sigma_2 = 1.5$.

When the effect of a variable, e.g., agitation intensity, on droplet size is studied, not only average size may vary, but also the relative width of the distribution, or even the distribution type. In such cases, the change of an average size (or the outcome of a measurement giving a particular average, such as sedimentation or turbidity) may not be sufficient as a characteristic, or even be misleading. For instance for log-normal distributions, increasing $\sigma$ from 0.5 to 1 causes the ratio $x_{32}/x_{10}$ to increase from 1.56 to 4.

Some authors have used the ratio of $\bar{x}$ obtained to the original size $\bar{x}_0$ as a measure of emulsification or homogenization results. But under most conditions droplets are broken down until a critical size $x_{cr}$ is reached, and $x_{cr}$ depends on conditions during emulsification and not, or only slightly, on $\bar{x}_0$. Hence, $\bar{x}/\bar{x}_0$ may be a misleading parameter.

## IV. Droplet Formation

### A. Film Formation

At the onset of emulsion preparation we often have two bulk phases. In some apparatus we may envisage that "lumps" of one phase (to become the disperse one) are "cut off" by, say, a stirrer blade, giving rise to coarse droplets. This means that a film of the future continuous phase is made in the other one. One may also envisage that parts of a film of disperse phase are cut off, giving rise to fairly small droplets.

Anyhow, such a film should exist for a little while, but if no surfactant is present, it is very unstable. It tends to drain rapidly under the influence of gravity. Moreover, the film must end somewhere, and at such a site the interface is strongly curved; consequently, interfacial forces tend to straighten the interface.

Films can only exist for some time if a surfactant is present. Film stability has been studied extensively for films of water in air, as in a foam. A recent review is by Lucassen [89]. For emulsions, the situation is somewhat different, and it has scarcely been studied, but some qualitative conclusions can be drawn from analogies with foams.

The main effect caused by a surfactant is that it allows an interfacial tension gradient $d\gamma/dz$ to exist. Such a gradient exerts a tangential stress on the liquid; or if a liquid streams along an interface with surfactant, an interfacial tension gradient develops. At equilibrium $-d\gamma/dz = \eta G_0$, where $G_0$ is the velocity gradient at the interface; i.e., $G_0 = (du_z/dy)_{y=0}$ if y is the distance from the surface. This causes a considerable resistance against drainage of a film. The gradient can support the weight of a film if $|2d\gamma/dz| \geq \rho_C gh$, where h is film thickness. This implies that films of the order of 0.1 mm thick or less would be fairly stable, assuming $d\gamma/dz \approx 1\ \mathrm{N\ m^{-2}}$. In emulsion formation the situation is more complicated. There will be no true equilibrium between viscous and surface forces during drainage of a film, and the viscosity ratio of both phases will also have an effect. Nevertheless, surface tension gradients must be essential.

The generation of a $\gamma$ gradient by streaming implies that the area of the interface is locally expanded or compressed. Surfactant molecules will diffuse toward the interface and try to restore the low interfacial tension. The extent to which $\gamma$ differs from its equilibrium value is expressed as the surface dilational modulus $E = d\gamma/d(\ln A)$. E depends on the nature of the surfactant, on the rate of transport toward the interface, and on the rate of extension or compression of the interface. In the case of a thin film, the transport toward the interface stops very soon, because diffusion over a very small distance is so rapid; the limited amount of surfactant present is then the factor determining E. In this case the Gibbs elasticity of the film $E_f$, which equals twice the surface dilational modulus, is given by [95,89]

$$E_f = -\frac{2\, d\gamma/d(\ln \Gamma)}{1 + (1/2)h\; dm/d\Gamma} \tag{9}$$

$\Gamma$ is the surface concentration or surface excess of the surfactant (mol $m^{-2}$). To be precise, the surfactant activity rather than m should be used, and Eq. (9) does certainly not apply above the critical micelle concentration. Equation (9) shows that $E_f$ is higher for a thinner film, and its drainage is impeded only if $E_f$ is rather high; if $E_f = 0$, a $\gamma$ gradient cannot develop.

The Gibbs elasticity plays another important role. The film will often be stretched, and in the presence of sufficient surfactant the thinnest part of the film will have the highest $E_f$, hence the greatest resistance against stretching. This implies a strong stabilizing mechanism, as already explained by Gibbs. But Lucassen [89] pointed out that $E_f = 0$ for $m = 0$ and goes through a maximum when increasing m. Consequently, when m is very low, stretching of the film will cause a decrease of $\Gamma$ and $E_f$ can be lowest for the thinnest part of the film; stretching will now cause the thinnest part to stretch fastest, and the film is unstable. This implies that surfactant concentration should not be too low, since the rapid film drawing (as may happen during emulsification) may lead quickly to depletion of surfactant in the films. Moreover, Prins has shown [24] that a thin film which is being stretched always breaks as soon as $\gamma$ exceeds a critical value $\gamma_{cr}$, which depends on the system. For rapid stretching and low m, $\gamma_{cr}$ will soon be reached.

In the reasoning given we have tacitly assumed two prerequisites to hold. First, the surfactant is in the continuous phase, i.e., in the film. If it is in the disperse phase, $E_f$ will always be very low and the film is unstable. This fits in with Bancroft's rule (Sec. I.A). Second, Gibbs equilibrium between surfactant in the bulk and in the interface is attained. This is not true for polymers as surfactants, and in that case $E_f$ is more difficult to predict; mostly the films appear to be fairly stable.

We may conclude that interfacial tension gradients are essential in emulsion formation, as originally pointed out by Van den Tempel [23], and that droplet formation is probably easier if films are drawn at not too high a speed. This would imply slow stirring at the beginning of emulsification.

### B. Disruption of a Plane Interface

We will now consider two bulk phases, separated by a plane interface which must be disrupted to produce drops. The phases have densities $\rho_1$ and $\rho_2$ (we will assume $\rho_1 \geqslant \rho_2$) and viscosities $\eta_1$ and $\eta_2$. The interfacial tension $\gamma$ can be low in the presence of surfactant, since the processes considered here are relatively slow and surfactant concentration is still high; thus, adsorption of surfactant can more or less keep up with the formation of new interface (see Sec. I.C.).

How then are droplets formed?

#### 1. *Turbulence*

In some cases we may envisage turbulent eddies to disrupt the interface; see, e.g., Davies [25]. Eddies with a shear-stress velocity $u_e$ may cause local pressures of the order of $\rho u_e^2$. At the other side of the interface there may also be turbulence, but on average pressure differences of $(\rho_1 - \rho_2)u_e^2$ may be expected, and these must surpass the Laplace pressure $2\gamma/R$. An additional condition is that the size of the eddy $l_e \approx 2R$; in that case the interface may be disrupted. The possibility is calculated easiest for isotropic turbulence (see Sec. VI.A). The highest $u_e$ is found in small eddies for which $u_e \approx \varepsilon^{1/4}\eta^{1/4}\rho^{1/2}$ and $l_e \approx \varepsilon^{-1/4}\eta^{3/4}\rho^{-1/2}$. Even if energy desnity is very high, these conditions can be met simultaneously only if either $\gamma$ is extremely low (say $10^{-5}$ N m$^{-1}$) or if $\eta$ is so high as to exclude turbulence. Turbulence will mostly not be isotropic on the scale considered, and prediction is then much more difficult [25]. Approximate calculations lead, however, to the same conclusion: mostly no disruption unless $\gamma$ is very low.

#### 2. *Capillary Ripples*

Any disturbance of the interface will cause surface waves (ripples) to develop, which may in some cases lead to shredding of droplets. The calculation of the properties of surface waves is a formidable problem and has been completely solved only for small amplitude [26]. For one-dimensional transverse waves the relation between wavelength ($\lambda = 2\pi/k$) and frequency ($\nu = \omega/2\pi$) is then given by Kelvin's equation:

$$\omega^2(\rho_1 + \rho_2) = g(\rho_1 - \rho_2)k + \gamma k^3 \tag{10}$$

The first term of the right-hand side is due to gravity, and can usually be ignored in our case. The second term is due to surface forces, hence the name capillary waves. The waves are usually damped, but if an external source keeps acting at a suitable frequency, standing waves are formed and the amplitude may increase until the interface is disrupted. Droplets of a diameter of roughly $\lambda/2$ would result. However, in experiments with standing waves on an air/water interface, droplet diameters of $\lambda/3$ to $\lambda/10$ were found [6]. Figure 3 shows that a stirrer with a high-revolution rate (e.g., $\nu$ = 200 Hz), and certainly an ultrasonic device (e.g., $\nu$ = 20 kHz), may be able to produce fairly small drops. Whether the two-dimensional waves needed are formed and whether the standing waves will exist long enough to produce droplets, will, however, depend on the actual conditions.

The damping of low-amplitude capillary waves is completely known [26]. It turns out that both the viscosity of the liquids and the presence of a surfactant enhance damping. But if surfactant concentration is locally different, as may occur in practice, the amplitude of the wave may actually increase at the site of lower $\gamma$. It has been reasoned [6] that this may be a cause of droplet formation during emulsification. Large $\gamma$ gradients will also cause interfacial turbulence and possible droplet shredding. (See Sec. VIII.A, point 5.)

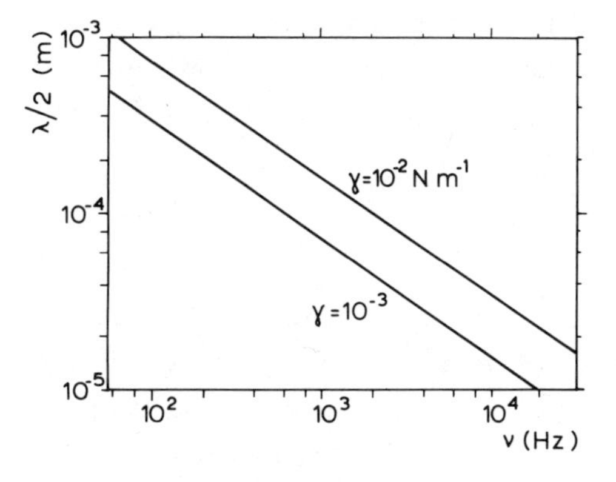


Figure 3. Relation between wavelength ($\lambda$) and frequency ($\nu$) of one-dimensional transverse waves on an interface, according to Eq. (10). $\rho_1 - \rho_2 = 100$, $\rho_1 + \rho_2 = 2000$ kg m$^{-3}$.

### 3. *Rayleigh-Taylor Instability*

Under some conditions, an interface is unstable, i.e., local perturbations tend to increase; a detailed treatment is given by Chandrasekhar [27]. The Rayleigh-Taylor instability occurs when the interface is accelerated perpendicular to its plane and directed from the lighter into the heavier phase. The acceleration due to gravity must be taken into account.

Gopal [6,28] has studied the significance of this instability for emulsification, particularly during shaking. For the case of inviscid liquids and constant $\gamma$ (constant in time and distance, hence no surfactant), the interface will be unstable if $k^2 < (\rho_1 - \rho_2)a/\gamma$ and if the acceleration a has the proper direction. Then all such ripples will be amplified, but maximum instability will occur for a certain wavelength $\lambda_{opt}$, given by

$$k_{opt}^2 = \frac{(\rho_1 - \rho_2)a}{3\gamma} \tag{11}$$

Assuming $\Delta\rho = 100$ kg m$^{-3}$, $\gamma = 0.01$ N m$^{-1}$, and a = 20 m s$^{-2}$ gives $\lambda_{opt} = 24$ mm; even if $\gamma = 10^{-3}$ and a = 100, we only have $\lambda_{opt} = 3.5$ mm. Hence, only coarse drops can be formed, unless a is very high.

A further analysis shows [6,28] that an asymmetric wave profile develops, since the heavy phase is thrown in the form of "spikes" or "fingers" into the lighter phase (Fig. 4). As long as the amplitude $\alpha$ is small, it is approximately given by

$$\alpha \approx \frac{(\rho_1 - \rho_2)at^2}{2(\rho_1 + \rho_2)} \tag{12}$$

The fingers could spontaneously break into globules, but this would take some time, at least a few tenths of a second (see Sec. IV.B.1). Hence, the importance of the Rayleigh-Taylor instability is doubtful. The effects of viscosity and variable surface tension will probably not make disruption any easier. Moreover, liquids of equal density can certainly be emulsified, though Eqs. (11) and (12) predict the absence of Rayleigh-Taylor instability.

### 4. *Kelvin-Helmholtz Instability*

This instability arises when the two phases move with different velocities $u_1$ and $u_2$ parallel to the interface [27]. It causes, for instance, waves to develop on a water surface over which the wind blows. The analysis given is valid, again, for inviscid liquids and constant $\gamma$. It is then found that any perturbation will be enhanced if

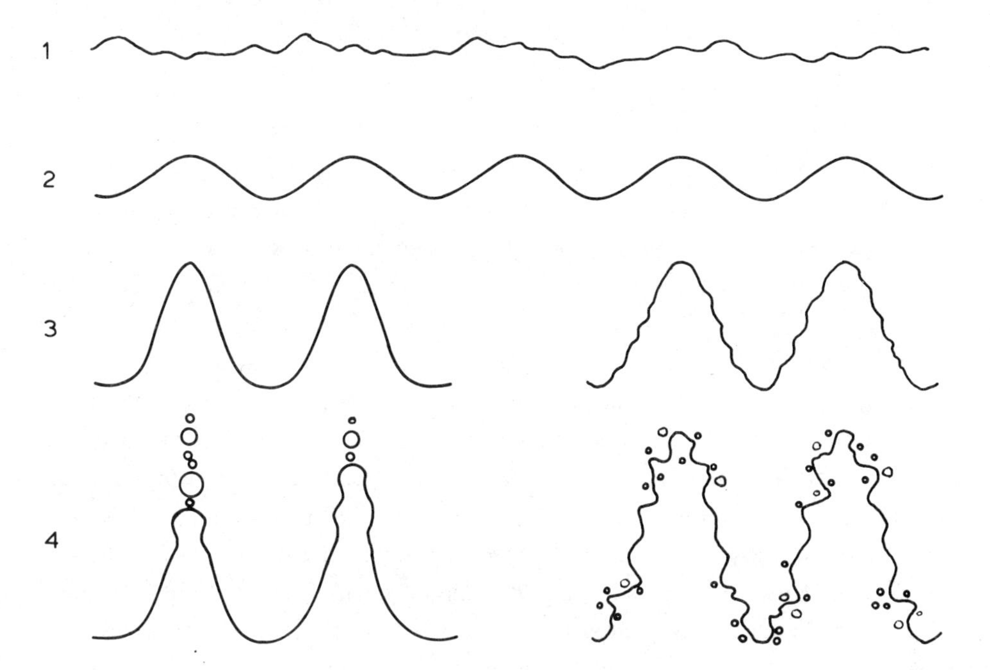


Figure 4. Rayleigh-Taylor instability of an interface; the denser phase is below; the interface is accelerated in the downward direction. At the left side pure Rayleigh-Taylor instability is depicted; at the right side Kelvin-Helmholtz instability develops at the flanks of the waves. (From Refs. 6 and 28.)

$$k > \frac{a(\rho_1^2 - \rho_2^2)}{\rho_1\rho_2(u_1 - u_2)^2} \tag{13a}$$

and

$$(u_1 - u_2)^2 > 2(\rho_1 + \rho_2)\,\frac{[a\gamma(\rho_1 - \rho_2)]^{1/2}}{\rho_1\rho_2} \tag{13b}$$

If condition (13b) is not fulfilled, surface forces will damp the waves. Assuming $\gamma = 10$ mN m$^{-1}$, $\Delta\rho = 100$ kg m$^{-3}$, and $a = g = 10$ m s$^{-2}$, we find that disturbances of practically any wavelength will be enhanced for $(u_1 - u_2) > 0.1$ m s$^{-1}$, and such velocity differences are not inconceivable. Whether this will lead to droplet shredding is not certain. The viscosity of the liquids probably enhances the effect, but $\gamma$ gradients may well diminish it.

### 5. *Other Aspects*

It is difficult to deduce which of the above-mentioned mechanisms will be mainly responsible for droplet formation. It will differ according to conditions, and it may often be a combination of mechanisms. It has been argued [6,28] that at the flank of the large ripples caused by the Rayleigh-Taylor instability, a Kelvin-Helmholtz instability can develop, since the phases move here with respect to each other. This would then lead to disruption (see Fig. 4). From Eq. (12) we infer that $u_1 - u_2 \approx at/40$, and it appears possible that velocity differences of $0.1 \text{ m s}^{-1}$ can be realized in some cases.

It may finally be mentioned that a liquid interface moving in a capillary is unstable if the displacing liquid would have a higher velocity than the other liquid, if either could move independently [29]. Since the permeability of the capillary will usually be equal for both liquids, it follows from Darcy's law that the interface is unstable if the displacing liquid has the lower viscosity. The result is a tendency for "fingering," i.e., the protrusion of threads of the displacing liquid into the displaced one. This is counteracted by the Laplace pressure, but under certain conditions droplets may be formed [30].

## C. Disruption of a Cylindrical Thread

In some cases cylindrical threads of disperse phase are formed during emulsification. Disruption of a plane interface may lead to fingering (Secs. IV.B.2 to IV.B.4). A jet of one liquid may directly be injected into the other one (Table 2, No. 3). A large drop may be drawn into a thread (Sec. V, Fig. 10). Break up of a thread results in droplets.

### 1. *Stationary Cylinders*

We consider a neutrally buoyant, infinite liquid cylinder of radius $\Theta$ in an unbound volume of another liquid. Such a thread is subject to small random deformations, whose properties have been studied by Rayleigh (see e.g., Ref. 27). Most deformations are damped and only varicose (i.e., axisymmetric) deformations of a wavelength $\lambda > 2\pi\Theta$ lead to instability. This means that they are amplified until the thread breaks into droplets (Fig. 5). If $\lambda > 2\pi\Theta$, the surface area of the deformed thread becomes smaller for increasing amplitude $\alpha$, hence the instability. The increase in $\alpha$ with time is given by

$$\alpha = \alpha_0 \exp(\beta t) \tag{14}$$

and as soon as $\alpha = \Theta$ the thread breaks. Hence the disruption time is given by

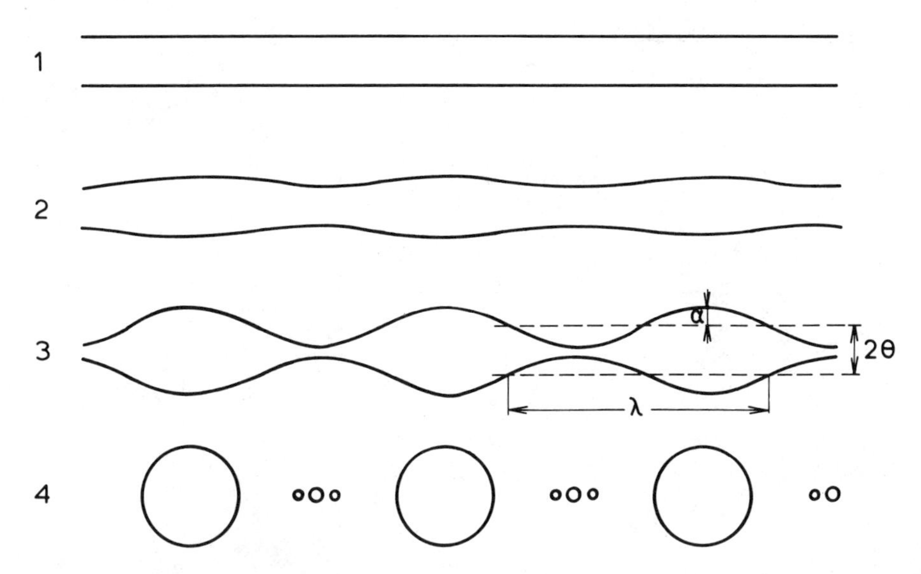


Figure 5. Instability of a stationary liquid cylinder. (From Ref. 32.)

$$\tau_{disr} = \frac{1}{\beta} \ln \frac{\Theta}{\alpha_0} \tag{15}$$

Assuming $\alpha_0 \approx 0.01\Theta$, we would have $\tau_{disr} \approx 5/\beta$.

The instability has been further analyzed by Tomotika [31] for the case of constant $\gamma$ and no inertial effect. He found that

$$\beta = \frac{\gamma}{2\eta_C \Theta} F(b,q) \tag{16}$$

where F is a complicated function of $b = k\Theta = 2\pi\Theta/\lambda$, and $q = \eta_D/\eta_C$. There is an optimum wavelength $\lambda_{opt}$ for which $\beta$ is maximum, and this deformation becomes dominant and causes final breakup. $F_{max} = F(b_{opt}, q)$ is given in Fig. 6. It follows that $\tau_{disr}$ is proportional to $\Theta$ and to $1/\gamma$. The radius of the resulting droplets follows from

$$r = \Theta \left( \frac{3\pi}{2b_{opt}} \right)^{1/3} \tag{17}$$

See Fig. 6.

Rumscheidt and Mason [32] have tested the theory for $q = 0.01$ to 10 and viscosities of either phase > 1 Pa s. Both deformation rate and droplet size agreed well with the theory. The drops were of almost equal size, except that three tiny satellite droplets formed between each pair of normal ones. Thus for $\gamma = 10^{-2}$ N m$^{-1}$ and $\Theta = 1$ mm, we have $1/\beta \approx 10$ ms, or somewhat larger for large $\eta_C$.

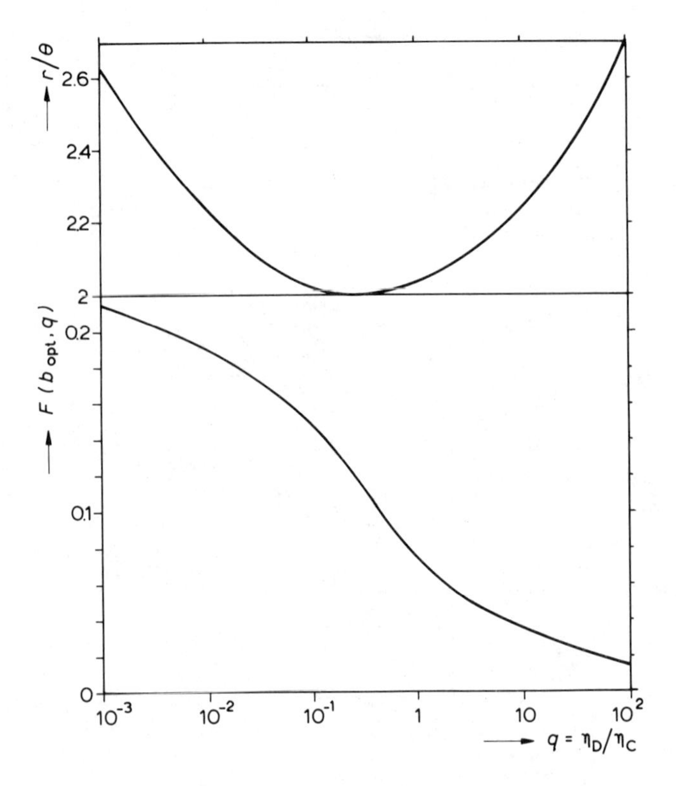


**Figure 6.** Maximum value for the instability function F(b, q) [Eq. (16)] of a stationary liquid cylinder of radius $\Theta$, and the corresponding drop radius after breakup, from Eq. (17). (Data from Refs. 31, 32.)

For $q < 0.1$, $F_{max} \approx 0.23(1 - b_{opt}^2)$. For very large q, we have [27] $F(b,q) \approx (1 - b^2)/3q$. Since $b_{opt}$ is now small [as follows from Fig. 6 in combination with Eq. (17)], there is no distinct maximum in $\beta$, hence no clear $\lambda_{opt}$. Consequently, a very viscous thread will not break up in such a regular way, but rather gives drops of variable size.

Up until now, we have neglected inertial effects. This is justified if Re << 1, which is often the case; Re may be defined as $Re = 2\Theta^2 \beta\rho_D/\eta_D$. Still, inertial effects often play some role, since the very regular deformation and breakup cannot be expected without the disturbances along the length of the thread influencing each other; hence a wave phenomenon; hence inertial effects.

If Re >> 1, viscous effects may be neglected, and now we obtain [27]

$$\beta_{max} = 0.343 \left(\frac{\gamma}{\Theta^3 \rho_D}\right)^{1/2} \tag{18}$$

and $b_{opt} = 0.7$, hence $r = 1.9\Theta$. This applies, for instance, to a water jet of a few millimeters radius. For intermediate Re, there appears to be no solution available.

The case of a stationary cylindrical thread contained in a cylindrical tube (coaxial, no inertia, $\gamma$ constant) has also been studied, both theoretically and experimentally [33]. It turns out that the containment scarcely influences $\lambda_{opt}$, but it may have a strong effect on $\beta$. Particularly for large q, $\beta$ is much decreased, the more so as $\Theta/\Theta_{tube}$ is larger.

### 2. *Restrained Jets*

If a cylinder of liquid moves from an orifice with average initial velocity u into another fluid, the stability of this jet is not appreciably different from that of a stationary cylinder, unless u is either very small or very large.

If u is very small, drops detach from the orifice, and very roughly their volume equals $2\pi\gamma\Theta_0/g\,\Delta\rho$, where $\Theta_0$ is the radius of the orifice. Many factors affect drop size, and complete solutions have been given [91,92], also for non-Newtonian (power-law) liquids [93]. When u increases, drop size mostly decreases, often after a slight initial increase. At a critical u, of the order of $(\gamma/\rho_D\Theta_0)^{1/2}$, a jet of liquid is formed before it breaks into drops [91], and now drop size is roughly as predicted for stationary cylinders.

If u becomes still higher, the thread begins to twist, i.e., sinusoidal deformations occur. One would assume the Kelvin-Helmholtz instability to be responsible. The stability is in any case decreased, i.e., smaller drops are formed in a shorter time [34]. There appears to be no good theory for the critical velocity; it is larger for a smaller orifice.

For very high u, the jet becomes turbulent. If we define $Re = 2\Theta_0 u \rho_D/\eta_D$, turbulence sets in for $Re \approx 8{,}000$ to $10{,}000$. $\Theta_0$ is the initial jet radius; radius increases with distance from the orifice. We now have a very different mechanism for disruption. The turbulent eddies inside the jet cause breakup and the viscous resistance of the continuous liquid hinders this [25]. Hence $\Theta_0$ and $\gamma$ have no influence. Droplet size shows quite a spread; while droplet-size distribution from a laminar jet may have $\sigma_0 \approx 0.1$, a turbulent jet gives, for instance, $\sigma_0 \approx 0.4$ [34]. The average droplet size is given by [25]

$$\bar{x} \approx 3100\left(\frac{\eta_D}{\rho_D}\right)^{1/4}\left(\frac{\eta_C}{\rho_C}\right)^{3/4} u^{-1} \qquad (19)$$

Hence for a W/O system with $\eta_D = 1$, $\eta_C = 10$ mPa s, and $u = 10$ m s$^{-1}$, we have $\bar{x} \approx 1.7$ mm. For a similar O/W system smaller drops would be formed, but a very high u is needed to produce turbulence. Consequently, it is mostly of little advantage to produce a turbulent jet.

*3. Extending Threads*

A drop placed in a liquid that undergoes axisymmetric elongational flow (see Sec. V.A) will be extended into a thread if the Weber number is large enough. For disruption by viscous forces,

$$\text{We} = G \frac{\eta_C r}{\gamma} \tag{20}$$

For elongational flow, G is defined as the elongation rate, i.e., the velocity gradient in the direction of flow: $G = du_z/dz$. This case has been investigated, both theoretically and experimentally, by Mikami et al. [35] for constant $\gamma$ and $\text{Re} = 2\Theta^2 G\rho_C/\eta_C << 1$, i.e., absence of inertial effects. This condition, combined with a not too small We, means that roughly $\eta_C^2 > \Theta\rho_C\gamma$ is needed; this can only be achieved for large $\eta_C$.

Assuming that we start with an undeformed cylindrical thread of radius $\Theta_0$, the thread becomes thinner according to

$$\Theta = \Theta_0 \exp\left(\frac{-Gt}{2}\right) \tag{21}$$

Equation (21) also assumes that G is the same inside and outside the thread or that the velocity is continuous across the thread boundary. But as soon as a $\gamma$ gradient develops, i.e., when even a little surfactant is present, this need not be true, and one would expect thinning rate to be less, particularly if $q >> 1$.

The extension has a strong damping effect on instabilities, the more so for larger We and larger q. This is owing to the fact that the deformations are extended as well, and only those that initially have a small $\lambda$ eventually lead to breakup. The theory is complicated, but one result is that for q around 1, $\tau_{disr} \approx 3/G$, increasing with q. The further the thread can be extended before it breaks, the smaller the resulting droplets. Figure 7 gives results for the expected droplet size, but these are little better than order-of-magnitude estimates. Moreover, breakup is fairly irregular, and besides the "major" drops formed, thin filaments remain which break eventually into much smaller droplets.

In practice, it is difficult to induce axisymmetric elongational flow for a sufficient length of time. However, Mikami et al. [35] have shown that plane hyperbolic flow (see Sec. V.A) gives about the same results, which implies that the flow around the thread becomes almost axisymmetric. If G alters during the process, which will often be the case in practice, prediction of results is difficult.

*4. Effect of Surfactants*

Apart from the obvious effect that presence of a surfactant lowers $\gamma$, it may have other effects. This has been studied by Carroll and Lucassen for the case of a

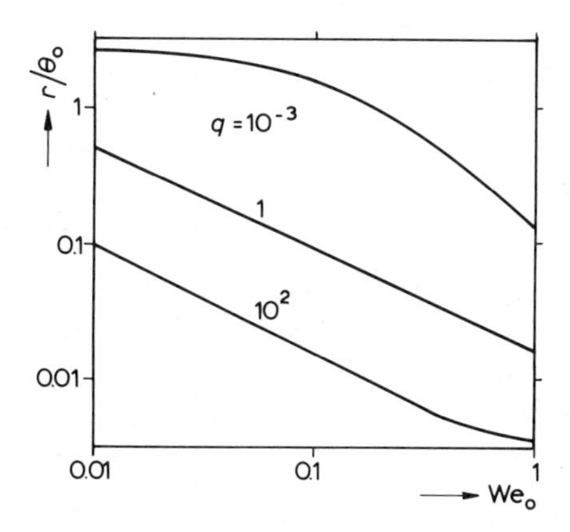


**Figure 7.** Approximate radius (r) of drops resulting from extending liquid threads of initial radius $\Theta_0$ as a function of initial Weber number $We_0$ and viscosity ratio q. (Data from Ref. 35.)

stationary liquid thread (supported by a coaxial solid thread) [36] and of a contained moving thread (unsupported, Re ≈ 1) [37]. The surfactant allows the development of interfacial tension gradients $d\gamma/dz$ (see Sec. IV.A) which impede the tangential motion of an interface, hence the flow of liquids along the interface. Mostly, the surfactant therefore tends to make the thread more stable, but it also slows down the thinning rate in extending threads. The overall effect turns out to be rather small. The quantity $r/\Theta$ is very little affected; $\beta/\gamma$ becomes smaller, at most by a factor 4, but in most cases the decrease is less pronounced.

It must be stressed that during emulsification, conditions (e.g., G) vary considerably and that time scales are very different from those in the mentioned experiments. Hence, it is difficult to predict what happens, though it is reasonable to assume that smaller drops should result if a surfactant is present.

Finally, it should be noted that "satellite drops" are not always observed if a surfactant is present [37].

## V. Disruption of Droplets in Laminar Flow

### A. Types of Flow and Their Effects

A general discussion of the effect of various flow patterns on deformation and breakup of droplets is given by Hinze [38]. We will here consider only laminar flow, i.e., absence of turbulence. We will also exclude inertial effects. These can be of two kinds.

The droplet may be accelerated with respect to the continuous phase. In that case the Weber number, which is the ratio of external stress over Laplace pressure, may be defined as

$$We^* = \Delta\rho(\Delta u)^2 \frac{r}{\gamma} \tag{22}$$

where $\Delta\rho$ and $\Delta u$ are the differences in density and velocity, respectively, between drop and continuous phase. Falling raindrops can be disrupted in this way [1], but in the case of emulsification we always have $We^* \ll 1$, since high $\Delta u$ cannot be realized for a small drop in a liquid. Hence we may exclude this inertial effect.

Inertial effects in the continuous liquid occur if the Reynolds number, which is the ratio of inertial to viscous forces, is high. In this case we have to consider the drop Reynolds number, and for laminar flow of velocity gradient G we have $Re = 2\,r^2G\,\rho_C/\eta_C$. Re(drop) $\ll 1$ is needed, and the flow is then called *creeping flow;* it is a much more severe condition than that for laminar flow through a tube. The Weber number is now as in Eq. (20), and roughly $We > 1$ is needed for breakup. Combination with Re leads to the condition $r \ll \eta_C^2/\gamma\,\rho_C$. Only in Sec. V.D will we consider inertial affects.

Another condition is mostly that the flow is steady, i.e., that G does not alter with time.

Figure 8 illustrates some types of flow and their effects on drops for small deformations. It considers plane or two-dimensional flow, i.e., flow pattern is assumed not to change in a direction perpendicular to the plane of drawing. Pure rotational flow (Fig. 8a) causes no deformation in a spherical drop placed at the origin; the drop only rotates. Pure elongational flow only causes deformation [into a prolate (cigar-shaped) ellipsoid] and no rotation; hence, it is sometimes called *irrotational flow.* Simple shear causes both.

In simple shear, the droplet is deformed into a prolate ellipsoid with its long axis at an angle of about $\pi/4$ to the z axis, and the liquid in the drop circulates around the origin. For greater deformation, the long axis of the ellipsoid turns towards the z axis, and when deformation becomes of the dumbbell type as shown in Fig. 10a, internal circulation stops [46]. Although the *shear rate,* commonly defined as $d(\Delta z/y)/dt$, equals G, the rate of pure deformation equals G/2 in simple shear, since half of the motion of a volume element is in fact rotation.

Plane hyperbolic flow is one example of elongational flow, i.e., flow with a velocity gradient in the direction of flow. G is also called *rate of elongation* or *rate of extension.* Elongational flow has a stronger effect on droplets than has shear flow (like Fig. 8b) [39]. For the same velocity gradient it exerts greater viscous stresses, owing to the absence of rotation. The *Trouton viscosity* $\eta_{el}$ is defined as the ratio of viscous stress over elongation rate, and we have

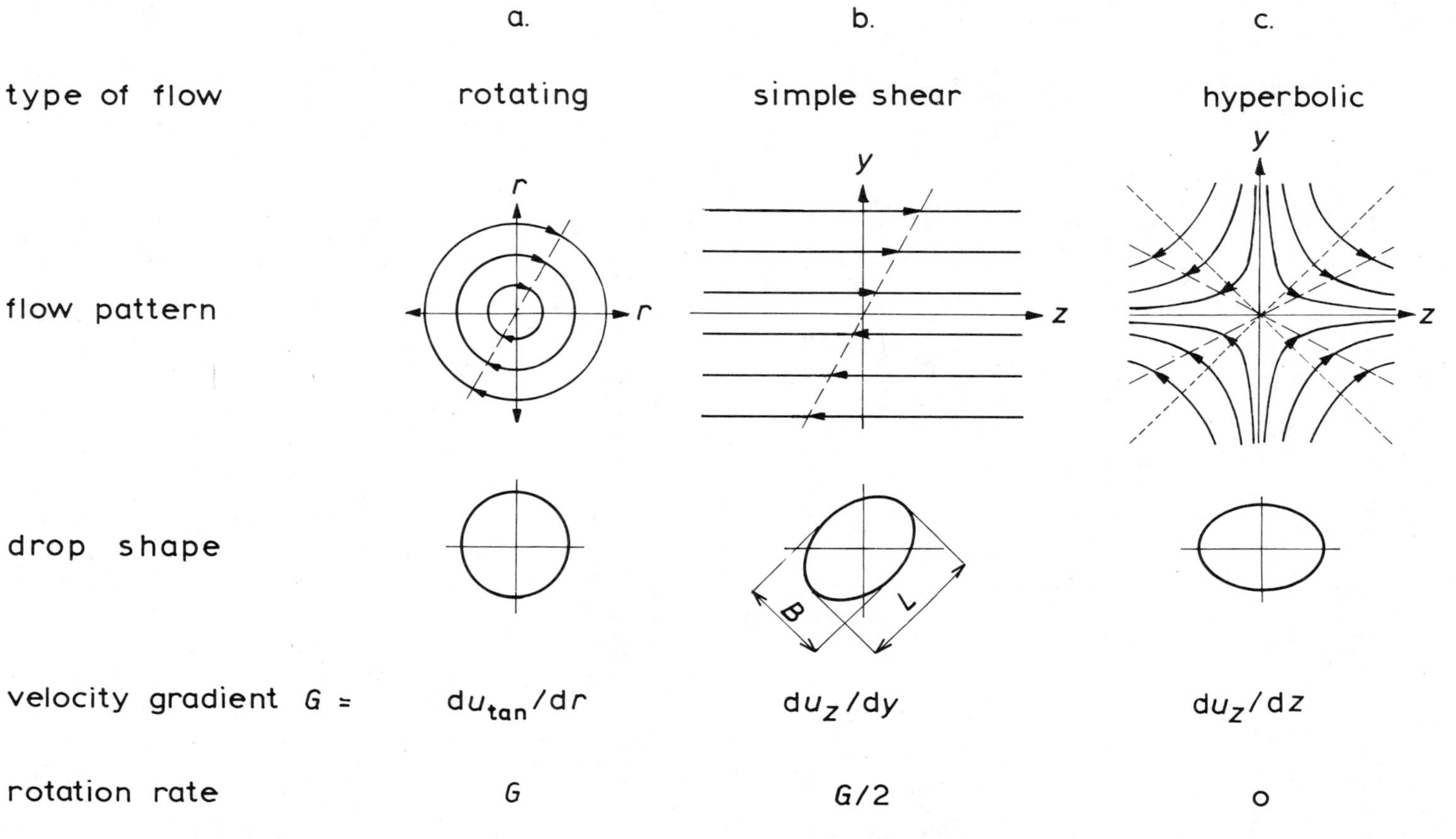


Figure 8. Various types of two-dimensional flow and its effect on droplet deformation and rotation.

$$\eta_{el} = \text{Tr}\ \eta \tag{23}$$

where for Newtonian fluids the Trouton number Tr = 3. Moreover, in the case of simple shear, deformation of a drop diminishes the stress on it since its dimension in the direction of G (y axis in Fig. 8) diminishes. For elongational flow, deformation causes an increase of dimension in the direction of G (now the z axis), hence the stress increases.

Other types of flow exist. Axisymmetric hyperbolic flow may be envisaged from Fig. 8c; if z is the axis of symmetry, the flow is converging and similar to that through a constriction in a tube (see also Fig. 12), and it deforms the drop into a prolate ellipsoid. If y is the axis of symmetry, the flow diverges radially, and the drop becomes an oblate (i.e., lenticular) ellipsoid; we will not consider this case.

Intermediate types of flow exist, of course, and during emulsification a drop will usually encounter these. The drop moves with the flowing liquid, having the same velocity as if it were a point body located at its center of gravity. In simple shear, the drop is nevertheless under the same conditions, wherever its position. But in most other flow types, the flow field around the drop depends on its position, and Figs. 8a and c are only valid for position in the origin.

A drop will eventually break in laminar flow, if its Weber number (We = $G\ \eta_C r/\gamma$) exceeds a critical value $We_{cr}$. This implies that for certain conditions (flow pattern, $G, dG/dt, \eta_C, \eta_D, \gamma$) there is a critical drop radius $r_{cr}$, so that drops with $r > r_{cr}$ will be broken up. $We_{cr}$ primarily depends on flow type and q, and this is treated below. A recent review on drop deformation and breakup is by Acrivos [40].

## B. Simple Shear

The classical analysis is by Taylor [41,42]; see also Ref. 43. Prerequisites of the theory are given in Table 3, and in view of the neglect of the effect of deformation on the stress (prerequisite 8), its success is remarkable. The equilibrium deformation, given by

$$D = \frac{L - B}{L + B} \tag{24}$$

is equal to We(19q + 16)/(16q + 16) for not too high q; see Fig. 8 for L and B. It is generally found that the drop breaks for D > 0.5, which corresponds to an extension of the drop to 2.1 times its original diameter.

Others have extended the theory (see Table 3) and we will follow Cox [99]; see also Torza et al. [2]. A small deformation develops with time approximately as

Table 3. Conditions in Theories on Droplet Deformation in Laminar Flow

| Theory by: | Taylor [41,42] | Cox [2,99] | B-B & A[a] [44] |
|---|---|---|---|
| Valid for flow types: | | | |
| Simple shear | + | + | + |
| Plane hyperbolic | + | | + |
| Axisymmetric extensional | | | + |
| Prerequisites: | | | |
| 1. Steady, unbounded flow | + | + | + |
| 2. Incompressible liquids | + | + | + |
| 3. Newtonian liquids | + | + | + |
| 4. Negligible inertial effects[b] | + | + | + |
| 5. Negligible thermal agitation[c] | + | + | + |
| 6. Steady state | + | | |
| 7. $\gamma$ constant | + | + | + |
| 8. Stresses for undeformed drop | + | | |
| 9. Deformation small | + | + | |
| 10. Tangential stress continuous[d] | + | + | + |
| 11. Tangential velocity continuous[d] | + | + | + |
| 12. Normal stress = Laplace pressure[e] | + | + | + |
| 13. Normal velocity continuous[f] | + | + | + |

[a] Barthès-Biesel and Acrivos; theory includes burst of drops.
[b] Hence small drop; no effect of density.
[c] Hence drop not very small.
[d] Follows from prerequisites 4 and 7.
[e] Follows from prerequisite 4.
[f] Follows from prerequisite 2 and the equation of continuity.

$$D_t \approx D\left[1 - \exp\left(\frac{-\gamma t}{r\eta_D}\right)\right] \tag{25}$$

hence Eq. (2). The situation is in fact more complicated, since the drop may tend to oscillate, particularly if $\gamma$ is small; see Ref. 2. The final deformation is given by

$$D = \frac{5(19q + 16)}{4(1 + q)[(19q)^2 + (20/We)^2]^{1/2}} \tag{26}$$

For $q \to 0$, $D \to We$, as in Taylor's theory, and for $q \to \infty$, $D \to 0$. For $q > 4$ and $We > 1$, $D \approx 5/4q$, irrespective of We. Consequently, very viscous drops can hardly be deformed. This is because the drop cannot deform as fast as the velocity gradient can cause deformation; by applying Eq. (2) we have $\eta_D r/\gamma << 1/G$, which corresponds to We >> 1/q. But the rotation rate is still G/2; so the drop merely rotates.

The theory has been extended to include large deformation and breakup of drops [44], but the resulting equations are very complicated. $We_{cr}$ is found to be a function of q only, and theory corresponds reasonably well with experimental results. Experiments with single large drops are usually performed in Couette flow (between coaxial cylinders), which can be made to be very close to simple shear flow. Figure 9 gives experimental results by Grace [45]. As expected, there is no break up for q > 4. The increase in $We_{cr}$ for decreasing q is not so much due to a larger We being needed for deformation [see Eq. (26)], but to breakup taking place at larger deformation (here given as $L_{cr}/2r$).

Others have similar, but not exactly equal, results [2,42,43,46]. Discrepancies are not surprising, since it has been shown [2] that the mode of breakup also depends on dG/dt, particularly for high G. This is illustrated in Fig. 10. It implies that, though $r_{cr}$ can be predicted with great accuracy, the size of the resulting droplets cannot.

Figure 9 also gives $n = d \log We_{cr}/d \log q$. This can be used to predict the effect of viscosity changes on breakup, since $(\partial \log r_{cr}/\partial \log \eta_D)_{\eta C} = n$ and $(\partial \log r_{cr}/\partial \log \eta_C)_{\eta D} = -(n + 1)$.

C. Elongational Flow

The effects of plane hyperbolic flow have been studied both theoretically and experimentally [42–45]. See Table 3 for theories. Experiments have mainly been performed with an apparatus consisting of four rotating cylinders causing a flow field as in Fig. 8c, in which a relatively large drop is kept at the origin. Results are shown in Fig. 9. It is seen that $We_{cr}$ is smaller than for simple shear, and that bursting of the drop is also possible for large q.

Bursting does not mean that the drop breaks in two; see Fig. 10, d and e. For small q the drop suddenly attains a slender, pointed shape and small droplets are detached from the pointed ends until the drop volume falls below $4\pi r_{cr}^3/3$. For moderate or high q the drop is suddenly pulled into a thread, which keeps thinning until Rayleigh instability causes it to break into droplets (see Sec. IV.C.3). Hence, a more viscous drop can be broken into smaller droplets.

In a flow field intermediate between simple shear and elongation, breakup is also possible for q > 4. $We_{cr}$ increases as the proportion of elongation in G diminishes, but only for almost completely simple shear does the drop not burst [47].

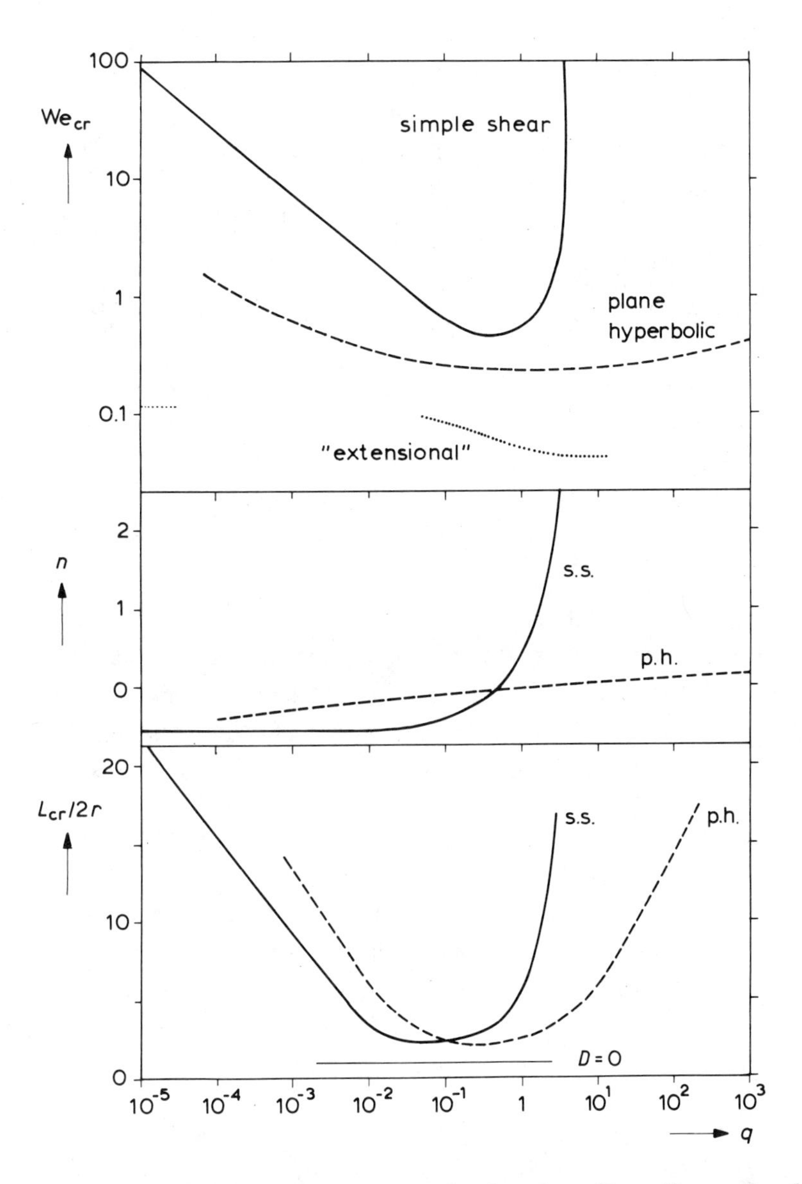


Figure 9. Critical Weber number for breakup $We_{cr}$ (i.e., the drop will break in the region above the curve); viscosity exponent $n = d \log We_{cr}/d \log q$; and the largest drop dimension at burst $L_{cr}$ relative to original drop diameter; for various types of steady flow as a function of the viscosity ratio q. (Data from Ref. 45. The result for axisymmetric extensional flow is purely theoretical, and is from Ref. 44.)

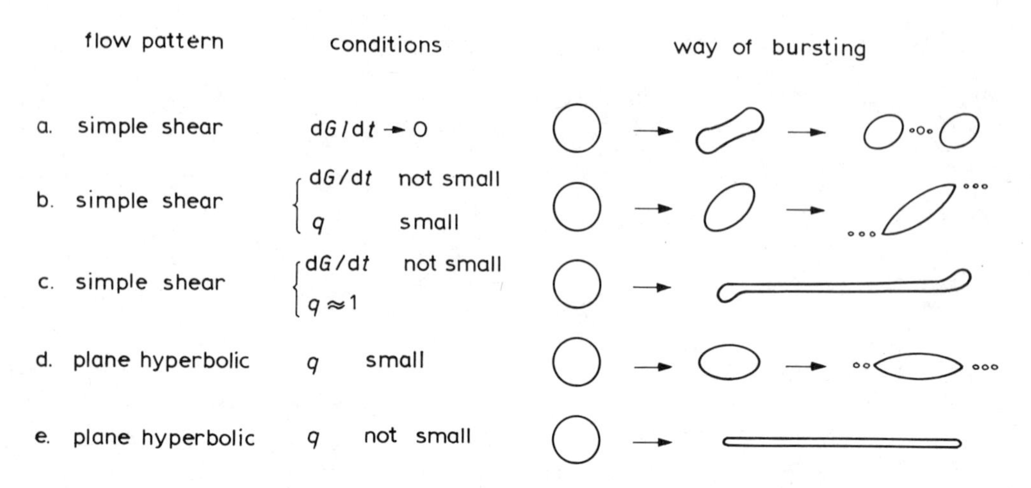


**Figure 10.** Droplet bursting under various conditions. (From Refs. 2 and 43.)

Three-dimensional elongating flow, e.g., converging axisymmetric hyperbolic flow, has apparently not been studied, since it is very difficult to realize for sufficient times. Some theoretical calculations have been made [44], and the result is shown in Fig. 9. It appears that a still smaller We is needed for breakup.

## D. Poiseuille Flow

In laminar flow in a tube or between parallel plates, i.e., in Poiseuille flow, a parabolic velocity profile develops. G is proportional to the distance from the center of the tube. If $r \ll h$ (see Fig. 11), the theory of Sec. V.B can be applied, taking G at the center of gravity of the drop. But if 2r approaches h, a very different situation arises. The drop moves to the center of the stream,

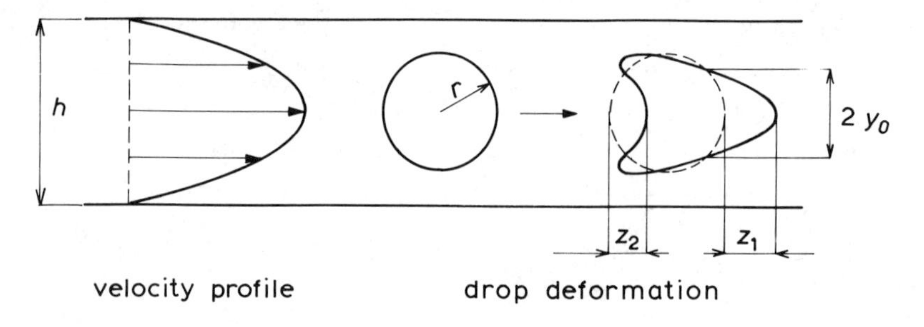


**Figure 11.** Droplet deformation in two-dimensional Poiseuille flow in a very narrow slit. (From Ref. 48.)

where $G = 0$. But since liquid velocity u varies considerably over distances equal to drop size, the drop may still be deformed if u is large (see Fig. 11).

Kiefer [48] has studied this situation for two-dimensional Poiseuille flow, hence between parallel plates. Considering the dynamic thrust of the liquid only (hence neglecting viscous stress), it can be derived from his analysis that the *final deformation*, defined as $(z_1 + z_2)/2r$, is approximately given by $(y_0/r)^2 \rho_C \bar{u}^2 (r/h)^4 (r/\gamma)$. In a first approximation, breakup would occur if

$$\frac{\rho_C \bar{u}^2 r^5}{h^4 \gamma} > 2 \tag{27}$$

where $\bar{u}$ is the average liquid velocity. Since necessarily $2r < h$, an additional condition to the validity of (27) is that

$$\frac{\rho_C \bar{u}^2 h}{\gamma} > 64 \tag{27a}$$

Moreover, Eq. (26) should predict a small deformation.

The theory is particularly applicable for large q, low $\eta_C$, and very small h. As an example, $h = 0.1$ mm, $u = 10\ \text{m s}^{-1}$ (hence $\bar{G} = 2 \times 10^5\ \text{s}^{-1}$), $\gamma = 10\ \text{mN m}^{-1}$, and $\rho_C = 1\ \text{mg m}^{-3}$ would give $r_{cr} = 29\ \mu\text{m}$.

Kiefer [48], using a somewhat more sophisticated approach, found reasonable agreement between theory and the observed maximum drop size ($\eta_C \approx 1$ mPa s, $2r/h = 0.3$ to $0.7$).

## E. Non-Newtonian Liquids

Two deviations from Newtonian behavior should be considered. First, viscosity depends on the velocity gradient, i.e., often shear rate. One commonly speaks of apparent viscosity $\eta'$ and for $G \to 0$, $\eta' = \eta'_0$. In our case, only shear-thinning liquids need be considered, i.e., $\eta'$ decreases with increasing G. The liquid may have a yield stress, which means that below a certain G no flow takes place; hence $\eta'_0 = \infty$. One consequence of shear-thinning can be that G differs more from place to place in an apparatus than would be the case for a Newtonian liquid. In extreme cases (strong dependence of $\eta'$ on G or a yield stress), very high shear rates develop in a very thin layer of the liquid (often near a wall), while elsewhere G is very low. This is called *plug flow;* it happens particularly in dispersions.

For some liquids, only shear-thinning need be considered, and these are called *purely viscous*. Often, a second deviation from Newtonian behavior becomes prominent, i.e., elastic deformation. Such liquids are called *viscoelastic* or *memory fluids;* they also show shear thinning. Characteristic is the relaxation time for disappearance of the elastic stress $\tau_{mem}$. Actually, the situation is more complicated, and more parameters are needed to describe flow behavior; see, e.g., Ref. 49. If $1/G >> \tau_{mem}$, viscoelastic effects are absent. $\tau_{mem}$ can be up to several seconds; in Newtonian liquids it is very small (e.g., $\sim 10^{-12}$ s for water). If a viscoelastic liquid is elongated, the Trouton number defined in Eq. (23) becomes >3, the more so as $G\tau_{mem}$ is larger; $Tr = 10^4$ is possible. Consequently, for high Tr, large elongation rates (G) cannot easily be realized, and the flow pattern becomes such that elongation rate is diminished at the expense of shear rate.

A review on motion, deformation, and burst of drops in viscoelastic liquids is given by Zana and Leal [50]. We will first consider a Newtonian drop in a non-Newtonian liquid. If the latter liquid is purely viscous, deformation and burst are not different from that in a Newtonian liquid, provided that $\eta_C$ is taken as $\eta'$ at the prevailing G. But purely viscous liquids are nearly always dispersions, and the effect of the presence of small particles in the continuous phase on droplet breakup has apparently not been studied, though many emulsions fall within this category (e.g., some ointments and margarine). The effects of plug flow are obvious: breakup only occurs at local sites.

For simple shear in viscoelastic liquids, deformation and burst of drops have been studied experimentally by Flumerfelt [51]. The manner of deformation is much as in a Newtonian liquid, although the ellipsoid aligns more in the direction of flow. The critical drop size for breakup is given by

$$\frac{r_{cr}\eta_D G}{\gamma} = C_1\tau_{mem}G + C_2 \tag{28}$$

$C_1$ and $C_2$ depend on the rheological parameters of both liquids; $C_1$ is of the order of $q_0 = \eta_D/(\eta'_C)_0$, and $C_2$ is generally 0.05 to 0.4, increasing with $q_0$. Note that when $\tau_{mem} \to 0$, Eq. (28) reduces to $We_{cr} = C_2/q$, corresponding to the result for Newtonain liquids. Note also that rearrangement of Eq. (28) and letting $G \to \infty$ leads to

$$r_{cr} \geqslant C_1 \frac{\tau_{mem}\gamma}{\eta_D} \tag{28a}$$

This implies that smaller drops can never be broken up, however large G is. For example, $C_1 = 0.01$, $\tau_{mem} = 1$ s, $\gamma = 10$ mN m$^{-1}$, and $\eta_D = 10$ mPa s would give $r_{cr} \geqslant 1$ cm; and this is certainly not an exceptional case [51].

Hence, breakup can become very difficult if the relaxation time $\tau_{mem}$ becomes appreciable. In practice, somewhat smaller drops can be formed (for instance, 2 to 5 times as small) if the flow is not steady, i.e., when dG/dt is large.

For extensional flow of viscoelastic liquids much the same results are often found [50]. But if $\gamma$ is very close to 0 and $\eta_D$ is large, a different mode of deformation and breakup may occur, the thread into which the drop has been drawn showing rocking motions and then buckling.

The breakup of a non-Newtonian drop in extensional flow in a Newtonian liquid was found to be roughly as for a Newtonian drop with $\eta_D = (\eta'_D)_0$ [51]. But if $\tau_{mem}$ is high, breakup can be more difficult [94].

### F. Application to Emulsification

The conditions under which the theories given above can be applied may be derived from Table 3. Many prerequisites will be more or less met during actual emulsification procedures, but particularly numbers 1, 4, and 7 may not.

To begin with, flow during emulsification will rarely be steady: the droplet will encounter different G and even different flow types during its deformation. But in some cases flow conditions may be fairly constant. It then remains to see whether flow is effectively unbounded.

Let us first consider Couette flow, i.e., flow between coaxial cylinders, in which the outer cylinder moves with a circumferential speed u, gap width being h. We have practically simple shear with $G = u/h$ and application of the theory of Sec. V.B gives as approximate condition for droplet breakup that $u\eta_C r/h\gamma \geqslant 1$. To ensure absence of turbulence, $Re = u\rho_C h/\eta_C$ should be smaller than about $10^3$. Using S.I. units throughout and taking $\rho_C = 10^3$, combination of both conditions gives $h/\eta_C \geqslant (r/\gamma)^{1/2}$. $r/\gamma = 10^{-4}$ is not an excessive condition for emulsions with low $\eta_C$; so $h \leqslant 10^{-2}\ \eta_C$; for $\eta_C = 10^{-3}$ (e.g., water) this implies $h \leqslant 10\ \mu m$. Such a condition is very difficult to realize, and if it can be done, the theory cannot be applied since the condition for unbounded flow $2r << h$ is not fulfilled. This does not mean that droplet breakup is impossible for small h; probably it is even easier. An analysis along the lines of Sec. V.D might lead to a suitable theory.

For other flow regimes, about the same conditions would apply. It follows then that simple-shear theory is not applicable for low $\eta_C$, though it may be useful for larger $\eta_C$.

Similar reasoning may be followed for elongational flow. As an example, axisymmetric converging flow is depicted in Fig. 12; some very approximate relations are also given. Also here, Re will usually be high for sufficient G, but the convergence tends to "squeeze out" turbulence. Another condition is, however, that the time during which elongational flow exists $\tau_{el}$ ($\approx 2/\overline{G}$) should be longer than $\tau_{def}$ ($\approx \eta_D r/\gamma$). Hence, $G_{cr}\eta_D r/\gamma < 2$ is needed, and for breakup roughly $G_{cr}\eta_C r/\gamma > 0.5$ (see Fig. 9). This implies $q < 4$. Hence, breakup will be difficult to realize in this particular configuration for emulsions with low $\eta_C$.

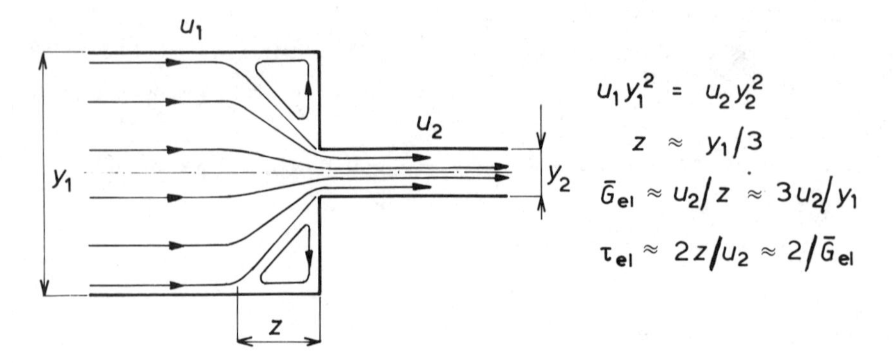


**Figure 12.** Axisymmetric elongational flow of a Newtonian liquid through a constriction in a tube. The relations given apply for high velocity u and $y_1^2 >> y_2^2$.

The condition for absence of inertial effects has been given in Sec. V.A: $r << \eta_C^2/\gamma\ \rho_C$. It follows, again, that $\eta_C$ should not be very low, say, $\eta_C > 0.01$ Pa s.

Usually, surfactants are present during emulsification. Hence $\gamma$ will not be constant (prerequisite 7), and it follows that tangential stress and velocity will not be continuous across the interface (prerequisites 10 and 11); see also Secs. IV.A and VIII.A. The overall effect of the surfactant is probably that disruption of droplets is easier.

It follows that the theory of droplet breakup in steady flow can with some reservations be applied to colloid mills (simple shear) for not too low $\eta_C$ and to flow through constrictions (axisymmetric extensional flow) for $q < 4$. It is disappointing that such studies have, to the author's knowledge, not been made. It can be derived from a few literature reports that the droplet size distribution indeed showed a fairly sharp upper limit, consistent with a well-defined $r_{cr}$ [52,53], and that the magnitude of $r_{cr}$ was probably not far off from what would be predicted. This relates to a study with a colloid mill, with $q = 0.6$ [52], and to flow through constrictions, with $q \approx 10^{-4}$ [53]; in both cases $r_{cr} \approx 10\ \mu m$. There is also one report on the colloid mill from which it follows that for $q \approx 30$, $We \approx 15$ did not lead to breakup [16]. But systematic studies, in which G, $\eta_C$, and $\eta_D$ have been varied and size distributions measured, appear to be lacking.

## VI. DISRUPTION OF DROPLETS IN TURBULENT FLOW

### A. Characterization of Turbulence

It will be clear from Sec. V.F that, particularly for low $\eta_C$, flow conditions during emulsification are mostly turbulent, and even intensely turbulent. Inertial forces are now predominant and can lead to droplet breakup. A general treatment of turbulent flow and its effects on particles and drops is in the books by Levich [1] and Davies [25]. The theory is mainly due to Kolmogorov. See Spielman [5] for collision rates.

Turbulent flow is usually characterized by the local flow velocity u in respect to its time-average value $\bar{u}$. The velocity u fluctuates in a chaotic way. The deviations are characterized by $u' = \langle (u - \bar{u})^2 \rangle^{1/2}$, hence a root-mean-square time average. In the direction of flow $\bar{u}$ equals the overall flow velocity; in directions perpendicular to the flow direction, $\bar{u} = 0$. Usually u' depends on direction; flow is isotropic if u' is the same in any direction.

Only isotropic flow can be characterized in a simple way, but it is difficult to realize practically. The large eddies of turbulent flow, comparable in size to, e.g., pipe diameter, have a fairly small u'; they transfer their kinetic energy to smaller eddies which have a higher u', and these to still smaller ones, and so on. Hence a spectrum of eddy sizes exist. Those that contain the largest amount of kinetic energy are called *energy-bearing eddies*, and have a size $\ell_e$. The smallest ones have such a high u' that their kinetic energy is dissipated into heat. They are called *energy-dissipating eddies*, and their size is $\ell_0$.

Though flow is hardly ever isotropic on the scale of the largest eddies, it is easier to realize this condition locally, on a scale comparable to $\ell_e$ and smaller; and these scales interest us. In pipe flow, Re > 50,000 is a sufficient condition. Hence, we may profitably apply the theory of isotropic turbulence.

In isotropic turbulence conditions are largely governed by the energy density per unit mass $\varepsilon/\rho$ and the kinematic viscosity $\eta/\rho$. The size of the smallest eddies or Kolmogorov scale is given by

$$\ell_0 = \eta^{3/4} \rho^{-1/2} \varepsilon^{-1/4} \tag{29}$$

This is a useful parameter since it gives the smallest size of particles that can conceivably be deformed by turbulent flow. Moreover it is needed to check the validity of other equations.

For distances in the range of $\ell_0$ to $\ell_e$ we have for the velocity in an eddy,

$$u'(\ell) = C \varepsilon^{1/3} \ell^{1/3} \rho^{-1/3} \tag{30}$$

and for the characteristic time of an eddy,

$$\tau(\ell) = \frac{\ell}{u'(\ell)} = C\, \ell^{2/3} \varepsilon^{-1/3} \rho^{1/3} \tag{31}$$

$1/\tau(\ell)$ is called the *eddy frequency*. C is a constant; the constants are usually of the order of unity. Note that $u'(\ell)$ depends on $\ell$, but not on $\eta$.

Turbulent flow strongly enhances the collision rate of small particles. The number of collisions (or encounters) per unit time of particles of radius r in concentration N with a central particle of radius $r_1$ is given by

$$J = 8N(r + r_1)^{7/3}\varepsilon^{1/3}\rho^{-1/3} \tag{32}$$

if $r + r_1$ is of the order of $\ell_e$. From Eq. (32) follows Eq. (4b).

For distances $x \ll \ell_e$, hence over part of an eddy, we have

$$u'(x) = C\varepsilon^{1/2}x\eta_C^{-1/2} \tag{33}$$

and for $(r + r_1) \ll \ell_e$,

$$J \approx 2.6N(r + r_1)^3\varepsilon^{1/2}\eta_C^{-1/2} \tag{34}$$

From Eq. (34) follows Eq. (3b).

## B. Conditions Governing Droplet Size

The theory outlined here is largely due to Kolmogorov, but several others have contributed (e.g., Refs. 1, 38, and 54).

A droplet with a diameter x of the order of $\ell_e$ will experience a velocity difference $u'(x)$ between opposite sides, when it is near an energy-bearing eddy. This leads to a pressure difference, according to Bernoulli's law, of about $\rho_C\{u'(x)\}^2$. For breakup, this should be larger than the Laplace pressure $4\gamma/x$, or in other words, We > 1 is required. It is a matter of debate whether a single meeting of the droplet with a suitable eddy suffices for breakup; it may be that several eddies have to pass the droplet at regular intervals so that the droplet is caused to oscillate. In any case, using Eq. (30) for $u'(x)$ leads to the diameter $x_{max}$ of a droplet that has just broken up

$$x_{max} = C\varepsilon^{-2/5}\gamma^{3/5}\rho_C^{-1/5} \tag{35}$$

This is one of the most useful equations for emulsion formation. It gives the order of magnitude of the largest drops that can be present although not the exact size, since the value of the constant is uncertain. It gives the precise dependence on $\varepsilon$ and $\gamma$; note that $\eta_C$ should have no effect, unless it becomes so high as to prevent isotropic turbulence. In turbulent flow, there is a whole spectrum of eddy sizes, and so a spread in droplet size will result. If flow remains isotropic, a change in $\varepsilon$ will not much alter the relative eddy spectrum; so the shape of the droplet size distribution will not alter much either. Consequently, Eq. (35) would also hold, albeit with a different constant, for average x.

**Table 4.** Sample Calculations for Droplet Disruption in Isotropic Turbulent Flow. Order of Magnitude Only

| Variable[a] | Unit | Eq. | Tank with stirrer | Ultra Turrax | Homogenizer |
|---|---|---|---|---|---|
| $\varepsilon$ | $W\ m^{-3}$ | — | $10^4$ | $10^8$ | $10^{12}$ |
| $\ell_0$ | $\mu m$ | (29) | 18 | 1.8 | 0.2 |
| $x_{max}$ (if $> \ell_0$) | $\mu m$ | (35) | 400 | 10 | (0.25)[b] |
| $x_{min}$ | $\mu m$ | (36) | 0.3 | 0.3 | 0.3 |
| $\tau(x_{max})$ | $\mu s$ | (31) | 2500 | 10 | (0.04)[b] |
| $\tau_{def}$ ($\eta_D = 10^{-3}$)[c] | $\mu s$ | (2) | 20 | 0.5 | 0.01 |
| $\tau_{def}$ ($\eta_D = 1$)[c] | $\mu s$ | (2) | $2 \times 10^4$ | 500 | 10 |
| $x_{max}$ (if $<< \ell_e$) | $\mu m$ | (37) | (3000)[b] | 30 | 0.3 |

[a]Other variables: $\gamma = 0.01\ N\ m^{-1}$, $\rho_C = 10^3\ kg\ m^{-3}$, $\eta_C = 10^{-3}$ Pa s.
[b]Theory does not hold here.
[c]For a globule of size $x_{max}$.

There are, of course, a number of conditions that must be fulfilled. Besides high Re, $x_{max} > l_0$ is needed. Moreover, droplet Reynolds number Re* = $xu'(x)\rho_C/\eta_C$ should be larger than, say 5; otherwise flow near the droplet becomes laminar, Eq. (30) would not hold locally, and breakup by inertial forces becomes impossible. Combining this condition with $We_{cr} = 1$ gives for the minimum drop size to which Eq. (35) applies,

$$x_{min} \approx \frac{3\eta_C^2}{\gamma\rho_C} \tag{36}$$

Moreover, $\tau_{def}$ should not be larger than the eddy time $\tau(x_{max})$ as given in Eq. (31), since otherwise there is not enough time for the drop to deform. $\tau_{def}$ is given in Eq. (2), although it is probably not quite applicable, since inertial effects are involved in the present case. Some examples are given in Table 4. It is seen that the conditions are usually fulfilled, unless $\varepsilon$ is very high or $\eta_D$ is high.

That does not mean that under these conditions drops cannot be broken up. Shinnar [54] has considered the viscous stress caused by eddies, and so for $x << l$. Now Eq. (33) applies and $G \approx u'(x)/x$. In the author's opinion, local flow pattern is probably near to plane hyperbolic flow; hence elongation is the mechanism and the theory of Sec. V.C would apply. Figure 9 shows $We_{cr} \approx 0.5$, with little dependence on q. The result is

$$x_{max} \approx C\gamma\varepsilon^{-1/2}\eta_C^{-1/2} \tag{37}$$

Examples are also given in Table 4. It follows that for very high $\varepsilon$ this mechanism will prevail, the more so when $\eta_C$ is larger.

Droplet size may be the resultant of disruption and coalescence, and if coalescence is the size-determining mechanism, the relations may be different. Shinnar and Church [55] supposed that the energy needed to pull flocculated drops apart before they can coalesce is essential, and that the adhesive energy is proportional to droplet diameter, as follows from the DLVO theory (see Chap. 3). In the author's opinion, $\eta_C$ would also come into play. Since the kinetic energy involved in the relative motion of two equal-sized droplets would be proportional to $\rho_C\{u'(x)\}^2x^3$, use of Eq. (30) leads to

$$x_{max} = \varepsilon^{-1/4} f(\rho_C, \eta_C, A) \tag{38}$$

where A is the Hamaker constant (cf. Chap. 3, Sec. IV.B). There are, however, many simplifications and uncertainties in this theory.

## C. Application to Various Apparatus

Several studies have been made to check Eq. (35), mostly in stirred tanks. Some workers did not add surfactants and used various methods to prevent droplet coalescence before examination. Others did use surfactants, which usually had little effect on the results, except on the constant in the equation. The value of this "constant" is variable in any case. This may be due to differences between average and local $\varepsilon$. For example, the size of a tank in which a stirrer is put has little effect on the local energy density near the stirrer blades, while average $\varepsilon$ is inversely proportional to tank volume. It has indeed been found [56,57] that tank size, if above a certain minimum, has little effect on droplet size, though the time needed to obtain these droplets increases, of course, with size. Hence, as the spread in $\varepsilon$ increases, the constant C will decrease.

Another problem is the definition of $x_{max}$. Some workers take the diameter corresponding to 90 or 95% of the cumulative volume distribution. Others use an average diameter, often $x_{32}$, or specific surface area.

*1. Stirrers*

For stirrers, $\varepsilon$ is usually proportional to $\rho\nu^3X^2$, where $\nu$ is the revolution rate and X stirrer diameter. Insertion in Eq. (35) gives the "Hinze-Clay relation"

$$x_{max} = C\nu^{-6/5}X^{-4/5}\rho_C^{-3/5}\gamma^{3/5} \tag{39}$$

For turbine mixers and $x_{max}$ of the order of 100 μm, this equation has been found to hold good [e.g., Refs. 56 to 59]. Various values for C are found, but it is usually about 0.1.

This holds only for small $\phi$. If $\phi$ is varied up to $\phi \approx 0.3$, $x_{max}$ is usually proportional to $1 + C\phi$; values for C of 2.5 to 5.4 have been reported [56,58, 59]. The increase in particle size with increasing $\phi$ is usually ascribed to increased coalescence, but it should be noted that dispersed particles tend to depress turbulence as well.

Others have found $x_{max}$ to be about proportional to $\nu^{-3/4}$ [54,60], at least below a certain $\nu$ and for very long stirring times. Shinnar [54] reasoned that below a certain critical $\nu$ prevention of coalescence would be dominating, and Eq. (38) indeed gave the correct exponent, though $\phi$ was only 0.05; above that $\nu$, breakup would determine drop size. This could only be true, in the author's opinion, if locally (e.g., very near the stirrer tips) $\varepsilon$ would be much higher than average, leading to small drops that would coalesce again immediately afterward, unless the average $\varepsilon$ would be sufficient to prevent it. Shinnar indeed observed coalescence at high and not at low $\nu$. Whether this situation occurs will probably depend on many conditions, and some other explanation may perhaps be needed.

For turbulent flow between coaxial cylinders, Eq. (35) has been found to hold reasonably well, with $C \approx 0.73$ [38].

In a rotor-stator apparatus, like an Ultra Turrax, much smaller $x_{max}$ can be found, e.g., 10 μm. The author measured $\varepsilon$ from temperature increase and found good agreement with Eq. (35), with $C \approx 1.3$. Wiedman and Blenke [9] investigated emulsification in an Ultra Turrax in great detail. The relation between $\varepsilon$ and $\nu$ is complicated and depends on constructional details, but $\varepsilon$ could be well predicted. Eq. (35) was well obeyed; the constant decreased with the number of passages through the active zone of the stirrer, as is to be expected.

From the results of some authors the width of the size distribution can be derived. It was widely different: $\sigma_2 = 0.3$ to 1.4. In general, $\sigma_2$ decreases with increasing stirring time (see Sec. IX.A).

*2. Pipe Flow*

For turbulent pipe flow Eq. (35) takes the form

$$x_{max} = C\theta^{2/5}\gamma^{3/5}\rho^{-3/5}\bar{u}^{-6/5} \tag{40}$$

where $\theta$ is the pipe radius. Near the wall of the pipe, however, u' is larger than average; so smaller droplets would result; the relation for $x_{max}$ should then also be different, but there is no agreement among authors [1,25]. Experimental results are conflicting also. According to Levich [1], Eq. (40) would predict $x_{max}$ reasonably well if Re is high and time of flow long enough. The width of the droplet size distribution appears to be very wide.

It should be realized that impingement of the liquid stream on baffles or protrusions may locally cause higher energy densities.

*3. Homogenizers*

In homogenizers pressure $p_h$ can be varied from, for instance, from 3 to 50 MPa, leading to $x_{32}$ values of, e.g., 1 to 0.2 μm. The net quantity of energy dissipated per unit volume is equal to $p_h$. Since flow velocity u in the valve is approximately proportional to $(p_h/\rho)^{1/2}$ and the time available for energy dissipation will be proportional to 1/u, Eq. (35) takes the form

$$x_{max} = Cp_h^{-3/5}\gamma^{3/5}z_d^{2/5} \tag{41}$$

where $z_d$ is the effective flow distance over which the energy is dissipated.

If the back pressure (i.e., the pressure in the liquid after leaving the valve) is small and the machine itself not very small (i.e., Re not too low), Eq. (41) agrees with the experiment very well, even for Re as low as $10^4$ [8,17]. This is somewhat surprising, since the condition $x_{max} \gg \ell_0$ is not fulfilled. $Cz_d^{2/5} \approx 0.25$ is S.I. units, which corresponds to a constant of about 2.5 in Eq. (35). Since for $p_h > 5$ MPa size distribution type and width are fairly constant, Eq. (41) also applies to average diameters. Between homogenizers, the constant (i.e., $Cz_d^{2/5}$) varied by a factor of about 2.3. Also $\sigma_2$ may vary; it was mostly between 0.7 and 1.1. Others have found that $x_{43}$ was approximately proportional to $p_h^{-0.7}$ (see Refs. 17, 48, and 61), but this concerned very small machines.

There are two more conditions in order for Eq. (41) to apply. One is that the initial droplet size should not be too large, say, <50 μm, since otherwise a very polydisperse emulsion results. The reason is presumably that insufficient time is available to break up such large drops. Moreover, a large drop may completely fill the valve slit. The other condition is that sufficient surfactant should be present to cover the newly formed droplets, since otherwise, again, a very wide size distribution is found.

When homogenizing milk and cream, $\bar{x}$ increased with $\phi$ when $\phi$ was above a certain value, say, 0.2 for a low and 0.05 for a high $p_h$ [61]. But the dependence on $\phi$ will probably depend strongly on conditions.

### D. Effect of Rheological Properties

Equation (35) predicts that viscosity of neither phase has an effect on droplet size resulting from emulsification in turbulent flow.

In one study with a stirred tank ($q = 0.01$ to 1, $x_{32} \approx 30\ \mu m$) little effect of $\eta_C$ on $x_{32}$ was found. Still lower q, hence larger $\eta_C$, gave an increase in $x_{32}$, but Re was too low to assure turbulence [62].

In another study with several homogenizers and an Ultra Turrax ($q = 5$ to 20, $x_{32} = 0.3$ to $7\ \mu m$) the effect of $\eta_C$ on $x_{32}$ was rather variable, but mostly $\bar{x}$ decreased with increasing $\eta_C$. Particularly in a small homogenizer $\sigma_2$ increased with increasing $\eta_C$ and $x_{32}$ was about proportional to $\eta_C^{-0.25}$ [63]. This may imply a mechanism of breakup that depends both on inertial forces [Eq. (35)] and viscous drag [Eq. (38)]. Higher $\eta_C$ moreover decreases Re, leading to less isotropic turbulence, hence a wider spread in eddy velocities, and so a larger $\sigma_2$. See also Fig. 13.

If $\eta_C$ is increased by adding a hydrophilic polymer, the result may be quite different, at least when the polymer concentration is not too low and when the ratio of the stretched length of the polymer molecule $L_{pol}$ to $\ell_0$ is above a certain

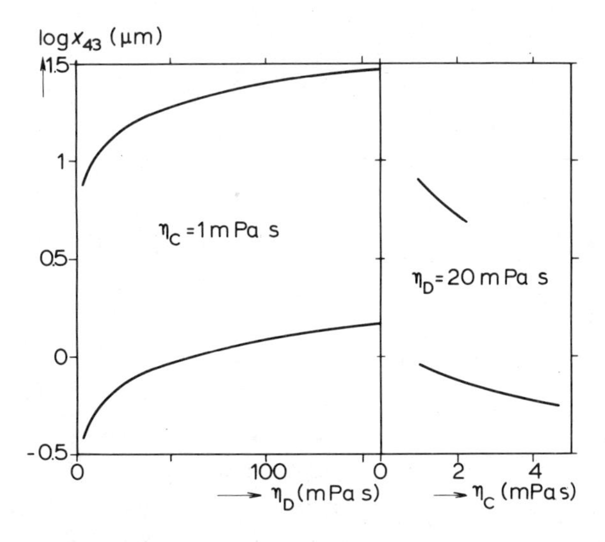


**Figure 13.** Effect of viscosity $\eta$ of either phase on average drop size $x_{43}$ of oil-in-water emulsions. The upper curves pertain to emulsification in an Ultra Turrax, the lower ones to a homogenizer. (Data from Ref. 63.)

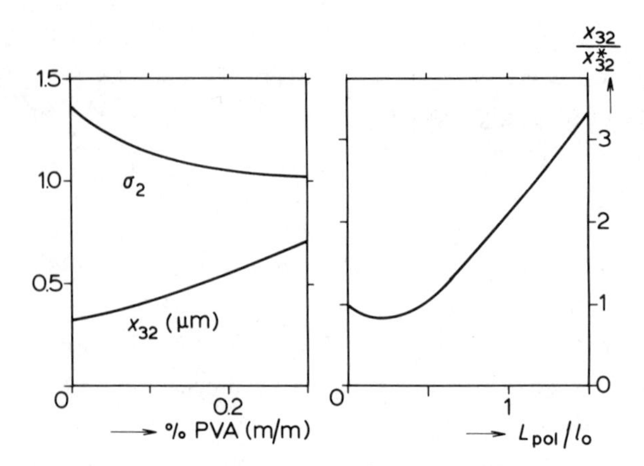


**Figure 14.** Effect of added poly(vinyl alcohol) on average droplet size $x_{32}$ and distribution width $\sigma_2$ of oil-in-water emulsions, produced under various conditions. $L_{pol}$ = stretched length of PVA molecules; $\ell_0$ = Kolmogorov scale; $x^*_{32}$ is drop size obtained for the same $\eta_C$ but without PVA being present. (From Ref. 64.)

value [64]. Some results are shown in Fig. 14. In a homogenizer, $x_{32}$ considerably increased and $\sigma_2$ decreased with increasing $\eta_C$. The explanation is presumably that the polymer depresses turbulence, i.e., it removes the smallest eddies from the spectrum; this would cause both an increase in average size and a decrease in distribution width of the droplets. If $L_{pol}/\ell_0 < 0.5$, the effect of the polymer was like that of any other substance increasing $\eta_C$.

According to Eq. (35), droplet size should be independent of the viscosity of the disperse phase, but often this is not so. In extensive studies with paraffin oil-in-water emulsions with various surfactants, q = 3 to 160, and with various machines (homogenizers, Ultra Turrax, ultrasonic device), resulting in $x_{43} = 0.5$ to 30 μm, $x_{43}$ was found to be proportional to $\eta_D^{0.37}$ for constant $\varepsilon$ [63,72]; see also Fig. 13. The explanation is probably that $\tau_{def}$ becomes increasingly larger than $\tau(\ell)$; see Table 4. This would imply that larger eddies become responsible, hence a higher $\sigma_2$, as is indeed found.

It is somewhat puzzling, however, that the dependence on $\varepsilon$ predicted by Eq. (35) is still obeyed so well. Perhaps we have a situation where viscous drag cannot be neglected, so that breakup would be best described by an intermediate of Eqs. (35) and (38); hence $x_{max}$ would be proportional to, say, $\varepsilon^{-0.45}$. The dependence, if slight, of $x_{max}$ on $\eta_C$ also points to this. But with increasing $\varepsilon$, the discrepancy between $\tau_{def}$ and $\tau(\ell)$ increases, which tends to increase $x_{max}$. The overall result may be that $x_{max}$ is just proportional to $\varepsilon^{-0.4}$. It would need painstaking studies to settle this point.

## VII. Disruption of Droplets by Cavitation

### A. Generation and Effects of Cavitation

Cavitation is the sudden formation and subsequent collapse of cavities, i.e., bubbles containing vapor and gas, in a liquid. It can happen if the local pressure falls below the vapor pressure. The formation of such bubbles is subject to nucleation, and the negative pressures needed for homogeneous nucleation at a reasonable rate are usually excessive. Consequently, heterogeneous nucleation must occur, but the nature of the catalytic impurities responsible is largely unknown. Anyhow, pressure must be much (say, 1 atm) below the vapor pressure for cavitation to occur; but number and formation rate of nuclei, hence cavities, appear to depend on largely unknown properties of the liquid. The lower the local pressure the more regular and reproducible the cavitation.

Low pressures can be invoked by imparting to the liquid a very high velocity or by moving an object at very high speed through the liquid; remember that $p + (1/2)\rho u^2$ = constant. Controlled cavitation is usually generated by sound waves, particularly of ultrasonic frequency. There is a vast amount of literature on this subject; a good review is given by Basedow and Ebert [65]. We will restrict ourselves to aqueous media.

We are concerned with longitudinal sound waves, with a velocity

$$u_s = \lambda\nu = (\kappa\rho)^{-1/2} \tag{42}$$

where $\kappa$ is the adiabatic compressibility. In water we have $u_s = 1500\ \mathrm{m\ s^{-1}}$; but the inclusion of some air or vapor bubbles greatly increases $\kappa$, hence decreases $u_s$. For emulsification $\nu = 20$ kHz is often used, hence $\lambda \approx 75$ mm, very much larger than droplet size. The peak pressure induced by the wave depends on its intensity I (in $\mathrm{W\ m^{-2}}$) according to

$$p_u = \pm(2u_s\rho I)^{1/2} \approx \pm 1730\ I^{1/2} \tag{43}$$

Hence $P_u = \pm 1$ atm ($10^5$ Pa) corresponds to $I = 3\ \mathrm{kW\ m^{-2}}$. Often, I should be around $10^5\ \mathrm{W\ m^{-2}}$ to produce good cavitation.

Cavities may be either oscillating or transient. Transient cavities, which collapse directly after formation and growth, are important for our case. Collapse will occur if $\nu$ is smaller than the natural frequency of the cavity, given by

$$\nu_{nat} \approx \left[\frac{4(p_0 + 2\gamma/r_0)}{\rho}\right]^{1/2} / r_0 \tag{44}$$

where $p_0$ and $r_0$ are pressure and radius of the bubble at the onset of collapse, hence at its maximum size. In water always $\nu_{nat} > 20$ kHz, hence cavities are transient at such ultrasound frequencies. Collapse proceeds extremely fast, over a time much shorter than $1/\nu$.

The fast collapse locally produces a high pressure, e.g., $10^8$ to $10^{11}$ Pa, increasing with decreasing $\nu$. A tiny shock wave (with a much smaller period than the generating ultrasound wave) of very high intensity is produced. If, near the place of collapse, something is present, say, a particle, which can be excited by the shock wave, a very high energy density exists in a very small region, and for a very short time. Cavitation is a way to concentrate the energy of the ultrasound locally. Nucleation determines the distribution of the regions of high $\varepsilon$ in space and time. Quantitative treatment is still imperfect.

Cavitation can produce many effects, and generally the shock waves are held responsible. Chemical reactions in the liquid may occur, erosion of metal surfaces, and disruption of small particles. Emulsion droplets can be disrupted, but, if standing waves occur, they can also be brought to coalescence, presumably in regions of fairly low amplitude. Polymer molecules can be broken, and this subject has been well studied [65]. Above a certain I, the breakup rate increases linearly with I; it is not affected by density differences; it decreases with increasing temperature, or more generally it decreases if the bubble can contain more vapor. It is concluded that the shock waves induce local velocity gradients, causing very rapid movement of parts of the polymer molecule with respect to the solvent, so that the molecule breaks.

### B. Ultrasonic Emulsification

Laboratory investigations have usually been done with piezoelectric or magnetostriction apparatus at $\nu = 10$ to 100, mostly 20, kHz. Typically, a metal probe emitting intense ultrasound from its tip is dipped into the liquid. The liquid may be forced to flow through a narrow slit directly below the tip of the probe.

In the beginning of emulsification, waves are produced at the interface and these give rise to fairly coarse drops, perhaps via the Rayleigh-Taylor mechanism [6,66]; see Secs. IV.B.2 and IV.B.3. Mercury-in-water emulsions, where there is a very large density difference, are formed by this mechanism, without cavitation being needed.

For the breakup of oil droplets cavitation must be responsible, since both phenomena run more or less parallel. Breakup is more efficient at lower $\nu$, in accordance with the higher shock-wave pressure this will produce. (Experiments with $\nu < 20$ kHz are difficult to perform; the sound would be intolerable for the experimenters.) The exact mechanism of droplet disruption is not known. It may be that the drop oscillates at a natural frequency until it bursts; the frequency of the shock waves (cavitation noise) shows a fairly broad spectrum

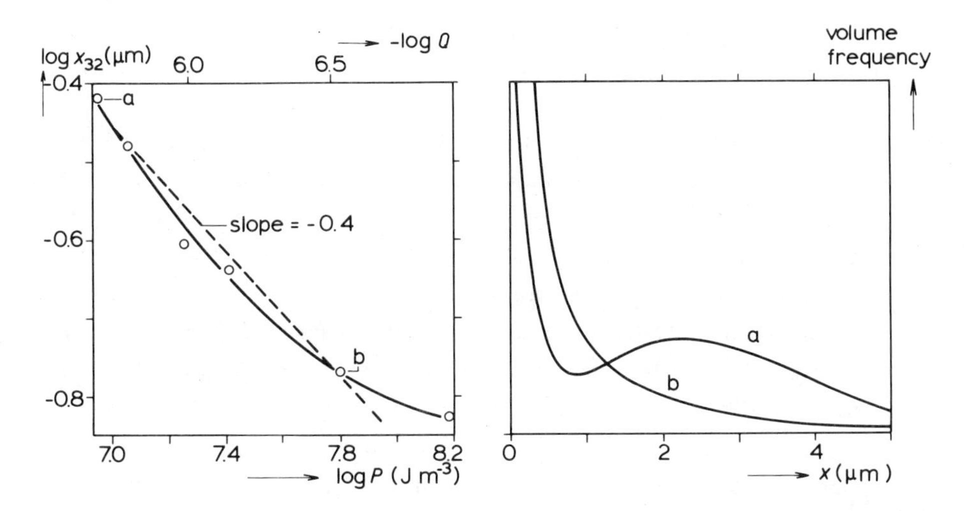


**Figure 15.** Average droplet size $x_{32}$ produced by ultrasonic emulsification as a function of the flow rate Q or of the energy dissipated per unit volume P. Two examples of the resulting volume-frequency distributions are given. (Data from Ref. 69.)

and some frequencies will fit, but it is difficult to decide whether the intensity and duration are sufficient. The imploding cavity may tear off parts from nearby drops, particularly if the cavity forms at the surface of the drops; but the number of cavities will usually be much less than the number of drops. Finally, local velocity gradients may cause effects as discussed in Sec. VI.B. In any case, the stresses acting on the droplets cannot be predicted.

An example of results is in Fig. 15. It is seen that the dependence on $\varepsilon$ is about as predicted by Eq. (35). It does not make much difference whether $\varepsilon$ is varied by varying flow rate or I [68; unpublished results by the author, 1976]. The occurrence of a bimodal size distribution may be an indication that two mechanisms act. If treatment is prolonged, average droplet size becomes very small [66,68,69]. This is in accordance with the fact that $\varepsilon$ is quite high, but only locally.

Results of ultrasonic emulsification are poorly reproducible. Both average size and distribution width are quite variable; $\sigma_2$ can be from 0.4 to 2.2 [63,68, 69]. Several factors may be important. For instance, preemulsification may sometimes cause larger droplets, not smaller [68]. In another investigation, adding glycerol to an O/W emulsion (0 to 47%, i.e., increasing $\eta_C$ 4.5 times) increased the resulting $x_{32}$ from 0.36 to 1.5 μm, and decreased $\sigma_2$ from 1.6 to 0.6. Note that the effect is quite different in the case of turbulence (Fig. 13). But vapor pressure $\kappa$ and possibly nucleating properties are altered by adding glycerol, as well as increasing $\eta_C$.

The effects of adding polymers to the continuous phase and of altering $\eta_D$ are comparable to the results shown in Figs. 13 and 14 [63]. This may point to the importance of rapid velocity fluctuations for droplet breakup.

Two aspects need mentioning. First, in some cases, stable emulsions can be made, even of fairly high $\phi$, without the use of surfactants [6]; the explanation can only be guessed. Second, "overprocessing" may occur, i.e., an increase in particle size with prolonged treatment, after an initial decrease in size [67,68]; the increase can be as much as a factor of 4. The explanation may be that the ultrasonic waves slowly degrade the surfactant, particularly when it is a protein. Alternatively, prolonged cavitation may lead to a loss of catalytic impurities required for nucleation of cavities. Further studies would be needed to settle this point.

It has been claimed that emulsification in a "liquid whistle" (see Fig. 2) is also caused by ultrasonic waves, but very few experimental results on droplet size are available. It appeared that in a small laboratory machine, droplet sizes were somewhat larger than obtained in a homogenizer operating at the same pressure drop (namely, 1.5 MPa). Moreover, removing the knife from the machine did not much alter the results [70]. Consequently, turbulence may still be an important mechanism in such machines, though some effect of cavitation cannot be ruled out.

### C. Cavitation in Homogenizers

It has often been argued that in the valve slit of a high-pressure homogenizer, negative pressures occur, so that cavitation can take place. Cavitation would then be responsible for droplet breakup. This aspect has recently been reviewed and studied by Kurzhals [71]. He showed that under particular conditions cavitation can indeed be intense. This is when there is some back pressure $p_{ex}$ at the exit of the homogenizing valve by use of a second valve, so that the Thoma number $Th = (p_{ex} - p_v)/(p_h - p_v)$ is about 0.2; here $p_h$ is homogenization pressure and $p_v$ is vapor pressure. For a valve with a very short passage time, the greater part of the energy dissipated in the homogenizing valve could then be ascribed to ultrasound energy. The proportion of ultrasound energy is correlated with a decrease in droplet size. It appears likely that in such cases droplet breakup is mainly caused by cavitation. However, for a normal valve and low $p_{ex}$ (e.g., $Th = 0.05$, as is common) ultrasound energy amounted to only a few percent of the total energy dissipated, and turbulence must be considered to be the main mechanism for breakup.

## VIII. Effects of Surfactants

### A. Factors Affected

Presence of surfactant may affect emulsion formation in many ways, and one aspect has been considered in Sec. IV.A. But experimental studies to elucidate the mechanisms are difficult. Most processes during emulsification take <0.1 ms (see Table 1), and interfacial properties can usually not be measured at such time scales. In other words, we need to know dynamic surface properties (see, e.g., Van den Tempel [96] for a review) and only the static properties are known. Moreover, most emulsifiers are mixtures of several components, or contain at least some impurities, and the more surface-active components tend to predominate in the adsorbed layer. This is precisely what will happen at the interface between two layers in a beaker. However, it may be different for adsorption at emulsion droplets, because of the then much larger surface-to-volume ratio, which has the effect that the minor components are soon depleted. Hence, the composition of the surface layer is different.

Purely from reasoning, the following effects may be expected (see also Sec. IV.A).

*1. Interfacial tension* will be lowered, and its effective value during droplet breakup affects droplet size. If viscous forces are predominant, x is proportional to $\gamma$ (Sec. V); if the inertial effects of turbulence are determining, x is proportional to $\gamma^{3/5}$ (Eq. 35). Such effects occur in wire drawing also (Sec. IV.C.4). Effective $\gamma$ will be partly determined by adsorption rate. Eq. (34) gives some help and examples are to be found in Table 1. The smaller the droplet, the slower the predicted adsorption rate, but experimental studies in turbulent streaming have apparently not been made. Surfactant may also be transported by spreading along the interface (see also Sec. I.C).

If emulsification is continued, equilibrium between the amount adsorbed ($\Gamma$, in mol m$^{-2}$) and the concentration in the continuous phase ($m_C$, in mol m$^{-3}$) may be expected. But then, effective $\gamma$ also depends on the surface dilational modulus $E = d\gamma/d(\ln A)$, because the drop surface is being enlarged during deformation and breakup. E depends on many factors. When one considers a plane interface between quiescent bulk phases in one of which a surfactant is soluble, E can be theoretically derived [89,90] from the properties of longitudinal waves in the interface, i.e., when the interface is periodically compressed and expanded:

$$E = \frac{-d\gamma/d(\ln \Gamma)}{(1 + 2\zeta + 2\zeta^2)^{1/2}} \tag{45}$$

$$\zeta = \frac{dm}{d\Gamma}\left(\frac{D}{2\omega}\right)^{1/2} \tag{45a}$$

D is the diffusion coefficient of the surfactant (usually $\sim 10^{-9}$ $m^2$ $s^{-1}$) and $\omega = 2\pi\nu$, $\nu$ being the frequency of the longitudinal wave; for transient expansion d(ln A)/dt may be used instead of $\omega$. It may be noted [89] that E equals the modulus in a thin film, i.e., half the Gibbs elasticity, for $h = (4D/\omega)^{1/2}$ [see Eq. (9)].

E depends much on the type of surfactant, since that determines the relations of $d\gamma/d(\ln \Gamma)$ and $dm/d\Gamma$ with m. Eq. (45) cannot be expected to be quite correct, particularly since activity rather than m should be used; above the critical micelle concentration $m_{cr}$, it is no longer valid.

In the case of emulsification the situation is much more complicated since the interface is not plane, transport of surfactant is by convection, liquid streaming along the interface, and depletion of surfactant in the film between closely approaching drops. Nevertheless, some qualitative conclusions can be drawn. For m = 0, E = 0. When m increases, E at first rapidly increases, since for most surfactants $-d\gamma/d(\ln \Gamma)$ and $d\Gamma/dm$ are already large for small m; moreover, d(ln A)/dt is large (of the order of $10^4$ $s^{-1}$); hence $\zeta$ is small. E can become quite high, at least of the order of 20 mN $m^{-1}$. For higher m, $dm/d\Gamma$ becomes large, so E decreases, and above $m_{cr}$ E is mostly small, even if d(ln A)/dt is high.

It may be concluded that E will mostly decrease during emulsification, since $m_C$ decreases (by depletion) and d(ln A)/dt increases (smaller drops are deformed more quickly). Hence effective $\gamma$ during breakup will, in the final stages of emulsification, be between the equilibrium value and $\gamma_0$, probably about midway ($\gamma_0 = \gamma$ in the absence of surfactant). When insufficient surfactant is present to give something like a fully covered interface, effective $\gamma$ will be nearer to $\gamma_0$.

When the surfactant is a polymer, molecules may unfold at the interface after adsorption. This implies that $\gamma$ may also be lowered by further adsorption of segments of one molecule. The rate of such a process is poorly known, and so is E.

2. *The surface free energy needed* for enlarging the drop surface is now $\gamma\,\Delta A + A\,\Delta\gamma$, instead of $\gamma\,\Delta A$ alone; $\Delta\gamma$ is determined by E. It implies that more energy is needed, though less than $\gamma_0\,\Delta A$. The effect is small, and probably insignificant. In addition, viscous resistance to surface enlargement may cost energy if the surface dilational viscosity $\Delta\gamma[d(\ln A)/dt]$ is large at the prevailing rate of extension. Nothing is known about this, but the effect is presumably insignificant too.

3. *An interfacial-tension gradient* arises when liquid streams along the interface, generating a tangential stress of $-d\gamma/dz$ (see Sec. IV.A). Assuming that E is high, i.e., negligible relaxation of the gradient due to adsorption and desorption of surfactant in the available time, the stress can be of the order of

$(\gamma_0 - \gamma)/x$, and for $\gamma_0 - \gamma = 10$ mN m$^{-1}$ and $x = 1$ μm this gives $\sim 10^4$ Pa. The result is that tangential stress and velocity are no longer continuous across the droplet boundary (prerequisites 10 and 11 of Table 3). This implies that internal circulation in the droplet is impeded or even prevented, and this facilitates droplet deformation, hence breakup.

4. *Coalescence* of newly formed drops is slowed down. Emulsification is mostly not possible without surfactants; remember that many natural substances contain some surfactants, such as fatty acids and monoglycerides in edible oils. See, for the effects of surfactants on coalescence, Chap. 3.

The stabilizing mechanism during emulsification is usually ascribed to the Gibbs-Marangoni effect, following Van den Tempel [23]. We consider the film between two approaching droplets; see Sec. IV.A for the Gibbs elasticity $E_f$ of such a film. During emulsification, adsorption of surfactant is mostly incomplete, so that $\gamma$ decreases with time. But the film will be rapidly depleted of surfactant. Hence $\gamma$ will be higher, and highest where the film is thinnest, because $E_f$ is highest there [see Eq. (9)]. Consequently, an interfacial-tension gradient develops. This causes surfactant molecules to move in the interface in the direction of highest $\gamma$, and this motion drags liquid along with it; this is the Marangoni effect. It can also be explained as a liquid motion caused by the tangential stress $-d\gamma/dz$. Small gradients may already cause considerable streaming. The liquid of the continuous phase thus forces its way into the gap between the approaching droplets, thereby driving them apart. Hence, a stabilizing mechanism, but only if the surfactant is in the continuous phase. If it is in the disperse phase, Gibbs elasticity is low, hence no $\gamma$ gradients and no liquid streaming. This may well be the explanation of Bancroft's rule (see Sec. I.A).

In the case of polymers with their high molecular weight, $E_f$ would be very high, since the molar concentration is low. But additional adsorption of polymer segments could take place, causing a low $E_f$. The actual situation is mostly unknown.

5. *Interfacial instability* may occur, which can facilitate emulsification. Interfacial-tension gradients cause Marangoni effects (see point 4) and if these are strong, the interface may be so distorted as to shred droplets.

The classical analysis is by Sternling and Scriven [97] recently improved by Gouda and Joos [98]. A prerequisite for interfacial instability is a gradient of the chemical potential of a solute across the interface. A plane interface is, for instance, unstable if the solute lowers $\gamma$ and is transported from a more viscous to a less viscous phase. This causes convection near the interface (the so-called roll cells) and a persistent $\gamma$ gradient. Hence the instability, which may have a small enough wavelength to cause fairly small droplets to be formed.

But now we come into the field of spontaneous emulsification, in which other factors may play a role too (e.g., Ref. 75), and which is outside the scope of this chapter.

But even without spontaneous emulsification, Marangoni effects may aid in droplet formation. If turbulent motion or mechanical vibration causes the interface to be locally curved with the concave side in the phase that provides the surfactant, the curved part will have a higher $\gamma$, since it receives the smallest quantity of surfactant molecules per unit surface area; hence transport of surfactant and streaming of liquid toward the point of strongest curvature; hence an unstable situation. This may well cause droplet shredding. The mechanism may explain the small droplets obtained with the nascent-soap method, or with some nonionic surfactants dissolved in the future disperse phase. However, the combined action of turbulence and dynamic interfacial effects forms a formidable problem that is far from being understood.

Droplet shredding naturally occurs easiest when $\gamma$ is very low, which may be the case for some surfactants if present at high concentration. Viscous drag exerted by the continuous phase on the surface of a droplet may cause local accumulation of surfactant (see point 3). This may cause (particularly for proteins) a very low $\gamma$ at one spot of the droplet. It is not known to what extent this facilitates its disruption. If, as is sometimes assumed, a negative interfacial tension [74] can occur, this will certainly promote droplet shedding [75]. Even in the absence of a viscous drag, $\gamma$ may temporarily be negative for some combinations of surfactants, but then a microemulsion is formed (see Chap. 4).

## B. Effect on Droplet Size

In several studies the surfactant concentration has been varied, and generally droplet size decreases with increasing concentration. But droplet size also depends strongly on the type of surfactant, and even for high $\varepsilon$ and sufficient surfactant to give a more or less complete coverage of the interface, droplet size may vary by a factor of, say, 3. It is poorly correlated with $\gamma$, as measured under static conditions. Width of the size distribution is also quite variable.

Interpretation of results is difficult since surfactant is depleted by adsorption, and it is the concentration in the continuous phase $m_C$, not the total concentration $m_E$ that determines $\gamma$ and $\Gamma$, and thereby emulsion properties. Studies in which $m_C$ was measured after emulsification have apparently not been made, except for certain polymers. We will treat various kinds of surfactants separately.

### *1. Anionic Surfactants*

Concentration in the continuous phase can be calculated for the case of sodium dodecyl sulfate from careful studies by Rehfeld [76] who emulsified benzene in

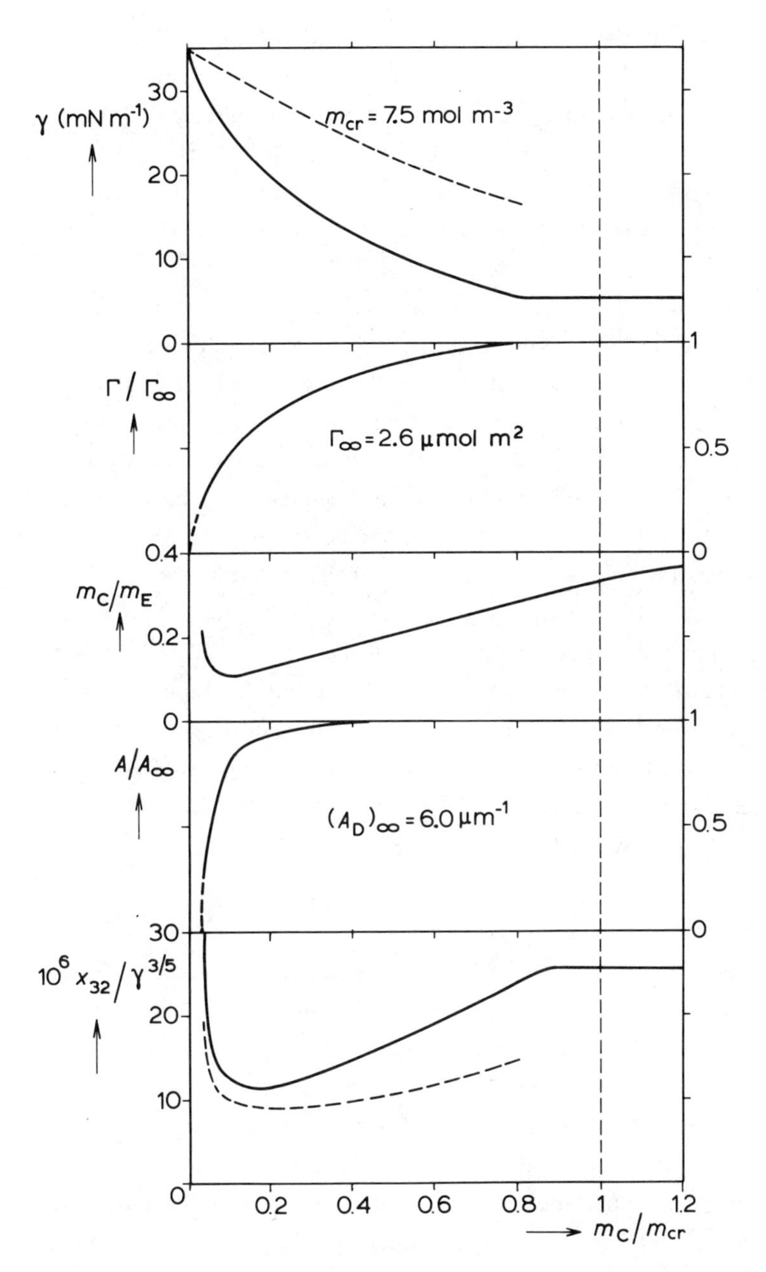


**Figure 16.** Benzene-in-water emulsions with sodium dodecyl sulfate as the surfactant produced with a high-speed mixer; $m_C$ = molar surfactant concentration in the aqueous phase after emulsification. Given are interfacial tension $\gamma$, surface excess $\Gamma$, fraction of surfactant not adsorbed $m_C/m_E$, interfacial area $A_D$, and average droplet size $x_{32}$ over $\gamma^{3/5}$. The broken lines are rough estimates for the situation during droplet breakup, taking the surface dilational modulus E into account, and for $\Delta A = A/2$. E is calculated from Eq. (45), assuming $D = 10^{-9}$ $m^2$ $s^{-1}$ and $2\omega = 10^5$ $s^{-1}$. (Data from Ref. 76.)

water, $\phi = 0.5$, with a high-speed mixer; he made sure to obtain equilibrium. Results are summarized in Fig. 16. For very low $m_C$ (or probably very low $\Gamma$), coalescence seems to be predominant, and emulsification is impossible. Note that the curve of specific surface area extrapolates to zero for finite $m_C$. At $m_C$ near $m_{cr}$, the parameters measured no longer change, except, of course, the fraction of surfactant adsorbed. The lowest x (highest $A_D$) obtained, namely, at high $m_C$, will probably be determined by Eq. (35). This implies that $x_{32}/\gamma^{3/5}$ should be constant. Taking $\gamma$ as measured, a noticeable minimum in the curve is found. But effective $\gamma$ will be higher (Sec. VIII.A, point 1), and in Fig. 16 an estimate is given that is probably on the high side. It is seen that the minimum persists. It is tempting to explain this by the Gibbs-Marangoni effect since $E_f \to 0$ for both $m \to 0$ and $m \to \infty$. For intermediate m, $E_f$ is high and a strong Marangoni effect, hence stability to coalescence, may be expected (see also Sec. IX.B). But other explanations cannot be ruled out.

Other anionic surfactants have been studied less well, but the trend is probably the same [77,78]. Realize that $m_{cr}$, the relations between $\Gamma$, $\gamma$, and $m_C$ and the minimum $\Gamma$ needed to prevent rapid coalescence will differ between surfactants. Realize also that $m_C$ cannot easily be predicted; it depends on the total amount of surfactant added, as well as on $\phi$ and the resulting $x_{32}$, hence on $\varepsilon$. Curves giving $\bar{x}$ as a function of total concentration may therefore be quite different; it also makes a difference whether m is expressed per unit volume of emulsion ($m_E$) or of disperse phase: $m_E/(1 - \phi)$.

### 2. *Nonionic Surfactants*

These are among the most common. The relation between $\bar{x}$ and $m_C$ is often different. There is mostly no clear $m_{cr}$, and a plateau value of A or $\bar{x}$ is not reached [72,78]; values for $\Gamma$ are unknown. Hence $\bar{x}$ decreases with increasing m, up to very high concentration. The higher $m_E$, the lower the energy needed to produce a finely dispersed emulsion, though this depends on surfactant type. We come near to spontaneous emulsification or solubilization.

Temperature may be important since it affects the solubility of the surfactant in both phases. Shinoda [79] introduced the concept of the *phase-inversion temperature* $T_{inv}$ for nonionic surfactants. Usually for $T < T_{inv}$ an O/W and for $T > T_{inv}$ a W/O emulsion is formed. Near $T_{inv}$ extensive solubilization occurs, and droplet size may become very small if much surfactant is present, even with gentle stirring. For example, $x_{30}$ may for emulsification at a few degrees below $T_{inv}$ be 0.4 times $x_{30}$ obtained at a much lower temperature [80]. The emulsions are, however, not very stable near $T_{inv}$; so they should be rapidly cooled. See Chap. 5 for additional comments on the phase-inversion temperature.

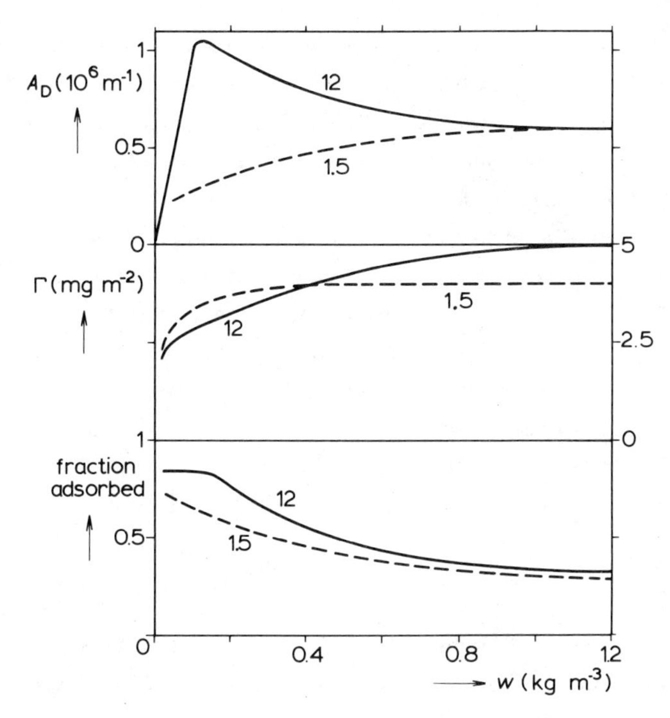


Figure 17. Oil-in-water emulsions; surfactant is poly(vinyl alcohol) (with 12 and 1.5% acetate groups, respectively; degree of polymerization 1600); produced with an Ultra Turrax; $\phi = 0.2$. Interfacial area $A_D$, surface excess $\Gamma$, and fraction of PVA adsorbed as a function of the concentration of PVA in the aqueous phase after emulsification w. (Data from Ref. 81.)

### 3. *Polymers*

When these are used as surfactants, the situation is more complicated, since adsorption is more or less irreversible, and a true equilibrium situation is not reached (see Chap. 1). Because of this, and because of the difficulty of predicting transport rate and polymer unfolding, $\gamma_{eff}$ is very difficult to estimate. We include proteins which are extensively used as surfactants in edible emulsions.

Extensive studies have been made by Lankveld and Lyklema [81] with poly(vinyl alcohol) (PVA) of various acetate content and various molecular weights. Some typical results are shown in Fig. 17. The behavior of PVA with 12% acetate groups is striking; some other polymers show roughly the same trend, e.g., sodium caseinate [81] and partially esterified poly(methacrylic acid), if highly ionized [82]. The authors assume that the molecules do not or hardly unfold at the interface during emulsification, so that $E_f$ can be high if $w_C$ is low. This implies a strong Marangoni effect. At higher $w_C$, $E_f$ is too low, and the coalescence rate increases. The PVA with 1.5% acetate groups would be far more flexible, hence rapid unfolding at the interface occurs, and so we obtain a low $E_f$ even at low $w_C$ and no Gibbs-Marangoni effect occurs. Consequently $A_D$ is lower ($x_{32}$ higher) despite the probably lower $\gamma$. The authors [81,82] give additional evidence to corroborate their explanation, and it is certainly an attractive hypothesis. But other effects may also play a role.

Most polymers show a behavior like either of the two types of Fig. 17, and that of the broken lines appears to be more common. Quantitatively, the results are quite variable [21,68,81–84]. They depend on polymer type, molecular weight, solvent quality, and, if it is a polyelectrolyte (e.g., a protein), pH and ionic strength. Moreover, emulsification "history" may affect results [21,68,81]: diluting the emulsion at a certain stage, or adding extra polymer, or preemulsification. As yet there is no adequate theory to provide a logical framework for the multitude of different observations.

### C. Formation of Surface Layers

For small-molecule surfactants, Gibbs's law for adsorption applies, and $\Gamma$ is determined by the equilibrium concentration in the bulk (see, e.g., Fig. 16). We will therefore restrict the discussion to polymers.

Polymer adsorption is more or less irreversible, though partly attached molecules can presumably be torn off if G is very large near the droplet. Polymer molecules can unfold at the interface, depending on the space (unoccupied interface) and time available. Moreover, the interface can be enlarged (see Fig. 1, lines 1 and 3), causing stretching and unfolding; and it can be compressed (Fig. 1, lines 3 and 4), possibly causing polymer molecules to fold up. They may also desorb, but often not very easily. Compression of an interfacial layer of some polymers may cause almost zero interfacial tension. Compression caused by coalescence is very variable.

Because of these effects, $\Gamma$ of an emulsion is mostly not equal to the $\Gamma$ found from adsorption experiments under quiescent conditions. In other words, there is no true adsorption isotherm. Mostly, $\Gamma$ is better predicted from $w_E/A_E$, i.e., the amount of polymer added per unit surface area of disperse phase formed, rather than from the equilibrium concentration $w_C$. $A_E = 6\phi/x_{32}$ and $x_{32}$ is a function of $\varepsilon$, $\phi$, $w_E$, etc. $\Gamma$ will, moreover, depend on the type of polymer and further composition of both phases, but if these are equal, $\Gamma$ is, within limits, determined by $w_E/A_E$ irrespective of how it has come about [68,83,84]. Coalescence does indeed favor high $\Gamma$ [81,83], as is suggested by the results in Fig. 17 for $w_C \approx 0.1$. But the conditions mentioned will also largely determine coalescence. If, however, emulsification is not yet complete, $\Gamma$ may be different.

From several literature results [68,81–84] it is deduced that roughly

$$\Gamma \approx C_1 \log \frac{w_E}{A_E} + C_2 \tag{46}$$

Taking both $\Gamma$ and $w_E/A_E$ in mg $m^{-2}$, we find that mostly $C_2 \approx 0$, while $C_1$ varies about from 1 to 6 for different polymers and compositions. There is mostly a minimum $\Gamma$ below which no emulsion can be made; it varies from 0.1 to 2 mg $m^{-2}$. Some polymers show a maximum $\Gamma$, i.e., the curves level off at this value; it mostly varies from 2 to 5 mg $m^{-2}$. $\Gamma$ values above 10 mg $m^{-2}$ have not been reported. It should be stressed that Eq. (46) is an empirical relation and only gives a trend: on close inspection, the curves often show deviations, though rarely over 20%.

If small-molecule surfactants are present in addition to the polymer, $\Gamma$ (polymer) often tends to be lower as the surfactant displaces the polymer from the interface [83]. Generally, $\bar{x}$ also becomes smaller. Remember that, for instance, many oils contain at least a little surfactant.

Polymers are usually mixtures, e.g., of different molecular weight. Protein preparations often contain different species, and there may be preferential adsorption [83]. As long as adsorption is irreversible, preference will depend on the rates of transport to and of unfolding at the interface. If transport is diffusion controlled, the smallest molecules arrive fastest. But it is usually controlled by convection, and often Eq. (34) will apply. Transport of molecules of radius $r_1$ to a droplet of radius r is then proportional to $m(r + r_1)^3$; for high shear rate, the same relation holds. Assuming that mass concentration w is proportional to $mr_1^3$, we obtain for $\Gamma$ in unit mass per unit area as a function of time,

$$\Gamma(t) = Cw_C r \left(1 + \frac{r_1}{r}\right)^3 F(t) \tag{47}$$

where $F(t) = t$ as long as $\Gamma(t) \ll \Gamma(t = \infty)$. Often, $r_1 \ll r$, so that $r_1$ has little effect on adsorption rate. But if $r_1$ becomes of the order of r, larger surfactant molecules or particles are preferentially adsorbed. This effect is mostly strongest for the smallest droplets in the emulsion since for them $r_1/r$ can be largest. Particularly in protein solutions, fairly large aggregates can be present, and these are "swept out" of the solution during emulsification. Smaller droplets have indeed a far higher $\Gamma$ than the larger ones if there is a large spread in surfactant particle size [85].

We are now coming to the field of adsorbing solid particles, instead of polymers, and $\Gamma$ can be much higher, e.g., 20 mg $m^{-2}$ or more [83].

If the interfacial area initially produced by droplet disruption is appreciably larger than the area that can be covered by the polymer available, aggregates of droplets may be formed (see Fig. 1, line 7). Two droplets now share one or more polymer molecules; the same applies when solid particles are the "surfactant." Usually, the aggregates are disrupted again (Fig. 1, line 8), and coalescence occurs (Fig. 1, line 4), until equilibrium is reached. This is because the newly disrupted droplets usually come into a region where $\varepsilon$ (hence disruptive force) is too low to achieve droplet breakup, but large enough to disrupt aggregates. However, if $\varepsilon$ suddenly drops by a large amount, aggregation may proceed while the aggregates cannot be broken down. This may happen in a homogenizer, just outside the valve, at least for certain polymers or protein aggregates as surfactants [86]. Very viscous emulsions result.

## IX. Formation Rate

### A. Disruption

The time needed for disruption of a drop is probably a few times that needed for deformation as given by Eq. (2): about $\eta_D r/\gamma$. Disruption must be repeated many times. Assuming a drop to be broken into two equal droplets, reduction of r by a factor 10 needs about 10 steps; hence several times 10 steps must normally occur. But every next disruption takes less time, and barring other restrictions, emulsification could mostly be complete within a second or so. However, it often takes minutes.

Part of the explanation may be that machines capable of disrupting small droplets may be less suitable for the beginning of emulsification and vice versa. Consequently, better results (i.e., smaller drops, narrower distribution) are often observed when preemulsifying the mixture by stirring; this applies, for instance, to colloid mills [18] and homogenizers [17]. These machines are able to reduce droplets from, say, 10 to 0.25 μm in a very short time, e.g., in a homogenizer in <0.1 ms. Also high-speed rotor-stator machines and ultrasonic emulsifiers can usually produce fine emulsions within seconds.

Much slower processes often occur in mixers, particularly in the final stages of emulsification. Figure 18 shows this and also that for a higher surfactant concentration it takes longer to obtain equilibrium. This seems to be a general rule, also for anionic surfactants near or above the critical micelle concentration (see Fig. 16). Mostly, more surfactant gives smaller droplets, hence a greater number of disruptions, and the increase in emulsification time is not surprising.

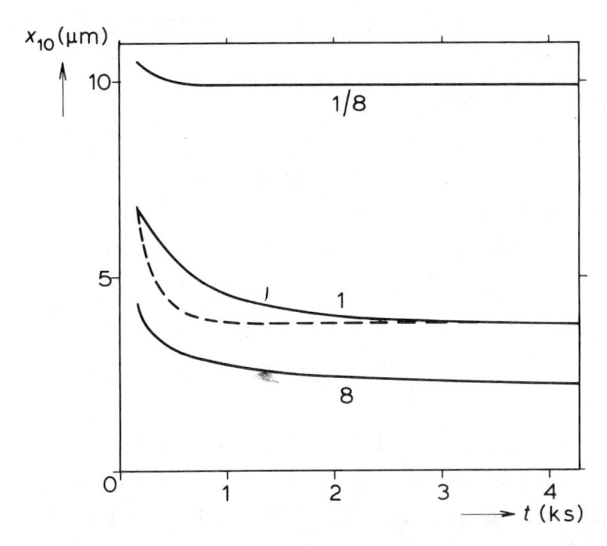


Figure 18. Oil-in-water emulsions; surfactant is Tween 80; produced in a high-speed mixer; $\phi = 0.5$. Average droplet size $x_{10}$ as a function of stirring time t; parameter is amount of surfactant added in kg/100 kg oil. The broken line is according to Eq. (48). (Data from Ref. 72.)

But if more surfactant does not cause smaller droplets, the time needed to obtain equilibrium droplet size also increases. See further Sec. IX.B.

Type of surfactant [15,72,82] and stirring rate [87] also effect emulsification rate; it increases with revolution number. The effect of volume fraction is not so clear; it is probably small if the concentration of surfactant in the continuous phase after emulsification is kept constant.

The variability of emulsification rate, which need not be correlated to final droplet size, is yet another reason for care in interpreting results of studies on the effect of process or product variables on droplet size or creaming rate.

The disruption of droplets is often treated as a first-order reaction: $dN/dt = KN$ [6,87,88]. Assuming a final number $N_\infty$ (since there is a final average diameter $\bar{x}_\infty$ which depends on $\varepsilon$, $\gamma$, etc.) leads to

$$\bar{x} = \frac{\bar{x}_\infty}{1 - \exp(-Kt)} \tag{48}$$

where K is the rate constant ($s^{-1}$). However, this equation does not fit the results (see, e.g., Fig. 18). There is probably a whole spectrum of reaction rates, depending, for instance, on droplet size. The energy density (or the velocity gradient) will vary from site to site in the apparatus, and it may take a long time before all the volume elements of the emulsion have passed the zone where conditions for breakup are most favorable.

The latter consideration implies that much energy may be dissipated at a rate where it cannot cause further breakup. This is, of course, a waste of energy, and continuous operation can be more efficient. If a certain average drop size is to be reached in batchwise operation, it is usually advantageous to increase the energy input per unit time, for instance by increasing stirring rate and shorten the duration. But this should not be overdone since otherwise $\bar{x}$ and particularly $\sigma$ (relative spread in x) tend to become very high. As a matter of fact, decrease of $\bar{x}$ and $\sigma$ also occur when a continuous emulsification process, like homogenizing, is repeated: during a single passage not all the volume elements pass the regions most favorable for breakup. These trends are illustrated in Fig. 19.

A slow decrease of $\bar{x}$ with continued stirring may also be caused by a steady increase in temperature which causes for instance $\eta_D$ to decrease, facilitating breakup (Sec. VI.D). Sometimes "overprocessing" is reported, i.e., an increase of $\bar{x}$ after some time. Except in the case of emulsification by cavitation (see Sec. VII.B), this can, in the author's opinion, only be due to changes in composition, particularly degradation of surfactants. A clear example is copious air inclusion, which may cause surface denaturation of such proteins as egg

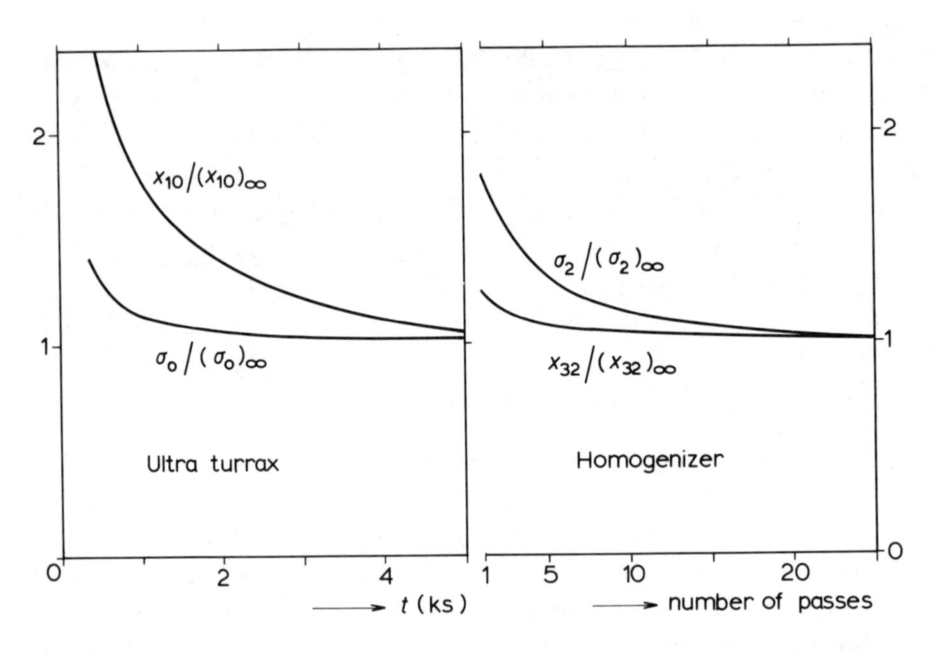


**Figure 19.** Average droplet size x and distribution width $\sigma$ divided by their final values as a function of stirring time and number of passes, respectively. (Data from Refs. 15 to 17.)

albumin, and we may then expect coalescence rate to increase with time. It was indeed found that in this case $\bar{x}$ went through a minimum [21].

### B. Coalescence

It is sometimes assumed [6,15] that the final value of $\bar{x}$ obtained is the result of a balance between coalescence and disruption. The idea is that coalescence rate would increase more rapidly with increasing N, based on the assumption that it is governed by collision rate. We would then obtain $dN/dt = K_1N - K_2N^2$, $K_2$ being the rate constant for coalescence, and $N_\infty = K_1/K_2$ [6]. This explanation is very different from that offered in Secs. V and VI. It would perhaps account for the slow attainment of equilibrium. In the author's opinion, however, both $K_1$ and $K_2$ would strongly depend on droplet size.

Coalescence is certainly possible and even very probably during emulsification. When two phases are stirred in a vessel until equilibrium is reached, and thereupon the conditions are altered, e.g., surfactant concentration is lowered [21] or stirring rate slowed down [59], drop size often increases, although slowly. The coalescence appears to be quite variable, and the relations among process conditions, composition, and coalescence rate are still obscure.

However, coalescence rate need not be a strong function of collision rate. When a slightly unstable emulsion is brought under high shear, this has negligible effect on coalescence rate, although the collision rate is increased by some orders of magnitude [73]. Coalescence is probably of importance only for droplets that have just been formed and have not yet acquired an equilibrium adsorption layer. Such drops are mostly formed from the breakup of one "parent" drop; so they are close to each other in any case, largely independent of $\phi$. One would assume Gibbs-Marangoni effects, as described in Sec. VIII.A, point 4, to be responsible for the prevention of coalescence.

This may be in agreement with the results presented in Fig. 16, particularly the effect of surfactant concentration (m) on droplet size. To this should be added, that although increasing $m_C$ beyond $0.5\ m_{cr}$ did not alter $x_{32}$, it did alter the time needed to obtain this value: the higher $m_C$, the longer it took. This fits in with a decrease in Gibbs elasticity, hence less Marangoni effect, and thus more coalescence.

It may be concluded that the quantitative importance of coalescence is unknown. It is probably quite variable, and may be an important cause of the differences in drop size found with different surfactants. Since coalescence probably takes place with freshly formed drops, this notion can be reconciled with the apparent existence of a critical drop size (larger ones are broken up) depending mainly on hydrodynamic conditions, and much less on $\phi$.

## X. Summing Up

As far as our *understanding* of emulsification is concerned, deformation and breakup of droplets in laminar flow has been studied best; the effect of surfactants might be further investigated. Practical studies to test the results obtained with single drops in emulsifying machines in which laminar flow predominates (e.g., colloid mills) are badly needed.

For the most part, emulsification takes place in turbulent flow. There is a reasonable understanding of important parts of the process, but a quantitative theory comprising all aspects has yet to be worked out. This will be difficult since there are so many variables which mutually influence each other's effect. Particularly the role of the surfactant, taking into account dynamic surface properties and coalescence rate, needs further investigation. The effect of $\eta_D$ is not yet explained either.

The mechanism of droplet breakup by cavitation is still to be elucidated. At the present we can at best qualitatively explain some trends.

With a view to *application* of our knowledge, we will summarize the effect of the most important variables, mainly on the resulting droplet size, though other aspects may be important too. A problem is that the variables often do not affect the result in the same way under different conditions.

i. *Type of apparatus*. Many considerations besides droplet size obtained will determine the choice of method or machine. But for high $\eta_C$, say, above 0.1 or 1 Pa s, machines like a colloid mill will usually be most efficient, and otherwise apparatus producing turbulence or cavitation. Figure 20 gives some results, but it should be noted that the stirrer was used batchwise and that continuous-flow machines are generally more efficient. If a very small $\bar{x}$ is needed, the high-pressure homogenizer is nearly always to be preferred. Generally, apparatus dissipating the available energy in the shortest time possible (hence giving high $\varepsilon$) and having, in addition, the most homogeneous flow conditions give best results. The size of the machine may also have effect (see Sec. II.B).

ii. *Energy input*. See Fig. 20 and Secs. VI.C and VII.B. In a colloid mill one would expect $\bar{x}$ to be proportional to $P^{-1}$. Remember that there may be quite a difference between total and net energy input, particularly in small machines.

iii. *Duration* for batchwise operation and *repetition* for continuous operation. Generally, both average droplet size and distribution width decrease, but the rate of decrease rapidly falls (see Fig. 19). Remember that, for instance, creaming rate is roughly proportional to $(x_{53})^2$, so that decreasing the distribution width may profoundly improve creaming stability. Nevertheless, prolonged or often-repeated treatments are seldom economic.

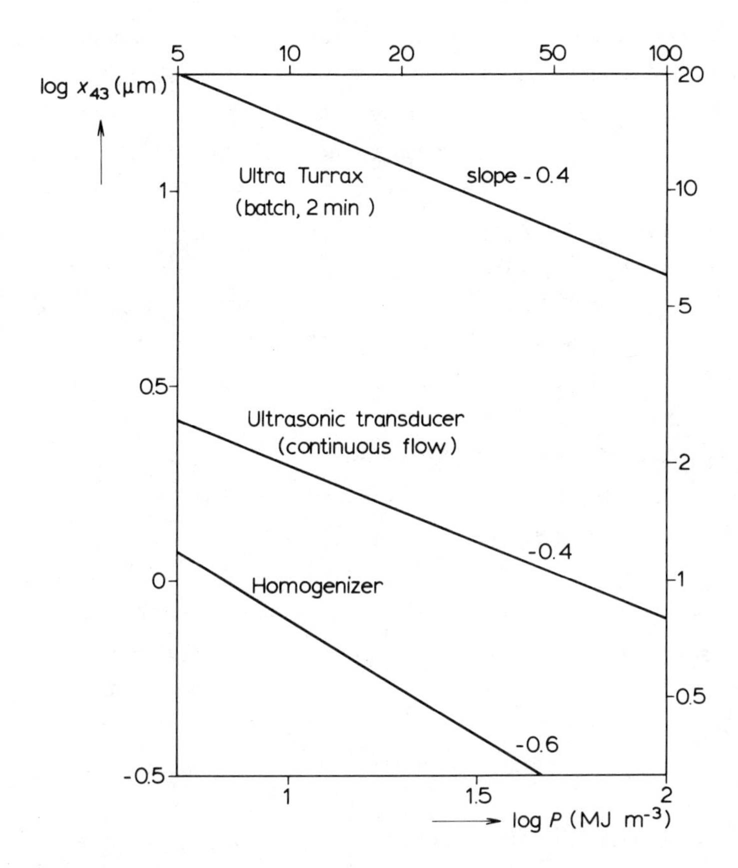


**Figure 20.** Average droplet size $x_{43}$ as a function of net energy input P (varied by varying intensity, not duration of treatment) for diluted paraffin oil-in-water emulsions produced in various machines. Approximate average results. (From author's experiments, 1976.)

iv. *Rheological properties* of either phase. $\eta_C$ first affects Re, hence type of flow, and consequently, the prevailing mechanism for droplet breakup: laminar or turbulent flow, with viscous or inertial forces, respectively.

In simple shear flow, disruption much depends on $q = \eta_D/\eta_C$. In elongational flow, q has little effect. In both types, $\bar{x}$ decreases with increasing $\eta_C$. See Fig. 9. In turbulent flow viscosities have less effect (see Fig. 13).

Relations are more complicated for non-Newtonian liquids. See Sec. V.E for laminar flow. Polymers in the continuous phase may depress turbulence, hence impair disruption in turbulent flow (Fig. 14). If the disperse phase contains solid particles, e.g., fat crystals in oil, droplet disruption can mostly not be achieved by forces acting through the continuous phase, and a ball or roller mill should be used.

v. *Surfactant* type and concentration. The surfactant is of primary importance for other properties of the emulsion, noticeably stability to coalescence or flocculation, but it also markedly affects droplet size. The higher $\varepsilon$, the smaller the effect of surfactant on $\bar{x}$. For most anionic surfactants, higher concentration gives lower $\bar{x}$ up to a plateau value, reached near the critical micelle concentration; for many nonionics decrease in $\bar{x}$ continues, and the more surfactant is used, the lower the energy needed to obtain a certain $\bar{x}$; for polymers, $\bar{x}$ also decreases at first with increasing concentration, but it sometimes increases again. See Sec. VIII.B, Fig. 17.

vi. *Volume fraction.* Mostly $\bar{x}$ increases with increasing $\phi$ if other conditions are kept constant, but often the effect is small. If the amount of surfactant per unit volume of disperse phase is kept constant, the equilibrium concentration of surfactant in the continuous phase increases with increasing $\phi$, and particularly with a nonionic surfactant, this may cause a decrease in $\bar{x}$ [72].

vii. *Method of preparation.* Several variants exist, for instance, in the way of adding both phases and the surfactant. It is often advantageous to add the disperse phase bit by bit to the continuous one; a practice similar to this is preemulsification. Since the energy consumed is mostly proportional to the volume treated, it is often advantageous to make the emulsion at high $\phi$ and subsequently dilute it to the desired $\phi$. See Sec. II.C for variants in the way of adding surfactant. If the surfactant consists of a mixture of polymers, the sequence of adding them may profoundly affect the composition of the surface layers since adsorption is more or less irreversible.

viii. *Temperature* affects many variables: $\eta_D$, $\eta_C$ (and along with them Re), solubility of the surfactant in either phase. Often, smaller droplets are formed at higher temperature. For polymers, temperature may affect solvent quality, and this may, in turn, influence polymer conformation, hence $\Gamma$. General rules cannot be given.

## Acknowledgments

The author is indebted to Drs. M. van den Tempel, A. Prins, and J. Lucassen for helpful comments; to Dr. J. Lucassen, Mr. J. E. Schaap, and Dr. E. Tornberg for the use of unpublished material; and to Mr. H. Stempher for the performance of hitherto unpublished experiments.

## Symbols

| | |
|---|---|
| $a$ | acceleration ($m\ s^{-2}$) |
| $A$ | area ($m^2$) |
| $A_D$, $A_E$ | specific surface area ($m^{-1}$) |
| $b$ | $kr$, $kR$, or $k\theta$ |
| $B$ | smallest dimension of deformed drop (m) |
| $C$ | undefined constant |
| $D$ | relative deformation; diffusion constant ($m^2\ s^{-1}$) |
| $E$ | surface dilational modulus ($N\ m^{-1} = kg\ s^{-2}$) |
| $E_f$ | Gibbs elasticity ($N\ m^{-1}$) |
| $f(x)$ | number frequency distribution ($m^{-4}$) |
| $F$ | a function |
| $g$ | acceleration due to gravity ($m\ s^{-2}$) |
| $G$ | velocity gradient ($s^{-1}$); strain rate ($s^{-1}$) |
| $h$ | gap or slit width (m) |
| $I$ | intensity of sound wave ($W\ m^{-2} = kg\ s^{-3}$) |
| $J$ | collision frequency ($s^{-1}$) |
| $k$ | wave number $= 2\pi/\lambda$ ($m^{-1}$) |
| $K$ | rate constant ($s^{-1}$ or $m^{-3}\ s^{-1}$) |
| $\ell$ | size of eddy (m) |
| $\ell_e$ | size of energy-bearing eddy (m) |
| $\ell_0$ | size of energy-dissipating eddy, Kolmogorov scale (m) |
| $L$ | largest dimension of deformed drop (m) |
| $m$ | molar concentration ($mol\ m^{-3}$) |
| $m_{cr}$ | critical micelle concentration ($mol\ m^{-3}$) |
| $N$ | number per unit volume ($m^{-3}$) |
| $p$ | pressure ($Pa = kg\ m^{-1}\ s^{-2}$) |
| $P$ | energy uptake ($J\ m^{-3} = kg\ m^{-1}\ s^{-2}$) |
| $q$ | viscosity ratio $= \eta_D/\eta_C$ |
| $Q$ | flow rate ($m^3\ s^{-1}$) |
| $r$ | radius of sphere (m); distance from center (m) |
| $R$ | radius of curvature (m) |
| $S_n$ | nth moment of size distribution ($m^{n-3}$) |
| $t$ | time (s) |
| $T$ | temperature (K) |
| $u$ | velocity ($m\ s^{-1}$) |
| $u'$ | root-mean-square velocity fluctuation ($m\ s^{-1}$); eddy velocity ($m\ s^{-1}$) |
| $w$ | mass concentration ($kg\ m^{-3}$) |
| $x$ | drop diameter (m) |

$\bar{x}$ unspecified average diameter (m)

$x_{nm}$ average diameter, derived from nth over mth moment (m)

$X$ diameter of impeller or stirrer (m)

$y$ distance in a direction perpendicular to z (m)

$z$ distance in the direction of flow or of the interface (m)

Re Reynolds number

Th Thoma number

Tr Trouton number

We Weber number

$\alpha$ amplitude (m)

$\beta$ growth factor of instability ($s^{-1}$)

$\gamma$ interfacial tension ($N\ m^{-1} = kg\ s^{-2}$)

$\Gamma$ surface excess ($mol\ m^{-2}$ or $kg\ m^{-2}$)

$\varepsilon$ energy density ($W\ m^{3} = kg\ m^{-1}\ s^{-3}$)

$\eta$ viscosity ($Pa\ s = kg\ m^{-1}\ s^{-1}$)

$\eta'$ apparent viscosity (Pa s)

$\theta$ radius of cylinder (m)

$\kappa$ adiabatic compressibility ($Pa^{-1} = m\ s^{2}\ kg^{-1}$)

$\lambda$ wavelength (m)

$\nu$ frequency (Hz); revolution rate ($s^{-1}$)

$\rho$ mass density ($kg\ m^{-3}$)

$\sigma_n$ size-distribution width, i.e., relative standard deviation of drop size, weighted with the nth moment of the distribution

$\tau$ relaxation time (s); characteristic time (s)

$\phi$ volume fraction or disperse phase

$\omega$ angular frequency = $2\pi\nu$ ($s^{-1}$)

Subscripts

cr critical

C (of the / per unit of) continuous phase

D (of the / per unit of) disperse phase

E (of the / per unit of) emulsion

el in elongational (extensional) flow

## References


1. V. G. Levich, *Physicochemical Hydrodynamics*, Prentice-Hall, Englewood Cliffs, 1962, Chaps. 5 and 8.
2. S. Torza, R. G. Cox, and S. G. Mason, *J. Colloid Interface Sci. 38*:395 (1972).
3. A. Prosperetti, in *Proc. Intern. Colloquium on Drops and Bubbles* (D. J. Collins, M. S. Plesset, and M. M. Saffren, eds.), Jet Propulsion Laboratory, Passadena, 1974, p. 357.
4. C. A. Miller and L. E. Scriven, *J. Fluid Mech. 32*(3): 417 (1968).
5. L. A. Spielman, in *The Scientific Basis of Flocculation* (K. J. Ives, ed.), Sijthoff & Noordhoff, Alphen a. d. Rijn, 1978, pp. 63-88.
6. E. S. R. Gopal, in *Emulsion Science* (P. Sherman, ed.), Academic, London and New York, 1968, Chap. 1.
7. P. Becher, *Emulsions: Theory and Practice*, 2nd ed., Reinhold, New York, 1965, Chap. 7.
8. H. Mulder and P. Walstra, *The Milk Fat Globule: Emulsion Science as Applied to Milk Products and Comparable Foods*, C. A. B., Farnham Royal and Pudoc, Wageningen, 1974, Chaps. 6, 9, and 10.
9. H. Wiedman and H. Blenke, *Chemie-Anlagen und Verfahren 1976*(4):82.
10. L. W. Phipps, *J. Phys. D: Appl. Phys. 8*:448 (1975).
11. S. Torza, R. G. Cox, and S. G. Mason, *Phil. Trans. Royal Soc. 269*(1198): 295 (1971).
12. W. R. Marshall, Jr., *Atomization and spray drying*, Chem. Eng. Progr. Monogr. Ser. 2 *50*:(1954).
13. K. Masters, *Spray Drying*, Leonard Hill, London, 1972, Chap. 6.
14. V. A. Jones, C. W. Hall, and G. M. Trout, *Food Technol. 18*:1463 (1964).
15. L. Djaković, P. Dokić, and P. Radivojević, *Kolloid-Z.Z. Polym. 244*:324 (1971).
16. L. H. Rees, *J. Soc. Dairy Technol. 21*:172 (1968).
17. P. Walstra, *Neth. Milk Dairy J. 29*:279 (1975).
18. S. Tsukiyama, A. Takamura, and M. Nakano, *Yakugaku Zasshi 93*:1131 (1973); *Chem. Abstr. 79*:138118 (1973).
19. C. E. Swift, C. Lockett, and A. J. Fryer, *Food Technol. 15*:468 (1961).
20. E. Tornberg and A. M. Hermansson, *J. Food Sci. 42*:468 (1977).
21. K. N. Pearce and J. Kinsella, *J. Agr. Food Chem. 26*:716 (1978).
22. P. Walstra and H. Oortwijn, *Neth. Milk Dairy J. 29*:263 (1975).
23. M. van den Tempel, *Proc. 3d Int. Congr. Surface Activity*, Cologne, 1960, Vol. 2, p. 573.
24. A. Prins, in *Foams* (R. J. Akers, ed.), Academic, London, 1976, Chap. 5.
25. J. T. Davies, *Turbulence Phenomena*, Academic, New York and London, 1972, Chaps. 8 to 10.
26. E. H. Lucassen-Reynders and J. Lucassen, *Advan. Colloid Interface Sci. 2*:347 (1969).
27. S. Chandrasekhar, *Hydrodynamic and Hydromagnetic Stability*, Clarendon, Oxford, 1961, Chaps. 10 to 12.
28. E. S. Raja Gopal, in *Rheology of Emulsions* (P. Sherman, ed), Pergamon, Oxford and London, 1963, pp. 15-25.
29. J. Bear, *Dynamics of Fluids in Porous Media*, American Elsevier, New York, 1972.
30. R. Raghavan, *Diss. Abstr. B. 31*:4734 (1971).
31. S. Tomotika, *Proc. Royal Soc. A. 150*:322 (1935).

32. F. D. Rumscheidt and S. G. Mason, *J. Colloid Sci. 17*:260 (1962).
33. T. Mikami and S. G. Mason, *Canad. J. Chem. Eng. 53*:372 (1975).
34. E. G. Richardson, in *Flow Properties of Disperse Systems* (J. J. Hermans, ed.), North-Holland, Amsterdam, 1953, Chapt. 2.
35. T. Mikami, R. G. Cox, and S. G. Mason, *Int. J. Multiphase Flow 2*:113 (1975).
36. B. J. Carroll and J. Lucassen, *J. Chem. Soc. 2. Faraday Trans. 1 70*:1228 (1974).
37. B. J. Carroll and J. Lucassen, in *Theory and Practice of Emulsion Technology* (A. L. Smith, ed.), Academic, London, 1976, Chap. 1.
38. J. O. Hinze, *Amer. Inst. Chem. Eng. J., 1*:289 (1955).
39. M. van den Tempel, *Chem. Eng. 1977*(2):95.
40. A. Acrivos, in *Proc. Intern. Colloquium on Drops and Bubbles* (D. J. Collins, M. S. Plesset, and M. H. Saffren, eds.), Jet Propulsion Laboratory, Passadena, Calif., 1974, p. 390.
41. G. I. Taylor, *Proc. Royal Soc. A. 138*:41 (1932).
42. G. I. Taylor, *Proc. Royal Soc. A. 146*:501 (1934).
43. F. D. Rumscheidt and S. G. Mason, *J. Colloid Sci. 16*:238 (1961).
44. D. Barthès-Biesel and A. Acrivos, *J. Fluid Mech. 61*:1 (1973).
45. H. P. Grace, *Eng. Foundation 3d Res. Conf. on Mixing*, Andover, 1971 (cited in Ref. 40).
46. H. J. Karam and J. C. Bellinger, *Ind. Eng. Chem., Fund.* 7:576 (1968).
47. S. V. Kao and S. G. Mason, unpublished results, 1976.
48. P. Kiefer, Ph.D. Dissertation, University of Karlsruhe, 1977.
49. S. Middleman, *The Flow of High Polymers*, Interscience, New York, 1968.
50. E. Zana and L. G. Leal, in *Proc. Intern. Colloquium on Drops and Bubbles* (D. J. Collins, M. S. Plesset, and M. M. Saffren, eds.), Jet Propulsion Laboratory, Passadena, Calif, 1974, p. 428.
51. R. W. Flumerfelt, *Ind. Eng. Chem., Fund. 11*:312 (1972).
52. A. Takamura, S. Noro, S. Ando, and M. Koishi, *Chem. Pharm. Bull. 25*:2617 (1977).
53. H. Mulder, F. C. A. den Braver, and T. G. Welle, *Neth. Milk Dairy J. 10*:214 (1956).
54. R. Shinnar, *J. Fluid Mech. 10*:259 (1961).
55. R. Shinnar and J. M. Church, *Ind. Eng. Chem. 52*:253 (1960).
56. J. W. van Heuven, Ph.D. Dissertation, Technical University of Delft, 1969.
57. S. Tsukiyama and A. Takamura, *Chem. Pharm. Bull. 23*:616 (1975).
58. T. Vermeulen, G. M. Williams, and G. E. Langlois, *Chem. Eng. Sci. 22*:1267 (1967).
59. Y. Mlynek and W. Resnick, *Amer. Inst. Chem. Eng. J. 18*:122 (1972).
60. W. A. Rodger, V. G. Trice, and J. H. Rushton, *Chem. Eng. Progr. 52*:515 (1956).
61. J. D. S. Goulden and L. W. Phipps, *J. Dairy Res. 31*:195 (1964).
62. S. Tsukiyama, A. Takamura, Y. Fukuda, and M. Koishi, *Chem. Pharm. Bull 24*:414 (1976).
63. P. Walstra, *Dechema Monogr. 77*:87 (1974).
64. P. Walstra, *Chem. Eng. Sci. 29*:882 (1974).
65. A. M. Basedow and K. H. Ebert, *Adv. Polymer Sci. 22*:83 (1977).
66. M. K. Li, *Diss. Abstr. B 37*:5237 (1977).
67. E. S. Rajagopal, *Proc. Indian Acad. Sci. 49A*:333 (1959).
68. E. Tornberg, *J. Food Sci. 45*:1662, 1980.

69. P. Walstra, *Neth. Milk Dairy J. 23*:290 (1969).
70. J. E. Schaap, unpublished results 1974.
71. H. A. Kurzhals, Ph.D. Dissertation, University of Hannover, 1977.
72. L. M. Djaković and P. Dokić, *Tenside Detergents 14*:126 (1977).
73. M. A. J. S. van Boekel and P. Walstra, *Colloids Surfaces 3*:99 (1981).
74. C. A. Miller and L. E. Scriven, *J. Colloid Interface Sci. 33*:360 and 371 (1970).
75. G. Smits, Ph.D. Dissertation, University of Groningen, 1977.
76. S. J. Rehfeld, *J. Colloid Interface Sci. 24*:358 (1967).
77. L. Jürgen-Lohmann, *Kolloid-Z. 124*:77 (1951).
78. E. L. Rowe, *J. Pharm. Sci. 54*:260 (1965).
79. K. Shinoda, *J. Colloid Interface Sci. 24*:4 (1967).
80. K. Shinoda and H. Saito, *J. Colloid Interface Sci. 30*:258 (1969).
81. J. M. G. Lankveld and J. Lyklema, *J. Colloid Interface Sci. 41*:475 (1972).
82. J. T. C. Böhm and J. Lyklema, in *Theory and Practice of Emulsion Technology* (A. L. Smith, ed.), Academic, London, 1976, Chap. 2.
83. H. Oortwijn and P. Walstra, *Neth. Milk Dairy J., 33*:212 (1979).
84. E. Tornberg, *J. Sci. Food. Agr., 29*:867 (1978).
85. P. Walstra and H. Oortwijn, *Neth. Milk Dairy J., 36*:103 (1982).
86. L. V. Ogden, P. Walstra, and H. A. Morris, *J. Dairy Sci. 59*:1727 (1976).
87. S. Tsukiyama, A. Takamura and N. Nakura, *Chem. Pharm. Bull. 22*:1902 (1974).
88. E. S. Rajagopal, *Kolloid-Z. 167*:17 (1959).
89. J. Lucassen, in *Physical Chemistry of Surfactant Action* (E. H. Lucassen-Reynders, ed.), Surfactant Sci. Series, Vol. 10, Marcel Dekker, New York, 1979.
90. J. Lucassen and M. van den Tempel, *Chem. Eng. Sci. 27*:1283 (1972).
91. G. F. Scheele and B. J. Meister, *Amer. Inst. Chem. Eng. J. 14*, 9 and 15 (1968).
92. R. Kumar, *Chem. Eng. Sci. 26*:177 (1971).
93. R. Kumar and Y. P. Saradhy, *Ind. Eng. Chem., Fund. 11*:307 (1972).
94. M. van den Tempel, personal communication, 1979.
95. A. Prins, C. Arcuri, and M. van den Tempel, *J. Colloid Interface Sci. 24*:84 (1967).
96. M. van den Tempel, *J. Non-Newtonian Fluid Mech. 2*:205 (1977).
97. C. V. Sternling and L. E. Scriven, *Amer. Inst. Chem. Eng. J. 5*:514 (1959).
98. J. H. Gouda and P. Joos, *Chem. Eng. Sci. 30*:521 (1975).
99. R. G. Cox, *J. Fluid Mech. 37*:601 (1969).

# 3

# Emulsion Stability

THARWAT F. TADROS / Imperial Chemical Industries, PLC Ltd., Bracknell, Berkshire, England

and

BRIAN VINCENT / University of Bristol, Bristol, England

## I. General Considerations

### A. Introduction

There are many examples one could quote of naturally occurring, apparently stable emulsion systems: milk and the oil-in-water (O/W) and water-in-oil (W/O) emulsions associated with oil-bearing rocks are just two examples. One could also list a wide range of technological products and processes which involve the production of stable emulsions, for example, the food industry (e.g., mayonnaise), detergency (removal of oil deposits), pharmacy (drug administration), cosmetics (skin creams), agricultural sprays (pesticides and herbicides), and the spraying of bitumen emulsions on road surfaces. This list serves to illustrate the wide range of technologies involved. In addition, there are a number of related systems which, although arguably should not strictly be classified as emulsion systems, do have many structural and physical features in common with "true" emulsions. Examples here would be biological cells (e.g., blood cells) and liposomes, as well as liquid aerosol droplets and foams. Over the last 30 years or so there has been a growing interest in the application of physics to these systems in order to understand the factors controlling their stability and breakdown. That this is an important facet is illustrated by the fact that in some cases the need is to produce emulsions of long-term stability, whereas in other cases emulsions of controlled, limited stability are required (particularly in multistage processes where it is essential that an initially stable emulsion eventually breaks, e.g., bitumen road emulsions). Finally, there are those cases where naturally occurring, unwanted stable emulsions have to be broken down. Examples of the last-mentioned group would be the W/O emulsions which build up in oil-storage tanks or in engine fuels and the O/W emulsions that arise in effluent waters.

It will be shown in later sections that emulsions can be stabilized against breakdown by one or more different mechanisms. Basically there are three of these to consider: charge stabilization, steric stabilization, and stabilization by adsorbed solid particles at the liquid/liquid interface.

With regard to charge stabilization, one important difference between particulate dispersions and emulsions is that the latter cannot usually be stabilized by this mechanism without the natural presence of, or the deliberate addition of, an ionic "stabilizer," e.g., an ionic surfactant or a polyelectrolyte. The reason for this is that the natural charge separation at, say, a pure oil/water interface is very small and only arises owing to the preferential *desorption* of $H^+$ ions relative to $OH^-$ ions on the aqueous side of the interface arising from image forces. With solid/liquid dispersions (e.g., Au sols or AgI sols), however, the natural charge separation, owing to preferential adsorption of potential-determining ions ($H^+/OH^-$ or $Ag^+/I^-$), may be sufficiently large to lead to charge stabilization in the absence of any added stabilizer. Other factors come in here as well, such as the fact that the larger part of the potential drop across the interface occurs on the oil side of a pure oil/water interface.

Adsorption of a macromolecular stabilizer also gives rise, as we shall see, to a steric mechanism of stabilization. Indeed, if a nonionic surfactant or polymer is present, or is added, the steric mechanism dominates. Many naturally occurring emulsions are stabilized by the presence of macromolecular materials. In addition, in the case of geologically occurring emulsions, such as those in oil beds, the presence of solid particles (e.g., silica, bentonite clays) gives rise to the third mechanism referred to above, and unraveling the physics of these systems is a complex matter.

The application of physics to these systems requires an understanding of the forces operating within them. The primary objective of this chapter is to elucidate these theoretical aspects. However, we do not aim at a comprehensive text on the subject; we only seek to highlight the main assumptions and conclusions. Thus, as far as possible, derivation of equations is avoided and the reader is referred instead to the original article or other reviews. In addition, we outline the various experimental techniques used and summarize the main experimental findings, again referring to the original or review literature where appropriate.

The remainder of this section is devoted to a review of the hierarchy of different systems falling within the broad classification of emulsions, followed by a general discussion of the types of breakdown that can occur. Consideration is then given in Sec. II to the different types of forces operating in emulsion systems, and a general thermodynamic description of stability and breakdown is presented. The remaining sections then deal in more detail with the different facets of emulsion stability and breakdown.

### B. Classification of Emulsion Types

Table 1 represents an attempt to classify the wide range of emulsion systems whose stability is to be considered. This particular classification subdivides the different types into (1) the nature of the "stabilizing moeities"; (2) the basic "structure" of the system. In both cases the lists represent a hierarchy of increasing complexity.

List a in Table 1 reflects the increasing degree of complexity of the liquid/liquid interface, and the discussion in this chapter draws heavily on the material presented in Chap. 1 in which the physical chemistry of these interfaces is discussed in detail. The one class of stabilizing moeities that perhaps falls outside the general group listed in a is that of solid particles. The basis of stabilization in this instance is somewhat different from the rest of the list, and a separate section of this chapter is therefore devoted to this subject. With regard to list b, the formation and properties of micellar emulsions and microemulsions will be dealt with in Chap. 4, the others in the list having been considered already in Chaps. 1 and 2. However, we shall aim in the next section to give a broad thermodynamic description of the whole range of emulsion types listed under b in Table 1.

**Table 1.** Classification of Emulsion Types

| a. Nature of stabilizing moeity | b. Structure of the system |
|---|---|
| None (i.e., two pure immiscible liquids) | Nature of the internal/external phases (e.g., O/W or W/O) |
| Simple molecules and ions | Micellar emulsions/microemulsions |
| Nonionic surfactants | Macroemulsions |
| Ionic surfactants | Bilayer droplets |
| Surfactant mixtures | Double and multiple emulsions |
| Nonionic polymers | Mixed emulsions |
| Polyelectrolytes | |
| Biopolymers | |
| Mixed polymers and surfactants | |
| Liquid crystalline "phases" | |
| Solid particles | |

### C. Classification of Emulsion-Breakdown Processes

We now turn to a general description of the various breakdown processes that may occur in emulsion systems. There are basically five ways in which the structure of a dispersion of liquid droplets in a continuous liquid medium can change. These are summarized below, and each is dealt with more fully in subsequent sections.

i. No change in droplet size (or droplet size distribution), but buildup of an equilibrium droplet concentration gradient within the emulsion. In limiting cases the result is a close-packed (usually random but in certain cases ordered) array of droplets at one end of the system,[†] with the remainder of the volume occupied by the continuous phase liquid. This phenomenon results from external force fields, usually gravitational, centrifugal, or electrostatic, acting on the system. "Creaming" is the special case in which the droplets collect in a concentrated layer at the top of an emulsion.

ii. Again no change in *basic* droplet size or distribution but the buildup of aggregates of droplets within the emulsion. The individual droplets retain their identity. This process is known as *flocculation* and results from the existence of attractive forces between the droplets. In cases where these

[†]In some cases this may result in distortion of the droplet shape from spherical to polyhedral, but there is no change in the *volume* of the individual droplets.

net attractive forces are relatively weak, an equilibrium degree of flocculation may be achieved (so-called weak flocculation), associated with the reversible nature of the aggregation process. The exact nature of the equilibrium state has not as yet been resolved in detail, but it will depend, one presumes, on the characteristics of the system. On the one hand, one can envisage the buildup of an aggregate size distribution (i.e., singlets, doublets, triplets, etc.). On the other hand, a process more akin to "phase separation" (similar to those which occur with simple molecular systems) may result [1], i.e., the attainment of an equilibrium between singlet drops and large aggregates—in the limit, one large aggregate. One then has, effectively, a dispersed "phase" in equilibrium with a flocculated or aggregated "phase" (cf., a vapor-condensed phase equilibrium). It should be noted that the terms *weak* and *strong* flocculation in this context are relative terms only; there is no real dividing line. By *strongly flocculated* one is usually referring to a system in which virtually all the droplets are present in aggregates or in an aggregated phase (the molecular analogy here would be a solid phase with a negligible equilibrium vapor pressure).

iii. In which flocculated droplets in an aggregate in the bulk of the emulsion (see item 2) or, alternatively, droplets within a close-packed array resulting from sedimentation or creaming, coalesce to form larger droplets. This results in a change in the initial droplet size distribution. The limiting state here is the complete separation of the emulsion into the two immiscible bulk liquids (i.e., into one large drop suspended in the middle of the continuous phase, in the absence of gravity). Coalescence thus involves the elimination of the thin liquid film (of continuous phase) which separates two droplets in contact in an aggregate or a close-packed array. The forces to be considered here, therefore, are the forces acting within thin-liquid-film systems in general. As we shall see, these can be multivarious and complex. In mixed emulsion systems, for example, in which there are droplets of liquid 1 and also liquid 2 dispersed in a continuous phase of liquid 3, coalescence between liquid 1 and liquid 2 droplets occurs only if liquids 1 and 2 are miscible. If they are immiscible, then either droplet adhesion or engulfment occurs (Fig. 1). In either case, however, the thin liquid film between the contacting droplets is eliminated. A more complete description of this behavior is presented below.

iv. An alternative way in which the average droplet size in an emulsion can increase, without the droplets coalescing, occurs if the two liquids forming the dispersed phase and the continuous phase, respectively, are not totally immiscible. So far we have implicitly assumed total immiscibility. This is strictly a hypothetical condition since *all* liquid pairs are mutually miscible

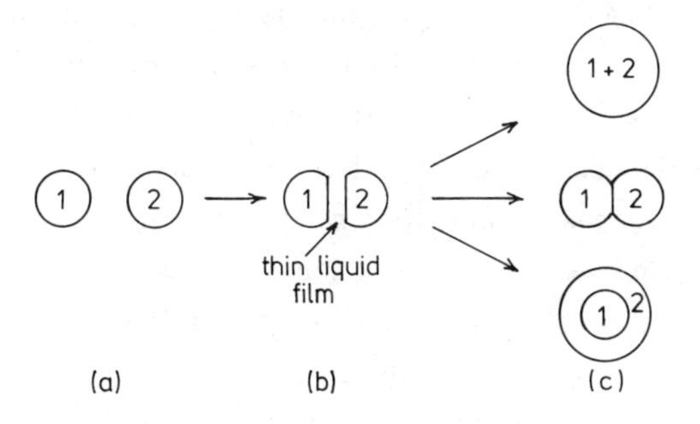


**Figure 1.** Droplet coalescence, adhesion, and engulfment.

to some finite extent. Even liquid pairs which are commonly referred to as *immiscible* often have mutual solubilities which are not negligible. For example, for benzene + water at 25° the saturation concentration of water in benzene is 0.072 wt% and that for benzene in water is 0.180 wt%. If one starts with a truly monodisperse emulsion system then no effects arising from this mutual solubility will arise. However, if the emulsion is polydisperse, larger droplets will form at the expense of the smaller droplets owing to the process known as *Ostwald ripening* (see Chap. 1). In principle, the system will tend to an equilibrium state in which all the droplets attain the same size (this may, of course, mean that one ends up with one single large drop). In essence, Ostwald ripening is associated with the difference in chemical potential for droplets of different size.

v. So far discussion has been limited to changes in emulsion structure associated with the buildup of droplet concentration gradients, the clustering of droplets in aggregates, or changes in droplet size distribution. A further way in which the structure may change is for the emulsion to "invert," e.g., for an O/W emulsion to change to a W/O emulsion. This may be brought about by a change in temperature or concentration of one of the components or by the addition of a new component to the system. It is debatable whether this should be included as a separate, distinct category of emulsion breakdown. First, it is not a unique process in itself but rather a composite of other processes, in particular iii and iv above. Second, one could argue that the emulsion does not undergo "breakdown" in the generally accepted meaning of the word, since a new emulsion is formed which may, moreover, be in a higher state of dispersion than the original emulsion. It is, however, convenient to mention phase inversion at this stage.

One major difficulty in the experimental study of emulsion breakdown is the isolation of the four main processes referred to in items i to iv above, i.e., sedimentation (creaming), flocculation, coalescence, and Ostwald ripening. Certainly, in practical systems all four processes may appear to occur simultaneously or sequentially in any order. This will depend on the relative rate constants for the four basic processes. The one exception is that coalescence, as such, may only follow flocculation or clustering arising from sedimentation (or creaming). Moreover, the four rate constants are not independent since, for example, that for sedimentation will depend on how much flocculation, coalescence, and Ostwald ripening has been achieved at any given stage in the overall breakdown processes. Even in "fundamental studies" the isolation of the four processes is experimentally difficult. For example, in order to eliminate sedimentation effects one should strictly work with two liquids having zero density difference, so that gravitational forces are absent, or to work under conditions of zero gravity. Indeed, some recent experiments on the stability of liquid dispersions in low gravity fields have recently been described [2].

A further experimental difficulty may arise, especially in flocculation or coalescence studies, associated with the presence of solid/liquid interfaces, e.g., the walls of the containing vessel, and possibly also the air/liquid interface if the system is an open one. Flocculation of droplets with (or "adsorption" of droplets at) these interfaces is always a possibility which should be considered. The resulting buildup of droplet concentration in these regions may well lead to increased rates for droplet flocculation and coalescence (so-called surface flocculation and coalescence). Indeed, this phenomenon is often exploited in emulsion breakdown, for example, in the use of solid filters to break emulsions [3].

## II. Forces in Emulsion Systems and the Thermodynamics of Emulsion Breakdown

The physics of emulsion systems is complex and in many aspects still presents major challenges to physicists or physical chemists working in the field. This complexity is due to the fact that one is dealing with physics applied at two levels, i.e., at the droplet (semimacro) level and also at the molecular level. For example, in discussing sedimentation and flocculation, the relevant forces to be considered are essentially long range, at least relative to forces on the molecular level. External field forces, such as gravity, centrifugal forces, or applied electrostatic forces, are effectively of infinite range. Interdroplet field forces, such as electrical double-layer forces and dispersion forces, may be effective at droplet separations of tens or hundreds of nanometers. In addition, the inertial (hydrodynamic) forces which are relevant are obviously those acting on the droplets themselves, and the diffusional (or entropic) forces that operate are those associated with *droplet* concentration gradients or differences within the emulsion system (and therefore with the configurational entropy of the *droplets* as a whole).

Details of local molecular structure are not of primary importance. It is sufficient, for example, in calculating the gravitational force on a droplet to know its size and density, and the density and viscosity of the continuous phase. Similarly, in calculating the dispersion force between two droplets across a liquid medium one needs to know their size and the bulk dielectric behavior (ideally as a function of frequency) of the liquids involved. For the electrical double-layer forces one needs, in addition to the size of the droplets, their surface (or some related) potential, and the bulk electrolyte concentration and dielectric constant of the continuous phase.

On the other hand, in considering processes such as coalescence, Ostwald ripening, and phase inversion, in addition to knowledge of droplet geometry and concentration, it is essential to have details of local molecular structure and composition, both in the droplets and at the interface. One is now concerned with the inter*molecular* forces, *molecular* diffusion, and *molecular* entropies.

It is not intended in this section to discuss the details of the different forces referred to above; these are dealt with in the subsequent sections. Rather the intention is to try to correlate the different phenomena referred to in Sec. I in terms of the overall physics of the system. As discussed in Sec. I, a general description of the overall, combined kinetics of sedimentation, flocculation, coalescence, and Ostwald ripening presents a formidable and, as yet, largely untackled problem (although the kinetics of the individual processes are considered separately in later sections). We can, however, make some progress in discussing the thermodynamics of emulsion breakdown as an integral process. This is best achieved by considering the free-energy changes associated with emulsion formation and breakdown.

Let us start with the simplest possible system and then gradually build up the degree of complexity. To this end we shall consider the formation and breakdown of emulsions of two (hypothetically) immiscible pure liquids, 1 and 2, such that Ostwald ripening effects are absent. We shall assume that external field forces are also absent. The only breakdown processes to consider, therefore, are flocculation and coalescence. The lowest free-energy state and, therefore, the equilibrium state is one liquid suspended as a spherical drop in the other (i.e., state I in Fig. 2). The total free energy $G^I$ of the system in state I may be expressed in the form

$$G^I = G_1^I + G_2^I + G_{12}^I + G_{s1}^I \tag{1}$$

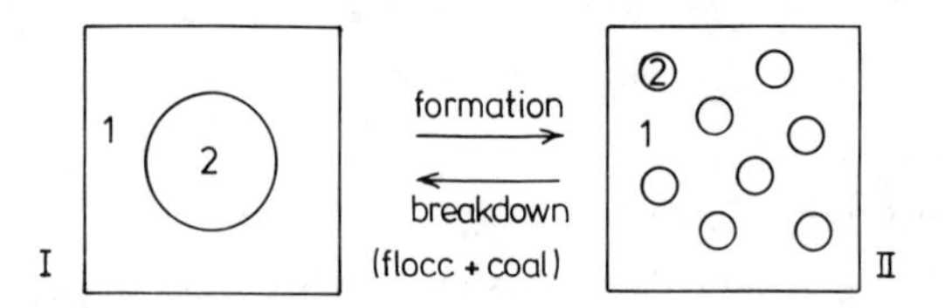


Figure 2. Emulsion formation and breakdown.

where $G_1^I$ and $G_2^I$ are the total bulk free energies of molecules of species 1 and 2, respectively. $G_{12}^I$ (strictly $G_{12}^{\sigma,I}$) is the excess interfacial free energy associated with the liquid/liquid interfaces. This is given by [see Chap. 1, Eq. (38)]

$$G_{12}^I = \gamma_{12} A^I \tag{2}$$

where $\gamma_{12}$ is the interfacial tension and $A^I$ is the total interfacial area, which is a minimum if the droplet is a sphere. The term in brackets in Eq. (1) is the excess free energy associated with the interface between liquid 1 and the solid container. $G_{s1}^I$ may be expressed in a form similar to $G_{12}^I$ [Eq. (2)]. In general, the solid/liquid interfacial area will be relatively small, and thus $G_{s1}^I$ can be neglected. It will be omitted from subsequent equations.

Suppose free energy is supplied to the system under isothermal conditions (i.e., work is done on the system in the form of stirring or ultrasonic irradiation with the system being immersed in an efficient thermostatic bath) such that an emulsion is formed (state II, Fig. 2). The free energy of this state is given by

$$G^{II} = G_1^{II} + G_2^{II} + G_{12}^{II} - T \cdot S_2^{II,\text{config}} \tag{3}$$

where $S_2^{II,\text{config}}$ is the configurational entropy of the droplets of liquid 2 formed. To a first approximation,[†]

$$S_2^{II,\text{config}} = -nk \left[ \ln \phi_2 + \left( \frac{1 - \phi_2}{\phi_2} \right) \ln(1 - \phi_2) \right] \tag{4}$$

Here k is the Boltzmann constant, n is the number of droplets formed, and $\phi_2$ is the volume fraction of liquid 2. Since $G_1^I = G_1^{II}$ and $G_2^I = G_2^{II}$ the free energy of emulsion formation $\Delta G_{\text{form}}$ is given by

$$\Delta G_{\text{form}} = G^{II} - G^I = G_{12}^{II} - G_{12}^I - T\,\Delta S^{\text{config}} \tag{5}$$

[†]Equation (4) may be derived using a simple lattice model for the emulsion, choosing as the element size the equivalent diameter of the droplets.

(note that $\Delta S^{config} \backsim S^{II, config}$, since $S^{I,config} \backsim 0$). If the difference in interfacial area between states I and II is $\Delta A$, then combination of Eqs. (2) and (5) gives

$$\Delta G_{form} = \gamma_{12} \Delta A - T \Delta S^{config} \tag{6}$$

From Eq. (16) it may be seen that $\Delta G_{form}$ can be either positive or negative, depending on the relative magnitudes of the two terms on the right-hand side of the equation, both of which are themselves positive. In general, $\gamma_{12} \Delta A >> T \Delta S^{config}$, and therefore $\Delta G_{form}$ is large and positive. However, if $\gamma_{12}$ is sufficiently small, then it is possible for $\Delta G_{form}$ to be negative, i.e., for spontaneous emulsification to occur. The limiting value of $\gamma_{12}$ at which this occurs is found by equating $\Delta G_{form}$ to zero is Eq. (6). Setting $\Delta A = n4\pi r^2$ and substituting Eq. (4) then leads to†

$$\gamma_{12} = -\frac{kT\{\ln \phi_2 + [(1 - \phi_2)/\phi_2] \ln (1 - \phi_2)\}}{4\pi r^2} \tag{7}$$

For example, for the case where $\phi_2 = 0.5$ and $r = 100$ nm, $\gamma_{12}$ must be $< \backsim 5 \times 10^{-5}$ mN m$^{-1}$ for $\Delta G_{form}$ to be negative and for spontaneous emulsification to occur. Similarly, for $\phi_2 = 0.1$ and $r = 25$ nm, $\gamma_{12}$ must be $< \backsim 1.7 \times 10^{-3}$ mN m$^{-1}$. Values of $\gamma_{12}$ of these orders of magnitude are barely determinable using current techniques, e.g., the spinning drop method (see Chap. 1, Sec. I.F.6). Such extremely low interfacial tensions are only rarely achieved, however, particularly under equilibrium conditions. They are more readily achieved under nonequilibrium conditions, for example, during molecular transport across an interface. This is relevant to the discussion of spontaneous emulsification (Chap. 2) and microemulsion formation (Chap. 4).

In most cases $\Delta G_{form}$ is positive; this implies that $\Delta G_{break}$ is negative. This means that an external supply of free energy is required for emulsions to form and that once they have formed they are thermodynamically unstable or metastable. The free energy path for the breakdown of an emulsion, from state II to state I (Fig. 2), is shown schematically in Fig. 3 by the full line. There are no free-energy barriers either to flocculation or to coalescence. The kinetics of both breakdown processes is diffusion controlled: in the case of flocculation, by the diffusion of the droplets, and in the case of coalescence, by diffusion of

†Reiss [4] has given a more detailed statistical mechanical analysis of this problem. The analysis presented here only serves as an illustration of the point. The two analyses do predict critical $\gamma_{12}$ values (for negative $\Delta G_{form}$) of the same order of magnitude, however.

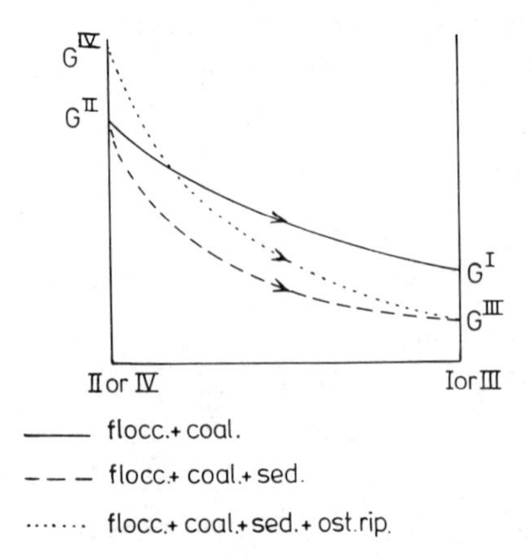


**Figure 3.** Free-energy path in emulsion breakdown.

molecules of liquid 1 out of the thin liquid film formed between two contacting droplets of liquid 2 (see Fig. 1b).

The dashed line in Fig. 3 corresponds to the case where sedimentation or creaming is superimposed upon the flocculation and coalescence processes. This occurs, for example, when there is a density difference between species 1 and 2 such that gravitational potentials have to be taken into account. The final state of the system (state III) is now the more familiar one of two liquid phases separated by a flat interface, the lower phase having the higher density. $G^{I} - G^{III}$ (Fig. 3) represents the net decrease in gravitational potential energy of the system associated with the sedimentation process.

The dotted line in Fig. 3 represents the situation if, in addition to the above effects, Ostwald ripening also has to be taken into account. This occurs if the initial state (IV) is polydisperse and the liquids have a finite mutual solubility. Additional free-energy changes occur which are associated with changes in the chemical potential of the molecules of species 2 as the size of the droplets changes [see the Kelvin equation, Chap. 1, Eq. (48)].

So far we have not discussed the role of interdroplet forces in the thermodynamics of emulsion formation/breakdown. These will be discussed more specifically in subsequent sections; we only introduce them in a general way here. Between all emulsion droplets there will be van der Waals forces. In general, there is also some contribution from electrostatic forces. For example, even a pure water/hydrocarbon interface may have a net surface charge owing to preferential adsorption of $OH^-$ ions. If other ions are present in the system (e.g., $K^+$ and $NO_3^-$), there may also be preferential adsorption or desorption of one of these species. Added surfactants or polymers, which adsorb strongly at the interface, give rise to steric interactions, as well as to electrostatic interactions if they are charged.

In general, since the presence of adsorbed species at an interface lowers the interfacial tension $\gamma_{12}$, then the free energies of both the initial, dispersed state (II or IV) and the final, coalesced state (I or III) will be lower than in a system of two pure liquids. In addition, the absolute value of $\Delta G_{break}$ will be smaller. This may be seen from Eq. (8), which is the analog of Eq. (6),

$$\Delta G_{break} = \gamma_{12}\,\Delta \mathcal{A} - T\,\Delta S^{config} \tag{8}$$

(Note that $\Delta \mathcal{A}$ and $\Delta S^{config}$ are now *negative*.) In the situation depicted in Figs. 2 and 3 for two pure liquids, the role of the interdroplet forces (for the moment considered to be only the van der Waals forces) in the flocculation and coalescence processes has been masked. They are taken account of in changes in $\gamma_{12}$ that occur during the breakdown process. The point is $\gamma_{12}$ is the same for the initial (II) and final states (I) in Fig. 2 (provided that the droplets in state II are sufficiently far apart so that their average separation is greater than the range of the interdroplet forces). In the flocculation process, however, droplets come into contact and are separated by a thin liquid film. The interfacial tension of each of the two interfaces forming this film is lower than that of an isolated interface. This is expressed by the following relation:

$$\gamma_{12}(h) = \gamma_{12}(\infty) + \frac{G_i(h)}{2} \tag{9}$$

where h is the thickness of the thin film and $G_i(h)$ the interaction free energy of the two droplets across the film. If only (attractive) van der Waals forces are operating, then G(h) is negative and $\gamma_{12}(h)$ is always less than $\gamma_{12}(\infty)$, the interfacial tension of an isolated interface. The film thus spontaneously thins until it ruptures, i.e., coalescence occurs. The value of $\gamma_{12}$ appearing in Eqs. (6) and (8) is $\gamma_{12}(\infty)$.

If repulsive forces are also operating during the flocculation and coalescence processes, then the simple picture shown in Fig. 3 may not be correct, i.e., the free energy of the system may not fall continuously as breakdown proceeds.

Consider first the case where, due to adsorbed ions, there is an electrostatic repulsion ($G_E$) between the droplets, as well as the van der Waals attraction ($G_A$). The form of the net G(h) curve is then the familiar DLVO [5] one, shown schematically in Fig. 4. Consideration of the form of G(h) and Eq. (9) suggests that $\gamma_{12}(h)$ in the film between two contacting droplets is greater than $\gamma_{12}(\infty)$ in the region of $G_{max}$ in Fig. 4. Indeed, unless any two approaching droplets have sufficient mutual kinetic energy to surmount this free-energy barrier $G_{max}$, then $\Delta G_{flocc}$ will be positive (since $\Delta\gamma$ is positive).

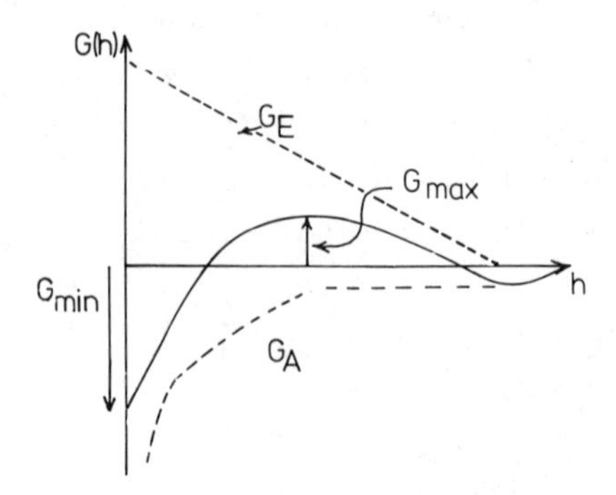


Figure 4. General form of G(h) versus h for DLVO-type system.

A similar situation holds with regard to the coalescence step if adsorbed species are present in the interfaces. In order for coalescence to occur, it is necessary either for the interface to expand in order to create free surface (i.e., $A$ to increase), and/or for desorption of stabilizer to occur (i.e., $\gamma$ to increase).[†] In either case, $\Delta G_{coal}$ will be positive.

Strictly speaking, the $\Delta G_{flocc}$ and $\Delta G_{coal}$ terms referred to here are activation free energies (i.e., $\Delta G^{A}_{flocc}$ and $\Delta G^{A}_{coal}$) since if either process does occur then the net free-energy change is negative. The situation is summarized in Fig. 5.

The intermediate state V is a metastable state and represents a flocculated emulsion that has undergone no coalescence. If $\Delta G^{a}_{coal}$ is sufficiently high, it may remain in this state indefinitely. Similarly, state II is also a metastable one, and if $\Delta G^{a}_{flocc}$ is sufficiently high, a "stable," dispersed system may persist indefinitely. However, states II and V in these cases represent states of kinetic stability rather than true thermodynamic stability (see below).

The dashed curve in Fig. 5 represents the situation if there is no free-energy barrier to flocculation, but if there is a very large barrier to coalescence. Such a situation would arise for neutral droplets stabilized, for example, by an adsorbed (neutral) polymer. In this case only the long-range van der Waals forces and the short-range steric repulsion forces are operating. If $\Delta G_{flocc}$ is not too large (say <10 kT per droplet), then the flocculation is reversible and an equilibrium is set up, as discussed in Sec. I.C.

In Fig. 5 it is seen that

---

[†]The so-called steric interaction in this case is then effectively associated with these two processes (see later).

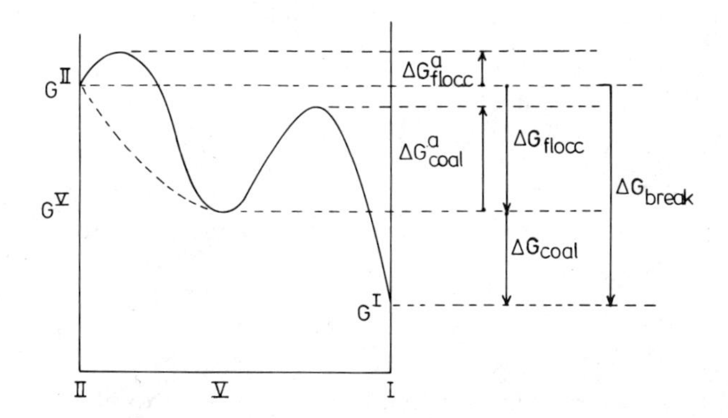


**Figure 5.** Free-energy path for breakdown (flocculation plus coalescence) for droplets having both attractive and repulsive (electrostatic plus steric) interactions.

$$\Delta G_{break} = \Delta G_{flocc} + \Delta G_{coal} \tag{10}$$

It is worth considering the individual contributions to $\Delta G_{flocc}$ and $\Delta G_{coal}$ in the light of Eqs. (8) and (10). In Chap. 1 [Eq. (37)] it was shown that the excess interfacial (Helmholtz) free energy $A^{\sigma}$ associated with the presence of an interface is given by†

$$A^{\sigma} = \gamma_{12}\mathcal{A} + \sum_i \mu_i n_i^{\sigma} \tag{11}$$

If an interface disappears owing to coalescence, the change in free energy ($\Delta G_{coal}$) is simply given by

$$\Delta G_{coal} = -\Delta(\gamma_{12}\mathcal{A}) \tag{12}$$

[Note that the $\Sigma\, \Delta\mu_i n_i^{\sigma}$ term disappears since the chemical potential of species i is the same in either bulk phase and in the interface.]

Inspection of Eqs. (8), (10), and (12) leads to the conclusion that

$$\Delta G_{flocc} = \mathcal{A}\, \Delta\gamma_{12} - T\Delta S^{config} \tag{13}$$

remembering $\Delta(\mathcal{A}\gamma_{12}) = \gamma_{12}\Delta\mathcal{A} + \mathcal{A}\,\Delta\gamma_{12}$

†An expression similar to Eq. (11) may be written in terms of the Gibbs free energy $G^{\sigma}$. In practice negligible error is introduced in replacing the Helmholtz free energy by the Gibbs free energy.

Thus, $\Delta G_{flocc}$ is made up of two terms. The $A \Delta\gamma_{12}$ term is associated with the change in interfacial tension in the contact region of two droplets (i.e., for the two surfaces in the thin film separating the droplets). The $T \Delta S^{config}$ term is the change of configurational entropy of the system discussed previously. Both terms are negative. Under most circumstances the $A \Delta\gamma$ term dominates, so that $\Delta G_{flocc}$ is negative (i.e., flocculation is thermodynamically spontaneous). However, if $A \Delta\gamma$ is less than $T \Delta S^{config}$, then $\Delta G_{flocc}$ is positive. The dispersion is then thermodynamically stable with respect to flocculation. This situation is illustrated schematically in Fig. 6.

Spontaneous flocculation will not occur, and the dispersion will have to be "concentrated" (e.g., by sedimentation) before coalescence can occur. The condition $|A \Delta\gamma_{12}| < |T \Delta S^{config}|$ may be realized if $\Delta\gamma_{12}$ is small, i.e., $G_{min}$ in Fig. 4 is small. Since $|\Delta S^{config}|$ decreases as the droplet number concentration increases, one can envisage that at some initial droplet concentration the condition $\Delta G_{flocc} = 0$ is reached, i.e., below this concentration the emulsion is thermodynamically stable ($\Delta G_{flocc}$ positive), but that beyond this concentration the emulsion becomes thermodynamically unstable ($\Delta G_{flocc}$ negative) and reversible flocculation occurs, as discussed above.

Returning to Fig. 2 the breakdown process (II → I) has certain broad features in common with nucleation and growth in molecular phase changes. For example, the growth of a liquid phase from a supersaturated vapor phase involves an aggregation step. The free energy of the final state is lower than the initial state. In this case, however, there is only one maximum in the free-energy profile of the process (corresponding to the formation of critical size nuclei), as opposed to two possible maxima for the emulsion breakdown process (Fig. 5). This point

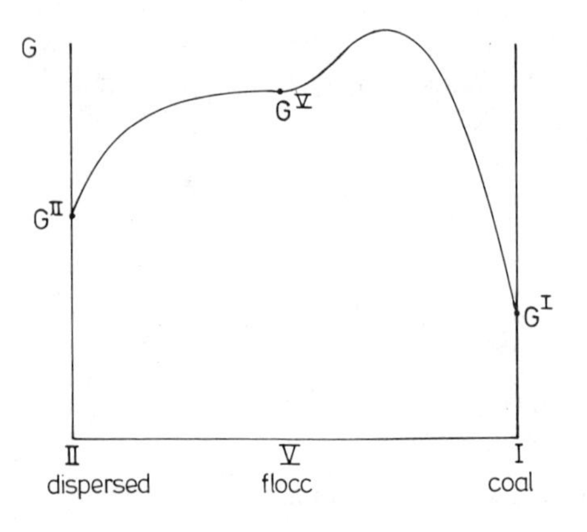


**Figure 6.** Free-energy path for an emulsion which is thermodynamically stable with respect to flocculation.

is made only to illustrate the parallel between phenomena at the molecular level and at the dispersion level and the growing use of classical thermodynamics to describe the behavior of dispersions and emulsions [6,7].

In the following sections the various breakdown process referred to will be discussed in more detail. In particular, more attention is paid to the kinetic aspects of emulsion breakdown.

## III. Sedimentation and Creaming

### A. Introduction

The buildup of a sediment or cream in emulsion systems is well known and is a direct consequence of the density difference between the droplets and the continuous phase. The word *cream* originates from the separation of the fat globules in milk into two layers, one containing ~35% (by volume) fat (the cream layer), the other ~8% fat. Although the density difference between milk fat and water is the primary factor in this separation process, other parameters are obviously important. For example, it has long been recognized that the creaming of milk may be prevented by homogenization, i.e., by passing the milk through a fine clearance valve under pressure to break the larger droplets into smaller ones. Thus average droplet size and droplet size distribution are important features.

Possible equilibrium states for an emulsion with respect to sedimentation are illustrated schematically in Fig. 7. Case (a) represents the case of small droplets where the gravitational force is opposed by a diffusional force (i.e., associated with the translational kinetic energy of the droplets). A Boltzmann-type equi-

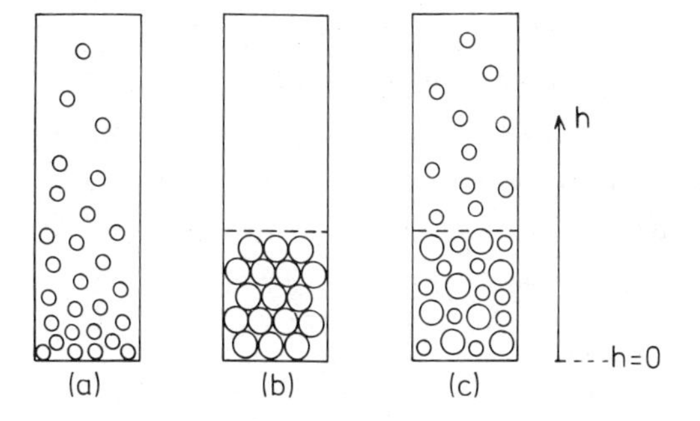


**Figure 7.** Possible equilibrium state for an emulsion with respect to sedimentation: (a) sedimentation and diffusion comparable (small particles); (b) sedimentation exceeds diffusion (large particles); (c) mixed-particle-size emulsion.

librium distribution is set up, the droplet concentration $C_d$ at height h being given by

$$C_d(h) = C_d(0) \exp\left(-\frac{mgh}{kT}\right) \tag{10a}$$

where m is the effective mass of a droplet, given by

$$m = \frac{4}{3}\pi a^3 \Delta\rho \tag{11a}$$

where a is the hydrodynamic radius of the droplets and $\Delta\rho$ is the density difference between the two liquid phases. Note that mgh is the potential energy of a droplet at height h. For no appreciable separation to occur [ i.e., $C_d(L) \simeq C_d(0)$], then [8]

$$kT \gg \frac{4}{3}\pi a^3 \Delta\rho \; gL \tag{12a}$$

where L is the length of the containing vessel.

Case (b) in Fig. 7 represents the situation for large droplets. Here the concentration forces are very much bigger than the opposing diffusional force, so that effectively "complete" sedimentation occurs, i.e., now

$$kT \ll \frac{4}{3}\pi a^3 \Delta\rho \; gL \tag{13a}$$

For polydisperse droplets the situation illustrated by case (c), i.e., a bimodal distribution, typifies what happens. Here again a lower, separate layer appears to form, but with some smaller droplets remaining in the upper layer. This is the situation encountered with unhomogenized milk, although in the reverse sense (creaming rather than sedimentation). A further complicating feature is when flocculation or coalescence or both occur simultaneously with sedimentation. Both processes lead to an increase in the effective, average droplet size and tend to widen the droplet size distribution. If weak, reversible flocculation occurs (see Sec. I.C ), then large flocs form in equilibrium with singlet droplets, and an equilibrium situation similar to case (c) in Fig. 6 can again result, where now the supernatant phase contains essentially only singlet droplets.

In certain industrial applications creaming or sedimentation is desirable, e.g., in the separation of isoprene droplets from water in the rubber latex industry, or in the collection of cream. However, in many cases it is undesirable, e.g., in many pharmaceutical preparations and agricultural sprays. Methods for controlling the rate of sedimentation or creaming are, therefore, desirable. Sedimentation/creaming rates are considered in the next section. The section following deals with the physical properties of creams and sediment layers.

B. Sedimentation/Creaming Rate

Assuming a steady state has been reached, the velocity v of uniform, nondeformable, noninteracting spheres of radius a in a liquid of viscosity $\eta_0$ may be determined by equating the gravitational force with the opposing hydrodynamic force, as given by Stokes' law [9], i.e.,

$$\frac{4}{3}\pi a^3 \Delta\rho\, g = 6\pi\eta_0 a v \tag{14}$$

Thus,

$$v = \frac{2\Delta\rho\, g a^2}{9\eta_0} \tag{15}$$

For example, for oil droplets (density 0.8 g $cm^{-3}$) the upward velocity depends on droplet size as follows: $v = 2 \times 10^{-3}$ cm $h^{-1}$ for droplets of diameter 0.1 μm; $v = 0.2$ cm $h^{-1}$ for droplets of diameter 1 μm. If a centrifuge is used

$$g = \omega^2 x \tag{16}$$

where x is the mean distance of the centrifuge tube from the axis of rotation and $\omega$ is the angular velocity ($2\pi f$, where f is the number of revolutions per second). Note that if the length of the centrifuge tube is not small in relation to x, then the applied centrifugal field cannot be considered to be uniform over the length of the tube.

An alternative way of increasing the rate of sedimentation for charged droplets is to apply an external electrostatic field X across the ends of the vessel containing the emulsion. The electrostatic force $F_e$ on a droplet having a uniform surface charge density $\sigma$ is given by

$$F_e = 4\pi a^2 \sigma X \tag{17}$$

This involves the assumption that the droplets are small enough so that the net charge may be considered as a point charge in an unperturbed electrostatic field. It is also assumed that the droplets are not polarized. In order to obtain v, $F_e$ now replaces the gravitational force on the left-hand side of Eq. (14). However, for droplets of the size normally encountered in emulsions the assumption of an unperturbed field is not applicable, and, particularly when water is the continuous phase, the effects of the electrical double layer have to be considered.

In many practical cases the objective is to decrease v rather than to increase it. Inspection of Eq. (15) indicates that this may be achieved by decreasing $\Delta\rho$ or increasing $\eta_0$. The former may be achieved by the addition of a suitable second solvent, the latter by the addition of "structuring" or "gelling" agents, such as high molecular-weight polymers or particles (silica, clay, etc.). The general application of Eq. (15) depends on the validity of Stokes' law. With liquid droplets the assumption of a smooth surface is reasonable, except perhaps when solid particles are present as stabilizers. These may give rise to surface irregularities (up to ~0.1 μm) [10].

The question of deformability is a more serious problem. A liquid droplet moving within a second liquid phase has an internal circulation imparted to it. As a result of this, its motion through the continuous phase has a "rolling" as well as a "sliding" component. This situation has been treated theoretically by Rybczynski [11] and Hadamard [12], who produced the following equation for v,

$$v = \frac{2\,\Delta\rho\, ga^2}{3\eta_0}\,\frac{\eta_0 + \eta}{3\eta_0 + 2\eta} \tag{17a}$$

where $\eta$ is the viscosity of the internal phase. For the case $\eta >> \eta_0$, Eq. (17a) predicts v to be 50% greater than that obtained from Stokes' law, whereas for two liquids having similar viscosities (i.e., $\eta \sim \eta_0$), v is only about 20% greater than the Stokes' law value. It would seem, however, that Eq. (17a) does not agree well with experimental data for v [13]. It has been suggested that this could be due to the neglect of the interfacial viscosity contribution; this is particularly the case when highly structured adsorbed or spread layers are present at the liquid/liquid interface. Frumkin and Levich [14] have considered this and their modification of the Rybczynski-Hadamard equation includes a term for the Gibbs surface excess. Paradoxically, if the interfacial viscosity contribution is very high, then the droplets again behave effectively as solid particles, and Stokes' law should again apply. Clearly, more accurate experimental data on well-defined systems is required to test the validity of the various theoretical treatments for v. In this connection, O'Brien et al. [15] have suggested a multiple-beam interferometric technique for the measurement of creaming rates for hydrocarbon droplets in water.

With large droplets shape distortion may occur owing to changes in pressure with "vertical height" of the droplet. The difference in pressure between the "top" and "bottom" of the droplet is $\Delta\rho$ gd, where d is the (distorted) vertical diameter. Any deviation from spherical geometry leads to an increase in surface area $\Delta A$. Thus the distorting force due to gravity is opposed by the work necessary to increase the surface area ($\gamma\,\Delta A$). For small deformities d ~ 2a, the fractional change in the radius is $\Delta\rho\, ga^2/\gamma$. For $\Delta\rho = 0.1$ g cm$^{-3}$ and $\gamma = 2$ mN m$^{-1}$, a 2-μm-diameter droplet would undergo a distortion in radius of ~$5 \times 10^{-5}$% whereas for a 200-μm-diameter droplet the distortion would be ~0.5%. Thus, the effect is really only significant for large droplets.

Another limitation to be considered in the application of Stokes' law is that it strictly applies only to noninteracting droplets, and thus it is only valid for very dilute emulsions. In most practical emulsions the droplet concentration is high. Moreover, there is normally a wide variation in droplet size. These problems have been discussed by Greenwald [16]. For concentrated emulsions one is usually interested in the mass rate of creaming, i.e., in the motion of the center of mass of the dispersed phase. In the absence of flocculation and coalescence an emulsion may be considered to consist of $n_i$ droplets of radius $a_i$, each having an effective mass $4\pi a_i^3 \Delta\rho/3$, and moving with a velocity $v_i$. The velocity v of the center of mass of the dispersed phase is given by

$$v = \frac{1}{V}\sum_i \frac{4\pi a_i^3 n_i v_i}{3} \tag{18}$$

where V is the total volume of the dispersed phase, i.e.,

$$V = \sum_i \frac{4\pi a_i^3 n_i}{3} \tag{19}$$

Substituting for $v_i$ from Eq. (15) yields

$$v = \sum_i \frac{8\pi g n_i a_i^5 \Delta\rho}{27\eta_0 V} \tag{20}$$

It should be remembered that Eq. (20) involves the same assumptions and approximations as Eq. (15).

## C. Properties of Creamed or Sedimented Layers

The structure depicted in Fig. 7b represents a sediment that has settled to an equilibrium volume, i.e., the mutual distances between droplets are determined by a balance between the external (gravitational, centrifugal, or electric) field and the mutual interdroplet forces (electrostatic and/or steric forces associated with an interfacial polymer layer). The structure shown in this figure represents a hexagonal close packed array, with an internal phase volume of 74%. Packing of this type may be observed, e.g., with monodisperse polystyrene or silica particles, and is responsible for the iridescence (light diffraction) colors often seen with these systems, but it is a somewhat idealistic picture for an emulsion system. Several complicating factors have to be taken into account. These may be listed as follows:

i. Polydispersity
ii. Deformation from spherical geometry

iii. Flocculation
iv. Coalescence

Even in the absence of all of these factors, it is unusual to obtain hexagonal close packing; random close packing is more generally the result. The concept of random packing of monodisperse spheres has been considered by several authors [17]. Experiments, aimed at supplementing the geometrical models of Bernal [18] for simulating the structures of liquids, were carried out in which 1/8-in.- diameter steel balls were allowed to settle and pack under conditions of mild agitation. In general, it was observed that the packing fell between the limits of 60 and 64%. These limits were designated as *loose* and *close* random packing, respectively, with corresponding average coordination numbers of 7 to 8 and 8 to 9. The effects of polydispersity are important in that the smaller droplets may fit into voids between the larger droplets in a packed sediment or cream. Packings of greater than 90% can be achieved in this way.

Deformation is frequently encountered and again leads to a reduced sediment or cream volume. The more rigid the interfacial film, the greater the resistance to deformation. On the other hand, the greater the size of the droplets and density difference between the two liquid phases, the greater the tendency for deformation to occur. In many cases, however, the droplets distort into polyhedral cells resembling a foam in structure, with a corresponding network of more-or-less planar thin films of one liquid separating cells of the other liquid. The stability of the system to coalescence then depends on the stability to rupture of these films.

Flocculation and coalescence of emulsions are described in later sections. Suffice to say here that if flocculation occurs in the emulsion, then, because flocs settle more rapidly in general, the overall rate of creaming will be faster. The final sediment or cream volume will be greater as well. This is analogous to the behavior of solid dispersions and suspensions. In highly stable dispersions the particles settle individually under gravity and move past each other to form a compact sediment which is difficult to redisperse. This phenomenon is usually described as *caking* or *claying*. On the other hand, if the particles are flocculated, they cannot easily move past each other, and a loose sediment forms, with a higher volume because of the more open structure. However, if the particles are only weakly flocculated, then this has the advantage that the sediment is eadily redispersed by mild agitation.

### D. Ultracentrifugal Studies of Creaming and Sedimentation

The first investigations in this area were carried out by Vold and Groot [19], by Garrett [20], and Rehfeld [21]. Vold and Groot found that the rate of settling decreased as settling progressed. Modern analytical ultracentrifuges allow one to follow the separation of emulsions in a quantitative manner. With a typical O/W emulsion three layers are generally observed: (1) a clear aqueous phase, (2) an opaque phase consisting of distorted polyhedral oil droplets, and (3) a clear, separated oil phase, resulting from coalescence of these polyhedra. The degree of emulsion stability has been expressed by Rehfeld [22] as the volume fraction of the opaque phase remaining after time t. Vold and Mittal [23] expressed this in the form,

$$\frac{t}{V} = \frac{1}{bV_{\infty}} + \frac{t}{V_{\infty}} \qquad (21)$$

where V is the volume of oil separated at time t, $V_{\infty}$ is the extrapolated volume at infinite time, and b is a constant. Thus, a plot of t/V versus t should give a straight line from which b and $V_{\infty}$ may be calculated. These two parameters have been taken as indexes of emulsion stability by Vold and Mittal [23].

Smith and Mitchell [24] used a centrifugal method to study the stability of paraffin oil/water emulsions stabilized with sodium dodecyl sulfate. Centrifugation produced not only the expected opaque cream layer and, at sufficiently high centrifuge speeds, a coalesced oil layer, but also an extra layer between the cream layer and the oil phase. This extra layer was found to be water continuous, and microscopic studies indicated that it contained oil droplets separated by thin aqueous films. Similar structures have been described by Sebba [25] and Lissant [26] as biliquid foams. On increasing the centrifuge speed, Smith and Mitchell [24] showed that the "foam"/cream boundary did not move, suggesting that there is no force barrier to be overcome in forming the foam layer from the cream layer. The interpretation gives was that in the foam layer the aqueous film separating two oil droplets thins to a "black" film under the action of van der Waals forces. The foam layer, therefore, resembles the "dispersed-phase/flocculated-phase" equilibrium of the type described by Long et al. [1] (for latex particles), except that the concentration in the dispersed, i.e., cream phase, is much higher than the concentration ranges studied by by these latter authors.

Smith and Mitchell showed that the boundary between the foam layer and the coalesced layer is associated with a force (or pressure) barrier. They noted the minimum centrifuge speed necessary to produce a visible amount of coalesced oil after 30 min, and defined a "critical pressure" in this way. They then compared experimental critical pressures at varying electrolyte concentrations with those calculated theoretically. Invariably the experimental values were lower than those calculated theoretically and this was accounted for in terms of thermal fluctuations leading to film rupture.

It was convenient to discuss these experiments at this point, in the context of creaming; a more complete discussion of coalescence, per se, is given in Sec. V.

## IV. FLOCCULATION

### A. Introduction

In Sec. I *flocculation* was defined as the aggregation of droplets to give three-dimensional clusters without coalescence occurring, i.e., the droplets retain their individual identities in the clusters (or flocs). In Sec. II the free-energy change $\Delta G_{flocc}$ accompanying flocculation was discussed in some detail. It was shown that for weakly interacting droplets $\Delta G_{flocc}$ may be positive or negative, depending on the droplet concentration. Also, for a given system, a critical droplet number concentration (or phase volume) exists below which the emulsion is *thermodynamically* stable with respect to flocculation, but above which reversible flocculation to an equilibrium extent occurs. For strongly interacting droplets this critical droplet concentration is so low as to be practically insignificant, and the flocculation may be classified as being irreversible. It should be remembered, however, that this is not strict terminology. No process is irreversible, as such, on an infinite time scale. The flocculation process is then sometimes preferably referred to as *coagulation*. However, even for a system which is thermodynamically unstable, flocculation may be effectively prevented if a large enough free energy barrier ($\Delta G^{a}_{flocc}$) exists between the dispersed (deflocculated) state and the flocculated state (cf. Fig. 5). The emulsion is then stable in the kinetic sense.

In the next section we consider the forces acting in an emulsion system which give rise to this behavior pattern. As will be seen, these forces are primarily the long-range London-van der Waals forces and electrostatic forces associated with the overlap of the electrical double layers around charged droplets. When added together this leads to the total mutual interaction free energy between two approaching droplets. This approach forms the basis of the classical DLVO theory [5] of colloid stability, first proposed independently by Derjaguin and Landau and by Verwey and Overbeek. In a subsequent section flocculation rates are discussed on the basis of this theory, together with some more recent modifications. In particular it will be shown that some care has to be exercised in applying this theory to the case of liquid droplets, as opposed to particulate systems for which it has been most used.

## B. Interdroplet Interactions

Some discussion of van der Waals interactions has already been given in Chap. 1 in the context of interfacial tensions. A more rigorous treatment is given here.

### 1. *London-van der Waals Interactions*

The van der Waals interactions between molecules are of three main types: dipole-dipole (Keesom) interactions, dipole-induced dipole (Debye) interactions, and dispersion (London) interactions. The dispersion forces arise from charge fluctuations within a molecule associated with the motion of its electrons. They therefore operate even between nonpolar molecules. Although the London force between two molecules is short range (being inversely proportional to the seventh power of their separation), it turns out that the net force between two colloidal particles or droplets is much longer range.

For two infinitessimal volume elements, $dV_1$ and $dV_2$ of material 1 separated by a distance h in vacuo, the interaction energy $dU_A$ is given by

$$dU_A = -\rho_1^2 \lambda_{11} \frac{dV_1\, dV_2}{h^6} \tag{22}$$

where $\lambda_{11}$ is the London interaction constant for two molecules of type 1 and $\rho_1$ is their number density. The constant $\lambda_{11}$ is related to the polarizability of the molecules concerned and hence may be derived from optical (refractive index) data. It has values generally in the range $10^{-78}$ to $10^{-76}$ J $m^6$ for different molecules, depending on their electronic structure. (For water $\lambda_{11} = 4.7 \times 10^{-78}$ J $m^6$.)

Two general approaches to the calculation of the interaction energy between condensed phases have been used. The first, due to de Boer [27] and Hamaker [28] assumes pairwise additivity of the forces and is sometimes referred to as the *microscopic approach.* The other general approach, initiated by Lifshitz [29], considers the interactions in terms of *macroscopic* properties of the media (e.g., dielectric data) and the assumption of pairwise additivity is implicitly overcome.

Hamaker [29] integrated Eq. (22) over the total volumes $V_1$ and $V_2$ of the two particles, i.e.,

$$U_A = -\frac{\lambda_{11}\rho_1^2}{h^6} \int_{V_1} dV_1 \int_{V_2} dV_2 \tag{23}$$

For two spheres of radius a and center-center separation R, the following result is obtained:

$$U_A = -\frac{\pi^2\rho_1^2\lambda_{11}}{6}\left[\frac{2a^2}{R^2 - 4a^2} + \frac{2a^2}{R^2} + \ln\left(\frac{R^2 - 4a^2}{R^2}\right)\right] \qquad (24)$$

which reduces to

$$U_A = -\pi^2\rho_1^2\lambda_{11}\frac{a}{12h} \qquad h \ll a \qquad (25)$$

where h (= R − 2a) is the surface-surface separation.

Alternatively, for two parallel plates of thickness $\delta$ separated by a distance d one obtains,

$$U_A = -\frac{\pi\rho_1^2\lambda_{11}}{48}\left\{\frac{1}{[(1/2)d]^2} + \frac{1}{[(1/2)d + \delta]^2} - \frac{2}{[(1/2)d + (1/2)\delta]^2}\right\} \qquad (26)$$

which reduces to

$$U_A = -\frac{\pi\rho_1^2\lambda_{11}}{12d^2} \qquad \delta \gg d \qquad (27)$$

or

$$U_A = -\frac{\pi\rho_1^2\lambda_{11}\delta^2}{2d^4} \qquad d \gg \delta \qquad (28)$$

The so-called Hamaker constant $A_{11}$ is defined by

$$A_{11} = \pi^2\rho_1^2\lambda_{11} \qquad (29)$$

and generally lies in the range $3 \times 10^{-20}$ to $5 \times 10^{-19}$ J.

Equations (24) and (26) refer to particles or droplets in vacuo. If they are suspended in a continuous phase medium of material 2, then the net interaction is reduced. This case was also considered by Hamaker [28]. The general form of Eqs. (24) and (26) is retained but the Hamaker constant must be replaced by an effective Hamaker constant A, i.e.,

$$A = A_{11} + A_{22} - 2A_{12} \qquad (30)$$

where $A_{22} = \pi^2\rho_2^2\lambda_{22}$

$A_{12} = \pi^2\rho_1\rho_2\lambda_{12}$

It is generally assumed that $\lambda_{12} \sim (\lambda_{11}\lambda_{22})^{1/2}$ and hence

$$A \sim (A_{11}^{1/2} - A_{22}^{1/2})^2 \qquad (31)$$

Inspection of Eq. (30) indicates that a small uncertainty in the estimation of either $A_{11}$ or $A_{22}$ leads to large uncertainties in the value of A.

The equations quoted above for $U_A$ do not allow for the finite time of propagation of the electromagnetic interaction at high values of h or d (i.e., ~20 nm or so). This so-called retardation effect was first included in the theory of intermolecular forces by Casimir and Polder [30]. Some calculations by Kitchener and Musselwhite [31] indicated that for typical emulsion droplets (1 μm diameter) the effect of retardation is to reduce their attraction energy at a separation of 100 nm or so to about 90% of the value predicted by Eq. (24). For practical purposes Schenkel and Kitchener [32] derived the following approximate equation for $U_A$ for two spheres

$$U_A = -Aa\left(\frac{2.45}{120\pi h^2} + \frac{2.17}{720\pi^2 h^3} - \frac{0.59}{3360\pi^3 h^4}\right) \tag{32}$$

This equation effectively replaces Eq. (25), but the latter is probably still the most useful and most simple form for calculating $U_A(h)$ curves for emulsion droplets, if one only requires approximate values.

Detailed discussion will not be given here of the Lifschitz approach for calculating $U_A(h)$ curves. It certainly leads to a more exact evaluation, but the necessary integrations involved are complex and require knowledge of the dielectric data over a wide frequency range for the materials involved. For recent treatments and developments of this approach the reader is referred to reviews by Winterton [33] and by Israelachvili [34], who have given very readable accounts of the subject. More advanced expositions are those by Israelachvili and Tabor [35], by Parsegian [36], and by Richmond [37].

Two of the important conclusions that emerge from the macroscopic approach are that the Keesom and Debye forces are explicitly included, and that, strictly speaking [38], the interaction is a free energy rather than a potential energy, i.e., $U_A$ should be replaced by $G_A$, as considered in Sec. II.

Discussion of the effects of adsorbed layers of the van der Waals interaction is postponed to the general discussion of the effects of adsorbed layers on interparticle interactions (Sec. VIII).

### 2. *Electrical Double-Layer Interactions*

As discussed in Chap. 1, Sec. II.C, charged droplets possess electrical double layers at the liquid/liquid interface. When two droplets, each carrying such a diffuse layer of counter-ions approach, for example in a Brownian encounter, repulsion results from an overlapping of these diffuse layers. A full account of electrical double-layer interactions, both for flat plates and for spheres, is given in the monograph by Verwey and Overbeek [56]. A more recent account has been given by Bell and Peterson [39], which includes a discussion of *un*symmetrical particles (i.e., shape, size, and surface potential differences). In this chapter we discuss only some of the more basic equations.

In developing equations for $G_E$, the electrostatic interaction free energy, two limiting cases may be distinguished: the constant potential and constant charge models. In the former it is assumed that thermodynamic equilibrium is maintained during particle collisions, such that the surface potential remains constant, but that there is a corresponding decrease in surface charge density. This condition only holds if the rate of particle approach, during a Brownian encounter, is slower than the rate at which ions can rearrange themselves in the electrical double layer. The other limiting condition, i.e., constant charge, corresponds to the case where there is no exchange of ions at the surface during the particle encounter; there is, however, a corresponding increase in surface potential. In practice the situation usually lies somewhere between these two limits. Fortunately differences between $G_E$ values calculated under these two limiting conditions only become significant at small separations; here the constant charge value is greater than the constant potential value.

$G_E$ may be calculated in two basic ways: (1) by calculating the free-energy change of the two double layers on overlap and (2) by deriving an expression for the mean force between the two particles and integrating to obtain the potential energy of mean force. Numerical solutions to the exact equations for flat plates have been evaluated and tabulated by Devereux and de Bruyn [40]. For spheres, on the other hand, which are more relevant in the context of emulsions, exact solutions are not available, and it is necessary to resort to approximate equations, each valid for different limiting conditions. One such approximate equation, derived by Derjaguin and Landau [5a], is

$$G_E = 2\pi a \varepsilon_r \varepsilon_0 \psi_0^2 \ln[1 + \exp(-\kappa h)] \tag{33}$$

where $\varepsilon_r$ is the dielectric constant of the medium, $\varepsilon_0$ is the permittivity of free space, $\psi_0$ is the surface potential, h is the surface-surface separation, and $\kappa$ is the reciprocal "thickness" of the electrical double layer, given by [see Chap. 2, Sec. II.C, Eq. (77)]

$$\kappa = \left( \frac{2z^2e^2c}{\varepsilon_r \varepsilon_0 kT} \right)^{1/2} \tag{34}$$

where z = valency of counter-ions

e = electronic charge

c = electrolyte concentration in bulk solution

k = Boltzmann constant

T = absolute temperature

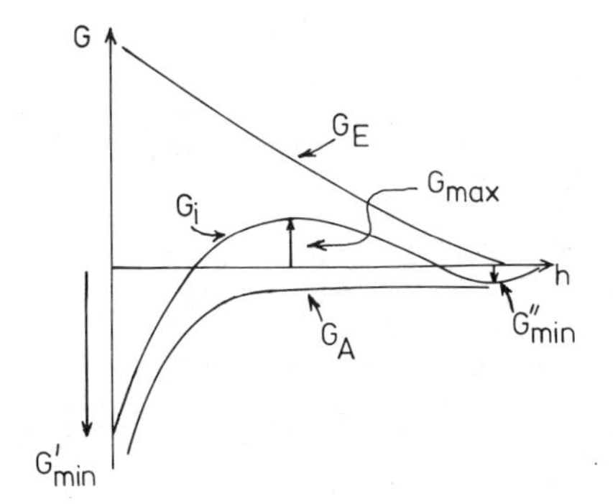


Figure 8. Schematic representation of G(h).

Equation (33) is valid only for low surface potentials (i.e., $z\psi_0 < 25$ mV), for spheres that are large in radius in comparison to $\kappa^{-1}$ (i.e., $\kappa a >> 1$), and for weak interactions between the spheres (i.e., $\kappa h > 1$). Verwey and Overbeek [5b] have given tables for $G_E$ for higher potentials, obtained using a more exact equation.

The application of Eq. (33) and similar equations is complicated, in the case of emulsion droplets, by the possible deformation of the droplets during collisions, and by the existance of an internal double layer within the droplets. For O/W droplets the latter effect is small for oils of low dielectric constant. Distortion is a more difficult problem. Some flattening of the droplets in the region of contact may well occur, and this leads to difficulty in assigning a value of a, the radius, in equations such as Eq. (33). It may be better to use the equations for flat plates. There is no satisfactory theoretical analysis for deformable droplets at present.

### 3. *Total Interaction Free Energy*

The total interaction free energy $G_i$ is obtained by summing the $G_A$ and $G_E$ contributions. A schematic representation of the $G_i(h)$ curve is given in Fig. 8. This curve is characterized by a primary minimum ($G'_{min}$) at close contact of the particles,[†] a maximum ($G_{max}$), and a secondary minimum ($G''_{min}$) at large interparticle or interdroplet separations. Weak, reversible flocculation may occur into this secondary minimum when it exists (see discussion below). For irreversible coagulation into the primary minimum to occur, the energy barrier has to be sur-

[†]Note that at very short distances between the particles, a strong repulsion appears due to the overlap of electron shells; this is the Born repulsion.

mounted. The height of the barrier is primarily controlled by the electrolyte concentration.

In order to illustrate these points the results of Srivastava and Haydon [41] are worth quoting. These authors calculated the $G_i(h)$ curve for paraffin oil/water emulsions stabilized by bovine serum albumin, using the following data: droplet radius = 1.24 $\mu$m; KCl concentration = 0.01 mol $dm^{-3}$; zeta potential (used in place of $\psi_0$, see below) = 20 mV; effective Hamaker constant $A = 1.72 \times 10^{-20}$ J. They found that the G(h) curve has a maximum $G_{max} > 50$ kT. Since this is considerably larger than the average thermal energy of the droplets [(3/2)kT], it may be safely assumed that no coagulation into the primary minimum occurs. However, at $h \sim 15$ nm, there is a secondary minimum, $G''_{min}$ a few kT in depth. This led to the observation of the formation of doublets in equilibrium with singlet droplets, but these doublets could be readily broken up by the application of weak mechanical forces, e.g., mild agitation. The existence and size of $G''_{min}$ depends on the four variables referred to above, i.e., droplet size, electrolyte concentration, Hamaker constant, and surface (or zeta) potential. Generally $G''_{min}$ is significant only for large droplets (>1 $\mu$m).

With smaller droplets (say, $\sim$0.1 $\mu$m), no secondary minimum appears in the $G_i(h)$ curves, and therefore flocculation is not observed. Moreover, for $\psi_0 \sim 50$ mV and $c \sim 10^{-3}$ mol $dm^{-3}$, $G_{max}$ is so high that coagulation does not occur. As $\psi_0$ is reduced or c increased, $G_{max}$ decreases. At a value of $\sim$20 kT, very slow coagulation will be observed (on a time scale of months). At sufficiently high values of c, $G_{max}$ is reduced to zero and rapid coagulation occurs, the process then being essentially diffusion controlled.

In establishing theoretical $G_i(h)$ curves for a given system the most difficult parameters to assign values are $\psi_0$ and A. The electrolyte concentration (c) is readily determined and the particle size may be determined by one of several techniques (see Sec. IV.D). The value for A is difficult to find, particularly if one is dealing with multicomponent liquid phases. However values have been tabulated for a number of pure liquids, and the reader is referred to the review articles by Gregory [42] and Visser [43].

With regard to $\psi_0$ the reader is referred to Chap. 1, Sec. II.C, where a discussion of the potential profile across a typical liquid/liquid interface is discussed (see Figs. 11 and 12 of Chap. 1, in particular). The surface potential $\psi_0$ may be identified with the potential at the plane containing the charged head groups of the surfactant molecules (Fig. 11 of Chap. 1) and thus $\psi_0 \sim \psi_1$ (Fig. 12 of Chap. 1). Surface potentials for liquid interfaces can be measured using, for example, vibrating disk electrodes [44]. On the other hand, $\psi_0$ may not be the appropriate potential to consider in equations such as Eq. (33) for $G_E$. If specific adsorption of counter-ions at the charge groups in the interface takes place, then the potential $\psi_d$, at the plane containing these adsorbed counter-ions (the so-called Stern plane) may be the more appropriate potential to use. In practice, what is frequently done is to use the zeta potential, i.e., the potential at the plane of shear.

This plane may be one or two solvent molecule diameters further out than the Stern plane. Clearly $\psi_0 > \psi_d > \zeta$. It is usually assumed that $\zeta$ is not too different from $\psi_d$, although the exact validity of this assumption is open to question. The advantage to be gained, however, lacking further insight, is that $\zeta$ is readily determined experimentally, whereas $\psi_d$ is not.

For emulsion droplets the most convenient method of obtaining $\zeta$ is by microelectrophoresis [44,45] which gives the electrophoretic mobility u of the droplets, where u is the droplet velocity under a uniform applied electric field. There are a number of equations available for calculating $\zeta$ from u [44,45], e.g.,

$$\zeta = \frac{\eta u}{\varepsilon_r \varepsilon_0} \qquad \kappa a > 100 \tag{35}$$

$$\zeta = \frac{3\eta u}{2\varepsilon_r \varepsilon_0} \qquad \kappa a < 1 \tag{36}$$

$$\zeta = \frac{3\eta u}{2\varepsilon_r \varepsilon_0} f(\kappa a) \qquad 1 < \kappa a < 100 \tag{37}$$

where $\eta$ is the viscosity of the medium.

Equation (35) is usually referred to as the *Smoluchowski* [46(b)] *equation*, Eq. (36) as the *Hückel* [47] *equation*, and Eq. (37) as the *Henry* [48] *equation*. The last-named takes into account the retardation effect of the electrical double layers in an applied electric field. More sophisticated theories, for example, that by Wiersema and Overbeek [49], take into account the relaxation effect also. Ottewill and Shaw [50] have tabulated values of u and $\zeta$ for various electrolyte types and concentrations.

One question that now arises in the use of equations relating u and $\zeta$, such as those referred to above, concerns the assignment of values to $\varepsilon_r$ and $\eta$. Strictly speaking, these are the values of the dielectric constant and viscosity, respectively, in the electrical double layer. Lyklema and Overbeek [51] have investigated this problem and concluded that $\varepsilon_r$ in the double layer is not too different from the bulk value, whereas $\eta$ may be significantly higher. However, lack of suitable data for electroviscous constants usually means that the bulk viscosity value is used in most cases.

## C. Flocculation Rates: Theory

It is convenient to split the discussions of the theory of flocculation kinetics into two parts: "fast" (i.e., diffusion-controlled) flocculation, where $G_{max} = 0$, and "slow" flocculation, where $G_{max} > 0$.

### 1. *Fast Flocculation*

The simplest model is that due to von Smoluchowski [46a] who considered the case where there is no interaction between two colliding particles until they come into contact, whereapon they adhere irreversibly. For such a system initially containing $n_0$ particles per unit volume, the number concentration n at time t was found to be

$$n = \frac{n_0}{1 + kn_0 t} \tag{38}$$

where k is the rate constant, given by

$$k = 8\pi Da \tag{39}$$

D is the diffusion coefficient, which, assuming Stokes' law to hold (see discussion in Sec. III), may be expressed as

$$D = \frac{kT}{6\pi\eta a} \tag{40}$$

which is independent of particle size. Thus, for oil droplets dispersed in an aqueous phase at 25°C ($\eta$ = 1 cP), $k = 5.5 \times 10^{-12}$ $cm^3$ $s^{-1}$.

Rearrangement of Eq. (38) yields the half-life $\tau_{1/2}$ of an emulsion (i.e., the time for the total number of droplets to fall to half the original value),

$$\tau_{1/2} = \frac{1}{kn_0} \tag{41}$$

Given the above value for k, it is instructive to evaluate $\tau_{1/2}$ for different values of $n_0$, or the equivalent, $\phi$ the volume fraction of the disperse phase. Some typical $\tau_{1/2}$ values are listed in Table 2 for different droplet size emulsions at different $\phi$ values, covering the range normally met with in practice. It can be seen that the time scales involved range over some 10 orders of magnitude. A dilute dispersion of large droplets may show no visible signs of flocculation over a day or so, whereas a high concentration of small droplets would appear to be instantly flocculated.

The Smoluchowski analysis also leads to an expression for the number of i-mer aggregates at time t, i.e.,

Table 2. Half-lives of Emulsions of Different Phase Volumes and Droplet Sizes

| $\phi$(%) / a ($\mu$m) | $10^{-3}$ | 1 | 10 | 50 |
|---|---|---|---|---|
| 0.1 | 76 s | 76 ms | 7.6 ms | 1.5 ms |
| 1 | 21 h | 76 s | 7.6 s | 1.5 s |
| 10 | 4 months | 21 h | 2 h | 25 min |

$$n_i = \frac{n_0(t/\tau_{1/2})^{i-1}}{(1 + t/\tau_{1/2})^{i+1}} \tag{42}$$

Note that in the derivation of Eqs. (38) and (42) it is assumed that flocculation has proceeded for a sufficient time to ensure that the necessary steady state conditions are established, i.e., $t > a^2/D$ (e.g., $\sim 5 \times 10^{-17}$ s for 1 $\mu$m radius droplets); so this is, in general, not a problem.

A number of modifications to the Smoluchowski fast rate constant need to be considered.

a. Reversibility. If reversible flocculation into a shallow energy well is occurring then any complete kinetic analysis must take into account the (backward) rate of deflocculation as well as the (forward) rate of flocculation, i.e.,

$$\frac{dn}{dt} = -k_f n^2 + k_b n \tag{43}$$

$k_b$, the rate constant for deflocculation, may well depend on floc size and the exact way in which the flocs break down (e.g., how many "contacts" are broken). This would mean that the second term on the right-hand side of Eq. (43) would need to be replaced by a summation over all possible modes of breakdown, making analysis of the kinetics complex.

Another complication in the analysis of the kinetics of reversible flocculation is that, as discussed in Sec. II, flocculation of this type is a critical phenomenon rather than a chain (or sequential) process. Thus, a critical droplet number concentration $n_{crit}$ has to be exceeded before flocculation occurs, i.e., flocculation becomes a thermodynamically favored process. The kinetics of weak, reversible flocculation has more in common, therefore, with nucleation kinetics than with chemical (e.g., polymerization) kinetics. That is not to say that doublets, triplets, etc., will not form transiently below $n_{crit}$. These are thermodynamically unstable, but their effective concentrations may be calculated from a suitable kinetic analysis.

b. Polydispersity. Most emulsion systems are not monodisperse. The flocculation rate for a polydisperse system is usually faster than that of the equivalent[†] monodisperse system. Analyses for the kinetics of flocculation of polydisperse systems have been given by Mueller [51a] and by Wiegner [51b].

c. Non-Stokesian Hydrodynamic Effects. Even with systems of monodisperse irreversibility coagulating spheres (e.g., aqueous polystyrene latex dispersions [52,60]) the rate constant for fast flocculation seems to depend on particle size and number concentration. The maximum value of $k$ found for these systems is $3 \times 10^{-12}$ cm$^3$ s$^{-1}$, compared to the Smoluchowski value of $5.5 \times 10^{-12}$ cm$^3$ s$^{-1}$. This is due to the fact that the hydrodynamic interactions for a diffusing particle obey Stokes' law only when that particle is isolated from others. In a real dispersion extra hydrodynamic forces have to be taken into account. Thus, for approaching particles, the assumption made in the Smoluchowski analysis that the relative diffusion coefficient $D_{12}$ is given by the sum of their two individual diffusion coefficients $D_1$ and $D_2$ is no longer valid. The extra hydrodynamic correction terms evaluated (by Spielman [53a] and by Honig et al. [53b]) do account, at least in part, for the observed dependence of $k$ on $n_0$ and a, although this problem is not yet totally resolved.

d. Non-Brownian Collisions. The so-called perikinetic flocculation discussed above assumes that particle collisions arise solely from Brownian diffusion of the particles, i.e., that the diffusion coefficient of the particles is given by the Einstein relation $D = kT/f$ where f is the frictional coefficient (Stokesian or otherwise) of the particles. If external energy is supplied to the system (e.g., shear, ultrasonic, or centrifugal) or the system is not at thermal equilibrium (so that convection currents arise), then the rate of droplet collisions is modified (usually increased), and the flocculation is referred to as *orthokinetic*. Although, for an irreversibly coagulating system, one might expect orthokinetic conditions to lead to an increased state of flocculation in a given time interval, with a weakly (reversibly) flocculating system the opposite is frequently the case, i.e., application of shear, for example, leads to deflocculation. The hydrodynamics is complex in these cases, but investigations of the orthokinetic flocculation of emulsions have been made by various authors, e.g., Swift and Friedlander [54a] and Hidy et al. [54b].

---

[†]By *equivalent* is meant that the particle radius is given by $a^3 = (1/n)\Sigma_i a_i^3 n_i$ and that $n = \Sigma_i n_i$.

*2. Slow Flocculation*

Smoluchowski [46] originally accounted for the effect of an energy barrier (i.e., $G_{max}$), arising from interparticle interactions, on the kinetics of coagulation by introducing a correction parameter $\alpha$, where $\alpha$ is the fraction of collisions which are "effective," i.e., where irreversible contacts are formed. This idea was based on the Arrhenius equation for chemical reactions, i.e.,

$$k = k_0 \exp\left(\frac{-G_{max}}{kT}\right) \tag{44}$$

The analogy with chemical reactions is not exact, however. In chemical kinetics, the rate of disappearance of a reactant is given by the absolute concentration of the transition-state species times the vibrational frequency of the bond in that species which has to break to form the products [55]. In coagulation kinetics the rate depends on the flux of particles to any one chosen particle [56]. Fuchs [57] showed that when particle interactions are present this flux is made up of two contributions, one due to the Brownian diffusion of the particles, the other due to the interactions. An account of Fuchs' analysis is given in standard texts, e.g., Kruyt [58] or Verwey and Overbeek's monograph [5b]. The resulting equation for the slow flocculation rate constant $k$ is

$$k = \frac{k_0}{W} \tag{45}$$

where W, the stability ratio, is given by

$$W = 2a \int_{2a}^{\infty} \exp\left(\frac{G_{max}}{kT}\right) R^{-2}\, dR \tag{46}$$

where R equals the center-center separation of two interacting particles or droplets. Note that W depends, therefore, on the extent of flocculation, and indeed on the morphology of the flocs that are formed in any given system. For this reason it is usual only to consider the early stages of flocculation ($t \to 0$) where, for the most part, doublets are the only aggregated structures present.

An approximate form of Eq. (46) for charge-stabilized dispersions has been given by Reerink and Overbeek [59]

$$W \sim \frac{1}{2} \kappa a \exp\left(\frac{G_{max}}{kT}\right) \tag{47}$$

where $\kappa$ is given by Eq. (34). These authors also derived the following theoretical relationship between W and the electrolyte concentration c for aqueous dispersions at 25°C:

$$\log W = \text{constant} - 2.06 \times 10^{9} \frac{a\gamma^2}{z^2} \log c \tag{48}$$

where

$$\gamma = \frac{\exp(ze\psi_0/2kT) - 1}{\exp(ze\psi_0/2kT) + 1} \tag{49}$$

Equation (48) was derived by combining Eq. (47) with the appropriate expressions for $G_A$ (Eq. 25) and $G_E$, i.e.,

$$G_E = 6.39 \times 10^{-9} \frac{a\gamma^2}{z^2} \exp(-\kappa h) \tag{50}$$

Note that Eq. (50) [5b] is an alternate form to Eq. (33) for $G_E$.

Equation (48) predicts that experimental plots of log W versus log c should be linear in the slow flocculation region. A number of workers have sought to test this relationship, the first definitive work probably being that of Ottewill and Shaw [60]. W was calculated from Eq. (45), where $k$ was taken to be the limiting ($t \to 0$) rate constant obtained from turbidimetric studies (see Sec. D). Unfortunately, although Eq. (48) predicts that the slope of the log W/log c plot should be proportional to particle size, these authors found no systematic variation in this slope with latex particle dispersions ranging in particle radius from 0.03 to 0.2 μm. No complete explanation at present exists for these observations, although Healy and Weise [61] have attempted to resolve the problem by taking into account secondary minimum effects.

In the fast coagulation regime ($G_{max} = 0$), W is independent of c, i.e., $d(\log W)/d(\log c) = 0$. Thus, an experimental plot of log W versus log c should consist of two linear portions, intersecting at the so-called critical flocculation (coagulation) electrolyte concentration (CFC) for that particular system at dispersion plus electrolyte. Typical theoretical curves are shown in Fig. 9. Note that in the fast regime W is actually less than 1. This occurs because there is then a net attraction due to van der Waals interactions between the particles. Further considerations of this effect have been given by Parfitt and McGowan [62]. The theoretical curves shown in Fig. 8 also illustrate the greater flocculating power of a two:two electrolyte as compared to a one:one electrolyte, in line with the well-known Schuze-Hardy rule. Experimental curves [60] are generally of the same form as those predicted theoretically.

## D. Flocculation Rates: Experimental

As discussed in Sec. IV.C, only the initial flocculation rates ($t \to 0$) are subject to theoretical analysis. Also, with emulsion systems, one is generally dealing with

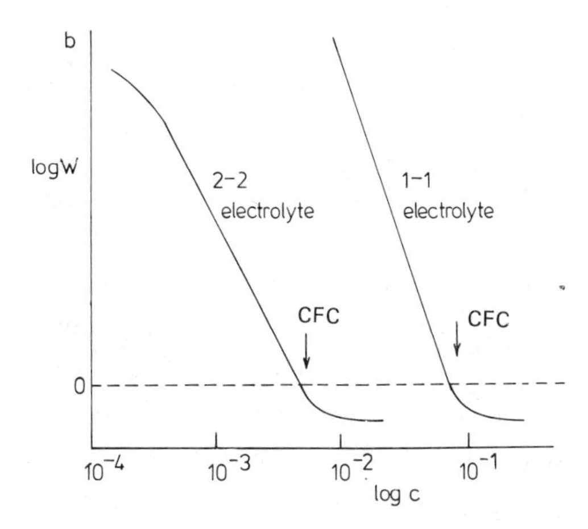


**Figure 9.** Plot of log W versus log c for 1:1 and 2:1 electrolytes.

polydisperse systems, and this renders any quantitative comparison with the theory outlined in the previous section of doubtful validity. However, three experimental techniques have been most widely used to monitor flocculation rates for emulsions: (1) light scattering, (2) particle counting, and (3) centrifugation methods. The first of these is restricted, in practice, to dilute systems (i.e., $\phi << 0.1\%$); the other two may be used for somewhat higher volume fractions ($\phi \sim 1-10\%$), depending on the droplet size and exact method used.

For concentrated emulsions no really satisfactory techniques have yet been developed for monitoring flocculation rates. Rheological techniques, of course, give information on the state of dispersion for a given system, but are not very suitable for following flocculation rates since the degree of flocculation is necessarily perturbed in a rheological experiment. Relaxation time measurements, in principle, contain information on both the forward (flocculation) and the backward (deflocculation) rate constants, but it is difficult to separate these in practice. Two general, practical points should be borne in mind: one should be careful in diluting any system that this in itself does not alter the state of dispersion; settling/creaming (e.g., in optical methods) should be avoided by gentle, end-over-end rotation of the containing vessel.

### *1. Light Scattering Methods*

In general, two types of light-scattering experiment are performed: (1) angular light scattering, where the intensity of the scattered light is measured as a function of the scattering angle, and (2) turbidity ($\tau$) or optical density (D) measurements, where the intensity (I) of the transmitted beam is measured.

$$\tau = \frac{1}{\ell} \ln \frac{I_0}{I} \tag{51}$$

and

$$D = \log_{10} \frac{I_0}{I} \tag{52}$$

where $\ell$ is the path length of the light through the emulsion and $I_0$ is the intensity of the incident beam. Although, in principle, method (1) contains more information, (2) is easier to carry out in practice (e.g., it is less sensitive to the presence of dust particles) using, say a twin-beam spectrophotometer or a stopped-flow apparatus. Also, it may be used at somewhat higher droplet number concentrations, i.e., before multiple scattering invalidates the simple theoretical treatments outlined below [63,64].

For reviews of the general theory of light scattering the reader is referred to the standard texts of van der Hulst [65] and Kerker [66]. A shorter review has been given by Kratohvil [67].

For droplets having radii appreciably less than the wavelength ($\lambda$) of the light used (i.e., $a < \lambda/20$; e.g., for $\lambda = 546$ nm, $a < 0.03$ μm) the Rayleigh theory of light scattering applies. Theoretically, therefore, this is only valid for microemulsion systems, although in practice it is often used for larger droplet size emulsions. In the Rayleigh theory, the turbidity $\tau_0$ of an initially monodisperse emulsion, containing $n_0$ singlet droplets each of volume $v_1$, is given by

$$\tau_0 = A' n_0 v_1 \tag{53}$$

where A' is an optical constant for the system,

$$A' = \frac{24\pi^3 \mu_0^4}{\lambda^4} \left( \frac{\mu^2 - \mu_0^2}{\mu^2 + 2\mu_0^2} \right)^2 \tag{54}$$

where $\mu$ and $\mu_0$ are the refractive indices of the droplet and continuous phase, respectively. By combining the Rayleigh theory with the Smoluchowski-Fuchs theory of flocculation kinetics, Oster [68] has derived the following expression for the turbidity of a flocculating dispersion as a function of time t:

$$\tau = A' n_0 v_1^2 (1 + 2n_0 kt) \tag{55}$$

Equation (55) predicts a linear dependence of $\tau$ on t. Experimental plots of $\tau$ (or D) versus t are usually curved, however, since, even if the singlet droplets are Rayleigh scatters, aggregates are almost certainly not. In addition, some of the other assumptions made in deriving Eq. (55) apply only to the early stages of flocculation. Nevertheless, initial rate constants obtained from the limiting slope ($t \to 0$) of $\tau$-t plots are useful, if only from a comparative viewpoint for most emulsion systems.

In order to extract absolute rate constants from turbidity-time plots the more general Mie theory of light scattering [65,66] must be used. This is applicable to droplets of any size and predicts for a flocculating dispersion

$$\tau = \sum_i n_i \pi a_i^2 K_{t,i} \tag{56}$$

where $n_i$ is the number of i-mer aggregates (i.e., containing i singlest) and is given by Eq. (43). $K_{t,i}$ is the total scattering factor for an i-mer. Tables of $K_{t,i}$ for different $\alpha$ and m values ($\alpha = 2\pi\mu_0 a_i/\lambda$ and $m = \mu/\mu_0$) have been compiled by Pangonis et al. [69].

$$\frac{\tau}{\tau_0} = \frac{\Sigma_i n_i a_i^2 K_{t,i}}{n_0 a_1^2 K_{t,1}} \tag{57}$$

Thus, by substituting for $n_i$, as given by Eq. (43), in Eq. (57) for various assumed values of $k$ (the rate constant), theoretical plots of $\tau/\tau_0$ may be constructed by taking the necessary number of terms on the right-hand side of Eq. (57) such that the sum converges [70]. Then, by comparing the computed $\tau/\tau_0$ curves with the experimental $\tau/\tau_0$ curves a best-fit value for $k$ may be established. In principle, it ought to be possible to compute theoretical $\tau/\tau_0$ plots for initially polydisperse systems having a known droplet size distribution, but as far as we are aware, this has not been tried.

Angular-light-scattering experiments have also been used [71,72] to follow flocculation kinetics, and again absolute rate constants may be determined. Lips et al. [71] have used the Rayleigh-Gans-Debye theory of light scattering to evaluate the light scattered at a given angle for systems where $\mu \backsim \mu_0$. Primary particles within an aggregate are assumed to scatter independently, the scattered light interfering with that from the other primary particles in the aggregate. Various model structures are assumed, and the Smoluchowski Eq. (43) again is used to predict $n_i(t)$. Angular light scattering is particularly useful at very low angles ($\backsim 1°$) [72] since in this region the change in intensity with degree of aggregation becomes more or less independent of the exact aggregate morphology. It was found [72] that the intensity of scattered light at these very low angles increases linearly with time for an aggregating system, and the rate constant can be obtained from the slope of this plot.

Photon correlation spectroscopy (PCS) [73-75], in which one determines the correlation function for laser photons arriving at the photomultiplier at some fixed scattering angle, is potentially a powerful method for following flocculation rates, since in principle the complete particle size distribution should be obtainable as a function of time, i.e., $n_i(t)$ for all i. This has yet to be achieved, however. Uzgiris and Costaschuk [76] were the first to report studies of flocculation using the PCS technique, in this case for polystyrene latex particles.

If one is interested in detecting the onset of low levels of flocculation, e.g., in determining stability-instability boundaries, rather than rate constants, then log $\tau$/log $\lambda$ plots can be useful. This has been discussed by Long et al. [71].

*2. Droplet Counting Methods*

These fall into two categories: (1) direct counting methods and (2) indirect counting methods. The direct counting method normally refers to optical microscopy, although electron microscopes have also been used with emulsion systems [77-79]. The latter technique includes freeze etching of the emulsion system; in principle, kinetic measurements could be carried out by freezing a flocculating emulsion at different time intervals.

Optical methods are, however, more straightforward and with modern optical microscopes one can normally resolve droplets as small as 0.5 μm diameter. By using, for example, a hemocytometer cell, total or singlet droplet counts can be made for a fixed number of graticule squares. The rate constant is then obtained from the linearized form of Eq. (38),

$$\frac{1}{n} = \frac{1}{n_0} + kt \tag{58}$$

An example of the use of an optical microscope method to determin k is given by the work of van den Tempel [80], who studied the flocculation rate of charged oil/water emulsions stabilized by Aerosol OT. The oil phase (chlorobenzene plus liquid paraffin) was chosen such that it had the same density as the aqueous phase, thus obviating problems due to creaming. At given time intervals, after adding an electrolyte as coagulant, the degree of flocculation was frozen by diluting the emulsion with a 1% solution of nonionic surfactant. The fast rate constants obtained at high electrolyte concentrations were generally faster than those predicted by the simple Smoluchowski theory, probably owing to polydispersity effects. The optical microscope method was later used by Srivastava and Haydon [81] for following both flocculation and coalescence kinetics.

Examples of indirect particle counting methods are the Coulter counter and various flow ultramicroscope particle counters, the most recent of which use laser light sources [82]. These indirect particle counting techniques are, in general, much faster than the direct optical microscope methods. The Coulter counter, which can count at ~$10^4$ particles $s^{-1}$, was originally developed for blood counts. It yields, in principle, droplet size distributions as a function of time, although because correction factors have to be applied to obtain $n_i(t)$, the method does not yield totally absolute values. The emulsion is made to flow through a narrow orifice (>30 μm diameter); when a droplet passes through, the resistance of the medium in the orifice changes and an electric pulse is generated, the size of which depends on the droplet volume. Using a multichannel analyzer, the droplet size distribution can be built up. There are, however, a number of limitations to the use of this instrument:

i. A correction factor is required to allow, statistically, for coincidence, i.e., when more than one droplet passes simultaneously through the orifice.
ii. The system should be free of dust.
iii. There is the possibility of floc breakdown when a floc is forced through the orifice.
iv. Only emulsion droplets >0.6 μm diameter can be conveniently studied.
v. High dilutions are necessary ($\sim 10^6$ to $10^7$ droplets per $cm^3$).
vi. A reasonably high background electrolyte concentration (up to 1 mol $dm^3$ NaCl) is necessary so that the conductance has a reasonable value.

Factors v and vi should be carefully considered in using this method, since dilution or addition of electrolytes may, of themselves, alter the state of dispersion of the emulsion. In spite of these limitations, the Coulter counter has been used by a number of workers [83-87] to follow the kinetics of both flocculation and coalescence.

In the flow ultramicroscope technique the emulsion is made to flow through a cell illuminated by a beam at 90° to the direction of flow. The particles appear as a speckle pattern in a dark background and may be counted manually [88,89] or by using a photoelectric counting device [82,90]. Walsh [91] has recently developed an instrument of this type, not only to count the total number of particles, but to measure the particle size distribution, again using a multichannel analyzer. One major advantage of optical counters of this type over the Coulter counter is that much smaller particle or droplet size systems can be studied. However, depending on the relative refractive indices of the particles and the dispersion medium, large particles (>1 μm) are difficult to study by this technique, unless the scattering angle is reduced from 90° to, say, <10°.

### 3. *Centrifugal Methods*

There are three basic techniques here: ultracentrifugation, disk centrifuges, and photosedimentometers. The ultracentrifuge may, in principle, be used to follow flocculation kinetics, and Bailey et al. [92,93] have described how the method could be applied to particle-size analysis for emulsions. However, the method is tedious and requires a method of analysis for the dispersed phase.

In photosedimentometers the optical density of a sedimenting dispersion is scanned as a function of height at various time intervals. Killmann [94] has recently described the construction of such an apparatus for use with dispersions. Since aggregates settle at different rates, in principle, information on the particle size distribution, as a function of time, should be extractable, assuming all the particles settle independently.

The disk centrifuge has been described by Allen [95]. The dispersion is placed in the center of a disk which is then rotated. Samples are then extracted manually at a given radial point at different time intervals. From a knowledge of the densities of the droplets and of the continuous phase the particle size distribution can be built up. In the Joyce-Loebl disk centrifuge, the manual sampling method is replaced by a light-intensity monitoring device, thus, in effect combining the attributes of the disk centrifuge and the photosedimentometer techniques.

## V. COALESCENCE

### A. Introduction

As illustrated in Fig. 1, when two droplets come in contact, e.g., in a cream or a floc, a thin liquid film or lamella forms between them. Coalescence results from the rupture of this film. This is on the assumption that the two droplets are formed of the same liquid or, at least, of miscible liquids. If not, then adhesion or engulfment may occur. Film rupture usually commences at a specified "spot" in the lamella, arising from thinning in that region. In order to understand the behavior of these films we need to consider two aspects of their physics: (1) the nature of the forces acting across the film—these determine whether the film is thermodynamically stable, metastable, or unstable—and (2) the kinetic aspects associated with local (thermal or mechanical) fluctuations in the film thickness.

The coalescence of emulsion droplets is thus governed by the behavior of the thin (LLL) film between the droplets. The physics is formally similar to that for other types of film, e.g., in soap bubbles and foams (VLV), or the spreading of a liquid on a solid or an immiscible substrate (VLS or VLL). Indeed, the coagulation of solid particles corresponds to the SLS case, but there we are only concerned with factor (1) above, since the local fluctuations in film thickness referred to in (2) can only arise with fluid interfaces.

Reviews of the stability of liquid films have recently been given by Buscall and Ottewill [96a], and by Aveyard and Vincent [96b]. The reader is referred to these for further description.

In this section we shall consider only symmetrical films, i.e., where the two interfaces forming the film are identical.

## B. Forces Across Liquid Films

Figure 10 shows the general features of the lamella between two droplets of the phase ($\alpha$) in a continuous phase ($\beta$). The thin film region, as discussed above, consists of two flat, parallel interfaces separated by a distance b. At the end of the film there is a border or transition region where the interfaces have a high curvature, i.e., compared to the curvature of the droplets themselves. Eventually, however, at larger values of b (effectively beyond the range of the forces operating across the film—see below) the curvature decreases to that of the droplets themselves, i.e., becomes effectively flat, on the scale considered here, for droplets in the 1-$\mu$m region. One may then define a macroscopic contact angle $\theta$ as shown in the diagram (Fig. 10).

A number of workers have attempted to measure the thickness of thin LLL films. For example, Sonntag et al. [97] have described an elegant light reflection technique for measuring the thicknesses of O/W/O lamellae, while Haydon et al. [98] have studied W/O/W systems in detail using lamellae stabilized by lecithin; these systems are excellent models for biological membranes.

In considering the forces acting across the film, two regions of separation b are of interest: $b > 2\delta$. In the first region, the forces acting are the long-range van der Waals forces and the electrical double-layer interactions. A

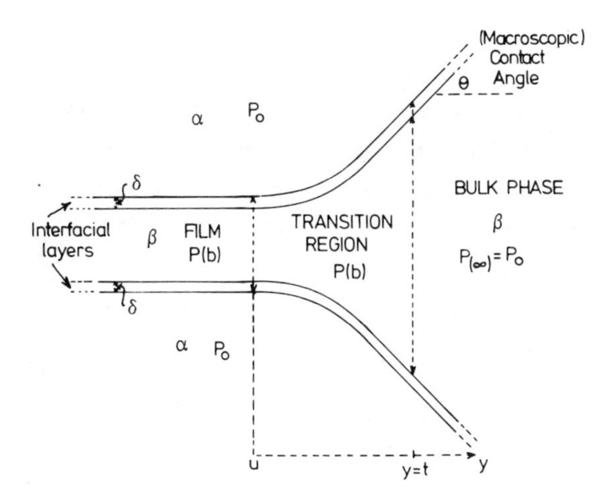


Figure 10. Diagrammatic representation of the thin film and border regions between two liquid droplets ($\alpha$) in a continuous phase ($\beta$). (From Ref. 96b.)

formal description of these has already been given in Sec. IV on flocculation. Suffice it to say here that the same type of interaction free-energy diagram exists as shown in Fig. 8, i.e., with a secondary minimum, a maximum, and a primary minimum. When the film is sitting in either the primary or secondary minimum, the net force on the film is zero (i.e., $dG_i/db = 0$). These two metastable states correspond with the so-called Newton† and common black films, respectively. The films appear black because b is, in general much smaller than the wavelength of the light being used, and interference between light reflected from the two interfaces is completely destructive.

When a film is in the primary minimum we then have to consider what happens if it tries to thin further (i.e., $b < 2\delta$). Here the so-called steric interactions come into play. Detailed discussion of these is postphoned to Sec. VIII. Suffice it to say at this point that these normally constitute a very steep repulsion force, associated with the distortion of the adsorbed polymer or surfactant in the interfacial layers. Film rupture requires, inter alia, the breakdown of these steric interactions.

The stability of thin films may be analyzed in terms of the relevant interaction free energies in the same way as in the section of flocculation. Various alternative (although essentially equivalent) descriptions are, however, possible. Examples of these are the disjoining pressure concept introduced by Derjaguin [100] and descriptions in terms of the interfacial tension of the film or of the chemical potential of the solvent [96b]. We shall briefly consider the first two.

### 1. *Disjoining Pressure*

The concept of disjoining pressure arose from Derjaguin's work on the structure of thin liquid films adhering to substrates. Its significance for emulsions was first pointed out by de Vries [101]. The disjoining pressure $\pi(b)$ is the net force, per unit area, acting in the direction normal to the two flat, parallel interfaces (Fig. 10). At any given value of b, $\pi(b)$ balances the excess normal pressure $p(b) - p(\infty)$ in the film. The quantity $p(b)$ is the normal pressure in a film of thickness b, and $p(\infty)$ is the normal pressure in a film of sufficiently thick so that $G_i(h) = 0$, i.e., the interaction free energy of the film is zero. Thus, $p(\infty)$ is equivalent, for a plane-parallel film, to the isotropic pressure in the adjacent bulk phase $p_0$. Thus,

$$\pi(b) = p(b) - p_0 \tag{59}$$

†Newton [99] was the first to record the observation of black films in soap bubbles.

As mentioned above, $\pi(b)$ will contain three contributions, i.e., from the London-van der Waals (dispersion) forces ($\pi_A$), from the electrostatic forces ($\pi_E$), and from short-range forces ($\pi_S$). It is assumed that gravitational forces can be neglected. The short-range forces include the steric interactions referred to earlier plus other van der Waals forces (Keesom and Debye), H-bonding, solvation forces associated with solvent structure at the interface, etc. Thus,

$$\pi(b) = \pi_A + \pi_E + \pi_S \tag{60}$$

$\pi_A$ is generally negative; $\pi_E$ and $\pi_S$ are positive. Thus, the net disjoining pressure may be positive (repulsive) or negative (attractive) at any given value of b. $\pi(b)$ is formally related to $\overline{G}_i(b)$ the specific excess free energy of the film, associated with these interactions through the relationship

$$\pi(b) = -\frac{d\overline{G}_i(b)}{db} \tag{61}$$

At the primary and secondary minima $\pi(b) = 0$, and then $p(b) = p_0$, Eq. (59), i.e., the normal pressure in the film equals the bulk hydrostatic pressure in the continuous phase. A film will tend to thin spontaneously, therefore, until it reaches a thickness where this condition holds. The film is then in mechanical equilibrium but is thermodynamically metastable.

Note that in the transition zone (Fig. 10), p(b) is always less than $p_0$ [i.e., $\pi(b)$ is negative] because of the high curvature of the interface [see the Laplace equation, Chap. 1, Eq. (46)]. Because p(b) here is also lower than p(b) in the flat film region, there is a tendency for liquid to be sucked into the transition region from the film, i.e., there is a capillary force acting in the direction parallel to the flat interfaces. This force is responsible for the draining of the film, but it does not enter into a discussion of the disjoining pressure, which acts in the direction normal to the interfaces as we have seen.

### 2. *Interfacial Tension of Liquid Films*

The concept of interfacial tension was discussed in some detail in Chap. 1. In particular it was shown there how the interfacial tension can be related to the variation in the tangential pressure tensor $p_t$ across an interface [see Chap. 1, Eq. (5)]. There will be a similar variation in $p_t$ across the liquid film at some thickness b. This is shown schematically in Fig. 11 [96]. We may thus define an interfacial tension for the whole film

$$\gamma(b) = \int_0^b (p_t - p_0)\, dx \tag{62}$$

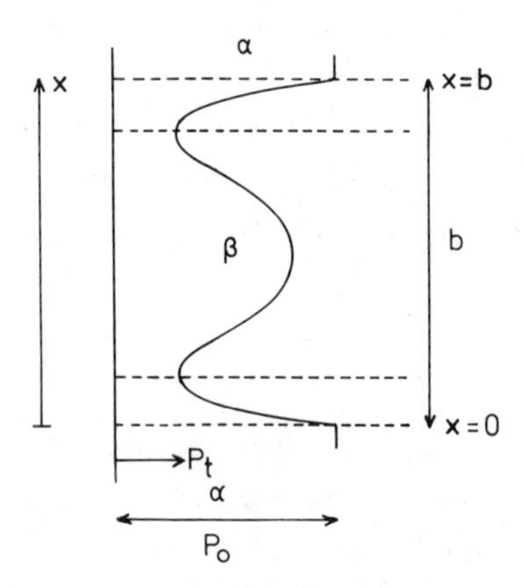


Figure 11. The variation in tangential pressure $P_t$ across a thin film.

By choosing some dividing plane in the middle of the film (conveniently at x = b/2, for a symmetrical film), we can arbitrarily divide $\gamma(b)$ into two contributions, one from the upper interface and one from the lower interface

$$\gamma(b) = \gamma^{\alpha\beta}(b) + \gamma^{\beta\alpha}(b) = 2\gamma^{\alpha\beta}(b) \tag{63}$$

where $\alpha\beta$ refers to the upper interface, and $\beta\alpha$ the lower interface, as depicted in Fig. 11.

Note that $\gamma^{\alpha\beta}(b) \neq \gamma^{\alpha\beta}(\infty)$, the interfacial tension of an isolated $\alpha\beta$ interface, i.e., that between the bulk liquids $\alpha$ and $\beta$ or, for an emulsion, in the region of the interface of the droplet away from the contact zone. $\gamma(b)$ is related to $\overline{G}_i(b)$ and $\pi(b)$ through the relations

$$\gamma(b) = \gamma(\infty) - \overline{G}_i(b) \tag{64}$$

$$\gamma(b) = \gamma(\infty) + \int_{\infty}^{b} \pi(b)\, db \tag{65}$$

Equation (65) is obtained by combining Eqs. (64) and (61).

## C. Film Rupture

The question of film rupture has been considered by various authors, e.g., Scheludko [102] and Vrij [103]. A detailed account has been given by Sonntag and Strenge [104] and the reader is referred to this work for further details.

Film rupture is a nonequilibrium effect and is associated with local thermal or mechanical fluctuations in the film thickness b. A necessary condition for rupture to occur, i.e., for a spontaneous fluctuation to grow, is that

$$\frac{d\pi_A}{db} > \frac{d\pi_E}{db} \tag{66}$$

However, this would assume that, at given value of b, there are no changes in $\gamma(b)$ due to the fluctuations. This is not so, however, since a local fluctuation is necessarily accompanied by a local increase in interfacial area, a consequent deficiency in surfactant or polymer adsorption in that region, and therefore, a local rise in interfacial tension. This effect, the so-called Gibbs-Marangoni effect, in fact opposes fluctuations. It also means that we need to add a term to the right-hand side of Eq. (66) to take account of this fluctuation effect in the local interfacial tension, i.e., Eq. (66) now reads

$$\frac{d\pi_A}{db} > \frac{d\pi_E}{db} + \frac{d\pi_\gamma}{db} \tag{67}$$

As a film thins locally due to fluctuations, if the condition expressed by Eq. (67) is met at some critical thickness, then the film becomes unstable and the fluctuation "grows" leading to rupture. It was Scheludko [102] who introduced this concept of a critical thickness. He gave the following relationship between the wave number K of the surface fluctuation responsible for rupture and the critical thickness $b_{cr}$ for a film in which only van der Waals forces are operating:

$$b_{cr} = \left(\frac{A\pi}{32\,K^2\gamma_0}\right)^{1/4} \tag{68}$$

where A is the net Hamaker constant for the film and $\gamma_0[=\gamma(\infty)]$ the interfacial tension of the isolated liquid/liquid interface. This equation neglects retardation corrections to $\pi_A$. K is in turn dependent on R, the radius of the (assumed) circular film zone.

Vrij [103] has derived alternate expressions for $b_{cr}$. For large thicknesses where $\pi_A << \pi_\gamma$,

$$b_{cr} = 0.268\left(\frac{A^2R^2}{\gamma_0\pi_\gamma f}\right)^{1/7} \tag{69}$$

For small thicknesses, where $\pi_\gamma << \pi_A$,

$$b_{cr} = 0.22\left(\frac{AR^2}{\gamma f}\right)^{1/4} \tag{70}$$

where f is a factor which depends on b.

Thus, Scheludko's and Vrij's equations (68) and (70) have the same form at small thicknesses. This equation has been checked experimentally by Scheludko [102] and by Sonntag and Netzel [105] for emulsion droplets. Equation (70) predicts that as $\gamma \to 0$ so $b_{cr} \to \infty$, i.e., the film should spontaneously rupture at large b values. However, this is certainly not observed; as $\gamma \to 0$ emulsion droplets become highly stable (cf. the discussion on microemulsions, Chap. 4). Moreover, Eq. (70) predicts that as $R \to 0$, $b_{cr} \to 0$, i.e., very small emulsion droplets should never rupture. Experiments on aqueous foam films suggest that one observes finite values for $b_{cr}$ as $R \to 0$. Moreover, experiments by Sonntag [104] on emulsion droplets showed no changes in $b_{cr}$ with changes in the size of the contact area. This is because the lamella formed between two oil droplets, at nonequilibrium separations, does not have the idealized, planar interface depicted in Fig. 11. Rather they have the structure shown, somewhat exaggerated, in Fig. 12, i.e., a dimpled structure. This does not arise from fluctuations, but rather is an effect produced by the draining of the solution from the film region and associated hydrodynamic effects. The hydrodynamic dimpling effect has been considered by several authors, e.g., by Frankel and Mysels [106]. The thinnest region of the film occurs at the periphery of the contact zone, and rupture will tend to occur here. With polymer-stabilized films, presumably because of the increased rigidity of the interfaces, dimpling is far less marked.

The dimpling effect also accounts for the fact that the interfacial tension $\gamma_0$ seems to have little effect on film rupture in emulsion systems. Sonntag [107] has demonstrated, for example, the independence of $b_{cr}$ on $\gamma_0$, using octane/water droplets stabilized by a nonylphenyl ethoxylate surfactant plus an oil-soluble surfactant. It would appear, however, that interfacial tension *is* important in the rupture of foam (VLV) films [103].

### D. Rate of Coalescence Between Droplets

Van den Tempel [80] derived an expression for the coalescence rate of emulsion droplets by assuming the rate to be proportional to the number of points of con-

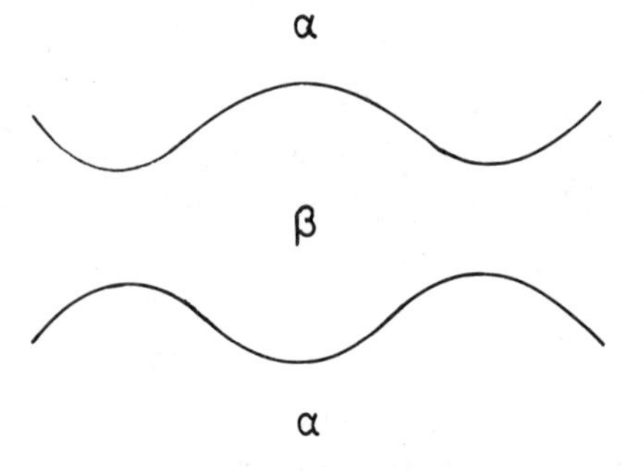


Figure 12. Schematic representation of the "dimple" between two emulsion droplets.

tact between particles in an aggregate. In this theory, both flocculation and coalescence are taken into account simultaneously. The average number of primary particles $n_a$ in an aggregate at time t is given by the Smoluchowski theory (see Sec. IV). The number of particles n which have not yet combined into aggregates at time t is given by

$$n = n_0(1 + kn_0)^{-2} \tag{71}$$

where $n_0$ is the number of singlet particles present initially.

The number of aggregates $n_v$ is given by

$$n_v = kn_0^2t(1 + kn_0t)^{-2} \tag{72}$$

The total number of primary particles in all aggregates is given by

$$n_0 - n_t = n_0\left[1 - \frac{1}{(1 + kn_0t)^2}\right] \tag{73}$$

Hence

$$n_a = \frac{n_0 - n_t}{n_0} = 2 + an_0t \tag{74}$$

If m is the number of separate particles existing in an aggregate, then $m < n_a$, as some coalescence will have occurred; m will be only slightly lower than $n_a$ if coalescence is slow, whereas $m \to 1$ if coalescence is very rapid.

The rate of coalescence is then proportional to $m - 1$, i.e., the number of contacts between the particles in an aggregate. Van den Tempel [80] observed that, in sufficiently dilute emulsions, small aggregates generally contain one large particle together with one or two small ones and are built up linearly. Thus, m decreases in direct proportion to $m - 1$, whereas m increases at the same time by adhesion to other particles. The rate of increase caused by flocculation is given [following Eq. (74)] by

$$\frac{dm}{dt} = an_0 - K(m - 1) \tag{75}$$

where K is the rate of coalescence.

Integrating Eq. (75), for the boundary condition m = 2 when t = 0

$$m - 1 = \frac{an_0}{K} + \left(1 - \frac{an_0}{K}\right) \exp(-Kt) \tag{76}$$

The total number of particles, whether flocculated or not, in a coagulating emulsion at time t is obtained by adding the number of unreacted primary particles to the number of particles in aggregates

$$n = n_t + n_v m = \frac{n_0}{1 + kn_0 t} + \frac{kn_0^2 t}{(1 + kn_0 t)^2}\left[\frac{kn_0}{K} + \left(1 - \frac{kn_0}{K}\right) \exp(-Kt)\right] \tag{77}$$

In Eq. (77) the first term on the right-hand side represents the number of particles which would have been present if each particle had been counted as one single particle. The second term gives the number of particles which arises when the composition of the aggregates is taken into account.

In the limiting case $K \to \infty$, the second term on the right-hand side of Eq. (77) is equal to zero, and the equation reduces to the von Smoluchowski equation (38). On the other hand, if $K = 0$, i.e., if no coalescence occurs, $n = n_0$ for all values of t. For the case $0 < K < \infty$, the effect of a change in the particle concentration on the rate of coagulation is given by Eq. (77). This clearly shows that the changes in particle number concentration with time depends on the initial particle number concentration. This illustrates the differences between emulsion breakdown and the coagulation of hydrophobic sols. In the latter case, the rate of increase of $1/n$ with time is independent of the initial particle number concentration. Some calculations of the change of $1/n$ with t, made by van den Tempel for reasonable values of k and K are shown in Fig. 13 for various $n_0$ values. It is clear that the rate of increase of $1/n$ with t rises more rapdily as $n_0$ increases. Van den Tempel [80] plotted $\Delta(1/n)$, i.e., the decrease of

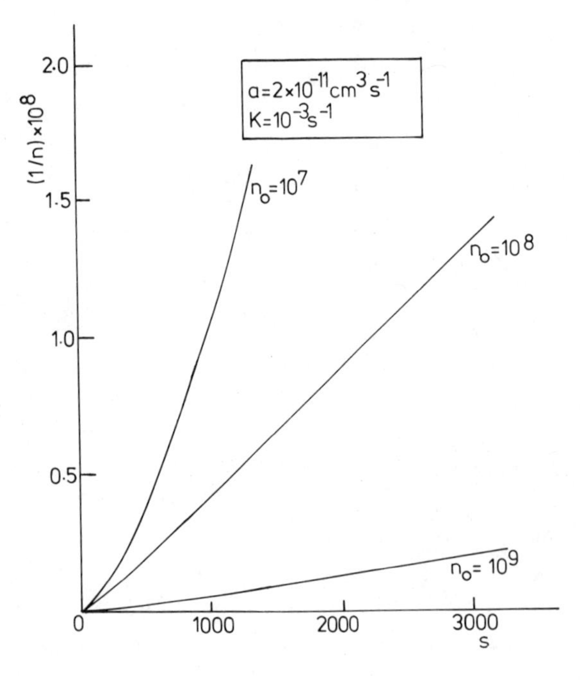


Figure 13. Variation of $1/n$ with time. (From Ref. 80.)

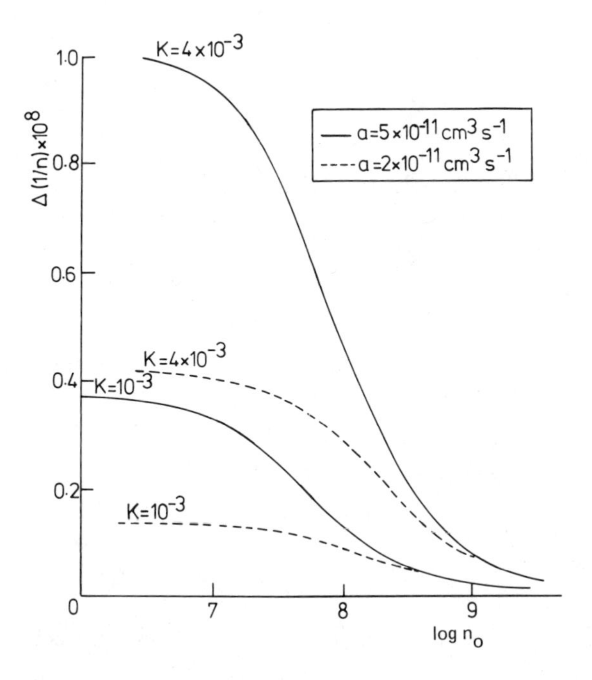


**Figure 14.** Plot of $\Delta(1/n)$ versus log $n_0$ for two values of K and k. (From Ref. 80.)

particle number concentration after 5 min, versus initial particle number concentration for two values of K and k. The results are shown in Fig. 14. This clearly shows that the rate of coagulation, as measured by the value of 1/n, does not change significantly with $n_0$, either for very dilute or for concentrated emulsions. However, in the region where $kn_0/K$ is of the order of unity, the rate of coagulation decreases sharply with increase of $n_0$.

To simplify Eq. (77), van den Tempel [80] made three approximations:

i. In a flocculating, concentrated emulsion $kn_0 \gg K$. In most real systems K is generally much smaller than unity and $kn_0 \geqslant 1$ is sufficient to satisfy this condition. Thus $kn_0$ rapidly becomes much larger than unity and the contribution from unreacted primary particles may be neglected. In this case Eq. (77) reduces to

$$n = kn_0^2 t\,(1 + kn_0 t)^{-2} \left\{ \frac{kn_0}{K}\,[1 - \exp(-Kt)] \right\} \tag{78}$$

since $kn_0 t \gg 1$, then $1 + kn_0 t \sim kn_0 t$, so that

$$n = \frac{n_0}{Kt}\,[1 - \exp(-Kt)] \tag{79}$$

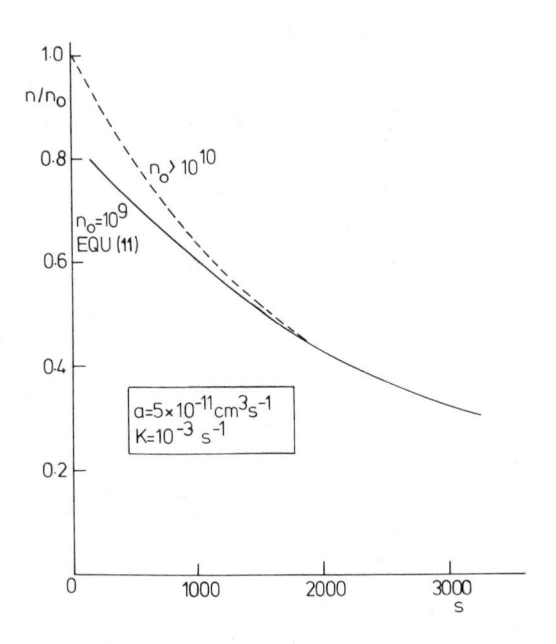


**Figure 15.** Change in particle number concentration with time for dilute and concentrated emulsions. (From Ref. 80.)

This means that the rate of coalescence no longer depends on the rate of flocculation in concentrated emulsions. Van den Tempel [80] calculated the change in particle number concentration with time for concentrated ($n_0 > 10^{10}\ cm^{-3}$) and dilute emulsions ($n_0 = 10^9\ cm^{-3}$) and for values of $k = 5 \times 10^{-11}\ cm^3\ s^{-1}$ and $K = 10^3\ s^{-1}$; the results are shown in Fig. 15.

It is seen that for concentrated emulsions Eqs. (77) to (79) yield similar results, whereas for dilute emulsions Eq. (78) gives rise to a serious divergence at values of t less than 1000 sec. Moreover, the particle number concentration is found to decrease approximately exponentially with time, until Kt becomes large compared with unity. This is confirmed by the experiments of Latzkar and McClay [108].

One other limitation in the application of Eqs. (77) to (79) lies in the assumption made in deriving Eq. (75), where the number of points of contacts between m particles was taken to be m − 1. This is obviously not the case for the flocculation of concentrated emulsions, where the aggregates consist of a large number of particles. In a regular, closely packed aggregate of spheres, all having the same size, each particle touches 12 other particles. The number of points of contact in such an aggregate will be proportional to m rather than m − 1. In a heterodisperse system one particle may even touch more than 12 other particles. This can be taken into account by rewriting Eq. (74) as

$$\frac{dm}{dt} = kn_0 - pKm \tag{80}$$

where p has a value between 1 and about 6. On integration, Eq. (80) gives

$$m = \frac{kn_0}{pK} + \left(2 - \frac{kn_0}{pK}\right)\exp(-pKt) \tag{81}$$

which replaces Eq. (76) for concentrated emulsions. This means that in concentrated emulsions the rate of coalescence increases with particle concentration in a manner dependent on the particle size distribution, the degree of packing, and the size of aggregates.

ii. In a very dilute emulsion, $kn_0/K$ can be much smaller than unity if coalescence occurs very rapidly. After coagulation has proceeded for sufficient time, such that $Kt \gg 1$, the second term on the right hand side of Eq. (77) may be neglected relative to the first term. This equation then reduces to Smoluchowski's equation, i.e., the rate of coagulation is independent of any coalescence.

iii. If the degree of coalescence is very small, the exponential term in Eq. (77) may be expanded in a power series, retaining the first two terms only when $K \ll 1$, such that the equation reduces to

$$n = n_0[1 - Kt(1 + kn_0t)^{-1} + Kt(1 + kn_0t)^{-2}] \tag{82}$$

This equation predicts only a very small decrease in particle number concentration with time, as expected.

iv. When coagulation has proceeded for a sufficiently long time, Kt may be much greater than unity. In this case, the exponential term may be neglected and since $kn_0/t \gg 1$ (in the denominator), then

$$n = \frac{n_0}{Kt} + \frac{1}{kt} \tag{83}$$

With large $n_0$, the first term on the right-hand side of Eq. (83) predominates, and hence, $1/kt$ can be neglected.

Davies and Rideal [109] discussed the problem of coalescence, incorporating an energy barrier term into the Smoluchowski equation, in order to account for the slow coalescence observed by Lawrence and Mills [110] for emulsions stabilized by sodium oleate. The Smoluchowski equation may be written in terms of the mean volume $\overline{V}$ of emulsion droplets, as follows [109],

$$\overline{V} = \frac{\phi}{n_0} + 4\pi DR\phi t \tag{84}$$

where $\phi$ is the volume fraction of the dispersed phase, and R is the collision radius. Equation (84) predicts that the mean volume of the droplets should be doubled in about 43 s, whereas experiments show that in the presence of sodium oleate, this takes about 50 days. To account for this an energy barrier term ($\Delta G_{coal}$) was introduced into Eq. (84) as follows:

$$\overline{V} = V_0 + 4\pi DR\phi t \exp(-\Delta G^a_{coal}/kT) \tag{85}$$

This predicts that an energy barrier of ~11 kT has to be overcome in the coalescence of oil/water emulsions stabilized by sodium oleate.

Substituting for D from Einstein's equation ($D = kT/6\pi\eta a$) and differentiating Eq. (85) with respect to t, the following relation results for the rate of coalescence for an oil/water emulsion:

$$\frac{d\overline{V}}{dt} = \frac{4\phi kT}{3\eta_w} \exp\left(\frac{-\Delta G^a_{coal}}{kT}\right) = C_1 \exp\left(\frac{-\Delta G^a_{coal}}{kT}\right) \tag{86}$$

where $\eta_w$ is the viscosity of the continuous phase (i.e., water for an oil/water emulsion), and $C_1$ is the collision factor defined by Eq. (86). For a water/oil emulsion, the corresponding relation would be

$$\frac{d\overline{V}}{dt} = 4(1 - \phi)\frac{kT}{3\eta_0} \exp\left(\frac{-\Delta G^{a,*}_{coal}}{kT}\right) = C_2 \exp\left(\frac{-\Delta G^{a,*}_{coal}}{kT}\right) \tag{87}$$

where $\eta_w$ is the viscosity of the oil-continuous phase, and $C_2$ is the corresponding collision factor.

In moderately sheared or stirred systems, the emulsion type (O/W or W/O), as well as the stability, depends on the coalescence rates, as given by Eq. (86) and (87). These equations also show that the emulsion type and stability depend, not only on the method of dispersion, but also on the phase volume and the viscosities of the two phases [111,112].

Davies and Rideal [109] considered the energy barrier in terms of the electrical potential at the surface of the oil drops, which arises for drops stabilized by ionic surfactants. The energy barrier was calculated on the assumption that the surface potential remains constant as two drops approach. This may not be true, unless the interfacial viscosity or interfacial compressional modulus is high. If $\psi_0$ is the surface potential at the drop/solution interface, then in the absence of specific adsorption of counter-ions, the energy barrier preventing coalescence should be proportional to $\psi_0^2$, according to the DLVO theory [see Eq. (33)]. Thus,

$$\Delta G^a_{coal} = B\psi_0^2 \tag{88}$$

where B is a constant that depends on the radius of curvature of the droplets. Since, as we have seen, two approaching droplets tend to flatten in the region of contact in a lamella, the radius of curvature to be used for emulsion drops may be considerably different from the actual drop radius. However, the degree of flattening is negligible for small emulsion droplets, i.e., $<1$ μm diameter. If there is specific adsorption of counter-ions, the electrical potential to be used in evaluating the electrical double-layer repulsion will be less than $\psi_0$. Indeed, it is then necessary to use the potential $\psi_d$ at the plane of specifically adsorbed ions, i.e.,

$$\Delta G^a_{coal} = B\psi_d^2 \tag{89}$$

A more recent theory for coalescence rate has been proposed by Hill and Knight [113] based on an analysis of the collision dynamics of slightly deformable spheres. The problem was treated as one of slow coagulation, i.e., as one in which only a small proportion of collisions leads to coalescence. Moreover, these authors have taken into account the polydispersity of emulsion systems and the dependence of coalescence probability on droplet size.

Basically, an expression was derived for the probability of coalescence of two droplets as a function of their radii and relative velocity. An average was then taken over all sizes of droplets, assuming the velocities to be distributed according to the Maxwell distribution law. The probability of coalescence was (arbitrarily) assumed to depend on the area of the common interface and on the pressure across this interface during an encounter, so that

$$\text{Chance of coalescence} \propto \int (\text{area} \times \text{pressure})\, dt \tag{90}$$

independent of the stage of coalescence reached. In this treatment, long range forces of repulsion and attraction were neglected.

The average probability g of coalescence between two droplets of radius p and q, approaching at relative velocity V and obliquity $\chi$, is given by the expression

$$g(p, q, V, \chi) = \beta\mu V \cos\chi \tag{91}$$

where $\beta$ is the constant of proportionality in Eq. (90) and $\mu$ is the reduced mass of the two droplets.

The average effect of all the individual coalescence processes was deduced from the probability of coalescence in a single collision by averaging the effects of all collisions between droplets of all sizes. The droplets are assumed to collide with a frequency given by kinetic theory. Assuming a Gaussian distribution of droplet radii, and that the available volume $\alpha$ per $cm^3$ is constant in any one experiment, i.e., independent of the degree of coalescence, Hill and Knight [113] derived the following expression for the rate of coalescence, expressed in terms of the variation in surface area A of the emulsion with time t,

$$A^{-1} = \frac{11\beta kT}{4\alpha} t + \text{constant} \tag{92}$$

This equation predicts a linear relationship between $A^{-1}$ and t, whereas Smoluchowski theory predicts a linear relationship between $A^{-3}$ (reciprocal mean volume) and t [see Eq. (84)]. A comparison between the two theories was attempted by Hill and Knight [113], using the experimental results of Lawrence and Mills [110], Lotzker and Maclay [108] and King and Mukherjee [114]. Some of the experimental data seemed to be better fitted using the Hill and Knight equation while others were better fitted using the Smoluchowski form. The authors also found that when the range of measurements is small, the results can be fitted by either theory. However, in view of the lack of precise measurements, and over a wide time range, it is difficult to decide, unambiguously, which one of the two approaches fits the experimental results best.

### E. Coalescence of Droplets at Planar Liquid/Liquid Interfaces

The study of the coalescence of droplets at a liquid/liquid interface has been carried out by several authors [115-121], not only as a means of studying stability using uniform, rigid, single droplets, but also to provide information about film rupture as a droplet approaches the interface, and hence to give information on the mechanism of coalescence. A review of the field has been given by Jeffreys and Davies [122]. In most of the experimental investigations, a liquid drop (phase $\alpha$) is released into an immiscible liquid (phase $\beta$) and allowed to come to rest at the planar $\alpha/\beta$ interface. The time interval between its arrival at the interface and its coalescence (i.e., the rest time) is recorded. The median rest time $\tau_{1/2}$ is regarded as a measure of drop stability. A typical experimental apparatus for such measurements is shown in Fig. 16, based on the work of Cockbain and McRoberts [115]. These authors conducted a systematic investigation of the stability of water and oil droplets at the O/W interface in the presence of different types of surface-active agents.

For a given system the stability of individual droplets was not constant and, hence, if thirty or more droplets were examined separately, a distribution curve is obtained as shown in Fig. 17, where N is the number of drops which do not coalesce within a time t. It can be seen that the distribution curve consists of two fairly well defined regions, one in which N decreases very slowly with t, followed by a region in which the decrease in N is much more rapid, and approximately exponential. This indicates that the lifetime of the droplets is determined by two distinct processes: first, drainage of the continuous phase from between

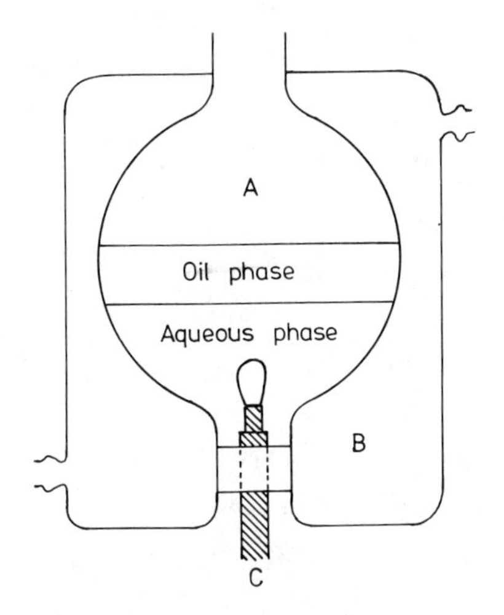


**Figure 16.** Apparatus for measurement of coalescence of liquid droplets at a liquid/liquid interface. (From Ref. 115.)

the drops and the plane interface, followed by rupture of the lamellar film. A rate constant k for the film rupture process (coalescence) may be obtained from the exponential portion of the curve, using the relation

$$\ln N = -kt + \text{constant} \tag{93}$$

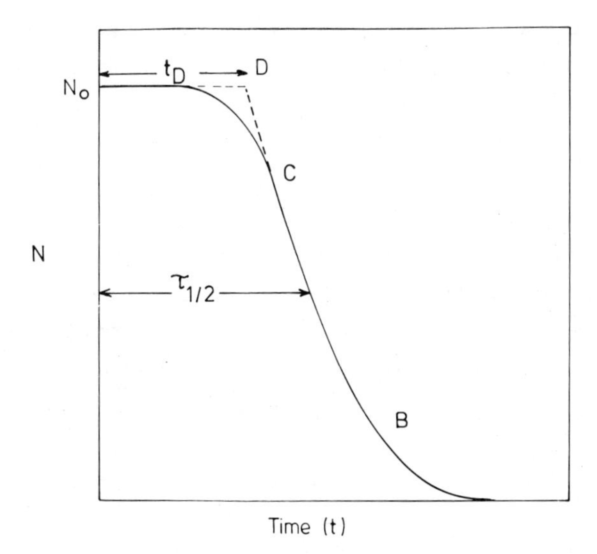


**Figure 17.** Typical distribution-time curve for the lifetimes of oil droplets. (From Ref. 115.)

The overall stability of the droplets is characterized by their half-life $\tau_{1/2}$. The difference $t_D$, as given by

$$t_D = \tau_{1/2} - \frac{1}{k} \ln 2 \qquad (94)$$

is assumed to be mainly determined by the rate of film drainage.

If the droplets are sufficiently stable, it is possible to follow the rate of coalescence with the interface of 5 to 30 drops formed at suitable intervals (e.g., 5 to 10 s). Cockbain and McRoberts [115] found that the rate constant $k$ was not significantly different from that obtained by studying the drops singly. Presumably this was due to the fact that the drops nearly always coalesced with the plane interface rather than with themselves and that the disturbance of the interface caused by the rupture of one drop did not markedly affect the coalescence of any other. Gillespie and Rideal [116] treated the coalescence of water droplets at the benzene/water interface in a semiquantitative manner by taking into account the various physical factors involved. As mentioned above, the time of coalescence of the droplets, for a given set of conditions, was found not to be constant, but to vary over a considerable range. The treatment by Gillespie and Rideal [116] considered, first, the case when stabilizing agents are absent.

If N is the number of droplets which do not coalesce in time t, it is convenient to discuss the variation of $N/N_0$ with time, where $N_0$ is the total number of droplets assessed. Figure 18 shows a plot of $N/N_0$ (on a log scale) versus t for water droplets of various size. When a droplet approaches an oil/water interface, it will be deformed and a thin film of radius $R_C$ will be formed between the droplet and the interface, as illustrated in Fig. 19. $R_c$ is given by the following approximate equation [123]:

$$R_c \sim a^2 \left( \frac{2 \Delta\rho g}{\gamma} \right)^{1/2} \qquad (95)$$

where a is the radius of the drop, $\Delta\rho$ is the difference in density between droplet and the medium, and $\gamma$ is the interfacial tension. Equation (95) was originally derived by Derjaguin and Kusakov [123] for small air bubbles at an interface.

Because of the difference in density between the droplet and the surrounding liquid, the film between the droplet and the interface will drain owing to gravity effects. The rate of drainage between approaching flat plates is given by the Raynold expression

$$-\frac{db}{dt} = \frac{2\pi F b^3}{3\eta A^2} \qquad (96)$$

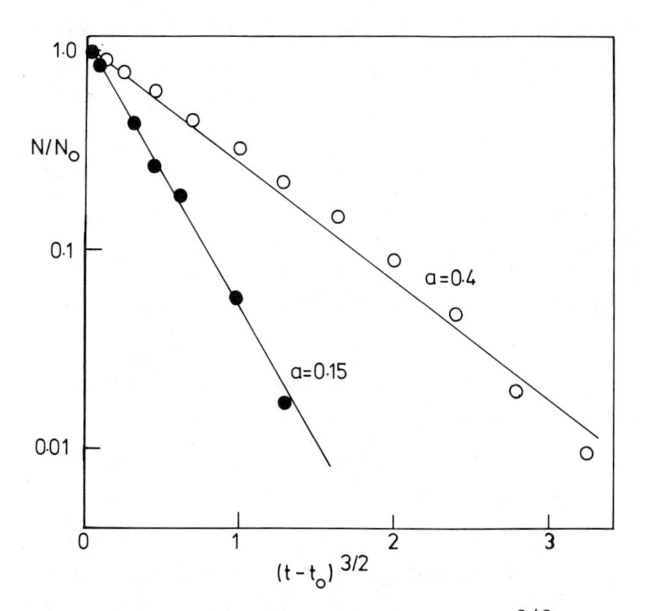


**Figure 18.** Plot of $N/N_0$ versus $(t - t_0)^{3/2}$ for water droplets of two different sizes. (From Ref. 116.)

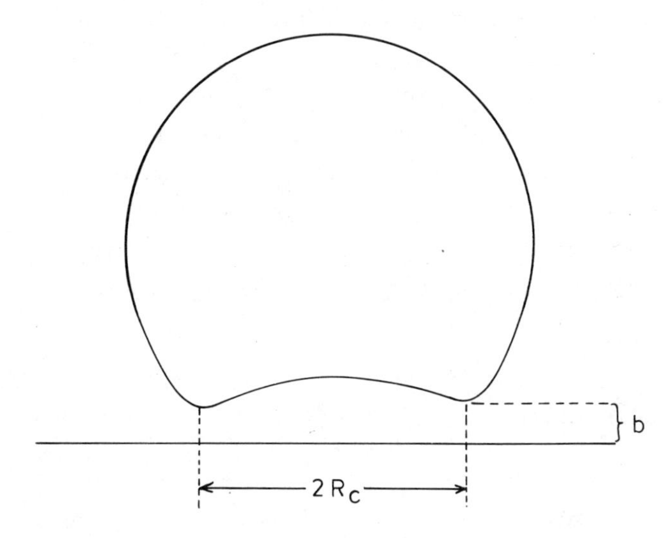


**Figure 19.** Approximate shape of a droplet at the oil/water interface. (From Ref. 116.)

where b is the distance between the plates of area A, and F is the force pressing one plate to the other. Using this expression and Eq. (95), the film thickness b at any time t is given by

$$\frac{1}{b} = \left( \frac{1}{b_0^2} + \frac{4\gamma^4 t}{\eta \rho g a^5} \right)^{1/2} \tag{97}$$

where $b_0$ is the film thickness at zero time, i.e., at the moment when the process of deformation of the drop is normally complete, $\eta$ is the viscosity of the liquid forming the film, and $\rho$ is the liquid density.

If film drainage is uniform, Eq. (97) suggests that it would take a very long time before coalescence by direct contact takes place. However, experimental observation indicates that the film does not drain uniformly, and hence rupture may take place before drainage is complete. This explains the wide distribution of lifetimes of drops at the liquid/liquid interface. Owing to the droplet deformation, the film thickness along the circle of circumference $2\pi R_c$ is thinner than it is anywhere else, and hence it is most likely that rupture takes place along this line. At any point on this line the thickness is likely to deviate from the mean value because of mechanical or thermal fluctuations. This may lead to a complex series of capillary waves (as first noted by Lord Rayleigh [124]), characterized by a frequency f and amplitude $A_0$. The probability of rupture P is related to f and $A_0$ by the equation

$$P = \frac{2\pi C_0 R_c A_0}{b} \sin \omega t \tag{98}$$

where $\omega = 2\pi f$, and $C_0$ is a constant determined by temperature fluctuations, the variation of $\gamma$ with temperature, mechanical shock, etc.

The experiments conducted by Cockbain and McRoberts [115] and Gillespie and Rideal [116] have clearly shown that rupture does not take place until the film has drained to a critical value $b_{cr}$, which takes a time $t_D$. This means that $P < 0.01$ up to time $t_D$. Hence the periodic changes in the drop shape at the interface apparently have little chance of resulting in rupture if $\sin \omega t < b/b_0$. Therefore, after time $t_D$, the probability of rupture per cycle is given by the expression [116]

$$P = \frac{2C_0 A_0 R_c}{b} \int_{\sin^{-1} b/b_{Cr}}^{\pi/2} \sin \omega t \, d(\omega t) \tag{99}$$

provided the mean thickness of the film b does not vary appreciably during the cycle. Thus

$$P = C_o A_0 R_c \left( \frac{1}{b^2} - \frac{1}{b_{cr}^2} \right)^{1/2} \tag{100}$$

From Eqs. (97) and (110), the probability of rupture per second (1/N) dN/dt is given by

$$-\frac{1}{N}\frac{dN}{dt} = \frac{2K}{3}(t - t_D)^{1/2} \tag{101}$$

and

$$\log \frac{N}{N_0} = -K(t - t_D)^{3/2} \tag{102}$$

where

$$K = fC_0 A_0 \left( \frac{6\gamma}{a\eta} \right)^{1/2} \tag{103}$$

K is effectively the coalescence rate constant. Equation (102) was tested by Gillespie and Rideal [116] for the coalescence of water droplets at the benzene/water interface. Linear plots of $\log(N/N_0)$ versus $(t - t_D)^{3/2}$ were obtained (see Fig. 18), and K was found to depend on the droplet radius; the larger the droplets, the more stable the emulsion. Over the range of droplet sizes investigated (1.5 < a < 4.5 mm), $Ka^{1/2}$ was found to be constant. The influence of temperature on coalescence rate is illustrated in Fig. 20 for water droplets at a paraf-

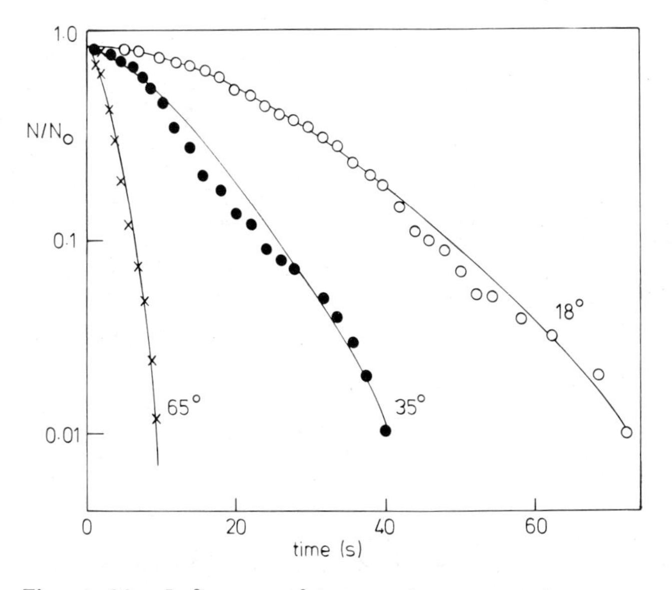


**Figure 20.** Influence of temperature on coalescence rate. (From Ref. 116.)

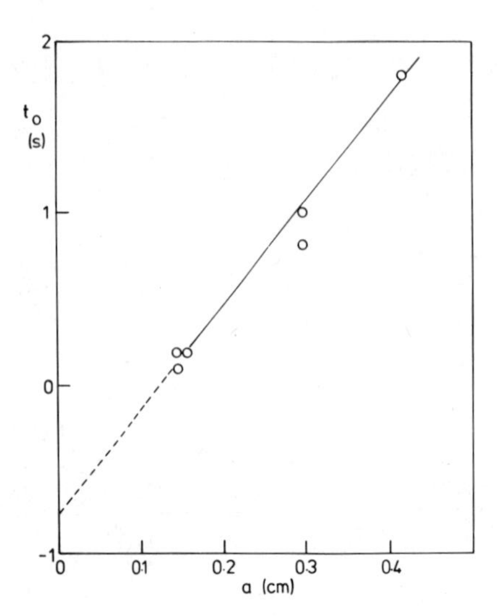


Figure 21. Variation of coalescence time with droplet radius. (From Ref. 116.)

fin/water interface. It is clear that, as the temperature increases, K also increases.

The time at which coalescence is first detected $t_D$ changes with the droplet radius a more or less linearly, as shown in Fig. 21. The time $t_D$ is given, following Eq. (97), by the expression

$$t_D = \frac{\eta \rho g a^5}{4\gamma^2}\left(\frac{1}{b_{cr}^2} - \frac{1}{b_0^2}\right) \tag{104}$$

where $b_0$ denotes the film thickness when droplet deformation is complete.

Since $Ka^{1/2}$ is constant, Eq. (103) indicates, that, with the exception of very small droplets (where measurements are difficult in any case), the product $fA_0C_0$, which describes the uneven film drainage, is independent of droplet size. Thus, under a given set of experimental conditions, $b_{cr}$ would be expected to depend only upon $R_c$. In fact,

$$b_{cr}^2 \propto R_c^2 \propto a^4$$

This, together with Eq. (103), indicates that

$$b_0 = -\frac{1}{2\gamma}\left(\frac{\eta \rho g a^5}{B}\right)^{1/2} \tag{105}$$

where B denotes the intercept of the straight line drawn through the experimental points. Calculated values of $b_0$ for water droplets of various sizes at

Table 3. Approximate Values of the Film Thickness at Various Stages in the Coalescence of Water Drops at the Benzene/Water Interface at 20°C

| a (mm) | $b_0$ (μm) | $t_D$ (s) | $b_{cr}$ (μm) | $t_f$ (s) | $b_f$ (μm) |
|---|---|---|---|---|---|
| 1.5 | 1.0 | 0.2 | 0.87 | 1.2 | 0.62 |
| 2.0 | 2.0 | 0.35 | 1.7 | — | — |
| 2.5 | 3.5 | 0.6 | 2.6 | — | — |
| 3.0 | 5.5 | 1.0 | 3.7 | 3.4 | 1.5 |
| 4.0 | 11.0 | 1.8 | 6.2 | 4.2 | 4.5 |

the benzene/water interface are given in Table 3. This table also gives the corresponding values of $b_{cr}$ calculated from Eq. (104). If $t_f$ is the maximum lifetime observed in a given experiment, the film would have a thickness $b_f$ at this time. Typical values of $t_f$ and $b_f$ are given in Table 3 for water droplets at the benzene/water interface. In the presence of surface-active agents at the oil/water interface, the interfacial tension $\gamma$ varies with the concentration of the stabilizing agent. Distortion of the film is, therefore, curtailed by the fact that, as the area changes, $\gamma$ increases and takes up the tangential stress [125]. In this case both $d\gamma/dc$ and $d\gamma/dT$ are important factors. However, the mechanism for film rupture, and hence, coalescence is similar to that in the absence of surfactants.

As mentioned above, Cockbain and McRoberts [115] found that after $t_D$, i.e., the time for complete drainage, N changes exponentially with t, and hence it is possible to obtain the coalescence rate constant K from the equation

$$\ln \frac{N}{N_0} = -Kt \tag{106}$$

Gillespie and Rideal [116] considered the time $t_D$ over which K was not constant. Equation (100) still holds, although the factor $(1/b^2 - 1/b_{cr}^2)^{1/2}$ is not given by Eq. (97) in all cases. Electric double-layer repulsion or electroviscous effects [126] will often slow the rate of drainage. The fact that $(1/N)dN/dt$ becomes constant for large t indicates either that b reaches an equilibrium value $b_{min}$ or that the change of film thickness with time is so slight that it may be regarded as constant. Thus, $t_D$ is the time taken for the film to drain to this value. However, the film thickness $b_0$ at time zero would be expected to be at least as great as that found in the absence of surfactant, since $\gamma$ is much smaller, and hence $R_c^2$ would be greater. Thus, from Eq. (97)

$$b_{min} \geqslant \frac{b_0}{(1 + 4b_0^2\gamma^2 t_D/\eta\rho g a^5)^{1/2}} \tag{107}$$

using $\gamma = 30$ mN m$^{-1}$ and the value of $b_0$ determined for water drops at the benzene/water interface (see Table 3), $b_{min} > 6 \times 10^{-6}$ cm.

If the double-layer thickness is very much less than the film thickness, the electroviscous effect may be accounted for using Elton's expression [126] for film drainage,

$$t = \frac{a^5 \rho g}{\gamma^2}\left[\eta\left(\frac{1}{b^2} - \frac{1}{b_0^2}\right) + \frac{2\pi^2\varepsilon^2\zeta}{16\kappa}\left(\frac{1}{b^4} - \frac{1}{b_0^4}\right)\right] \tag{108}$$

where $\kappa$ is the average specific conductance of the film, $\zeta$ is the zeta potential, and $\varepsilon$ is the dielectric constant of the liquid film. Equation (108) predicts that $b_{min}$ is greater than the value calculated neglecting the electroviscous effect. Thus, in the presence of surfactant, a film of considerable thickness prevails between the drop and the interface and coalescence occurs by rupture of this film.

Equation (108) predicts that, when double-layer repulsion is not significant, log $(N/N_0)$ is proportional to $t^{5/4}$, at large values of t. However, in view of the difficulty in measuring the lifetime accurately, it is difficult to decide whether log $(N/N_0)$ is actually proportional to $t^{5/4}$ or to t. When double-layer repulsion is significant, the film drains to a value $b_{min}$, and

$$\frac{1}{N}\frac{dN}{dt} = fA_0C_0R_c\left(\frac{1}{b_{min}^2} - \frac{1}{b_{cr}^2}\right)^{1/2} \tag{109}$$

using Langmuir's expression [127] for the interaction between two double layers, and at 20°C,

$$\frac{1}{b_{min}^2} = \frac{2 \times 10^8 \gamma}{\varepsilon a} \tag{110}$$

In this case a plot of log $N/N_0$ versus t would be linear after $b_{min}$ has been reached. The rate constant for rupture of the film in this case is given by

$$K = fA_0C_0R_c\left(\frac{1}{b_{min}^2} - \frac{1}{b_{cr}^2}\right)^{1/2} \tag{111}$$

If mechanical and thermal fluctuations, leading to rupture, are sufficiently large to make $b_{cr}^2 >> b_{min}^2$, then

$$K \simeq 2fC_0A_0 \left(\frac{\rho g}{3\varepsilon}\right)^{1/2} a^{3/2} \times 10^4 \tag{112}$$

This means that small drops would be more stable than large ones, as found by Cockbain and McRoberts [115].

It is interesting to compare the values of $fA_0C_0$ (i.e., the probability of rupture per unit time of film thickness a and unit radius) in the presence and absence of surfactants. For paraffin and benzene drops in aqueous solutions of cetyltrimethyl ammonium bromide, Cockbain and McRoberts [115] found values of K between $3 \times 10^{-2}$ and $3 \times 10^{-3}$ $s^{-1}$, for a = 0.1 cm. This corresponds to a value of $fA_0C_0$ between $0.7 \times 10^{-4}$ and $0.7 \times 10^{-5}$. The value for water droplets in benzene, in the absence of surfactants, was of the order of $2 \times 10^{-4}$. Thus, as expected, the probability of rupture is considerably less in the presence of surfactants.

It must be mentioned that the Gillespie-Rideal model [116] assumes the interface to be nondeformable; this is obviously unrealistic. This point was stressed by Elton and Pickett [128], who studied the coalescence of oil droplets at the oil/water interface in the presence of various concentrations of electrolytes. These authors considered the interface to be deformable, while the droplets were assumed to be essentially rigid and nondeformable. No attempt was made to derive a theoretical relationship for the distribution of coalescence times. Rather, the authors expressed their data in the form of an empirical relationship, namely,

$$\frac{N}{N_0} = \frac{1}{1 + at^n} \tag{113}$$

where a and n are constants which depend on droplet size, electrolyte concentration, and temperature, for a given system.

It is clear that a quantitative analysis of the drainage process requires analysis of the exact shape of the droplet as it approaches the interface. A complete theory should consider the deformability of both the droplet and the interface [129]. A sophisticated analysis of droplet shape was made by Charles and Mason [119]. These authors suggested that coalescence begins with the formation of a hole in the intervening aqueous phase which expands with a resulting decrease of the interfacial area and interfacial free energy. The velocity of hole expansion v was related to interfacial tension $\gamma$ and film thickness b by the expression

$$v = \frac{\sqrt{4\gamma}}{\Delta\rho\, b} \tag{114}$$

where $\Delta\rho$ is the difference in density between the disperse and continuous phases.

Princen [130] has solved the problem of the equilibrium shape of droplets at a liquid/liquid interface prior to complete coalescence. The existence of an equilibrium shape presupposes the presence of a repulsive force between the droplet and interface preventing coalescence. It was assumed that double-layer repulsion is the major contribution here. If the double-layer force is strong enough, the film will have a uniform thickness in the "contact area," at least at equilibrium. This corresponds to the secondary minimum in the potential energy-distance curve in the DLVO theory. It was also assumed that the film thickness in the contact area is small, compared with the droplet dimension, and also that the interfacial tension of each interface in the contact area is equal to that of the corresponding free interface. A schematic representation of the model used in the analysis is given in Fig. 22.

In this model, the droplet consisting of phase 1 and surrounded by an immiscible phase 2 rests at the interface separating phase 1 and 2. The lowest point of the droplet is chosen as the origin of the coordinate system, i.e., a horizontal x axis and a vertical z axis, which is the axis of revolution. The shape of the droplet and the interface may be described in terms of the function Z(x). Any point (x, z) may be associated with an angle $\phi$ measured between the normal at (x, z) and the negative z direction. The slope of the curve at

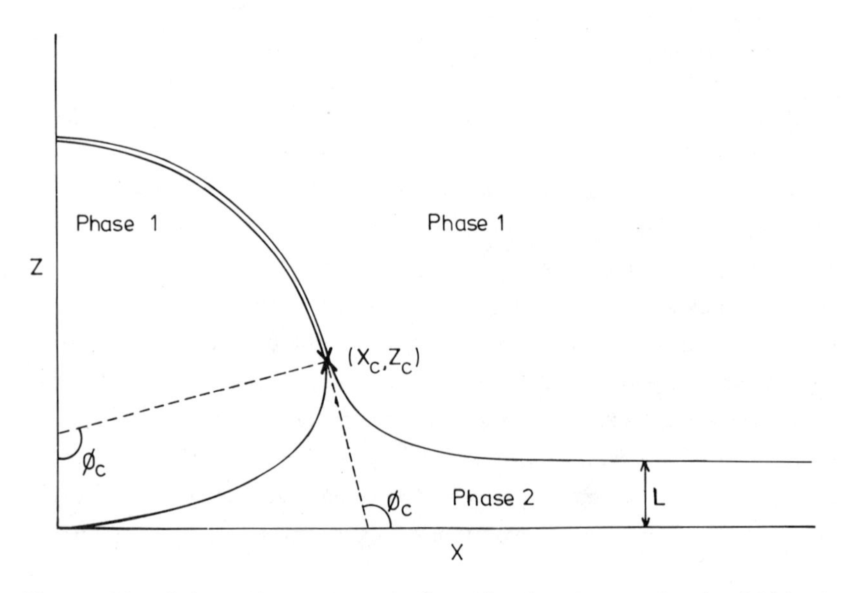


**Figure 22.** Schematic representation of a droplet at the liquid/liquid interface at equilibrium. (From Ref. 130.)

(x,z) is equal to tan $\phi$. Three regions may be distinguished: (1) the "free" droplet interface ($z < z_c$); (2) the region of "contact" between the droplet and the interface ($z > z_c$ and $|x| < |x_c|$); (3) the bulk interface ($x > x_c$) the so-called tails. At ($x_c$, $z_c$) the tail bends away from the drop and becomes horizontal far from the droplet.

For the free droplet interface, the shape may be obtained by equating the surface and gravity forces, leading to the following differential equation:

$$\frac{1}{\rho_1} + \frac{1}{\rho_2} = CZ + \frac{2}{b} \tag{115}$$

where b is the radius of curvature in the lowest point of the droplet, and $\rho_1$ and $\rho_2$ are the principal radii of curvature. For the area of contact between the droplet and the interface,

$$\frac{1}{\rho_1} + \frac{1}{\rho_2} = \frac{2}{R} \tag{116}$$

where R is the radius of curvature of the top of the droplet. For the tails,

$$\frac{1}{\rho_1} + \frac{1}{\rho_2} = C(Z - L) \tag{117}$$

Thus, the problem is reduced to solving the above differential equations using the appropriate boundary conditions. The properties of the droplet can then be described in terms of a dimensionless quantity $\beta$, given by

$$\beta = cb^2 \tag{118}$$

where b is the radius of curvature of the lowest part of the drop and c is given by the expression

$$c = g \frac{\Delta\rho}{\gamma} \tag{119}$$

where $\Delta\rho$ is the density difference between the two immiscible phases, $\gamma$ is the interfacial tension, and g is the acceleration due to gravity.

Princen has calculated [130] the drop shape as a function of $\beta$ over a wide range of values. The results of these calculations are shown in Fig. 23 in which q, the radius of the maximum vertical dimension of the droplet, p, the ratio of the radii of the deformed and undeformed droplets ($p = R/r$), and $\phi_c$ are each plotted as a function of $\beta$. As expected, q approaches unity for small values of $\beta$ and the limiting value of p is 2. The results also show that, with decreasing $\beta$, the shape of the free droplet approaches a perfect sphere; while that of the tail approaches a perfect plane. In the limit of infinitely large droplets, $\phi_c = 90°$, i.e., the droplet can be presented as a hemisphere with radius R. Thus, for the limiting values,

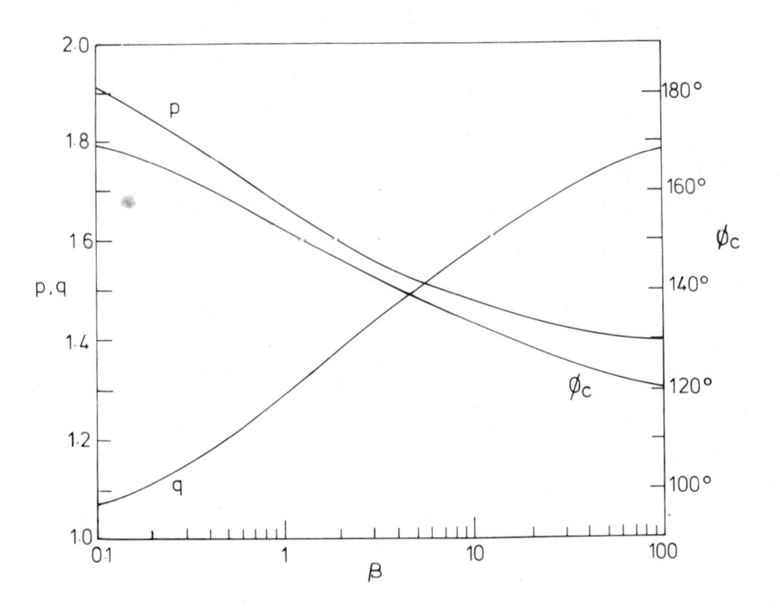


Figure 23. Dependence of $\phi_c$, p, and q on $\beta$ after Princen. (From Ref. 130.)

$$(\phi_c)_{r\to 0} = 180° \qquad q_{r\to 0} = 1 \qquad p_{r\to 0} = 2$$

$$(\phi_c)_{r\to \infty} = 90° \qquad q_{r\to \infty} = 2 \qquad p_{r\to \infty} = 1.260$$

The above theory may be applied directly to the problem of the rate of approach of a rising droplet to a horizontal liquid/liquid interface [130]. As mentioned above, the rate of drainage of the film between the approaching droplet and the interface is generally accepted to be given by the Reynolds equation, i.e.,

$$-\frac{db}{dt} = \frac{2\pi F}{3\eta A^2} b^3 \tag{120}$$

Equation (120) was originally derived to describe the drainage of two parallel, rigid disks with area A at a distance b, F being the force pressing the two disks together, and $\eta$ the viscosity of the draining fluid. However, in applying Eq. (120) to the case of a droplet approaching a liquid/liquid interface, a number of complications arise. First, the approaching interfaces are not resistant to the normal stresses which exist in the draining film because of the radial pressure gradient. As a result, the interfaces will not be quite parallel. Second, the interfaces are not generally resistant to tangential stresses which also exist during drainage. This means that there is a finite velocity at the interfaces which accelerates the rate of drainage. Third, at small film thicknesses, the van der Waals forces tend to increase, and double-layer interactions tend to reduce, respectively, the predicted rate of drainage. However, these complications are significantly reduced, and become unimportant, for systems with a high interfacial tension, small bulk density difference between the two phases, a high interfacial rigidity, and film thicknesses greater than ~200 nm. Under these conditions, Eq. (120) may be expected to represent adequately the rate of drainage. In the case of small droplets, both the droplet and the bulk interface are deformable; the area of contact between the droplet and the interface is given by,

$$A = \frac{4\pi}{3} cr^4 \tag{121}$$

where r is the droplet radius and $c = \Delta\rho\, g/\gamma$.

The force F on the droplet is simply the buoyancy force, i.e.,

$$F = \frac{4}{3} \pi r^3 \Delta\rho\ \rho \tag{122}$$

Substituting Eq. (121) and (122) into (120) gives

$$-\left(\frac{db}{dt}\right)_{r \to 0} = \frac{\gamma}{2\eta c r^5} b^3 \tag{123}$$

For larger droplets, A can be calculated using the droplet profile analysis given by Princen [130]. In that case, the buoyancy force is given by

$$F = \Delta\rho\ g[V\phi_c - \pi x_c^2(z_c - L)] \tag{124}$$

Since this force is exactly balanced by the downward force exerted by the curved interface in the contact area, a more convenient expression for F is given by

$$F = 2\pi x_c \gamma \sin \phi_c \tag{125}$$

In the limit of infinitely large droplets, where the droplet is a hemisphere with radius $R = r \sqrt[3]{2}$,

$$A = 2\pi R^2 \tag{126}$$

and

$$F = 2\pi R\gamma \tag{127}$$

Substituting Eqs. (126) and (127) in Eq. (120) then gives the following expression for the rate of drainage:

$$-\left(\frac{db}{dt}\right)_{r \to \infty} = \frac{\gamma}{3\eta R^3} b^3 = \frac{\gamma}{6\eta r^3} b^3 \tag{128}$$

In this case, the rate of drainage is independent of $\Delta\rho$. As shown above (see Fig. 18), the draining film is not uniform, but will have a maximum thickness ($b_{max}$) at the center and a minimum thickness ($b_{min}$) at the periphery. This was first observed by Derjaguin and Kusakov [123]. Frankel and Mysels [106] have calculated the rate of drainage in such a dimple, when $b_{min} << b_{max}$, neglecting both double layer and van der Waals interactions. The rate of drainage is given by the two following expressions, for the case where both interfaces are resistant to tangential stress.

$$-\frac{db_{min}}{dt} = \frac{\omega_{min}\gamma}{\eta c r^5} b_{min}^3 \tag{129}$$

and

$$-\frac{db_{max}}{dt} = \frac{\omega_{max}\gamma}{\eta c^3 r^{11}} b_{min}^5 \tag{130}$$

with $\omega_{min} = 2.09$ and $\omega_{max} = 21.8$.

For a uniform film, the Reynolds equation leads to

$$-\frac{db}{dt} = \frac{\omega_R\gamma}{\eta c r^5} b^3 \tag{131}$$

with $\omega_R = 2$. Using the same procedure, Princen [130] has calculated the rate of drainage for two other systems:

i. Two approaching small droplets of equal size which form a dimple on approach (Fig. 24), i.e., two truncated spheres. In this case, F is given by Eq. (122)

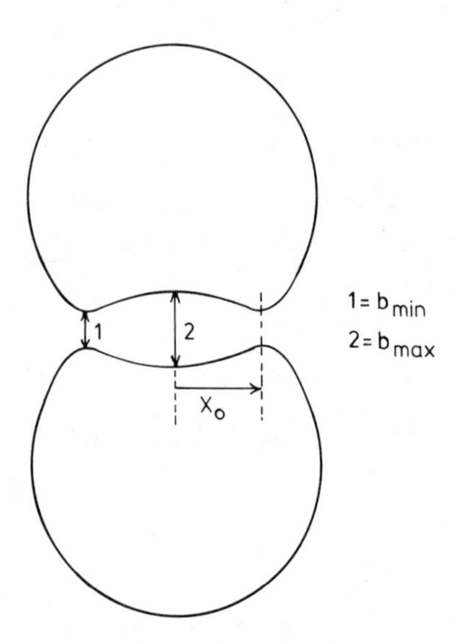


**Figure 24.** Picture of a "dimple" between two approaching droplets according to Princen. (From Ref. 130.)

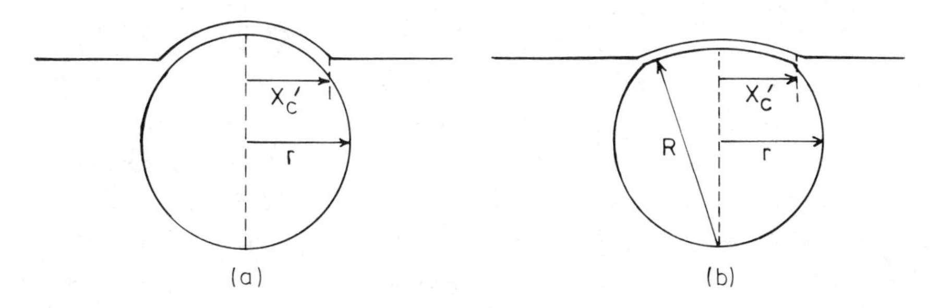


Figure 25. Approach of a droplet to a deformable liquid/liquid interface according to (a) Allen et al. [119] and (b) Elton et al. [128,129].

and $x_c = r^2\sqrt{2c/3}$. Assuming $x_c = x_0$, $\omega_{min} = 2.09$ and $\omega_{max} = 5.49$. Reynolds equation leads to Eq. (131) with $\omega_R = 2$. This case is obviously important for the coalescence of liquid droplets.

ii. A droplet approaching a horizontal deformable liquid/liquid interface. Dimpling was observed for this system by Allen, Charles, and Mason[119]. This is represented in Fig. 25a. If the droplet is also deformable [128,129], then the profile given in Fig. 25b will be observed. In that case $x_c = 2r^2\sqrt{2/3}$ and $A = 4/3\pi cr^4$. After a suitable transformation of coordinates and assuming $x_0 = x_c$, application of Frankel and Mysel's analysis [106] leads to $\omega_{min} = 0.524$ and $\omega_{max} = 1.37$ (Eqs. (129) and (130) respectively), while Reynolds equation would predict $\omega_R = 0.5$ [Eq. (131)].

A summary of the coefficients in the drainage Eqs. (129) to (131) for the three cases considered is given in Table 4.

Table 4. Numerical Coefficients in the Drainage Equations [Eqs. (129) to (131)]

| | Droplet approaching a solid nondeformable plane interface | Two approaching droplets of equal size | Droplet approaching a horizontal deformable liquid/liquid interface |
|---|---|---|---|
| $x_0$ | $r^2\sqrt{2c/c}$ | $r^2\sqrt{2c/3}$ | $2r^2\sqrt{c/3}$ |
| $\omega_{min}$ | 2.09 | 2.09 | 0.524 |
| $\omega_{max}$ | 21.8 | 5.49 | 1.37 |
| $\omega_R$ | 2 | 2 | 2 |

F. Surfactant Association and Enhanced Stability to Coalescence

It has long been known that mixed surfactants can have a synergistic effect on emulsion stability, with respect to coalescence rates. As early as 1940, Schulman and Cockbain [131] found that the stability of Nujol/water emulsions increases markedly using amphiphilic compounds (e.g., cetyl alcohol, cholesterol, etc.) in combination with sodium cetyl sulfate. The enhanced stability was attributed to the formation of intermolecular complexes at the oil/water interface. This complex formation was assumed to be associated with the formation of a densely packed interfacial layer. It would seem that the maximum effect is obtained when a water-soluble surfactant and an oil-soluble surfactant, capable of interacting at the oil/water interface, are used in combination. The interaction of two surfactant species of this type has been studied by Alexander and Schulman [132], who found that suitable combinations lead to greately enhanced stability as compared to the individual surfactants and that these films are generally associated with a very low interfacial tension, $\sim 0.1$ mN $m^{-1}$. This reduction in interfacial tension suggests some sort of cooperative adsorption, although quantitative measurements of the adsorption of the individual components have not, in general, been attempted. Moreover, it has also been claimed that for maximum stability of O/W emulsions, the interfacial film should be charged. The ability of surfactant molecules to complex at the O/W interface has been related to the packing geometry of the molecules at the interface. The implication is that those combinations which form a closely packed condensed film, such as cetyl sulfate and cholesterol, lead to a stable emulsion. In this case the condensed interfacial film consists of equal numbers of molecules of each species. It is found that if cholesterol is replaced by another molecule, such as cholesteryl ester or oleyl alcohol, then a condensed interfacial film did not form, because of some sort of steric hinderance, and enhanced stability was not observed. Thus, enhanced stability would seem to be directly related to specific interactions between the molecules of the two surfactant species.

Since the publication of Schulman's paper, other investigators have examined various mixtures of surfactants at the oil/water interface, using interfacial viscosity measurements. For example, Davies and Meyers [133] showed that simultaneous adsorption of sodium dodecyl sulfate and cetyl alcohol at the benzene/water interface results in a very high interfacial viscosity, although each component alone gives a low viscosity. The high interfacial viscosity was attributed to complex formation between the two components at the interface. Indeed, King and Goddard [134] suggested that a 1:2 complex forms between lauryl alcohol and sodium lauryl sulfate, or myristyl alcohol and sodium myristyl sulfate. However, such complex formation between the surfactant and cosurfactant may not, in itself, be the actual reason for enhanced stability [135,136]. An increase in the interfacial viscosity cannot be taken unambiguously as a criterion for complex formation. In fact Carless and Hallworth [137] have measured the interfacial viscosity in a wide variety of combinations of surfactants and stearyl alcohol, concluding that none of these formed complexes at the oil/water interface and that the interfacial viscosity was not enhanced (although the corresponding mixed monolayer at the air/water interface may show enhanced viscosity).

In a recent study Tadros [138] systematically investigated the effect of the addition of cetyl alcohol to cetyl trimethyl ammonium bromide on the adsorption at the xylene/water interface using interfacial tension measurements. The results showed that addition of the alcohol enhances the adsorption of the cationic surfactant at the interface. Although no measurement of the alcohol adsorption at the O/W interface has been attempted, it is likely that coadsorption of cetyl alcohol takes place. This has the effect of increasing the packing of the surfactant molecules at the interface. Moreover, addition of the alcohol resulted in an increase in the charge on the oil droplets in the region of saturation adsorption. It was thus concluded that a coherent strong interfacial film could develop at the O/W interface between the cationic surfactant and alcohol molecules. This film should act as a barrier preventing coalescence by virtue of its rheological properties, e.g., its high dilational viscoelasticity (see below). Indeed, measurements of the rate of coalescence of xylene/water emulsions (50% by volume at 25°C), have shown that addition of cetyl alcohol does result in very stable emulsions, relative to those prepared in the absence of the alcohol. A similar effect was also observed in the presence of high concentrations of electrolyte, i.e., 1.0 mol $dm^{-3}$ NaCl, or 0.5 mol $dm^{-3}$ $CaCl_2$. This means that the electrical double-layer barrier is not the dominant factor in determining the stability of the emulsions formed from these surfactant-cosurfactant mixtures.

Although numerous studies have been made of enhanced stability using mixed surfactants, the mechanism for the enhancement is not yet totally understood. However, the importance of this concept in industrial emulsions (e.g., in pharmaceutical, agricultural, cosmetic, and other formulations), where it is the rule rather than the exception to use mixed surfactants, would seem to justify further research in this area. Some of the ideas developed so far to explain enhanced stability in the presence of mixed films are given below.

### 1. *Interfacial Tension and Gibbs Elasticity*

It has long been known that minor constituents (often impurities) in surfactants have a marked effect on the surface tension of solutions. For example, the presence of minute quantities of lauryl alcohol in sodium lauryl sulfate is well known to cause a substantial reduction in the air/water interfacial tension for the surfactant solution below the critical micelle concentration (CMC). This has been attributed to the fact that although the alcohol molecules themselves are not strongly surface-active they can fit between the surfactant ions at the air/water interface, and this is the cause of the lowering of the interfacial tension. In other words, the presence of the main surfactant molecules renders the alcohol impurity more strongly surface-active, leading to the observed synergistic effect. Above the CMC, the amphiphilic molecules are solubilized into the micelles and the interfacial tension increases again. This explains the minima observed in the $\gamma$-log C plots for impure surfactant solutions.

Although ample data exist for the air/water interfacial tensions of mixed surfactant solution, corresponding results for the oil/water interface are scarce. The recent results by Tadros [138] for the xylene/water interface have shown that addition of cetyl alcohol to cetyl trimethyl ammonium bromide results in a lowering of the interfacial tension of the solution and a shift of the CMC to lower concentrations. The reduction in interfacial tension, on the addition of the alcohol, continues *above* the CMC of the surfactant. Thus, provided the amphiphilic compound does not partition into the oil phase, its presence in the mixed film should have an effect on the Gibbs' elasticity of that film. Prins and van den Tempel [139] have applied the theories of Gibbs' elasticity, developed by van den Tempel et al. [140], to the surfactant mixture sodium laurate plus lauric acid. In the presence of laurate ion, lauric acid has an extremely high surface activity. At half coverage, the interface contains 1.3 mol $dm^{-3}$ laurate ions and $4.8 \times 10^{-7}$ mol $dm^{-3}$ lauric acid (as laurate). Thus, under these conditions, the minor constituent can contribute more to the Gibbs' elasticity than does the major constituent. In fact, Prins and van den Tempel [139] have calculated a Gibbs' elasticity coefficient as high as $10^3$ mN $m^{-1}$. The high values of the Gibbs' elasticity observed by Mysels et al. [141] for sodium laurate foam films must be due to the presence of an impurity (probably lauric acid formed by hydrolysis of the sodium laurate). The high Gibbs' elasticity obtained in the presence of a second component could account, at least in part, for the enhanced stability of foam films and emulsions which is observed.

Prins et al. [142] have measured the dilational elasticity $\varepsilon$ defined by

$$\varepsilon = A \frac{d\gamma}{dA} \tag{132}$$

for sodium lauryl sulfate in the presence of lauryl alcohol at the air/water interface. They found that $\varepsilon$ increases markedly in the presence of the alcohol. A correlation between film elasticity and the coalescence rate of oil/water emulsions has been observed by Srivastava and Haydon [143], and also by Boyd et al., [144]. It seems likely that the high surface dilational elasticity obtained in the presence of the alcohol prevents lateral displacement of molecules in the thin film between approaching oil droplets; this would also account (in part) for the observed increased stability. However, most measurements of surface elasticity have been made for the air/water interface; measurements for the oil/water interface are at present scarce.

### 2. *Interfacial Viscosity*

It has long been assumed by many authors (see Chap. 1, Sec. II.G) that a high interfacial viscosity could account for the stability of liquid films. It is clear that this must play a role under dynamic conditions, e.g., when two droplets approach each other. A decrease in the rate of expulsion of liquid from the film results from the enhanced viscosity. Under static conditions the interfacial viscosity does not play a direct role. However, a high interfacial viscosity is often accompanied by a high interfacial elasticity (i.e., a high Gibbs-Marangoni effect), and therefore there may well be an indirect contribution to the increased stability of the corresponding emulsion arising from the increased viscosity of the interfacial film. Prins and van den Tempel [140] argued strongly against there being any role played by the interfacial viscosity. Two observations seem to support their argument. First, changes in temperature have only a small effect on the stability of foam films, even though they may change from slow draining to fast draining [145]. The slow drainage of films containing minor components has been observed by Miles et al. [146]; this was accounted for in terms of an increased interfacial shear viscosity. Second, the interfacial shear viscosity appears to drop suddenly on slightly increasing the concentration of the major component [147] (the film becoming fast draining), whereas their stability increases significantly. Based on these arguments, Prins and van den Tempel [139] discounted any effect of increased interfacial shear viscosity on enhanced emulsion stabilization resulting from the presence of a minor component; they attributed the effect solely to an increase in interfacial elasticity.

Most of the above discussion has been concerned with foam films. The situation with interfacial films at the oil/water interface is different in certain aspects. First, as mentioned, a high interfacial viscosity is rarely encountered [137,147]. Second, the adsorption of surfactants immobilizes the surface; the droplets are so small that any tangential shearing stresses to which they are subjected are immediately opposed by surface tension gradients, which need only to be very small. That the surfactant film is immobilized is reflected in the sedimentation rate of such droplets: Stokes' law appears to be obeyed (see Sec. III on creaming and sedimentation).

### 3. *Hindrance to Diffusion*

Another possible explanation for enhanced stability in the presence of mixed surfactants could be connected with the hindered diffusion of the surfactant molecules in the condensed film. This would imply that desorption of surfactant chains is hindered on the approach of two emulsion droplets, and hence thinning of the film is prevented. However, Hallworth and Carless [148] suggested that this cannot be the case, since saturation of the continuous aqueous phase with surfactant, without any addition of the cosurfactant, does not prevent demulsification, although desorption would be prevented. Moreover, they indicated that the contact time during droplet collisions is probably too short for desorption to be important [149] and the presence of nearby droplets should displace the aqueous diffusion medium [150].

### 4. *Liquid Crystalline Phase Formation*

Friberg and co-workers [151] have attributed the enhanced stability of emulsions formed with mixtures of surfactants to the formation of three-dimensional association structures, namely, liquid crystals. These structures can form, for example, in a three-component system of surfactant, alcohol, and water, as illustrated in the phase diagram shown in Fig. 26.

The presence of liquid crystalline phases in emulsion systems has been known for some time [152-154]. However, it is mainly due to the work of Friberg and collaborators [151] that the relationship between emulsion properties and the equilibrium phase diagrams of the constituents has been established. An interfacial layer containing the mixed surfactant and water forms at the oil/water interface. The presence and location of this liquid crystalline phase in emulsions have been studied by optical and electron microscopy. That formation of liquid crystalline phases leads to enhanced stability has been verified recently by Boyd et al. [154] for food emulsions. They studied two emulsifier systems, one of which formed a liquid crystalline phase with water and a second which did not. That

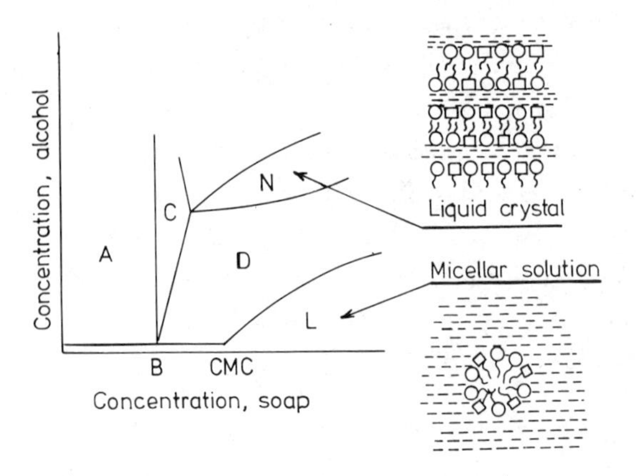


**Figure 26.** Phase diagram of surfactant-alcohol-water system. (From Ref. 151.)

which formed a liquid crystalline phase gave rise to an interfacial layer with pronounced viscoelastic behavior. On the other hand, the emulsifier system which did not give a liquid crystalline phase did not show this viscoelasticity. Moreover, the first emulsifier system gave rise to a more stable emulsion.

Friberg et al. [151] have stressed the importance of the interaction of the two surfactants with water for the formation of lamellar liquid crystalline phases. It seems that the formation of a liquid crystalline phase can only take place above a minimum water content. This demonstrates the importance of establishing the phase diagram of the two surfactants with water and the identification of the regions where liquid crystalline phases form.

In comparison with the surfactant-water interaction, the influence of the corresponding surfactant-oil interaction is less pronounced, although studies by Davies [155] have demonstrates some dependence of emulsion stability on oil/emulsifier interactions. Such a dependence has also been pointed out by Shinoda and Arai [156]. These authors have used the value of the phase inversion temperature (PIT) or hydrophile-lipophile balance (HLB) as a guide to emulsifier selection (cf. Chapter 6). The emulsifier-hydrocarbon interaction determines the PIT range for nonionic surfactants, since the association conditions are extremely sensitive to these interactions; for example, to the dipole-induced dipole interaction between aromatic hydrocarbons and polar surfactant molecules. This has been illustrated by Friberg [151] for emulsions of hexadecane or *p*-xylene in water plus nonylphenol ethoxylate (with an average of 9 moles of ethylene oxide). When hexadecane was used, two phases coexisted over certain regions of the phase diagram: one was essentially aqueous, the other essentially hydrocarbon. When the hexadecane was replaced with an equivalent amount of *p*-xylene, a single, liquid crystalline phase was formed.

There seems to be little doubt from experimental evidence that the presence of liquid crystalline phases has a marked effect on the stability of emulsions. However, the basis for this enhanced stability requires an explanation. This is not an easy task in view of the complexity of the system. Fribert et al. [151] have given an explanation in terms of the reduced attractive potential energy between two oil droplets, each surrounded by a layer of a liquid crystalline phase. They have also considered changes in hydrodynamic interactions in the interdroplet region; this affects the aggregation kinetics.

The hydrodynamic conditions in the interdroplet region have been treated by various investigators, including Derjaguin and Kusakov [157] and Frankel and Mysels [106]. This led to the concept of "dimpling," discussed previously. Some experimental observations have been made by Platikanov [158]. Further consideration of the hydrodynamic forces has been given more recently by Brenner [159], Spielman [160], Ivanov et al. [161], and Brenstein et al. [162]. In order to evaluate the hydrodynamic conditions in the interdroplet film, an estimation of the distortion of the two droplets from symmetry is essential. To this end, it is sufficient to estimate the displacement of the point of intersection between the straight line connecting the droplet centers and the tangent orthogonal to this line; this is clearly the point at which the maximum force acts. By calculating the Laplace pressure, before and after distortion, and taking reasonable values for the drop radius ($R = 10^{-6}$ m), Hamaker constant ($A = 10^{-19}$ J), and interfacial tension ($\gamma = 5$ mN m$^{-1}$), Friberg et al. [151] estimated that in fact only small changes from spherical symmetry occurred. This justified treating two approaching emulsion droplets as rigid spheres, even at close separations. However, at such short separations the hydrodynamic conditions in the liquid between the two droplets results in a reduced rate of approach [158-161]. Deviations from Stokes' law also become pronounced. An estimation of the reduction in flocculation kinetics may be obtained from Spielman's theory [160].

As indicated above, Friberg et al. [151] have calculated the effect on the van der Waals interaction of the presence of a liquid crystalline phase surrounding the particles. A schematic representation of the flocculation and coalescence of droplets with and without a liquid crystalline phase layer is given in Fig. 27, following Friberg et al. [151]. The upper part of Fig. 27 (A to F) represents the flocculation process when the emulsifier is adsorbed as a monomolecular layer. The distance d between the water droplets decreases to a distance m at which the thin film ruptures and the droplets coalesce; m is chosen to correspond to the thickness of the hydrophilic layers in the liquid crystalline phase. This simplifies the calculations and facilitates comparison with the case in which the liquid crystalline layer is adsorbed around the particles.

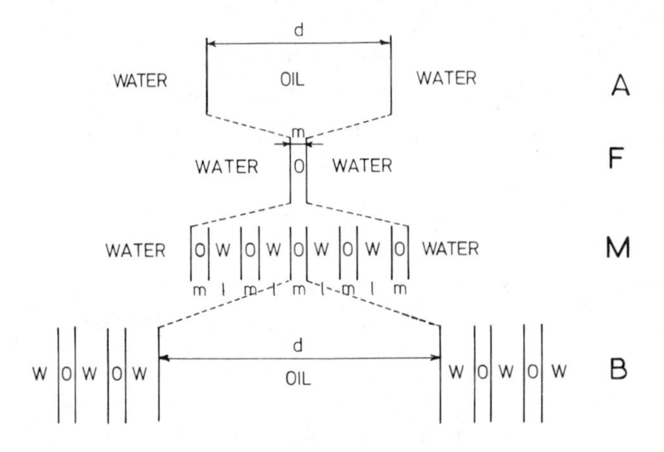


Figure 27. Schematic representation of flocculation and coalescence in the presence and absence of liquid crystalline phases. (From Ref. 151.)

The flocculation process for the case of emulsified droplets covered with liquid crystalline layers is illustrated in the lower part of Fig. 27 (B to M). The oil layer between the droplets thins to thickness m. The coalescence process which follows involves the removal of successive layers between the droplets until a thickness of one layer is reached (F); the final coalescence step occurs in a similar manner to the case for a monomolecular layer of adsorbed surfactant.

For case A, the van der Waals attraction is given by the expression

$$G_A = -\frac{A}{12\pi d^2} \tag{133}$$

For case B, $G_B$ can be obtained from the algebraic summation of this expression for the aqueous layer on each side of the central layer. The ratio $G_B/G_A$ is then given by the expression

$$\frac{G_B}{G_A} = d^2 \left\{ \sum_{p=0}^{n} \sum_{q=0}^{n} [d + (p + q)(\ell + m)]^{-2} + \sum_{p=0}^{n-1} \sum_{q=0}^{n-1} [d + 2\ell + (p + q)(\ell + m)]^{-2} - \sum_{p=0}^{n} \sum_{q=0}^{n-1} [2(d + \ell) + (p + q)(\ell + m)]^{-2} \right\} \tag{134,135,136}$$

where $\ell$ and m are the thicknesses of the water and oil layers, n is the number of water layers (which is equal to the number of oil layers), and p and q are integers.

The free-energy change associated with coalescence (i.e., $M \rightarrow F$) is calculated from the variation of the van der Waals interaction across the droplet walls. This treatment reflects the energy change associated with the layers squeezed out from the interdroplet region. The problem is circumvented by assuming that these displaced layers adhere to the enlarged droplets so that their free energy is not significantly changed in the process. In this manner, the ratio of the interaction energies in the states F and M is obtained from a summation of the van der Waals interactions from the individual layers on the water parts, i.e.,

$$\frac{G_M}{G_F} = m^2 \left\{ \sum_{p=0}^{n} [(m + p)(m + \ell)]^{-2} - \sum_{p=0}^{n} (p + \ell)(m + \ell)]^{-2} \right\} \tag{137}$$

To illustrate the relative importance of the van der Waals attraction energy, calculations were made using the above expressions, for the case of flocculation (i.e., A and B) and for the case of coalescence (i.e., F and M). The results are given in Fig. 28a and b, respectively.

Figure 28a shows that the influence of a liquid crystalline layer around the droplets on the flocculation process is insignificant. However, the effect on the

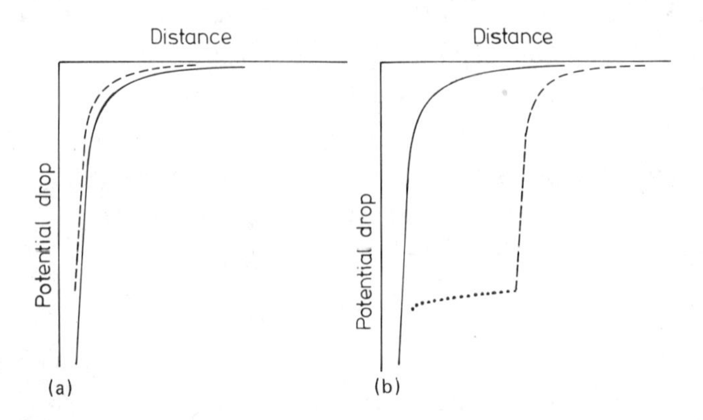


**Figure 28.** Van der Waals potential energy-distance curves for flocculation (A) and coalescence (B) between two droplets in the presence (---) and absence (——) of liquid crystalline phases. (From Ref. 151.)

free energy change due to coalescence is quite significant. For example, with nine layers on each droplet and a layer thickness of reasonable magnitude ($\ell$ = m = 5 nm) the total van der Waals interaction is reduced to only 10% of its original value for the coalescence process in the case of the layer structure. This is to be compared with 98% in the case of two droplets at the same distance, but separated by the oil phases instead of by the liquid crystalline phase. Even more important are the extremely small changes in attraction energy after removal of the first layers. The first layers give a drop corresponding to 1.5% of the total van der Waals interaction energy (A to F). The last layer, before the state F is reached, corresponds to 78% of the total van der Waals energy. It would seem, therefore, that the presence of a liquid crystalline phase has a pronounced influence on the distance dependence of the van der Waals energy, leading to a drastic reduction in the force of attraction between the emulsion droplets. It is important to note that the influence of the liquid crystalline layer is not simply a distance effect in the manner described by previous investigators [163,164] who have discussed the Vold effect [165]. The distance between the liquid crystalline covered droplets is of the order of 50 nm. The influence of the liquid crystalline layer is to change the distance dependence of the van der Waals interaction energy between the droplets, as illustrated by the dotted curve in Fig. 27b.

In the light of the above analysis, Friberg et al. [151] attributed the stabilizing action of liquid crystalline layers on emulsion stability to this reduction in the van der Waals energy. A necessary prerequisite for such stabilizing action is the localization of this liquid crystalline layer at the O/W interface. Although this has been demonstrated for several systems, it is by no means a universal occurrence.

## G. The Self-Bodying Action of Mixed Surfactant Systems

The use of a mixture of an anionic, cationic, or nonionic surfactant with an amphiphilic compound, such as a fatty alcohol, as cosurfactant, has long been used in the preparation of semisolid oil/water emulsions in many industrial applications, e.g., pharmaceutical and cosmetic preparations. These commercial preparations were in use long before any mechanism for the interactions was suggested. The cosurfactant itself acts as a relatively weak emulsifying agent for water/oil emulsions. The surfactant may well be a strong oil/water emulsifier (i.e., dispersing agent), but emulsions formed with it alone are somewhat unstable and fluid [166]. With the surfactant plus cosurfactant mixture, emulsions may be prepared which remain mobile at low total concentrations of the surfactants, but which become semisolid at higher concentration (~10% or less of the total weight of the emulsion). This process, whereby the agent which stabilizes the emulsion also controls the consistency between wide limits, is referred to as the *self-bodying* action. The essential feature of this self-bodying action is the introduction of a significant viscoelastic component into the rheological behavior [166]. This is distinct from the simple increase in viscosity of an emulsion with increase in surfactant concentration which is commonly observed, without the emulsion becoming viscoelastic [167]. Since the concentration of mixed emulsifier in these semisolid creams is generally far in excess of that required to form a close-packed layer at the interface, it seems unlikely that interfacial film rheology is of prime importance in determining the consistency of such emulsions.

In a series of papers Barry and co-workers [168] have examined various systems consisting of a fatty alcohol and an anionic, cationic, or nonionic surfactant. The mechanism of the self-bodying action of the anionic-mixed sodium dodecyl sulfate/cetyl alcohol emulsifier and of the cationic mixed emulsifier cetrimide (a mixture of alkyl trimethyl ammonium bromides) has been investigated using rheological measurements. For emulsions containing different fatty alcohols, it was reported that the alcohol chain length had an appreciable influence on the rheological properties, and that commercial cetostearyl alcohol (mainly a mixture of cetyl and stearyl alcohol) formed more stable emulsions than pure alcohols or a combination of the pure alcohols. In many of these investigations, the ternary system (i.e., surfactant plus alcohol plus water) was examined prior to the four component system (i.e., with the oil added). For the ternary system, the effect of changes in the amphophile-surfactant ratio on the structure and rheology of the systems formed, after the mixed emulsifier disperse in water, has been examined. The effect of increasing the proportion of mixed surfactant to water, when the ratio was kept constant at 9:1, which is an ideal ratio for stable emulsions [169], was investigated. To examine the self-bodying action of mixed surfactants when used in emulsions, the alcohol-surfactant ratio was also kept constant at nine to one, while the total concentration of the mixed emulsifier was varied. Most of the work was carried out with straight chain alcohols of the homologous series $C_{12} - C_{18}$ and commercial cetostearyl alcohol. The surfactants were mostly sodium dodecyl sulfate, alkytrimethyl ammonium bromides or a nonionic surfactant Cetomacrogol, $CH_3—(CH_2)_m—(OCH_2—CH_2)_n—OH$ (where m = 15 to 17 and n = 20 to 40).

In the above studies, the alcohol was usually melted and dispersed in the anionic or cationic surfactant solution. In this manner the molten alcohol disperses into the droplets and the surfactant and water penetrates to form an isotropic phase ($L_2$) of inverted, mixed micelles containing water (see Fig. 29). On further dilution with water, this phase rapidly transforms into a highly viscous liquid crystalline phase (LC) with a lamellar structure. Individual globules of molten cetyl alcohol stream through the system to produce elongated threads of the liquid crystalline phase [166]. The fluidity of this liquid crystalline phase increases as more water enters. Because of interfacial tension effects spherulites are formed. The final stage in the process is dissolution to form an

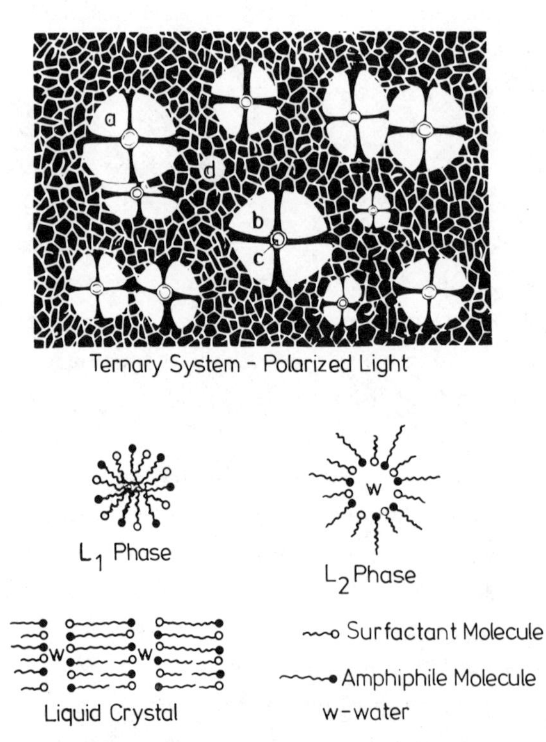


Figure 29. Schematic diagram showing various structures in surfactant-amphiphile mixtures. (From Ref. 166.)

isotropic, mixed-miceller solution ($L_1$). If the temperature is rapidly diminished by forced cooling, the alcohol solidifies and further interaction with the surfactant ceases, so that no liquid crystalline phase forms. No attempt was made to ensure that phase equilibrium occurs before cooling since, particularly with amphiphiles of long chain length, equilibration may require protracted heat treatment. This is not the usual practice in emulsion processes. The liquid crystalline phase freezes to a metastable (pseudomesomorphic) state [170], in which the intimate mixture of alcohol, surfactant, and water present in the smectic phase is maintained, with the system still showing optical anisotropy at room temperature.

With increasing alcohol concentration, a submicroscopic network structure builds up, linking solid alcohol and frozen spherulites. The result is a typical gel-like structure which exhibits solid properties. This structure has been verified by electron microscopy for the system containing a mixture of cetostearyl sulfate/cetostearyl alcohol [171]. A schematic picture of this system is given in Fig. 29. With time, the gel structure becomes weaker, and eventually it breaks down forming a crystalline residue. These crystals have a higher density and lower melting point than that of pure cetyl alcohol; they are mainly thin, flat, and polyhedral. When the components of the ternary system are impure, e.g., when cetostearyl alcohol is used instead of cetyl alcohol, the gels are more stable.

Oil/water emulsions may be prepared using ternary systems of the type discussed above with paraffin oil as the dispersed phase. For example, a mixture of cetyl alcohol and paraffin oil at 65°C is added to aqueous sodium dodecyl sulfate at the same temperature. The mixture is then agitated while the temperature drops to ambient. As mentioned above, usually the alcohol-surfactant ratio is kept constant at 9:1, while the total concentration of the surfactant is gradually increased. Since the total concentration of the surfactant mixture is much greater than that required to form a complex condensed film at the interface, the excess surfactant modifies the physical properties of the emulsion. A gel network builds up progressively in the continuous phase as the concentration of the mixed surfactant increases. The excess surfactant remains in the aqueous phase, either as monomer or as micelles depending on its concentration. Below the CMC mobile unstable emulsions form, whereas above the CMC a micellar pseudophase is present during the cooling process and some of the alcohol diffuses from the oil into the aqueous phase. The higher the concentration of the alcohol in the oil phase, the more molecules diffuse into the continuous phase.

However, the amount of water-insoluble alcohol which can be dispersed in a nonmicellar aqueous environment is small, although molten long-chain monohydric alcohols are unusual in that they stabilize in surfactant solutions and form a smectic phase below the normal CMC [172]. When the surfactant concentration is greater than the CMC, more alcohol is solubilized into the micelles. In the surfactant-plus-alcohol-plus-water system, the site of interaction is the molten alcohol/surfactant solution interface. In the equivalent emulsion, as a rule the site of interaction is at the oil (containing the alcohol)/water interface, where owing to the relatively high alcohol concentration, $L_2$ phase forms and transforms into ternary smectic phase. This elongates into threads along the flow lines of the mixture, breaks up into spherulites and may finally form a small amount of the isotropic phase $L_1$. A small amount of liquid paraffin is also solubilized into these phases. As the temperature falls below the penetration temperature of the alcohol, such solubilization is no longer possible [173], and a frozen metastable structure forms. This results in the formation of a viscoelastic gel which entraps oil droplets, and hence a marked increase in consistency is observed. A schematic diagram of this structure is given in Fig. 30, following Barry [168].

When the emulsions are diluted and examined microscopically, aggregates of droplets are visible. These often appear to be held together by a membrane sheath, i.e., frozen smectic phase. Some of the larger droplets are distorted and appear flattened or polyhedral in shape. They contain anisotropic crystals which are usually acicular, but occasionally flat and hexagonal, lying at the edges of the

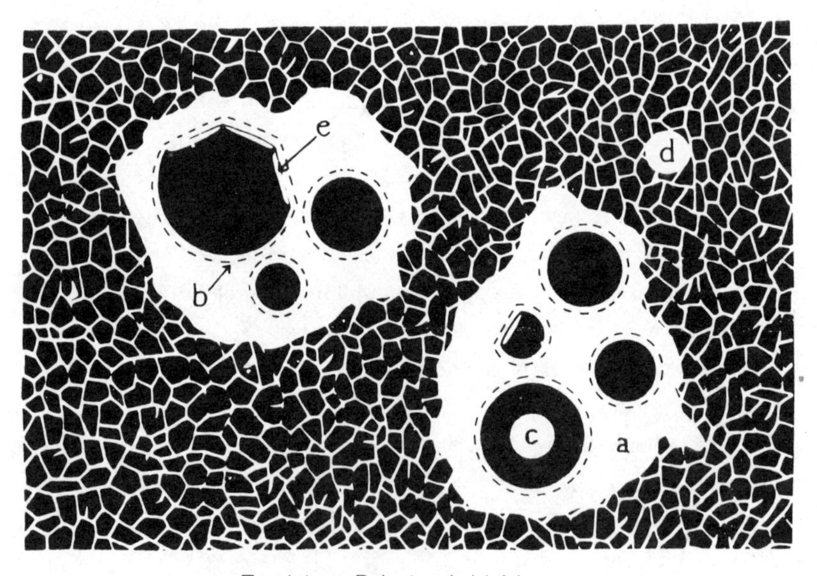


**Figure 30.** Schematic representation of structures formed in emulsions containing alcohol/surfactant mixtures. (From Ref. 166.)

droplets. On cooling, the amount of alcohol in the oil phase can exceed its solubility limit, resulting in the formation of a precipitate which deforms the droplets [168,174]. With time, the emulsion slowly becomes more mobile forming anisotropic crystalline deposits in the continuous phase. As mentioned above, replacing cetyl alcohol by the less pure cetostearyl alcohol results in more rheologically stable emulsions.

Rheological investigations using continuous shear measurements (e.g., a Ferranti-Shirley cone-and-plate viscometer) on the ternary system: sodium dodecyl sulfate plus cetyl alcohol plus water [168], have shown considerable hysteresis in the shear-rate/shear-stress curves, the effect increasing with increase in alcohol content (see Fig. 31). As there is negligible structure rebuilding on standing, the systems undergo irreversible shear breakdown or irreversible softening. They are not thixotropic. Figure 30 shows also spur points which could be attributed to the gel network in the system [166].

The influence of temperature on the rheological properties of these ternary systems has also been investigated by Barry and Shotton [168] and by Davis [175]. Barry and Shotton [168] measured the relative viscosity $\eta_{rel}$ using a Ferranti-Shirley viscometer at a shear rate of 816 $s^{-1}$, whereas Davis [175] measured the total compliance $J_T$ after an arbitrary time of 16 min, as a function of temperature. The results are shown in Fig. 32. A maximum in the apparent relative viscosity and a minimum in $J_T$ occur in the region of 42.5°C. This temperature represents the transition temperature from frozen to liquid smectic phase. Similar rheological behavior has been reported by Barry and Saunders [168] for ternary systems consisting of a cationic (or nonionic) surfactant plus alcohol plus water.

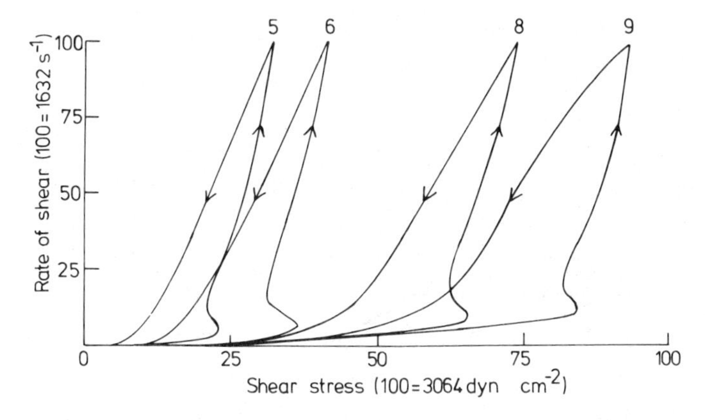


Figure 31. Rheological curve for the ternary system sodium dodecyl sulfate/cetyl alcohol/water at various molar ratios of alcohol/surfactant. (From Ref. 166.)

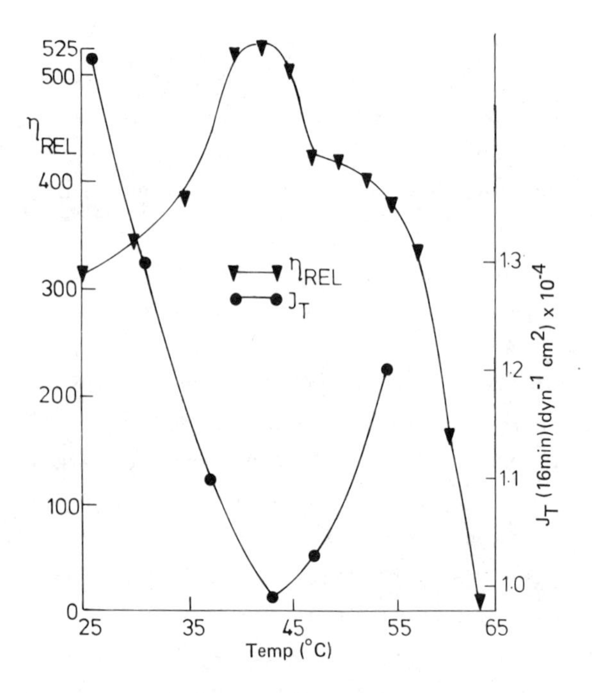


**Figure 32.** Influence of temperature on the relative viscosity and compliance of the ternary system sodium dodecyl sulfate/cetyl alcohol/water. (From Ref. 166.)

In the presence of the oil phase, the flow curves also show hysteresis loops, but spur points are either absent or much less prominent. During shearing, aggregates of droplets disrupt releasing entrapped continuous phase. Again, the area of the hysteresis loops increases with increase of emulsifier concentration, and a similar increase in the extent of gel formation is observed as in the absence of the oil. However, with the corresponding emulsion some structure rebuilding has been observed, indicating that the system is partly thixotropic. Moreover, the consistency and rheological stability of these emulsions depends on the chain length of the alcohol [168]. In fact, emulsions prepared using commercially available cetostearyl alcohol show higher consistency and rheological stability compared with those prepared using the pure alcohol components or even a mixture of these. Indeed, a mobile emulsion of low stability forms when pure stearyl alcohol is used as the oil-soluble component of the mixed surfactant system. Such differences are related to the relatively long chain length and high melting point of the stearyl alcohol; these lead to a reduction in the strength of the viscoelastic network joining the droplets together.

Owing to the complex rheological behavior of the systems discussed above, i.e., their viscoelasticity, the information obtained using continuous shear experiments (e.g., using the Ferranti-Shirley instrument) is obviously limited. Moreover, experimental difficulties encountered using the cone-and-plate viscometer, such as slippage, evaporation, etc., raise questions as to the validity of these measurements. For this reason Barry and co-workers [166,168] have carried out well-controlled experiments, under conditions of small strain, i.e., creep and oscillatory tests. In these experiments the system is examined in its rheological ground state, i.e., where the method of testing does not interfere with the static structure [176]. The results of measurements of this type may be quantitatively analyzed using the established theory of viscoelasticity [177].

For the ternary systems, sodium dodecyl sulfate plus cetyl alcohol plus water and alkyl trimethyl ammonium bromides plus cetostearyl alcohol plus water, studied by Barry and co-workers [166,168], the creep compliance $J_1$ decreases and the viscosity $\eta_1$ increases as the surfactant concentration increases. For paraffin oil/water emulsions prepared using the mixed surfactant, sodium dodecyl sulfate plus cetyl alcohol, $J_1$ was found to decrease and $\eta_1$ to increase sharply over a narrow surfactant concentration range (2–4 per cent). Thereafter, the

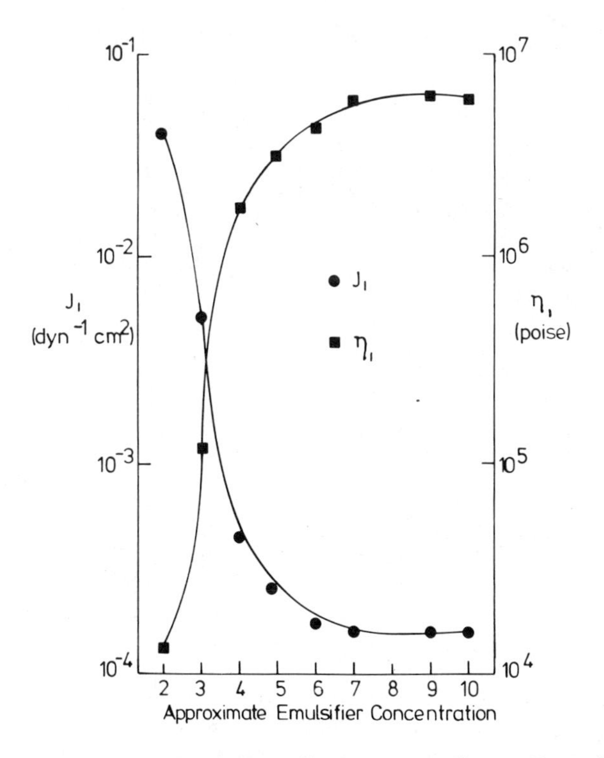


Figure 33. Variation of creep compliance $J_1$ and viscosity $\eta_1$ with emulsifier concentration for paraffin oil/water emulsions, stabilized by the emulsifier mixture: sodium dodecyl sulfate plus cetyl alcohol. (From Ref. 166.)

change was smaller and, eventually, both $J_1$ and $\eta_1$ reached a limiting value. These results are shown in Fig. 33. For the emulsions stabilized with the alkyl trimethyl ammonium bromide ("cetrimide") plus cetostearyl alcohol mixtures, the total creep compliance $J_t$ continues to decrease with increasing emulsifier concentration, up to a concentration of 16%. Moreover, this system shows an aging effect, whereby $J_t$ (above 5% emulsifier) shows a decrease on storage over 4 months, indicating strengthening of the gel network with time. The results are shown in Fig. 34.

As with continuous shear experiments, the variation in compliance with temperature shows a minimum at 42.5°C for both types of surfactant systems. Such a minimum is thought to be due to something like the melting of the gel network. It occurs at a lower temperature for the emulsion relative to that of the ternary system. This may be due to the inclusion of oil into the gel network, which lowers the melting point of the smectic phase. Similar information on the viscoelastic properties of such systems may be obtained using oscillatory measurements (see Sec. IV.C.2). In these measurements the stress (or strain) is varied periodically at a fixed frequency. If the viscoelastic behavior is linear (at small amplitudes), the amplitude of the strain is proportional to the amplitude of the stress, and the strain alternates sinusoidally out of phase with the stress.

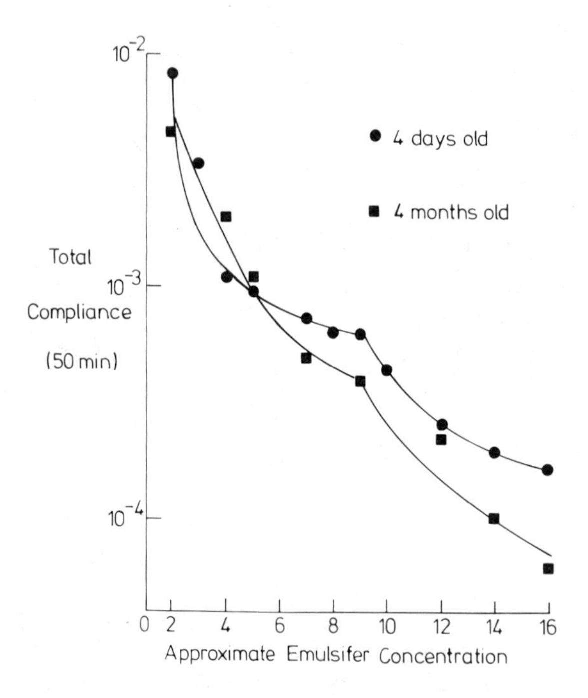


**Figure 34.** Variation of creep compliance $J_t$ with emulsifier concentration for the mixture cetrimide/cetostearyl alcohol. (From Ref. 166.)

The tangent of this phase angle ($\tan\phi$) is known as the loss tangent and is a measure of the ratio of energy lost to energy stored in a cyclic deformation. $\phi$ is zero for a Hookeian solid and 90° for a Newtonian liquid. $\tan\phi$ was measured by Barry and Shotton [168] for the ternary system sodium dodecyl sulfate plus cetyl alcohol plus water, over a frequency range of 2.5 decades, using a Weisenberg rheogonimeter. $\tan\phi$ is low (corresponding to $\phi$ varying approximately from 12.5 to 4°) with the value rising slightly as the frequency decreases. These features indicate that the system exhibits mainly elastic properties, confirming the gel network structure.

## VI. The HLB Concept and Emulsion Stability

### A. Introduction

The selection of different surfactants in the preparation of either O/W or W/O emulsions is often still made on an empirical basis. However, in 1949, Griffin [178] introduced a semiempirical scale for selecting an appropriate surfactant or blend of surfactants. This scale, termed the *hydrophile-lipophile balance* (HLB), is based on the relative percentage of hydrophilic to hydrophobic groups in the surfactant molecule(s). Surfactants with a low HLB number normally form W/O emulsions, whereas those with a high HLB number form O/W emulsions.

A summary of the HLB range required for various purposes [179] is given in Table 5. The relative importance of hydrophilic and hydrophobic groups was first recognized when mixtures of surfactants were used with varying proportions of surfactants having a low and high HLB number. The efficiency of any combination, as judged by disperse phase separation, was found to pass through a maximum when the blend contained a particular concentration of the surfactant with the higher HLB number. The original method for determining HLB numbers, developed by Griffin [178] is quite laborious and requires a number of trial and

**Table 5.** A Summary of HLB Ranges and Their Application

| HLB range | Application |
|---|---|
| 3 to 6 | W/O emulsifier |
| 7 to 9 | Wetting agent |
| 8 to 18 | O/W emulsifier |
| 13 to 15 | Detergent |
| 15 to 18 | Solubilizer |

error procedures. Later Griffin [179] developed a simple equation which permits the calculation of the HLB number for certain types of nonionic surfactants, particularly the alcohol ethoxylates $R(CH_2CH_2O)_xOH$ and the polyhydric fatty acid esters. For example, for the polyhydric fatty acid esters, the HLB number is given by the expression

$$\mathrm{HLB} = 20\left(1 - \frac{S}{A}\right) \tag{138}$$

where S is the saponification number of the ester and A the acid number of the acid. Thus a glyceryl monostearate with S = 161 and A = 198, will have an HLB value of 3.8. However, in many cases it is difficult to determine the saponification number accurately, e.g., esters of tall oil, rosin, beeswax, and lanolin, and for these a relationship based on the weight per cent of oxyethylene E and the polyhydric alcohol P has been given by Griffin [179], namely

$$\mathrm{HLB} = \frac{E + P}{5} \tag{139}$$

If the surfactant contains only poly(ethylene oxide) as the only hydrophilic group, e.g., the alcohol ethoxylates, the HLB number is simply

$$\mathrm{HLB} = \frac{E}{5} \tag{140}$$

However, the above equations cannot be used for nonionic surfactants containing propylene oxide or butylene oxide, nor can it be used for ionic surfactants. The ionization of the hydrophilic group tends to make them even more hydrophilic in character, so that the HLB number cannot be determined from the weight percent of the hydrophilic groups. In that case, the laborious procedure suggested by Griffin [178] must be used.

Davies [180] has devised a method for calculating the HLB values for surfactants directly from their chemical formulas, using empirically determined group numbers. Thus, a group number is assigned to various emulsifier component groups and the HLB is then calculated from these numbers using the following empirical relation:

$$\mathrm{HLB} = 7 + \sum (\text{hydrophilic group numbers}) - \sum (\text{lipophilic group numbers}) \tag{141}$$

The group numbers for various component groups are given in Table 6. Davies [180] has shown that the agreement between HLB numbers calculated using the above empirical relation and those determined experimentally is quite satisfactory.

Table 6. HLB Group Numbers

| Groups | Group number |
|---|---|
| Hydrophilic | |
| —$SO_4Na^+$ | 38.7 |
| —$COO^-H^+$ | 21.2 |
| —$COO^-Na^+$ | 19.1 |
| N (tertiary amine) | 9.4 |
| Ester (sorbitan ring) | 6.8 |
| Ester (free) | 2.4 |
| —COOH | 2.1 |
| —O— | 1.3 |
| CH (sorbitan ring) | 0.5 |
| Lipophilic | |
| —CH— ) | |
| —$CH_2$— ) | 0.475 |
| —$CH_3$— ) | |
| —CH— ) | |
| Derived | |
| —($CH_2$-$CH_2$—O— | 0.33 |
| —($CH_2$-$CH_2$—$CH_2$—O)— | −0.15 |

*Source:* Ref. 186.

B. Experimental Determination of HLB

Various procedures have later been developed to determine the HLB of different surfactants. For example, Griffin [171] found a correlation between the HLB number and the cloud points of polyoxyethylene derivatives. This is illustrated in Fig. 35, in which the cloud point of a 5% aqueous solution is plotted against the HLB number. This means that for such a class of surfactants measurements of the cloud point should at least give a rough estimate of the HLB number. Greenwald et al. [182] developed a titration procedure for estimating the HLB

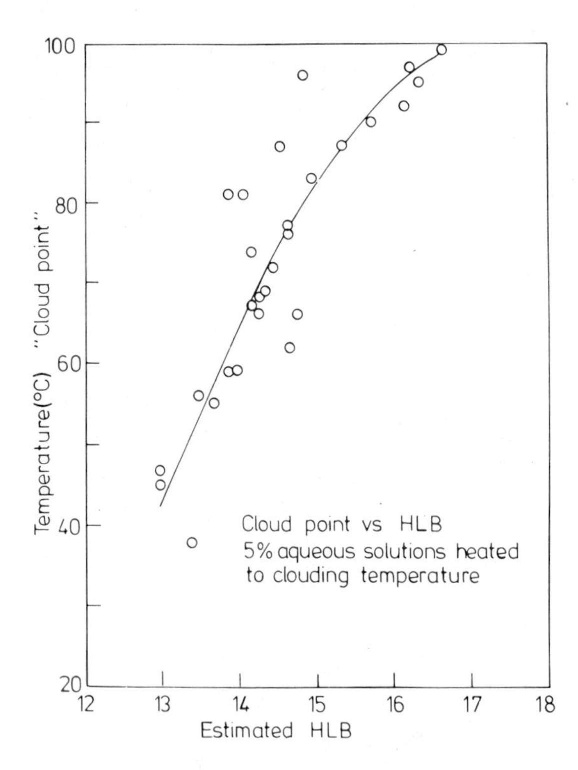


**Figure 35.** Relationship between cloud point and HLB.

number. In this method, a 1% solution of surfactant in benzene plus dioxane is titrated with distilled water at constant temperature until a permanent turbidity appears. Greenwald [182] found a good linear correlation between the HLB number and the water titration value for polyhydric alcohol esters. This is shown in Fig. 36. However, the slope of the line depends on the class of material involved.

Racz and Orban [183] developed a calorimetric method for estimating the HLB value based on the measurement of the heat of hydration of ethoxylated surfactants. For liquid hydrophilic surface-active agents, the enthalpy of hydration Q is related to the HLB value through the empirical relation

$$\mathrm{HLB} = 0.42Q + 7.5 \tag{142}$$

For solid surfactants, a much more complex relationship would be required since the solid-to-liquid phase change also contributes to the observed enthalpy change.

Becher and Birkmeier [184] suggested that gas liquid chromatography (GLC) could be used to determine the HLB value. Since in GLC the efficiency of separation depends on the polarity of the substrate with respect to the components of the mixture, it should be possible to determine the HLB directly by using the

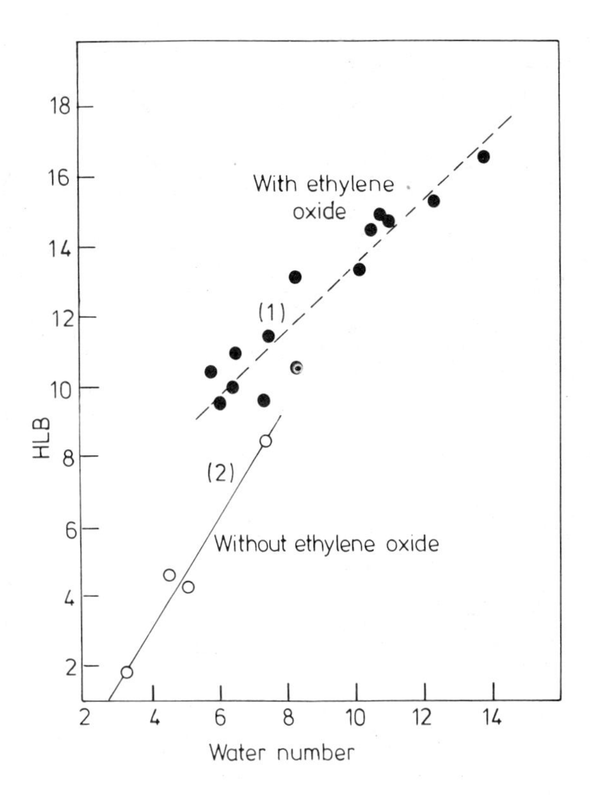


**Figure 36.** Correlation of HLB with the water number. (From Ref. 182, reprinted with permission from H. L. Greenwald, G. L. Brown, and M. N. Fineman, Anal. Chem. *28*:1693. Copyright 1956 American Chemical Society.)

surfactant as the substrate and passing an oil phase down the column. Thus, when a 50:50 mixture of ethanol and hexane is passed down a column of a simple nonionic surfactant, such as sorbitan fatty acid esters and polyoxyethylated sorbitan fatty acid esters, two well-defined peaks, corresponding to hexane (which appears first) and ethanol, appear on the chromatograms. A good correlation was found between the retention time ratio $R_t$ (ethanol/hexane) and the HLB value. Such a relationship is shown in Fig. 37. Statistical analysis of the data gave the following empirical relationship between $R_t$ and HLB,

$$\mathrm{HLB} = 8.55\,R_t - 6.36 \tag{143}$$

where

$$R_t = \frac{R_t^{\mathrm{EtOH}}}{R_t^{\mathrm{hexane}}}$$

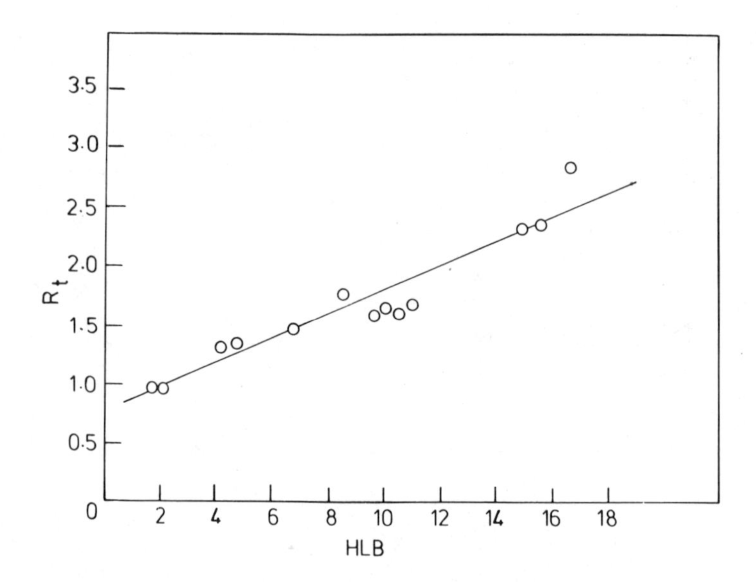


**Figure 37.** Correlation between retention time and HLB of sorbitan fatty acid esters and polyoxyethylated sorbitan farry acid esters. (From Ref. 184.)

C. Relationship Between HLB and Emulsion Coalescence

Davies [180,186] has attempted to relate HLB values to the relative coalescence rates of oil/water and water/oil emulsions, and to the partition of the surfactant between the oil and water phases. If the droplets have a surface potential $\psi_0$ the repulsive potential energy is proportional to $\psi_0^2$ [see Eq. (33)] and hence, the coalescence rate $K_{O/W}$ of the oil droplets in O/W emulsions is given by

$$K_{O/W} = A_1 \exp\left(\frac{-B\psi_0^2}{RT}\right) \tag{144}$$

Here $A_1$ is the hydrodynamic collision factor, given by

$$A_1 = \frac{4\phi RT}{3\eta_0} \tag{145}$$

where $\eta_0$ is the viscosity of the medium, $\phi$ is the volume fraction of the oil phase, and B is a constant with a value of 0.24. This was derived empirically from the results of Lawrence and Mill [111] on coalescence kinetics. In the presence of nonionic surfactants, the repulsive force is not only generated by the electrical potential, but also by a hydration barrier $E_h$ due to the strong bonding of water molecules to the ethylene oxide units of the chains. To a first approximation, this energy barrier is equal to $\theta\Sigma E_h$, where $\theta$ is the fraction of surface covered and $\Sigma E_h$ is the total energy barrier due to hydration. In this case, the rate of coalescence is given by the equation

$$K_{O/W} = A_1 \exp\left(-\frac{B\psi_0^2 - \theta\Sigma E_h}{RT}\right) \tag{146}$$

For the W/O emulsion case, coalescence occurs through water bridging across the oil film between the droplets. Hence, the rate of coalescence $K_{W/O}$ depends on hydrodynamic factors, including the hydrodynamic collision factor $A_2$, the number m of $CH_2$ groups in each hydrocarbon chain, and the fraction of surface covered by emulsifier molecules $\theta$, i.e.,

$$K_{W/O} = A_2 \exp\left(-\frac{2m\omega\theta}{RT}\right) \tag{147}$$

where $\omega$ is the energy barrier involved in the transfer of a $CH_2$ group into water, i.e., 1.3 kJ $mol^{-1}$. Thus, the relative rate of coalescence is given by the equation

$$\frac{K_{W/O}}{K_{O/W}} = \frac{A_2}{A_1} \exp\left(\frac{B\psi_0^2 + \theta\Sigma E_h - 2m\omega\theta}{RT}\right) \tag{148}$$

Therefore, an O/W emulsion will form if $K_{W/O}/K_{O/W} \gg 1$ and a W/O forms if $K_{W/O}/K_{O/W} \ll 1$.

Equation (148) may be rewritten as follows (substituting the necessary numerical values for B and W):

$$\frac{RT}{600\theta}\, m\, \frac{A_1K_{W/O}}{A_2K_{O/W}} = \frac{0.24\psi_0^2}{600\theta} + \frac{\Sigma E_h}{600} - m \tag{149}$$

From Eq. (141), and substituting the group number for —$CH_2$—, namely 0.475 (see Table 5),

$$HLB - 7 = \sum \text{hydrophilic group numbers} - m \times 0.475$$

or

$$\frac{HLB - 7}{0.475} = \frac{\Sigma \text{ hydrophilic group numbers}}{0.475} - m$$

Since

$$\text{Hydrophilic group number for a single hydrated group} = \frac{E_h}{1260} \tag{150}$$

and

$$\text{Hydrophilic group number for a charged group} = \frac{1.9 \times 10^{-4}\psi_0^2}{\theta} \tag{151}$$

by combining Eqs. (148) and (149) one obtains

$$\ln \frac{(A_1K_{W/O})}{(A_2K_{O/W})} = 2.2\theta\,(HLB - 7) \tag{152}$$

This relates the relative rate constants to the HLB number. If the oil and water have the same viscosity and the volume fraction $\phi = 0.5$ then $A_1 = A_2$. Equation (152) predicts that for an HLB value of 7, neither a W/O or O/W emulsion is preferentially formed. For a surfactant of HLB value = 11, which fully covers the surface of the droplets (i.e., $\theta \sim 1$), $K_{O/W} = 10^{-4} K_{W/O}$ and, therefore, an O/W emulsion results.

Equation (150) indicates that the HLB group number is proportional to the energy barrier to coalescence set up by the water which is firmly bound to the hydroxyl or ester groups on the surface-active agent molecule. This also explains the correlation between HLB and the cloud point of nonionic surfactants (see Sec. B). For an ionic surfactant the HLB group number for a group such as sulfate or sulfonate depends on $\psi_0$ and $\theta$. Thus, strictly speaking, the contribution of a charged group to the HLB number is not constant and varies with $\psi_0$ or $\theta$ or both.

Davies [186] also tried to relate the distribution of a surfactant between the water and oil phases with the HLB number. For example, the distribution of an alcohol between water and oil is determined by the free energy of transfer of the molecule from water to oil, i.e.,

$$-\Delta G_{W\to 0} = RT \ln \frac{C_O}{C_W} \tag{153}$$

where $C_O$ and $C_W$ are, respectively, the equilibrium emulsifier concentrations in the oil and in the water phases. The free energy of transfer is made up of two terms corresponding to the hydrophilic (i.e., the OH group) and hydrophobic parts (i.e., the alkyl groups) of the molecules. The free energy of transfer of an OH group from water to oil is equal to +13.4 kJ $mol^{-1}$, whereas that of an alkyl group is $-3.36m$ kJ $mol^{-1}$, where m is the number of $-CH_2-$ groups in the chain.

Therefore,

$$\Delta G_{W\to 0} = +13.4 - 3.36\ m \tag{154}$$

If the molecule contains more than one hydrophilic group, then

$$\Delta G_{W\to 0} = \Delta G_{W\to 0}\ \text{(hydrophilic groups)} - 3.36m \tag{155}$$

Combining Eqs. (153) and (155) gives

$$\frac{RT}{3.36} \ln\left(\frac{C_W}{C_O}\right) = \frac{\Sigma\, \Delta G_{W\to 0}\ \text{(hydrophilic groups)}}{3.36} - m \tag{156}$$

Comparing Eqs. (156) and (148),

$$\frac{A_1 K_{W/O}}{A_2 K_{O/W}} = \left(\frac{C_W}{C_O}\right)^{0.75/\theta} \tag{157}$$

where

$$E_h = 0.75\ \Delta G_{W \to 0}\ \text{(uncharged hydrophilic group)} \tag{158}$$

$$\frac{0.32\psi_0^2}{\theta} = \Delta G_{W \to 0}\ \text{(uncharged hydrophilic group)} \tag{159}$$

Equation (157) gives a rationalization of the Bancroft rule [187] which states that the phase in which the stabilizating agent is more soluble will be the continuous phase. Thus, surfactants which are preferentially soluble in water stabilize O/W systems and vice versa. Equation (156) predicts that $E_h$ (the work done in removing a hydrophilic group from water) must be of the same magnitude as the total work done in transferring the group from water to oil. This is indeed true for nonpolar hydrocarbons such as decane, where the interaction between hydrophilic groups and the hydrocarbon is small. Moreover, for nonionic surfactants, $\psi_0 = 0$, and the Bancroft rule should be obeyed, as long as the oil phase is nonpolar and $A_1 = A_2$. For ionic surfactants, on the other hand, Bancroft's rule will also be obeyed if Eq. (159) is still valid. This will be the case for reasonable values of $\psi_0$. For low values of $\psi_0$, i.e., at small surfactant concentrations and/or high electrolyte concentration, Eq. (120) is no longer valid and the Bancroft rule is not obeyed.

It is useful to combine Eq. (156) with that for the HLB value, Eq. (149); this gives

$$\text{HLB} - 7 = 0.36 \ln \frac{C_W}{C_O} \tag{160}$$

and

$$\text{Hydrophilic group number for single hydrated group} = \frac{\Delta G_{W \to 0}\ \text{(uncharged hydrophilic group)}}{1680} \tag{161}$$

$$\text{Hydrophilic number for a charged group} = \frac{\Delta G_{W \to 0}\ \text{(charged hydrophilic group)}}{1680} \tag{162}$$

Again, if Eqs. (161) and (162) are correct, Eq. (160) will hold, i.e., it is possible to relate the distribution characteristics of the surfactant with its HLB number. Thus, the Bancroft rule and HLB number are equivalent; i.e., if the surfactant is preferentially soluble in water, its HLB number is greater than 7 and vice versa. A test of Eq. (160) has been made by Davies [186] who found a linear relationship between ln $(C_W/C_O)$ and HLB value for a number of nonionic emulsifiers.

It should be pointed out, however, that the method of dispersing the oil into the water and vice versa used in practice has a marked effect on the emulsion type. Hence the above theory for emulsions prepared by simple mixing techniques has to be modified. In practice, oil and water are usually forced between two shearing plates in a variety of emulsifying machines. Davies [186] developed a laboratory apparatus for studying the tendency to form an emulsion of certain oil by measuring the phase volume of oil $\phi$ at which the emulsion inverts from oil continuous to water continuous, or vice versa. When $\phi$ is near unity the emulsion will always be oil continuous irrespective of other factors, whereas if $\phi$ is small the emulsion will always be water continuous. Between these two extremes lies the inversion point $\phi_i$. In this manner, the effect of other factors such as oil viscosity, HLB, etc., can be studied. For example, at a specific rotor speed of the stirrer, and using petroleum ether as the oil, Davies [188] found that $\phi_i$ increased linearly with HLB in the range 6.7 to 16. At lower HLB values, $\phi_i$ increased again, with decreasing HLB because the water-continuous systems were no longer homogeneous emulsions. The dependence of $\phi_i$ on HLB also depends upon other factors such as rotor speed, the nature of the oil and its viscosity, the concentration of the emulsifier and even on the material of the plate of the rotor. Oil-wetted plates such as Perspex or hard rubber strongly favor oil-continuous systems. This is illustrated in Fig. 38 which shows the variation of $\phi_i$ with HLB for various plate materials.

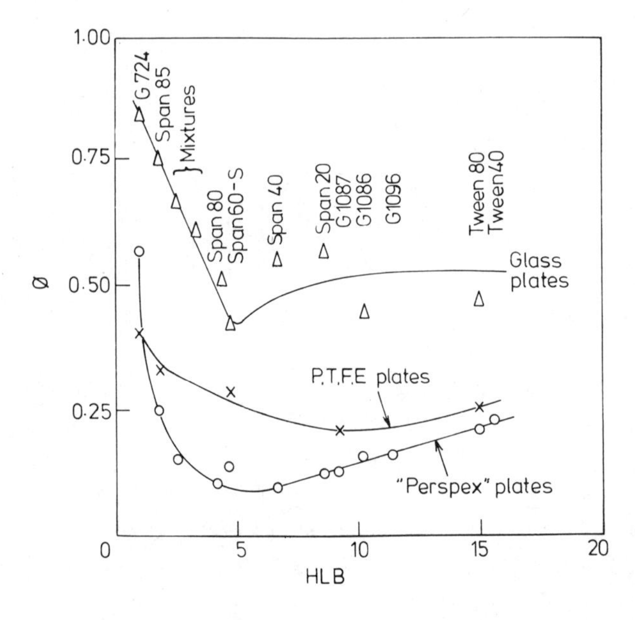


Figure 38. Variation of $\phi_i$ with HLB according to Davies [188].

The rise in $\phi_i$ at low HLB is due to the release of free oil, incompletely emulsified. These results also show that the effect of HLB on emulsion type is not in accord with Eqs. (148) and (152), when using this laboratory machine. In that case, the rate of coalescence is no longer important, and the formation of the emulsion depends much more strongly on the hydrodynamics of mixing. Indeed, the marked dependence of $\phi_i$ on the materials of the plates confirms the importance of unequal hydrodynamic effects, to which the surface active agents contribute possibly by altering the relative energies and the relative rates of wetting of the solid plates by the oil and the water.

Thus, in general, emulsification depends on the rates of formation of the two types of emulsions, as well as the rates of their breaking once formed. In continuous flow machines the first is dominant, owing to the high shear gradients. Moreover, under these conditions, the mixing regime of the two phases may be very important. For example, Sherman [189] found that by carefully adding water to an oil phase containing the emulsifier, while vigorously stirring by hand and ensuring that no additional water was added until the previous addition was effectively dispersed, $\phi_i$ for W/O to O/W inversion was greater than when using mechanical stirring with continuous slow addition of water to the oil phase.

Thus, at best, the dependence of emulsion type and stability on the HLB is only qualitative and cannot be generalized. This is not surprising since the equations discussed above are based on a simplified picture of coalescence and do not take into account the important role of liquid-film thinning between droplets and the importance of shear and the dilational properties of the adsorbed layer of surfactant around the droplets. It is not unexpected, therefore, that many authors have cast some doubt on the validity of the HLB concept. For example, Riegelmann and Pichon [190] claimed that stable O/W emulsions could be obtained in the range of HLB number of 2 to 17, indicating that a low HLB number does not always ensure the formation of a W/O emulsion. This was also supported by Peterson et al. [191] who found no correlation between the HLB value of several classes of nonionic surfactants and the type of emulsion produced. Moreover, Sherman [192] showed that the dependence of emulsion type on HLB value depends not only on emulsifier concentration, but also on the phase volume of the oil. For example, sorbitan monolaurate with an HLB number of 8.6 stabilizes paraffin oil/water emulsions at low concentration, but at 5 to 6% w/w, the O/W emulsion inverts to a W/O emulsion when the volume fraction of the oil is high. Thus, at one HLB number, both types of emulsions can be obtained, depending on the conditions. Similar behavior has been observed with other nonionic surfactants, e.g., polyoxyethylene derivatives. In spite of the above shortcomings, the HLB index can be very useful in the practical formulation of emulsions. A list of HLB numbers has been given by Becher [193] for commercial emulsifiers and similar lists may also be found in the technical literature for many other commercially available emulsifiers. In practice, in formulating a given emulsion, one may take any pair of emulsifying agents, which fall at opposite ends of the HLB scale, e.g., Tween 80 (sorbitan monooleate with 20 ethylene oxide units HLB = 15) and Span 80 (sorbitan monooleate, HLB = 5), using them in various proportions to cover a wide range of HLB number and emulsions are made in the same way, with a few percent of the emulsifying blend. The stability of the emulsions is then evaluated at each HLB number from the rate of droplet coalescence (or qualitatively by measuring the rate of oil separation) to find the most effective HLB value, i.e., the HLB value providing optimum stability. Having found the most effective HLB value, various other surfactant pairs are compared at this HLB number, until the most effective pair is found.

## VII. PHASE INVERSION

### A. Introduction

Phase inversion, as the name implies, is the process whereby the internal and external phase of an emulsion suddenly invert, e.g., O/W to W/O or vice versa. This subject was briefly discussed earlier (see Sec. VI.C) in connection with the relationship between HLB and emulsion type. As mentioned there, Davies [186] defined the phase volume of oil $\Phi_i$ as the volume at which an emulsion inverts from oil continuous to water continuous, or vice versa. A detailed investigation of the effect of phase volume on emulsion inversion has been carried out by Sherman [194].

It was shown that the viscosity of an emulsion, at a given emulsifier concentration, gradually increases with increase in phase volume, but that at a certain critical volume fraction $\Phi_c$ there is a sudden decrease in the viscosity. This critical volume fraction corresponds to the point at which the W/O emulsion inverts to an O/W emulsion. $\Phi_c$ was found to increase with increasing emulsifier concentration. Similar observations were also made by Ross [195]. The sharp decrease in viscosity observed at the inversion point is due to the sudden reduction in disperse phase volume fraction. For example, if one starts with a W/O emulsion, then, on increasing the volume fraction of the water phase (disperse phase), the viscosity of the emulsion increases gradually until a maximum value is obtained, generally when $\Phi \sim 0.74$. When inversion takes place to an O/W emulsion, the volume fraction of the disperse phase (the oil) will now be 0.26; hence the dramatic decrease in viscosity.

In the early theories of phase inversion, it was postulated that inversion takes place as a result of the difficulty in packing emulsion droplets above a certain volume fraction. For example, according to Ostwald [196], an assembly of spheres of equal radii should occupy 74% of the total volume of the system. Thus, at a phase volume $\Phi > 0.74$, the emulsion droplets would have to be packed more densely than is possible. This means that any attempt to increase the phase volume of the emulsion beyond that point should result in distortion, breaking, or inversion. A number of investigations have clearly indicated the invalidity of this argument, inversion being found to take place at phase volumes much greater or smaller than this critical value. For example, Pickering [197] has shown that paraffin oil/aqueous surfactant solution emulsions containing 99% by volume of oil can be obtained. It has been argued [198] that these are not strictly emulsions but dispersions in soap gels. On the other hand, Shinoda and Saito [199] showed that inversion of olive oil/water emulsions takes place at $\Phi = 0.25$. Moreover, as discussed above, Sherman [194] has shown that the volume fraction at which inversion takes place depends to a large extent on the emulsifier. Ostwald's theory applies only to the case of packing of rigid, nondeformably spheres of equal size. Emulsion droplets are neither resistant to deformation nor are they, in general, of equal size. The wide distribution of droplet size makes it possible to achieve a higher internal phase volume fraction by virtue of the fact that the smaller droplets can be fitted into the interstices between the larger ones. If one adds to this the possibility that the droplets may be deformed into polyhedra, even denser packing is possible.

At present, there does not seem to be any theory available which can explain phase inversion in a quantitative manner. However, location of the inversion point is of practical importance, particularly when emulsions are stored over a long period of time. As will be discussed below, in the case of emulsions stabilized with nonionic surfactants, the emulsion sometimes inverts at a certain critical temperature, termed the *phase inversion* temperature, which in some cases may be correlated with certain properties of the system, such as the HLB value of the emulsifier or even the rate of coalescence of the emulsion.

### B. Phase Inversion Temperature

In a series of papers Shinoda and co-workers [198-200] (cf. also Chap. 5) have shown that the phase inversion temperature (PIT) of nonionic emulsifiers is influenced by surfactant HLB number. The following conclusions were drawn: (1) the size of emulsion droplets depends on the temperature and the HLB of emulsifiers; (2) the droplets are less stable toward coalescence close to the PIT; (3) relatively stable O/W emulsions are obtained when the PIT of the system is some 20 to 65°C higher than the storage temperature; (4) a stable emulsion is obtained by rapid cooling after formation at the PIT; and (5) the optimum stability of an emulsion is relatively insensitive to changes of HLB value or PIT of the emulsifier, but instability is very sensitive to the PIT of the system. Later Shinoda et al. [201] studied the effect of the molar mass and the molar mass distribution of the hydrophilic chain lengths of alkyl or alkyl aryl polyoxyethylene ethers and emulsifiers (having the same PIT) on the stability of O/W and W/O emulsions. The authors found that stability against coalescence increases markedly as the molar mass of the lipophilic and hydrophilic groups increase. Moreover, the emulsions showed maximum stability when the distribution of the hydrophilic groups was fairly broad. It was also found that in those cases where the distribution of the hydrophilic chains is broad, the cloud point is lower and the PIT is higher, than in the corresponding case for narrow size distributions. Thus, the PIT and HLB number are directly related parameters, provided the distribution of the oxyethylene chains of the emulsifiers is similar. The HLB number of an emulsifier of broad distribution is lower than that of the pure emulsifier having the same PIT. A plot of HLB value and PIT for cyclohexane/water emulsions (1:1 by volume), stabilized by a number of nonionic surfactants [202], is shown in Fig. 39. This figure is useful for the estimation of the change of the HLB value and the optimum hydrophilic chain length from the change of

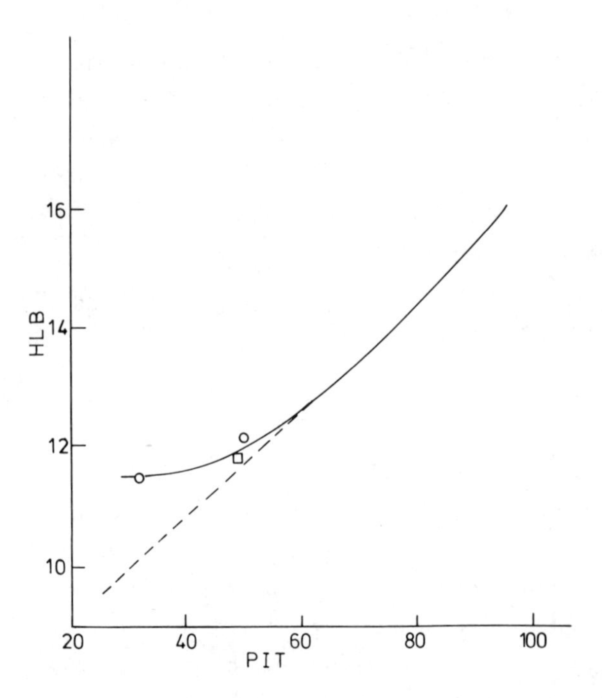


**Figure 39.** Relationship between HLB and PIT for cyclohexane/water emulsions stabilizers with nonionic surfactant (five per cent).

the PIT at various temperatures. Addition of salts [202] reduces the PIT, and therefore an emulsifier with a higher PIT value is required in the presence of electrolytes in order to obtain a more stable emulsion because the optimum PIT of the emulsifier is fixed if the temperature is fixed [199].

In view of the above correlation between PIT and HLB and the known dependence of the kinetics of droplet coalescence on HLB (see Sec. VI) Sherman and co-workers [144,203-204] suggested the use of the measured PIT value as a rapid method for assessing emulsion stability. Figure 40 shows the relationship between the rate of coalescence for Nujol/water emulsions prepared using blends Tween and Span surfactants of various HLB numbers. The data suggest that there is indeed a relationship between PIT and the rate of droplet coalescence. Thus, on this evidence it does seem that measurement of the PIT may provide a quick method for assessing the stability of an emulsion. However, it should be stressed that the correlation between PIT and coalescence rate has so far been observed only with a limited number of emulsions and emulsifier systems. Further work is needed to test the correlation (if any) with other systems.

### C. Measurement of Phase Inversion Temperature

Various methods may be used to determine the PIT of an emulsion. The simplest method is by direct visual assessment [156,206]. Techniques involving dilution

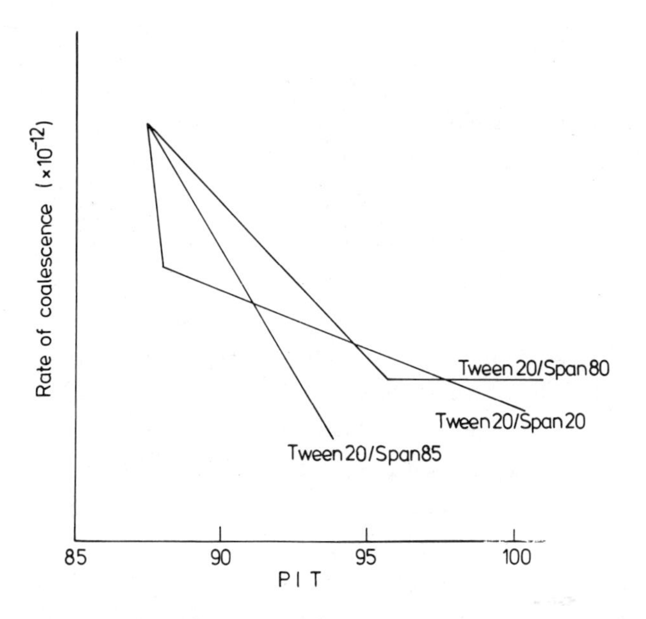


**Figure 40.** Relationship between emulsion stability and PIT. (From Ref. 203.)

[206] suffer from the disadvantage that this may itself lead to phase inversion. A sensitive method for measuring the PIT is to follow the conductivity of the emulsion as a function of temperature. This method has been commonly used by many investigators [207-210]. However, it is not applicable to emulsion systems containing high concentrations of electrolytes. Matsumoto and Sherman [205] suggested a differential thermal analysis technique for measuring the PIT. At the PIT, a sharp peak is observed. Differential scanning calorimetry may also be used for locating the PIT. This method is fairly rapid and may be used for measurements of high PITs, provided precautions are taken to minimize evaporation of the sample during the measurements.

Since, at the PIT, there is usually a large change in the viscosity of the system, rheological measurements on the bulk emulsion also provide a sensitive method for locating the PIT and have been used by many investigators, e.g., Florence and Rogers [210], Lin and Lambrechts [211], and Sherman [194]. Suncerland and Enever [212] combined a programmed heating unit with a Ferranti-Shirley cone-and-plate viscometer to measure the PIT accurately. A sharp change in the apparent viscosity of the emulsion (paraffin oil/water, stabilized by nonionic surfactant mixtures) was observed at the PIT. This increase in viscosity may be attributed to the aqueous continuous phase being replaced by the more viscous oil phase. This method was capable of resolving the PIT to within ±1°C.

## VIII. Stabilization by Macromolecules

### A. Introduction

The use of naturally occurring macromolecular substances, such as gums and proteins, for the stabilization of emulsions has long been used in practice, for example, in food and pharmaceutical emulsions [213,214]. The fat globules in milk are stabilized against coalescence by adsorbed proteins. In an early investigation, [215] the stabilization of such emulsions was attributed to the formation of "tough skins" formed at the interface between an oil droplet and an aqueous solution of the macromolecular substance. Various other investigators [216] have used microscopy to demonstrate the formation of such thick skins at various interfaces. However, most of this work has been carried out with ill-characterized systems, without any purification either of the oils or the macromolecular substances. Indeed, it is highly probable that the oils used contain substantial amounts of surface-active compounds. Moreover, such qualitative observations cannot distinguish between the formation of simple adsorption layers, interfacial precipitates, or some other form of complex film formed by the interaction between the macromolecules and surfactant impurities [217].

As described in detail in Chap. 1, the adsorption of macromolecules at the liqud/liquid interface is a slow process and equilibrium is only reached after long times (hours or days), depending strongly, for example, on the molar mass distribution of the adsorbed molecule. At equilibrium, adsorbed homopolymer molecules form an adsorbed layer comprised of loops and tails extending into the solution phase. Thus, when two oil droplets approach each other, e.g., through a Brownian encounter, a repulsive force is generated owing to the presence of these adsorbed layers. This stabilizing effect is usually referred to as *steric stabilization* [218]. Several theories have now been developed for these steric interactions, which describe the stabilizing effect (with respect to flocculation) of adsorbed polymers or nonionic surfactants on particulate dispersions. These theories may, in principle, also be applied to the case of the stabilization of emulsions against flocculation. Moreover, the macromolecular film formed at the oil/water interface can, in view of its viscoelastic properties, in certain circumstances, provide a "mechanical" barrier to coalescence. This is particularly the case with many adsorbed protein molecules, which denature on adsorption and hence collapse at the interface, forming a fairly rigid film. A brief summary of the viscoelastic properties of such macromolecular films and the correlation with stability towards coalescence will be given later. First, however, we discuss stabilization toward flocculation.

## B. Steric Interactions

It is convenient to consider three groups of interactions between two droplets separated by distance b (Fig. 41): (1) the van der Waals term $G_A$; (2) for charged droplets, the electrical double-layer interactions $G_E$; and (3) steric interactions $G_S$.

For $b > 2\delta$, only the first two sets of interactions arise. Both the van der Waals and the electrical double layer interactions are modified by the presence of the adsorbed layer; a detailed discussion of this is presented later.

For $b < 2\delta$, the steric interactions, which result from the interference of the two adsorbed layers, in general dominate the total interaction. Indeed, the

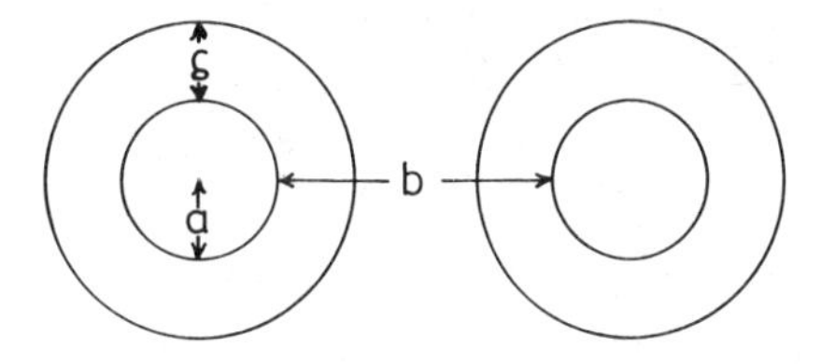


Figure 41. Interaction of two emulsion droplets with adsorbed polymer layers.

basic mechanism of the steric interaction may be described as keeping the droplets far enough apart (i.e., b sufficiently large) such that the van der Waals attraction is minimal.

As with the theories of electrical double-layer interactions, two limiting cases for steric interactions may be considered:

i. Interaction under equilibrium adsorption conditions
ii. Interaction under constant adsorption conditions

With case i, equilibrium is maintained between the stabilizing molecules adsorbed at the two interfaces concerned and those present in bulk solution, i.e., some desorption of molecules occurs on approach. This case is equivalent to the constant potential case in the DLVO theory. It is obviously restricted to weakly adsorbed molecules, such as nonionic surfactants, and does not, therefore, apply in general to polymeric adsorbents. With case ii, constant adsorption is maintained during particle collision; this is equivalent to the constant charge case in the DLVO theory. Constant adsorption also implies that the fraction of segments p actually adsorbed at the surface remains constant, although the segments protruding into the bulk solution (loops or tails) and those in trains may redistribute themselves during contact.

For the constant adsorption case, again two limiting cases may be visualized [219]: (1) interpenetration of adsorbed layers with no compression (this situation can only apply for separations greater than one adsorbed layer thickness, i.e., $b > \delta$); (2) compression of adsorbed layers without any interpenetration. These two cases are illustrated in Fig. 42. One may consider situations where each of these limiting cases may occur. For example, if the segment concentration in the adsorbed layer is relatively high, then compression (rather than interpenetration) will tend to occur. This is usually the case with low molecular weight or

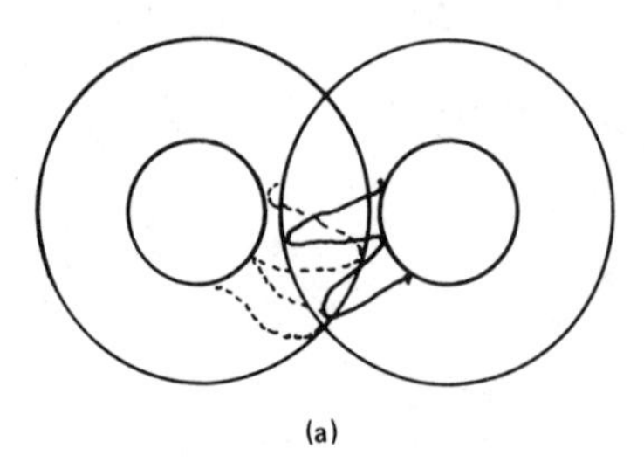


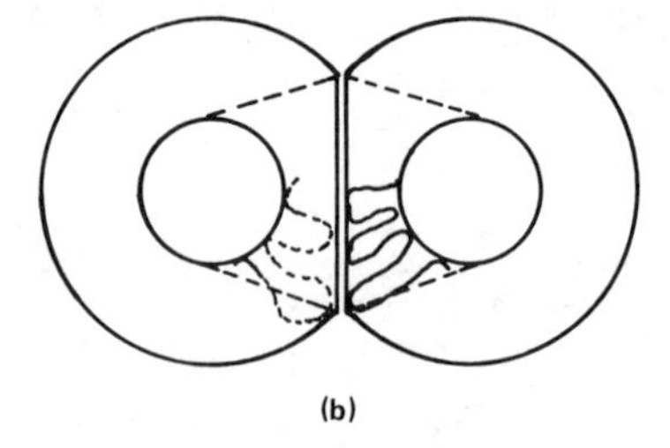


Figure 42. Two limiting (constant adsorption) models for steric stabilization: (a) interpenetration of adsorbed layers without compression; (b) compression without interpenetration.

branched chain molecules. However, with high-molecular-weight substances, where the segment concentration at the periphery of the adsorbed layer is small, and also at low surface coverage, interpenetration tends to predominate.

For intermediate situations some interpenetration and some compression simultaneously occur. It is useful here to consider the case of the buildup of polymer concentration c in a good solvent. At a certain concentration (c*), which will be strongly molar mass dependent, the polymer coils just begin to overlap with their neighbors in some form of close-packed array. Beyond c*, interpenetration occurs but the coils also collapse back toward their theta (unperturbed) dimensions. At concentration c** the coils have their theta dimensions, and moreover, the segment concentration throughout the whole solution is now more or less uniform. The concentration region between c** and the amorphous solid/melt is sometimes referred to as the *diluted melt region*. The concentration of polymer in an adsorbed layer around a particle, at monolayer coverage, is generally in the region of c*.

There are two contributions to the steric interactions. If compression occurs, the configuration entropy of the chains is reduced since the total volume available to each chain is reduced. This term is known as the *volume restriction* $G_{VR}$ or *elastic* term $G_{EL}$. In addition, whether compression and/or overlap occurs there will be a buildup in segment concentration in the interaction zone and this leads to an increase in the local osmotic pressure and free energy. This latter term is referred to as the *mixing* term $G_M$.

### 1. *Volume Restriction Term*

Mackor [220] was the first to analyze this term in an attempt to explain the experimental data of van der Waarden [221] on the stabilization of carbon black in hydrocarbons by adsorbed alkyl benzenes. He assumed a model based on inflexible rods, terminally adsorbed but freely jointed to a flat surface. The number of configurations available to the rod was assumed to be proportional to the surface area swept out by the free end. A schematic picture showing the effect

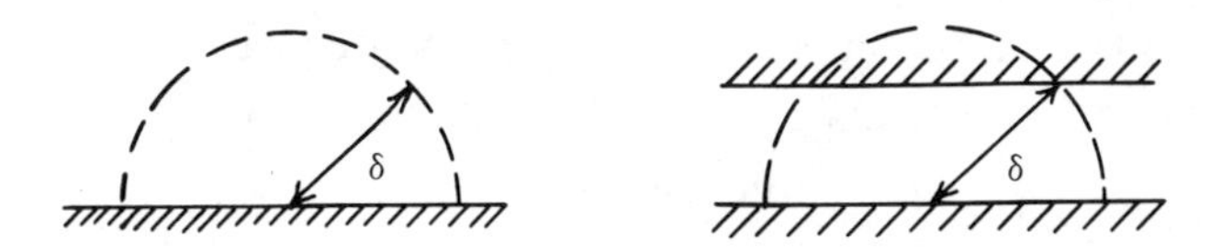


Figure 43. Schematic representation of the effect of the approach of a second surface on the configuration of a terminally adsorbed rod, according to Mackor [220].

of approach of a second surface on the surface area swept by a rod is shown in Fig. 43.

The number of configurations $W_\infty$ prior to the approach of the second surface is proportional to $2\pi\delta^2$, where $\delta$ is the length of the rod; whereas on approach of the second surface to a distance b, $W_b$ is proportional to $2\pi\delta^2(b/\delta)$. Thus, the ratio of the number of configurations, after and before interaction, is given by

$$\frac{W_\infty - W_b}{W_\infty} = 1 - \frac{b}{\delta} \tag{163}$$

Using the Boltzmann relation, the change in configuration entropy $\Delta S$ is given by

$$\Delta S = k \ln \Omega = k \ln \left(1 - \frac{b}{\delta}\right) \tag{164}$$

where k is the Boltzmann constant. The free energy of repulsion $G_{VR}$ is then given by

$$G_{VR} = N_s kT \Theta_\infty \left(1 - \frac{b}{\delta}\right) \tag{165}$$

where $\Theta_\infty$ is the fraction of surface covered by an adsorbed layer with the surfaces at infinite separation, and $N_s$ is the number of adsorbed rods per unit area of surface; the other terms have their usual meeting.

It is clear from Eq. (165) that $G_{VR} = 0$ when $b = \delta$, i.e., just at the point where the particles are no longer in contact. Some calculations were made by Ottweill and Walker [222] for the nonionic surfactant molecule $C_{12}H_{15}$—$(CH_2CH_2O)_6OH$, adsorbed on polystyrene latex particles. The length of the molecule was taken as 4 nm. Putting $N_s = 2 \times 10^{14}$ cm$^{-2}$, corresponding to one site per 0.5 nm$^2$ (i.e., the value predicted from the adsorption isotherm) and $\Theta = 0.1$, which just allows free gyration with no lateral interactions, the variation of $\Delta G_{VR}$ with b was calculated and is shown in Fig. 44.

In this treatment, the lateral interaction between neighboring rods was not taken into account and hence Mackor's theory only applies at very low coverages. This theory was later extended by Mackor and van der Waals [223] to higher surface coverages, using a quasi-lattice model for the solvent. The treatment was restricted to the case of dilute solutions, where the interaction between adsorbed and nonadsorbed molecules may be neglected. Using a statistical-mechanical approach, Mackor and van der Waals [223] derived the following adsorption equations for rigid solute molecules having n segments.

$$\frac{\Theta_1}{\Theta_2} = \frac{m}{\ell}\left\{\frac{(1 - \Theta_1)[1 - 2(n-1)\Theta_1/\ell z]}{(1 - \Theta_1 - n\Theta_2)(1 - \Theta_1/mz)}\right\}^{n-1} \tag{166}$$

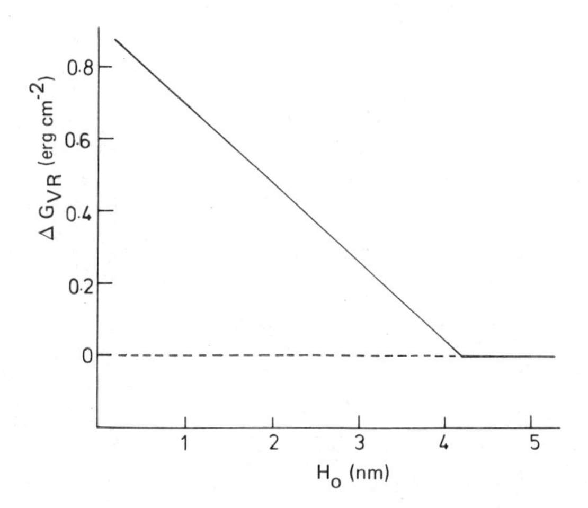


**Figure 44.** Free energy of repulsion as a function of distance of approach $H_0$ for interaction between two parallel plates, with a low surface coverage of adsorbed rod-shaped molecules.

$$\frac{a_B}{a_A^n} \exp\left(\frac{-\chi}{kT}\right) = \frac{\theta_2}{\ell z} \frac{[1 - 2(n-1)\theta_2/\ell z]^{n-1}}{(1 - \theta_1 - n\theta_2)^n} \tag{167}$$

where

$$\theta_1 = \frac{N(B_1)}{N}$$

$$\theta_2 = \frac{N(B_2)}{N}$$

are the fractional coverages of vertically and horizontally oriented molecules, occupying $N(B_1)$ and $N(B_2)$ sites of the total N sites on the surface, respectively. The quantities $\ell$ and m are the fractional coordination numbers of sites in the same layer and the two adjacent layers, respectively (i.e., $\ell + 2m = 1$); $\chi$ is the net free energy of adsorption per active site; $a_A$ and $a_B$ are the activities of the solvent and solute, respectively, in bulk solution. Equations (166) and (167) enable one to determine the adsorption isotherm as a function of $a_A$ and $\chi$. Moreover, the ratio of vertically to horizontally oriented molecules may be evaluated.

Mackor and van der Waals [223] also derived the following equation for the interfacial tension $\gamma$ (i.e., the free energy per unit area of surface):

$$\frac{\gamma}{N^s}kT = -n \ln a_A + \ln(1 - \Theta_1 - n\Theta_2) + (n-1)\ln(1-\Theta_2)$$

$$- mz(n-1)\ln\left(1 - \frac{\Theta_1}{mz}\right) - \frac{\ell z}{2}\ln\left[\left(1 - 2(n-1)\frac{\Theta_2}{\ell_2}\right)\right] \quad (168)$$

Having obtained equations for the adsorption and interfacial tension for an isolated interface, Mackor and van der Waals were then able to consider the effect of the approach of a second interface. They first considered the case of dumbbell-shaped solute molecules B (i.e., each occupying two segments) and spherical solvent molecules A. A system containing $n_A$ and $n_B$ molecules of A and B, with chemical potentials $\mu_A$ and $\mu_B$, respectively, was considered. When the number d of lattice planes separating the two planes $\alpha$ and $\beta$ (each with surface area $\mathbb{A}$) is large, no interaction between the adsorbed layers occurs and the adsorption isotherm and interfacial tension are as given by Eqs. (166) to (168). However, when the planes are brought together to a distance $d < 4$, interaction occurs and a repulsive force is generated. To calculate the free energy of repulsion, three states of the system corresponding to $d = 3$, $d = 2$, and $d = 1$, (see Fig. 45) were considered.

The difference in free energy between any of the states illustrated in Fig. 45, G, and that of large separation $G_\infty$ may be expressed as follows:

$$\Delta G = G - G_\infty = 2\mathbb{A}(\gamma - \gamma_\infty)_{n_A, n_B} + \sum_{i=A,B} n_i(\mu_i - \mu_i^\infty) \quad (169)$$

The first term is the variation in interfacial tension associated with desorption, whereas the second term is the change in bulk chemical potential arising from this desorption process. For the case whre the volume of bulk solution is effectively infinite, the second term is negligible and thus

$$\Delta G = 2\mathbb{A}(\gamma - \gamma_\infty)_{\mu_A^\infty \mu_B^\infty} \quad (170)$$

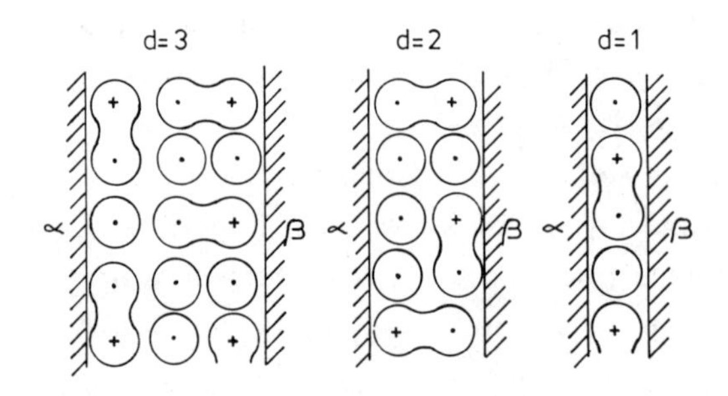


**Figure 45.** Interaction between parallel plates with adsorbed dumbbell-shaped solute molecules. (From Ref. 223.)

The problem is then reduced to the determination of $\gamma$ and the adsorption isotherm for each of the cases illustrated in Fig. 45. Proceeding in this way, Mackor and van der Waals evaluated $G_{VR}$ for each of the cases illustrated in Fig. 45, using the equation

$$\frac{G_{VR}}{N_s kT} = \frac{2}{N_s kT}(\gamma - \gamma_\infty) \tag{171}$$

for a simple cubic lattice ($z = 6$, $\ell z = 4$, $mz = 1$) and a hexagonal lattice ($z = 12$, $\ell z = 6$, $mz = 3$). The results of the calculations are shown in Fig. 46.

The value obtained for $G_{VR}$ depends on the nature of the lattice, the adsorption energy of the adsorbed molecules, and, of course, on d. As long as d exceeds twice the length of the adsorbed molecules, no repulsion is felt, but as soon as d falls below this limit $G_{VR}$ increases rapidly with decrease in d.

To assess the physical consequences of the repulsion, $G_{VR}$ should be compared with $G_A(b)$,

$$G_A = -\frac{A}{12\pi b^2} \tag{172}$$

where b (= rd) is the distance between the planes, r is the lattice dimension, and A is the Hamaker constant. For convenience, the case for the hexagonal lattice is considered, where the distance between nearest neighbors is a. The distance between two adjacent layers of sites is $(1/3)a\sqrt{6}$ (i.e., the height of a tetrahedron with side a) and

$$b = \frac{1}{3}\sqrt{6}\,(ad) \tag{173}$$

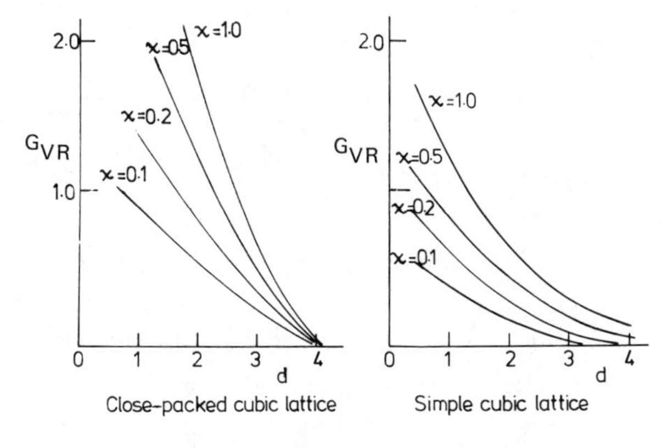


**Figure 46.** $G_{VR}$ versus d at various values of the adsorption energy $\chi = (a_B/a_A^n)\exp(\chi/kT)$. (From Ref. 223.)

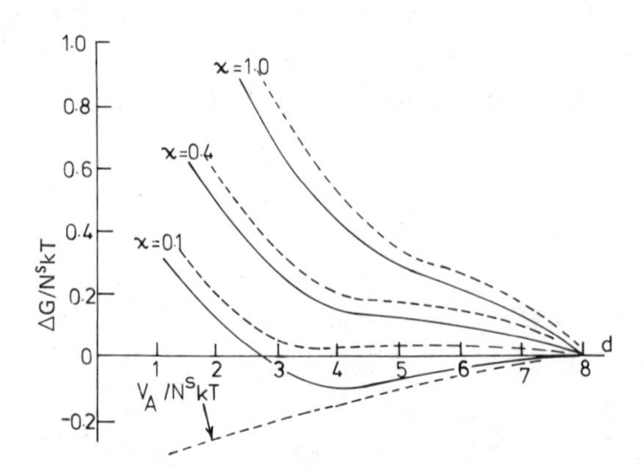


Figure 47. Total free energy of interaction $\Delta G$ between two adsorbing planes as a function of d, for the adsorption of rigid linear tetramers. (From Ref. 223.)

The number of sites per unit area $N^S$ is given by

$$N^S = \left(\frac{1}{2}\sqrt{3}\,a^2\right)^{-1} \tag{174}$$

This leads to

$$G_A = -\frac{N^S A\sqrt{3}}{16\pi d^2} \tag{175}$$

$$\frac{\Delta G_{VR}}{N^S kT} = \frac{0.0344\,6\ A}{kT}\,\frac{1}{d^2} \tag{176}$$

Taking A to be 50 kT (i.e., $A = 2 \times 10^{-19}$ J), the results of the calculations made by Mackor and van der Waals [223] are shown in Fig. 47. The results given in Fig. 47 show that even at low coverage ($\chi = 0.1$) stabilization occurs. However, the Mackor and van der Waals theory is too crude to be applied to the case of high molar mass, flexible macromolecules. For the latter, several theories have been developed in recent years and these are discussed below.

Clayfield and Lumb [224] developed a computer simulation method for calculating the reduction in configurational entropy on compressing a terminally anchored chain on the approach of a second surface. Solvent effects were not considered. A Monte Carlo technique was used, together with a cubic lattice, implying a 90° bond angle for the simulated polymer chains. Note more than one monomer unit was allowed to occupy any lattice site, thus excluded-volume effects are taken into account. The results were expressed as the proportion $W(\ell_1)$ of the total number of configurations for which the distance of furthest extremity of the chain from the surface is less than $\ell_1$. The decrease in configurational entropy $\Delta S$ resulting from the reduced freedom of the chain is derived from the Boltzmann expression, leading to

$$\Delta S = k \ln \frac{1}{W(\ell_1)} \tag{177}$$

Thus, the interaction free energy per polymer molecule $G_{VR}$ is given by

$$\frac{G_{VR}}{kT} = - \ln W(\ell_1) \tag{178}$$

A plot of $-\ln W(\ell_1)$ versus L ($=\ell_1/\ell_r$, where $\ell_r$ is the root mean square value of $\ell$), for various numbers of segments per chain, is shown in Fig. 48.

Random chains adsorbed on a surface have their extremities at a variety of distances $\ell$ from the surface; the root-mean-square (rms) $\ell_r$ value represents the thickness of the unperturbed adsorbed layer. On average, a random chain in free solution has the same diameter in any direction and this is more or less true also for an unconfined adsorbed chain as well. The surface coverage $\Theta$ may be taken (arbitrarily) to be unity when the chains occupy an average area of $\ell_r^2$. In the case of very strong adsorption, packing closer than this can be obtained (i.e., $\Theta > 1$); this results in some lateral compression of the chains. The free energy of compression per chain is $G_{VR}$ and per unit area it is $G_{VR}\Theta/\ell_r^2$. Two cases were considered by Clayfield and Lumb [224]: sphere-plate and sphere-sphere geometries. For the former case, the free energy of repulsion is given by the expression

$$\frac{G_{VR}}{kT} = 2\pi\Theta \int_0^{\pi/2} \left\{ \frac{2a/\ell r + (b/\ell r)^2}{(1 + \cos\phi)^2} - \sin\phi \cos\phi \, [-\ln W(L)] \right\} d\phi \tag{179}$$

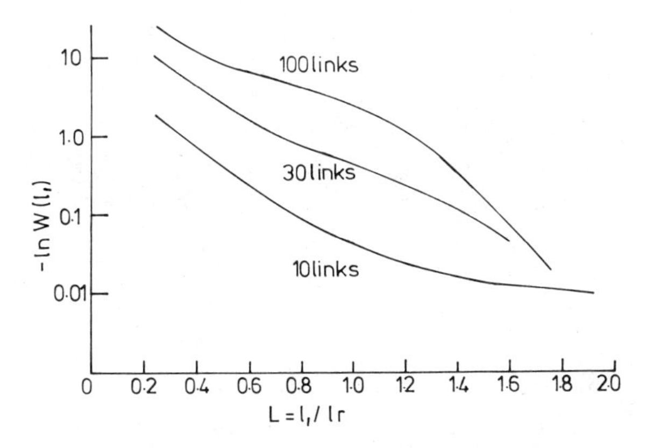


**Figure 48.** Free energy of compression (in kT units) for various random chains. (From Ref. 224.)

where a is the sphere radius, b is the distance between the surface of the sphere and that of the plate, and $\phi$ is the angle the molecules on the sphere make with the axis of the sphere perpendicular to the plane. Equation (179) may be integrated numerically to obtain $\Delta G_{VR}/kT\Theta$ as a function of $b/2\ell_r$ for various values of $a/\ell_r$ and for various numbers of segments in the chain; the integration is carried out until $d\,G_{VR}/d\phi$ becomes negligible.

For sphere-sphere interactions,

$$\frac{G_{VR}}{kT} = -2\pi\Theta \int_0^{\pi/2} \left(\frac{a}{\ell_r} + \frac{b}{2\ell_r}\right) \tan\phi \sec\phi \ln W(L)\, d\phi \tag{180}$$

Equation (180) can also be integrated numerically. By way of illustration, some of the calculations made by Clayfield and Lumb [224] (combining $G_{VR} + G_A$) are given in Fig. 49. Some of the consequences of Clayfield and Lumb's calculations are

i. $G_{max}$ increases with a, at a given $\ell_r$; this is an unexpected result in view of the increase of $G_A$ with a.
ii. A reduction of $\Theta$ lowers $G_{max}$ with little or no change in $G_{min}$.
iii. With increasing values of $a/\ell_r$, $G_{max}$ and $G_{min}$ both increase.
iv. Increasing the length of the polymer molecule with the objective of increasing the adsorbed layer thickness may actually lead to a decrease in stability.

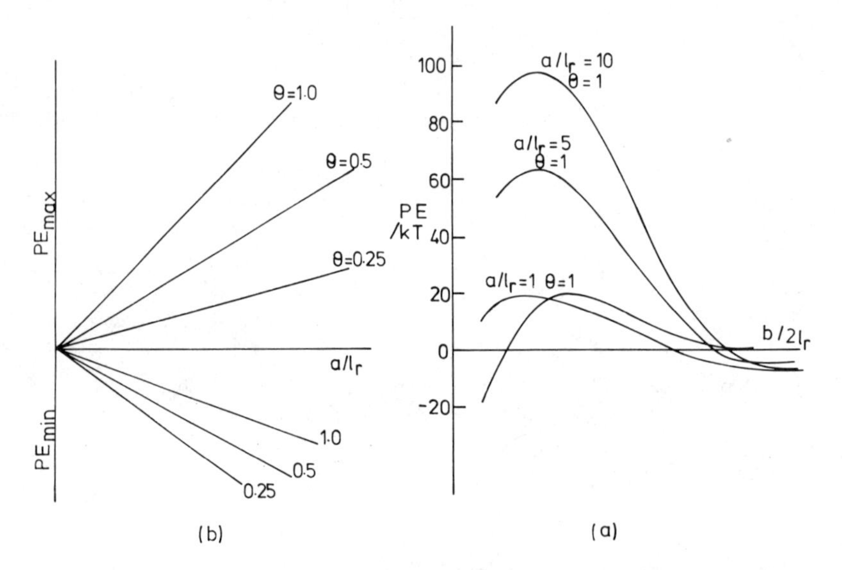


**Figure 49.** (a) Potential energy of interaction between a sphere and a plane coated with a 30-link chain; $A/kT = 25$. (b) Potential energy maxima and minima as a function of $a/\ell_r$ for 30-link chain; $A/kT = 12.5$. (From Ref. 224.)

v. There is a lower limit to the size of particles stabilized by a given polymer chain length. This occurs because larger molecules, having a higher surface area per molecule, provide only a small number of interacting molecules on close approach of a particle and a plane. On the other hand, smaller molecules, although preventing irreversible flocculation, may allow reversible flocculation (associated with $G_{min}$) to occur.

Meier [225] has also developed a theory for the interface between two surfaces covered with terminally anchored flexible polymers. The theory was developed for the case of parallel plates only. The excluded volume contribution to the free energy of repulsion was evaluated explicitly using random flight statistics. The basis of Meier's method is to obtain solutions to the diffusion equation with the appropriate boundary conditions in order to evaluate the configurational free energy.

The diffusion equation for polymer molecules may be written in the form

$$\frac{\partial (W_{\bar{r}})}{\partial n} = \frac{\ell^2}{6} \nabla^2 W(\bar{r}) \tag{181}$$

where $W_{\bar{r}}$ is the probability of end-to-end distance $\bar{r}$ of a chain of n elements of length $\ell$. The volume restriction term was obtained again using the Boltzmann equation

$$G_{VR} = -2\nu kT \ln P_N(d) \tag{182}$$

where $P_N(d)$ is the probability that all the segments of a given molecule are within a distance d of the surface to which the chain is anchored, with the other surface at infinity, and $\nu$ is the number of molecules per unit area.

As two surfaces, each having a layer of terminally anchored chains, are brought together, the volume available to the chains necessarily decreases with the result that some available configurations are lost. The new value for $P_N(d)$ can be determined from the diffusion equation. This leads to an expression of the form

$$P_N(d) = \sum_{m=-\infty}^{m=+\infty} \exp\left(\frac{-6m^2 d^2}{N\ell^2}\right) - \exp\left(\frac{-3(2m - 1)^2 d^2}{2N\ell^2}\right) \tag{183}$$

for relatively low coverages, so that

$$G_{VR} = -2\nu kT \sum_{m=-\infty}^{m=+\infty} \exp\left(-\frac{6m^2 d^2}{N\ell^2}\right) - \exp\left(\frac{-3(2m - 1)^2 d^2}{2N\ell^2}\right) \tag{184}$$

$(N\ell^2)^{1/2}$ is the root-mean end-to-end distance of the chains, where N is the total number of elements of the chain. Note that

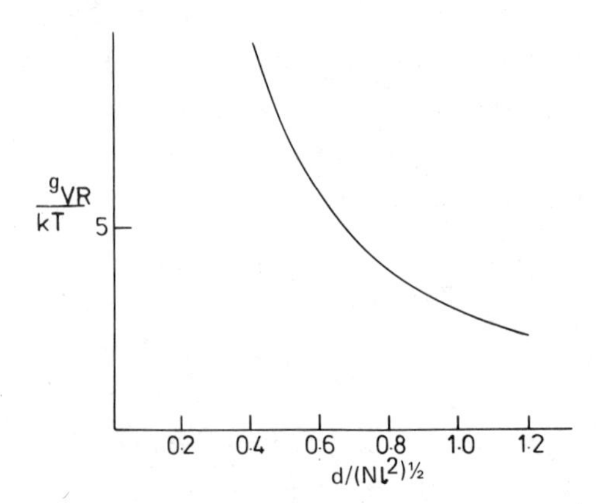


Figure 50. Variation of volume restriction term with distance of separation. (From Ref. 225, reprinted with permission from D. J. Meier, J. Phys. Chem. *71*:1861. Copyright 1967 American Chemical Society.)

$$\Delta G_{VR}(d) = 2\nu g_{VR}(d) \tag{185}$$

where $g_{VR}$ is the free energy per chain. Meier [115] plotted $g_{VR}(d)$ versus $d/(N\ell^2)^{1/2}$, as shown in Fig. 50. The results given in Fig. 50 show that the volume restriction term is quite high, being of the order of 5 kT per chain when the separation is one-half the rms end-to-end distance of the free chain. However, there is an error in Meier's theory [16] associated with an incorrect derivation of the segment density distribution which arises by allowing configurations which penetrate the particle surface. This error was subsequently corrected by Hesselink, as discussed below.

Hesselink [226] derived the segment density distributions for adsorbed macromolecules using Hoeve's equation [227] for the loop size distribution. He then considered the influence of a second parallel interface on the configurational free energy of a loop, tail, and bridge. He also derived the density distribution for a loop and a tail in the presence of a second surface. Hesselink [226] used a six-choice cubic lattic for these calculations. The segment density distribution $\rho_2(i,x)$ of a single loop of size i on a six-choice cubic lattice in the direction x normal to the adsorbing surface is given by

$$\rho_2(i,x) = 12x(i\ell^2)^{-1} \exp\left(-\frac{6x^2}{i\ell^2}\right) \tag{186}$$

where $\ell$ is the length of a segment and

$$\int_0^\infty \rho_2(i,x)\, di = 1 \tag{187}$$

Equation (186) was derived by summing the probabilities, $\rho(i,k,x)$ of finding the kth segment at a distance x from the interface for all values of k, where $\rho(i,k,x)$ is the probability of finding the terminal segment of a chain of k segments at x times the probability of finding the $i - k$ segments at $x > 0$.

Hoeve's equation [227] for the loop size distribution is

$$n_i = A_i^{-3/2} \exp (i\lambda) \tag{188}$$

where $n_i$ is the number of loops of i segments, $\lambda$ is the free energy of adsorption per segment (in units of kT) and A is a normalizing constant. Integration of Eq. (187) leads to

$$n_i = n_a \pi^{-1/2} (\bar{i}) i^{-3/2} \exp\left( - \frac{ia^2}{\bar{i}^2} \right) \tag{189}$$

For a copolymer,

$$n_i = n(\bar{i})^{-2} \exp\left( - \frac{i}{\bar{i}} \right) \tag{190}$$

The following equation for the segment density distribution of an adsorbed homopolymer was derived

$$\rho_h = 2a\sqrt{6}\ (\bar{i}\ell)^{-1} \exp\ (-2ax\sqrt{6}\ \bar{i}\ell) \tag{191}$$

and for a copolymer,

$$\rho_c = 12x(i\ell)^{-2} \int_0^\infty \exp\left( \frac{-i}{\bar{i}} - \frac{6x^2}{i\ell^2} \right) di \tag{192}$$

Hesselink [226] considered that, on the approach of a second particle, the loop size distribution of an adsorbed homopolymer is continuously rearranged. Since long loops are more hindered than short loops, long loops will then be converted to shorter loops. However, during a Brownian encounter of two colloical particles insufficient time is available for any rearrangement of the loop size distribution. Thus, the loop size distribution may be assumed to be constant. However, this is not the case when considering the distribution of segments within any given loop or tail. Calculations by Hesselink showed that, during a Brownian encounter of two particles, a loop with ~480 segments has sufficient time to continuously readjust to its time-average configuration in the region between the two interfaces, without making contacts with the adsorbing interface.

Hesselink [226] then calculated the reduction in number of configurations for a tail and a loop on approach of a second impenetrable interface. This leads to

$$R_1(i,d) = \sum_{v=-\infty}^{\infty} \exp\left(-\frac{6v^2d^2}{i\ell^2}\right) - \exp\left(-\frac{3d^2(2v+1)^2}{2i\ell^2}\right) \tag{193}$$

for tails, and

$$R_2(i,d) = \sum_{v=-\infty}^{\infty} \left(1 + \frac{2vd}{\ell}\right) \exp\left(-\frac{6vd(vd+1)}{i\ell^2}\right) \tag{194}$$

for loops. Equation (194) is similar to that of Meier [16]. The increase in free energy for a tail and a loop associated with configurational restriction due to an approaching second interface is given by

$$G_{VR} = -kT \ln R_j(i,d) \tag{195}$$

The results of calculations using Eq. (195) are given in Fig. 51. When two interfaces, both covered by v (= n/i) tails (or loops) per unit area, approach each other, the resulting increase in free energy $G_{VR}$ per unit area of surface due to volume restriction effects is given by

$$G_{VR} = -2kT \sum_i n_i \ln (R_i, d) = -2VkTV_{i,d} \tag{196}$$

where

$$V_{i,d} = -in^{-1} \int_0^{\infty} \ln \sum_{v=-\infty}^{\infty} \left(1 - \frac{12v^2d^2}{i\ell^2}\right) \exp\left(-\frac{6v^2d^2}{i\ell^2}\right) di \tag{197}$$

For equal loops and $d/\sqrt{i\ell^2} \geqslant 1$,

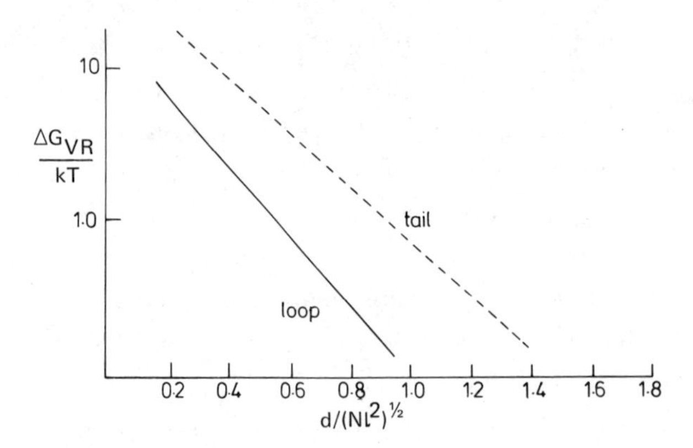


**Figure 51.** Volume restriction term for a tail and a loop. (From Ref. 226, reprinted with permission from F. T. Hesselink, J. Phys. Chem. 75:2096. Copyright 1971 American Chemical Society.)

$$V_{i,d} = -2\left(1 - \frac{12d^2}{i\ell^2}\right)\exp\left(-\frac{6d^2}{i\ell^2}\right) \tag{198}$$

*Mixing Term*

Fischer [228] was the first to analyze the interpenetration or mixing term $G_M$. The following basic assumptions were made in order to evaluate $G_M$ for two spherical particles: (1) the segment concentration in the adsorbed layer is uniform, (2) the segment concentration in the overlap region is the sum of the individual concentrations from both adsorbed layers, and (3) the Flory-Krigbaum theory [229] of dilute polymer solutions is applicable. A schematic representation of the overlap of adsorbed layers is given in Fig. 52.

On overlap, the chemical potential of the solvent $\mu_1^\beta$ in the overlap region $\delta V$ is less than that in the rest of the adsorbed layer $\mu_1^\alpha$, since its activity or concentration is less. Solvent molecules will, therefore, tend to diffuse into this region from the bulk solution, tending to force the particles apart. This difference in chemical potential of the solvent in the overlap zone and the rest of the adsorbed layer $\mu\Delta_1$ may be related to the excess osmotic pressure $\pi_E$ through the relation

$$\Delta\mu_1 = \mu_1^\alpha - \mu_1^\beta = -\overline{V}_1(\pi^\alpha - \pi^\beta) = -\pi_E\overline{V}_1 \tag{199}$$

where $\overline{V}_1$ is the molar volume of the solvent, and $\pi_E$ is related to the second virial coefficient B of the polymer solution by

$$\pi_E = RTBc^2 \tag{200}$$

where c is the concentration of the polymer in the adsorbed layer. The free energy of overlap is then

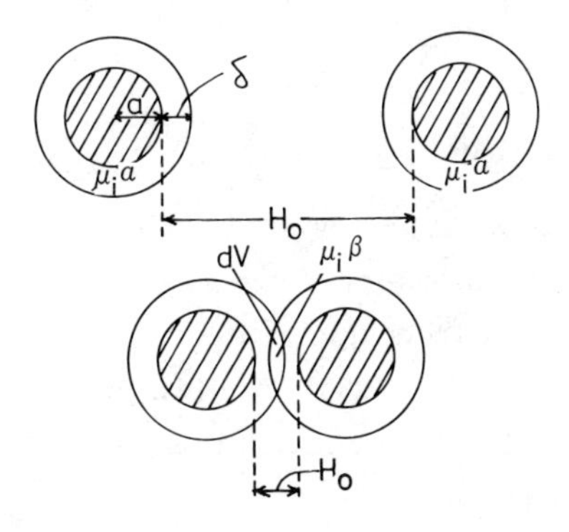


**Figure 52.** Schematic representation of the overlap of adsorbed layers.

$$G_M = 2\int_0^{dV} \pi_E \, \delta V = 2\pi_E \, \delta V \tag{201}$$

For two spherical particles,

$$G_M = 2RTc^2B \frac{4\pi}{3}\left(\delta - \frac{H_0}{2}\right)\left(3a + 2\delta + \frac{H_0}{2}\right) \tag{202}$$

or

$$\frac{G_M}{kT} = N_{av}c^2B \frac{4\pi}{3}\left(\delta - \frac{H_0}{2}\right)\left(3a + 2\delta + \frac{H_0}{2}\right) \tag{203}$$

It is possible to arrive at Eqs. (202) and (203) directly from the Flory-Krigbaum theory of polymer solutions [229]. The free energy of mixing of polymer segments with solvent molecules in a small volume element $\delta V$ is given by

$$\delta(\Delta G_M) = kT(\delta n_1 \ln \phi_1 + \delta n_2 \ln \phi_2 + \chi \, dn_2 \, \phi_2) \tag{204}$$

where $\delta n_1$ is the number of polymer molecules contained in $\delta V$, $\phi_1$ and $\phi_2$ are the volume fractions of solvent and polymer, respectively, and $\chi$ is the Flory-Huggins interaction parameter. ($\chi$ is related to the second virial coefficient B.) The total change in the free energy of mixing for the whole interaction zone $\delta V$ is, therefore, obtained by summing over all the volume elements comprising $\delta V$. This leads to

$$G_M = \frac{4\pi kT}{3V_1} \phi_2^2 \left(\frac{1}{2} - \chi\right) \delta V \tag{205}$$

or

$$\frac{G_M}{kT} = \frac{4\pi}{3V_1} \phi_2^2 \left(\frac{1}{2} - \chi\right)\left(\delta - \frac{H_0}{2}\right)^2\left(3a + 2\delta + \frac{H_0}{2}\right) \tag{206}$$

Equations (205) and (206) are identical to equation (203), except that the virial coefficient B is now replaced by the interaction parameter $\chi$, which may in fact be split into two terms, namely,

$$\chi = \chi_h + \chi_s \tag{207}$$

where $\chi_h$ is the original Flory-Huggins enthalpy of mixing parameter (as measured calorimetrically) and $\chi_s$ is an additional entropy parameter which is associated with excess volume of mixing effects, structural effects due to hydrogen bonds, etc. Some calculations of $G_M$ by Ottewill and Walker [222], for two polystyrene spherical particles of radius 16.2 nm, covered with the nonionic surfactant $C_{12}H_{25}(CH_2CH_2O)_6OH$, are given in Fig. 53 for various values of $\chi$. In

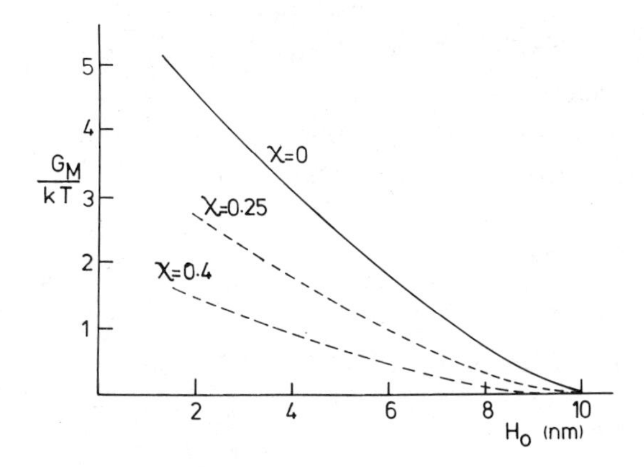


**Figure 53.** Variation of the free energy of mixing with distance of separation between two polystyrene latex particles with adsorbed surfactant $C_{12}E_6$ for various values of $\chi$. (From Ref. 222.)

these calculations $\delta$ was taken to be 5 nm (as measured from the sedimentation coefficient using an ultracentrifuge). $c_2$ was calculated from the adsorption isotherm to be 0.26 g $cm^{-3}$.

Although, for high molar mass polymers, the assumption that the segment concentration in the overlap region is the sum of the individual concentrations may be valid, at least for small degrees of interaction, the assumption of uniform segment density in the adsorbed layer is invalid. In order to obtain an expression for the mixing term for particles with adsorbed or anchored macrommolecules, the segment density distribution should be taken into account. This was considered both by Meier [225] and by Hesselink et al. [226,230]. Meier [225] considered that as the surfaces containing adsorbed polymer chains are brought together, the segment density in the space between the surfaces increases. Again, the Flory-Krigbaum theory [229] was used to estimate the free energy of mixing. It was assumed that molecules adsorbed on the two opposing surfaces are positioned directly opposite one another, but that the surface coverage is sparse enough so that overlap between adjacent polymer molecules on the same surface may be neglected.

The theory considers the jth molecule adsorbed on a planar surface at x = 0 and the kth molecule adsorbed directly opposite on the second surface at x = d. The free energy of mixing $\delta G_M$ of polymer segments from k and j with solvent in the volume element $\delta V$ is given by

$$\delta G_M = \frac{kT\ \delta V}{V_1}[\ln (1 - \rho_k V_s - \rho_j V_s) + \chi_1 V_s(\rho_k + \rho_j)](1 - \rho_k V_s - \rho_j V_s) \tag{208}$$

where $V_1$ is the molar volume of the solvent, $V_s$ is the segment volume, and $\rho_j$ and $\rho_k$ are (number) densities of segments from the two chains, j and k. The free energy of mixing per pair of molecules on opposing surfaces is obtained by integrating Eq. (208) over an $\delta V$. As the opposing surfaces are brought from infinity to d, the change in free energy of mixing $G_M(d)$ is given by

$$G_M(d) = \int_V (dG_M)_d - \int_V (dG_M)_\infty \tag{209}$$

Expanding the logarithms in Eq. (208) and retaining terms up to $\rho^2 V_s^2$, one obtains

$$G_M(d) = \frac{kTV_s^2}{V_1}\left(\frac{1}{2} - \chi\right)\left[\int_V (\rho_j + \rho_k)_d^2 \, dV - \int_V (\rho_j + \rho_k)_\infty^2 \, dV\right] \tag{210}$$

The quality of the solvent is incorporated through the $\chi$ parameter, which is related to the expansion factor $\alpha$ of the polymer in solution through

$$\alpha^5 - \alpha^3 = \frac{27\, i^2 V_s^2 (1/2 - \chi)}{(2\pi)^{3/2} V_1 \langle i\ell^2\rangle_0^{3/2}} \tag{211}$$

where $\langle i\ell^2\rangle_0^{1/2}$ is the root-mean-square end-to-end distance under $\Theta$ conditions (i.e., the "unperturbed" dimensions). Since $\int \rho_i^2 \, dx = \int \rho_k^2 \, dx$ and $\int (\rho_i \rho_k)_\infty \, dx = 0$, then

$$G_M(d) = \frac{2kTV_s^2}{V_1}\left(\frac{1}{2} - \chi\right)\left[\int_V (\rho_j)_d^2 \, dV - \int_V (\rho_j)_\infty^2 \, dV + \int_V (\rho_j \rho_k)_d \, dV\right] \tag{212}$$

combining Eqs. (211) and (212), Hesselink et al. [230] arrived at the expression

$$G_M = 2\left(\frac{2\pi}{9}\right)^{3/2} (\alpha^2 - 1)\, kT\, \nu^2 \langle r^2\rangle M_{i/d} \tag{213}$$

where $\langle r^2\rangle = \alpha^2 i\ell^2$ is the root-mean-square end-to-end distance in the solvent under question, and

$$M_{(i,d)} = \langle r^2\rangle^{1/2}\left[\int_0^d (\hat{\rho}_j)_d^2 \, dx - \int_0^\infty (\hat{\rho}_j)_\infty^2 \, dx + \int_0^d (\hat{\rho}_j \hat{\rho}_j)_d \, dx\right] \tag{214}$$

The caret indicates that $\hat{\rho}$ is normalized such that $\int \hat{\rho}_d = 1$.

The normalized segment density distribution for tails and loops were obtained using the equations given for the volume restriction term. By substitution of these in Eq. (214), followed by numerical integration, $\Delta G_M$ was obtained. An analytical solution was possible for homopolymers with equal loops, i.e.,

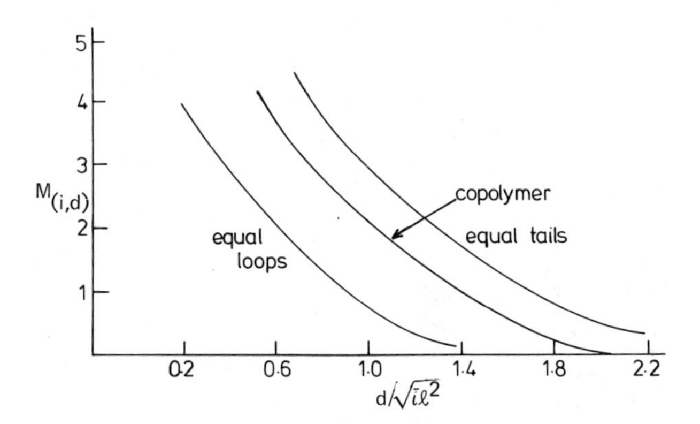


**Figure 54.** Variation of $M_{(i,d)}$ with $d/\sqrt{i\ell^2}$. (From Ref. 226, reprinted with permission from F. T. Hesselink, J. Phys. Chem. 75:2099. Copyright 1971 American Chemical Society.)

$$G_M = 2\left(\frac{2\pi}{9}\right)^{3/2} \nu^2 \, kT(\alpha^2 - 1)\langle r^2\rangle M_{(i,d)} \tag{215}$$

with

$$M_{(i,d)} = 3\pi^{1/2}\left(\frac{6d^2}{i\ell^2} - 1\right) \exp\left(-\frac{3d^2}{i\ell^2}\right) \tag{216}$$

Figure 54 shows the variation of $M_{(i,d)}$ with $d/\sqrt{\bar{i}\ell^2}$ for particles covered by equal tails, equal loops, and copolymers.

### 3. *Total Interaction*

The total interaction between two polymer-covered particles is given by

$$G_i = G_E + G_A + G_s \tag{217}$$

where $G_E$ is the double-layer repulsion, $G_A$ is the van der Waals attraction, and $G_s$ is the steric repulsion. It is assumed that the two contributions to $G_s$ are additive, i.e.,

$$G_s = G_{VR} + G_M \tag{218}$$

## C. Modification of $G_E$ Owing to the Presence of an Adsorbed Polymer Layer

The adsorption of macromolecules usually leads to the redistribution of ions in the double layer around a charged particle, as well as displacement of the plane of shear. These effects are schematically shown in Fig. 55 for a negative particle with an adsorbed homopolymer. Moreover, the presence of an adsorbed nonionic macromolecule may affect the surface charge density by changing either the ion-

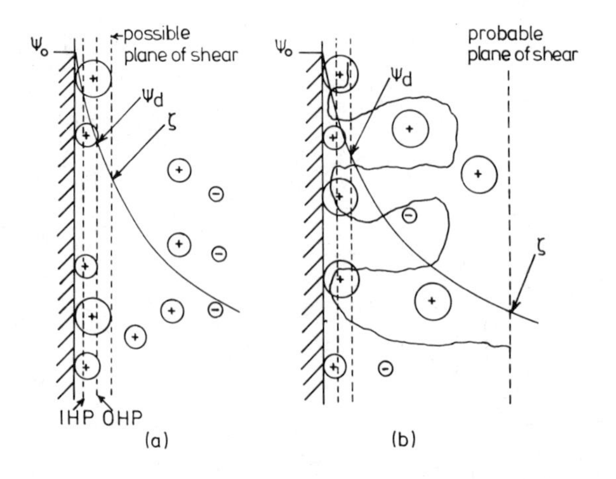


**Figure 55.** Schematic representation of the structure of the electrical double layer in the (a) absence and (b) presence of adsorbed homopolymer.

ization of any surface groups or the adsorption of potential-determining ions. There could also be some displacement of both specifically adsorbed counter-ions or oriented water dipoles. This will lead to a change in the thickness of the inner layer, and the permittivity of the diffuse (outer) layer. Thus, to evaluate $G_E$ under these conditions requires knowledge of how all these effects vary with polymer coverage. Not all of these effects vary independently and, hence, to account for them in calculating $G_E$ becomes a formidable task for which no comprehensive theory is as yet available. However, as an illustration, we will show below how the adsorbed layer leads to a change in $\psi_d$ and hence in $G_E$.

The double-layer interaction for two spheres may be expressed in an approximate form using Derjaguin's expression, Eq. (33). In the absence of an adsorbed polymer layer, when $\kappa a >> 1$,

$$G_E = 2\pi \varepsilon\varepsilon_0 a\psi_0^2 \ln [1 + \exp(-\kappa H_0)] \tag{219}$$

or allowing for the thickness of the inner layer,

$$G_E = 2\pi \varepsilon\varepsilon_0 a\psi_d^2 \ln [1 + \exp(-\kappa(H_0 - d)] \tag{220}$$

where $\psi_d$ is the potential at the Stern or outer Helmholtz plane (OHP), d is the thickness of the inner layer, $\varepsilon$ is the dielectric constant of the diffuse double layer, and $\varepsilon_0$ the permittivity of free space. In Eq. (220), it is the common practice to replace $\Psi_d$ by the zeta potential ($\zeta$), since the shear plane is very close to the OHP. However, when there is an adsorbed polymer layer this cannot be justified, since the shear plane will now be related to the thickness of the adsorbed layer and to whether this behaves as a free-draining or non-free-draining layer. Thus, in this case, it is necessary to establish a relationship between $\psi_d$ and $\zeta$ in order to calculate $G_E$. In the case where the presence of polymer does not cause any distortion of the diffuse double layer, $\psi_d$ and $\zeta$ are related through the Gouy-Chapman equation

$$\tanh \frac{ze\zeta}{4kT} = \tanh \frac{ze\psi_d}{4kT} \exp[-\kappa(\delta - d)] \tag{221}$$

where $\delta$ is the thickness of the adsorbed layer. Thus, if $\delta$ is known, e.g., from hydrodynamic measurements, and an estimate for d can be made, it is possible to calculate $\psi_d$ from the measured $\zeta$. However, for those adsorbed polymers where the segment density in the adsorbed layer is large, distortion of the diffuse layer takes place and Eq. (221) is no longer valid. In this case, a new model of the double layer is required for the evaluation of $\psi_d$. Such a theory is at present not available.

### D. Modification of $G_A$ Owing to the Presence of an Adsorbed Layer

An analysis of the effect of adsorbed layer on $G_A$ was first given by Vold [231], who derived the following expression for particles covered with an adsorbed layer of thickness $\delta$ (see Fig. 56):

$$G_A = -\frac{1}{12}[H_s(A_{22}^{1/2} - A_{33}^{1/2})^2 + H_p(A_{33}^{1/2} - A_{11}^{1/2})^2 + 2H_{ps}(A_{22}^{1/2} - A_{33}^{1/2})(A_{33}^{1/2} - A_{11}^{1/2})] \tag{222}$$

where $H_s$ is the Hamaker function for two spheres of radius $a + \delta$ with their surfaces (of adsorbed layers) separated by a distance $\Delta$. $H_p$ is the Hamaker function for spheres of radius a with their surfaces separated by a distance $b = \Delta + 2\delta$. $H_{ps}$ is the Hamaker function for a sphere of radius a and one or radius $a + \delta$ separated by a distance $\delta + \Delta$. Using the Hamaker equation (see Sec. II.B.1),

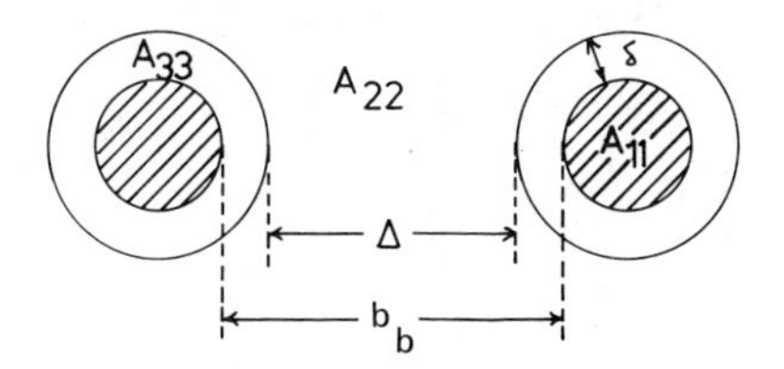


**Figure 56.** Model for the calculation of $G_A$ in the presence of adsorbed layers after Vold [231].

$$G_A = -\frac{1}{12}\left[(A_{22}^{1/2} - A_{33}^{1/2})^2\left(\frac{a+\delta}{\Delta}\right) + (A_{33}^{1/2} - A_{11}^{1/2})\left(\frac{a}{\Delta + 2\delta}\right) + \frac{4a(A_{22}^{1/2} - A_{33}^{1/2})(A_{33}^{1/2} - A_{11}^{1/2})(a+\delta)}{(\Delta+\delta)(2a+\delta)}\right] \tag{223}$$

As an illustration for the effect of adsorbed layer thickness on $G_A$, Otewill [232] calculated $G_A$ for thickness of 0 to 4 nm, taking $A_{11} = 2 \times 10^{-19}$J, $A_{22} = 6 \times 10^{-20}$J and $A_{33} = 5.6 \times 10^{-20}$J. The results are shown in Fig. 57. Vincent [233] has extended Vold's treatment for other geometries, taking retardation effects into account. A schematic diagram showing the effect of $A_{33}$ and $G_A$ is shown in Fig. 58 for the case $A_{11} > A_{22}$. It can be seen that, depending on the value of $A_{33}$, the net effect of adding the adsorbed layer can be to decrease or increase $G_A$. However, in this approach the distance which is fixed is the distance b and $\Delta$ between the outermost surface of the bare and covered particles, respectively. However, if one keeps the distance r between the particle centers constant, then one should consider changes in $G_A$ relative to the value of $G_A$ rather than $G_A^0$ (see Fig. 57). In this case, the effect of adding adsorbed polymer is always to increase $G_A$, except for a very small region of $A_{33}$ values close to and slightly less than $A_{22}$, where $A_{22} < A_{11}$.

For an adsorbed polymer layer, two additional features must be considered. First, the adsorbed layer will be composite, in that it will consist of both polymer segments and solvent molecules, and second, the segment density will not be

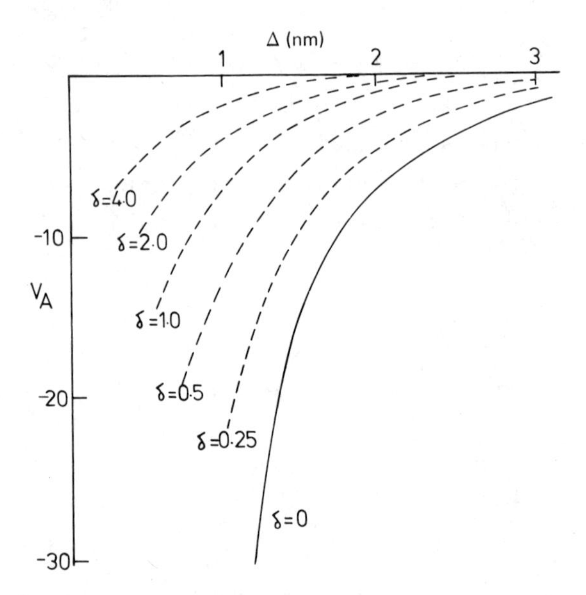


Figure 57. Variation of $V_A$ with $\Delta$ at various values of $\delta$.

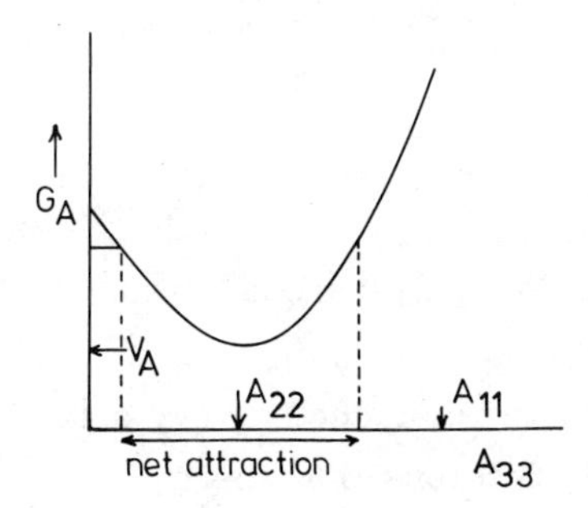


Figure 58. The effect of the Hamaker constant of the adsorbed layer on the van der Waals attraction.

uniform throughout the layer. Both of these effects have been discussed by Vincent [233] and some theoretical $G_A$ curves in the presence of an adsorbed homopolymer were given, assuming an exponential (Hoeve) type distribution of segments. For comparison, the calculations based on a uniform segment distribution were also shown.

## E. Total Interaction Free Energy

For simplicity, we will consider the case where $G_E$ is negligible and also where the adsorbed polymer layer has no effect on $G_A$, i.e., $A_{33} = A_{22}$. In this case $G_A$ for two flat plates is given by

$$G_A = -\frac{A}{12\pi b^2} \tag{224}$$

and

$$G_i = G_{VR} + G_M - \frac{A}{12\pi b^2} \tag{225}$$

Substituting for the value of $G_{VR}$ and $G_M$ from Eqs. (196) and (215),

$$G_i = 2\nu kTV(\bar{i},d) + 2\left(\frac{2\pi}{9}\right)^{3/2} \nu^2 kT\bar{i}(\alpha^2 - 1)M_{(\bar{i},d)} \quad \frac{A}{12\pi b^2} \tag{226}$$

Thus, the main parameters determining the free energy of interaction $G_i$ on the approach of two particles having an adsorbed polymer layer are

i. $\bar{i}$, the average number of segments per loop (or tail), or rather the mean square loop size $\bar{i}\ell^2 = \langle r^2\rangle_0\alpha^2$, where $\langle r^2\rangle_0$ is proportional to the size of the loops

ii. $\nu$, the number, or $\omega$, the mass, of adsorbed loops (or tails) per unit area ($\omega = \nu M/N_a$)

iii. $\chi$, the "quality" of the solvent expressed in terms of the expansion factor $\alpha$ [see Eq. (211)]

iv. The mode of attachment of the macromolecules

v. A, the Hamaker constant

Some calculations were made by Hesselink et al. [230] for polystyrene adsorbed on a flat surface. The results are shown in Fig. 59.

At large values of b, $G_A$ predominates, but at shorter distances a very steep repulsion prevents further approach. Unlike the DLVO interaction energy-distance curve, only one minimum $G_{min}$ is found even at very low coverages ($\omega = 10^{-10}$ g cm$^{-2}$) over the whole range of b. $G_M$ starts at higher values of b than does $G_{VR}$. This is logical since osmotic repulsion starts as soon as the polymer layers overlap, whereas the volume restriction effect starts only after some degree of interaction between the adsorbed layers has taken place. The role of $G_{min}$ is paramount, as we shall see in the next section, in controlling the stability-flocculation behavior of sterically stabilized dispersions and emulsions.

## F. Flocculation of Sterically Stabilized Emulsions

In order that a dispersion or emulsion, stabilized by adsorbed polymer, remain deflocculated, the following criteria must be fulfilled:

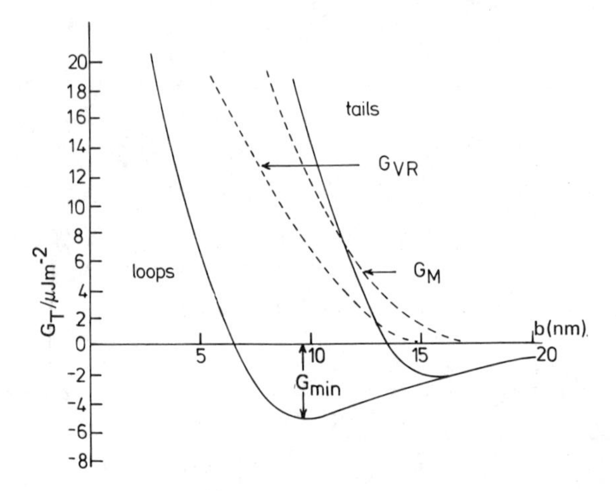


Figure 59. Total interaction free energy between two flat plates having adsorbed tails or loops: $A = 10^{-20}$ J, $\alpha = 1.2$, $\omega = 2 \times 10^{-8}$ g cm$^{-2}$, M = 6000, $\langle r^2\rangle_0^{1/2} = 52$ nm, area per chain = 50 nm$^2$. (From Ref. 226, reprinted with permission from F. T. Hesselink, J. Phys. Chem. 75:2100. Copyright 1971 American Chemical Society.)

i. There should be full coverage (by the adsorbed polymer) of the particle or droplet interface. If this is not the case, bridging flocculation may then arise, i.e., polymer molecules may become simultaneously adsorbed on the interfaces of both droplets or particles. Alternatively, "bare patches" on the two surfaces may come into direct contact, leading to coagulation (or even coalescence).

ii. The polymer must be firmly anchored at the interface. Homopolymers are not usually the most effective structures in this respect. For emulsions, AB block copolymers are particularly suitable, where A is soluble in one liquid phase and B is soluble in the other. If the polymer is not strongly adsorbed at the interface, desorption may occur during a droplet collision.

iii. The polymer layer must be thick enough that $G_{min}$ (see Sec. VIII.E) is negligibly small. A fuller discussion of the relationship between stability and $G_{min}$ is given later in this section.

iv. The stabilizing moiety (e.g., that part of an AB block copolymer which is soluble in the external phase) must be in a good solvent environment. If $\chi > 0.5$, then $G_M$ [see Eq. (215)] becomes *negative*, i.e., the steric interaction becomes a net attraction.

We now turn to a discussion of the role of $G_{min}$ in determining the stability-flocculation behavior of dispersions and emulsions. It is convenient to divide the discussion into three sections: (1) *high* molar mass, terminally anchored polymer chains; (2) *low* molar mass, terminally anchored chains; (3) multipoint anchored chains (e.g., adsorbed homopolymer having a loop and train type configuration at the interface).

### 1. *High Molar Mass, Terminally Anchored Chains*

Here $G_M$ dominates the steric interaction, at least at low degrees of interpenetration. Indeed, Napper [234] has argued that one may neglect any contribution from $G_{VR}$ until $b < \delta$ (Fig. 59), i.e., when the polymer layer on one interface comes into direct contact with the second interface itself. This argument should hold for droplets as well as particles, although one has to remember that droplets may well become distorted (flattened) in the interaction region.

For high-molar-mass chains, the polymer adsorbed layer is usually sufficiently thick that one can safely also neglect any contribution from $G_A$ to the total interaction. Thus, in terms of Eq. (225), $G_i$ is simply given by $G_M$. Inspection of Eq. (215) shows that $\chi$ is the main parameter controlling the stability. For a given polymer-solvent system, $\chi$ may be conveniently changed by varying the temperature or adding a nonsolvent to the external phase.

Napper [235] has made extensive studies of the flocculation of sterically stabilized dispersions, using high molar mass stabilizers (generally, $M > 10^4$), and

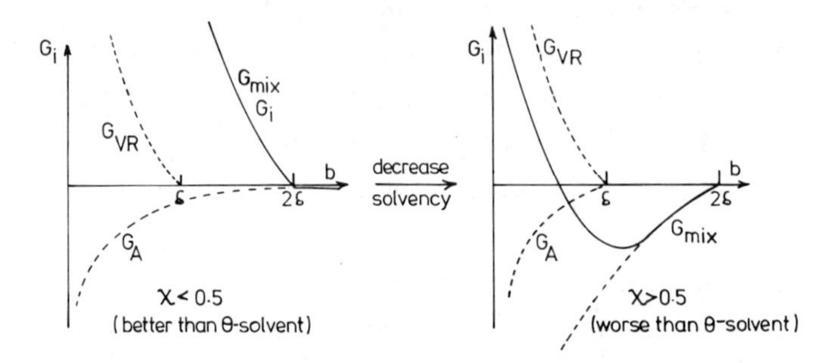


**Figure 60.** The interaction free energy-separation curves for two droplets stabilized by high-molar-mass polymer.

evaluating the effect of changing $\chi$. This is shown schematically in Fig. 60. A significant value of $G_{min}$ is attained when $\chi > 0.5$, i.e., just beyond $\Theta$ conditions for the stabilizing polymer. In this way Napper was able to demonstrate a close correlation between the critical flocculation temperature (CFT) of a dispersion and the corresponding $\Theta$ temperature of the stabilizing polymer, and also between the critical flocculation volume fraction (CFV) of a dispersion and the volume fraction of added nonsolvent required to give a $\Theta$-solvent mixture for the stabilizing polymer.

With regard to the CFT, some dispersions flocculate on cooling and others on heating. Generally speaking (but *not* always), the former occurs when a nonaqueous solvent is the external phase, while the latter occurs when water is the external phase. This behavior mirrors the temperature phase diagrams for the polymer plus solvent system in question. Everett and Stagemann [236] have recently given one example of a system which shows a CFT both on heating *and* on cooling. This was a poly(methylmethacrylate) dispersion in various *n*-alkanes, stabilized by a polystyrene/poly(dimethylsiloxane) block copolymer. The phase diagrams for the corresponding poly(dimethylsiloxane) plus alkane systems show both an upper and a lower consolute temperature.

There has not been a great deal of work to date on the application of these ideas to emulsion systems. One recent example, however, is given in a paper by March and Napper [237]. These authors have investigated the stability-flocculation behavior of O/W emulsions stabilized by block and graft copolymers. In particular, they studied toluene droplets dispersed in 0.39 mol $dm^{-3}$ aqueous $MgSO_4$ solutions and stabilized by poly(ethylene oxide). They monitored the stability of the emulsion in terms of the volumes of the cream layer. A stable emulsion would show little sign of creaming within a given period of time (typically 2 h), whereas when flocculation occurred a cream layer rapidly formed. Fig-

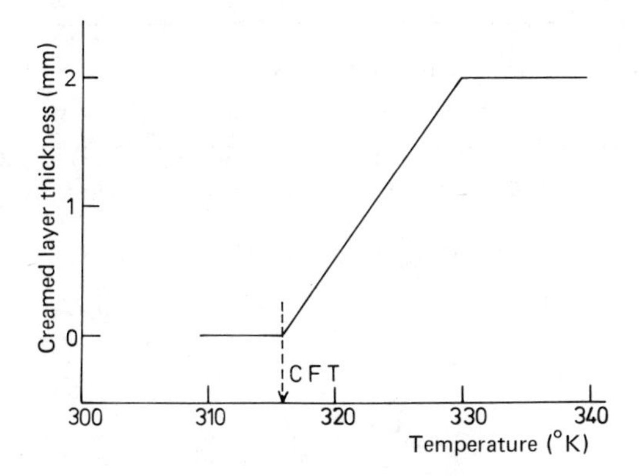


**Figure 61.** Location of the CFT for a sterically stabilized emulsion [257].

61 shows typical data obtained and the location of the CFT, which was taken to be 318K in this particular case. It seems that the observed flocculation was reversible, i.e., the emulsion could be readily redispersed on cooling below the CFT. This suggests the absence of any coalescence in the cream layer. This was confirmed by microscopic studies; these showed that the droplets retained their spherical shape in the cream layer, although there was some evidence of flattening in the contact zones between the droplets.

Various AB block copolymers of poly(ethylene oxide) (PEO), poly(vinylacetate) (PVAc), poly(vinylalchohol) (PVA), polyacrylamide (PAM), poly(acrylic acid) (PAA), poly(laurylmethacrylate) (PLM), and polystyrene (PS), in various combinations were tried. In all the cases studied, a close correspondence between the CFT and the $\Theta$ temperature was found. The results are set out in Table 7.

### 2. *Low Molar Mass, Terminally Anchored Chains*

In this case one cannot neglect the roles played by $G_{VR}$ and $G_A$, since the segment concentration in the adsorbed layer is now significantly high. Thus, the modified equations for $G_A$ allowing for the presence of the adsorbed layer, as discussed in Sec. VIII.D, must be used.

Because $G_{VR}$ now tends to come into play almost at the same place as $G_M$ (i.e., at $b \sim 2\delta$, Fig. 62), the changes in $G_M$ resulting from change in solvency (i.e., the $\chi$ parameter) for the stabilizing polymer tend to be somewhat masked. However, changes in solvency also lead to changes in $G_{VR}$ and $G_A$. This occurs because, generally speaking, a decrease in solvency leads to a contraction in the thickness of the adsorbed layer (and consequently an increase in average seg-

Table 7. CFT Values for Emulsions[a]

| Type | Stabilizing Moiety | Anchor polymer | Internal phase | External phase | CFT (K) | Θ(K) |
|---|---|---|---|---|---|---|
| O/W | PEO | PVAc | Toluene | 0.39 mol $dm^{-3}$ $MgSO_4$ | 318 ± 2 | 318 ± 2 |
| O/W | PVA | PVAc | Toluene | 2.0 mol $dm^{-3}$ NaCl | 297 ± 2 | 299 ± 2 |
| O/W | PAM | PVAc | Toluene | 2.1 mol $dm^{-3}$ $(NH_4)_2SO_4$ | 297 ± 2 | 298 ± 2 |
| O/W | PAA | PVAc | Toluene | 0.2 mol $dm^{-3}$ HCl | 298 ± 2 | 287 ± 5 |
| W/O | PLM | PAA | Water | *n*-Pentanol | 301 ± 3 | 303 ± 5 |
| W/O | PS | PEO | Water | Cyclohexane | 308 ± 3 | 307 ± 1 |

[a]For key to polymers see text.
*Source:* Ref. 237.

ment density). The result of this is that there is an increase in $G_{min}$ on decreasing the solvency, but the change is more gradual than for high-molar-mass polymers, and, moreover, the correlation of the CFT with the Θ temperature may no longer hold. A typical example of the behavior of dispersions stabilized by low-molecular-mass, terminally anchored chains is the work of Cowell, Li-In-On, and Vincent, [242] who studied the stability-flocculation behavior of (neutral) polystyrene particles with terminally attached poly(ethylene oxide) chains of molar mass 750 and 2000, dispersed in aqueous $MgSO_4$ solutions. Here the CFT values were found to be significantly lower than the corresponding Θ temperatures for poly(ethylene oxide) in these aqueous solutions. A schematic representation of the effect of solvency (e.g., increasing temperature in the case of poly(ethylene oxide) in aqueous electrolyte solutions) on $G_{min}$ is shown in Fig. 62. This figure should be contrasted with Fig. 60 for *high*-molar-mass, terminally anchored chains. These two figures, of course, represent extreme conditions, and are only meant to be schematic. There is no sudden change in the situation on going from low- to high-molar-mass chains. The main point we wish to emphasize is that, whereas $G_{min}$ always tends to increase in depth on decreasing the solvency for the stabilizing polymer, the lower the molar mass of the stabilizing terminally

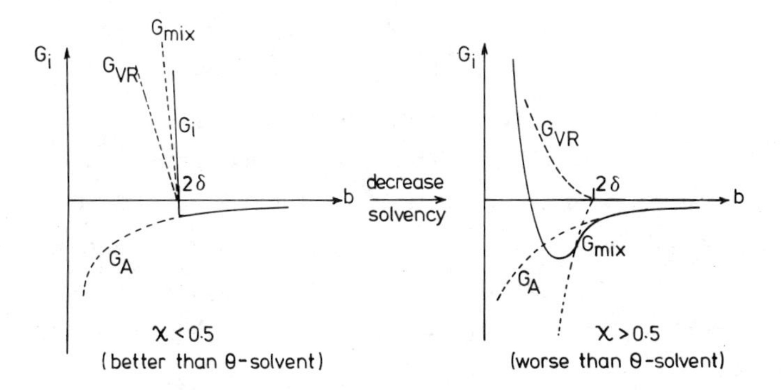


Figure 62. The interaction free energy-separation curves for two droplets stabilized by *low*-molar-mass polymer.

anchored chains the more gradual the change in $G_{min}$, and the less likely is there to be any correlation between the CFT and the Θ temperature of the corresponding polymer in solution.

### 3. *Adsorbed Homopolymers/Multipoint Anchored Polymers*

Here the situation is even more complex, and it is difficult to make general predictions of what may be expected. In a sense, the situation now lies somewhere between the two extremes depicted in Figs. 60 and 62. The average segment density in the adsorbed layer may be quite low, but because most of the segments are likely to be in loops rather than in tails significant interpenetration without compression is unlikely. Thus, although the effect of $G_A$ may again be small, $G_{VR}$ may still be important, and, therefore a close correlation of the CFT with the Θ temperature may not occur. One example of this was demonstrated by Dobbie et al. [238], who found CFT values well in excess of the corresponding Θ-temperature for aqueous polystyrene lattices stabilized by adsorbed polystyrene/poly(ethylene oxide) block copolymers at low coverages, where it was supposed that the poly(ethylene oxide) chains were adsorbed in a loop/train configuration.

In all that has been discussed so far it has been implicitly assumed that the concentration of free polymers in the continuous phase is negligible small. Where this is not the case then flocculation may arise as a result of either of two different effects. First, if the CFT, in the absence of free polymer, lies above the lower consulute temperature (in the case of an aqueous continuous phase) or below the upper consolute temperature (in the case of a non-aqueous continuous phase), then flocculation may occur if the cloud point curve (i.e., the temperature-concentration phase diagram for the polymer solution) is crossed on increasing the polymer concentration. An example of this is the work of Lambe et al. [239] for aqueous polystyrene lattices stabilized by adsorbed poly(vinyl alcohol). Second, even in a good solvent environment, far from the cloud-point curve, flocculation may occur when the polymer concentration in the continuous phase reaches a value where the polymer chains are overlapping with each other, i.e., around c*. This effect was demonstrated by Vincent and co-workers [240-242], and an explanation has been offered by Luckham et al. [241]. These authors have also shown that beyond c** the system tends to stabilize again. A complete explanation for this phenomenon has not yet been given, however.

Another general feature of sterically stabilized dispersions is that where $G_{min}$ is not too deep (less than a few kT, say) then the CFT may be dependent on the particle or droplet concentration. This feature has been discussed by Cowell, Li-In-On, and Vincent [242] following some earlier work by Long et al. [1]. Cowell et al. [242] discussed the stability of dispersions containing weakly interacting particles in terms of the free energy of flocculation $\Delta G_f$. It was suggested that $\Delta G_f$ may be conveniently split into two contributions

$$\Delta G_f = -T\ \Delta S_{hs} + \Delta G_i \tag{227}$$

Here $\Delta S_{hs}$ is the decrease in configurational entropy of the whole system accompanying flocculation. $\Delta S_{hs}$ has been evaluated for hard-sphere systems by Percus and Yevick [243] and their equation has been adopted in the case of microemulsion systems by Overbeek [244]. Per particle or droplet, $\Delta S_{hs}$ decreases as the volume fraction or number concentration of particles or droplets increases. $\Delta G_i$ is the change in free energy, per particle or droplet, resulting from its interaction with its neighbors in a floc. $\Delta G_i$ is, therefore, a function of $G_{min}$. Because both $\Delta S_{hs}$ and $\Delta G_i$ are negative, it is the subtle interplay of these two terms which determines the overall sign of $\Delta G_f$, as given by Eq. (227). When $\Delta G_f$ is positive the dispersion will be thermodynamically stable, whereas when it is negative, weak reversible flocculation will be observed. This analysis predicts, for an emulsion or dispersion at a given temperature, a critical droplet or particle number concentration (or volume fraction), below which the system is stable, but above which flocculation occurs. When $G_{min}$ becomes large, then $\Delta G_i$ tends to dominate $\Delta G_f$ and the critical number concentration becomes so low as to be insignificant. The system is then thermodynamically unstable at all practical concentrations.

## G. Direct Measurement of Particle Interactions

One of the first direct measurements of the magnitude of steric repulsive forces came from the work of Andrews et al. [245]. They studied, using a capacitance technique, the thickness of nonaqueous black films, formed from solutions of glycerol monooleate of varying chain length. The films were supported between bulk aqueous solutions of sodium chloride. The adsorption of glycerol monooleate was calculated from interfacial tension measurements. A dc potential applied across the films resulted in their compression and thinning. The equilibrium thickness of the films decreased with increasing hydrocarbon chain length of the solvent from $C_9$ to $C_{16}$, i.e., as the chain length of the stabilizing monooleate chains was approached. At zero applied potential the film thickness in *n*-decane was found to be 4.4 nm, and this decreased to 3.6 nm on applying a potential of 0.4 V. On increasing the potential, the film took a period of minutes to hours to reequilibrate. This time was longer, the greater the hydrocarbon chain length of the solvent. At equilibrium, the net force on the film is zero, so that

$$F_R + F_A + F_S = 0 \tag{228}$$

where $F_S$ is the steric repulsion force, $F_A$ the London-van der Waals attraction force, and $F_R$ is the applied electrical compression force, given by

$$F_R = \frac{CV^2}{2b_h} \tag{229}$$

where C is the specific capacitance, V the applied voltage, and $b_h$ is the thickness of the hydrocarbon core, that is, equal to the measured thickness b minus twice the thickness of a layer comprising the polar groups; the latter was taken to be $2 \times 0.45$ nm.

$F_A$ is related to the thickness b of the film by the Hamaker equation

$$V_A = -\frac{A}{6\pi b^3} \tag{230}$$

where A is the Hamaker constant. Combining Eqs. (228) to (230) leads to the following expression for $F_S$:

$$F_S = \frac{A}{6\pi b^3} + \frac{CV^2}{2b_h} \tag{231}$$

A plot of $F_S$ versus b is shown in Fig. 63. The change in free energy $\Delta G$ as the film thins is given by

$$\Delta G_T = \frac{A}{12\pi b^2} - \int_{\infty}^{b} F_S \, db \tag{232}$$

Variation of $\Delta G$ with b is also shown in Fig. 63. Both the change in $F_S$ and $\Delta G$ with b follow the trends predicted by the Hesselink et al. [230] theory. The

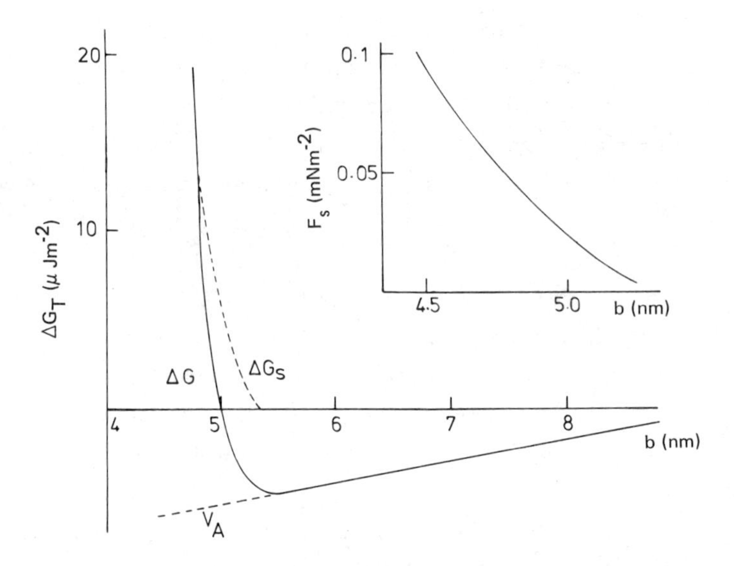


**Figure 63.** Variation of $F_S$ (insert) and $\Delta G_T$ with equilibrium film thickness of glycerol monooleate/decane films between two bulk aqueous saturated sodium chloride solutions, ($A = 3.48 \times 10^{-21}$J). (Data from Ref. 245.)

thicker films formed with the lower molar mass solvents, in the absence of an externally applied potential, contain a larger volume fraction of solvent and are more easily compressible. Calculations of the steric repulsion using Fischer's model [228] showed that the segment density in the overlap region must be very small compared to the average segment density in the film. Since for the thick film, the measured values of b are roughly twice the fully extended stabilizer chain dimensions, it must be the case that at any given time, only a small fraction of the oleate chains are in fact in the fully extended state. This result implies that even for relatively small surfactant molecules the assumption of uniform segment density in the adsorbed layer is invalid.

Recently, van Vliet and co-workers [246,247] have measured the equilibrium thickness and rate of thinning of films stabilized by poly(vinyl alcohol) (PVA) as a function of the hydrostatic pressure on the films, by measuring the intensity of reflected light from the film. From a knowledge of the hydrostatic pressure, $F_H$, and the van der Waals attraction $F_A$, it is possible to calculate the variation of the steric force $F_S$ with film thickness from the relation

$$F_A + F_H + F_S = 0 \tag{233}$$

Some results obtained by van Vliet and co-workers are shown in Fig. 64 for two PVA samples (PVA 205 and PVA 217) with molar masses of 27,000 and 86,000,

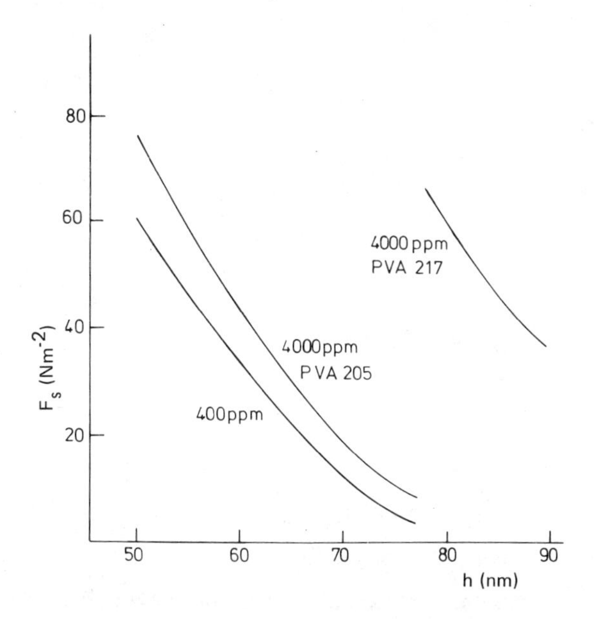


**Figure 64.** Steric repulsion of PVA-stabilized films as a function of thickness. (From Refs. 246 and 247.)

respectively, and a degree of hydrolysis in each case of 87 to 89%. As expected, the steric repulsion becomes stronger when the two adsorbed polymer layers approach each other. There is a pronounced effect of the molar mass of the PVA on the distance at which the adsorbed layers start to interact, but the rate of change of $F_S$ with distance is not strongly dependent on the molar mass. The dependence of b, at a given $F_S$, on $M_V$ is of the order $b \sim M_V^{0.3}$. The thickness $b_0$ at which the two adsorbed layers start to interact can be obtained by extrapolating the curves in Fig. 64 to $F_S = 0$. A linear extrapolation is possible when $b^{-2}$ or $b^{-3}$ is plotted against $F_S$. The thickness $b_0$ was found to be of the order of 76 to 77 nm for the PVA sample with $M_v = 27{,}000$ (independent of PVA concentrations over the range 400 to 4000 ppm) and 100 nm for $M_v = 86{,}000$. However, these values of $b_0$ are lower than those obtained from ellipsometric measurements.

The free energy of repulsion $G_S$ between two adsorbed layers versus distance curve can be calculated by graphical integration of the force-distance curve. The results are shown in Fig. 65. These free energy-distance curves were compared with those computed from the theory of Hesselink et al. [230], with the additional consideration of the presence of long tails (see Fig. 66).

The interaction between long tails is particularly important at the periphery of the adsorbed layers. It does not hold nearer to the interface itself, because of the lateral crowding by loops. $G_S$ was calculated by considering the case where a significant proportion of polymer segments are adsorbed in tails. The

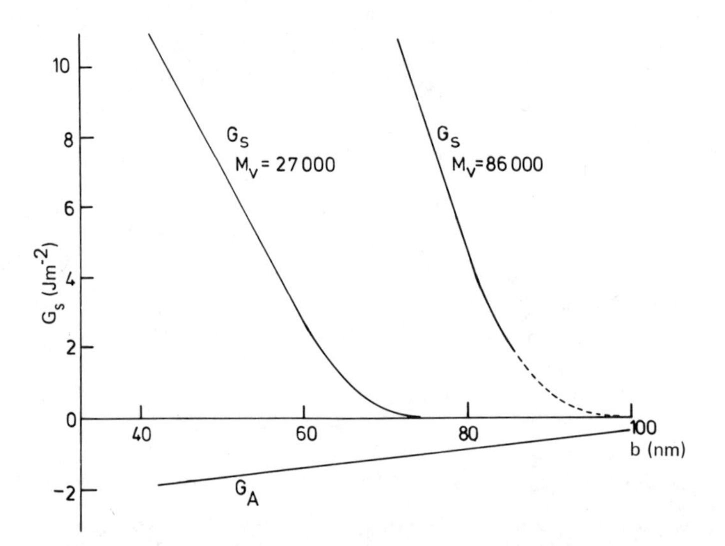


**Figure 65.** Free energy of repulsion and van der Waals attraction energy-distance curves for PVA-stabilized liquid films ($A = 4.38 \times 10^{-20}$ J). From Ref. 246, 247.)

results are shown in Fig. 67. These calculations illustrate the importance of tails in determining the steric repulsion in the outer part of the adsorbed layer. They also clearly show why steric repulsion between two adsorbed polymer layers already starts at large distances of separation and increases relatively slowly with decreasing separation.

Although the above investigations on liquid films give a great deal of information on the steric repulsion between two polymer layers, they do not represent

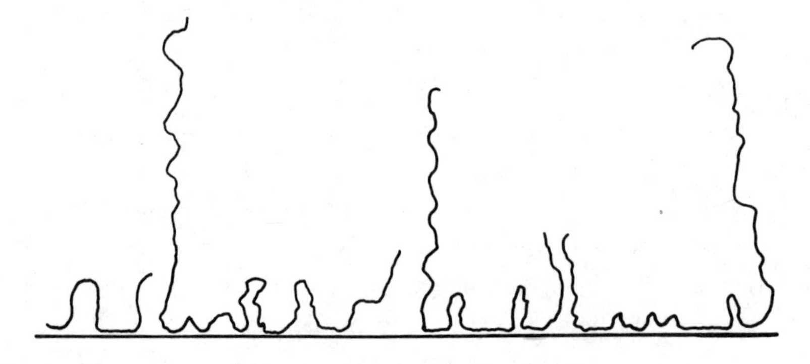

**Figure 66.** Possible model of a PVA layer adsorbed at an aqueous solution/air interface, illustrating the presence of long tails.

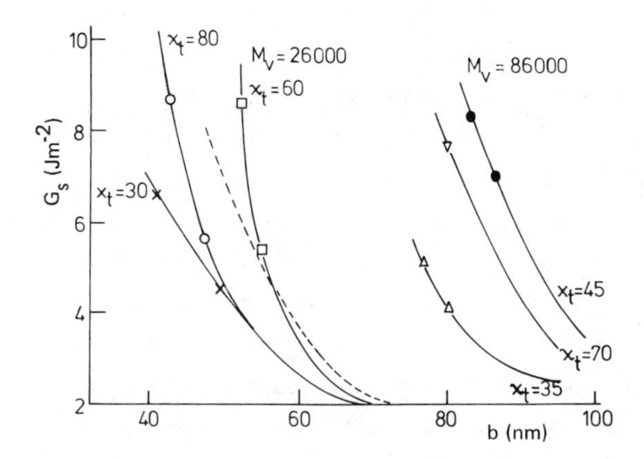


Figure 67. $G_s$-b curves. Dashed curves experimental results. Full curves calculated according to Hesselink et al. [230] at various percentages of polymer segments $x_t$ in tails. x□△● one tail per molecule. ○▽ two tails per molecule. (From Ref. 247.)

the real situation in a disperse system consisting of particles or droplets. The first attempt to measure the repulsive force between particles having adsorbed polymer layers was made by Doroszkowski and Lambourne [248] who measured the compression force for latex particles spread on a heptane/water interface using a surface balance. The latex particles were stabilized with an AB graft copolymer where the A (anchoring) group was poly(methylmethacrylate) and the B (soluble) group was poly(hydroxystearic acid) (PHS) with a number-average molecular weight of 1600. All free unadsorbed polymer in the continuous phase of the latex was removed, and it was ensured that the particles spread at the interface were on the heptane side. A series of latices of different sizes (0.14-0.49 μm) were studied. The average area per adsorbed stabilizer molecule was of the order of 5.4 $nm^2$. The interparticle repulsion-distance curves showed that repulsion starts at a distance of separation of 20 to 29 nm, indicating an adsorbed layer thickness of the order of 10 to 14.5 nm. This is higher than the fully extended chain length of the PHS chain (~10 nm), indicating some distribution of chain lengths.

The work done in compressing the adsorbed layers of the latex particles could be calculated by integrating the experimental pressure-area curves, assuming that the adsorbed layer on each particle interacts with that from each of its six nearest neighbors in the same plane. The results were then compared with theoretical calculations based on the simple Fischer model [228], i.e., assuming uniform segment density in the adsorbed layer. The agreement between theory and experiment was remarkably good, in spite of the neglect of the volume restriction term. This would support the argument that in this particular system the mixing term contributes more strongly to the free energy of repulsion.

The above experiments give the repulsive force between the particles in two dimensions. Measurements in three dimensions have been made using a compression cell apparatus by Ottweill and coworkers [249-251]. Latex dispersions (average diameter 155 nm) in a hydrocarbon medium (dodecane) and containing a stabilizer layer of poly(12-hydroxy stearic acid) were compressed in a specically designed osmotic pressure cell and the change in the volume fraction of the dispersion was measured as a function of the applied pressure, at room temperature. Initially, at a volume fraction of 0.47, osmotic equilibrium occurred slowly. However at a volume fraction of about 0.55, the resistance to compression increased considerably with small changes of volume fraction, and at 0.566 very considerable resistance to compression was experienced. Indeed the system started to behave as a material with a large elastic modulus. Thus, a strong repulsive force is found at a volume fraction of 0.566, the onset of repulsive force being between 0.53 and 0.548. Electron microscopy showed the particles to be in a hexagonally close packed array. It was thus possible to calculate the distance between the particle cores and from a knowledge of the bare particle radius, the shell thickness was found to be 7.25 nm. Since the length of the fully extended poly(12-hydroxystearic acid) is about 9.00 nm, it was suggested that some penetration or compression of these layers mush have taken place during the compression process. It was evident from these measurements that the strong repulsive force is caused by the interaction of the poly(12-hydroxystearic acid) chains in the outer shells, the repulsive force being of short range. The results also suggest that the free energy of repulsion becomes finite at about the point of shell contact, increases gradually during interpenetration or compression of the shells and then, as the shells become resistant to compression, the repulsive force rises very rapidly with further small increases in volume fraction. These results are in qualitative agreement with the simple theory of steric stabilization discussed above. The fact that the steric barrier will support a strong compressive force over a long interval of time suggests that, on the short time scale involved in a Brownian encounter, the repulsive force is more than adequate to ensure a stable colloidal dispersion in the medium in question.

Recently, Israelachvili et al. [252] measured the force between two curved mica surfaces with adsorbed poly(ethylene oxide) ($M_w$ = 148.000) layers in 0.04 mol $dm^{-3}$ $MgSO_4$ (a good solvent) as a function of surface separation. The system exhibited large irreversibilities as well as short-term and long-term time-dependent effects. After an arbitrary equilibration time (2 h), these authors could distinguish three interaction regimes, at relatively large (150 to 300 nm), medium

(10 to 150 nm), and short (~5 nm) distances of separation, which they designated as the α, β, and γ regimes, respectively. The repulsive force in the α regime rises sharply and roughly exponentially with distance. It may possibly be associated with the "diffuse interpenetration" domain described by Napper [253]. The repulsive force in the β regime depends on the rate of approach, and it was observed that the surfaces can undergo sudden jumps toward each other, suggestive of the yield characteristic of polymeric gels possessing a network structure undergoing partial collapse. This behavior may be attributed to slow adsorption-desorption processes, including bridging of tails during compression/ decompression, and to slow reconformation of the polymeric network. In the γ regime, the repulsive force increases very steeply with decrease of distance of separation; the polymer is desorbed and is forced out between the surfaces. When significant overlap has occurred, adhesion between the plates takes place, probably as a result of polymer bridging.

## H. Mechanical Properties of Macromolecular-Stabilized Emulsions and Their Resistance to Coalescence

It has long been recognized that the remarkable stability of emulsions containing adsorbed layers of protein, gums, and other macromolecular substances towards coalescence can be attributed to their mechanical properties, namely, the viscosity and elasticity of the interfacial film [254-257]. Biswas and Haydon [258-261] have systematically investigated the rheological characteristics of various proteins, namely, albumin, pepsin, poly(α-L-lysine), and arabinic acid, at the oil/water interface and correlated these measurements with the stability of the oil droplets at a planar oil/protein solution interface.

The viscoelastic properties of the adsorbed films were studied using two-dimensional creep and stress relaxation measurements in a specially designed rheometer. In the creep experiments, a constant torque (expressed in millinewtons per meter) is applied, and the resulting deformation γ (in radians) is recorded as a function of time. The creep recovery is determined by recording the deformation when the torque is withdrawn. In the stress-relaxation experiments, a certain deformation γ is produced in the film by applying an initial stress, and the deformation is maintained constant by decreasing the stress.

Figure 68 represents a typical creep curve for bovine serum albumin films. The curve shows an initial, instantaneous deformation, characteristic of an elastic body, followed by a nonlinear flow that gradually declines and approaches the steady flow behavior of a viscous body. After 30 min, when the external force is withdrawn, the film tends to revert to its original state, with an initial instantaneous recovery followed by a slow one. The original state, however, is not

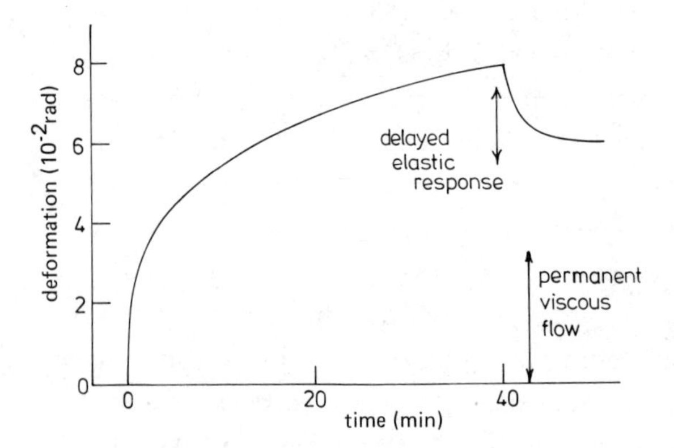


**Figure 68.** Creep behavior of an adsorbed bovine serum albumin film at pH 5.2 at a petroleum ether/water interface under a constant stress of 0.116 dyn $cm^{-2}$.

attained even after 20 h and the film seems to have undergone some flow. This behavior illustrates the viscoelastic property of the bovine serum albumin film. Linear viscoelasticity has been observed over moderate ranges of shear. Moreover, the dynamic shear modulus measured by oscillation experiments was generally higher than that obtained from the ideal region of the stress-strain curves. However, the magnitude of oscillations in the dynamic method is very small and lies appreciably in the regions of very high $d\sigma/d\gamma$.

Biswas and Haydon [258] found a striking effect of the pH on the rigidity of the protein film. This is illustrated in Fig. 69, where the shear modulus

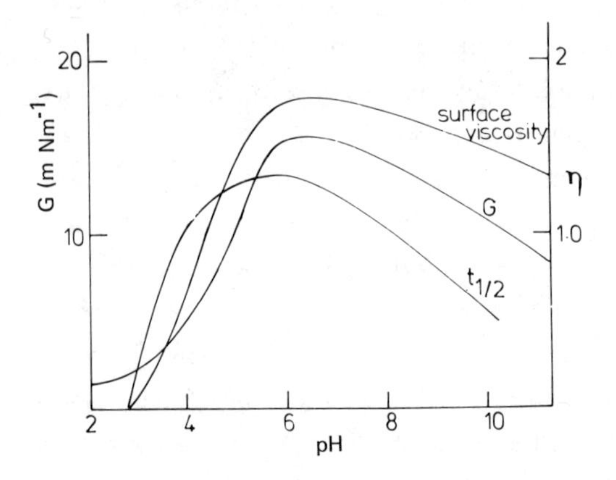


**Figure 69.** Shear modulus, surface viscosity, and half-life of petroleum ether drops resting beneath a plane petroleum ether/0.1 mol $dm^{-2}$ aqueous KCl solution interface.

G and surface viscosity $\eta$ is plotted as a function of pH. The elasticity is seen to be a maximum at the isoelectric point of the protein. Biswas and Haydon [260] then measured the rate of coalescence of petroleum ether drops at a planar oil/water interface by measuring the lifetime of a droplet resting beneath the interface. The half-life of the droplets was plotted against pH, as shown in Fig. 69, which clearly illustrates the correlation with G or $\eta$. In fact, Biswas and Haydon [258] were able to derive an equation relating the time for coalescence $\tau$ with the viscoelasticity and the thickness of the adsorbed film b and the critical distortion of the plane interface under the weight of the drop

$$\tau = \eta \left[ 3C' \frac{b^2}{A} - \frac{1}{G} - \phi(t) \right] \tag{234}$$

where G is the (instantaneous) elasticity, $\eta$ the long-term viscosity (i.e., for infinite time of retardation), $\phi(t)$ is the retarded elastic deformation per unit stress, and 3C' is a critical deformation factor.

Equation (234) predicts that

i. The lifetime of the drop ($\tau$) increases with increase of viscosity of the protective film.
ii. The rate process leading to coalescence is not influenced by the instantaneous elasticity, but this quantity is likely to set a limit on this process through the critical deformation factor 3C'.
iii. The lifetime should depend on the film thickness and vary linearly with $b^2$ if the retarded elasticity $\phi(t)$ is neglected.
iv. $\tau$ should be a fixed and not a fluctuating quantity.

The results obtained by Biswas and Haydon [260] indicate clearly that no significant stabilization occurred in the case of nonviscoelastic films. However, the presence of viscoelasticity is not sufficient to confer stability, particularly when drainage is rapid. For example, it was found that the highly viscoelastic films of bovine serum albumin or pepsin could not stabilize water/oil emulsions; the same was found for pectin and gum arabic films. It was clear that in these cases the drainage of the solvent was rapid even from between two rigid films, e.g., water/oil droplets in the case of bovine serum albumin. In fact, as expected, it is only after solvent drainage has taken place, and the disperse phases are still separated by a film of high viscosity, that enhanced stability occurs. It was concluded from these investigations that, in agreement with the prediction of theory, the experimental requirements for stability to coalescence are the presence of a film of high viscosity (the elasticity being of less importance) and also a film of appreciable thickness. It is also necessary that the main part of the film thickness should be located on the continuous-phase side of the interface.

Nielsen et al. [262] have also correlated the mechanical properties of various macromolecular-stabilized interfaces with droplet stability. In nearly all cases where a large interfacial rigidity or interfacial viscosity was found, the droplets were very stable. However, there were a number of cases where stable droplets could be made but no interfacial viscosity or rigidity could be measured. Polyelectrolytes, such as poly(methacrylic acid) and carboxymethylcellulose, were found to stabilize droplets better in their acid form than as salts; this behavior correlates with the interfacial viscosity behavior but not with the bulk solutions viscosity. It was thus concluded that a stabilizing agent stabilized an emulsion against coalescence most effectively when it was near the point of percipitation from solution.

## IX. Stabilization by Solid Particles

### A. Introduction

It has long been known that very effective stabilization against coalescence can be obtained by using very finely divided solids as emulsifying agents [263]. These emulsions are sometimes referred to as Pickering emulsions; the type of emulsion produced depends on which phase preferentially wets the solid particles [264]. If the particles are wetted more by the water phase, a water/oil emulsion is produced; where the aqueous phase preferentially wets the solid particles an oil/water emulsion is produced. The first attempt to explain the stability of solid stabilized emulsions, in a quantitative manner, was made by van der Minne [265], who showed that under thermodynamic equilibrium conditions, the adsorption of solid particles at the liquid/liquid interface is governed by their contact angle against the solid; an obtuse angle against the solid phase facilitates the stabilization. All subsequent theories for solid-stabilized emulsions centered on the importance of the contact angle at the interface. For example, Schulman and Leja [266] indicated that, in the presence of surface-active agents, oil/water emulsions are obtained with solid powders when the contact angle (as measured through the water phase) is slightly less than 90°. When the contact angle is slightly in excess of 90°, water/oil emulsions are produced. Thus, the contact angle at the three phase (oil/water/solid) boundary is critical in controlling the stabilization by given solid particles. These particles have to be very finely divided, i.e., very small in comparison with the emulsion droplet size. Thus an optimum contact angle is required to collect the particles at the liquid/liquid interface by capillary forces. Solids that are too strongly hydrophilic, such as silica or alumina, will pass into the aqueous phase too quickly, whereas if the solid particles are either made too hydrophobic (e.g., by the adsorption of a very long chain flotation collector), or are intrinsically hydrophobic (e.g., certain carbon blacks), then water/oil emulsions will be produced. This was illustrated, for example, in studies by Takakuwa and Takamori [267].

The mechanism of stabilization of emulsion droplets by the presence of finely divided solids at the liquid/liquid interface is far from being understood at present. Presumably, the presence of solid particles at the liquid/liquid interface plays an important role in preventing the thinning of the liquid film between the droplets. For the solid particles to be effective they should form a continuous monoparticulate film. Contact angle hysteresis is probably important in preventing displacement of the meniscus. For that reason "rough" asymmetric particles such as bentonite clays are more efficient than "smooth" spherical particles. The particles should also form a coherent film at the interface. This is achieved by capillary forces which tend to bring the particles close together at the interface. The smaller the radius of curvature of the meniscus between the particles, the larger the force of attraction; this explains the need for very fine solids.

Whatever the mechanism is, it is obviously necessary for the solid particles to be located at the interface, and this should be governed by the magnitude of the contact angles formed with the two liquid phases. Thus, it is necessary to consider the balance of forces for a solid sphere at the liquid/liquid interface. This process is analogous to froth flotation [6,7], where it is essential for a particle which is completely wetted by the solution around it to be attached to an air bubble of sufficient size to carry it from bulk solution to the surface of the solution. In practice, the particles are small in comparison with the air bubbles and hence one air bubble may have many particles attached to it. Hence the analogy with emulsion droplets stabilized by finely divided solid particles.

## B. Location of Solid Particles at the Liquid/Liquid Interface

Let us assume for simplicity a simple model of spherical particles of radius r, having the same density as the two liquids 1 and 2 (again of equal density), i.e., there are no effects from gravitational forces (see Fig. 70). If the particle is entirely in the liquid L, it has a free energy $G_1$ given by

$$G_1 = 4\pi r^2 \gamma_{1S} \tag{235}$$

where $\gamma_{1S}$ is the corresponding interfacial tension. Similarly, if the particle is completely immersed in liquid 2, it has a surface free energy $G_2$:

$$G_2 = 4\pi r^2 \gamma_{2S} \tag{236}$$

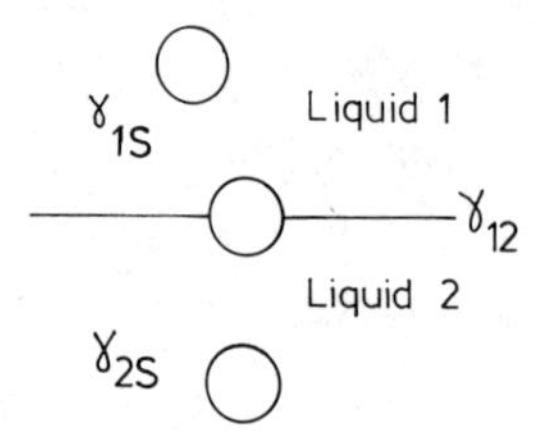


**Figure 70.** Schematic picture of a spherical solid particle at the liquid/liquid interface.

If the particle stays at the interface, it will be partly immersed in liquid 1 and partly in liquid 2. If $A_{1S}$ and $A_{2S}$ are the areas of the sphere immersed in liquids 1 and 2, respectively, then the free energy for the sphere at the interface is given by

$$G = A_{1S}\gamma_{1S} + A_{2S}\gamma_{2S} \tag{237}$$

When the sphere is located at the interface it displaces an area $A_{12}$ of the interface with an interfacial tension of $\gamma_{12}$. At equilibrium, the total surface free energy must be at a minimum, i.e.,

$$dG = A_{1S}\gamma_{1S} + A_{2S}\gamma_{2S} + A_{12}\gamma_{12} = 0 \tag{238}$$

The areas $A_{1S}$, $A_{2S}$, $A_{12}$ can be calculated from simple geometry as shown in Fig. 71. Thus

$$2\pi r dh\gamma_{2S} + (-2\pi r dh)\gamma_{1S} - \pi(2r - 2h)dh\gamma_{12} = 0 \tag{239}$$

which simplifies to

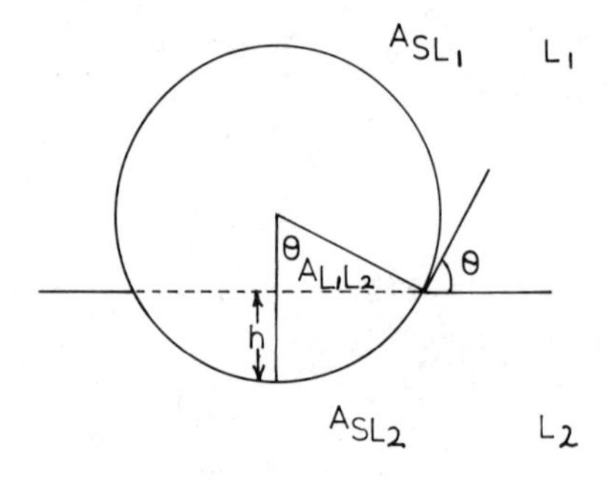


**Figure 71.** Solid sphere at the liquid/liquid interface.

$$\gamma_{1S} - \gamma_{1S} = \left(1 - \frac{h}{r}\right)\gamma_{12} \tag{240}$$

Since

$$\cos\Theta = 1 - \frac{h}{r} \tag{241}$$

then

$$\gamma_{1S} - \gamma_{2S} = \gamma_{12}\cos\Theta \tag{242}$$

which is essentially the Young-Dupré equation for contact angle equilibrium. Thus, a particle will seek a stable position in the interface such that the angle $\Theta$ becomes the equilibrium contact angle. In practice, owing to the influence of gravity, there will be a slight displacement from this position until the net force due to the interfacial tension effects just equals the gravitational force. Particles forming a finite contact angle at the liquid/liquid interface remain there, whereas those completely wetted by either of the liquids will be displaced into that liquid phase.

## C. Mechanism of Stabilization by Solid Particles

It is perhaps possible to understand why solid particles provide a barrier impeding coalescence, if one considers the free energy changes involved when a particle at equilibrium in the interface is displaced to either phase. If the particle is removed into the water phase, the area of the O/W interface is increased by $\pi a^2 \sin^2\Theta$ and an area $2\pi a^2(1 - \cos\Theta)$ of the solid would be transferred from oil to water. The surface-free-energy increase would be given by

$$\Delta G = \pi a^2 \sin\Theta\, \gamma_{OW} + 2\pi a^2(1 + \cos\Theta)(\gamma_{WS} - \gamma_{OS}) \tag{243}$$

However, since

$$\gamma_{OW}\cos\Theta = \gamma_{WS} - \gamma_{OS} \tag{244}$$

then

$$\Delta G = \pi a^2 \sin^2\Theta\,\gamma_{OW} + 2\pi a^2(1 + \cos\Theta)\gamma_{OW}\cos\Theta$$

$$= \pi a^2\gamma_{OW}[\sin^2\Theta + 2\cos\Theta\,(1 + \cos\Theta)] = \pi a^2\gamma_{OW}(1 + \cos\Theta)^2 \tag{245}$$

Similarly, the free-energy increase on transfer of the sphere from its equilibrium position to the oil phase would be

$$\Delta G = \pi a^2\gamma_{OW}[\sin^2\Theta - 2\cos\Theta\,(1 + \cos\Theta)] = \pi a^2\gamma_{OW}\,(1 - \cos\Theta)^2 \tag{246}$$

The difference, being the energy increase on transfer from oil to water, is

$$4\pi a^2 \gamma_{OW} \cos\Theta = 4\pi a^2 (\gamma_{WS} - \gamma_{OS}) \tag{247}$$

Thus, if $\Theta = 90°$, i.e., when the equilibrium position for the particle is half in the water and half in the oil phase, the free energy necessary to take the sphere from the interface to either liquid is the same and $\gamma_{WS} - \gamma_{OS} = 0$; the free-energy change required is provided entirely by the elimination of some O/W interface. Thus, even if the spheres form a close packed layer with $\Theta = 90°$, not all the O/W interface can be eliminated. The maximum proportion which could be eliminated is $\pi/2\sqrt{3} = 0.91$. Thus, although the movement of spheres to an existing interface from either bulk phase lowers the free energy, the creation of new interface to accommodate more spheres involves an increase of free energy. The spheres can considerably decrease the extra free energy which the emulsion must store, compared to the state in which the particles are present in the liquid phases, but they cannot gain for the emulsion a net free-energy advantage. However, once situated at the interface, the solid particles do make coalescence of the droplets more difficult, essentially by keeping the droplets from coming into close contact. For two adjacent droplets, each carrying a close packed array of solid spheres, to coalesce some or all of the solid particles have to escape into either of the bulk liquid phases or, alternatively, the droplets will need to be distorted sufficiently to provide free surface for contact. Each of these processes requires an increase in free energy, providing a barrier to coalescence.

## Symbols

| | |
|---|---|
| $A_1$ | hydrodynamic collision factor |
| A | Helmholtz free energy, also Hamaker constant, also acid number |
| $A_0$ | amplitude |
| C | collision factor |
| D | diffusion coefficient |
| $E_h$ | hydration barrier |
| E | percent of ethylene oxide |
| F | force |
| $F_e$ | electrostatic force |
| G | Gibbs free energy |
| $G_E$ | electrostatic interaction free energy |
| I | light intensity |
| J | creep compliance |
| K | coalescence rate |

| | |
|---|---|
| L | length of container |
| N | number of particles, also number of sites |
| P | probability of film rupture, percent of polyhydric alcohol |
| R | center-to-center separation between two spheres; also radius of curvature |
| $R_C$ | film radius |
| S | entropy; also saponification number |
| T | absolute temperature |
| $U_A$ | van der Waals attraction |
| V | volume of oil phase |
| W | stability ratio; also number of configurations |
| X | electrostatic field |
| Z | valency |
| A | interfacial area |
| a | sphere radius |
| b | film thickness, distance between particles |
| c | concentration |
| d | distance between plates |
| f | frequency |
| g | acceleration due to gravity |
| h | height; also distance |
| i | number of segments |
| K | Boltzmann constant; also rate constant of flocculation |
| l | distance of chain from the surface |
| m | mass; also number of particles; also number of $CH_2$ groups |
| n | number of moles; also number of particles |
| p | pressure |
| r | radius |
| t | time |
| u | mobility |
| v | velocity |
| w | energy barrier |
| x | distance from axis of rotation |
| $\gamma$ | interfacial tension |
| $\sigma$ | thickness of a plate, thickness of interfacial layer, length of a rod |
| $\varepsilon$ | permittivity (dielectric constant); also dilational elasticity |
| $\eta$ | viscosity |
| $\theta$ | fraction of surface covered; also contact angle |
| $\kappa$ | reciprocal "thickness" of the electrical double layer |
| $\mu$ | chemical potential, also refractive index |
| $\pi$ | disjoining pressure |

| | |
|---|---|
| $\pi_E$ | excess osmotic pressure |
| $\rho$ | density; also segment density distribution |
| $\rho_1$, $\rho_2$ | principal radii of curvature of a liquid film |
| $\sigma$ | surface charge density |
| $\tau$ | turbidity; also lifetime of a drop |
| $\tau_{1/2}$ | half life |
| $\chi$ | polymer-solvent interaction parameter |
| $\lambda$ | London interaction constant; also free energy of adsorption per segment |
| $\psi$ | surface potential |
| $\zeta$ | zeta potential |
| $\omega$ | angular velocity |
| $\phi$ | volume fraction |
| $\phi_{(t)}$ | retarded elastic deformation |

## References


1. J. Long, D. W. J. Osmond, and B. Vincent, *J. Colloid Interface Sci. 42*:545 (1973).
2. L. L. Lacey and O. H. Guenther, *Progr. Astronaut. Aeronaut. 52*:495 (1977).
3. B. J. Carroll, in *Surface and Colloid Science* (E. Matijević, ed.), Vol. 9, Wiley-Interscience, New York, 1977, p. 52.
4. H. Reiss, *J. Colloid Interface Sci. 53*:61 (1975).
5. (a) B. V. Derjaguin and L. Landau, *Acta Physiochem. USSR 14*:633 (1941); (b) E. J. W. Verwey and J. Th. G. Overbeek, *Theory of the Stability of Lyophobic Colloids*, Elsevier, Amsterdam, 1948.
6. See, e.g., *Faraday Disc. Chem. Soc. 65*:(1978).
7. T. L. Hill, *Thermodynamics of Small Systems*, Benjamin, New York, 1963.
8. J. T. Davies, in *Progress in Surface Science* (J. F. Danielli, K. G. A. Pankhurst, and A. C. Riddiford, eds.), Vol. 2, Academic Press, New York, 1964.
9. G. G. Stokes, *Phil. Mag, 1*:337 (1851).
10. P. Walstra and H. Cortwijn, *Neth. Milk Dairy J. 29*:263 (1975).
11. W. Rybczynski, *Bull. Acad. Sci. Cracovie*, Ser. A *1911*:40.
12. J. Hadamard, *Compt. Rend. 154*:1735 (1911).
13. V. G. Levich, *Physicochemical Hydrodynamics*, Prentice-Hall, New York, 1962.
14. A. N. Frumkin and V. G. Levich, *Zh. Fiz. Khim. 21*:1183 (1947).
15. R. N. O'Brien, A. I. Feher, and J. Leja, *J. Colloid Interface Sci. 42*:218 (1973).
16. H. L. Greenwald, *J. Soc. Cosmetic Chemists 6*:164 (1955).
17. G. D. Scott, *Nature 188*:908 (1960); S. Yerazunis, S. W. Cornell, and B. Winter, *Nature 207*:835 (1965).
18. J. D. Bernal, *Nature 183*:141 (1959); *185*:68 (1960); J. D. Bernal and J. Mason, *Nature 188*:910 (1960).
19. R. D. Vold and R. C. Groot, *J. Phys. Chem. 66*:1969 (1962); *J. Soc. Cosmetic Chemists 14*:2 (1963); *J. Colloid Sci., 19*:384 (1964); *J. Phys. Chem. 68*:

3477 (1964); *Proc. Int. Congr. Surface Activity (4th)*, Gordon and Breach, London, 1967, Vol. 2, p. 1233.
20. E. R. Garrett, *J. Pharm. Sci. 51*:35 (1962).
21. S. J. Rehfeld, *J. Phys. Chem. 66*:1966 (1962).
22. S. J. Rehfeld, *J. Colloid Interface Sci. 46*:448 (1974).
23. R. D. Vold and K. L. Mittal, *J. Colloid Interface Sci. 38*:451 (1972); *J. Soc. Cosmetic Chemists 23*:171 (1972); *J. Colloid Interface Sci. 42*:436 (1973).
24. A. L. Smith and D. P. Mitchell, in *Theory and Practice of Emulsion Technology* (A. L. Smith, ed.), Academic, New York, 1976, p. 61.
25. F. Sebba, *J. Colloid Interface Sci. 40*:468 (1972).
26. K. J. Lissant and K. G. Mayhan, *J. Colloid Interface Sci. 42*:201 (1972).
27. J. H. de Boer, *Trans. Faraday Soc. 32*:21 (1936).
28. H. C. Hamaker, *Physica 4*:1058 (1937).
29. E. M. Lifshitz, *Zh. Eksper. Teoret. Fiz. 29*:94 (1955).
30. H. B. G. Casimir and D. Polder, *Nature 158*:787 (1946); *Phys. Rev. 73*:360 (1948).
31. J. A. Kitchener and P. R. Musselwhite, in *Emulsion Science* (P. Sherman, ed.), Academic, New York, 1968, Chap. 2.
32. J. H. Schenkel and J. A. Kitchener, *Trans. Faraday Soc. 56*:161 (1960).
33. R. H. S. Winterton, *Contemp. Phys. 11*:559 (1970).
34. J. N. Israelachvili, *Contemp. Phys. 15*:159 (1974).
35. J. N. Israelachvili and D. Tabor, in *Progress in Surface Membrane Science* (J. F. Danielli, M. D. Rosenberg, and D. A. Cadenhead, eds), Academic, New York, 1973, Vol. 7, p. 1.
36. V. A. Parsegian, in *Physical Chemistry: Enriching Topics from Surface and Colloid Science* (H. van Olphen and K. J. Mysels, eds.), Thorex, La Jolla, Calif., 1975, Chap. 4.
37. P. Richmond, *Colloid Science* (Specialist Periodical Reports), The Chemical Society, London, 1975, Vol. 2, p. 130.
38. J. N. Israelachvili, *J. Chem. Soc. Faraday Trans. II, 69*:1729 (1973).
39. G. D. Bell and G. C. Peterson, *J. Colloid Interface Sci. 42*:542 (1972).
40. O. F. Devereux and P. L. de Bruyn, *Interaction of Plane Parallel Double Layers*, MIT Press, Cambridge, Mass. 1963.
41. S. N. Srivastava and D. A. Haydon, *Proc. Int. Congr. Surface Activity (4th)*, Gordon & Breach, London, 1967, Vol. 2, p. 1221.
42. J. Gregory, *Adv. Colloid Interface Sci. 2*:396 (1969).
43. J. Visser, *Adv. Colloid Interface Sci. 3*:331 (1972).
44. Cf. D. J. Shaw, *Introduction to Colloid and Surface Chemistry*, 2d ed., Butterworths, London, 1970, pp. 83 and 148.
45. D. J. Shaw, *Electrophoresis*, Academic, New York, 1969.
46. M. V. Smoluchowski, (a) *Z. Phys. Chem. 92*:129 (1917); (b) *Bull. Acad. Sci. Cracovie 1903*:182.
47. E. Hückel, *Physik. Z. 25*:204 (1924).
48. D. C. Henry, *Proc. Roy. Soc. (London) A133*:106 (1931).
49. P. H. Wiersma, A. L. Loeb, and J. Th. G. Overbeek, *J. Colloid Interface Sci. 22*:78 (1966).
50. R. H. Ottewill and J. N. Shaw, *J. Electroanal. Chem. 37*:133 (1972).
51. (a) G. Wiegner, *Kolloid Z., 58*:167 (1932); (b) H. Müller, *Kolloid Z. 38*:1 (1926); *Kolloid Beih. 26*:257 (1928).
52. A. Watillon and A. Joseph-Petit, Disc. Faraday Soc. *42*:143 (1966).
53. (a) L. A. Spielman, *J. Colloid Interface Sci. 36*:562 (1970); (b) P. Honig, G. J. Roebersen, and P. H. Wiersma, *J. Colloid Interface Sci. 36*:97 (1971).

54. (a) D. L. Swift and S. K. Friedlander, *J. Colloid Sci. 19*:621 (1964); (b) G. M. Hidy, *J. Colloid Sci. 20*:123 (1965); G. M. Hidy, and D. K. Lilly, *J. Colloid Sci. 20*:863 (1965).
55. H. Erying, *J. Chem. Phys. 3*:107 (1935).
56. A. S. C. Lawrence and O. S. Mills, *Disc. Faraday Soc. 18*:98 (1964).
57. N. Fuchs, *Z. Physik 89*:736 (1934).
58. H. R. Kruyt, *Colloid Science*, Vol. I, Elsevier, Amsterday, 1952.
59. H. Reerink and J. T. G. Overbeek, *Disc. Faraday Soc. 18*:74 (1954).
60. R. H. Ottewell and J. N. Shaw, *Disc. Faraday Soc. 42*:154 (1966).
61. G. R. Weise and T. W. Healy, *Trans. Faraday Soc. 66*:490 (1970).
62. D. N. L. McGowan and G. D. Parfitt, *J. Phys. Chem. 71*:449 (1967).
63. S. W. Churchill, G. C. Clark, and C. M. Slieperich, *Disc. Faraday Sco. 32*:192 (1960).
64. D. H. Napper and R. H. Ottewill, *J. Colloid Sci. 19*:72 (1964).
65. H. C. van de Hulst, *Light Scattering by Small Particles*, Wiley-Interscience, New York, 1957.
66. M. Kerker, *The Scattering of Light and Other Electromagnetic Radiation*, Academic, New York, 1969.
67. J. Kratohvil, *Anal. Chem. (Ann. Rev.) 38*:517R (1966).
68. G. Oster, *J. Colloid Sci. 15*:512 (1960).
69. W. J. Pangounis, W. Heller, and A. Jackson, *Tables of Light Scattering Functions for Spherical Particles*, Wayne University Press, Detroit, Mich., 1957.
70. B. Vincent, *Thesis*, Bristol (1969).
71. A. Lips, C. Smart, and E. Willis, *Trans. Faraday Soc. 67*:2979 (1971).
72. A. Lips and E. Willis, *J. C. S. Faraday Trans. I 69*:1226 (1973).
73. B. R. Ware and W. H. Flygare, *Chem. Phys. Lett. 12*:81 (1971).
74. P. N. Pusey, in *Industrial Polymers: Characterisation by Molecular Weight* (J. H. S. Green and R. Dietz, eds.), Transcripts Brooks, London, 1973.
75. B. J. Bernal and R. Pecora, *Dynamic Light Scattering*, Wiley-Interscience, New York, 1976.
76. E. E. Usgiris and F. M. Costachuk, *Nature 242*:77 (1973).
77. K. J. Lissant and K. J. Mayhan, *J. Colloid Interface Sci. 42*:201 (1973).
78. D. R. Eley, M. J. Hey, J. D. Symonds, and J. H. R. Willison, *J. Colloid Interface Sci. 54*:462 (1976).
79. S. S. Davis, T. S. Purewal, and A. S. Burbage, *J. Pharm. Pharmacol. 28*(Supp.):60 (1976).
80. M. van den Tempel, *Rec. Trav. Chim. 72*:433, 442 (1953).
81. S. N. Srivastava and D. A. Haydon, *Proc. Int. Congr. Surface Activity (4th)*, Vol. 2, Gordon and Breach, London, 1967, p. 1221.
82. P. McFadyen and A. L. Smith, *J. Colloid Interface Sci. 45*:573 (1973).
83. R. E. Wachtel and V. K. La Mer, *J. Colloid Sci. 17*:531 (1962).
84. W. I. Higuchi, R. O. Koda, and A. P. Lemberger, *J. Pharm. Sci. 51*:683 (1962); *52*:49 (1963).
85. E. L. Rowe, *J. Pharm. Sci. 54*:260 (1965).
86. H. Mima and N. Kitamore, *J. Pharm. Sci. 55*:441 (1966).
87. P. Becher, S. E. Trifiletti, and Y. Machida, in *Theory and Practice of Emulsion Technology* (A. L. Smith, ed.), Academic, New York, 1976, p. 271.
88. A. Watillon and F. van Grunderbeek, *Bull. Soc. Chim. Belge 63*:115 (1964).
89. R. H. Ottewill and D. J. Wilkins, *J. Colloid Sci. 15*:572 (1960).
90. B. V. Derjaguin and G. J. Valsenko, *J. Colloid Sci. 17*:605 (1962).
91. D. Walsh, private communication.
92. E. D. Bailey, J. B. Nichols, and E. O. Kraemer, *J. Phys. Chem. 40*:1149 (1936).

93. J. B. Nichols and E. D. Bailey, in *Physical Methods of Organic Chemistry* (A. Weissberger, ed.), 2d ed., Interscience, New York, 1949.
94. E. Killmann and J. Eisenlauer, *J. Colloid Polymer Sci.*
95. T. Allen, *Particle Size Measurement*, Chapman and Hall, London, 1975.
96. (a) R. Buscall and R. H. Ottweill, *Colloid Science* (Specialist Periodical Reports), Vol. 2, The Chemical Society, London, 1975, p. 191; (b) R. Aveyard and B. Vincent, *Progr. Surface Sci. 2*:59 (1977).
97. H. Sonntag, J. Netzel, and K. Klare, *Kolloid Z. Z. Polym. 211*:121 (1966).
98. J. Taylor and D. A. Haydon, *Disc. Faraday Soc. 42*:51 (1966); *Nature 217*:739 (1968); D. R. Andrews, E. D. Maner, and D. A. Haydon, *Spec. Disc. Faraday Soc. 1*:46 (1971); D. F. Billett and D. A. Haydon, *Proc. Roy. Soc. (London) A347*:141 (1975).
99. I. Newton, *Opticks*, London, 1704.
100. B. V. Derjaguin and E. Obuchar, *J. Colloid. Chem. 1*:385 (1935); *Acta Physiochem. USSR 1*:5 (1936); B. V. Derjaguin and H. Kussakov, *Bull. Acad. Sci. USSR 1936*:471; Acta *Physiochem. USSR 10*:25 (1939); B. V. Derjaguin and R. L. Scherbakar, *Kolloidn. Ah. 23*:33 (1961).
101. A. J. de Vries, *Foam Stability*, Rubber Stichting, Delft, 1957; *Rec. Trav. Chim. 77*:81, 209, 283, 383 (1958); *Proc. Int. Congr. Surface Activity (3d)*, Vol. 2, 1960, p. 566.
102. A. Scheludko, *Proc. Kon. Ned. Akad. Ventensch. B56*:87 (1962); *Adv. Colloid Interface Sci. 1*:391 (1967); A. Scheludko and E. Maner, *Trans. Faraday Soc. 64*:1123 (1968).
103. A. Vrij, *Disc. Faraday Soc. 42*:23 (1966).
104. H. Sonntag and K. Strenge, *Coagulation and Stability of Disperse Systems*, Halstead-Wiley, New York, 1969.
105. H. Sonntag and J. Netzel, *Tenside 3*:296 (1966).
106. S. P. Frankel and K. J. Mysels, *J. Phys. Chem. 66*:190 (1962).
107. H. Sonntag, *Colloid J. USSR 33*:440 (1971).
108. H. Lotzkar and W. D. McClay, *Ind. Eng. Chem. 35*:1294 (1943).
109. J. T. Davies and E. K. Rideal, *Interfacial Phenomena*, Academic, New York, 1961; J. T. Davies, in *Progress in Surface Science* (J. F. Danielli, K. G. A. Pankhurst, and A. Riddiford, eds.), Vol. 2, Academic, New York, 1964.
110. A. S. C. Lawrence and O. S. Mills, *Disc. Faraday Soc. 18*:98 (1954).
111. W. C. Griffin, *Kirk-Othmer Encyclopedia of Chemical Technology* 3d ed., Vol. 8, Wiley, New York, 1979, p. 900.
112. W. Clayton, *The Theory of Emulsions and Their Chemical Treatment* (C. G. Summer, ed.), 5th ed., Churchill, London, 1954.
113. R. A. W. Hill and J. T. Knight, *Trans. Faraday Soc. 61*:170 (1965).
114. H. C. King and J. Mukherjee, *J. Soc. Chem. Inc. 58*:243 (1939); *59*:185 (1940).
115. E. G. Cockbain and T. S. McRoberts, *J. Colloid Sci. 8*:440 (1953).
116. T. Gillespie and E. K. Rideal, *Trans. Faraday Soc. 52*:173 (1956).
117. L. E. Nelsen, R. Wall, and C. Adams, *J. Colloid Sci. 13*:441 (1958).
118. T. Watanabe and M. Kusui, *Bull. Chem. Soc. Jap. 31*:236 (1958).
119. G. E. Charles and S. G. Mason, *J. Colloid Sci. 15*:105 (1960); *15*:236 (1960); R. S. Allan and S. G. Mason, *Trans. Faraday Soc. 57*:2027 (1961).
120. G. V. Jeffreys and J. L. Hawksley, *J. Appl. Chem. 12*:239 (1962).
121. T. D. Hodgson and J. C. Lee, *J. Colloid Interface Sci. 30*:94, (1969).
122. G. V. Jeffreys and G. A. Davies, *Recent Advances in Liquid-Liquid Extraction*, Oxford, 1971, p. 495.
123. B. V. Derjaguin and M. Kusakov, *Acta Physiochem. USSR 10*:25 (1939).
124. Lord Rayleigh, *Proc. Roy. Soc. 29*:71 (1879).

125. V. G. Levich, *Acta Physiochem. USSR 14*:321 (1941).
126. G. A. H. Elton, *Proc. Roy. Soc. (London) A194*:259, 298 (1941).
127. I. Langmuir, *J. Chem. Phys. 6*:893 (1938).
128. G. A. H. Elton and R. G. Picknett, *Proc. Int. Congr. Surface Activity (2nd)*, Vol. 1, Blackwells, London, 1957, p. 288.
129. D. C. Chappelear, *J. Colloid Sci. 16*:186 (1961).
130. H. M. Princen, *J. Colloid Sci. 18*:178 (1963).
131. J. H. Schulman and E. G. Cockbain, *Trans. Faraday Soc. 36*:661 (1940).
132. A. E. Alexander and J. H. Schulman, *Trans. Faraday Soc. 36*:960 (1940).
133. J. T. Davies and G. R. A. Meyers, *Trans. Faraday Soc. 56*:691 (1960).
134. H. C. King and E. D. Goddard, *J. Phys. Chem. 67*:1965 (1963).
135. R. D. Vold and K. D. Mittal, *J. Colloid Interface Sci. 38*:541 (1972).
136. S. Friberg, *Kolloid Z. Z. Polym. 244*:333 (1971).
137. J. E. Carless and G. Hallworth, *J. Colloid Interface Sci. 26*:75 (1968).
138. Th. F. Tadros, *Colloids and Surfaces 1*:3 (1980).
139. A. Prins and M. van den Tempel, *Proc. Int. Congr. Surface Activity (4th)*, Vol. II, Gordon and Breach, London, 1967, p. 1119.
140. M. van den Tempel, J. Lucassen, and E. H. Lucassen-Reynders, *J. Phys. Chem. 69*:1978 (1965).
141. K. J. Mysels and M. C. Cox, *J. Phys. Chem. 65*:1107 (1961).
142. A. Prins, C. Arcuri, and M. van den Tempel, *J. Colloid Interface Sci. 24*:84 (1967).
143. S. N. Srivastava and D. A. Haydon, *Proc. Int. Congr. Surface Activity (4th)*, Vol. 2, Gordon and Breach, London, 1967, p. 1221.
144. J. Boyd, C. Parkinson, and P. Sherman, *J. Colloid Interface Sci. 41*:359 (1972).
145. T. G. Jones, K. Durham, W. P. Evans, and M. Camp, *Proc. Int. Congr. Surface Activity (2nd)*, Vol. I, Butterworths, London, 1957, p. 225.
146. G. D. Miles, J. Ross, and L. Shedlovsky, *J. Amer. Oil Chemists' Soc. 27*:268 (1950).
147. A. S. C. Lawrence and B. C. Blakely, *Disc. Faraday Soc. 18*:268 (1954).
148. G. Hallworth and J. E. Carless, *J. Pharm. Pharmacol. 245*:71 (1972).
149. P. H. Elworthy and A. T. Florence, *J. Pharm. Pharmacol. 215*:71 (1972).
150. E. G. Cockbain and T. S. McRoberts, *J. Colloid Sci. 8*:440 (1953).
151. S. Friberg, P. O. Jansson, and E. Cederberg, *J. Colloid Interface Sci. 55*:614 (1976); S. Friberg, *J. Colloid Interface Sci. 37*:291, (1971); S. Friberg, L. Mandell, and M. Larson, *J. Colloid Interface Sci. 29*:155 (1969); S. Friberg and I. Walton, *Amer. Soap Perfumer 85*:27 (1970); S. Friberg and P. Solyom, *Kolloid Z. Z. Polym. 236*:173 (1970); S. Friberg and L. Mandell, *J. Pharm. Sci. 59*:1001 (1970); S. Friberg and L. Rydhag, *Kolloid Z. Z. Polym. 244*:233 (1971).
152. B. W. Burt, *J. Soc. Cosmetic Chemists 16*:465 (1965).
153. R. Salisbury, E. E. Leuellen, and L. T. Chavkin, *J. Amer. Pharm. Ass., Sci. Ed. 43*:117 (1954).
154. J. Boyd, P. Sherman, and N. Krog, in *Theory and Practice of Emulsion Technology* (A. L. Smith, ed.), Academic, New York, 1974, p. 37.
155. J. T. Davies, *Rec. Prog. Surface Sci. 2*:129 (1964).
156. K. Shinoda and H. Arai, *J. Phys. Chem. 68*:3485 (1964).
157. B. V. Derjaguin and M. Kusakov, *Acta Physiochim. USSR 30*:25 (1939); B. V. Derjaguin, *Disc. Faraday Soc. 42*:317 (1966).
158. D. Platikanov, *J. Phys. Chem. 68*:3619 (1964).
159. H. Brenner, *Adv. Chem. Eng. 6*:328 (1966).
160. L. A. Spielman, *J. Colloid Interface Sci. 33*:562 (1970).

161. I. B. Ivanov, B. P. Radoev, T. Tr. Traykov, D. St. Dimitrov, and Chr. St. Vassilliett, *Proc. Conf. Colloid Surface Sci. (2nd)*, Budapest, Elsevier, Amsterdam, 1975, p. 583.
162. D. F. Bernstein, W. I. Higuchi, and N. T. H. Ho, *J. Colloid Interface Sci. 39*:439 (1972).
163. P. Becher, *J. Colloid Interface Sci. 42*:645 (1973).
164. D. W. J. Osmond, B. Vincent, and F. A. Waite, *J. Colloid Interface Sci. 42*:262 (1973).
165. M. J. Vold, *J. Colloid Sci. 16*:1 (1961).
166. B. W. Barry, *Rheol. Acta 10*:96 (1971).
167. P. Sherman, *J. Pharm. Pharmacol. 16*:1 (1964).
168. B. W. Barry, *J. Colloid Interface Sci. 28*:82 (1968); B. W. Barry and G. M. Saunders, *J. Colloid Interface Sci. 34*:300 (1970); B. W. Barry and G. M. Saunders, *J. Pharm. Pharmacol. 22S*:139 (1970); B. W. Barry, *J. Colloid Interface Sci. 32*:551 (1970); B. W. Barry and E. Shotton, *J. Pharm. Pharmacol. 19S*:110 (1967); B. W. Barry and G. M. Saunders, *J. Colloid Interface Sci. 35*:689 (1971); *36*:130 (1971); G. M. Eccleston, B. W. Barry, and S. S. Down, *J. Pharm. Sci. 62*:1954 (1973); B. W. Barry and G. M. Eccleston, *J. Pharm. Pharmacol. 25*:244, 294 (1972); B. W. Barry and G. M. Saunders, *J. Colloid Interface Sci. 41*:331 (1972); B. W. Barry and G. M. Eccleston, *J. Texture Studies 4*:53 (1973); B. W. Barry and G. M. Saunders, *J. Colloid Interface Sci. 38*:616, 626 (1972).
169. L. M. Spalton and R. W. White, *Pharmaceutical Emulsions and Emulsifying Agents*, 4th ed., Chemist and Druggist, London, 1964.
170. G. W. Gray, *Molecular Structure and Properties of Liquid Crystals*, Academic, New York, 1962.
171. F. D. Gstirner, D. Kottenberg, and A. Maas, *Arch. Pharm. 302*:340 (1969).
172. P. Ekwall, L. Mandell, and K. Fontell, in *Liquid Crystals* (G. H. Brown, ed.), Part 2.
173. A. J. Hyde, D. H. Langbridge, and A. S. C. Lawrence, *Disc. Faraday Soc. 18*:239 (1954).
174. F. A. J. Talman and E. M. Rowan, *J. Pharm. Pharmacol. 20*:810 (1968).
175. S. S. Davis, *J. Pharm. Sci. 58*:418 (1969).
176. B. Warburton and B. W. Barry, *J. Pharm. Pharmacol. 20*:255 (1968).
177. J. D. Ferry, *Viscoelastic Properties of Polymers*, Wiley, New York, 1961.
178. W. C. Griffin, *J. Soc. Cosmetic Chemists 1*:311 (1949).
179. W. C. Griffin, *J. Soc. Cosmetic Chemists 5*:249 (1954).
180. J. T. Davies, *Proc. Int. Congr. Surface Activity (2d)*, Vol. 1, Butterworths, London, 1959, p. 426.
181. W. C. Griffin, *Off. Dig. Fed. Paint Varn. Prod. Clubs 28*:466 (1956).
182. H. L. Greenwald, G. L. Brown, and M. N. Fineman, *Anal. Chem. 28*:1693 (1956).
183. I. Racz and E. Orban, *J. Colloid Sci. 20*:99 (1965).
184. P. Becher and R. L. Birkmeier, *J. Amer. Oil Chemists' Soc. 41*:169, (1964).
185. A. E. Anacker and V. K. LaMer, *Ann. N.Y. Acad. Sci. 58*:807 (1954).
186. J. T. Davies and E. K. Rideal, *Interfacial Phenomena*, Academic, New York, 1961.
187. W. D. Bancroft, *J. Phys. Chem. 17*:501 (1913).
188. J. T. Davies, *Proc. Int. Congr. Surface Activity (3d)*, Vol. 2, 1960, p. 585.
189. P. Sherman, in *Emulsion Science* (P. Sherman, ed.), Academic, New York, 1968, Chap. 3.
190. S. Riegelman and G. Pichon, *Amer. Perfum. 77*(4):31 (1962).

191. R. Peterson, R. D. Hamil, and J. D. McMahon, *J. Amer. Pharm. Ass.* *53*:651 (1964).
192. P. Sherman, in *Rheology of Emulsions* (P. Sherman, ed.), Pergamon, London, 1963, p. 73.
193. P. Becher, *Emulsions: Theory and Practice*, Reinhold, New York, 1965, p. 235.
194. P. Sherman, *J. Soc. Chem. Inc. (London) 69* (Suppl. No. 2):570 (1950).
195. S. Ross, *J. Soc. Cosmetic Chemists 6*:184 (1955).
196. W. O. Ostwald, *Kolloid Z. 6*:103 (1910); 7, 64 (1910).
197. S. U. Pickering, *J. Chem. Soc. (London) 91*:2002 (1907).
198. K. Shinoda, *J. Colloid Interface Sci. 25*:396 (1967).
199. K. Shinoda and H. Saito, *J. Colloid Interface Sci. 30*:2, 258 (1969).
200. K. Shinoda, *Proc. Inter. Congr. Surface Activity (5th)*, Vol. 2, Butterworths, London, 1968, p. 275.
201. K. Shinoda, H. Saito, and H. Arai, *J. Colloid Interface Sci. 35*:624 (1971).
202. K. Shinoda and H. Takeda, *J. Colloid Interface Sci. 32*:642 (1970).
203. C. Parkinson and P. Sherman, *J. Colloid Interface Sci. 41*:328 (1972).
204. C. J. Parkinson and P. Sherman, *Colloid Polym. Sci. 255*:172 (1977).
205. S. Matsumoto and P. Sherman, *J. Colloid Interface Sci. 33*:294 (1970).
206. F. R. Newman, *J. Phys. Chem. 18*:34 (1914).
207. S. S. Bhatnagar, *J. Chem. Soc. 117*:542 (1920).
208. M. Aoki and T. Matsuzaki, *J. Pharm. Sco. Jap. 83*:761 (1967).
209. T. Shimamota, *J. Pharm. Soc, Jap. 82*:1237 (1962).
210. A. T. Florence and J. A. Rogers, *J. Pharm. Pharmacol. 23*:233 (1971).
211. T. J. Lin and T. C. Lambrecht, *J. Soc. Cosmetic Chemists 20*:180 (1969).
212. V. B. Sunderland and R. P. Enever, *J. Pharm. Pharmacol. 24*:804 (1972).
213. S. Friberg (ed.), *Food Emulsions* Marcel Dekker, New York, 1976.
214. T. W. Schwartz, *Amer. Perfum. Cosmet. 77*:85 (1962).
215. J. A. Serralach and G. Jones, *Ind. Eng. Chem. 23*:1016 (1931); J. Serralach and R. J. Owen, ibid. *25*:816 (1933).
216. H. N. Holmes and D. N. Cameron, *J. Amer. Chem. Soc. 44*:66 (1922); P. A. Rehbinder and E. K. Wenström, *Kolloid Z. 53*:145 (1930); A. B. Taubman and A. F. Koretskii, *Colloid J. USSR 20*:613 (1958); J. M. Martinez Moreno, C. Gomez Herrera, E. Marquez Delgado, C. Janar del Valle, and L. Duran Hildago, *J. Amer. Oil Chemists' Soc. 37*:582, (1960); E. Shotton, K. Wibberley, and A. Vaziri, *Proc. Intern. Congr. Surface Activity (4th)*, Vol. II, Gordon and Breach, London, 1967, p. 1211.
217. Th. F. Tadros, in *Theory and Practice of Emulsion Technology* (A. L. Smith, ed.), Academic, New York, 1976, p. 281.
218. W. Heller and T. L. Pugh, *J. Chem. Phys. 22*:1778 (1954).
219. B. Vincent, *Adv. Colloid Interface Sci. 4*:193 (1974).
220. E. L. Mackor, *J. Colloid Sci. 6*:490 (1951).
221. M. van der Waarden, *J. Colloid Sci. 5*:317 (1950); *6*:443 (1951).
222. R. H. Ottewill and J. Walker, *Kolloid Z. 227*:108 (1968).
223. E. L. Mackor and J. H. van der Waals, *J. Colloid Sci. 7*:535 (1952).
224. E. J. Clayfield and E. C. Lumb, *J. Colloid Interface Sci. 22*:269, 285 (1966).
225. D. J. Meier, *J. Phys. Chem. 71*:1861 (1967).
256. F. Th. Hesselink, *J. Phys. Chem. 73*:3488 (1969); *75*:65 (1971).
257. C. A. J. Hoeve, E. A. Di Marzio, and P. Peyser, *J. Chem. Phys. 42*:2558 (1965); C. A. J. Hoeve, *J. Chem. Phys. 43*:3007 (1965); *44*:1505 (1966).
228. E. W. Fischer, *Kolloid Z. 160*:120 (1958).
229. P. J. Flory and W. R. Krigbaum, *J. Chem. Phys. 18*:1086 (1950).
230. F. Th. Hesselink, A. Vrij, and J. Th. G. Overbeek, *J. Phys. Chem.* *75*:2094 (1971).

231. M. J. Vold, *J. Colloid Sci. 16*:1 (1961).
232. R. H. Ottweill, in *Nonionic Surfactants* (M. J. Schick, ed.), Marcel Dekker, New York, 1967, Chap. 19.
233. D. W. J. Osmond, B. Vincent, and F. A. Waite, *J. Colloid Interface Sci. 42*:262 (1973); B. Vincent, *J. Colloid Interface Sci. 42*:270 (1973).
234. D. H. Napper, *Trans. Faraday Soc. 64*:1701 (1968).
235. D. H. Napper, *J. Colloid Interface Sci. 58*:390 (1977).
236. D. H. Everett and J. Stageman, *Disc. Faraday Soc. 65*:230 (1978).
237. G. C. March and D. H. Napper, *J. Colloid Interface Sci. 61*:383 (1977).
238. J. W. Dobbie, R. Evans, D. V. Gibson, J. B. Smitham, and D. H. Napper, *J. Colloid Interface Sci. 45*:557 (1973).
239. R. Lambe, Th. F. Tadros, and B. Vincent, *J. Colloid Interface Sci. 66*:77 (1978).
240. F. K. R. Li-In-On, B. Vincent, and F. A. Waite, ACS Symposium Series *9*, p. 165, American Chemical Society, Washington, 1975.
241. P. F. Luckham, B. Vincent, and F. A. Waite, *J. Colloid Interface Sci. 73*:508 (1980).
242. C. Cowell, F. K. R. Li-In-On, and B. Vincent, *J. Chem. Soc. Faraday Trans. I 74*:332 (1978).
243. J. K. Percus and J. Yevick, *Phys. Rev. 110*:1 (1958).
244. J. Th. G. Overbeek, *Faraday Disc. Chem. Soc. 65*:1 (1978).
245. D. M. Andrews, E. D. Manev, and D. A. Haydon, *Spec. Disc. Faraday Soc. 1*:46 (1970); D. M. Andrews, Ph.D. Thesis, Cambridge (1970).
246. T. van Vliet, *Mededel. Landbauwhogeschol,* Wageningen 77-1 (1977).
247. T. van Vliet, and K. Lyklema, *Disc. Faraday Soc. 65*:25 (1978).
248. A. Doroszkowski and R. Lambourne, *J. Polym. Sci. Part C 34*:253 (1971).
249. L. M. Barclay and R. H. Ottewill, *Spec. Disc. Faraday Soc. 1*:138 (1970).
250. L. M. Barclay, A. Harrington, and R. H. Ottewill, *Kolloid Z. Z. Polym. 250*:655 (1972).
251. R. J. R. Cairns, R. H. Ottewill, D. W. J. Osmond, and I. Wagstaff, *J. Colloid Interface Sci. 54*:45 (1976).
252. J. N. Israelachvili, R. K. Tandon, and L. R. White, *Nature 277*:120 (1979).
253. D. H. Napper, in *Colloid and Interface Science* (M. Kerker, A. C. Zettlemoyer, and R. L. Rowell, eds.), Vol. 1, Academic, New York, 1977, pp. 413-430.
254. J. J. Bikerman, *Surface Chemistry,* Academic, New York, 1958.
255. P. Rehbinder and A. A. Trapezhnikov, *Acta Physicochim. USSR 62*:257 (1938).
256. C. W. N. Cumber and A. E. Alexander, *Trans. Faraday Soc. 46*:235 (1950).
257. W. Clayton, *The Theory of Emulsions,* Churchill, London, 1954, pp. 92-98.
258. B. Biswas and D. A. Haydon, *Proc. Royal Soc. A271*:296 (1963).
259. B. Biswas and D. A. Haydon, *Proc. Royal Soc. A271*:317 (1963).
260. B. Biswas and D. A. Haydon, *Kolloid Z. 185*:31 (1962).
261. B. Biswas and D. A. Haydon, *Kolloid Z. 186*:57 (1962).
262. L. E. Bielsen, R. Wall, and G. Adams, *J. Colloid Sci. 13*:441 (1958).
263. S. U. Pickering, *J. Chem. Soc. 1934*:1112.
264. A. J. Scarlett, W. L. Morgan, and J. H. Hildebrand, *J. Phys. Chem. 31*:1566 (1927).
265. L. J. van der Minne, *Over Emulsions,* Amsterdam, 1928, p. 66.
266. J. H. Schulman and L. Leja, *Trans Faraday Soc. 50*:598 (1954).
267. T. Takakuwa and T. Takamori, *Proc. 6th Mineral Processing Cong.*, Cannes, Pergamon, Oxford, 1963.

# 4

# Microemulsions

STIG E. FRIBERG and RAYMOND L. VENABLE / University of Missouri-Rolla, Rolla, Missouri



## I. Introduction and Historical Review

The types of systems which are now frequently called *microemulsions* were brought to the attention of the scientific community beginning in the 1940s by the late J. H. Schulman and a series of collaborators [1]. The use of such systems in technology was established about 10 years earlier; wax formulations [2] and fuels [3] being two conspicuous applications. However, the earliest usage of the term *microemulsion* as such appears to be in 1959 [4]. Prior to that time the spherical particles that were obtained were referred to as hydrophilic oleomicelles or oleophilic hydromicelles, depending on whether the continuous phase was water or oil, respectively. Whatever they are called, the systems typically contain relatively large amounts of oil and water along with a surfactant and cosurfactant, except in the case of certain hydrophobic surfactants where no cosurfactant may be required. These systems spontaneously form at contact between the components, are frequently essentially transparent to the eye, are of low viscosity, and may be thermodynamically stable—all quite in contrast with the properties of normal or macroemulsions.

Schulman and his collaborators used low-angle x-ray diffraction [5], light scattering [6], ultracentrifugation [7], electron microscopy [8], nuclear magnetic resonance (NMR) [9,10], and viscosity [11] measurements to investigate various aspects of those so-called microemulsions. Others have also used NMR [12,13,14], conductivity [12], light scattering [15], and electron microscopy [15] measurements. Even with the great amount of information gathered in the above studies, as well as others to be mentioned later, there still seems to be a divergence of opinion as to the mechanism of formation and stabilization of microemulsions.

The typical mode of preparing solutions used by Schulman and co-workers was to take an emulsion of a hydrocarbon and water, stabilized by a soap such as potassium oleate, and add an alcohol such as *n*-hexanol to the above mixture until it suddenly became transparent. One could conversely take a mixture of hydrocarbon, alcohol, and surfactant and add water as was done by Shah [12]. In either case, as the system goes from very turbid to transparent, there must be a great upheaval on the molecular scale as the structure-forming elements of the solution rearrange themselves.

Thus it was quite logical to look for the driving force behind this drastic rearrangement of molecules. A brief historical review of the ideas advanced to explain the experimental observation is now presented. All of the systems studied contained a surfactant or soap and a cosurfactant which was capable of hydrogen bonding. The interface was considered to be liquid and to have a two-dimensional pressure given by $\pi = \gamma_{O/W} - \gamma_i$, where $\gamma_{O/W}$ is the interfacial tension between the oil and water (O/W) with no surface-active materials present, and $\gamma_i$ is the interfacial tension of the interfacial phase or interphase. Schulman and co-workers [8,16,17] suggested that a sufficiently large surface pressure could make $\gamma_i$ negative during the mixing stages, resulting in dispersion to smaller droplets. The surface pressures that would characteristically be required were of the order of 50 dyn/cm.

After a time it was realized by Prince [18] that in fact the hydrogen-bonding cosurfactant, most frequently an alcohol, would be distributed between the aqueous and hydrocarbon phases and that the value of $\gamma_{O/W}$ which should be used was a $(\gamma_{O/W})_a$, a much lower value reflecting the presence of the alcohol at the oil-water interface. This gave relief from the need for unrealistically high film pressures required by the original theory, while still allowing a negative interfacial tension to be vitally important. However, for this negative interfacial tension to produce the tiny spherical droplets encountered in the so-called microemulsions, the existence of a duplex film with different tensions at the film/oil and film/water interfaces, respectively, was required. Again Prince [19] suggested that only a flat film would be duplex and the formation of the curved surface would equalize the two surface pressures and tensions. It is not possible to test this model directly since it is not possible to measure interfacial tensions of tiny droplets such as are involved here.

To represent adequately the phase equilibria in a four-component system would require a tetrahedral phase diagram of the type illustrated in Figs. 4, 12, and 13. However, this is cumbersome, and it is possible to present a great deal of information about the phase equilibria by taking planes from such a tetrahedron and presenting these on three-component phase diagrams as illustrated in Fig. 14. Such diagrams can be used to place emphasis on the micellar nature of the so-called microemulsions as espoused by Winsor [20], Palit [21], Ekwall [22], Adamson [23], Gillberg [14], Shinoda [24], Ahmad [25], and many others. Rance and Friberg [26] briefly discuss such diagrams, pointing out two main regions on the phase diagrams: one forming the W/O microemulsions as a direct continuation of the cosurfactant solutions containing inverse micelles, and another that extends from the aqueous micellar solution in a more complicated manner. The latter region encompasses the O/W microemulsion composition which may not be thermodynamically stable. This reinforces the idea that one should be careful about jumping to conclusions as to what phases are present, or what the structure of the particles in the solution is, on the basis of limited types of information from parts of the phase diagrams.

In addition to these presentations of the experimental aspects of micellar solutions, Murphy [27] has very thoroughly discussed the relationships of various components of the interfacial free energy. Other factors [28-30] have been shown to be equally important as interfacial free energy in stabilizing these micellar solutions.

Much, if not all, of the work mentioned thus far has dealt with micellar solutions or so-called microemulsions prepared using ionic surfactants or soaps. Temperature effects have not been mentioned because temperature effects at the room temperature level are not particularly prominent in such systems. This brings out one of the major differences between ionic and nonionic surfactants in the so-called microemulsions because temperature effects at the room temperature level are very pronounced in the latter case. One other major difference between ionic and nonionic surfactants is that with nonionic surfactants no cosurfactant is needed. The commonly used alkyl carboxylates, sulfates and sulfonates are too hydrophilic to form microemulsions alone; a more hydrophobic cosurfactant must be added.

When a solution of a nonionic surfactant and hydrocarbon in water is heated, the solution becomes visibly turbid at a temperature known as the *cloud point*. At this temperature or at a slightly higher one, the solution separates into a surfactant-rich phase and a water-rich phase [31-35]. Conversely, if a hydrocarbon solution of a surfactant and water is cooled, there exists a temperature, sometimes called the *haze point*, below which an oil-rich phase and a surfactant-rich phase form. This demonstrates that the behavior of oil-continuous solutions is often just the reverse of water-continuous ones.

In addition to these temperature effects, Saito and Shinoda [31] report that the cloud point of a polyoxyethylene-type surfactant in aqueous solution is raised as the average oxyethylene chain length is increased. Shorter oxyethylene chains enhanced solubilizing power for a given length of hydrocarbon chain of the surfactant. The effect of varying the length of the hydrocarbon chain of the surfactant had a relatively small effect on solubilizing ability. In general the opposite effects can be expected for inverse micellar phenomena in an oil-continuous medium.

As an aqueous solution of a nonionic surfactant which contains large amounts of solubilized hydrocarbon is heated, a variety of things occur. The solubilization of oil increases rapidly (along with micelle size) as the cloud point is approached [36]. As mentioned previously, the system separates into two phases at or slightly above the cloud point for an aliphatic hydrocarbon. One phase contains very small amounts of surfactant and hydrocarbon and is appropriately referred to as the *aqueous phase*. The other phase, or surfactant phase, therefore has a higher concentration of the surfactant and contains large amounts of hydrocarbon and water. As the temperature is further increased, the surfactant phase coalesces with the hydrocarbon phase to form an oil-continuous solution with inverse surfactant micelles and solubilized water. The temperature range over which this phase reversal occurs is known as the *phase inversion temperature* (PIT) [31], (cf. also Chap. 5).

In the following sections some of the problems with the microemulsion structure and stability will be treated. The theoretical approach to microemulsion stability will be briefly illustrated followed by a section on the chemical aspects of the phenomenon emphasizing the influence of different chemical components on the microemulsion regions in the systems.

The sections on chemical aspects of the microemulsion phenomena is intended to be of assistance in formulation activities; it is followed by a brief treatment on applications of microemulsions.

## II. Microemulsion Structure and Stability

The attempts to explain the microemulsion systems are an illustrative example of the traditional conceptual differences between a chemist and a physicist. A physicist is interested in the interaction between particles, and the question of stability of a dispersed system is answered by finding the total free-energy minimum. The microemulsion is treated as a particulate system and the total free energy may be divided into contributions. The chemist, on the other hand, is interested in the molceular interaction among the chemical components, and his or her interpretation may be in the form of chemical equilibrium constants for different association states of the chemical compounds.

It is obvious that both approaches are useful in order to obtain a complete understanding of the phenomena and that both conceptions have advantages and disadvantages. The physicist's approach is useful in obtaining information about the basis for the stability of the system at maximum solubilization of water or oil For these conditions the tendency of the chemist to evaluate phenomena in terms of molecular ratios may lead to simplifications not based on reality, such as the notion of "complex formation" between sodium dodecyl sulfate and dodecanol at an emulsion droplet interface. The physicist's approach is, on the other hand, of little help in practical formulation activities since the relation between the amounts of chemical components and the free-energy terms is not simple. The chemical approach has obvious merits for formulation purposes. The following two sections may be independently studied and the reader specifically interested in formulations may directly proceed to Sec. II.B.

### A. Stability Theory

The question of microemulsion stability has been an intriguing problem for a long time. Schulman and his collaborators [1,4-12,16-19,37-39] intuitively emphasized the importance of a low interfacial free energy. In fact, suggestions were made of a negative interfacial free energy; a conclusion later shown to be premature [40].

#### 1. *Thermodynamic Stability*

The first attempt to evaluate the importance of different free-energy components was made by Ruckenstein [28] who summarized calculations of enthalpic (van der Waals potential, potential from the compression of the diffuse electric double layer, and the interfacial free energy) and entropic components. An evaluation of the entropic contribution to the free energy of a dispersed system has also been made by Reiss [29]; a useful compilation of different theories up to that stage is found in the book by Shah and Schechter [41].

Ruckstein found the van der Waals interaction between particles to be negligible compared to other free energy contributions and the interparticle potential was restricted to effects arising from the compression of the electrical double layer. The calculation of this repulsion potential followed Overbeek [42], with a model of one droplet under consideration surrounded by 12, neighboring droplets located anywhere on a shell of a sphere with radius R and by the remaining droplets homogeneously distributed in the space outside a sphere with radius $R_2$. The potential for one droplet located anywhere in the sphere with radius $R_1 - 2R$ was obtained.

$$U = \frac{18\varepsilon\psi_0^2 R^2}{K^3(R_1 - 2R)^3 R_1}\left[1 + \frac{1.48(1 + 1.3KR_1)e^{-0.3KR_1}}{(KR_1)^2}\right]$$

$$\times \{[K(R_1 - 2R) - 1] + [K(R_1 - 2R) + 1]\, e^{-2K(R_1 - 2R)}\} \qquad (1)$$

The calculation of dispersity entropic contributions was filled with difficulties since the geometrical probabilities were not available. Ruckenstein calculated the upper and lower limits of the entropy term with the following expression as a reasonable average of the two:

$$\Delta S_M = km \ln \frac{N_t}{m}$$

The interfacial free energy was separated into terms for the formation of the electric double layer and for the interfacial free energy of uncharged particles. The first was calculated using the Debye-Hückel approximation.

$$f_0 = \frac{\psi_0^2}{8/R\Pi}[\varepsilon_1(1 + K_1 R) + \varepsilon_2(K_2 R \coth K_2 R - 1)] \qquad (2)$$

The second term was taken as the rest from the zero free-energy difference between components in equilibrium with the microemulsion.

The results of the calculations are instructive because of the estimation of the level of interfacial free energy needed in order to obtain a stable microemulsion. Equation (2) exemplifies [28] this phenomenon: the interfacial tension stays at the level of $10^{-2}$ mN $m^{-1}$ for radii such as are found in microemulsions.

It should be observed that these calculations use one term for the interfacial free energy [27]

$$\gamma = \int_{Z_1}^{Z_2} p(1 - 2zh + z^2 k)\, dz \qquad (3)$$

The terms within brackets are not individually analyzed; the parentheses is used as a scaling factor. However, the pronounced influence of the cosurfactant-surfactant ratio (vide infra) makes a further discussion of the expression justified and useful.

The term $\int_{Z_1}^{Z_2} p\,dz$ is the pure stretching part of the interfacial tension; the change in surface free energy with change in surface area at constant mean and Gaussian curvature in the transition zone. The second term $\int_{Z_1}^{Z_2} pzH\,dz$ measures the negative of the change in interfacial free energy with the change in mean curvature at constant interfacial area and Gaussian curvature. The third term, the torsion stress, measures the change in interfacial free energy with change in Gaussian curvature at constant area and mean curvature.

For the common macrodisperse systems, terms two and three are without importance, since the interfacial tension (the first term) is high and the droplet curvature large compared to the thickness of the interface zone. Murphy [27] has estimated the limit at which the second term will be important and found the interfacial zone to be of the magnitude of half the radius of the droplet. For some microemulsion systems the interfacial zone is certainly of this magnitude and the first moment of the interfacial stress is important for their stability. To our knowledge no measurements separating the two terms of the interfacial tension have so far been made.

Robbins [43] treated the phase behavior of nonionics using the bending energy approach, ignoring the interparticle interactions; the stability was inferred from the conditions in a duplex film of nonionic surfactant. The volume ratios of hydrophobic and hydrophilic parts of the surfactant could adequately be related to the solubilization of water in W/O microemulsions and the corresponding phenomena in an O/W microemulsion.

A more complete treatment, including the chemical potential of the two phases, the interparticle potentials, the two terms of the interfacial free energy, and the entropic contribution was recently accomplished by Miller [44].

Equalizing the chemical potential of the dispersed phase with its value for the pure compound and combining the result with the equation for full balance at the interface [27], the simple condition

$$f^{\sigma} \approx \frac{\overline{H}}{R} \tag{4}$$

was obtained after approximations. (Cf. Ref. [43].)

Breaking the total free energy of the system into the free energy of the continuous phase, of the material dispersed within the droplets, of the interface, and of the entropic contribution; the total equilibrium condition may be expressed

$$\frac{4\pi R\overline{H}}{kT} + \frac{5}{2} - \ln\left[\left(\frac{2\pi \bar{g} kT}{h^2}\right)^{3/2} v_2\right] + \frac{2B_2}{v_2} + \frac{M}{v_2}\left(\frac{\partial B_2}{\partial m}\right)_{T,P,n_v} = 0 \tag{5}$$

Limiting the interparticle interaction to the van der Waals interaction, implying that there is no charge on the droplets, relations between the surfactant concentration and the droplet size may be calculated.

The results demonstrated the insignificance of the van der Waals interaction. The equilibrium droplet radius was found approximately the same for the Hamaker constant both equal to zero and to $3 \times 10^{-20}$. This is in agreement with the Ruckenstein [28] results. Another result of the model was a change in the trend of the free energy-radius relationship, with increasing surfactant concentration. The minimum in the free-energy function versus the radius M became more shallow with increasing concentration and disappeared completely for the highest values. This means that aggregates with an infinite number of molecules possess the lowest free energy; a prediction that agrees well with experimental results both for water-in-oil (W/O) and oil-in-water (O/W) microemulsions with nonionic surfactants [45].

These considerations are valid also for ionic systems with the proper inclusion of the electrostatic interactions. The relation to the chemical composition is, however, not simple; especially to the influence of the cosurfactant. The cosurfactant/surfactant ratio has the most pronounced influence on the properties (vide infra) and a special treatment is justified.

Eicke [46] has approached the cosurfactant problem by considering the thermodynamic stability of a binary oil/water system in the presence of a third component; the additive. The presence of such an additive will increase or decrease the miscibility of the two-component system.

The representation of the free energy of the system using temperature plus either the mole fraction of the binary O/W system or the mole fraction of the additive was shown to be interchangeable. The particular intricacy arising from the presence of a fourth component, the cosurfactant, was overcome by the following consideration [47]. Since most of the surfactant molecules in a microemulsion are assimilated at the interface, the polar part of the amphiphilic molecule was assigned to the water volume while the polar tail was envisioned as belonging to the nonpolar volume.

Thermodynamic considerations of such a system, limited to small concentrations of the additive, give a relation between the solubility of the additive and the Gibbs free energy of the system. An essential feature of the resulting relation is that the surfactant and cosurfactant do not act independently at the interface. The influence of the cosurfactant was seen as a modifier of the action of the surfactant. It is essential to realize that this result by no means implies the formation of a "complex" between the surfactant and the cosurfactant.

The generalization of the problem by this treatment was demonstrated by using atypical "cosurfactants" such as benzene, cyclohexane, dodecyl benzene, and carbon tetrachloride. In principle the treatment is applicable to all types of surfactants with the limitation of small concentrations of additive. This condition is, of course, an obstacle to its use for ionic systems, for which the cosurfactant concentration is of the order of 10 percent by weight.

Talmon and Prager [48,49] have developed a statistical thermodynamics of the microemulsion phenomenon by applying the Voronoi tesselation [50] to the volume. The Voronoi polytopes were randomly filled by oil or water for which no molecular miscibility was visualized. The surfactant was assumed entirely absorbed to the interfaces between the polytopes.

This means that composition-dependent free energy terms are absent and the free energy is calculated from entropic contributions only. The basic condition arises from the size of the surface generated by the concentration of the surfactant $C_s$ and the volume fraction of oil $\phi_2$. The results showed the presence of two-phase areas, for example, a W/O and O/W microemulsion in equilibrium with each other, but with no three-phase areas.

Inclusion of curvature effects and corrections for overcounted configurations made it possible to achieve phase diagrams with a three-phase area bordering three two-phase areas.

The different theoretical approaches have been useful in pointing out the essential terms for thermodynamic stability of a microemulsion. Future development may be characterized by a closer approach to models which take the spatial influence on the association structures into account. The importance of this influence is found in the formation of lyotropic liquid crystals when maximum water stability is exceeded. The thermodynamic treatment of such equilibria certainly offers a great challenge.

### 2. *Kinetic Stability*

Gerbacia and Rosano [51,52] experimentally found some W/O microemulsions not to be stable on a long-term basis in spite of the fact that the water content was only 2.6% on a surfactant-plus-cosurfactant concentration of 16.1%. A theory was developed to describe the kinetic stability using microemulsion droplets as a model, a choice that probably should have been further justified in view of what is known about the association conditions at such low water content. The theory was essentially limited to the energy changes occurring in the layers of surfactant and cosurfactant lining the microemulsion water droplet during coalescence. The free energy changes in the film were calculated using regular solution theory [53], both for the enthalpic component and for the entropic component. In addition, the van der Waals attraction potential was added using the Hamaker approach modified by Vold for droplets with surface layers [54].

The kinetic aspects of microemulsion stability has also been treated by Eicke [55,56] who used fluorescence measurements to detect a rapid exchange of electrolyte solutions and water between inverse micelles in iso-octane arising from collisions.

Ruckenstein [57] has given a general treatment of kinetic and static stability applicable to colloidal systems. The approach follows Cahn's [58] evaluation of the kinetics of spinodal decomposition with the necessary modification for unequal volumes.

The diffusional flux

$$J_1 = -J_2 = -\frac{1}{kT}\left(\frac{D_1}{V_1}\nabla\mu_1 - \frac{D_2}{V_2}\nabla\mu_2\right) \tag{6}$$

is modified for a concentrated system by a change of the chemical potential due to perturbation from the uniform state. After introducing mass conservation and a small perturbation, the growth coefficient is obtained

$$\alpha(\beta) = -\frac{\phi_{10}\beta^2}{kT}\left[v_2 D_1 \phi_{20}\left(\frac{\partial^2 f}{\partial \phi_2^2}\right)_{\phi_{20}} + 4\Pi\int_0^\infty r^2\left(\frac{\sin \beta r}{\beta r} - 1\right) U(r)g(r)\, dr\right] \tag{7}$$

Since $(\sin \beta r)/\beta r - 1 \leqslant 0$ and $U(r) < 0$, it is obvious that if $(\partial^2 f/\partial \phi_2^2)_{\phi_{20}} > 0$, the growth coefficient $\alpha < 0$ for all values of $\beta$ by definition, e.g., the system is stable for all perturbations.

However, the opposite $(\partial^2 f/\partial \phi_2^2)_{\phi_{20}} < 0$ does not always lead to instability if

$$\left|\left(\frac{\partial^2 f}{\partial \phi_2^2}\right)_{\phi_{20}} v_2 D_1 \phi_{20}\right| < -4\Pi\int_0^\infty r^2 U(r)g(r)\, dr \tag{8}$$

sufficiently large values of $\beta$ will give $\alpha(\beta) < 0$ and stability. Instability is encountered for small values of $\beta$.

The value of $\alpha$ has the dimension $\beta^{-1}$ and is a measure of the inverse of the time of perturbation. A series of approximations gave

$$\alpha \approx 10^{-5}\left(\frac{\partial^2 f}{\partial \phi_2^2}\right)^2_{\phi_{20}} s^{-1}$$

demonstrating the fact that for sufficiently small values of $(\partial^2 f/\partial \phi_2^2)_{\phi_{20}}$ an unstable system can survive a small perturbation for a long time.

## B. The Chemical Approach

The microemulsions may be stabilized by a single hydrophobic ionic surfactant such as di-octyl sulfosuccinate, a single nonionic surfactant of the polyethylene glycol alkyl ether type or a combination of a hydrophilic surfactant such as alkyl sulfate, alkyl soap, or a cationic alkyl ammonium halide together with a hydrophobic amphiphile of medium chain length (the cosurfactant), such as pentanol. Recently combinations of ionic surfactants with nonionic ones also have shown interesting properties for special applications.

The essential features for the different kinds of surfactants will be treated in the following sections. The surfactant-cosurfactant combination will be introduced first with regard to its historical interest; microemulsions stabilized by a single ionic surfactant will be included in this section, while the nonionic surfactants will receive a separate treatment in view of their special properties.

### 1. *Ionic Surfactants*

Microemulsions stabilized by ionic surfactants commonly have four chemical components: water, hydrocarbon, ionic surfactant, and a cosurfactant, the latter usually a medium-chain-length alcohol such as pentanol. Of these four components the three polar compounds form several colloidal association structures such as normal and inverse micelles, in addition to different structures of lyotropic liquid crystals. These phenomena have been extensively studied by Ekwall and co-workers [59], and although Ekwall never investigated microemulsions per se, the careful research into phase equilibria of amphiphilic substances by his group during the 1960s forms much of the basis for the following treatment of association phenomena in microemulsion systems. The description will cover normal and inverse micelles; the lamellar liquid crystal will be mentioned only in connection with the phases in equilibrium with microemulsions.

a. Normal Micelles: Micellization has been extensively covered in the literature from its first introduction [60-62]. A large number of review articles are available [63-64]. In spite of all these efforts progress in the field is still rapid [65-67]. The solubilization of the cosurfactant and the hydrocarbon in normal micelles is modest (Fig. 1). The thermodynamics of solubilization by micelles has not been extensively investigated in the literature. However, in a recent paper [68] it was shown that the free energy of solubilization in different homologous series of detergents was linearly related to the number of carbon atoms in the alkyl chain of the detergent.

Figure 2 is a good illustration of the difference between normal micellar solutions and microemulsions. The microemulsion space close to the water-hydrocarbon axis is marked in the figure; neither the solubilization of cosurfactant (A) nor hydrocarbon (B) are even of a magnitude needed in order to justify the term microemulsion (C). The microemulsion, to deserve the name, should contain water and hydrocarbon well in excess of the amount of surfactant. The distinction made [12] between micellar systems and microemulsions is certainly justified for this case.

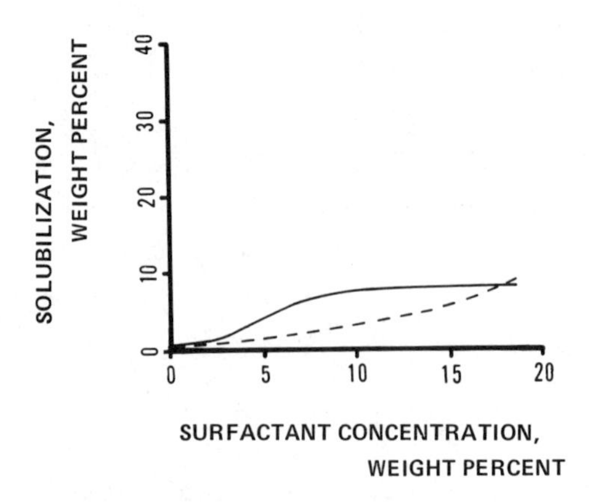


Figure 1. Typical solubilization for a cosurfactant (——) and a hydrocarbon (---) in an aqueous micellar solution. (CMC ≈ 2.5%)

Information about the direct relation between micellar association and microemulsion system is found instead in the solubilization of water and surfactant in the cosurfactant. This phenomenon will be treated in the next section.

b. Inverse Micelles and W/O Microemulsions: The cosurfactant, a medium-chain-length alcohol, dissolves water only to a small degree. Figure 3 [69] shows an example for which the solubility of water is 6% by weight (A). The surfactant displays an even lower solubility in pure form; the diagram shows a solubility of 2.5% by weight (B). However, the mutual influence of two compounds on the solubility is profound. Addition of a moderate amount of water gives rise to a pronounced

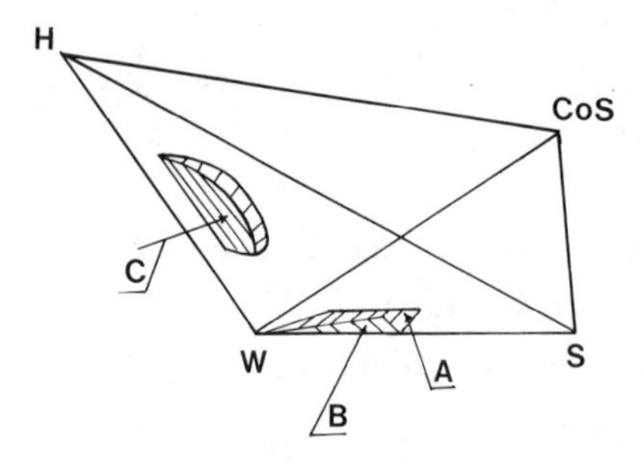


Figure 2. Neither the solubilization of cosurfactant (A) nor of hydrocarbon (B) in an aqueous micellar solution deserves to be called microemulsion formation. The microemulsions (C) should contain high amounts of both water and hydrocarbon: W = water; S = surfactant; CoS = cosurfactant; H = hydrocarbon.

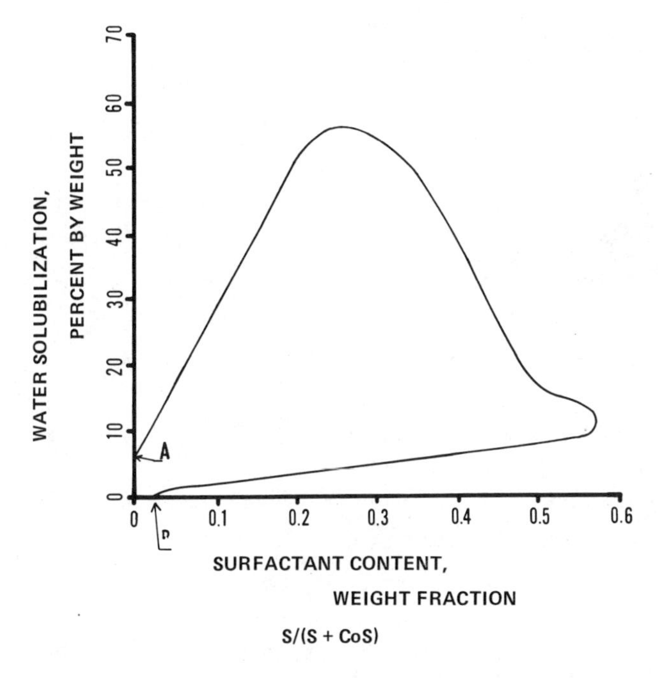


Figure 3. The solubility in pentanol (CoS) of potassium oleate (surfactant, S) and of water show a strong mutual dependence.

increase in the solubility of the soap. At a level of 10% water the solubility of the surfactant has increased to 58%, an order of magnitude enhancement. A corresponding increase of the water solubilization is found at a certain cosurfactant-surfactant ratio; for the combination pentanol/potassium oleate, illustrated in Fig. 3 a cosurfactant-surfactant weight ratio of 2.85 resulted in a maximum solubilization of water of 56%.

These alcohol solutions are hence characterized by high water solubility in an organic solvent (pentanol), and an attempt to prepare thermodynamically stable W/O microemulsions by addition of the fourth component by dissolving a hydrocarbon into the pentanol is rational. The solubility regions at different hydrocarbon contents form a direct continuation of the pentanol solution at zero content of hydrocarbon (Fig. 4). The figure also shows the overlap between these inverse micellar solutions and the microemulsion space. The inverse micellar solutions with high water and hydrocarbon content are in fact identical to the thermodynamically stable W/O microemulsions discussed by Schulman in his classical contributions.

Microemulsions other than these also exist and will be treated later in this chapter. The thermodynamically stable W/O microemulsions are, however, of such importance that they justify a more extensive treatment of the characteristics of inverse micellar solutions.

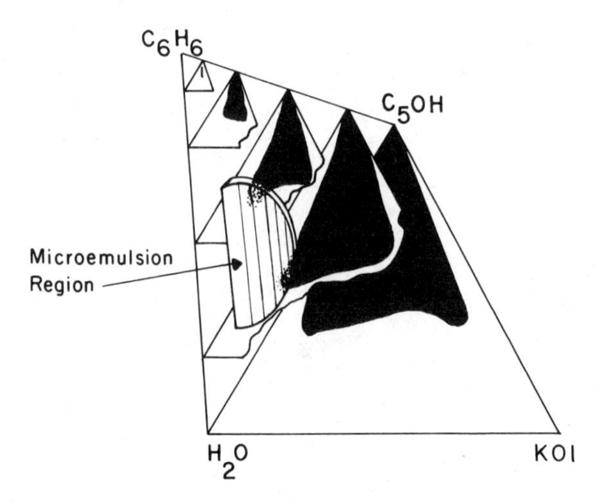


Figure 4. At the addition of hydrocarbon the W/O microemulsions form a direct continuation of the pentanol inverse micellar solutions.

*(1) Structure of the inverse micellar solution* The solubility region in Fig. 3 and base plane of Fig. 4 is traditionally referred to as an *inverse micellar solution* [39], but in reality also contains other species with systematic transitions between different structures. Several investigations have been performed [15,30,69-72,55, 28] on these solutions by different methods. The results agree with the following interpretation.

At low water content no interassociation of the surfactants takes place. The surfactant molecules exist in a monodispersed state with water and cosurfactant molecules both associated as ligands to the metal ion and/or hydrogen-bonded to the polar group of the surfactant organic part. This model is supported by the results from early investigations by Shah on such systems [12], using NMR and electric conductivity. The results were interpreted in the form of a model with monomolecular dispersion of the surfactant.

Determination of the dielectric properties by Clausse [70] and by Eicke [71,72] shed further light on the associations. The absence of the expected dielectric relaxation of the Cole-Cole type was conspicuous at water weight fractions of less than 0.2 in a hexadecane/water/potassium oleate/hexanol microemulsion. This dielectric relaxation is directly related to interfacial polarization and its absence is a rational consequence of the model with no surfactant interassociation, e.g., no inverse micelles and hence no interface at low water concentrations (cf. also Chap. 10). Light scattering [15] showed extremely low scattering intensity from solutions of low water contents. Figure 5 shows the scattering of a pentanol solution of soap with a water contant between 5 and 35% by weight to be less than that of a pure

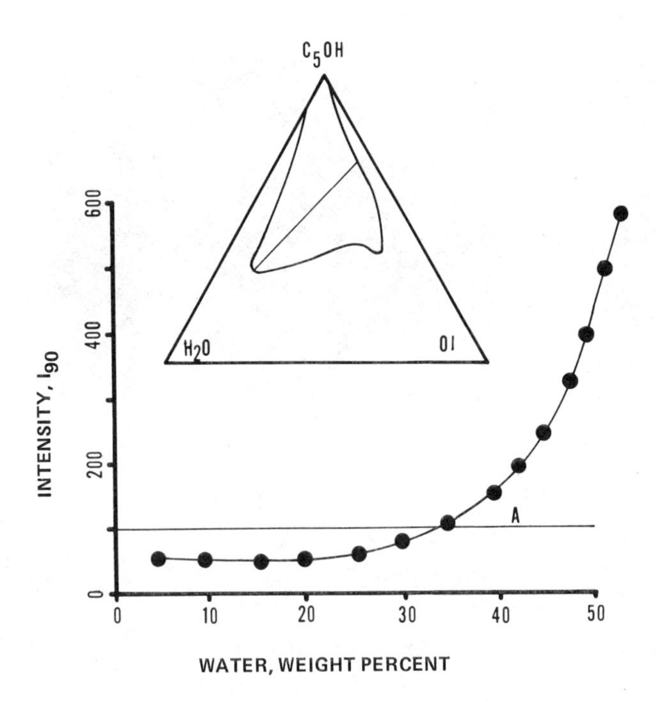


Figure 5. Light-scattering data show the pentanol solutions not to contain micelles at low water concentrations. The transition to micellar solution is gradual from about 20% water to about 45% when the increase of scattering becomes linear.

hydrocarbon (benzene = 100 scale units, A, Fig. 5). The association to larger aggregates at high concentrations of water is obvious from the data. Electron microscopy of replicas from freeze fractured samples [15] (Fig. 6) showed no pattern on replicas from the area of low scattering intensity (Fig. 5), while the samples with higher water contents (corresponding to the concentration range with pronounced light scattering) showed the presence of particles of a reasonable size for inverse micelles.

The results demonstrate the existence of two regions in the alcohol solution of ionic surfactant and water: one region at low water content with no inverse micelles, and one at higher water content in which the inverse micelles have formed.

The transition between the two is not sudden, as in aqeuous solution; the light scattering data [15] (Fig. 5) show the gradual transition from the low-scattering molecular dispersion to the straight-line increase of intensity with water concentration for the micellar solution. The concept of critical micellization concentration is obviously not applicable to this kind of solution. The transition is of importance for the properties of microemulsions (vide infra), and the stability factors for the two aggregates will be treated.

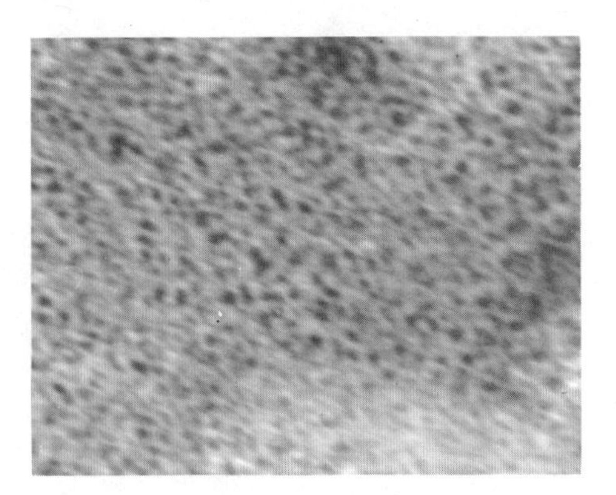

Figure 6. Electron microscopy of replicas from freeze-fractured samples showed a pronounced difference between samples with low (right) and high (left) light-scattering intensity (cf. Fig. 5).

The definition of inverse micelles used in this section is in accordance with the concepts of Ekwall and collaborators. In the terminology of Eicke, the small aggregates are called *inverse micelles* and the inverse micelles called *microemulsion droplets*. The rationale behind this terminology is the presence of an interface which is characteristic of the microemulsion state. Until a consensus on nomenclature is reached, the reader has to accept the confusion.

*(2) Stability of small aggregates* The water/surfactant interaction appears as the decisive stability factor for these aggregates; a hypothesis based on two conditions. The first one is the fact that the *minimum* water content to form the solution (Fig. 3, lower limit) is a straight line. This means that the increase of soap solubility above its solubility in pure pentanol is proportional to the amount of added water; a minimum molecular water-soap ratio of 3.0 is required to dissolve the potassium oleate in pentanol. The second condition supporting the assumption is the result [15] showing a constant water-soap ratio for the beginning of the association to larger aggregates. The initiation of further aggregation took place when each soap molecule had eight water molecules associated. These experimental results were judged of sufficient interest to justify investigations on bonding energies in small aggregates of this kind.

The aggregates are obviously too large to permit exact ab initio calculations; instead the semiempirical CNDO/2 program† was used [73]. The CNDO/2 program

†A description of the approximations and deviations from exact results in the CNDO/2 program is found in the Appendix.

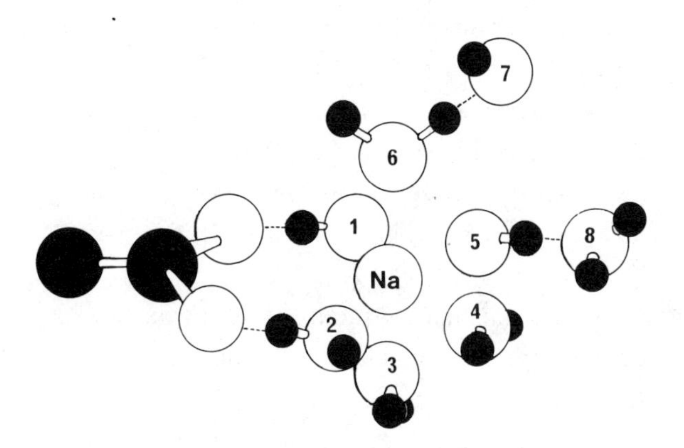


Figure 7. In the model for molecularly dispersed soap with associated water, two water molecules (1 and 2) are both hydrogen bonded to the carboxylic group and also served as ligands to the sodium ion. Four water molecules (3 to 6) are employed as ligands and two molecules (7 and 8) are hydrogen bonded to the ligand water.

cannot accommodate potassium atoms; the calculations were instead performed on a sodium soap system. The conditions for solutions with sodium soaps are similar to those found with potassium soaps; the modification is a higher water-soap ratio for minimum water content to dissolve the soap.

The model used (Fig. 7) was a sodium soap with two water molecules bound to the carboxylic group with a hydrogen bond and also acting as ligands to the sodium ion. The next four water molecules were organized as ligand molecules to the sodium ion; further water molecules were placed in a second sphere, hydrogen-bonded to the inner water molecules.

The results showed the energy for the bonding of the inner water molecules to be extremely high. The two water molecules (1 and 2 in Fig. 7), with both a hydrogen bond to the ionized carboxylic group and a ligand bond to the sodium ion, gave a binding energy of ≈ 50 kcal/mole which is approximately five times the value for water molecules in liquid water. The four water molecules (3 to 6) with only a ligand bond to the sodium ion experienced a binding energy of 25 kcal/mole. Water molecules in the spheres outside this inner one (7 and 8) showed a binding energy at a slightly lower energy level than the one in liquid water.

The model used for these calculations is admittedly schematic and the highly approximate results cannot be used for quantitative evaluations of the binding energies. The importance of the results is instead found in the trend of binding energy for the water molecules with their position. Water molecules close to the two ions have high binding energy; much in excess of that in pure water. This high energy confers stability to the small aggregates. The water molecules outside the inner layers display about the same binding energy as those in liquid water, and hence the stability is not extended to aggregates with a high number of water molecules. This result is in excellent agreement with the light-scattering data (Fig. 6), showing association when the number of water molecules exceed 8. In that concentration range inverse micelles are formed; their stability will be treated in the next section.

*(3) Stability of inverse micelles* The inverse micelles containing a high number of water molecules do not easily lend themselves to a treatment using quantum chemistry. Instead a model with continuous layers has to be used [51,52,71].

Eicke [71] has treated inverse micelles summarizing the interactions from

i. Solvent-solvent dispersion interactions.
ii. Dispersion interaction between the hydrocarbon chain of the amphiphile and the solvent molecules.
iii. Dispersion interaction between the hydrocarbon chains in the micelle.
iv. Coulombic interactions in the form of
   a. Repulsion between charges of identical sign
   b. Attraction between charges of opposite sign
   c. Electrostatic energy between opposite charges in the monomers

The results gave evidence of maximum stability for a certain size of micelles and agreed well with experimental results.

These calculations were performed on inverse micelle models with a continuous central aqueous core lined by a layer of charged surfactants, with part of the counter-ions forming an inner diffuse double layer. Eicke applied his results to inverse micelles stabilized by a single surfactant. The microemulsions used for practical purposes usually use combinations of a more hydrophilic surfactant with a strongly hydrophobic amphiphile, commonly a medium-chain-length alcohol, pentanol. The results are in principle also applicable to such systems.

However, such systems of four components possess one additional stability factor whose influence is less directly amenable to evaluation. This factor is the critical cosurfactant-surfactant ratio. Figure 4 demonstrates the maximum water solubilization occurring [69] at a surfactant-cosurfactant *ratio* of 0.27, with a steep reduction in solubilization capacity at both sides of the maximum.

In order to understand the basis for this maximum, limiting the study to the conditions in the inverse micellar solution is not sufficient; the conditions in the phase in equilibrium with the inverse micellar solution must also be taken into account. The maximum of water solubilization may not depend on conditions in the inverse micellar solution. Instead, structural changes in the phase(s) in equilibrium with the inverse micellar solution may be decisive. For such a case, calcula-

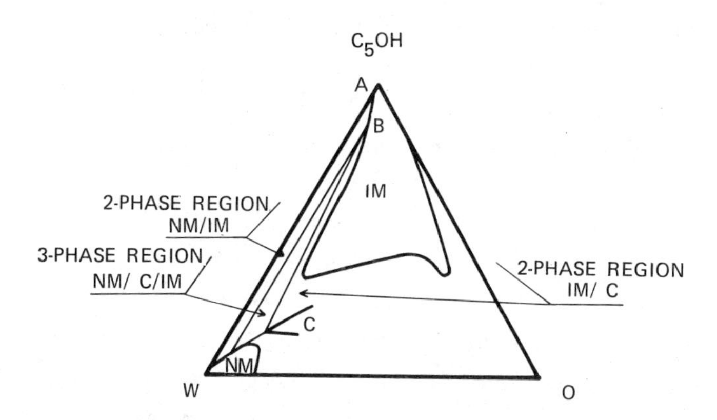


**Figure 8.** At sufficiently low surfactant concentrations (A to B) (cf. Fig. 5) the pentanol solution (IM) does not contain inverse micelles and the equilibrium body is an aqueous micellar solution (NM). At higher surfactant concentration, where inverse micelles are formed the equilibrium is with a lamellar liquid crystalline phase (LC).

tion on internal stabilization factors for the inverse micellar solution are obviously insufficient. The pertinent phase equilibria in the system (water, pentanol, and potassium oleate) are shown in Fig. 8. The figure shows the pentanol solution (IM) to be in equilibrium with the aqueous micellar solution (NM) only at low concentrations of the surfactant (A to B). The two-phase region NM/IM (left in the figure) does not include the part of the pentanol solution where inverse micelles are found (Fig. 5). Inverse micelles are found at surfactant concentrations in excess of B. In this range the equilibrium is not with the aqueous solution with normal micelles, but instead with the lamellar liquid crystal (LC).

The conditions in microemulsions are similar [69] and two conclusions are immediately evident. There is no phase change taking place in the phase in equilibrium with the W/O microemulsions when the latter reach the optimal surfactant/cosurfactant ratio for water solubilization (cf. Figs. 4 and 8). Furthermore [69], there is no extremum in the structure of the liquid crystalline phase at the composition in equilibrium with the inverse micellar solution of maximum water solubilization.

Hence, the reason for the variation of water solubility with the cosurfactant/surfactant ratio and its maximum must be found in the conditions in the pentanol solution, and the analysis by Eicke [46,71] is useful also for the four-component systems. The analysis is facilitated by the results from determinations of counter-ion binding. These results showed a minimum of counter-ion binding at maximum water solubilization, indicating a role for the absorbed cosurfactant as a moderator of the electric properties at the inverse micellar interface. The increased surfactant-cosurfactant ratio leads to an enhanced counter-ion binding giving an increased repulsion (term d, described above) and instability to the inverse micelle. The lamellar arrangement in the liquid crystalline phase is energetically advantageous.

*(4) Influence of electrolytes* Several applications of microemulsions involve solutions with high electrolyte content and the solubility regions for W/O microemulsions or inverse micelles in the presence of a salt is of considerable interest.

The variation of water solubilization with added electrolyte [69] in *soap* is systematic with two noticeable effects. The first one is an enhanced *minimum* water necessary to obtain solubility and the second effect is an increased surfactant-cosurfactant ratio for maximum water solubilization. In addition, a more pronounced dependence on the ratio between the two compounds (Fig. 9) is observed.

The enhanced minimum water concentration is a reflection of the fact that the small aggregates at low water concentrations do not form droplets, but are molecularly dispersed soap with attached water and alcohol molecules. In the region (Fig. 5) in which the soap molecules are not associated, the aqueous aggregates are not sufficiently large to accommodate an electrolyte. For a *soap* system [69] preliminary calculations indicated each NaCl ion pair to require an additional 30-35 water molecules over the minimum amount to dissolve the soap.

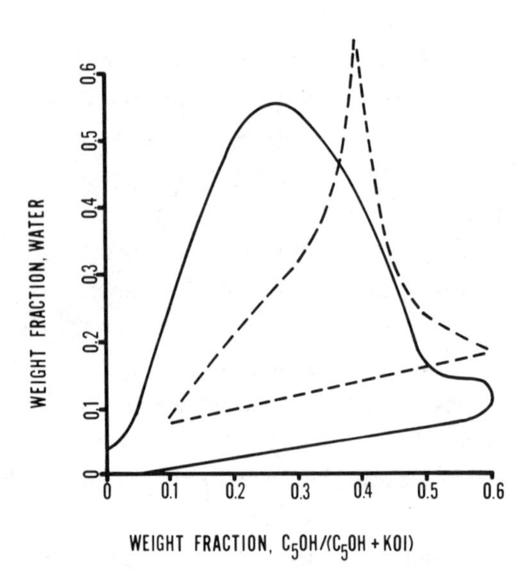


**Figure 9.** The presence of an electrolyte (NaCl 0.5 M) (---) changes the solubility of potassium oleate (KOl) and water in pentanol ($C_5OH$). Curve (——) shows solubility for pure water (cf. Fig. 3).

The enhanced surfactant/cosurfactant ratio necessary for maximum water solubilization is caused by the changes in the inner electric double layer. It is important for the stability of the inverse micelles; its compression [28,30,71] means a reduction in the free counter-ions and an enhancement of the bound ones. The effect is a reduction of the free-energy component from the diffuse electric double layer and an increase of the repulsion between dipoles at the inverse micellar surface.

These modifications of the properties should lead to less stable aggregates and it is necessary to evaluate the properties of the liquid crystal in equilibrium with the inverse micellar solution in order to understand the stability change.

The stability of the liquid crystal is critically dependent on the properties of the electric double layer [74,75]; addition of an electrolyte will cause a compression of the electric double layer, reducing the stability at large interplanar distances. Microemulsions and inverse micellar solutions with maximum water content are in equilibrium with a liquid crystal with high water content. Addition of electrolyte will destabilize the lamellar arrangement, the stability region of the liquid crystal will be shifted toward higher content of surfactant and be replaced by the inverse micellar solution that will expand toward higher ratio of surfactant. The reason for the shift to higher surfactant-cosurfactant ratio for maximum water solubilization is a good example of properties of the phase in equilibrium determining the stability of the inverse micellar solution. Theoretical treatment of this case is still lacking.

*(5) Influence of hydrocarbon* Thermodynamically stable W/O microemulsions are formed by addition of hydrocarbon to the pentanol solutions (Fig. 4) described in the preceeding section. The addition essentially implies a dilution of the pentanol matrix by the hydrocarbon, but in addition the hydrocarbon has some effects of its own. These effects concern the maximum water solubilization and the water concentration at which surfactant association is initiated. They will be discussed in the following section.

The maximum water solubilization is proportional to the amount of surfactant and cosurfactant for hydrocarbon contents less than 50% (Fig. 4) and the surfactant-cosurfactant ratio also remains similar. At higher hydrocarbon content the maximum water content tapers off (Fig. 10). The water solubilization at 90% hydrocarbon is little shifted toward the monomers. It is easy to realize the coupling between this equilibrium and the adsorption of the alcohol at the interface. An adsorption of Y molecules at the micellar interface will give

$$XC_5OH_{monom} \underset{}{\overset{K_1}{\rightleftarrows}} \frac{X}{2}(C_5OH)\ \frac{X}{Y}\ (C_5OH)_{Y,\ mic\ interf}$$

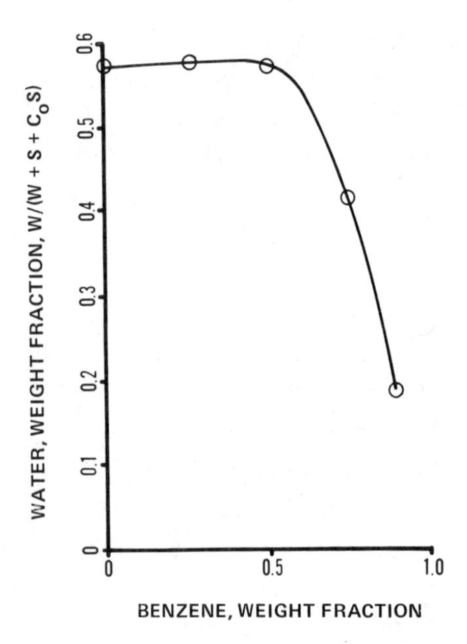


**Figure 10.** The maximum solubility of water (counted less hydrocarbon) in the inverse micellar pentanol solution is drastically reduced when benzene was added in excess of 50% by weight.

The straight-line portion at high water concentrations in the curves from light-scattering data [15] such as those in Fig. 5 can be extrapolated to give approximate lower limits for micellization in inverse micellar solutions. The results (Fig. 11) show that aromatic hydrocarbons give rise to a considerable reduction of the extrapolated concentration for micellization; the corresponding reduction for the aliphatic hydrocarbon is small.

The small changes of the "micellization concentration" with aliphatic hydrocarbons may be interpreted in a qualitative manner, along the lines indicated by Muller [76]. A reduction in the dielectric contstant of the medium should give rise to smaller aggregates, associating at a lower concentration. The strong influence of the aromatic hydrocarbons and their difference from the effect of the aliphatic ones is obviously outside this small influence. One interpretation would be a more specific interaction between the benzene nucleus and the polar part of the surfactant/water structure. The nature of this interaction, if any, has not yet been clarified.

*(6) W/O microemulsions with a polyethylene glycol ether cosurfactant* The medium chain-length alcohols suffer a disadvantage in their biological aggressivity, making them less useful for pharmaceutical and cosmetic applications. A suitable replacement may be found in a combination of an ionic and a nonionic surfactant [77].

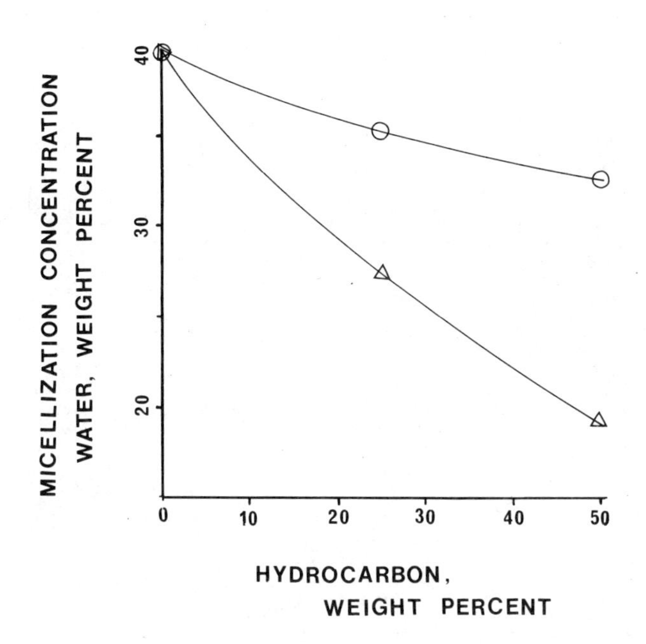


Figure 11. The presence of benzene (Δ) causes a more pronounced reduction of the onset of micellization in a pentanol solution of potassium olease and water than does addition of decane (o).

Such microemulsions show properties different from those formed with the usual combination of ionic surfactant and medium chain length alcohol as cosurfactant. A comparison of Figs. 12 and 4 reveals an obvious change when the pentanol is replaced by the oxyethylene ether. The microemulsion regions are not a direct extension of the solubility area for the basic system without hydrocarbon; the addition of hydrocarbon leads to a pronounced increase of water solubilization capacity. In fact the system with no hydrocarbon (base plane, Fig. 12) shows maximum solubility for water with nonionic surfactant only in the inverse micellar solution. The typical initial increase with addition of the ionic surfactant is found only when hydrocarbon is present.

Another difference is found in the fact that the amount of ionic surfactant at maximum water content is considerably smaller for the combination with oxyethylene ether (Fig. 12) than with pentanol (Fig. 4).

c. O/W Microemulsions: These emulsions have been treated by several investigators; the contributions by Prince [18,78] should be noted for their discussion on the importance of lowering the interfacial tension. Rosano and collaborators [38, 51,42,67,76,79], actually working with W/O microemulsions with low water content, were the first ones to point out that microemulsions need not be thermodynamically

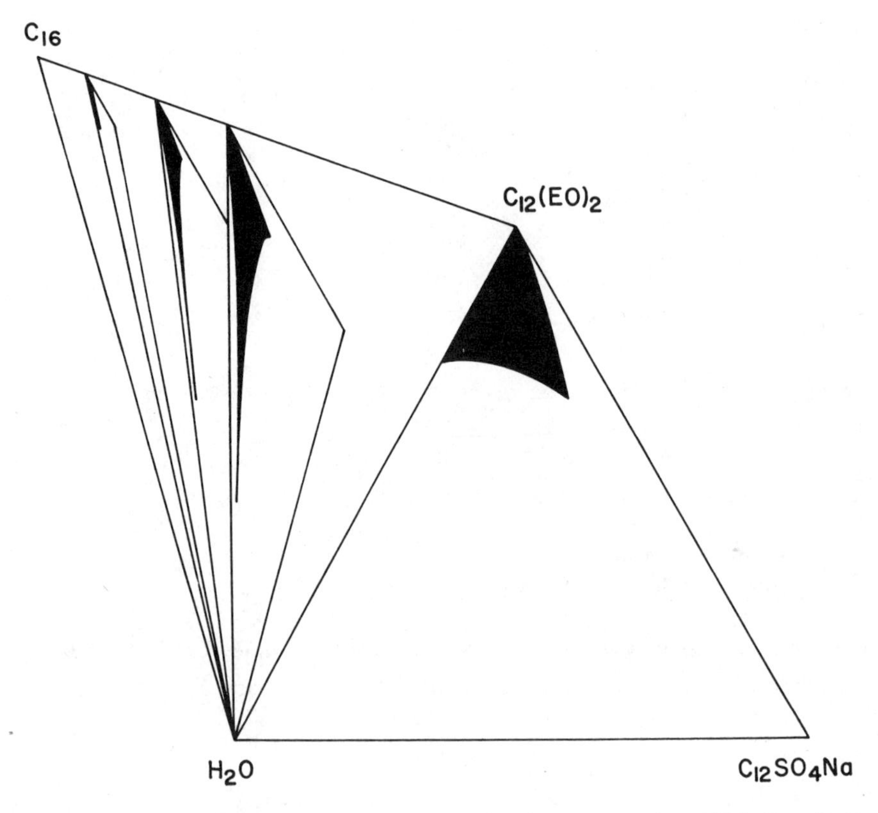


Figure 12. W/O microemulsions with bis-(ethylene glycol) dodecyl ether as a cosurfactant do not form as a direct continuation of the cosurfactant solution of water and surfactant at zero hydrocarbon content. The addition of hydrocarbon results in a structural change in the system giving rise to inverse micelles and a pronounced water solubilization. (From Ref. 77.)

stable. For the O/W microemulsions this is an important question and the experimental results are of sufficient interest for formulation work to merit some elucidation. The following sections will treat the connection with micellar solutions, the influence of order of addition for the components, influence of the hydrocarbon nature and some experience of long-term stability.

*(1) Relation to micellar solutions* The aqueous micellar solutions (Figs. 1 and 2) solubilize only modest amounts of cosurfactant or hydrocarbons. The level of approximately 10% by weight generally observed does not deserve the designation microemulsion. For this case, as has been mentioned earlier [36], the free energy of solubilization of water-insoluble organic compounds in micelles has been reported to be linearly related to the number of carbon atoms in the alkyl chain of the detergent.

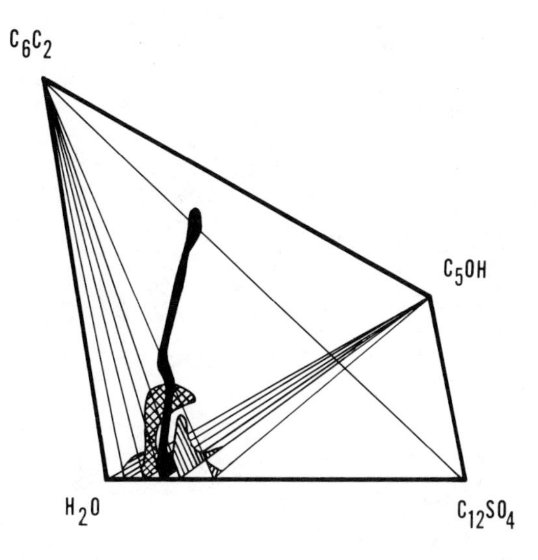


Figure 13. The O/W microemulsions emanate from the aqueous micellar solution, being critically dependent on the concentration of surfactant in water: $C_{12}SO_4$ = sodium dodecyl sulfate; $C_5OH$ = pentanol; $C_6C_2$ = p-xylene. (From Ref. 26.)

However, a *combination* of the hydrocarbon and the cosurfactant at an *optimum* level of surfactant in the aqueous solution will produce a microemulsion arising from the aqueous micellar solutions (Fig. 13). A comparison with the inverse micellar solutions (Fig. 4) is instructive; in that case a *combination* of water and soap will give mutually high solubility. For the O/W microemulsion a combination of hydrocarbon and cosurfactant will give the enhanced solubility.

The usual representation [39] of such systems is in a triangular diagram of water, hydrocarbon, and a surfactant/cosurfactant mixture at the three corners. Figure 13 makes it obvious that such a choice is not optimal; the representation in a triangle aqueous solution of (1) surfactant, (2) cosurfactant and (3) hydrocarbon facilitates the high solubilization region. A triangle from the water and hydrocarbon corners towards a point on the surfactant/cosurfactant axis would transform the large microemulsion area originating at $H_2O$ 85%/$C_{12}SO_4$15% to a small ellipse, which would be difficult to find and the relation to the micellar solutions would be lost.

*(2) Order of addition of components* In the system water, sodium dodecyl sulfate, pentanol, and *p*-xylene, the water with dissolved surfactant represents one component and a phase map of the solubility areas is illustrated by Fig. 14. The C and B areas belong to the W/O microemulsions (Fig. 4) and will not be treated. The region A is the O/W microemulsion illustrated in black in Fig. 13. This region was obtained by addition of the hydrocarbon as the last component. Addition of the co-

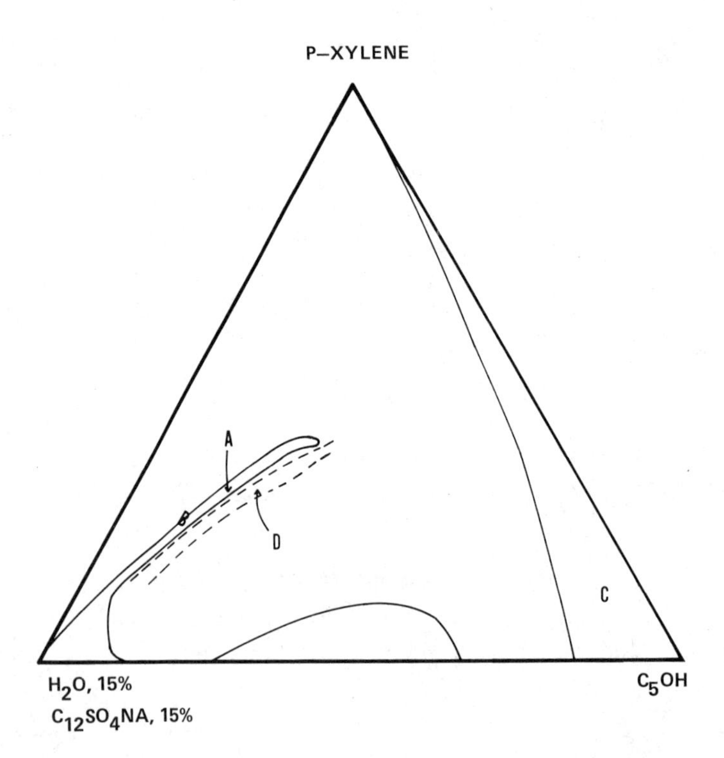


Figure 14. The solubility area for O/W microemulsions depends on the order of addition for the components. Addition of the hydrocarbon as the last component (A) gives solutions with more longevity than those (D) to which the cosurfactant was added last. The regions B and C are parts of the W/O microemulsions.

surfactant as the last component, which is the traditional manner of preparation [1,4-11,18,19,51,52,79], resulted in a different region; the area D surrounded by a dashed line. Microemulsions in both areas were transparent at preparation but those in area D separated rapidly, within 1 or 2 weeks. Of the microemulsions in area A, the ones with highest amount of hydrocarbon separated within days, and the ones with highest amount of water remained stable for several months.

The common preparation of microemulsions by adding the cosurfactant as the last component derives historically from Schulman's view of the role of the cosurfactant, and it is natural that it is still the preferred method, but the formulator should evaluate different methods to obtain maximum stability.

*(3) Influence of the nature of the hydrocarbon* W/O microemulsions are only insignificantly influenced by the nature of the hydrocarbon as to the region of stability (Fig. 4). The O/W microemulsions display an opposite behavior; the influence of the nature of the hydrocarbon is pronounced as far as maximum and min-

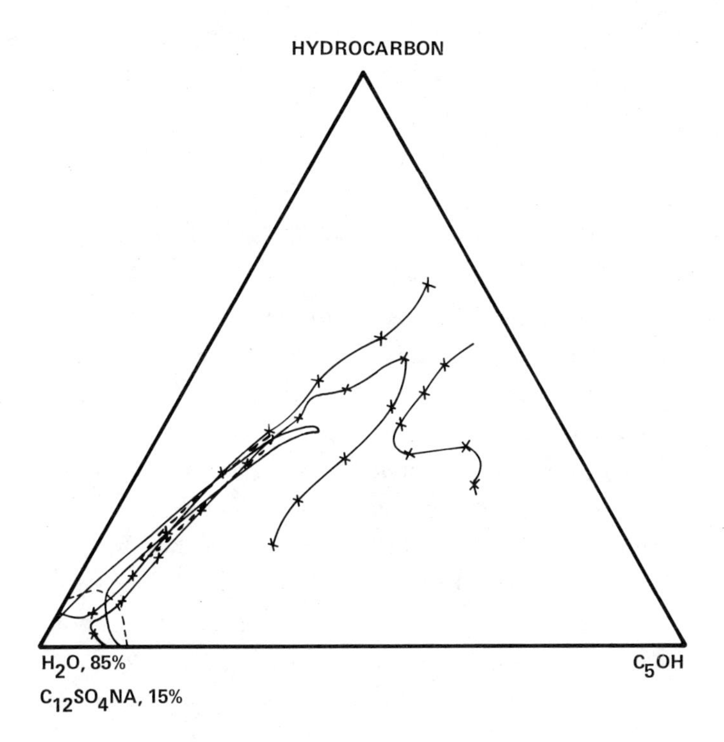


Figure 15. Different hydrocarbons give different amounts of solubilization, but the hydrocarbon/cosurfactant ratio remains identical: (——) *p*-xylene, (---) benzene, and (XXX) *n*-hexane.

imum water solubilization pattern (Fig. 15). The aliphatic *n*-hexane forms a continuous solubility area from the aqueous solution of the surfactant to the cosurfactant-hydrocarbon solution. The solubility region of the *p*-xylene is continuous from the aqueous solution, but does not connect to the hydrocarbon-cosurfactant solution. The microemulsion with benzene formed a microemulsion area with connection neither to the aqueous solution nor to the cosurfactant/hydrocarbon solution.

The nature of the hydrocarbon influenced only the level of solubilization; the hydrocarbon/cosurfactant ratio for the solubilization is constant. The hydrocarbon-surfactant weight ratio was constant for all three hydrocarbons, varying but little with the amount of water. The reason for this behavior is not yet clarified; further discussions must await increased knowledge about the structure in this part of the system.

*(4) Long-term stability and structural changes* The only samples for which no long-term changes could be found were those with low content (30% by weight) of the aqueous solution in the system containing *n*-hexane. The solutions in the

central part of the narrow solubility region either separated into several phases or were transformed to anisotropic phases. A more complete description of these changes is as yet not available.

### 2. *Nonionic Surfactants*

The systems presented for microemulsions with ionic surfactant were characterized by their response to the surfactant-cosurfactant ratio for W/O microemulsions and by the corresponding behavior, depending on the water-surfactant ratio, for O/W systems. The temperature effects at room temperature and the influence of the nature of the hydrocarbon have little significance and have not attracted significant research interest. The conditions for microemulsions stabilized by nonionics are entirely different: the nature of the hydrocarbon and temperature play dominant roles in the microemulsion formulation. A good example is provided by Fig. 16 [35] demonstrating the profound dependence on both temperature and nature of hydrocarbon. The conditions have been described in detail by Shinoda and coworkers [31,80-84]. The following treatment will be limited to the conditions which give rise to microemulsion states. The examples given are from investigations on aliphatic hydrocarbons; the conversion to other hydrocarbons has earlier been described [84].

The microemulsion phenomenon with nonionic surfactants is intimately related with the PIT phenomenon (the HLB temperature) [84] (Chap. 6) and the following section will describe the variation of solubilization when the temperature is increased through this range.

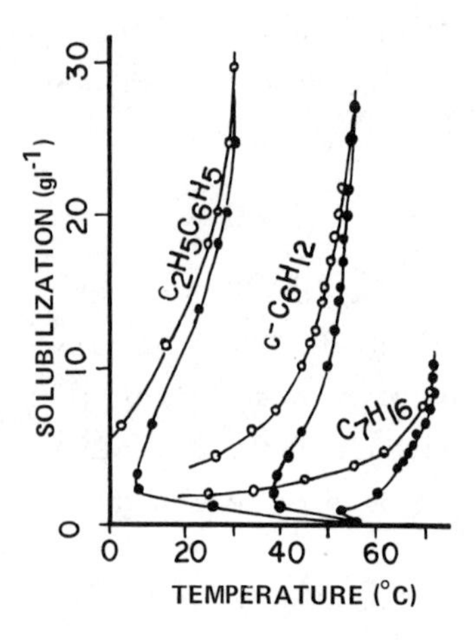


**Figure 16.** Solubilization by nonionic surfactants strongly depends on the temperature and the nature of the hydrocarbon. (From Ref. 83.)

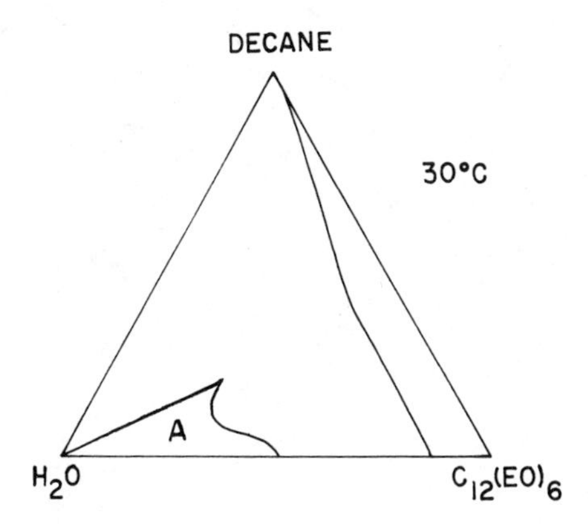


**Figure 17.** At temperatures below the HLB temperature, the surfactant is soluble in water forming normal micelles which solubilize hydrocarbon (A).

At low temperatures, far below the HLB level, the nonionic surfactant is soluble in water and the solubilization of hydrocarbon takes place in normal micelles. Figure 17 [85] shows the solubilization area (A) for decane at 30°C in an aqueous solution of hexaethylene glycol dodecyl ether. This is a normal solubilization and the term *microemulsion* is unwarranted.

An increase of the temperature in excess of the cloud point results in the surfactant not being soluble in the water; the solubilization area with normal micelles begins at a minimum concentration of surfactant $\beta$ in Fig. 17. In addition to the normal micelles, a new solubilization phenomenon is observed, characterized by both a maximum and a minimum hydrocarbon concentration for stability. This

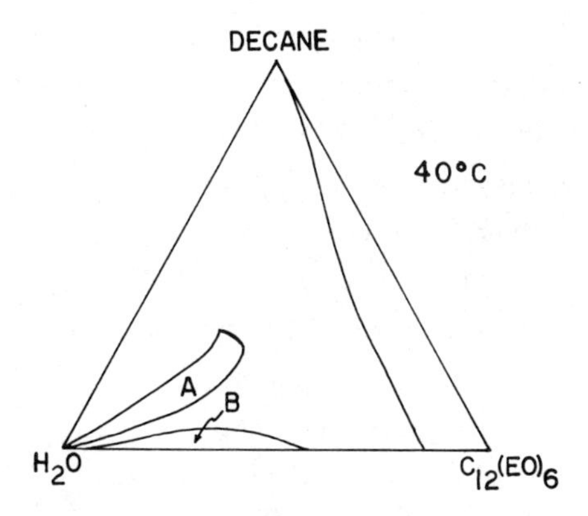


**Figure 18.** When the temperature approaches the HLB region, separate solubility areas (A) and (B) are formed from the aqueous micellar solution (t = 40°C).

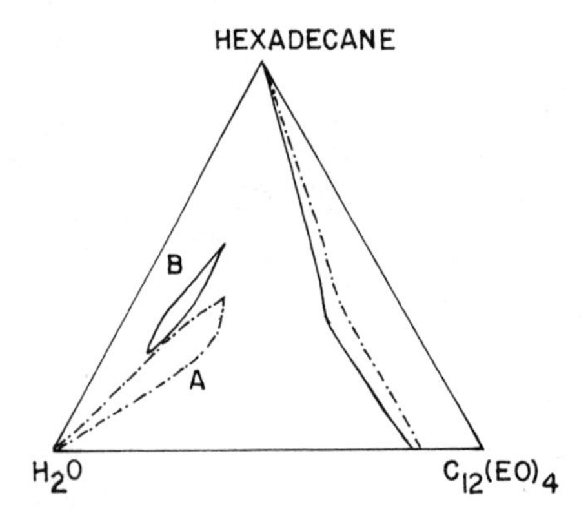


Figure 19. At temperatures close to the HLB temperature, the sectorial solubility region from the water is broken and the surfactant phase (B) appears an an isolated area.

region A in Fig. 18 certainly deserves the name microemulsion, and although its structure has not yet been experimentally clarified, the model used by Robbins [43] and others appears useful. The surfactant forms an interfacial layer between a central core of hydrocarbon and the surrounding aqueous continuum.

Further increase of the temperature leads to a separation of the microemulsion from the aqueous phase and a three-phase area is experienced. Figure 19 shows this development for the system water, tetraethylene glycol dodecyl ether, and hexdecane. The isolated phase was called the *surfactant phase* by Shinoda; it forms microemulsion with minimum amount of surfactant. This kind of microemulsion cannot be diluted with water nor with hydrocarbon. Its structure has not yet been determined with exactness; several suggestions involving bicontinuous structures have been made [83,86,87].

At temperatures in excess of the HLB value [88], the composition of the surfactant phase will be shifted toward higher surfactant concentration (Fig. 20), and the surfactant phase will coalesce with the hydrocarbon-emulsifier solution to the right in the diagram. The solubility area after coalescence permits pronounced solubilization of water and the solutions may be defined as W/O microemulsions. In some limited temperature ranges, solubility areas like those in Fig. 21 [85] may be encountered; the water-rich area has earlier been observed and named *second solubilization* [89]. Such systems may deserve the name *microemulsion*.

The microemulsions with nonionic surfactants are obviously extremely temperature sensitive. Attempts to improve this by the use of mixtures of nonionic and ionic surfactants give systems with improved temperature stability, but higher concentrations of surfactant were needed [86]. Mixtures [90] with extremely small amounts of ionic surfactant have shown interesting properties.

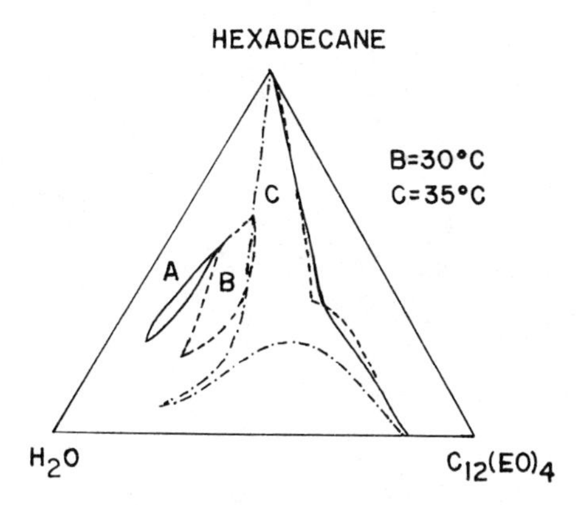


Figure 20. At temperatures above the HLB temperature, the surfactant phase (A) is at first transferred toward higher emulsifier concentrations and finally coalescence with the emulsifier/hydrocarbon takes place: (——) t = 25°C; (---) t = 30°C; (·—·—·—) t = 35°C.

### 3. *Detergentless Microemulsions*

Recently [91] the existence of microemulsions was demonstrated in systems with no surfactant present. In an assembly of water, *n*-hexane and 2-propanol a region (A, Fig. 22) was found to contain microdroplets of water in a hexane-rich organic matrix. The presence of aggregates of colloidal size was proved using ultracentrifugation and determination of electrical conductivity, in addition to visual examination. These microdroplets distinguished area A from area B, the properties of

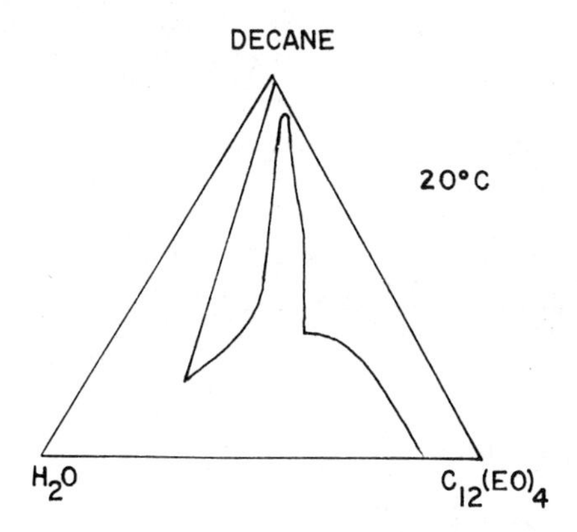


Figure 21. For some systems, in excess of but close to the HLB temperature, the water solubilization takes place in a sector emanating from the hydrocarbon corner.

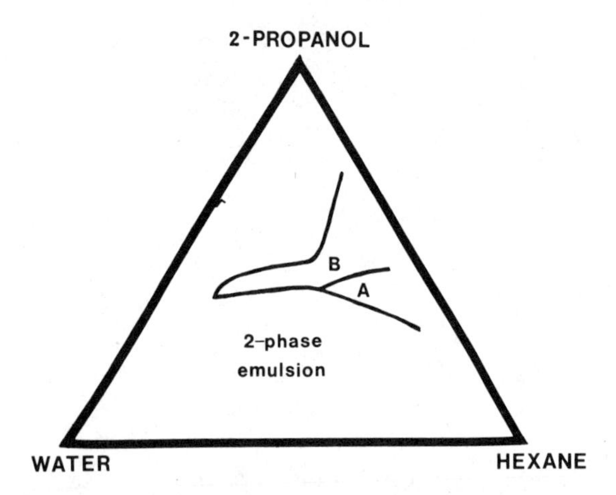


Figure 22. Detergentless microemulsions (A) are found in an area between the 2-propanol solution B and the two-phase emulsion. (From Ref. 91.)

which resembled a solution with molecular associations only. These microemulsions typically contained 8 to 10 wt % water with 45 to 46% each of the hexane and isopropanol.

The detergentless microemulsions offer promise of becoming a versatile and reasonably priced solvent for chemical reactions between water- and hydrocarbon-soluble chemicals. With this in mind, Holt and co-workers [92,93] have launched a program to determine the influence on the stability region by electrolytes and bases.

## III. APPLICATIONS OF MICROEMULSIONS

The microemulsions have reached a wide variety of applications; the chapter by Prince in Ref. 94 is enjoyable because of the personal involvement in the whole development by the author. A highly specialized volume on the problems encountered in tertiary oil recovery has recently been published [86,95]. In addition, some information may be found in the volumes by Lissant [96].

The following section will relate applications to the structure of microemulsions and show applications in which the ultrafine dispersity of the system is an advantage. It will emphasize the future applications somewhat to the disadvantage of the historical perspective.

### A. Fuels

One typical example of an application in which the size of the dispersed droplets is of importance is the use of microemulsions as fuels for combustion engines. There are numerous claims in patents and the literature [3,97-100] concerning improvement of the performance of gasoline engines by using microemulsions. The effects are similar to those observed in general combustion [92,101]; lower generation of nitrous oxides and less soot are found.

The possibility that microemulsions should reach a general application in the gasoline motor appears less probable. The use of gasoline is at present at such a level that a surfactant use of 5% by weight would involve tens of millions of tons of chemicals per year; a figure that obviously would require a development of chemical industry to a volume one magnitude higher than at present.

A more reasonable application would be as fuel for diesel engines. The use of this motor is considerably more limited and its specialized usage in highly controlled environments, such as in underground localities, make a more expensive fuel a possibility if the enhanced costs may be countered by reduced capital equipment, such as air purification installations. Experience so far with the diesel engine [102,103] show that the common emulsions with a drop size of 1 to 50 $\mu$m show little if any improvement in fuel performance. The greatest problem is the delay of ignition time, which in turn causes an increased fuel consumption, with no change in the nitrous oxide level in the exhaust gases and insignificant reduction of the smoke content.

These problems have been overcome by the use of microemulsions with an optimal choice of emulsifiers [102,103]. A cetane number in excess of 40 could be retained at a water content of 20% without addition of conventional cetane improvers such as amyl nitrate. The emulsifiers are long-chain polyethylene glycol dodecyl alkyl ethers, solubilizing up to 30% water in a standard diesel oil with a cetane number 43.

The results showed no improvement in the performance of the fuel; the use of fuel increased slightly based on the content of oil and emulsifiers. Improvement was found in the quality of the exhaust gases, especially at low speeds and high loads. Figure 23 shows the relative emission in g/hph of NO against smoke content for different injection timings. The upper part of the figure demonstrates the dilemma with the diesel motor for low speed and high load and a conventional fuel. A reduction of the injection time from 28 to 8° BTDC brings down the NO content from 15 to 5 g/hph, but gives a ten-fold increase in the smoke content. The advantage of the microemulsion fuel is obvious; the reduction in NO content is not countered by a similar increase of the smoke content even at low values of BTDC.

Continued investigation of these phenomena are at present in progress in order to determine the specific influence of droplet size.

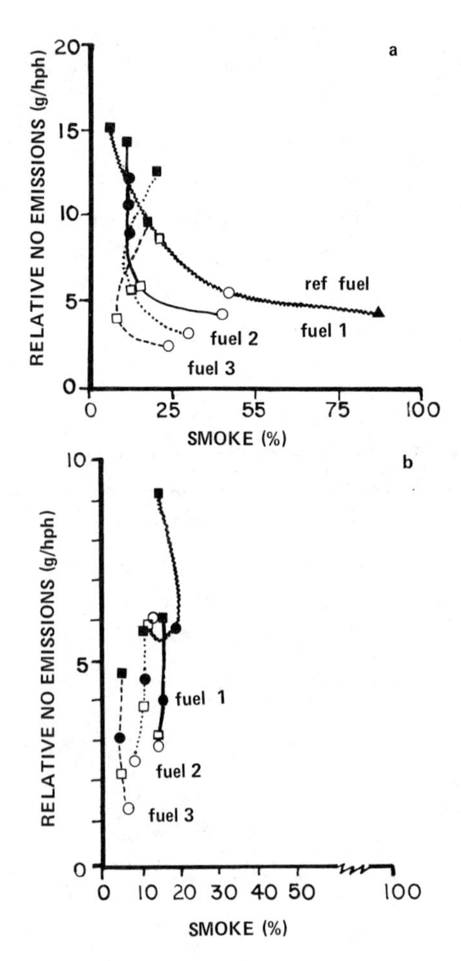


Figure 23. Amounts of emitted smoke and NO at various injection timings; ■ 28, ● 23, ★ 18, ○ 13, and ▲ 8° BTDC. The reference fuel (▀▄▀▄) is a diesel oil with cetane number 43. Fuel 1 ( ▬▬ ), fuel 2 (★★★), and fuel 3 (---) are microemulsion fuels with 10, 20, and 30% water, respectively. (a) Speed, 1380 rpm and 100% load ($P_e$ = 8.4 kp/cm$^2$). (b) Speed, 2300 rpm and 75% load ($P_e$ = 5.4 kp/cm$^2$). (From Ref. 103, G. Gillberg, and S. Friberg, in *Evaporation-Combustion of Fuels*, Advances in Chemistry Series No. 166 (J. T. Zung, ed.), 1978, p. 228.

B. Microemulsification of Waxes for Polishes

The first commercial use of a microemulsion was for the preparations with carnauba wax by George Rodowald in 1928 [94]. He found that boiling water, added to molten wax, gave a translucent emulsion. This wax emulsion turned out to give special properties to leather finish and to floor polish; after drying, the surface was glossy without buffing. The reason for the glossy surface is immediately evident; the particle size in the emulsion was less than 1500 Å.

The "emulsion" at room temperature was in reality a suspension of semisolid wax particles, stabilized by electric double-layer forces. The formation at high temperatures took place in a liquid system; the necessary reduction in interfacial free energy being facilitated by the fact that the carnauba wax contains a large number of —OH groups. These kind of microemulsions belong to the O/W ones (Fig. 13). A typical composition is shown in Table 1, and it is obvious that the emulsifier content at the level of 3.5% is less than that needed for pure hydrocarbons (Fig. 13). The explanation is found in the higher preparation temperature which extends the areas in Fig. 13 and in the high oxygen content of the wax. The action of the –OH groups of the wax replaces the cosurfactant in hydrocarbon compositions, reducing the total emulsifier content by a factor of 2 to 3.

### C. Tertiary Oil Recovery

The traditional production methods for crude oil leave part of the oil content in the reservoir, and the predicted shortage of petroleum fuels has stimulated an interest in the chemical recovery of the remaining oil. To discuss the subject completely of necessity reaches widely into different fields, such as the economic conditions, the political situation in the Middle East, and the taxation level decided by the U.S. Congress, the possibility of increasing the production facilities for petroleum sulfonates and polymers, as well as the scientific and technical knowledge necessary for a technically satisfactory process. These factors are treated in numerous publications, the reader with an interest covering the whole area may consult recent review articles [95].

**Table 1.** Composition of a Commercial Microemulsion

| Component | Wt % |
|---|---|
| Carnauba wax | 13.8 |
| Oleic acid | 1.7 |
| Potassium hydroxide | 0.5 |
| Borax | 1.0 |
| Water | 83.0 |

The following treatment will concentrate on the properties of the microemulsions used in the process, after a brief description of the conceptual problems involved in the recovery.

### *1. The Fundamentals of the Recovery Problem*

The remaining oil is trapped in pores; Fig. 24 gives a view of the conformation observed in the porous rock. The local conditions vary to a great extent and the oil may be trapped discontinuously for nonwetting conditions or may form a continuous layer if the oil is wetting the rock surface. The former case is, of course, less demanding. As is easily recognized, one requirement of flooding is a pronounced reduction in interfacial tension; this is a necessity in order to lower the pressure so as to remove the trapped oil against capillary forces. The capillarity problem alone has resulted in a large number of publications on the subject [104-108].

The reduction of interfacial free energy is accomplished by the use of surfactants, mainly petroleum sulfonates. Two main approaches have been employed, one using a low surfactant concentration solution filling a large portion of the pore volume, between 15 and 60%, the other one utilizing a high concentration of surfactant, but with the volume restricted to a small part of the pore volume.

In spite of the rich literature on the subject [95] direct comparisons are relatively few. Gogarty [109] compared results from Sankvik [110], Hill [111], and Davis [112] finding the efficiency ratios (oil recovered/surfactant injected) in turn equal to 0.70 (2), 2.62 (1.3), and 4.09 (10.4). The numbers within brackets show the surfactant concentrations and indicate some superiority for the high concentration systems.

Before the microemulsions used for flooding are further described, a short description of the total flooding system will be given.

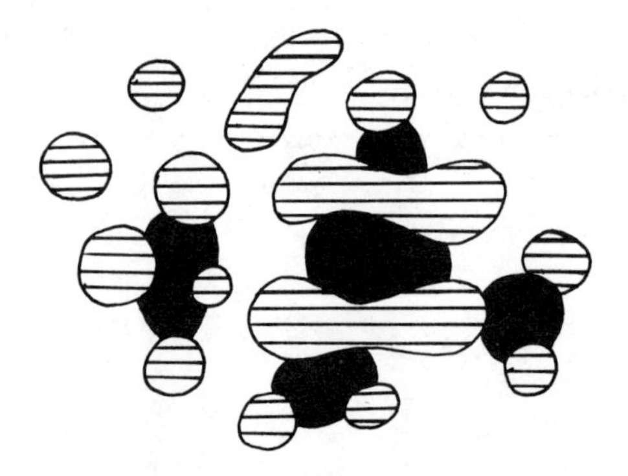

**Figure 24.** Porous rock with trapped oil.

*2. The Total Flooding System*

The removal of the residual oil from its reservoir was early described [113] and includes the following regions in order as they move [114] through the reservoir (Fig. 25). The surfactant solution, the microemulsion, is preceeded by the oil bank of released oil. It is succeeded by a polymer solution the mission of which is to prevent uneven diffusion of the surfactant (mobility control). The polymer solution is followed by water flooding. The properties of the polymer solution are economically important and the development in this area is intense and the knowledge of their interaction with porous rock [115-117] and with the microemulsion [118] is extensive. In spite of this, authorities [114] gave the verdict "All polymers we have studied suffer from one or more of the following problems: mechanical [119], chemical [120], thermal or bacterial [121] degradation, injection face plugging [122], excessive adsorption and inaccessible pore volume [123], or undesirable phase behavior when mixed with surfactants [124]."

*3. Microemulsions for Tertiary Oil Recovery*

The term *surfactant* in tertiary oil recovery includes several components: (a) the surfactant itself, usually a petroleum sulfonate [110,125,126], (b) a cosurfactant which may be ethoxylated alcohols or sulfated ethoxylated alcohols [127,128], and (c) alcohols, ethers, glycols, etc., which are called *cosolvents*.

These components are commercial mixtures and it is obvious that the number of components is too large for the determination of a traditional phase diagram. The phase conditions are instead presented as a map with water, hydrocarbon, and "surfactant" as the three components. In principal, this approach [125,129,130] will give a phase diagram (Fig. 26) with a single phase reaching from the oil corner to aqueous solutions. This is an expected behavior considering the phase

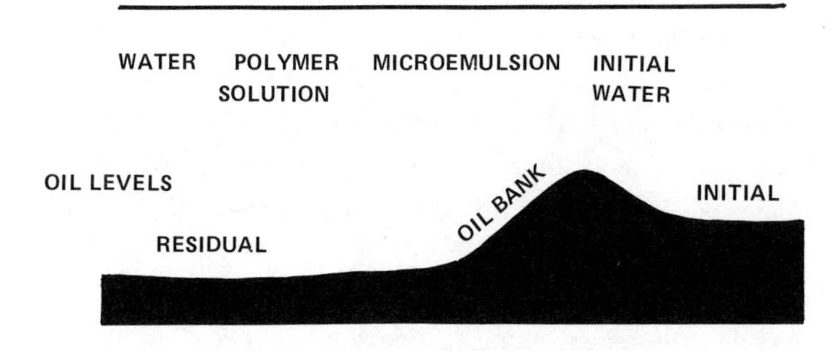


**Figure 25.** Microemulsion flooding involves the passage of several zones through the rock.

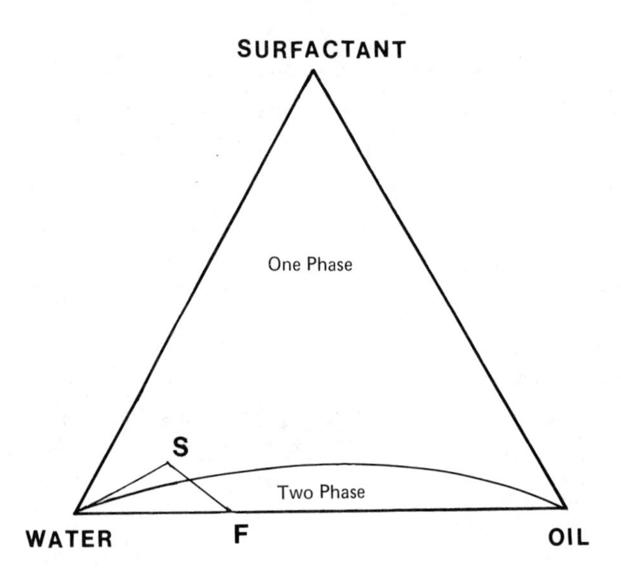


Figure 26. The conditions before microemulsions reaches the site may be modeled by the composition at F. When the microemulsion passes, the composition resembles S. After the microemulsion, the composition moves toward the aqueous corner.

regions in diagrams of pure compounds. Figure 15 clearly shows the continuous region obtained also with pure compounds.

According to Healy and Reed [125,129,130], the extension of the one-phase area toward the water-oil axis is considered of importance for tertiary oil recovery as reflecting conditions at the front and the rear of the moving microemulsion bank. At the beginning, water and oil only are present, and the point F in Fig. 26 [116] describes the state. The presence of the microemulsion will increase the concentration of "surfactant" and the composition of the system will move from the oil-water axis into the one-phase area (F $\rightarrow$ S, Fig. 26). When the microemulsion has passed the zone, the concentration of water will increase and the total composition will be shifted toward the aqueous corner in the diagram.

During the one-phase state, all oil present will be dissolved and recovered; for the two-phase area, it is essential that the interfacial tension has a sufficiently low value. This fact has given rise to an intense interest in low-tension systems, and numerous publications have been concerned with interfacial properties during the last decade [131-141].

These contributions give evidence that microemulsions are not a necessity for obtaining low interfacial tensions. However, several theoretical approaches indicate extremely low interfacial tensions between a microemulsion and a solution.

Ruckenstein's contributions [28,57] contain detailed theories for the stability of microemulsion systems in which the interfacial free energy is calculated.

Miller and collaborators [142] demonstrated the influence of association aggregate size in one of the phases for low interfacial tensions by an adaptation of the Cahn and Hilliard [143-145] theory for interfacial tension, modified to be useful for compositions far from critical points. The interfacial tension was found to be given by

$$\gamma = 2 \int_{\phi\alpha}^{\phi\beta} \{ \kappa(\phi)[F(\phi) - F^*(\phi)] \}^{1/2} \, d\phi \tag{9}$$

$\kappa$ was evaluated by summation of the intermicellar potential, using a cell theory with a uniform distribution of micelles in excess of a distance

$$\kappa = -\frac{\Pi}{3}\left(\frac{3}{4\Pi a_0^3}\right)^2 \int_D^\infty u(r - 2a_0) r^4 \, dr \tag{10}$$

In order to make a comparison with experimental results, the intermicellar Hamaker constant was adjusted and the micellar size at phase separation to a surfactant phase was used.

The comparison with experimental results was good; Fig. 27 clearly shows the predicted tensions both between the brine and the surfactant phase to be of the same magnitude as those obtained experimentally.

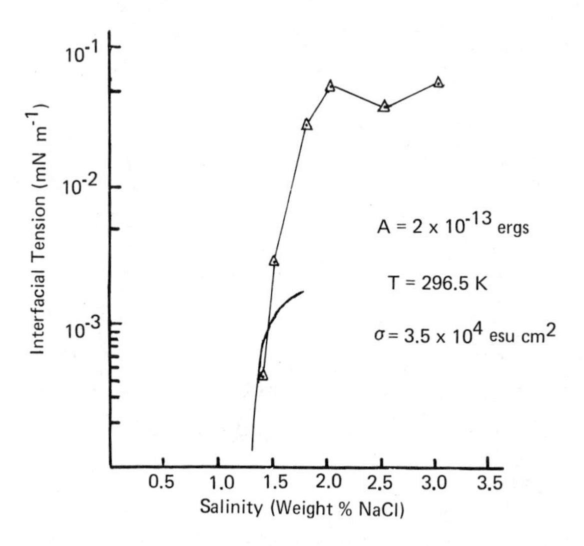


**Figure 27.** Comparison between theoretical [118] and experimental results for interfacial tension.

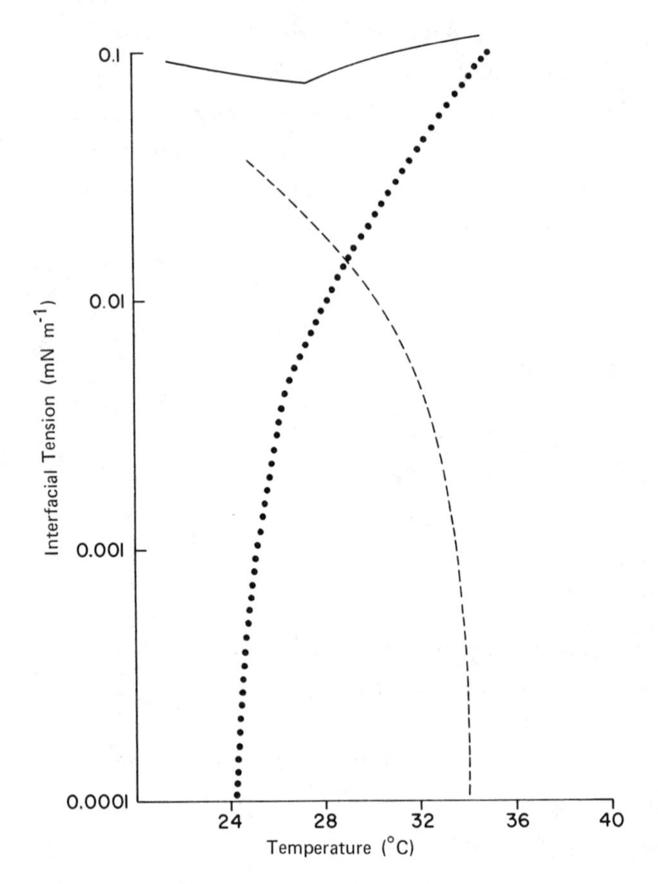


Figure 28. Interfacial tension in a system water, nonionic surfactant, and hydrocarbon: (——)$\gamma_{W/O}$, (· · ·)$\gamma_{W/\text{surfactant phase}}$, and (---)$\gamma_{O/\text{surfactant}}$.

The relative merit of two- and three-phase systems for ultra-low interfacial tensions is well demonstrated by the results from nonionic surfactants [146]. Water and hydrocarbon form a surfactant phase with a single nonionic surfactant in a limited temperature region [31].

The surfactant phase is formed by phase separation from the aqueous phase at increasing temperatures and will, at higher temperatures, coalesce with the hydrocarbon phase. Determination of the three interfacial tensions [146] gives a reasonable picture of the importance of the microemulsion (the surfactant phase) for the interfacial tension. The aqueous and the oil phase show a minimum for the interfacial free energy at the PIT value, in accordance with the views of Shinoda [83]. This minimum reaches to a value of $10^{-2}$ mN $m^{-1}$. The values of the interfacial tension against the surfactant phase, on the other hand, could be as low as possible; experimental limitations on temperature and vibration control limited the results to the values in Fig. 28. The results are instructive as to the difference between associated and nonassociated systems and illustrate the importance of microemulsion formation for ultra-low interfacial tension.

## D. Reactions in Microemulsions

The catalytic effect of association structures has been known since 1959 [147]. The interest up to 1970 mainly focused on reactions involving effects by normal micelles [148,149]; the catalytic effects by inverse micelles containing water were demonstrated in 1971 [134,150], followed by a publication on water-free systems [151]. The latter development has been described in a book [152]. The reactions in what may truly be described as microemulsions have not attracted corresponding attention.

Microemulsions are useful for the study of the biologically important porphyrine-metalporphyrine systems. The porphyrine is chosen to be soluble in the oil phase and the combination with metal ion from the aqueous phase takes place at the interface giving a reasonable model for biological systems.

Letts and Mackay [153] used O/W microemulsions to study the interaction between metal ions and tetraphenyl porphine. Of Mg(II), Mn(II), Zn(II), Co(II), and Cu(II), only Cu(II) was incorporated into the porphine in the presence of anionic surfactants, while cationic ones prevented the process. The authors suggested the existence of a copper-porphine complex at the interfacial region. The complex was stabilized by the presence of triphenyl phosphine. Smith and Barden used the detergentless microemulsions [91] to study the same system, finding a rapid incorporation of metal ions into protoporphyrine IX [154].

The detergentless microemulsions were also used for investigations on the interactions between the side chains of $N^{\alpha}$-dodecanoyl-amino-alcohols and Cu(II) ions [155]. The chelation of Cu(II) through an imidazole nitrogen and the amide carbonyl oxygen of $N^{\alpha}$-dodecanoyl-L-distinidol was proved by comparison with EPR spectra from simpler compounds. The agreement with the EPR data for Cu(II)-substituted human carbonic anhydrase [156] was an interesting and noteworthy feature.

The usefulness of the microemulsion media for modeling of biological systems has further been emphasized by Mackay [157] in his contributions on the photodegradation of chlorophyll. The results clearly showed the influence of the interface on the process and encourage further explanation of these sytems.

Although the data on reactions in microemulsions so far are not very extensive it appears obvious that this kind of dispersion promises an interesting challenge for the future.

## SUMMARY

Microemulsions have been described, emphasizing the thermodynamic basis for their stability, their connection with micellar solutions, and some of their applications.

It appears obvious that this kind of system offers a great potential for both scientific contributions, perhaps foremost in their use as reaction media, and technical development.

## APPENDIX: FUNDAMENTAL APPROXIMATIONS IN CNDO/2

### Introduction

Molecular orbital theory has its basis in quantum mechanics. Molecules are "built" from first principles; atomic nuclei are placed at bonding distances, and electrons are added to molecular orbitals, the lowest energy orbital being filled first. The molecular orbitals are constructed from a linear combination of atomic orbitals (LCAO). The molecular orbitals are made self-consistent by adjusting the coefficients of the atomic orbitals until the total energy does not vary to within a specified tolerance limit.

Molecular orbital calculations involve a large amount of variables and many-center integrals to describe electron-electron, electron-nuclei, and nucleus-nuclei interactions. Also, since the electronic motion is not independent, but correlated, the most accurate calculations take this physical reality into account, increasing the complexity even further.

Thus, application of ab initio molecular orbital calculations involving no empirical parameters have been limited to small molecules. Advances are currently being made using ab initio methods also for intermolecular forces, yet for the present the semiempirical molecular orbital techniques are preferred being more efficient and accurate for larger molecular systems including intermolecular interactions.

In the semiempirical approach, the mathematical approximations used in the ab initio method are retained, and the integral terms are approximated by semiempirical parameters. The following sections will describe the approximations and parameters of the CNDO/2 method [158-163].

### CNDO/2

Molecular orbital calculations using CNDO/2 treat only valence shell electrons explicitly (Table 2). The inner shell electrons are treated as part of a nonpolarizable core. Slater type orbitals (STOs) are used for the atomic orbital basis set.

Table 2. Molecular Energy and Geometry: A Comparison between CNDO/2 Calculations and Experimental Results

| Molecule | Bond | Bond length Calc. | Bond length Exp. | Bond angle Calc. | Bond angle Exp. | Dipole moment Calc. | Dipole moment Exp. |
|---|---|---|---|---|---|---|---|
| HF | HF | 1.000 | .92 | | | 1.86 | 1.82 |
| $H_2O$ | HO | 1.03 | .98 | 104.7 | 104.3 | 2.08 | 1.85 |
| $H_3N$ | HN | 1.07 | 1.02 | 106.4 | 109.1 | 2.08 | 1.47 |
| $H_2CO$ | HC | 1.12 | 1.12 | 115.0 | 116.0 | 1.92 | 2.37 |
| | CO | 1.25 | 1.21 | | | | |

*Approximation 1*

The first approximations of CNDO/2 is zero differential overlap (ZDO). Here the electron repulsion integrals between different atomic orbitals is set equal to zero, given by

$$(\mu\nu|\lambda\sigma) = (\mu\mu|\lambda\lambda)\delta_{\mu\nu}\delta_{\lambda\sigma}$$

where $\delta_{ij}$ is the Kronecker delta, which equals unity when i = j and zero for i ≠ j. The symbols in parenthesis represent a Dirac notation for the electron repulsion term.

$$(\mu\nu|\lambda\sigma) = \iint \phi_\mu(1)\frac{1}{r_{12}}\phi_\lambda(2)\phi_\sigma(2)\, dy_1\, dy_2$$

where $\phi_\mu(1)$ represents an atomic orbital containing electron (1).

This approach solves the three- and four-center integral problem by neglecting them all, meaning that the atomic orbitals in question ($\phi_i$ and $\phi_j$) do not overlap anywhere in space.

*Approximation 2*

The overlap integrals $\delta_{\mu\nu}$ are set equal to the Kronecker delta $\delta_{ij}$. The physical interpretation of the overlap integral [e.g., $(\mu\mu|\lambda\lambda)$] is an electron repulsion term between the electron cloud of $\phi_\mu^2$ and $\phi_\lambda^2$, with one electron in the atomic orbitals $\phi_\mu$ and $\phi_\lambda$.

The overlap integrals are also neglected when normalizing the molecular orbitals.

*Approximation 3*

As a result of approximation 1, the integral terms which remain of the form $(\mu\mu|\lambda\lambda)$ are given parameter values $\gamma_{\mu\lambda}$. These electron interaction integrals $\gamma_{\mu\lambda}$ depend only on the nature of the atoms to which $\phi_\mu$ and $\phi_\lambda$ belong. This results in a set of integrals $\gamma_{AB}$ which describe the average repulsion between a valence electron on atom A and valence electron on atom B.

*Approximation 4*

When $\phi_\mu$ and $\phi_\nu$ are both centered on atom A, the integrals $(\mu|V_b|\nu)$ are set equal to zero if $\mu \neq \nu$; when $\mu = \nu$, the integral is the same for all valence orbitals on atom A,

$$(\mu|V_B|\nu) = V_{AB}$$

$\Sigma_{B\neq A} V_{AB}$ gives the interaction of an electron in $\phi_\mu$ with the cores of all remaining atoms B,

$$V_{AB} = Z_B \gamma_{AB}$$

where $Z_B$ is the charge of atom B.

*Approximation 5*

Bonding between atoms is calculated from the core integrals $H_{\mu\nu}$.

$$H_{\mu\nu} = \beta^\circ_{AB} S_{\mu\nu} \qquad \mu \text{ on atom A}, \ \nu \text{ on atom B}$$

Here the overlap integral B is not neglected, as it contributes to the bonding. The proportionality constant $\beta^\circ_{AB}$ is called the *resonance integral.* It has no real physical significance, but B is purely an empirical number. Each atom has its characteristic $\beta^\circ$ value. $\beta^\circ_{AB}$ is a weighted average of $\beta^\circ_A$ and $\beta^\circ_B$,

$$\beta^\circ_{AB} = \frac{K(\beta^\circ_A + \beta^\circ_B)}{2}$$

K equals 0.75 when A or B is a second-row element and 1 otherwise.

The final parameter to be discussed is the one center integral $U_{\mu\mu}$ appearing in the diagonal matrix elements of the core Hamiltonian,

$$H_{\mu\mu} = U_{\mu\mu} - \sum_{B(\neq A)} (\mu|V_B|\nu)$$

$U_{\mu\mu}$ is an atomic quantity representing the energy of atomic orbital $\phi_\mu$ in the bare field of the core of its own atom. It is calculated from experimental atomic energy levels,

$$-\frac{1}{2}(I_\mu + A_\mu) = U_{\mu\mu} = \left(Z_A - \frac{1}{2}\right)\gamma_{AA}$$

where $I_\mu$ is the ionization potentidal of the orbital $\phi_\mu$ on atom A, and $A_\mu$ is the electron affinity of $\phi_m$ on atom A.

## Acknowledgments

The authors acknowledge the influence Professors Per Ekwall and Kozo Shinoda have had on their concept of microemulsions, as well as mnay stimulating discussions with Hans-Fredrik Eicke, Leon Prince, Henri Rosano, and Dinesh O. Shah. Clarence Miller kindly made available a manuscript prior to publication.

## Symbols

$a_0$ micellar radius

$B_2$ second virial coefficient

$D_\nu$ diffusional coefficient for specimen $\nu$

$f$ free energy per unit volume

$f^\sigma$ interfacial free energy

$F(\phi)$ free energy for a mixture of *uniform* composition $\phi$

$F^*(\phi)$ free energy for a mixture of bulk phases with a micelle volume fraction $\phi$

$\bar{g}$ mass of one droplet

$h$ Planck's constant

$\bar{H}$ bending stress

$J_\upsilon$ diffusional flux in a dispersed system

$k$ Boltzmann constant

$K_\upsilon$ inverse Debye-Hückel distance for the medium $\upsilon$

$\bar{K}$ torsion stress

$m$ number of droplets

$n_\upsilon$ number of moles of compound $\upsilon$

$N_t$ total number of sites in the system, $V/v_1$

$\bar{p}$ excess pressure in the interfacial transition zone

$P$ applied pressure on a droplet

$r$ distance between the centers of two particles

$R$ radius of the droplet under consideration

$R_1$ average distance between droplets $= 2R(0.74\phi_2)^{1/3}$

$R_2$ $1.3\ R_1$

$S$ entropy per unit area of interfacial layer

$S^\upsilon$ maximum solubility of the cosurfactant in the $\nu$ phase

$T$ absolute temperature

| | |
|---|---|
| U | interaction potential between molecules and/or particles |
| $v_1$ | volume of a molecule of the continuous phase |
| $v_\upsilon$ | volume of one molecule (or one particle) of $\nu$ |
| $x^\upsilon$ | mole fraction of species $\nu$ |
| z | coordinate perpendicular to the interface |
| $\alpha$ | growth coefficient (time dependence) for perturbations |
| $\beta$ | wave number of the perturbation |
| $\Gamma_\upsilon$ | amount of component at the interface |
| $\Delta^\sigma_\alpha$ | fractional increase of the volume forming a unit interfacial area by taking the components 1 and 2 from the continuous phases |
| $\Delta^p_\alpha$ | fractional increase of the volume forming the polar phase by taking the components 1 and 2 from the apolar phase |
| $\varepsilon_\upsilon$ | dielectric constant of the medium $\upsilon$ |
| $\mu_\upsilon$ | chemical potential per molecule or particle |
| $\nu = 1$ | continuous phase |
| $\nu = 2$ | dispersed phase |
| $\nu = a$ | apolar phase |
| $\nu = p$ | polar phase |
| $\xi$ | mole fraction of cosurfactant |
| $\tau$ | thickness of the interfacial zone (approximately the length of the surfactant molecule |
| $\phi_\upsilon$ | volume fraction of phase $\upsilon$ |
| $\phi_\nu$ | volume fraction of component $\nu$ |
| $\phi_{\nu 0}$ | volume fraction of phase $\nu$ in unperturbed state |
| $\psi_0$ | electric surface potential |

## References


1. T. P. Hoar and J. H. Schulman, *Nature 152*:102 (1943).
2. D. Bowden, and J. Holmstine, U.S. Patent 2,045,455 (1936).
3. V. R. Kokatnur, U.S. Patent 2,111,100 (1935).
4. J. H. Schulman, W. Stoeckenius, and L. M. Prince, *J. Phys. Chem. 63*:1677 (1959).
5. J. H. Schulman and D. P. Riley, *J. Colloid Sci. 3*:383 (1948).
6. J. H. Schulman and J. A. Friend, *J. Colloid Sci. 4*:497 (1949).
7. J. E. Bowcott and J. H. Schulman, *Z. Electrochem. 11*:117 (1955).
8. W. Stoeckenius, J. H. Schulman, and L. M. Prince, *Kolloid-Z. 169*:170 (1960).
9. D. F. Sears and J. H. Schulman, *J. Phys. Chem. 68*:3529 (1964).
10. I. A. Zlochower and J. H. Schulman, *J. Colloid Interface Sci. 24*:115 (1967).
11. C. E. Cooke and J. H. Schulman, in *Surface Chemistry* (P. Ekwall, ed.), Munksgaard, Copenhagen, Denmark, 1965, p. 231.
12. D. O. Shah and R. M. Hamlin, *Science 171*:483 (1971).
13. J. R. Hansen, *J. Phys. Chem. 78*:256 (1974).

14. G. Gillberg, H. Lehtinen and S. Friberg, *J. Colloid Interface Sci. 33*:40 (1970).
15. E. Sjōblom and S. Friberg, *J. Colloid Interface Sci. 67*:16 (1978).
16. J. H. Schulman and J. B. Montague, *Ann. N.Y. Acad Sci. 92*:366 (1961).
17. H. L. Rosano, H. Schiff, and J. H. Schulman, *J. Phys. Chem. 66*:1928 (1962).
18. L. M. Prince, *J. Colloid Interface Sci. 23*:165 (1967).
19. L. M. Prince, *J. Colloid Interface Sci. 29*:216 (1969).
20. P. A. Winsor, *Trans. Faraday Soc. 44*:376 (1948); *46*:762 (1950).
21. S. R. Palit, V. A. Moghe, and B. Biswas, *Trans. Faraday Soc. 55*:463 (1959).
22. P. Ekwall, I. Danielsson, and L. Mandell, *Kolloid-Z. 169*:113 (1960).
23. A. W. Adamson, *J. Colloid Interface Sci. 29*:261 (1969).
24. K. Shinoda and H. Kunieda, *J. Colloid Interface Sci. 42*:381 (1973).
25. S. I. Ahmad, K. Shinoda, and S. Friberg, *J. Colloid Interface Sci. 47*:32 (1974).
26. D. G. Rance and S. Friberg, *J. Colloid Interface Sci. 60*:207 (1977).
27. C. L. Murphy, Ph.D. Thesis, University of Minnesota, Minneapolis, Minnesota (1966).
28. E. Ruckenstein and J. C. Chi, *J. Chem. Soc. Faraday Trans. II 71*:1690 (1975).
29. H. Reiss, *J. Colloid Interface Sci. 53*:61 (1975).
30. S. Levine and K. Robinson, *J. Phys. Chem. 76*:876 (1972).
31. H. Saito and K. Shinoda, *J. Colloid Interface Sci. 24*:10 (1967).
32. T. Nakagawa and K. Tori, *Kolloid-Z. 168*:132 (1960).
33. T. Nakagawa, K. Kuriyama, and H. Inove, *J. Colloid Interface Sci. 15*:168 (1960).
34. K. Kunyama, *Kolloid-Z. 180*:55 (1962).
35. H. Saito and K. Shinoda, *J. Colloid Interface Sci. 32*:647 (1970).
36. K. S. Birdi, *Kolloid-Z. Z. Polym. 250*:731 (1972).
37. J. E. L. Bowcott and J. H. Schulman, *Z. Electrochem. 59*:283 (1955).
38. H. L. Rosano, *J. Soc. Cosmetic Chemists 25*:609 (1974).
39. L. M. Prince, *J. Colloid Interface Sci. 52*:182 (1975).
40. C. A. Miller and L. E. Scriven, *J. Colloid Interface Sci. 33*:360 (1970).
41. J. P. O'Connel and R. J. Brugman, in *Improved Oil Recovery by Surfactant Flooding* (D. O. Shah and R. S. Schechter, eds.) Academic, New York, 1977, p. 339.
42. W. Albers and J. Th. G. Overbeek, *J. Colloid Sci. 14*:510 (1959).
43. M. L. Robbins, in *Micellization, Solubilization and Microemulsions* (K. L. Mittal, ed.), Plenum, New York, 1977, p. 713.
44. C. A. Miller and P. Neogi, *J. Amer. Inst. Chem. Eng. 26*:212 (1980).
45. S. Friberg, L. Rydhag, and T. Doe, in *Lyotropic Liquid Crystals* (S. Friberg, ed.), *Adv. Chem. Series No. 152*, American Chemical Society, Washington, 1976, p. 28.
46. H.-F. Eicke, *J. Colloid Interface Sci. 68*:440 (1979).
47. C. Wagner, *Z. Physik. Chem. 132*:273 (1928).
48. Y. Talman and S. Prager, *Nature 267*:233 (1977).
49. Y. Talman and S. Prager, *J. Chem. Phys. 69*:2984 (1978).
50. J. L. Finney, *Proc. Roy. Soc. A319*:479 (1970).
51. W. E. Gerbacia and H. L. Rosano, *J. Colloid Interface Sci. 44*:242 (1973).
52. W. E. Gerbacia, H. L. Rosano, and M. Zajac, *J. Amer. Oil Chemists Soc. 53*:101 (1976).
53. G. Scatchard, *Chem. Rev. 8*:321 (1931).

54. M. J. Vold, *J. Colloid Sci. 16*:1 (1961).
55. H.-F. Eicke, *J. Colloid Interface Sci. 52*:65 (1975).
56. H.-F. Eicke, J. C. W. Shephard, and A. Steinman, *J. Colloid Interface Sci. 53*:678 (1978).
57. E. Ruckenstein, in *Micellization, Solubilization and Microemulsions* (K. L. Mttal, ed.), Plenum, New York, 1977, p. 755.
58. J. W. Cahn, *J. Chem. Phys. 42*:93 (1965).
59. P. Ekwall, in *Advances in Liquid Crystals* (G. H. Brown, ed.), Academic, New York, 1974, p. 7.
60. J. W. McBain, M. E. Laing, and A. F. Titley, *J. Chem. Soc. 115*:1279 (1919).
61. E. R. Jones and C. R. Bury, *Phil. Mag. 4*:1 (1927).
62. P. Ekwall, *Acta. Acad. Aboensis (Math. Phys.) 4*:1 (1927).
63. K. Shinoda, *Colloidal Surfactants*, Academic, New York, 1963.
64. P. Mukerjee, *Adv. Colloid. Interface Sci. 7*:241 (1967).
65. E. Ruckenstein and R. Nagarajan, *J. Phys. Chem. 79*:2622 (1975).
66. J. N. Israelachvili, D. J. Mitchell, and B. W. Ninham, *J. Chem. Soc. Faraday Trans. II 72*:1525 (1976).
67. G. Anianson, S. Wall, M. Almgren, H. Hoffman, I. Kielmann, W. Ulbricht, R. Zana, J. Lang, and C. Tondre, *J. Phys. Chem. 80*:805 (1976).
68. K. S. Birdi and T. Magonisson, *Colloid Polym. Sci. 254*:1059 (1976).
69. S. Friberg and I. Buraczewska, *Progr. Colloid Polym. Sci. 63*:1 (1978).
70. M. Clausse and R. Rayer. in *Colloid and Interface Science* (M. Kerker, ed.), Vol. II, Academic, New York, 1976, p. 217.
71. H.-F. Eicke and H. Christen, *J. Colloid Interface Sci. 46*:417 (1974); *48*:281 (1974).
72. H.-F. Eicke and H. Shepherd, *Helv. Chim. Acta. 57*:1951 (1974).
73. J. A. Pople, D. P. Santry, and G. A. Segal, *J. Chem. Phys. 43*:129 (1965).
74. S. Engstrom and H. Wennerstrom, *J. Phys. Chem. 82*:2711 (1978).
75. E. Barouch, E. Matijevic, T. A. Ring, and J. M. Finlan, *J. Colloid Interface Sci. 67*:1 (1977).
76. N. Muller, *J. Colloid Interface Sci. 63*:383 (1977).
77. H. Sagitani and S. E. Friberg, *J. Dispersion Sci. Technol. 7*:151 (1980).
78. L. M. Prince, *J. Soc. Cosmetic Chemists 21*:193 (1970).
79. W. Gerbacia, H. L. Rosano, and J. H. Whittam, in *Colloid and Interface Science* (M. Kerker, ed.), Vol. II, Academic, New York, 1976, p. 245.
80. K. Shinoda, *J. Colloid Interface Sci. 24*:4 (1967).
81. K. Shinoda, *J. Colloid Interface Sci.*, *34*:278 (1970).
82. K. Shinoda and T. Ogawa, *J. Colloid Interface Sci.*, *24*:56 (1967).
83. K. Shinoda and S. Friberg, *Adv. Colloid Interface Sci. 4*:281 (1975).
84. K. Shinoda and H. Arai, *J. Phys. Chem. 68*:3485 (1964).
85. S. E. Friberg, I. Buraczewska, and J. C. Ravery, in *Micellization, Solubilization and Microemulsions* (K. L. Mittal, ed.), Plenum, New York, 1977, p. 901.
86. S. Friberg, I. Lapczynska, and G. Gillberg, *J. Colloid Interface Sci. 56*:19 (1976).
87. L. E. Scriven, *Nature 263*:123 (1976).
88. T. A. Bostock, M. P. McDonald, G. J. T. Tiddy, and L. Warring, in *Surface Active Agents*, Society of Chemical Industry, 1980.
89. S. G. Frank and G. J. Zografi, *J. Colloid Interface Sci. 29*:27 (1969).
90. G. Gillberg, L. Eriksson, and S. Friberg, in *Emulsions, Lattices, and Dispersions* (P. Becher and M. N. Yudenfreund, eds.), Marcel Dekker, New York, 1978, p. 201.
91. G. D. Smith, C. E. Donelan, and R. E. Barden, *J. Colloid Interface Sci. 60*:488 (1977).

92. B. A. Keisen, D. Varie, R. E. Barden, and S. L. Holt, *J. Phys. Chem.* *83*:1276 (1979).
93. N. F. Borys, S. L. Holt and R. E. Barden, *J. Colloid Interface Sci.* *71*:526 (1979).
94. L. Prince, in *Microemulsions* (L. Prince, ed.), Academic, New York, 1977, Chap. 2.
95. D. O. Shah and R. S. Schechter (eds.), *Improved Oil Recovery by Surfactant and Polymer Flooding*, Academic, New York, 1977.
96. K. Lissant (ed.), *Emulsions*, Marcel Dekker, New York, 1976.
97. M. L. Robbins and J. H. Schulman, U.S. Patent 3,346,494 (1967).
98. D. W. Brownaweil and M. L. Robbins, U.S. Patent 3,527,581 (1970).
99. F. C. McCoy and G. W. Eckert, U.S. Patent 3,876,391 (1975).
100. S. Friberg and L. Lundborg, SW. Patent 368,898 (1974); U.S. Patent 3,902,869 (1975).
101. J. M. Pariel, R. Helion, and G. Robic, *Rev. Gen. Therm.* *II*:979 (1972).
102. E. Waldmanns and D. E. Wulfhorst, S.A.E. Report 700736, Society of Automotive Engineers, New York, 1970.
103. G. Gillberg and S. Friberg, in *Evaporation—Combustion of Fuels*, (J. T. Zung, ed.), Adv. Chem. Series No. 166, American Chemical Society, Washington, 1978, p. 221.
104. T. F. Moore and R. L. Slobod, *Producers Monthly* *20*:207 (1956).
105. N. R. Morrow and C. C. Harris, *Soc. Petrol. Eng. J.* *5*:15 (1965).
106. J. C. Melrose, *Can. J. Chem. Eng.* *48*:638 (1970).
107. J. C. Slattery, *Amer. Inst. Chem. Eng.* *20*:1145 (1974).
108. F. A. L. Dullien, *Amer. Inst. Chem. Eng.* *21*:299 (1975).
109. W. B. Gogarty, in *Improved Oil Recovery by Surfactant and Polymer Flooding* (D. O. Shah and R. S. Schechter, eds.), Academic, New York, 1977.
110. W. W. Gale and E. I. Sandvik, *Soc. Petrol. Eng. J.* *13*:199 (1973).
111. H. J. Hill, J. Reisberg, and G. L. Stegemeyer, *J. Petrol. Technol.* *25*:186 (1973).
112. J. A. Davis, Jr. and S. C. Jones, *J. Petrol. Technol.* *20*:1415 (1968).
113. J. J. Taber, I. S. K. Kamath and R. L. Reed, *Soc. Petrol. Eng. J.* *1*:195 (1961).
114. R. L. Reed and R. N. Healy, in *Improved Oil Recovery by Surfactant and Polymer Flooding* (D. O. Shah and R. S. Schechter, eds.), Academic, New York, 1977, p. 383.
115. A. B. Metzner, *ibid.*, p. 439.
116. B. B. Sandiford, *ibid.*, p. 487.
117. G. P. Willhite and J. G. Dominguez, *ibid.*, p. 511.
118. S. P. Trushenski, *ibid.*, p. 555.
119. J. M. Meacker, *Soc. Petrol. Eng. J.* *15*:311 (1975).
120. B. L. Knight, *J. Petrol. Technol.* *25*:618 (1973).
121. H. E. Gilliland and F. R. Conley, *Oil Gas J.* *43* (1976).
122. M. E. Yost and O. M. Stokke, *J. Petrol. Technol.* *27*:161 (1975).
123. S. P. Trushenski, D. L. Douben, and D. R. Parrish, *Soc. Petrol. Eng. J.* *14*:633 (1974).
124. H. Al-Rikibi and J. S. Osaba, *Oil Gas J.* 87 (1973).
125. R. N. Healy, R. L. Reed, and D. G. Stermark, *Soc. Petrol. Eng. J.* *16*:147 (1976).
126. L. W. Holm, *J. Petrol. Technol.* *23*:1475 (1971).
127. H. J. Hill, U.S. Patent 3,638,728 (1972).
128. H. R. Froming and W. S. Askew, U.S. Patent 3,714,062 (1973).
129. R. N. Healy and R. L. Reed, *Soc. Petrol. Eng. J.* *14*:491 (1974).

130. R. N. Healy, R. L. Reed, and C. W. Carpenter, *Soc. Petrol. Eng. J. 15*:87 (1975).
131. D. T. Wasan, L. Gupta, and M. K. Vora, *AIChE J. 17*:1287 (1971).
132. L. Gupta and D. T. Wasan, *Ind. Eng. Chem. Fundam. 13*:26 (1974).
133. R. L. Cash, J. L. Cayias, M. Hayes, and D. J. MacAllister, T. Schares, and W. H. Wade, *J. Petrol. Technol. 28*:985 (1976).
134. J. L. Cayias, R. S. Schechter, and W. H. Wade, *Soc. Petrol. Eng. J. 16*:351 (1976).
135. P. H. Doe, W. H. Wade, and R. S. Schechter, *J. Colloid Interface Sci. 59*:525 (1977).
136. R. S. Schechter, W. H. Wade and J. A. Wingrave, *J. Colloid Interface Sci. 59*:7 (1977).
137. R. L. Cash, J. L. Cayias, M. Hayes, D. J. MacAllister, T. Schares, W. H. Wade, and R. S. Schechter, *J. Petrol. Technol. 16*:985 (1976).
138. J. L. Cayias, R. S. Schechter, and W. H. Wade, *J. Colloid Interface Sci. 59*:31 (1977).
139. L. Cash, J. L. Cayias, G. Foureier, D. MacAllister, T. Schares, R. S. Schechter, and W. H. Wade, *J. Colloid Interface Sci. 59*:39 (1977).
140. P. H. Doe, M. El-Emary, W. H. Wade, and R. S. Schechter, *J. Amer. Oil Chem. Soc. 55*:505, 513, 570 (1978).
141. W. H. Wade, J. E. Morgan, R. S. Schechter, J. K. Jacobson, and J. L. Sologer, *Soc. Petrol. Eng. J. 18*:242 (1978).
142. C. A. Miller, R. Hivan, W. J. Benton, and T. Fort, *J. Colloid Interface Sci. 61*:554 (1977).
143. J. W. Cahn and J. E. Hilliard, *J. Chem. Phys. 28*:258 (1958).
144. H. T. Davis, *J. Chem. Phys. 62*:3412 (1975).
145. S. J. Salter and H. T. Davis, *J. Chem. Phys. 63*:3295 (1975).
146. A. Denss, personal communication, 1980.
147. E. F. J. Duynstee and E. Grunwald, *J. Amer. Chem. Soc. 81*:4540 (1959).
148. E. H. Cordes and R. B. Dunlap, *Accounts Chem. Res. 2*:329 (1969).
149. E. J. Fendler and J. H. Fendler, *Adv. Phys. Org. Chem. 8*:271 (1969).
150. S. E. Friberg and S. I. Ahmad, *J. Phys. Chem. 75*:2001 (1971).
151. E. J. Fendler, J. H. Fendler, R. T. Medacy, and W. A. Woods, *Chem. Commun.* 1497 (1971).
152. J. H. Fendler and E. J. Fendler, *Catalysis in Micellar and Macromolecular Systems*, Academic, New York, 1975.
153. K. A. Letts and R. A. Mackay, *Inorg. Chem. 14*:2990, 2993 (1975).
154. B. A. Keiser, R. E. Barden, and S. L. Holt, *J. Colloid Interface Sci. 73*:290 (1980).
155. G. D. Smith, B. B. Garrett, S. L. Holt, and R. E. Barden, *Inorg. Chem. 16*:558 (1977).
156. G. D. Smith, R. E. Barden, and S. L. Holt, *J. Coord. Chem. 8*:157 (1978).
157. C. E. Jones and R. A. Mackay, *J. Phys. Chem. 82*:63 (1978).
158. P. Schuster, in *The Hydrogen Bond,* (P. Schuster, G. Zundel and C. Sandorfy, eds.), North Holland, New York, 1976, p. 27.
159. J. N. Murrel and A. J. Harget, *Semi-Empirical Self-Consistent-Field Molecular Orbital Theory of Molecules*, Wiley-Interscience, New York, 1972.
160. J. A. Pople and G. A. Segal, *J. Chem. Phys. 43*:136 (1965).
161. J. A. Pople and G. A. Segal, *J. Chem. Phys. 44*:3289 (1966).
162. J. A. Pople and D. L. Beveridge, *Approximate Molecular Orbital Theory*, McGraw-Hill, New York, 1970.
163. J. A. Pople and D. L. Beveridge, *Approximate Molecular Orbital Theory*, McGraw-Hill, New York, 1970, pp. 89, 99, and 100.

# 5

# Phase Properties of Emulsions: PIT and HLB

KŌZŌ SHINODA and HIRONOBU KUNIEDA / Yokohama National University, Yokohama, Japan



## I. Introduction

The most important theme in the study of emulsions, of solubilization, and in many other applications, is the selection of a suitable surfactant which will satisfactorily emulsify or solubilize the chosen ingredients at a given temperature. For this reason, the hydrophile-lipophile balance (HLB) of the surfactant is a useful index. Clayton [1] drew attention to the concept of balanced surfactants in a series of patents dating back to 1933. What is involved here is the effect of the relative hydrophile character on the surface-active properties of the molecule. In a given

---

homologous series there is a point or range in which the hydrophile-lipophile balance is optimal for the particular application. However, the concept of HLB in its early stages was qualitative. Griffin [2], in his monumental paper, said:

> Any emulsion chemist who works with surfactants for a few years soon recognizes that there is a correlation between their behavior and their solubility in water. For example, he will use a water-soluble surfactant or blend to make an O/W emulsion. He will also use a water-soluble surfactant for solubilization and an almost completely water-soluble surfactant as a detergent. All the products of these applications may be said to exhibit aqueous characteristics; that is, they dilute readily with water and conduct electricity. For these purposes the emulsion chemist would under no circumstances use an oil-soluble surfactant. However, to make a water-in-oil emulsion, to couple water-soluble-soluble materials into an oil, or to make a dry-cleaning detergent, all of which are nonaqueous systems, he would choose an oil-soluble surfactant.
>
> This relationship of behavior and water solubility that is followed by most experienced emulsion chemists is so inexact in its usual form that it is only of value as a basis of thinking.

Schemes designed to put this concent on a quantitative basis have been advanced: these are the HLB numbers of Griffin [2,3] and Davies [4], the H/L numbers of Moore and Bell [5], the water number of Greenwald et al. [6], the phase inversion temperature (PIT, HLB temperature) of Shinoda [7-10], and, more recently, the emulsion inversion point (EIP) of Marszall [11-13].

In contrast to the other approaches, the PIT (HLB-temperature) method introduced by Shinoda employs a characteristic property of the *emulsion* (rather than of the surfactant molecule considered in isolation) as an indicator of the HLB of the surfactant, and utilizes the information obtained from a study of the effect of temperature on the properties of the system and on the HLB of the emulsifier at the oil/water interface. Since the PIT is a characteristic property of the emulsion, the effect of additives on the solvent, the effect of mixed emulsifiers or of mixed oils, et., are all reflected in the PIT, and thus tells us how the HLB of the emulsifier at the interface is changed. In other words the true HLB of the emulsifier is a function of the types of oils employed, the temperature, the composition of the surfactant, etc.

The HLB number and HLB temperature are interrelated and convertible one to the other, but the specifications of the system, i.e., the kinds of oils, the temperature, and so forth, must be known. Thus, study of the phase diagram, dispersion type, and emulsion stability in the water/oil/surfactant system as a function of temperature gives us a clear understanding of emulsions and will be described in Sec. II. The HLB-temperature method will be described in Sec. III.A, while the

HLB-number method will be briefly described in Sec. III.B, taking into account the concepts developed in Secs. II and III.A.

It should be emphasized that the HLB-number method and the HLB-temperature method complement each other. Both methods are applicable to nonionic emulsifiers, but not directly to nonionics, where the HLB number or temperature must usually be assigned empirically. A hydrophilic ionic emulsifier may have a very large HLB number. Emulsion studies reveal, however, that ionic emulsifiers and nonionic emulsifiers of the same HLB number do not behave similarly.

In order to evaluate the HLB of an ionic surfactant or to evaluate the optimum mixing ratio of an ionic surfactant and cosurfactant, a new method is described in Sec. IV. Since the estimation is based on the actual system, it is necessary to specify the kinds of oil, etc. The following scheme illustrates the relations among the various approaches.[†]

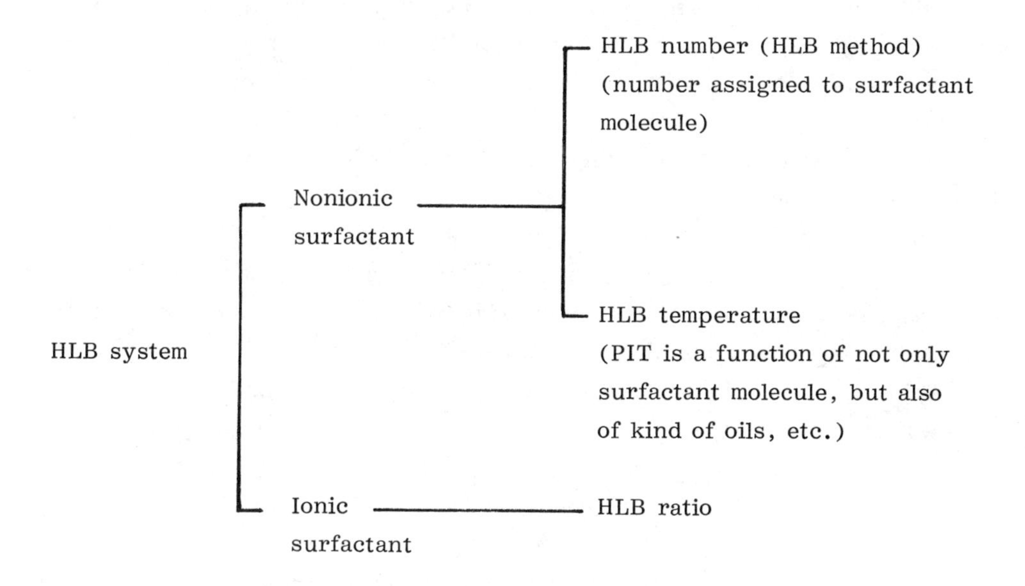


## II. Effect of Temperature on Phase Equilibria and Dispersion Type of a Ternary System Water/Cyclohexane/Nonionic Surfactant

The knowledge of phase equilibria in ternary systems composed of water, hydrocarbon, and nonionic surfactant is necessary in order to understand the dissolution behavior, the types of dispersion in emulsions, and the mutual solubility of

---

[†]Since the hydrophile-lipophile balance concept is fundamental, the senior author prefers to use the designation HLB system for the whole concept.

water and oil brought about the action of surfactants. Among saturated hydrocarbons, cyclohexane may well represent typical behavior in these systems. Although the composition of water versus oil has to be explored over the entire volume fraction range, the concentration of the surfactant may be limited to 1 to 10% from a practical viewpoint [14]. The complete phase diagram of a three-component system often clarifies the delicate difference among the types of oils and of the phases which separate [15-19].

The effect of temperature in particular has to be thoroughly explored in systems containing nonionic surfactants, since the effect is so remarkable and important [7,8]. Little attention was paid to the effect of temperature until 1963 [20]. However, since the change of temperature in solutions of nonionic surfactants corresponds to a change of HLB (or an effective change in oxyethylene chain length), studies of the effect of temperature readily provides a great deal of information [21,22]. The change in either the hydrophilic or lipophilic chain length of the surfactant shifts the phase equilibria and dispersion types to higher or lower temperatures, respectively, but the pattern is similar. Thus, a typical surfactant such as poly(oxyethylene) (9.7) nonylphenyl ether can represent typical behavior in these ternary systems. Hence, we shall restrict our discussion to phase equilibria and dispersion types for water-cyclohexane systems containing 3 and 7 wt% of this surfactant, as a function of temperature [14]. The study of these systems is useful in understanding the mutual relations among solubilization of oil in aqueous surfactant solubions; solubilization of water in nonaqueous surfactant solutions; the types, inversion, and stability of emulsions; and practical applications, such as detergency, dry cleaning, and emulsification.

The phase diagram of a water-cyclohexane system containing 7 wt% of poly(oxyethylene) nonylphenyl ether is shown in Fig. 1. The left-hand side of the figure corresponds to an aqueous surfactant solution containing a small amount of cyclohexane. The region $I_W$ is the oil-swollen micellar solution. The solubilization curve as a function of temperature is observed at a relatively low temperature. Solubilization of cyclohexane in an aqueous surfactant solution increases markedly close to the cloud point, but above this temperature a surfactant phase and a water phase separate. A large amount of water and cyclohexane dissolves in the surfactant phase, and the two phases (water and surfactant) coexist above the cloud-point curve, region $II_{D-W}$. If the amount of oil in the system is increased at this temperature, a separate oil phase appears. The central region indicated by III represents a three-phase region composed of water, surfactant, and oil phase. As the volume fraction of the surfactant phase is large (about 80% in a 7 wt% solution), the water or oil will disappear with a small change of composition or temperature. Thus, the three-phase region is narrow and small. It becomes larger in more dilute solution.

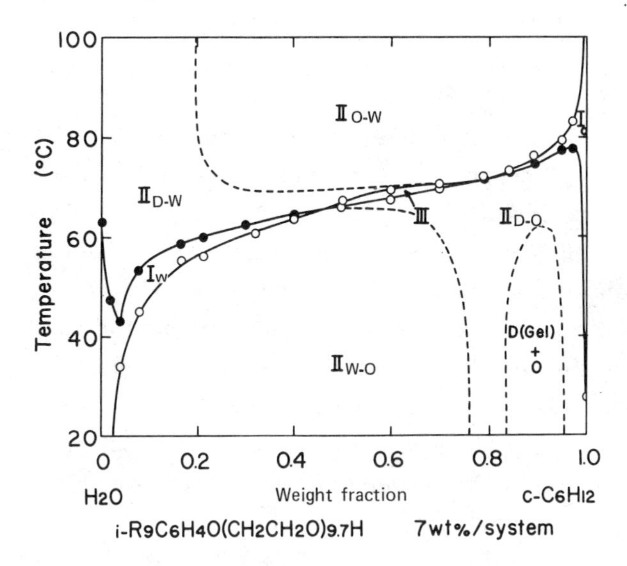


Figure 1. The phase diagram of the water-cyclohexane system containing 7 wt% of poly(oxyethylene) (9.7) nonylphenyl ether as a function of temperature. Cloud-point curve is indicated by •.

The region $II_{D-O}$, on the right of the three-phase region, is a two-phase region consisting of surfactant and oil phases. Since the solubility of water in a surfactant phase decreases, and that of oil increases, with increasing temperature, the result is an increase in the volume of water phase and a decrease in the surfactant plus oil phases.

The right-hand side of Fig. 1 corresponds to a nonaqueous solution of a nonionic surfactant containing a small amount of water. Region $I_0$ is a water-swollen micellar solution of cyclohexane. The solubilization curve of water in cyclohexane is observed at a relatively high temperature. Solubilization of water increases with decreasing temperature, particularly near the cloud point in a nonaqueous surfactant solution, but a surfactant phase separates from the cyclohexane below the cloud point.

Above the $I_W$, III, and $I_0$ regions two phases coexist. As the solubility of the surfactant in water is very small in these regions (or in this temperature range), the aqueous phase is nearly pure water. Accordingly, practically all the surfactant is in the oil phase. The concentration of surfactant in the oil phase increases with the change of composition from the right-hand side to the left-hand side, and finally, the surfactant becomes continuous in the nonaqueous phase. This tendency is strong around the phase inversion temperature. The right-hand side of the dotted line indicates the two-phase system consisting of water and oil phases, and the left-hand side corresponds to the two-phase system consisting of water and surfactant phases.

Similarly, two phases coexist below the $I_W$, III, and $I_O$ regions. The surfactant dissolves in water in these regions, but does not dissolve well in oil. At the left side of the two-phase region, excess oil separates from the oil-swollen micellar solution, region $II_{W-O}$. However, the relative concentration of the surfactant in the water phase increases with the change of composition from left to right, and, finally, the surfactant becomes continuous. Hence, the two phases on the right-hand side of the dotted line consist of surfactant (liquid crystal) and oil phases.

No three-phase region, i.e., surfactant phase, exists in a solution composed of pure (homogeneous) nonionic surfactant, water, and cyclo-hexane, because the transition from the oil-swollen aqueous micellar solution $I_W$ to the water-swollen nonaqueous micellar is sharp and solubilization of water and cyclo-hexane is so large [22].

### A. Characteristic Temperature for Mutual Dissolution of Oil and Water

The important feature of solutions of a nonionic surfactant is the notable increase in the solubility of oil in a aqueous surfactant solution [14,18,23], as well as that of water in a nonaqueous surfactant solution [24], at the cloud point. This tendency is shown also in the three-phase region by the fact that a large amount of oil and water dissolves into the surfactant phase.

Although the oil-swollen micellar solution $I_W$, the surfactant phase in region III, and the water-swollen micellar solution $I_O$ are three separate regions in the phase diagram, the changes in composition and structure of these phases are quite continuous. Region $I_W$ is a hydrophilic micellar solution, the surfactant phase is considered to be a micellar solution of a flat sandwichlike structure, whereas region $I_O$ is an oleophilic micellar solution. The curvature of the surfactant monolayer against oil (or water) seems to change continuously with temperature, since the very small interfacial tension between the surfactant phase and the water phase (smaller than 0.1 dyn/cm) increases with rise in temperature; whereas the tension between the surfactant and oil phases (which is also very small; less than 0.1 dyn/cm) increases with decreasing temperature [25]. This finding suggests that the adsorbed surfactant monolayer at the oil-water interface has a concave tendency towards oil at lower temperatures, is flat at medium temperatures, and has a convex tendency towards oil at higher temperatures. This results in an O/W-type emulsion at lower temperatures and a W/O type at higher temperatures. Although there is no solubilization at the temperature of region III, the solubility of oil in the surfactant plus water phases, or that of water in the surfactant plus oil phases, is high, and detergent action may well proceed at this temperature.

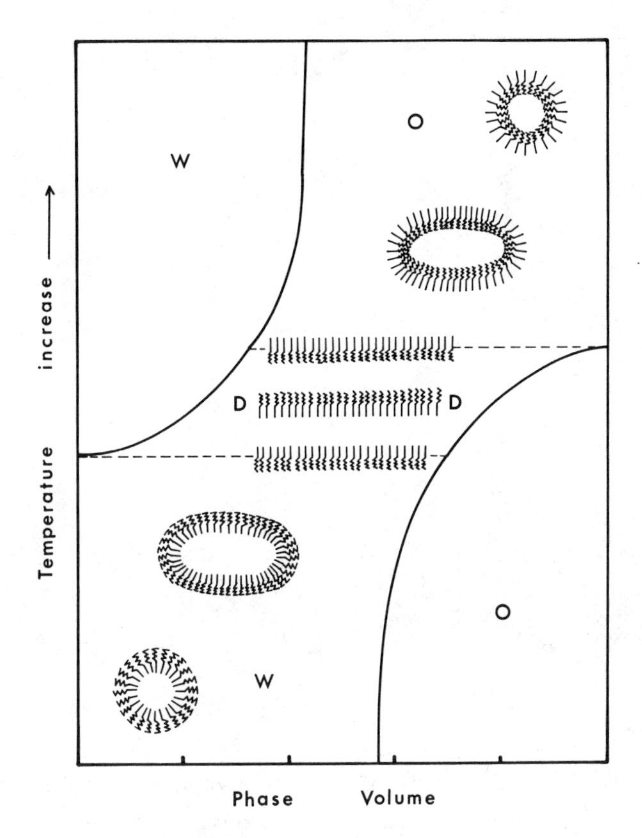


Figure 2. Nonionic surfactant dissolves in water forming normal micelles at low temperature and solubilizes oil. The aggregation number and solubilizing power of the micelles increase with temperature rise. At the three-phase region (or close to the phase inversion temperature) in the emulsion surfactant aggregates infinitely and dissolves a large amount of oil and water in the surfactant phase. Coexisting water and oil phases contian only a small amount of surfactant. At higher temperatures more oil dissolves and water separates from the surfactant phase, finally to oil phase.

The change in the aggregation number of the micelles, solubilizing power, the curvature of the surfactant aggregates, and the change of volume fractions of water, oil, and surfactant phases as a function of temperature in the neighborhood of the three-phase region is illustrated in Fig. 2.

## B. Effect of Oxyethylene Chain Length of Nonionic Surfactants on Phase Equilibria

If the oxyethylene chain length of nonionics is increased or decreased, similar phases diagrams are obtained, but shifted to higher or lower temperatures, re-

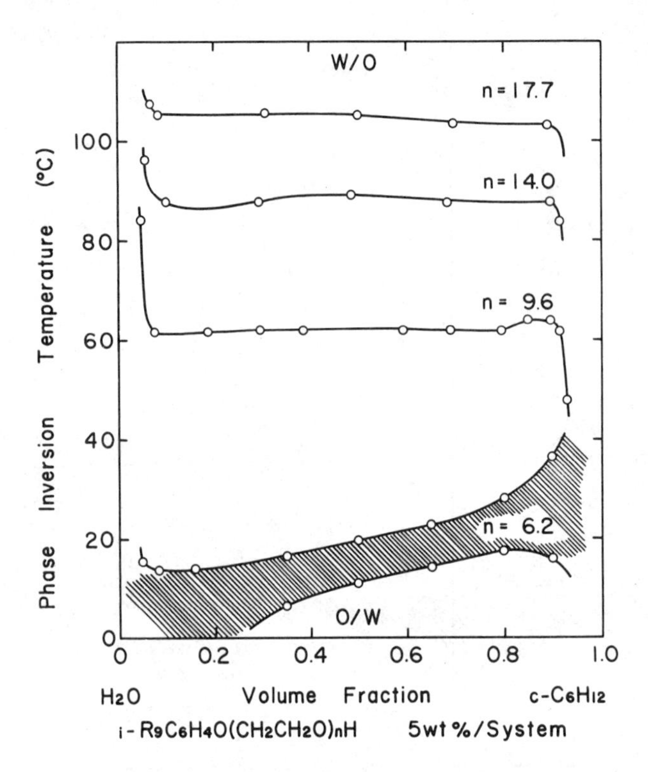


Figure 3. The effect of the hydrophilic chain length of nonionic surfactants on the PIT versus phase volume curves of cyclohexane/water emulsions. n is the mean oxyethylene chain length. (From Ref. 26.)

spectively, as shown in Fig. 3 [26]. If the temperature of the system is raised, the interaction between water and hydrophilic moiety of the surfactant decreases. Thus, the effect of temperature increase or a decrease in the oxyethylene chain length will be similar. This reasoning is confirmed by the phase diagram of non-ionic surfactants in water plus cyclohexane as a function of the average oxyethylene chain length of surfactant (Fig. 4) [22]. Here the oxyethylene chain length of the surfactant (ordinate) decreases, rather than the temperature, as in Fig. 3. The phase inversion temperature (PIT) in the emulsion (as well as the cloud point [27]) may be depressed or raised by the addition of salts to the aqueous phase, in analogy to the shortening or lengthening of the oxyethylene chain [28-30].

By varying amphiphiles, additives, hydrocarbons, and compositions, Winsor [31,32] was able to define the limits of completely solubilized systems and the nature of the equilibria of the solubilized phases with other phases. He classified equilibria as type I, solubilized phase in equilibrium with dilute hydrocarbon; type

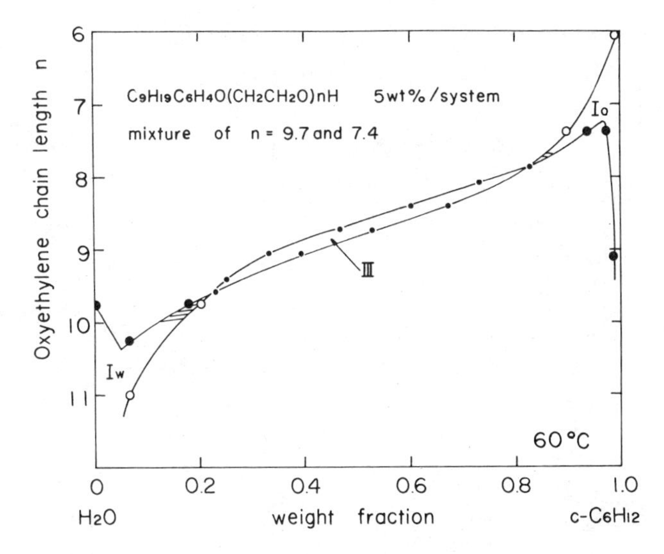


Figure 4. The effect of the average oxyethylene chain length of nonionics on the phase diagram of water-cyclohexane containing 5 wt% system of a mixture of $C_9H_{19}C_6H_4O(CH_2CH_2O)_{7.4}H$ and $C_9H_{19}C_6H_4O(CH_2CH_2O)_{9.7}H$. (From Ref. 22.)

II, solubilized phase in equilibrium with dilute aqueous phase; type III, solubilized phase in equilibrium with dilute hydrocarbon *and* dilute aqueous phases; and type IV, solubilized phase only. Furthermore, type IV systems were subdivided into various isotropic sol and birefringent gel phases. Gradual changes in composition led to conversion of one system into another. Regions $I_W$ and $I_O$ in the present study correspond to type IV, region III to type III, region $II_{W-O}$ to Type I, and region $II_{O-W}$ to type II of Winsor's classification. Similar solubilized systems have been studied by Schulman and co-workers [33-38], who coined the name *microemulsions* for these systems [cf. Chap. 4]. The surfactant phase and the $I_W$ and $I_O$ phases close to region III correspond to Schulman's so-called microemulsions. The remarkable difference between the present system under discussion and Schulman's systems is that the concentration of surfactant required to obtain a microemulsion is so much smaller (~ 5 wt%) in the present case than in Schulman's systems [14,22,47].

## C. Effect of Temperature on Dispersion Type

Excess water separates from a nonaqueous micellar solution at a high temperature. The dispersion type of this two-phase system is W/O over a wide range of volume fractions, as shown in Fig. 5. In the region where the volume fraction of oil is less than 0.2, the concentration of the nonionic surfactant is fairly high and the

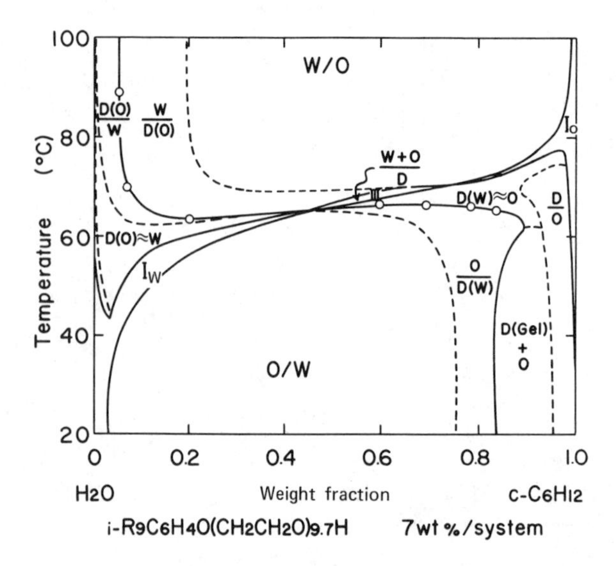


Figure 5. The effect of temperature and composition on the dispersion types of the system composed of water, cyclohexane, and poly(oxyethylene) (9.7) nonylphenyl ether (7 wt% system). (From Ref. 14.)

nonaqueous phase may be considered to be a surfactant phase in which hydrocarbon is dissolved. At this volume fraction, the solution is viscous. If the volume fraction of oil further decreases, the water phase (which occupies a very large volume fraction) finally becomes a continuous phase, i.e., phase inversion from a W/D(O) to a D(O)/W type occurs.

At the intermediate temperature three phases coexist, so that water, oil, and surfactant phases are more clearly distinguished in region III. The type of dispersion here is the (W + O)/D type; the oil phase disappears on the left side of region III, owing to the decrease of the volume fraction of hydrocarbon, so that the dispersion is either a W/D or a D/W type above or below the phase inversion temperature. A region exists between the phase inversion temperature and the cloud point curve in which water and surfactant phases are both continuous (W ≈ D).

On the other hand, the water phase disappears on the right side of region III, owing to the decrease of the volume fraction of water, so that the type of dispersion is either a D/O or an O/D type, above or below the phase inversion temperature, respectively. The change from surfactant phase to water proceeds gradually with temperature decrease or with an increase in the hydrophilic nature of the surfactant, as shown by the lower dotted line in Fig. 5. The dispersion in the $II_{D-O}$ region is not always a D/O type above the phase inversion temperature, but both phases are continuous in one part of the region. The type is clearly D/O in the region where the volume of oil phase is large, as shown in the region D/O of Fig. 5.

At low temperatures the surfactant dissolves in water and some hydrocarbon is solubilized in the aqueous micellar solution. Beyond the solubilization limit, excess hydrocarbon disperses as an O/W-type emulsion. Because of the change of composition from the left-hand side to the right-hand side, the concentration of surfactant in the water phase increases, since the amount of surfactant in the system is fixed. Finally, anisotropic stiff gel, liquid crystal, and oil phases coexist in region D(gel) + O. If the volume fraction of water further decreases, the stiff gel becomes a sol, and a two-phase system consisting of oil and surfactant phases is obtained. Similar phase equilibria and dispersion types are obtained in the cyclohexane-water system containing 3 wt% of poly(oxyethylene) (9.7) nonylphenyl ether, and are shown in Figs. 6 and 7.

It may be concluded from Figs. 1 to 7 that the oil phase is the continuous medium at high temperatures (or when the emulsifier is lipophilic), while the surfactant phase is the continuous medium at an intermediate temperature close to the phase inversion temperature (the HLB-temperature, or when the hydrophile-lipophile properties of the surfactant just balance), and that a water phase is the continuous medium at low temperature (or when the emulsifier is hydrophilic) in systems consisting of water, oil, and surfactant. The emulsion inversion point (EIP) proposed by Marszall [11-13] may correspond to the composition at which the dispersion type inverts from D/O to W/O on the right side of the diagram, from D/W to W/D on the

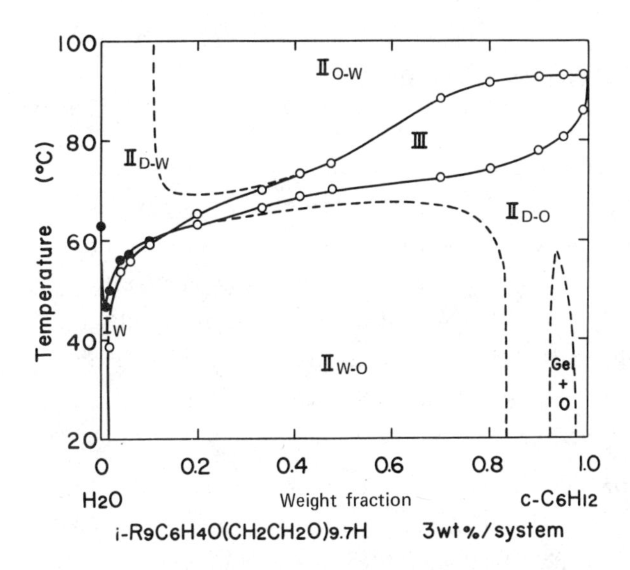


Figure 6. Phase diagram of the water-cyclohexane system containing 3 wt% of poly-(oxyethylene) (9.7) nonylphenyl ether as a function of temperature. (From Ref. 14.)

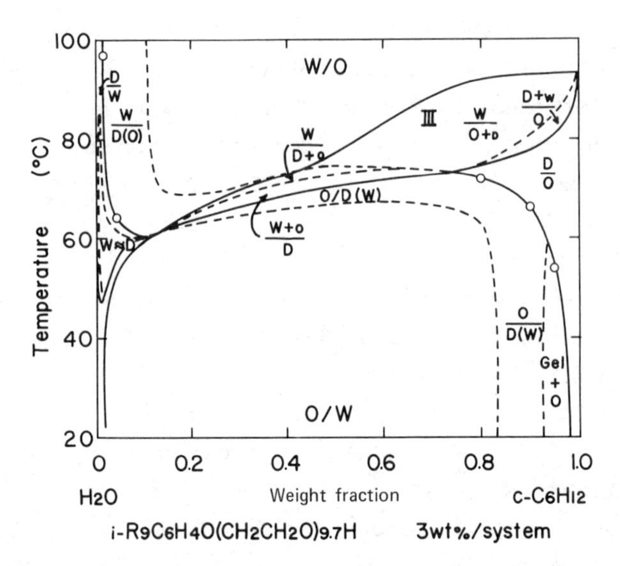


Figure 7. Effect of temperature and composition on the dispersion types of the system composed of water, cyclohexane, and poly(oxyethylene) (9.7) nonylphenyl ether (3 wt% system). (From Ref. 14.)

left, at about 20° C and 80° C, respectively, in Figs. 1 and 3. Since the boundary between D/O and O/W (or D/W and W/D) is a steep line or a flat curve at the PIT, the estimation of emulsifier HLB from the change in EIP may not be accurate.

## III. Hydrophile-Lipophile Balance of Nonionic Emulsifiers

It is clear from Figs. 1 to 7 that the HLB on a nonionic surfactant is a function of temperature, owing to the fact that the interaction between water and the hydrophilic group of the surfactant and the lipophilic group of the oil changes with temperature [23,24,39]. The required HLB numbers of oils and the experimental HLB numbers of surfactants are empirically determined from the maximum stability of emulsions as a function of the HLB of the surfactants. However, it is not easy to determine exact HLB numbers of particular surfactants, because the emulsion stability is not a sensitive function of the change of HLB numbers (Fig. 8).

The situation was described by Griffin as follows [2]:

> In its present form, the HLB-number system lacks exactness. A suitable simple laboratory method of measuring HLB values of surfactants accurately is needed. We have tried a variety of methods including solubility in water or various solvents, ratio solubility in two solvents, solubilization behavior for both oils and

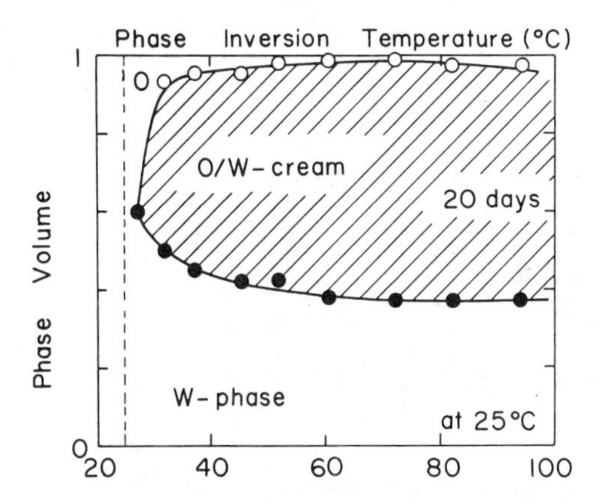


Figure 8. Effect of phase inversion temperature (PIT) of emulsifiers on the volume fraction of oil, cream, and water phases 20 days after emulsification. (Emulsified and stored at 25° C using a series of poly(oxyethylene) nonylphenyl ethers, the PITs of which vary from 27 to 94° C.) (From Ref. 9.)

dyes, surface and interfacial tension data, cloud-point behavior, and many other properties. Of these, the most promising is the determination of the cloud point of an aqueous solution of the surfactant. However, this test still possesses severe limitations.

## A. HLB Temperature (PIT)

The cloud point of an aqueous solution of surfactant surely reflects the hydrophilic property of the surfactant. If we adopt some characteristic property of emulsions, such as phase inversion temperature (PIT), as a measure of the HLB, it naturally reflects any given experimental condition, such as, for example, the effect of the sizes and types of the hydrophilic and lipophilic groups of the surfactants, and supplies information on the effect of the concentration of emulsifier, the phase volume, additives in both water and oil phases, the temperature, the effect of blending oils and surfactants, etc.

The comparison between the HLB temperature and the HLB number in regard to the amount and type of information supplied is summarized in Table 1. It is evident that the PIT is a most suitable parameter for the measurement of the HLB of nonionic surfactants [7,8,23]. An HLB number alone assigned to a surfactant molecule cannot, in principle, give information on the effect of any variable, but empirical corrections, such as the required HLB numbers of oils, required HLB numbers of water containing salts, etc., may be obtained by supplementing the stability measurements with the aid of the PIT data. On the other hand, the PIT

Table 1. Comparison between the PIT (HLB Temperature) and the HLB Number in Regard to the Information on Various Factors

| PIT (HLB temperature) | Factors | HLB number |
|---|---|---|
| ○[a] | Hydrophilic-liphophile balance of surfactant | ●[b] |
| ○ | Types of hydrophilic moiety of surfactant | ▲[c] |
| ○ | Types of lipophilic moiety of surfactant | ▲ |
| ○ | Types of oils | ● |
| ○ | Additives in water and/or oil phase | ▲ |
| ○ | Concentration of emulsifier | * |
| ○ | Phase volume | * |
| ○ | Temperature | * |
| ○ | Emulsion types | ● |
| ○ | Correlation with other properties | ▲ |
| *[d] | In the case of ionic surfactant | ▲ |

[a]○: Information is available accurately.
[b]●: Information is available less accurately.
[c]▲: Crude information is available.
[d]*: Almost no information is available.

is not observed if the surfactant is too lipophilic. In such a case, the PIT has to be evaluated by extrapolation from more hydrophilic homologues, or a low HLB number may be arbitrarily assigned.

Although the HLB temperature supplies all information about the experimental conditions in nonionics, the addition of an optimum amount of salt and cosurfactant is required in order to observe the HLB temperature of ionic surfactants since the HLB of ionic surfactants alone does not change appreciably with temperature, and the PIT cannot be observed. The effect of temperature on the emulsion type in the case of solutions of ionic surfactants is opposite to that for nonionics, i.e., O/W at the higher temperature and W/O at lower temperatures [39]. The optimum mixing ratios of ionics with cosurfactants to form lamellar liquid crystals also yields relatively accurate information concerning the HLB of ionic surfactants and will be discussed later.

The fact that emulsion stabilized with nonionic agents are W/O type at high temperatures, but change to O/W type at low temperatures, is evidence that the hydrophile-lipophile balance of nonionic surfactants becomes appreciably more hydrophilic as the temperature is depressed. The hydration forces between the hydrophilic moiety of the surfactant and water are stronger at the lower temperature (so that the surfactant is more hydrophilic), and the adsorbed monolayer at the oil/water interface may have convex curvature toward the water phase. When the hydration between the hydrophilic moiety of the surfactant and water diminishes at the higher temperature, the surfactant is more lipophilic, and the adsorbed monolayer at the oil/water interface may have concave curvature towards the water [8], and inversion of emulsion type occurs. Since the PIT is considered to be the temperature at which the hydrophilic-lipophilic property of the surfactant just balances, the selection of a suitable emulsifier by the HLB temperature is similar to the HLB number, in the sense that both depend on the hydrophile-lipophile balance.

### B. Comparison between HLB Temperature and HLB Numbers

In view of the above discussion, it is to be expected that there should be correlation between the HLB temperature and the HLB number. Such a correlation (for cyclohexane) is shown in Fig. 9. We can determine the HLB number from the HLB temperature of a surfactant with the aid of this calibration curve. As the PIT changes markedly with the type of oil employed, the correlation differs for different oil/water emulsions. As a result of the fact that the HLB of a surfactant is different at different oil/water interfaces, a correction is required in the original Griffin system. This correction is, in effect, supplied by the concept of required HLB number for oils.

The HLB number is defined at 25° C, but it is a function of temperature. Since the effect of temperature is in the opposite sense for the HLB numbers of ionic and nonionic surfactants, the same HLB numbers at 25° C would imply *higher* HLB numbers for ionics and *lower* HLB numbers for nonionics at higher temperatures, and vice versa. Thus, the equality of HLB numbers at 25° C is, in a sense, an artifact. On the other hand, the HLB number of a nonionic surfactant whose PIT is equal to 25° C is found to 9.5 for liquid paraffin and 11.1 for cyclohexane. Figure 10 may be used to find the required HLB numbers of oils at 25° C, because the HLB numbers of surfactants whose HLB temperatures are 25 to 70° C higher than the storage temperature are found to be the required HLB numbers [9]. The required HLB numbers estimated from the HLB temperature system and those

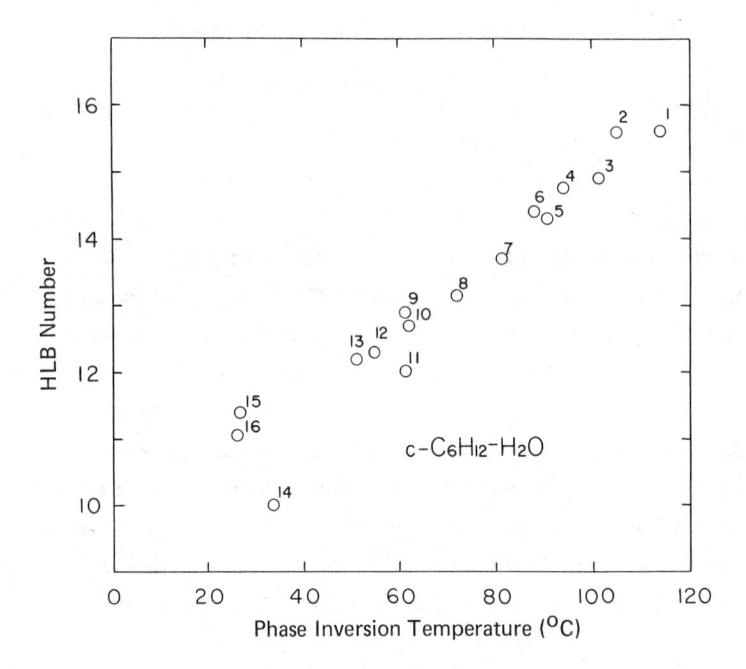


**Figure 9.** Correlation between the HLB numbers of nonionic surfactants and the phase inversion temperatures (PIT, HLB temperature) in cyclohexane/water emulsions stabilized with the surfactants (3 wt% system).

1. Tween 40
2. $i\text{-}R_9C_6H_4O(CH_2CH_2O)_{17.7}H$
3. Tween 60
4. $i\text{-}R_9C_6H_4O(CH_2CH_2O)_{14.0}H$
5. $i\text{-}R_{12}C_6H_4O(CH_2CH_2O)_{15}H$
6. $R_{12}O(CH_2CH_2O)_{10.8}H$
7. $i\text{-}R_8C_6H_4O(CH_2CH_2O)_{10}H$
8. $i\text{-}R_9C_6H_4O(CH_2CH_2O)_{9.7}H$
9. $i\text{-}R_8C_6H_4O(CH_2CH_2O)_{8.5}H$
10. $i\text{-}R_{12}C_6H_4O(CH_2CH_2O)_{9.7}H$
11. $R_{12}O(CH_2CH_2O)_{6.3}H$
12. $i\text{-}R_{12}C_6H_4O(CH_2CH_2O)_{9.4}H$
13. $i\text{-}R_9C_6H_4O(CH_2CH_2O)_{7.4}H$
14. $R_{12}O(CH_2CH_2O)_{4.2}H$
15. $i\text{-}R_8C_6H_4O(CH_2CH_2O)_6H$
16. $i\text{-}R_9C_6H_4O(CH_2CH_2O)_{6.2}H$

recommended earlier [40] are compared in Table 2. The agreement between the two methods is seen to be reasonable.

It is evident the systems are complementary. Information obtained from both systems can be utilized. Griffin's method is used to determine the HLB number of the surfactant, while the modified method is based on the solution properties and inversion of emulsion type in surfactant/water/oil systems.

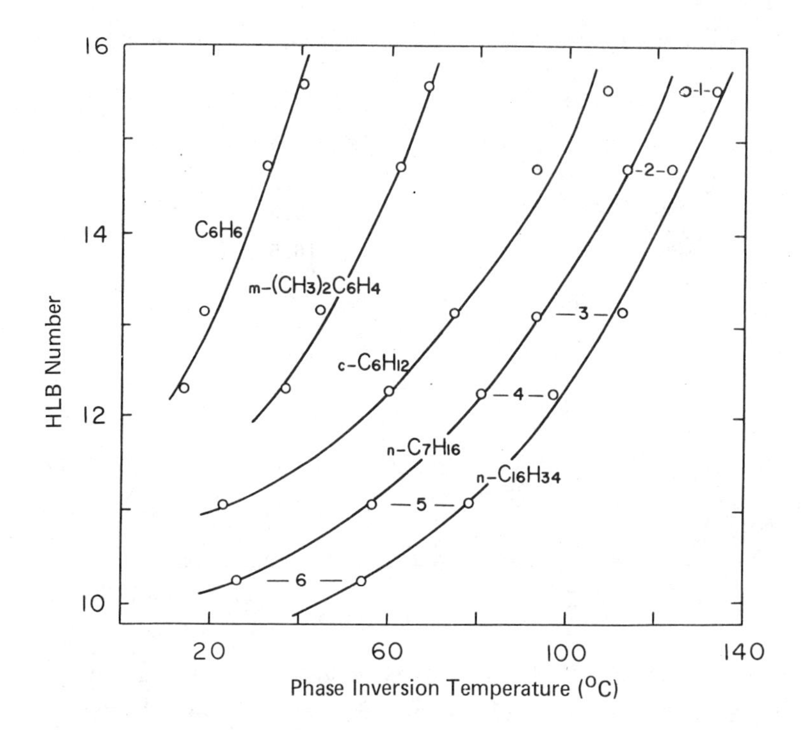


**Figure 10.** Correlation between HLB numbers and PITs in various oil/water (1:1) emulsions stabilized with nonionic surfactants (1.5% system).

1. $i\text{-}R_9C_6H_4O(CH_2CH_2O)_{17.7}H$
2. $i\text{-}R_9C_6H_4O(CH_2CH_2O)_{14.0}H$
3. $i\text{-}R_9C_6H_4O(CH_2CH_2O)_{9.6}H$
4. $i\text{-}R_9C_6H_4O(CH_2CH_2O)_{7.4}H$
5. $i\text{-}R_9C_6H_4O(CH_2CH_2O)_{6.2}H$
6. $i\text{-}R_9C_6H_4O(CH_2CH_2O)_{5.3}H$

[From K. Shinoda, *J. Chem. Soc. Jap.* *89*:435 (1968).]

### C. HLB Number

Becher has reviewed the literature on the definition and determination of hydrophile-lipophile balance through 1965 [40], and has briefly summarized the additional literature through 1972 [41]. Becher and Griffin [42] have prepared a bibliography of papers on HLB, covering the period 1949 to 1972. The relationship between and HLB and emulsion stability is discussed in Chap. 3.

Recently, Shinoda et al. [43] have used data on PIT for cyclohexane emulsions (cf. Table 3) to evaluate the relative contributions of the hydrophilic and lipophilic moieties to the HLB. It may be seen from Table 3 that the hydrophile-lipophile properties of the surfactant just balance if 2 to 2.7 methylene groups are added per oxyethylene group at 25° C. Similarly, for a surfactant whose PIT is equal to

Table 2. Comparison between the Griffin Required HLB Numbers and HLB Numbers Estimated from PIT

| Types of oils | HLB numbers [45] | HLB numbers (PIT) [10] | PIT (°C) |
|---|---|---|---|
| Mineral oil (paraffinic) | 10 | 10 | 110 |
| Propene tetramer | 14 | 12 | — |
| Kerosene | 14 | 12 | 94 |
| Trichlorotrifluoroethane | 14 | 12.5 | — |
| Cyclohexane | 15 | 13 | 70 |
| Carbon tetrachloride | 16 | 13.5 | 53 |
| Xylene | 14 | 14.5 | 46 |
| Toluene | 15 | 15.5 | 38 |
| Benzene | 15 | 16.5 | 21 |

25° C, and whose HLB is 10, the PIT will be unchanged if 3.15 methylene groups are added (or subtracted) per oxyethylene added (or subtracted). On the other hand, for a surfactant with the same PIT, but whose HLB is equal to, for example, 11.6, one oxyethylene is equivalent to 2.3 methylene groups.

Table 3. Molecular Formula of Emulsifiers and PIT in Emulsions (Water Cyclohexane) [43]

| Emulsifier | PIT (°C) | HLB number[a] |
|---|---|---|
| $R_6C_6H_4O(CH_2CH_2O)_{7.5}H$ | 52 | 13.0 |
| $R_9C_6H_4O(CH_2CH_2O)_{8.6}H$ | 50 | 12.4 |
| $R_{12}C_6H_4O(CH_2CH_2O)_{9.7}H$ | 51 | 12.2 |
| $R_{16}C_6H_4O(CH_2CH_2O)_{12.4}H$ | 48 | 12.6 |
| $R_8O(CH_2CH_2O)_{4.3}H$ | 25 | 11.9 |
| $R_{12}O(CH_2CH_2O)_{5.8}H$ | 25 | 11.6 |
| $R_9C_6H_4O(CH_2CH_2O)_{6.2}H$ | | 11.1 |
| $R_9C_6H_4O(CH_2CH_2O)_{4.5}H$ | 25 | 9.5 (liquid paraffin) |

[a]These HLB numbers are assigned to emulsifier at room temperature. Actual HLB number at 50°C is about 1 unit smaller.

*Source:* Ref. 43.

Shinoda and co-workers have recently [44] addressed the problem of the algebraic additivity of emulsifier HLB, implicit in the Griffin system. Evidence for a certain degree of nonadditivity has been presented [40], although the effect was not considered to be large. Shinoda and co-workers have pointed out that a more direct way of studying the HLB of surfactant blends as a function of composition would be to determine the PIT of a series of blends and compare these with the PIT of a series of single surfactants spanning the same range of HLB. Such studies reveal that, in a blend of strongly lipophilic and strongly hydrophilic surfactants, the individual emulsifiers may individually dissolve in the oil and water phases, so that the effective HLB of the blend could deviate from the weight average. If, for example, the lipophilic emulsifier is very soluble or the cmc in the oil phase is fairly high, the HLB of the *adsorbed* mixed monolayer could have a quite different composition from that suggested by the stoichiometry. In the work reported, the effect was not large, in agreement with the earlier work. In the case of W/O emulsions, a lower HLB than predicted was found.

Griffin [45] has recently updated his lists of emulsifier HLB numbers and required HLB for various oils (Tables 4 and 5).

## IV. HLB of Ionic Surfactants

The basis of the HLB system resides in the fact that a surfactant whose HLB is optimum for a particular application is more efficient than another surfactant of different HLB or different oxyethylene chain length. In the case of ionic surfactants, however, changing the HLB significantly is difficult, but it is possible to do so by changing the valency and concentration of counterions and by replacing some ionic surfactant with a lipophilic cosurfactant such as fatty acid, fatty alcohol, or poly(oxyethylene) glycol alkyl ether [46,47]. The addition of (or partial replacement by) a cosurfactant increases the solubilizing power of an ionic surfactant in a remarkable way [46-50], and emulsion stability also increases [51]. It is known that an emulsion may be stabilized by a condensed mixed film, as witness the classic case of the mixed film of cholesterol and sodium hexadecyl sulfate. Further addition of multivalent ions depresses the surface charge at the interface, which is equivalent to a depression of the HLB of the adsorbed film, and emulsion inversion may occur [52]. These processes, then, can be interpreted as a change in the HLB of adsorbed surfactant.

Evaluation of the HLB of an ionic surfactant may be performed by studying the phase diagram of the four-component system composed of ionic surfactant, cosurfactant, water, and oil [53,54]. For example, the phase diagram consisting of octylammonium chloride, octylamine, water, and xylene is shown in Fig. 11 [54]. In the

Table 4. HLB Numbers for Surfactants [45]

| Chemical designation and CAS registry number | Type | HLB number |
|---|---|---|
| Oelic acid (*112-80-1*) | N | 1.0 |
| Lanolin alcohols (*61788-49-6*) | N | 1.0 |
| Acetylated sucrose diester | N | 1.0 |
| Ethylene glycol distearate (*627-83-8*) | N | 1.3 |
| Acetylated monoglycerides | N | 1.5 |
| Sorbitan trioleate (*26266-58-6*) | N | 1.8 |
| Glycerol dioleate (*25637-84-7*) | N | 1.8 |
| Sorbitan tristearate (*26658-19-5*) | N | 2.1 |
| Ethylene glycol monostearate (*111-60-4*) | N | 2.9 |
| Sucrose distearate (*27195-16-0*) | N | 3.0 |
| Decaglycerol decaoleate (*11094-60-3*) | N | 3.0 |
| Propylene glycol monostearate (*1323-39-3*) | N | 3.4 |
| Glycerol monooleate (*25496-72-4*) | N | 3.4 |
| Diglycerine sesquioleate | N | 3.5 |
| Sorbitan sesquioleate (*8007-43-0*) | N | 3.7 |
| Glycerol monostearate (*31566-31-1*) | N | 3.8 |
| Acetylated monoglycerides (stearate) | N | 3.8 |
| Decaglycerol octaoleate (*66734-10-9*) | N | 4.0 |
| Diethylene glycol monostearate (*106-11-6*) | N | 4.3 |
| Sorbitan monooleate (*1333-68-2*) | N | 4.3 |
| Propylene glycol monolaurate (*10108-22-2*) | N | 4.5 |
| High-molecular-weight fatty amine blend | C | 4.5 |
| POE (1.5) nonyl phenol (ether) (*9016-45-9*) | N | 4.6 |
| Sorbitan monostearate (*1338-41-6*) | N | 4.7 |
| POE (2) oleyl alcohol (ether) (*25190-05-0*) | N | 4.9 |
| POE (2) stearyl alcohol (ether) (*9005-00-9*) | N | 4.9 |
| POE sorbitol beeswax derivative | N | 5.0 |
| PEG 200 distearate (*9005-08-7*) | N | 5.0 |
| Calcium stearoxyl-2-lactylate (*5793-94-2*) | A | 5.1 |
| Glycerol monolaurate (*27215-38-9*) | N | 5.2 |
| POE (2) octyl alcohol (ether) (*27252-75-1*) | N | 5.3 |
| Sodium-Ostearyllactate (*18200-72-1*) | A | 5.7 |
| Decaglycerol tetraoleate | N | 6.0 |
| PEG 300 dilaurate (*9005-02-1*) | N | 6.3 |
| Sorbitan monopalmitate (*26266-57-9*) | N | 6.7 |

Table 4. (Continued)

| Chemical designation and CAS registry number | Type | HLB number |
|---|---|---|
| *N*,*N*-Dimethylstearamide (*3886-90-6*) | N | 7.0 |
| PEG 400 distearate (*9005-08-7*) | N | 7.2 |
| High-molecular-weight amine blend | C | 7.5 |
| POE (5) lanolin alcohol (ether) (*61790-91-8*) | N | 7.7 |
| Polyethylene glycol ether of linear alcohol | N | 7.7 |
| POE octylphenol (ether) (*9002-93-1*) | N | 7.8 |
| Soya lecithin (*8020-84-6*) | N | 8.0 |
| Diacetylated tartaric acid esters of monoglycerides | N | 8.0 |
| POE (4) stearic acid (monoester) (*9004-99-3*) | N | 8.0 |
| Sodium stearoyllactylate (*18200-72-1*) | A | 8.3 |
| Sorbitan monolaurate (*1338-43-8*) | N | 8.6 |
| POE (4) nonylphenol (ether) (*9016-45-9*) | N | 8.9 |
| Calcium dodecylbenzene sulfonate (*26264-06-2*) | A | 9 |
| Isopropyl ester of lanolin fatty acids | N | 9.0 |
| POE (4) tridecyl alcohol (ether) (*24938-91-8*) | N | 9.3 |
| POE (4) lauryl alcohol (ether) (*9002-92-0*) | N | 9.5 |
| POP/POE condensate | N | 9.5 |
| POE (5) sorbitan monooleate (*9005-65-6*) | N | 10.0 |
| POE (40) sorbitol hexaolate (*9011-29-4*) | N | 10.2 |
| PEG 400 dilaurate (*9005-02-1*) | N | 10.4 |
| POE (5) nonylphenol (ether) (*9016-45-9*) | N | 10.5 |
| POE (20) sorbitan tristearate (*9005-71-4*) | N | 10.5 |
| POP/POE condensate (*9003-11-6*) | N | 10.6 |
| POE (6) nonylphenol (ether) (*9016-45-9*) | N | 10.9 |
| Glycerol monostearate-self emulsifying (*31566-31-1*) | A | 11.0 |
| POE (20) lanolin (ether and ester) | N | 11.0 |
| POE (20) sorbitan trioleate (*9005-70-3*) | N | 11.0 |
| POE (8) stearic acid (monoester) (*9004-99-3*) | N | 11.1 |
| POE (50) sorbitol hexaoleate (*9011-29-4*) | N | 11.4 |
| POE (6) tridecyl alcohol (ether) (*24938-91-8*) | N | 11.4 |
| PEG 400 monostearate (*9004-99-3*) | N | 11.7 |
| Alkyl aryl sulfonate | A | 11.7 |
| Triethanolamine oleate soap (*2717-15-9*) | A | 12 |
| POE (8) nonylphenol (ether) (*9016-45-9*) | N | 12.3 |
| POE (10) stearyl alcohol (ether) (*9005-00-9*) | N | 12.4 |
| POE (8) tridecyl alcohol (ether) (*24938-91-8*) | N | 12.7 |

Table 4. (Continued)

| Chemical designation and CAS registry number | Type | HLB number |
|---|---|---|
| POP/POE condensate | N | 12.7 |
| POE (8) lauric acid (monoester) (*9004-81-3*) | N | 12.8 |
| POE (10) cetyl alcohol (ether) (*9004-95-9*) | N | 12.9 |
| Acetylated POE (10) lanolin | N | 13.0 |
| POE (20) glycerol monostearate (*53195-79-2*) | N | 13.1 |
| PEG 400 monolaurate (*9004-81-3*) | N | 13.1 |
| POE (16) lanolin alcohol (ether) (*61790-81-6*) | N | 13.2 |
| POE (4) sorbitan monolaurate (*9005-64-5*) | N | 13.3 |
| POE (10) nonylphenol (ether) (*9016-45-9*) | N | 13.3 |
| POE (15) tall oil fatty acids (ester) | N | 13.4 |
| POE (10) octylphenol (ether) (*9002-93-1*) | N | 13.6 |
| PEG 600 monostearate (*9004-99-3*) | N | 13.6 |
| POP/POE condensate | N | 13.8 |
| Tertiary amines: POE fatty amines | C | 13.9 |
| POE (24) cholesterol (*27321-96-6*) | N | 14.0 |
| POE (14) nonylphenol (ether) (*9016-45-9*) | N | 14.4 |
| POE (12) lauryl alcohol (ether) (*9002-92-0*) | N | 14.5 |
| POE (20) sorbitan monostearate (*9005-67-8*) | N | 14.9 |
| Sucrose monolaurate (*25339-99-5*) | N | 15.0 |
| POE (20) sorbitan monooleate (*9005-65-6*) | N | 15.0 |
| POE (16) lanolin alcohols (ether) (*8051-96-5*) | N | 15.0 |
| Acetylated POE (9) lanolin (*68784-35-0*) | N | 15.0 |
| POE (20) stearyl alcohol (ether) (*9005-00-9*) | N | 15.3 |
| POE (20) oleyl alcohol (ether) (*25190-05-0*) | N | 15.3 |
| PEG 1000 monooleate (*9004-96-0*) | N | 15.4 |
| POE (20) tallow amine (*61790-82-7*) | C | 15.5 |
| POE (20) sorbitan monopalmitate (*9005-66-7*) | N | 15.6 |
| POE (20) cetyl alcohol (ether) (*9004-95-9*) | N | 15.7 |
| POE (25) propylene glycol monostearate (*37231-60-0*) | N | 16.0 |
| POE (20) nonylphenol (ether) (*9016-45-9*) | N | 16.0 |
| PEG (1000) monolaurate (*9004-81-3*) | N | 16.5 |
| POP/POE condensate | N | 16.8 |
| POE (20) sorbitan monolaurate (*9005-64-5*) | N | 16.9 |
| POE (23) lauryl alcohol (ether) (*9002-92-0*) | N | 16.9 |
| POE (40) stearic acid (monoester) (*9004-99-3*) | N | 16.9 |
| POE (50) lanolin (ether and ester) (*61790-81-6*) | N | 17.0 |

Table 4. (Continued)

| Chemical designation and CAS registry number | Type | HLB number |
|---|---|---|
| POE (25) soyasterol (*68648-64-6*) | N | 17.0 |
| POE (30) nonylphenol (ether) (*9016-45-9*) | N | 17.1 |
| PEG 4000 distearate (*9005-08-7*) | N | 17.3 |
| POE (50) stearic acid (monoester) (*9004-99-3*) | N | 17.9 |
| Sodium oleate (*143-91-1*) | A | 18.0 |
| POE (70) dinonylphenol (ether) (*9014-93-1*) | N | 18.0 |
| POE (20) castor oil (ether, ester) (*61791-12-6*) | N | 18.1 |
| POP/POE condensate | N | 18.7 |
| Potassium oleate (*143-18-0*) | A | 20 |
| *N*-cetyl-*N*-ethyl morpholinium ethyl sulfate (35%) (*78-21-7*) | C | 30 |
| Ammonium lauryl sulfate (*2235-54-3*) | A | 31 |
| Triethanolamine lauryl sulfate (*139-96-8*) | A | 34 |
| Sodium alkyl sulfate | A | 40 |

*Source:* Ref. 45.

region $L_1$ the octylammonium and octylamine are dissolved in water and xylene is solubilized. In the region $L_2$ the octylamine and octylammonium chloride are dissolved in xylene and form reverse micelles, solubilizing water. Point P, at which the ratio of octylammonium chloride to octylamine is 0.55:0.45, is the optimum ratio of surfactant for the solubilization of oil (xylene). Point Q, at which the ratio of octylammonium chloride to octylamine is 0.40:0.60, is the optimum ratio of surfactant for the solubilization of water in nonaqueous solution. We can thus conclude a balanced (in the HLB sense) mixed layer is formed when the mixing ratio is around 1.22 to 0.66. If we take into account the concentrations of surfactants which are singly dispersed, these estimates of the mixing ratio at the interface may be approached. In any case, we can determine the optimum mixing ratio of surfactant from Fig. 11.

However, this determination of the mixing ratio at which the HLB of the mixed adsorbed layer is optimum involves time-consuming experimental procedures, in just the same way as finding the optimum HLB number of an emulsifier (or emulsifier blend) for a particular oil.

A rough approximation to the HLB may be obtained from the water solubility of surfactants. In the case of nonionic surfactants, for example, the cloud point is an important index of the HLB. In the case of ionic surfactants, however,

Table 5. Required HLB Numbers for Emulsification of Oils and Waxes [45]

| Compound | CAS registry number | HLB number | Compound | CAS registry number | HLB number |
|---|---|---|---|---|---|
| | | | *O/W emulsion* | | |
| Acetophenone | *98-86-2* | 14 | Dimethyl silicone | *9016-00-6* | 9 |
| Dimer acid | *61788-89-4* | 14 | Ethylaniline | *103-69-5* | 13 |
| Isostearic acid | *2724-58-5* | 15-16 | Ethyl benzoate | *93-89-0* | 13 |
| Lauric acid | *143-07-7* | 16 | Fenchone | *1196-79-5* | 12 |
| Linoleic acid | *60-33-3* | 16 | Isopropyl myristate | *110-27-0* | 12 |
| Oleic acid | *112-80-11* | 17 | Isopropyl palmitate | *142-91-6* | 12 |
| Ricinoleic acid | *141-22-0* | 16 | Kerosene | *8008-20-6* | 12 |
| Cetyl alcohol | *36653-82-4* | 16 | Lanolin, anhydrous | *8006-54-1* | 12 |
| Decyl alcohol | *25339-17-7* | 15 | Lard | *61789-99-9* | 5 |
| Hexadecyl alcohol | *36653-82-4* | 11-12 | Laurylamine | *124-22-1* | 12 |
| Isodecyl alcohol | *25339-17-7* | 14 | Menhaden oil | *8002-50-4* | 12 |
| Lauryl alcohol | *112-53-8* | 14 | Methyl phenyl silicone | *42557-10-8* | 7 |
| Oleyl alcohol | *143-28-2* | 14 | Methyl silicone | *9076-37-3* | 11 |
| Stearyl alcohol | *112-92-5* | 15-16 | Mineral oil, aromatic | *8012-95-1* | 12 |
| Tridecyl alcohol | *112-70-9* | 14 | Mineral oil, paraffinic | *8012-95-1* | 10 |
| Arlamol E | *25231-24-4* | 7 | Mineral spirits | *8030-30-6* | 14 |
| Beeswax | *8012-89-3* | 9 | Mink oil | *8023-74-3* | 9 |
| Benzene | *71-43-2* | 15 | Nitrobenzene | *98-53-3* | 13 |

| | | | | | |
|---|---|---|---|---|---|
| Benzonitrile | *100-47-0* | 14 | Nonylphenol | *25154-52-3* | 14 |
| Bromobenzene | *108-86-1* | 13 | *ortho*-Dichlorobenzene | *95-50-1* | 13 |
| Butyl stearate | *123-95-5* | 11 | Palm oil | | 7 |
| Carbon tetrachloride | *56-23-5* | 16 | Paraffin wax | *8002-74-2* | 10 |
| Carnauba wax | *8015-86-9* | 15 | Petrolatum | *8009-03-8* | 7-8 |
| Castor oil | *8001-79-4* | 14 | Petroleum naphtha | *8030-30-6* | 14 |
| Ceresine wax | | 8 | Pine oil | *8002-09-3* | 16 |
| Chlorinated paraffin | *8029-39-8* | 12-14 | Polyethylene wax | *9002-88-4* | 15 |
| Chlorobenzene | *108-90-7* | 13 | Propylene tetramer | *9003-97-0* | 14 |
| Cocoa butter | | 6 | Rapeseed oil | *8002-13-9* | 7 |
| Corn oil | *8001-30-7* | 8 | Safflower oil | | 7 |
| Cottonseed oil | *8001-29-4* | 6 | Soybean oil | | 6 |
| Cyclohexane | *110-82-7* | 15 | Styrene | *100-42-5* | 15 |
| Decahydronaphthalene | *91-17-8* | 15 | Tallow | *61789-97-7* | 6 |
| Decyl acetate | *112-30-1* | 11 | Toluene | *108-88-3* | 15 |
| Diethylaniline | *91-66-7* | 14 | Trichlorotrifluoroethane | *76-13-1* | 14 |
| Diisooctyl phthalate | *27554-26-3* | 13 | Tricresyl phosphate | *1330-78-1* | 17 |
| Diisopropylbenzene | *25321-09-9* | 15 | Xylene | *1330-20-7* | 14 |
| | | | *W/O emulsion* | | |
| Gasoline | | 7 | Mineral oil | | 6 |
| Kerosene | *8008-20-6* | 6 | Stearyl alcohol | *112-92-5* | 7 |

*Source:* Ref. 45.

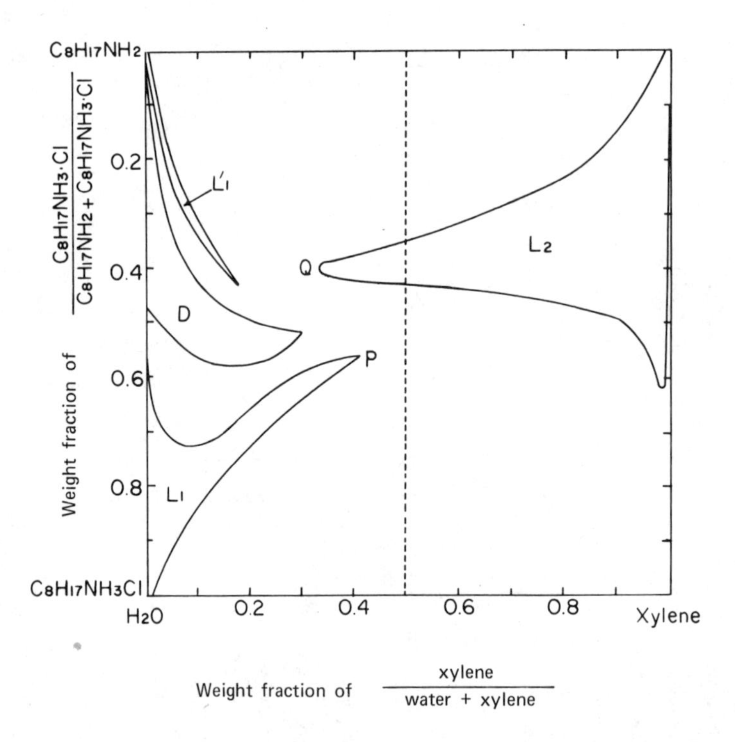


**Figure 11.** Phase equilibria of octylamine, octylammonium chloride, water, and *p*-xylene at 22° C. Octylamine + octylammonium chloride, and *p*-xylene + water were kept 20 wt% and 80 wt% of the system, respectively. (From Ref. 54.)

surfactant HLB does not change with temperature and hence there occur no phase separation phenomena comparable to the cloud point.

As shown on the water axis in Fig. 11, the solution behavior of octylammonium chloride + octylamine changes with composition. Mixed surfactant dissolved in water forms micelles in the region rich in ionic surfactant. If the ratio of cosurfactant increases, liquid crystalline phase separates, and the whole system becomes liquid crystalline. In this way, the dissolution behavior of ionic surfactant + cosurfactant in water may be used to estimate the HLB of the ionic surfactant or its mixtures with a cosurfactant.

Ekwall et al. have extensively studied the triangular phase diagrams of surfactant systems [15,16]. In most systems consisting of water, ionic surfactant, and alcohol they obtained more or less similar phase diagrams. One of the most intensively studied systems is reproduced in Fig. 12. An aqueous micellar solution is observed in the system containing a relatively small amount of alcohol. A lamellar liquid crystalline phase appears in solutions containing comparable amounts of ionic surfactant and alcohol, and a reversed micellar solution is observed in the

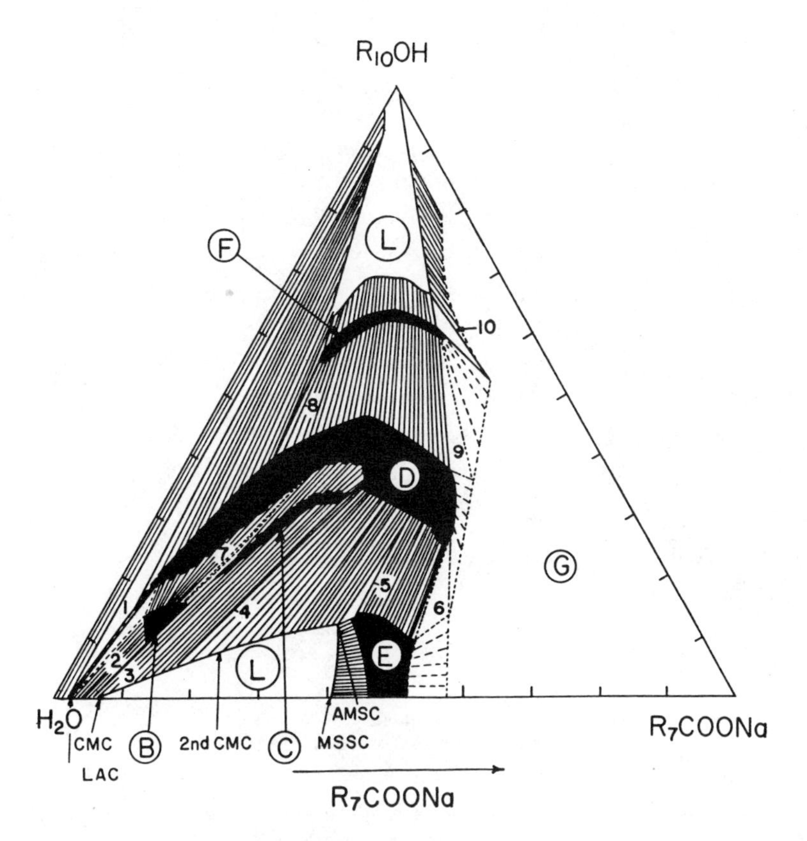


Figure 12. Phase diagrams for ternary systems containing sodium caprylate/decanol/water at 20° C. (From Ref. 16.)

alcohol corner of the diagram. We consider that the hydrophile-lipophile property of the mixed monolayer is balanced in a lamellar liquid crystal. The water axis in Fig. 11 corresponds to the solution behavior of an aqueous solution containing 20 wt% of ionic surfactant + cosurfactant in Fig. 12, i.e., we can observe the change in HLB of the surfactant mixture as a function of composition from the solution behavior of the mixture in water.

As pointed out, ordinary ionic surfactants are usually too hydrophilic and the triangular phase diagrams are similar. However, bivalent salts of anionic surfactants, such as $C_{12}H_{25}OCH_2CH_2OSO_3 \cdot Mg_{0.5}$ (or $Ca_{0.5}$) [46,47], and ionic surfactants which have two comparable hydrocarbon chains [55] appear to be well-balanced surfactants, and the solution behavior of dioctadecyl dimethylammonium (DODMACl) chloride in water actually exhibits such a tendency (Fig. 13). A lamellar liquid crystalline phase is readily formed in solutions containing only this surfactant. The phase diagram of water/DODMACl/decanol (upper portion of Fig.

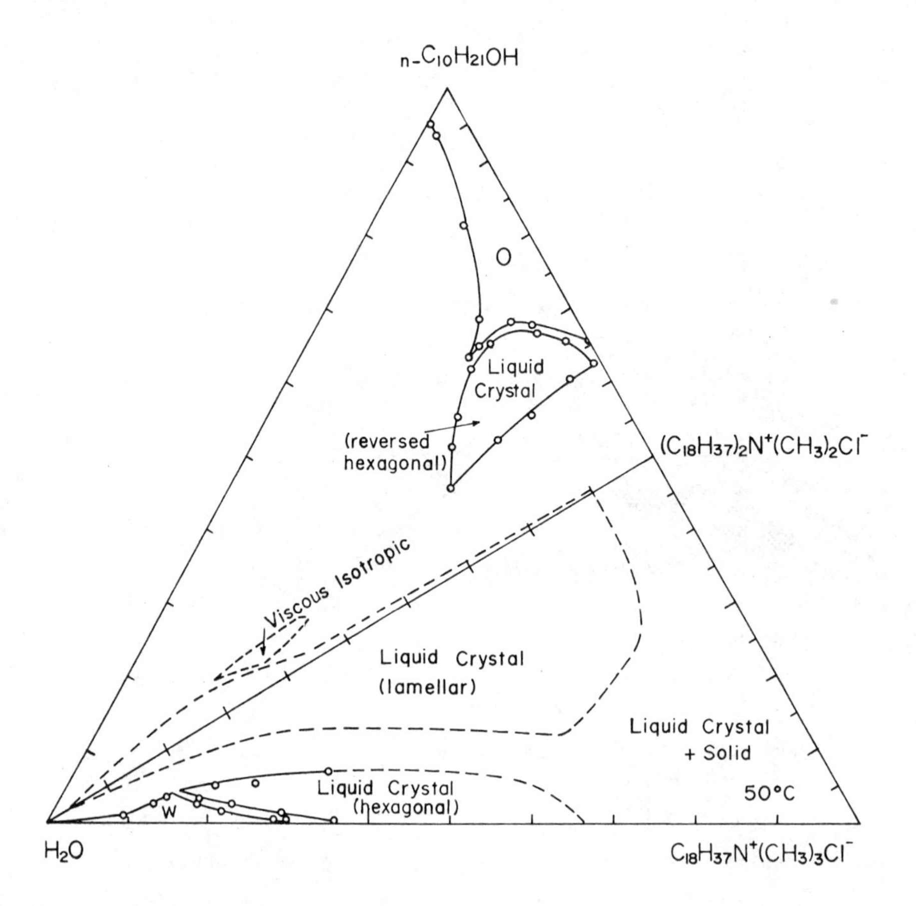


Figure 13. Combined phase diagram of two triangular phase diagrams, $(C_{18}H_{37})_2N(CH_3)_2Cl$/decanol/water and $(C_{18}H_{37})_2N(CH_3)_2Cl/C_{18}H_{37}N(CH_3)_2Cl$/water at 50° C. [From Ref. 55, reprinted with permission from H. Kunieda and K. Shinoda, *J. Phys. Chem. 82*:1710 (1977). Copyright by the American Chemical Society.]

13) resembles that of the alcohol-rich region in Fig. 12 and the phase diagram of water/DODMACl/dioctadecyl trimethylammonium chloride (lower half of Fig. 13) resembles that of the surfactant-rich region in Fig. 12.

The HLB of ionic surfactants is estimated from the mixing ratio of ionic surfactant and cosurfactant (alcohol) at which a lamellar liquid crystalline phase is formed in the ternary phase diagram (ionic surfactant/decanol/water) [16,17,55]. The semiquantitative order of the HLB of various ionic surfactants are plotted in Fig. 14.

Lin and Marszall have proposed an equation which relates the cmc and the HLB of ionic surfactants [56]. However, there is the objection that the cmc will de-

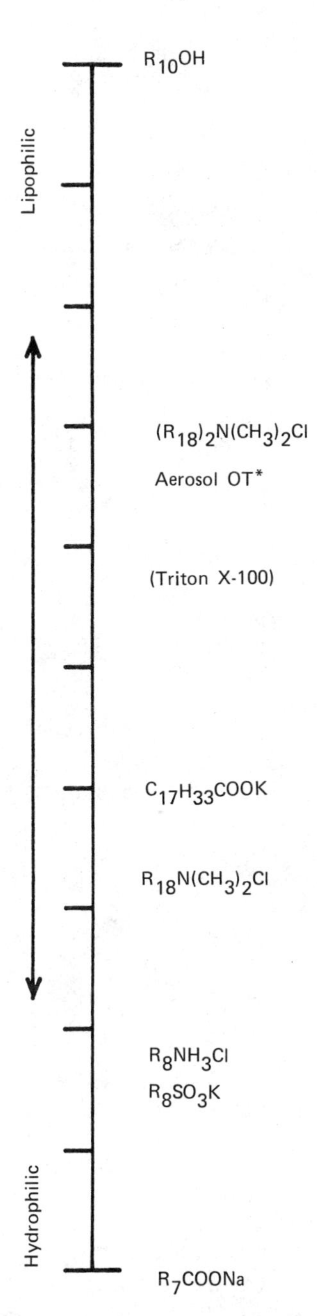


**Figure 14.** Semiquantitative order of HLB of the various ionic surfactants estimated from the mixing ratio of ionic surfactant and cosurfactant (decanol) at which a lamellar liquid crystal is formed [Aerosol OT: sodium 1;2-bis(2-ethylhexyloxycarbonyl)-1-ethanesulfonate].

crease geometrically with the size of the hydrophile and lipophile groups under conditions in which the HLB will be kept constant.

Little [57] has found a correlation between the HLB numbers and solubility parameters of surfactants. However, the solubility parameter is not effective in evaluating the solubility of surfactants. On the other hand, a correlation between the solubility of surfactant aggregates (inverse micelles) in hydrocarbon solvents may hold.

## References


1. W. Clayton, *Theory of Emulsions*, 4th ed., Blakiston, Philadelphia, Pa., 1943, p. 127.
2. W. C. Griffin, *J. Soc. Cosmetic Chemists 1*:311 (1949).
3. W. C. Griffin, *J. Soc. Cosmetic Chemists 5*:249 (1954).
4. J. T. Davies, *Interfacial Phenomena*, Academic, New York, 1963, pp. 371-383.
5. C. D. Moore and M. Bell, *Soap, Perfumery Cosmetics 29*:893 (1956).
6. H. L. Greenwald, G. L. Brown, and M. N. Fineman, *Anal. Chem. 28*:1963 (1956).
7. K. Shinoda and H. Arai, *J. Phys. Chem. 68*:3485 (1964).
8. K. Shinoda, *J. Colloid Interface Sci. 24*:4 (1967).
9. K. Shinoda and H. Saito, *J. Colloid Interface Sci. 30*:258 (1969).
10. K. Shinoda, *Proc. 5th Int. Congr. Surface Active Substances*, Barcelona *3*:275 (1968).
11. L. Marszall, *Acta Pol. Pharm. 32*:397 (1975).
12. L. Marszall, *Colloid Polymer Sci. 254*:674 (1976).
13. L. Marszall, *Fette, Seifen, Anstrichmittel 80*:289 (1978).
14. K. Shinoda and H. Saito, *J. Colloid Interface Sci. 26*:70 (1968).
15. P. Ekwall, I. Danielsson, and L. Mandell, *Kolloid-Z. 169*:113 (1960).
16. P. Ekwall, L. Mandell, and K. Fontell, *Mol. Cryst. Liq. Cryst. 8*:157 (1969).
17. G. Gillberg, H. Lehtinen, and S. Friberg, *J. Colloid Interface Sci. 33*:40 (1970).
18. S. Friberg and I. Lapczynska, *Progr. Colloid Polymer Sci. 55*:614 (1976).
19. S. Friberg, I. Lapczynska, and G. Gillberg, *J. Colloid Interface Sci. 56*:19 (1976).
20. P. Becher, *Emulsions: Theory and Practice*, 2d ed., Reinhold, New York, 1965, p. 164.
21. K. Shinoda and H. Sagitani, *J. Colloid Interface Sci. 64*:68 (1978).
22. K. Shinoda and H. Kunieda, *J. Colloid Interface Sci. 42*:381 (1973).
23. H. Saito and K. Shinoda, *J. Colloid Interface Sci. 24*:10 (1967).
24. K. Shinoda and T. Ogawa, *J. Colloid Interface Sci. 24*:56 (1967).
25. H. Saito and K. Shinoda, *J. Colloid Interface Sci. 32*:647 (1970).
26. K. Shinoda and H. Arai, *J. Colloid Interface Sci. 25*:429 (1967).
27. W. N. Maclay, *J. Colloid Interface Sci. 11*:272 (1956).
28. K. Shinoda and H. Takeda, *J. Colloid Interface Sci. 32*:642 (1970).
29. R. P. Enever, *J. Pharm. Pharmacol. 26*:128 (1974).
30. R. P. Enever, *J. Pharm. Sci. 65*:517 (1976).
31. P. A. Winsor, *Trans. Faraday Soc. 44*:376 (1948).
32. P. A. Winsor, *Solvent Properties of Amphilphilic Compounds*, Butterworths, London, 1954, Chap. 3-6.

33. J. H. Schulman and D. P. Riley, *J. Colloid Sci.* *3*:383 (1948).
34. J. H. Schulman and J. A. Friend, *J. Colloid Sci.* *4*:497 (1949).
35. J. H. Schulman and J. B. Montagne, *Ann. N.Y. Acad. Sci.* *92*:366 (1961).
36. J. E. Bowcott and J. H. Schulman, *Z. Elektrochem.* *59*:283 (1955).
37. J. H. Schulman, W. Stoeckenius, and L. M. Prince, *J. Phys. Chem.* *63*:1677 (1959).
38. J. Zlochower and J. H. Schulman, *J. Colloid Interface Sci.* *24*:115 (1967).
39. H. Kunieda and K. Shinoda, *J. Colloid Interface Sci.* *75*:601 (1980).
40. P. Becher, *Emulsions: Theory and Practice.* 2d ed., Reinhold, New York, 1965, pp. 231-235.
41. P. Becher, in *International Review of Science, Physical Chemistry, Ser.* 2, Vol. 7 (M. Kerker, ed.), Butterworths, London, 1975, pp. 251-253.
42. P. Becher and W. C. Griffin, *McCutcheon's Detergents and Emulsifiers,* MC Publishing, Glen Rock, N.J., 1974, p. 227.
43. K. Shinoda, H. Saito, and H. Arai, *J. Colloid Interface Sci.* *35*:624 (1971).
44. K. Shinoda, T. Yoneyama, and H. Tsusumi, *J. Dispersion Sci. Technol.* *1*:1 (1980).
45. W. C. Griffin, *Kirk-Othmer Encyclopedia of Chemical Technology,* 3d ed., Vol. 8, Wiley-Interscience, New York, 1979, pp. 913-916.
46. K. Shinoda and T. Hirai, *J. Phys. Chem.* *81*:1842 (1977).
47. K. Shinoda, *Pure Appl. Chem.* *52*:1195 (1980).
48. H. B. Klevens, *J. Chem. Phys.* *17*:1004 (1949).
49. H. B. Klevens, *J. Amer. Chem. Soc.* *72*:3581, 3780 (1950).
50. K. Shinoda and H. Akamatsu, *Bull. Chem. Soc. Jap.* *31*:497 (1958).
51. F. Z. Saleeb, C. J. Cante, T. K. Streckfus, J. R. Frost, and H. L. Rosano, *J. Amer. Oil Chemists Soc.* *52*:208 (1975).
52. J. H. Schulman and E. G. Cockbain, *Trans. Faraday Soc.* *36*:661 (1940).
53. S. Friberg, *Kolloid-Z.* *244*:333 (1971).
54. S. I. Ahmad, K. Shinoda, and S. Friberg, *J. Colloid Interface Sci.* *47*:32 (1974); K. Shinoda, H. Kunieda, N. Obi, and S. Friberg, *ibid.,* *80*:304 (1981).
55. H. Kunieda and K. Shinoda, *J. Phys. Chem.* *82*:1710 (1977); H. Sagitani, T. Suzuki, M. Nàgal, and K. Shinoda, *J. Colloid Interface Sci.* *87*:11 (1982).
56. I. J. Lin and L. Marszall, *J. Colloid Interface Sci.* *57*:85 (1976).
57. R. C. Little, *J. Colloid Interface Sci.* *65*:587 (1978).

# 6

# Emulsion Droplet Size Data

CLYDE ORR[†] / Georgia Institute of Technology, Atlanta, Georgia



## I. Introduction

The size distribution of its droplets is a most important parameter in characterizing any emulsion. Two emulsions may have the same average droplet diameter and yet exhibit quite dissimilar behavior because of differences in their distribution of diameters. Stability and resistance to creaming, rheology, chemical reactivity, and physiological efficiency are but a few of the phenomena influenced by both relative size and size distribution. This section includes neither the obtaining of size data nor the influence of size parameters on properties. It attempts, instead, to cover

---

[†]Present address: Micromeritics Instrument Corporation, Norcross, Georgia.

the presentation and manipulation of sizing data to yield the maximum of useful information in a concise and meaningful form so that once such size information is attained, it can be employed in correlations with production and properties as covered elsewhere.

Size is an individual droplet property, but any property of one droplet is hardly a value of interest in overall emulsion evaluation. It is the entire distribution of sizes within an emulsion that is of real concern. Practically, only a limited number of droplets are ever examined, but the resulting measurements will be treated as if they constituted a continuous distribution of sizes. As is usual when dealing with multivalued (i.e., distributed) variables, statistical methods will be utilized to limit the number of variables. Nevertheless, care should be taken to assure appropriate representation for all sizes in every measurement effort. A few hundred size measurements may justify the continuum approach if the distribution is narrow, while several thousand may barely be adequate if the distribution is wide. The only really satisfactory means for determining if sufficient numbers of droplets have been examined is to examine more and then determine if there has been a significant shift in the data. The following assumes adequate sampling and sufficient measurements for representation as continuous distributions.

Continuous distributions of sizes can be represented by any one of several graphical plots and by analytical and empirical functions. Any property of interest relating to size, such as total droplet surface, volume, or mass, can, in principle, be obtained once the distribution is adequately characterized.

There are many techniques for obtaining size information. Emulsion droplet data may be gathered in terms of the number N of droplets of specific length, namely the diameter D, by means of optical or electron microscopy [1-4]; as diameter versus droplet projected area S by application of turbidity techniques [5-8]; as diameter versus surface area A by light-scattering measurements [4,9,10-12]; as number versus volume V by electrical resistance counting [13,14,15]; and in terms of diameter versus mass M by x-ray transmission coupled with sedimentation [16], or perhaps, by hydrodynamic chromatography [17,18]. Thus the evaluation of an emulsion for size can involve measures of its droplet number, length (diameter), area, volume, and mass.[†]

## II. Mean Diameters

If the droplets of an emulsion were equally sized, all measurement methods should give the same values for diameter and number (concentration) provided measure-

---

[†]Both volume and mass need not be considered separately in many instances since they are related through an independent constant, the droplet density.

ments yielding projected area, surface area, and volume data were converted into the appropriate terms by means of the geometric relationships $S/N = \pi D^2/4$, $A/N = \pi D^2$, or $V/N = \pi D^3/6$. Monodisperse emulsions are very special cases and the far more likely situation of unequally sized droplets must be treated. Mean diameters need then to be utilized, the appropriate one or ones depending on the method of data determination, and the intended use to be made of the result.

Mean diameters are devices for representing as one value a system of unequal size droplets by focusing on particular parameters from among those of number, length, area, and volume (or mass). Any mean representation may involve two and only two characteristic parameters of the droplet system. By way of illustration, consider a system consisting of 10 droplets having diameters of 1, 2, 3, . . . , 10.

When these diameters, or lengths, are summed, the result is 55. This number divided by 10, since there are 10 droplets, gives a number-length mean diameter $D_{NL}$ of 5.5. This reveals that, in so far as number and length are concerned, the distribution of unequally sized droplets is equivalent to one of 10 equally sized droplets each having a diameter of 5.5. As another example, the total surface area of the 10 unequal droplets, as obtained by summing $\pi(1)^2$, $\pi(2)^2$, $\pi(3)^2$, . . . , $\pi(10)^2$, is $385\pi$. Ten equal size droplets having this total surface are found from $\sqrt{385\pi/10\pi}$ to have diameters of 6.21, which is the number-surface mean diameter $D_{NA}$.

In the above illustrations one initial parameter (number) was retained, but any two parameters may be utilized. Continuing with the 10 unequal droplets having a total surface of $385\pi$ and a total length of 55, the length-surface mean diameter $D_{LA}$ is the diameter of equal sized droplets having these two properties but, as will be seen, the number of them can no longer be 10. This mean diameter is $385\pi/55\pi$ which is 7, and the number is 55/7 or 7.857. In other words, 7.857 droplets each of diameter 7 have the same total length and the same total surface as the unequally sized set of 10. The definitions of these and other mean diameters are summarized in Table 1. Unfortunately, many different names have been applied to mean diameters over the years. Some, but not all, of them are indicated in the table. When using results from the technical literature, it is prudent to establish which mean is actually being employed. Still other means will be introduced subsequently.

The value of mean diameters is several fold. First, they provide a mechanism for reducing a mass of data to a conveniently handled form. Second, they serve as a guide for property correlation. If, for example, the initial reaction rates of two emulsions were being compared, a logical basis for the comparison might be a mean diameter involving surface area. Finally, mean diameters permit direct computation of product quantities. Should an emulsifier operating under certain conditions be known to produce emulsion droplets with a surface-volume mean diameter

Table 1. Mean Diameter Definitions

| Descriptive name (alternate found in literature) | Symbol | Mathematical expression[a] |
|---|---|---|
| Number-length mean (arithmetic mean) | $D_{NL}$ | $\frac{\Sigma D \Delta N}{\Sigma \Delta N}$ |
| Number-surface mean (surface mean, diameter of average surface) | $D_{NA}$ | $\left(\frac{\Sigma D^2 \Delta N}{\Sigma \Delta N}\right)^{1/2}$ |
| Number-volume mean (volume mean, diameter of average volume) | $D_{NV}$ | $\left(\frac{\Sigma D^3 \Delta N}{\Sigma \Delta N}\right)^{1/3}$ |
| Length-surface mean (linear mean, length diameter mean) | $D_{LA}$ | $\frac{\Sigma D^2 \Delta N}{\Sigma D \Delta N}$ |
| Length-volume mean (volume-diameter mean) | $D_{LV}$ | $\left(\frac{\Sigma D^3 \Delta N}{\Sigma D \Delta N}\right)^{1/2}$ |
| Surface-volume mean (surface mean, Sauter) | $D_{AV}$ | $\frac{\Sigma D^3 \Delta N}{\Sigma D^2 \Delta N}$ |
| Volume-(or weight-) moment mean [volume-(or weight-) mean, weight average particle size, De Brouckere] | $D_{VM}$ or $D_{WM}$ | $\frac{\Sigma D^4 \Delta N}{\Sigma D^3 \Delta N}$ |

[a]Could be expressed in terms of number percentages ($\Delta P$) as well as in actual numbers ($\Delta N$).

$D_{SV}$ of, say, 10 μm (= $1 \times 10^{-3}$ cm) and it was desired to know the surface area created by emulsifying 1000 $cm^3$ of liquid, the result is immediately calculated as 1000 $cm^3/1 \times 10^{-3}$ cm = $10^6$ $cm^2$. The other mean diameters can be similarly employed to relate number and total surface area, length (diameter) and volume, and the like.

## III. Data Classification

Once size measurements are available, it greatly facilitates the treatment to consolidate the information. This is called classifying the data. The task is accomplished by arranging the measured values into a number of intervals and tabulating the numbers that fall within each interval. Each interval is termed a class, and the droplet diameter limits ($D_1$ to $D_2$) of each class are known as *class boundaries*. The choice of class boundaries is quite important. Equal differences within class boundaries (arithmetic progression) will usually be adequate when the spread of the data is small. If the spread is wide and there is reason to believe droplet diameters may be logarithmically distributed, then equal differences between the logarithms of the diameters (geometric progression) may be better employed. As a general guide, the boundaries can be chosen so that the class interval divided by the average class diameter is approximately constant as demonstrated in a tabulation to follow. Selection of 10 to 20 classes is recommended since fewer can result in the discarding of valuable information and more creates excessive computation. The number of droplets in each class is called the frequency $\Delta N$ and the total number as employed here is $\Sigma \Delta N$. Of course, number percent $\Delta P$ can be used as well as actual numbers in most instances.

Suppose a microscopic examination of 1000 emulsion droplets yielded the classified data given in the two left-hand columns of Table 2. The other tabulated numbers are then readily calculated. These data can be graphed to show the percent by numbers within each size range as on Fig. 1, which is called a *histogram*, or as cumulative presentations represented by the curves of Fig. 2. The latter curves relate the percent of droplets by number equal to and less than, or equal to and greater than, the corresponding diameter. One of the curves is obtained by plotting accumulated percentages against the lower value of each class, while the other employs the upper value of each diameter interval. Either set of data can be employed in calculating mean diameters, for example, and will be so employed herein later. The arithmetic average class diameter rather than either limit is also useful for certain purposes as will be demonstrated.

Tabel 3 presents the calculations for mean diameters as defined in Table 1 using the data of Table 2. Table 4 shows conversions of the number-diameter data into surface area and volume (or mass) distribution, the results being presented also as cumulative plots on Fig. 3. Obviously, the limits of these curves must be identical, but their differing shapes emphasize the influence of surface and volume as opposed to diameter (length).

## IV. Conversion of Base

So far, this treatment has been based on experimental data relating numbers of droplets to specific diameter ranges, but, as noted previously, data might be acquired in terms of droplet surface, volume, or mass versus diameter. Geometrical factors are now required for conversion to another base. As an example, assume a turbidimetric analysis of an emulsion gave the results in the two left-hand columns of Table 5. The projected area distribution data can be converted into droplet volume data as carried out in the table or into distribution information on

Table 2. Classification of Data and Initial Calculations

| Experimental data | | | | | Percentages (number basis) | | |
|---|---|---|---|---|---|---|---|
| Droplet diameter range $D_1$ to $D_2$ (μm) | Number in range $\Delta N$ | Diameter interval $\Delta D$ (μm) | Average diameter D (μm) | Class boundary check[a] $\Delta D/D$ | Within range $\Delta P$ (%) | Equal to or less than maximum ($D_2$) diameter (%) | Equal to or greater than minimum ($D_1$) diameter (%) |
| 0.71-1.00 | 1 | 0.3 | 0.9 | 0.333 | 0.1 | 0.1 | 100 |
| 1.01-1.40 | 4 | 0.4 | 1.2 | 0.333 | 0.4 | 0.5 | 99.9 |
| 1.41-2.00 | 22 | 0.6 | 1.7 | 0.353 | 2.2 | 2.7 | 99.5 |
| 2.01-2.80 | 69 | 0.8 | 2.4 | 0.333 | 6.9 | 9.6 | 97.3 |
| 2.81-4.00 | 134 | 1.2 | 3.4 | 0.353 | 13.4 | 23.0 | 90.4 |
| 4.01-5.60 | 249 | 1.6 | 4.8 | 0.333 | 24.9 | 47.9 | 77.0 |
| 5.61-8.00 | 259 | 2.4 | 6.8 | 0.353 | 25.9 | 73.8 | 52.1 |
| 8.01-11.20 | 160 | 3.2 | 9.6 | 0.333 | 16.0 | 89.8 | 26.2 |
| 11.21-16.00 | 73 | 4.8 | 13.6 | 0.353 | 7.3 | 97.1 | 10.2 |
| 16.01-22.40 | 21 | 6.4 | 19.2 | 0.333 | 2.1 | 99.2 | 2.9 |
| 22.41-32.00 | 6 | 9.6 | 27.2 | 0.353 | 0.6 | 99.8 | 0.8 |
| 32.01-44.80 | 2 | 12.8 | 38.4 | 0.333 | 0.2 | 100 | 0.2 |
| | 1000 | | | | 100 | | |

[a]Approximately constant.

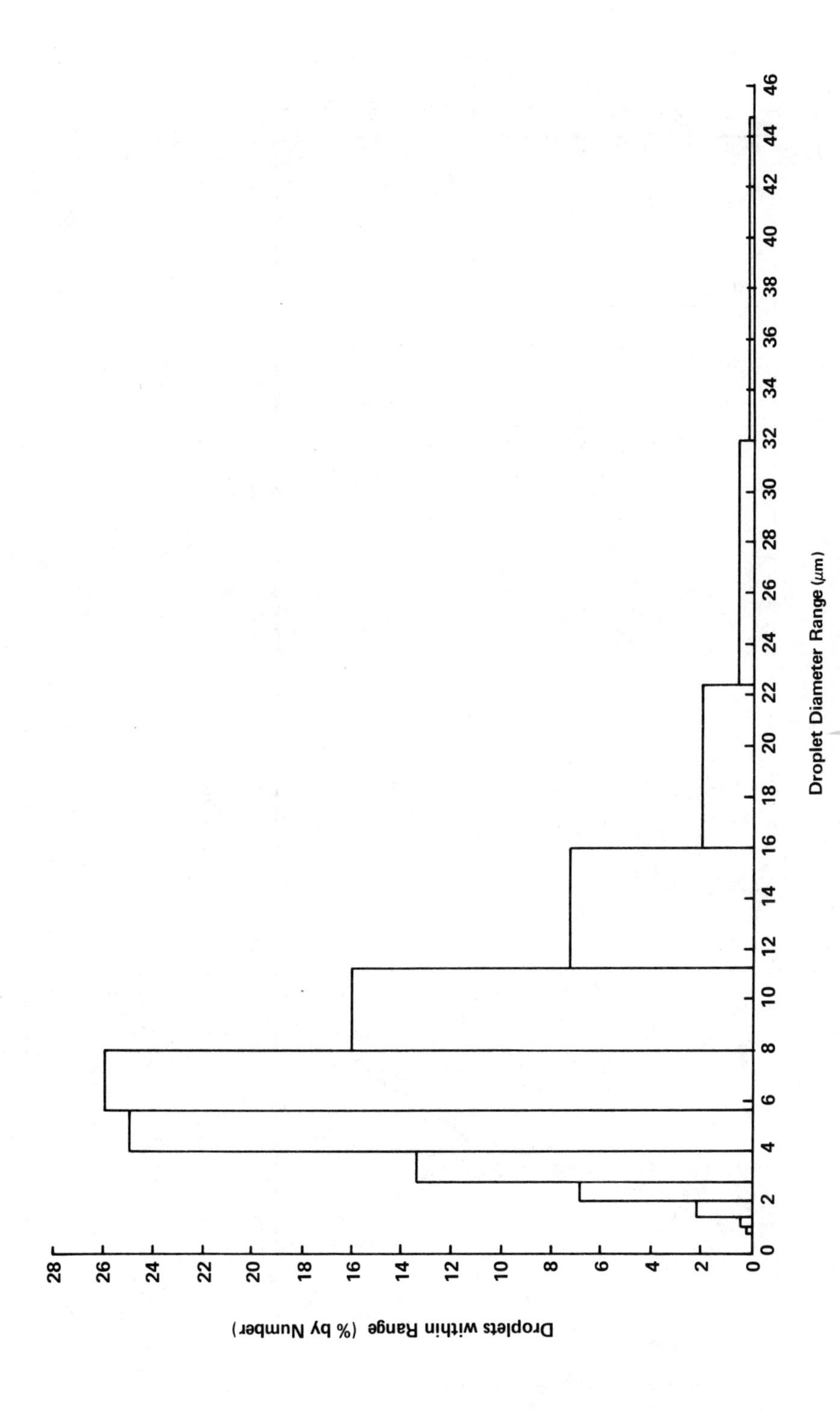


Figure 1. Histogram (data from Table 2).

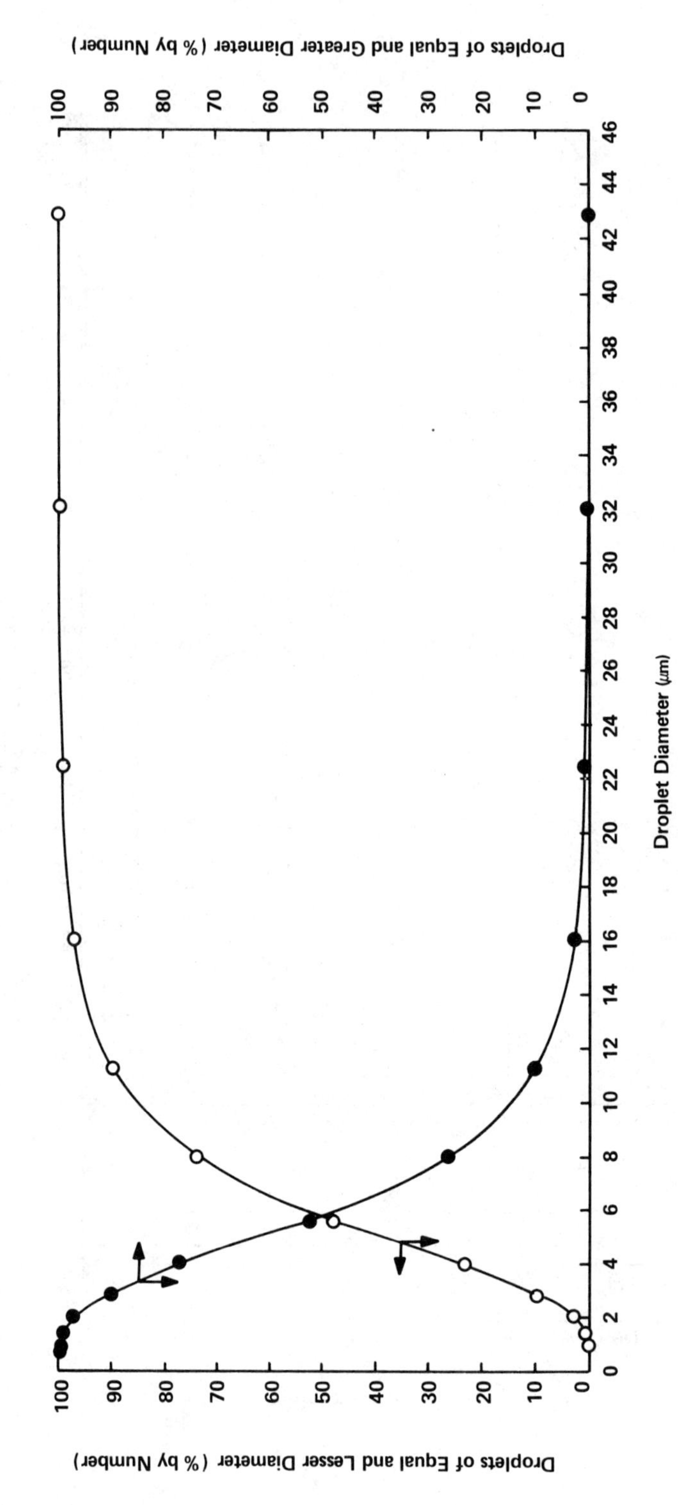


Figure 2. Cumulative plots of data from Table 2.

Table 3. Calculation of Mean Diameters

| Distribution data[a] | | Calculated terms | | | | | | |
|---|---|---|---|---|---|---|---|---|
| Average diameter D (μ) | Number percentage $\Delta P^b$ (%) | $D^2$ | $D^3$ | $D^4$ | $D\Delta P$ | $D^2\Delta P$ | $D^3\Delta P$ | $D^4\Delta P$ |
| 0.9 | 0.1 | 0.8 | 0.7 | 0.7 | 0.1 | 0.1 | 0.1 | 0.1 |
| 1.2 | 0.4 | 1.4 | 1.7 | 2.0 | 0.5 | 0.6 | 0.7 | 0.8 |
| 1.7 | 2.2 | 2.9 | 4.9 | 8.4 | 3.7 | 6.4 | 10.8 | 18.4 |
| 2.4 | 6.9 | 5.8 | 13.8 | 33.2 | 16.6 | 39.7 | 95.4 | 228.9 |
| 3.4 | 13.4 | 11.6 | 39.3 | 133.6 | 45.6 | 154.9 | 526.7 | 1790.7 |
| 4.8 | 24.9 | 23.0 | 110.6 | 530.8 | 119.5 | 573.7 | 2753.7 | 13218.0 |
| 6.8 | 25.9 | 46.2 | 314.4 | 2138.1 | 176.1 | 1197.6 | 8143.8 | 55377.8 |
| 9.6 | 16.0 | 92.2 | 884.7 | 8493.5 | 153.6 | 1474.6 | 14155.8 | 135895.4 |
| 13.6 | 7.3 | 185.0 | 2515.5 | 34210.2 | 99.3 | 1350.2 | 18362.8 | 249734.5 |
| 19.2 | 2.1 | 368.7 | 7077.9 | 135895.5 | 40.3 | 774.1 | 14863.6 | 285380.4 |
| 27.2 | 0.6 | 739.8 | 20123.6 | 547363.2 | 16.3 | 443.9 | 12074.2 | 328417.9 |
| 38.4 | 0.2 | 1474.6 | 56623.1 | 2174327.2 | 7.7 | 294.9 | 11324.6 | 434865.4 |
| | 100 | | | | 679.3 | 6310.7 | 82312.2 | 1504928.3 |

$D_{NL} = 679.3/100 = 6.8\ \mu m$

$D_{NA} = (6310.7/100)^{1/2} = 7.9\ \mu m$

$D_{NV} = (82312.2/100)^{1/3} = 9.4\ \mu m$

$D_{LA} = 6310.7/679.3 = 9.3\ \mu m$

$D_{LV} = (82312.2/679.3)^{1/2} = 11.0\ \mu m$

$D_{AV} = 82312.2/6310.7 = 13.0\ \mu m$

$D_{VM}$ or $D_{WM} = 1504928.3/82312.2 = 18.3\ \mu m$

[a] From Table 2.

[b] Actual numbers or percentages can be used interchangeably.

Table 4. Distribution of Number, Surface Area, and Volume (or Mass) as Percent Equal to and Less Than Corresponding Diameter

| Diameter (maximum in range)[a] D (μm) | Distribution | | |
|---|---|---|---|
| | By number[a] $\Delta P/\Sigma\ \Delta P$ (%) | By surface area $D^2\Delta P/\Sigma(D^2\Delta P)$ (%) | By volume (or mass) $D^3\Delta P/\Sigma(D^3\Delta P)$ (%) |
| 1.00 | 0.1 | Negligible | Negligible |
| 1.40 | 0.5 | Neglibible | Negligible |
| 2.00 | 2.7 | 0.1 | Negligible |
| 2.80 | 9.6 | 0.7 | 0.1 |
| 4.00 | 23.0 | 3.2 | 0.8 |
| 5.60 | 47.9 | 12.3 | 4.1 |
| 8.00 | 73.8 | 31.3 | 14.0 |
| 11.20 | 89.8 | 54.6 | 31.2 |
| 16.20 | 97.1 | 76.0 | 53.5 |
| 22.40 | 99.2 | 88.3 | 71.6 |
| 32.00 | 99.8 | 95.3 | 86.2 |
| 44.80 | 100 | 100 | 100 |

[a]From Table 2.

any other basis. The resulting volume-based data, for example, are shown converted to a number basis in Table 6 and then back into projected areas in Table 7. Comparing the results of Table 7 with the initial data of Table 5 is informative in that it shows multiple conversions using average class diameters and dropping certain digits in the conversion steps still results in the recovery of a distribution that differs only slightly from the original.

## V. Distribution Functions

While mean diameters and distribution curves have their utility, they do not extract all the value contained in a distribution of sizes. The spread, or scatter, of diameters (or surface or volume) about the mean contains much useful information. A distribution function whereby all the size data can be summarized as a precise mathematical expression is required to reveal the maximum of information. In those instances where measuring instrument limitations do not permit gathering size information above or below certain diameters, such an expression adds considerable confidence to any extrapolation of data. If the expression can be linearized

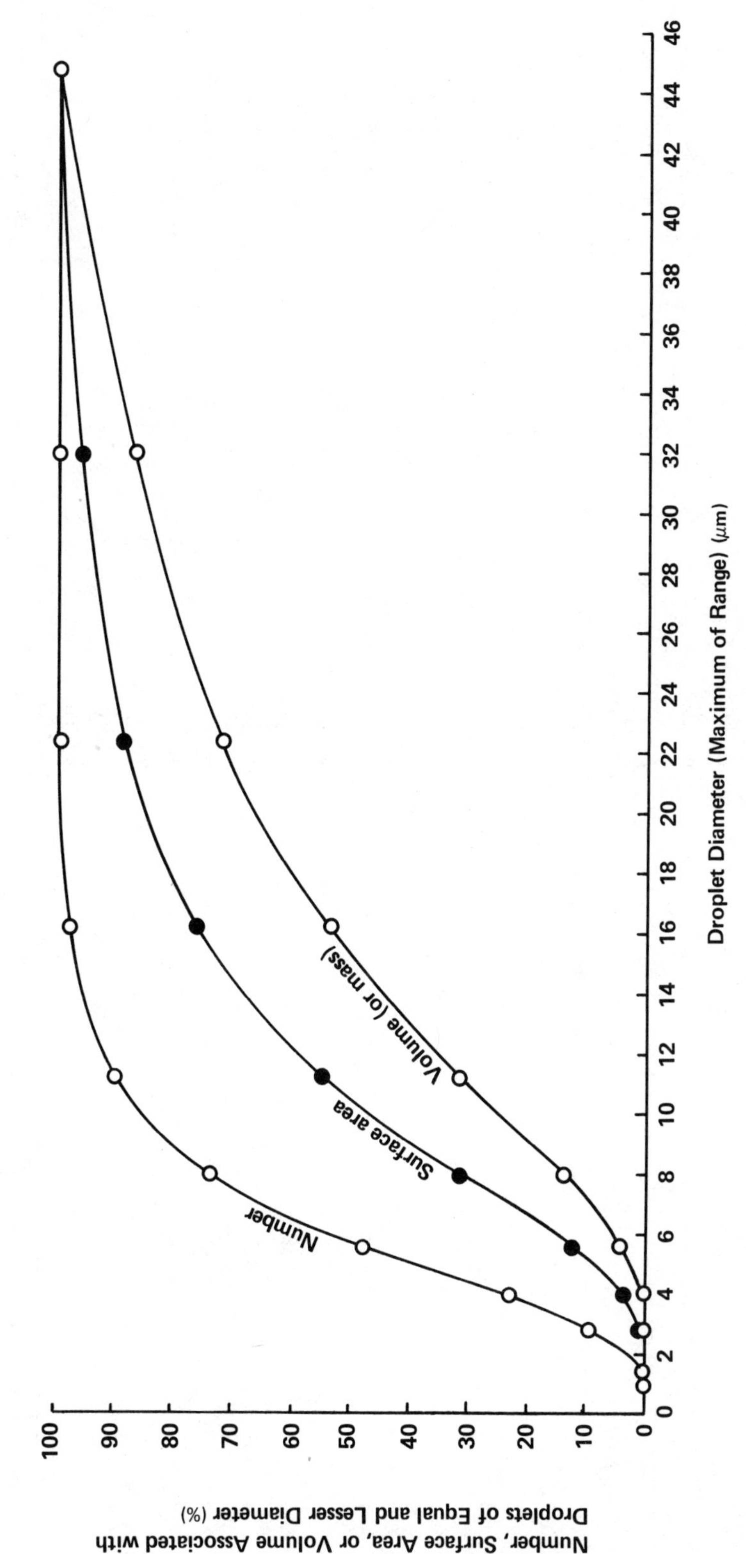


Figure 3. Cumulative distributions by number, surface area, and volume (or mass).

Table 5. Conversion of a Projected Area to a Volume (or Mass) Distribution

| | Experimental data | | | | |
|---|---|---|---|---|---|
| Diameter range (μm) | Percent by projected area less than smaller diameter (%) | Percent by projected area within range (%) | Relative volume within range[a] ($D_{avg}$ × % area within range) | Volume fraction [rel. vol. ÷ Σ (rel. vol.)] | Percent by volume less than smaller Diameter[b] (%) |
| 13-12 | 99.5 | 0.5 | 6.25 | 0.0148 | 98.5 |
| 12-11 | 96.2 | 3.3 | 37.95 | 0.0899 | 89.5 |
| 11-10 | 92.8 | 3.4 | 35.70 | 0.0846 | 81.1 |
| 10-9 | 89.2 | 3.6 | 34.20 | 0.0810 | 73.0 |
| 9-8 | 85.1 | 4.1 | 34.85 | 0.0826 | 64.7 |
| 8-7 | 80.5 | 4.6 | 34.50 | 0.0817 | 56.5 |
| 7-6 | 74.8 | 5.7 | 37.05 | 0.0878 | 47.8 |
| 6-5 | 67.8 | 7.0 | 38.50 | 0.0912 | 38.6 |
| 5-4 | 58.1 | 9.7 | 43.65 | 0.1034 | 28.3 |
| 4-3 | 47.7 | 10.4 | 36.40 | 0.0862 | 19.7 |
| 3-2 | 31.4 | 16.3 | 40.75 | 0.0965 | 10.0 |
| 2-1 | 4.8 | 26.6 | 39.90 | 0.0945 | 0.6 |
| 1-0 | 0 | 4.8 | 2.40 | 0.0057 | 0 |
| | | | 422.10 | | |

[a]Ratio of volume to projected area is $\pi D^3/6 \div \pi D^2/4 = 2D/3$; hence, multiplying this ratio by percent of projected area in range gives relative volume in range. The numerical constant is unimportant because it does not affect relative values. The average diameter is used.

[b]100-100[Σ (volume fraction)].

Table 6. Conversion of a Volume (or Mass) to a Number Distribution

| | From Table 5 | | | | |
|---|---|---|---|---|---|
| Diameter range ($\mu$m) | Percent by volume less than smaller diameter (%) | Percent by volume within range (%) | Relative number within range[a] $\left(\frac{N}{1000} = \frac{\%\ \text{vol. in range}}{D^3_{avg}}\right)$ | Number percent within range (%) | Percent by number less than greater diameter (%) |
| 13-12 | 98.5 | 1.5 | 1 | 0.01 | 100 |
| 12-11 | 89.5 | 9.0 | 6 | 0.07 | 99.99 |
| 11-10 | 81.1 | 8.4 | 7 | 0.08 | 99.92 |
| 10-9 | 73.0 | 8.1 | 9 | 0.10 | 99.84 |
| 9-8 | 64.7 | 8.3 | 14 | 0.16 | 99.74 |
| 8-7 | 56.5 | 8.2 | 30 | 0.35 | 99.58 |
| 7-6 | 47.8 | 8.7 | 32 | 0.37 | 99.23 |
| 6-5 | 38.6 | 9.2 | 55 | 0.63 | 99.86 |
| 5-4 | 28.3 | 10.3 | 113 | 1.30 | 98.23 |
| 4-3 | 19.7 | 8.6 | 201 | 2.32 | 96.93 |
| 3-2 | 10.0 | 9.7 | 621 | 7.16 | 94.61 |
| 2-1 | 0.6 | 9.4 | 2785 | 32.11 | 87.45 |
| 1-0 | 0 | 0.6 | 4800 | 55.34 | 55.34 |
| | | | 8674 | | |

[a]The volume of each droplet is $\pi D^3/6$. Droplet volume multiplied by the number of droplets in range is the relative (or %) volume within range. Hence $N(\pi D^3/6)$ = % volume in range. The numberical constant is unimportant because it does not affect relative values. The average diameter is used. Multiplication by 1000 prevents the numbers from appearing ridiculous; any multiplying factor could be used.

Table 7. Conversion of a Number to a Projected Area Distribution

| | From Table 6 | | | | |
|---|---|---|---|---|---|
| Diameter range ($\mu m$) | Percent by number less than greater diameter (%) | Number percent within range (%) | Relative projected area within range[a] ($D^2_{avg} \times N\%$) | Projected area fraction within range [rel. area ÷ Σ (rel. area)] | Percent by projected area less than smaller diameter[b] (%) |
| 13-12 | 100 | 0.01 | 1.56 | 0.0056 | 99.4 |
| 12-11 | 99.99 | 0.07 | 9.26 | 0.0330 | 96.1 |
| 11-10 | 99.92 | 0.08 | 8.82 | 0.0315 | 93.0 |
| 10-9 | 99.84 | 0.10 | 9.03 | 0.0322 | 89.8 |
| 9-8 | 99.74 | 0.16 | 11.56 | 0.0413 | 85.6 |
| 8-7 | 99.58 | 0.35 | 19.69 | 0.0703 | 78.6 |
| 7-6 | 99.23 | 0.37 | 15.63 | 0.0558 | 73.0 |
| 6-5 | 98.86 | 0.63 | 19.06 | 0.0680 | 66.2 |
| 5-4 | 98.23 | 1.30 | 26.33 | 0.0940 | 56.8 |
| 4-3 | 96.93 | 2.32 | 28.42 | 0.1014 | 46.7 |
| 3-2 | 94.61 | 7.16 | 44.75 | 0.1597 | 30.7 |
| 2-1 | 87.45 | 32.11 | 72.25 | 0.2579 | 4.9 |
| 1-0 | 55.34 | 55.34 | 13.84 | 0.0494 | 0 |
| | | | 280.20 | | |

[a]The project area of a droplet is $\pi D^2/4$. This value multiplied by the number percent in range gives the relative projected area in range. The average diameter is used.

[b]100-100Σ (projected area fraction).

in addition, application of it is simplified, and fitting of data to it becomes possible with a degree of independence not otherwise achievable. Furthermore, a linearized function permits ready graphical representation and affords maximum opportunities for interpolation, extrapolation, and comparison among systems. Finally, if the parameters of the distribution function can be related to the emulsification process or the properties of the resulting emulsion, still more useful information can result. There are many distribution functions [19-28], and obviously, they cannot all be covered here. Besides, the preponderance of emulsions appear to be adequately covered by a relatively few of them.

## A. Normal and Log-Normal

The normal, or Gaussian, distribution is perhaps the best known one dimensional distribution function. It serves as a convenient introduction. The function, written

$$\frac{dP}{dD} = \frac{100}{\sigma\sqrt{2\pi}} \exp\left(-\frac{(D - \overline{D})^2}{2\sigma^2}\right) \tag{1}$$

contains two adjustable parameters, $\overline{D}$ and $\sigma$. The first parameter $\overline{D}$ is called the *arithmetic mean diameter*, and should the function accurately describe droplet diameter-versus-number data, it would be the number-length mean diameter $D_{NL}$ of Table 1. The squared argument involving $\overline{D}$ in the exponent shows the function is symmetric about the mean. The second parameter $\sigma$ is a measure of the uniformity of diameters and is termed the standard deviation; it is defined by

$$\sigma^2 = \frac{\Sigma\,(D - \overline{D})^2}{\Sigma\, dP} \tag{2}$$

In practice the normal distribution relationship is unlikely to be applicable to emulsion size data for the simple reason that actual distributions are rarely symmetric; they tend to be skewed. If data from Table 2, for example, are rearranged as in the third and fourth columns of Table 8 and plotted as in Fig. 4, the skewing is evident. On the other hand, if D is replaced by log D and the data recalculated also as in Table 8 and replotted as in Fig. 5, a very nearly symmetrical curve results. This suggests [19] that an equation of the form

$$\frac{dP}{d(\log D)} = \frac{100}{\log \sigma_g \sqrt{2\pi}} \exp\left(-\frac{(\log D - \log \overline{D}_g)^2}{2 \log^2 \sigma_g}\right) \tag{3}$$

called the *log-normal distribution function*, should offer a more useful representation of most emulsion droplet diameters. This it generally does. It does not accurately represent all data, however, and persistence in its application to data not

Table 8. Calculations for Plots of Figs. 4 and 5

| From Table 2 | | | | | | |
|---|---|---|---|---|---|---|
| D(μm) | ΔP (%) | ΔD (μm) | ΔP/ΔD | log D | Δ(log D) | ΔP/Δ(log D) |
| 0.9 | 0.1 | 0.3 | 0.3 | −0.046 | 0.125[a] | 0.8 |
| 1.2 | 0.4 | 0.4 | 1.0 | 0.079 | 0.151[a] | 2.6 |
| 1.7 | 2.2 | 0.6 | 3.7 | 0.230 | 0.150 | 14.7 |
| 2.4 | 6.9 | 0.8 | 8.6 | 0.380 | 0.151 | 45.7 |
| 3.4 | 13.4 | 1.2 | 11.2 | 0.531 | 0.150 | 89.3 |
| 4.8 | 24.9 | 1.6 | 15.6 | 0.681 | 0.152 | 163.8 |
| 6.8 | 25.9 | 2.4 | 10.8 | 0.833 | 0.149 | 173.8 |
| 9.6 | 16.0 | 3.2 | 5.0 | 0.982 | 0.152 | 105.3 |
| 13.6 | 7.3 | 4.8 | 1.5 | 1.134 | 0.149 | 49.0 |
| 19.2 | 2.1 | 6.4 | 0.3 | 1.283 | 0.152 | 13.8 |
| 27.2 | 0.6 | 9.6 | 0.1 | 1.435 | 0.149 | 4.0 |
| 38.4 | 0.2 | 12.8 | Negligible | 1.584 | 0.150 | 1.3 |

[a]For example, 0.079 − (−0.046) = 0.125; 0.230 − 0.079 = 0.151.

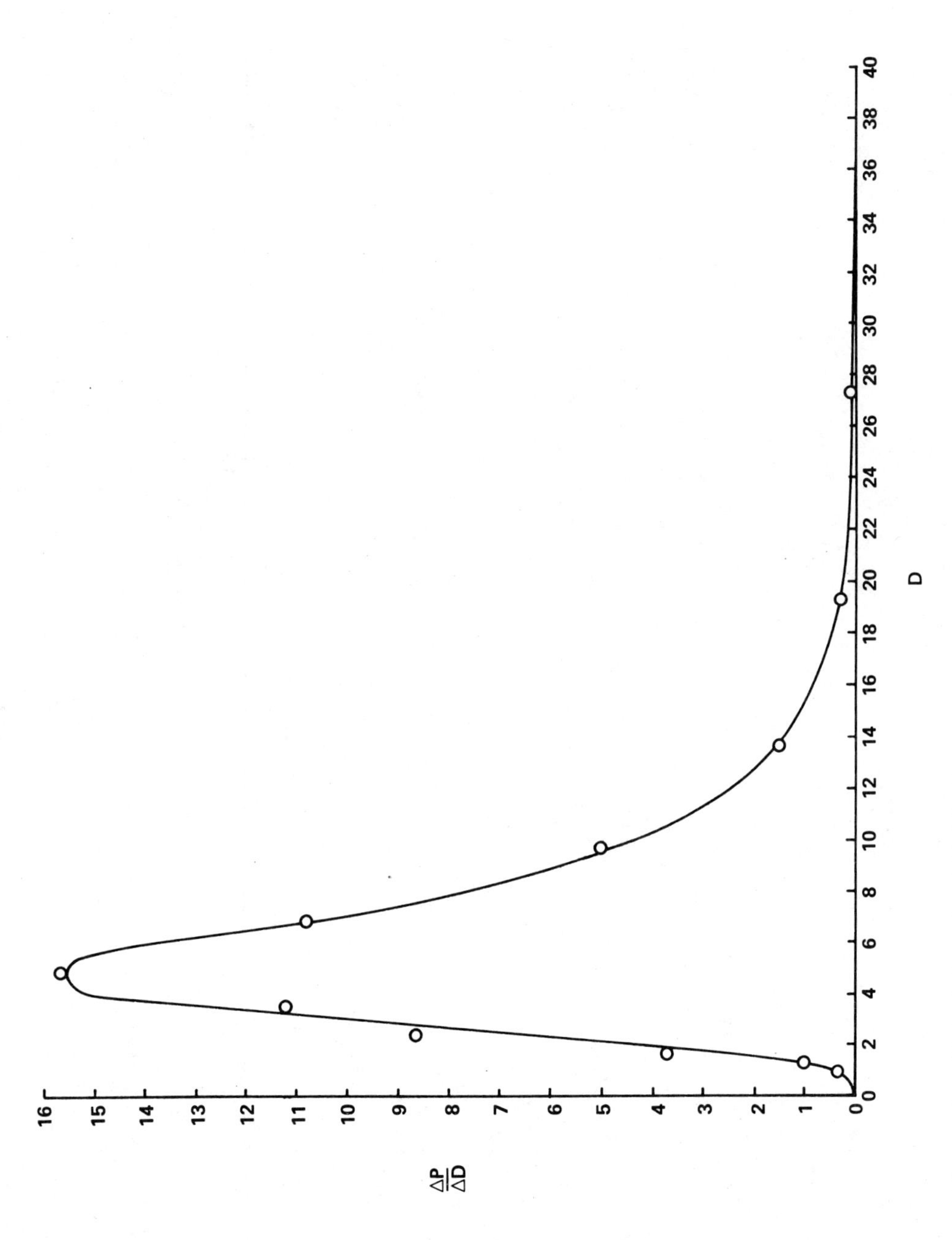


Figure 4. Plot of data from Table 8.

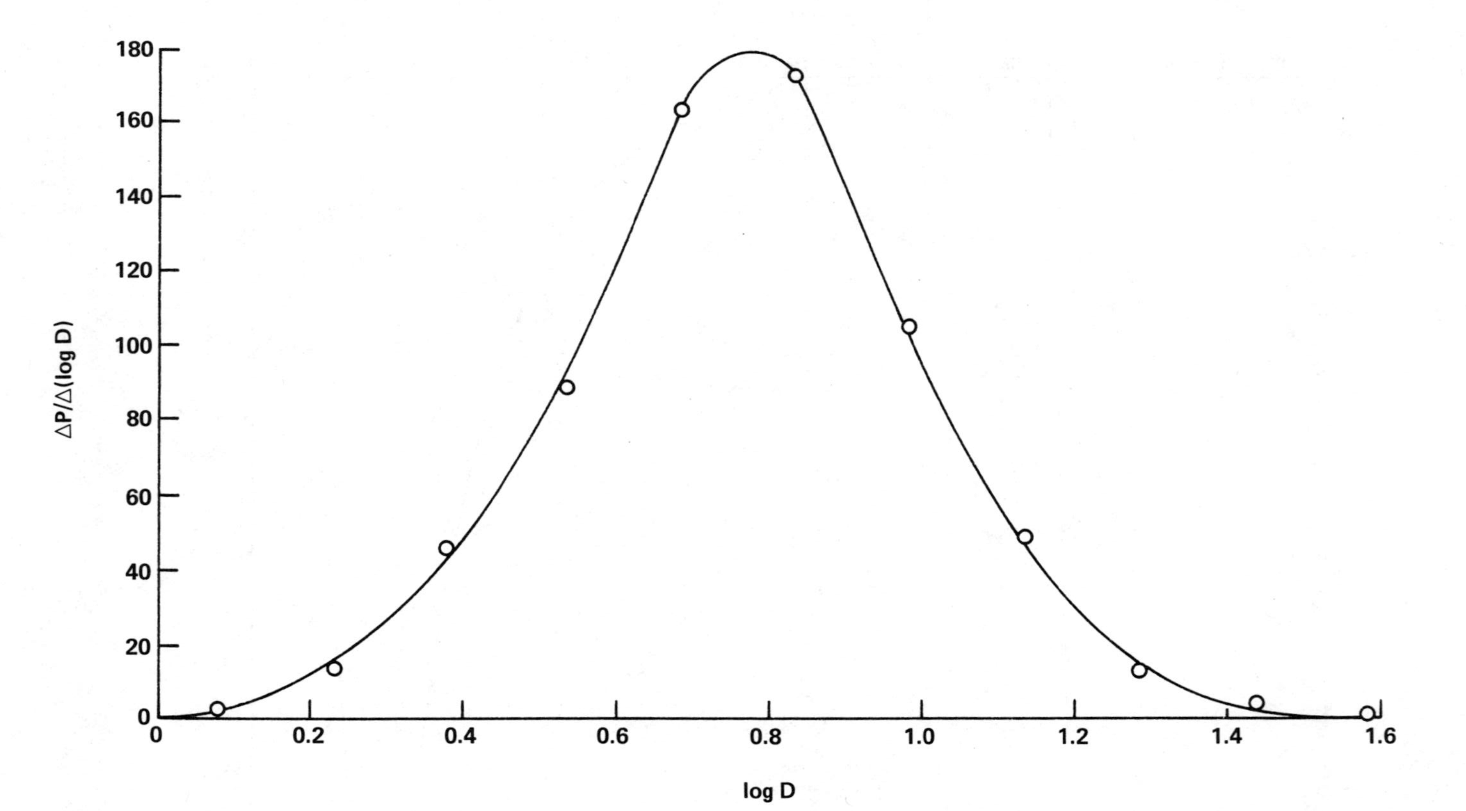


Figure 5. Plot of data from Table 8.

giving a good fit can lead to erroneous conclusions, which is equally true for other functions as well. There is some evidence that physical processes leading to dispersions should be expected to produce log-normally distributed components [29] (see also Chap. 2), but at this time the primary justification for using the log-normal function is merely that it is adequately representative in a great many instances. As will be shown below, the function is easy to apply and permits ready calculation of the several mean diameters. The latter attributes account for its frequent use even in situations of questionable validity. What constitutes valid and invalid application is unanswerable rigorously. The function is least likely to apply at the diameter extremes and most likely to fit in the midrange. Perhaps a satisfactory guide for applicability is a good fit over 80% of the diameter range.

The parameter $\overline{D}_g$ is termed the geometric mean diameter and is that corresponding to the apex in a plot such as Fig. 5. It is also sometimes called the count mean diameter when number frequency versus diameter data are employed as in the figure. To designate clearly the basis for the geometric mean diameter another suffix will be used subsequently; in the case of number-based data the designation will be $\overline{D}_{gN}$.

The parameter $\sigma_g$ is designated the geometric standard deviation and relates, as does its counterpart for the normal function, to the spread of diameters about the geometric mean diameter. Corresponding to Eq. (2), is is given by

$$\sigma_g^2 = \frac{\Sigma\,(\log D - \log \overline{D}_{gN})^2}{\Sigma\ d(\log P)} \tag{4}$$

When data are available in terms of droplet volume as a function of diameter, for example, Eq. (3) may be written

$$\frac{dV}{d(\log D)} = \frac{\pi D^3}{6 \log \sigma_g \sqrt{2\pi}} \exp\left(-\frac{(\log D - \log \overline{D}_{gV})^2}{2 \log^2 \sigma_g}\right) \tag{5}$$

Where $\overline{D}_{gV}$ is the volume-based geometric mean diameter and is the peak value as before. Area- and mass-based data can likewise be treated by substituting simple geometric relationships to give area and mass geometric mean diameters, $\overline{D}_{gA}$ and $\overline{D}_{gM}$, respectively. The geometric standard deviation $\sigma_g$ is unchanging with the basis of measurement because the spread of the distribution remains fixed regardless of measurement base.

Application of the normal and log-normal distribution functions is much more conveniently made by using normal or log-normal probability graph paper than by following a calculation procedure. Both types of paper are commercially available. The grid of the former has been prepared by noting any continuous cumulative distribution of data—for present purposes, say one listing the percentage of droplets

P smaller than diameter D—would plot a straight line on a graph of P versus D when represented by an equation of the form P = f(D). If the diameters were normally distributed, an expression of this form could be obtained by integrating the normal function, Eq. (1), written

$$P = \frac{100}{\sigma\sqrt{2\pi}} \int_0^D \exp\left(-\frac{(D - \overline{D})^2}{2\sigma^2}\right) dD \tag{6}$$

Unfortunately, the integral of Eq. (6) cannot be evaluated in closed form, but it is a standard form (the error function), and its value is readily obtained from tabulations in numerous handbooks. Normal probability paper is thus prepared with D on a linear scale and P on a percentage probability scale such that a straight line results when the integral of Eq. (6) is satisfied. Normally distributed, cumulative data must therefore plot a straight line on normal-probability graph paper. If a straight line is not obtained, the data are not normally distributed.

Log-probability paper is similarly prepared except that the integral form of Eq. (3)

$$P = \frac{100}{\log \sigma_g \sqrt{2\pi}} \int_0^D \exp\left(-\frac{(\log D - \log \overline{D}_g)^2}{2 \log^2 \sigma_g}\right) d(\log D) \tag{7}$$

is utilized. The diameter scale on log-probability paper is logarithmic. Cumulative data following the log-probability function are thus made to plot as a straight line on log-probability graph paper. As noted above, practical considerations may allow a good fit to a straight line over 80% of the data range. Figure 6 shows a log-probability grid with the cumulative number-diameter data of Table 2 (also Table 8) plotted on it. A satisfactory straight line is seen to result.

The midpoint, or peak diameter, on Fig. 5 is also the midpoint, i.e., the 50% point, on Fig. 6. The two corresponding diameter values 5.9 μm (= log 0.77) on Fig. 5 and 5.8 μm on Fig. 6 are seen to correspond within plotting limitations; they identify the geometric mean diameter $\overline{D}_{gN}$[†]. That the midpoint on Fig. 6 indeed represents the geometric mean diameter can be verified by evaluating Eq. (7) between the limits D = 0 and $D = \overline{D}_g$ which will give the result 0.5 or 50%.

The geometric standard deviation can be calculated from Eq. (4), but it is also readily obtained from the log-probability plot of data. If Eq. (7) is evaluated for the fraction lying within one standard deviation of the mean, i.e., between $D = \overline{D}_g$ and $D = \sigma_g \overline{D}_g$, the result will be 0.3413. This shows that points corresponding to 0.5 ± 0.3413 and 0.5 establish the value. On a straight-line plot as in Fig. 6 the geometric standard deviation is thus established as the ratio of the 84.13% diameter

[†]Subscript N is included since data are number based.

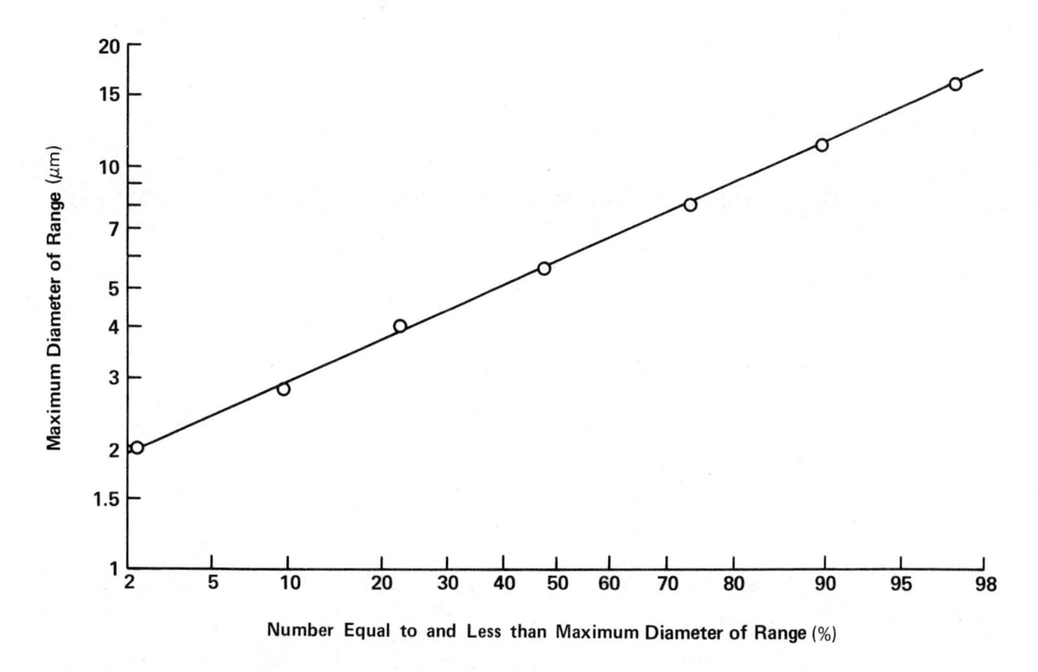


Figure 6. Log-probability plot of data from Table 2.

to the 50% diameter or as the ratio of the 50% diameter to the 15.87% diameter. Its value, as obtained from Fig. 6, is 10.0/5.8 = 5.8/3.37 = 1.72.

All mean diameters listed in Table 1 are readily calculated once the geometric mean diameter and the geometric standard deviation are determined for any set of emulsion droplet data, the necessary relationships deriving from the log-probability function and the definition of the particular mean. The number-length mean diameter $D_{NL}$, for example, is obtained from a combination of its definition and Eq. (3), written[†]

$$D_{NL} = \frac{\Sigma\, D\, dN}{\Sigma\, dN} = \frac{1}{\log \sigma_g \sqrt{2\pi}} \int_{-\infty}^{+\infty} \exp\left(-\frac{(\log D - \log \overline{D}_{gN})^2}{2 \log^2 \sigma_g}\right) D\, d(\log D) \qquad (8)$$

Rearrangement of this expression with the aid of integral functions yields

$$\log D_{NL} = \log \overline{D}_{gN} + 1.1513 \log^2 \sigma_g \qquad (9)$$

The other mean diameters are likewise related to the number-length geometric mean diameter $\overline{D}_{gN}$ and the geometric standard deviation $\sigma_g$ by the relations

[†]Number is more appropriate here than percentage.

$$\log D_{NA} = \log \overline{D}_{gN} + 2.3026 \log^2 \sigma_g \tag{10}$$

$$\log D_{NV} = \log D_{LA} = \log \overline{D}_{gN} + 3.4539 \log^2 \sigma_g \tag{11}$$

$$\log D_{LV} = \log \overline{D}_{gN} + 4.6052 \log^2 \sigma_g \tag{12}$$

$$\log D_{AV} = \log \overline{D}_{gN} + 5.7565 \log^2 \sigma_g \tag{13}$$

$$\log D_{VM} = \log D_{WM} = \log \overline{D}_{gN} + 8.0591 \log^2 \sigma_g \tag{14}$$

When the numerical values determined from Fig. 6 are inserted into Eq. (9), for example, there is obtained

$$\log D_{NL} = \log 5.8 + 1.1513 \log^2 1.72 = 0.8273$$

or $D_{NL} = 6.7\ \mu m$ which compares well with the value 6.8 μm as calculated in Table 3. The other mean diameters calculated by Eqs. (10) through (14), are, respectively, 7.8, 9.0, 9.0, 10.5, 12.1, and 16.3 μm which also represent acceptable agreement with values presented in Table 3. Drawing other lines parallel (the one value of $\sigma_g$ is common to all) to the one already plotted on Fig. 6 but through the above values located on the 50% probability line gives the distribution of diameters within the emulsion according to the particular definitions for surface, volume, etc.

Equations (9) through (14) were derived for and are appropriate for number-based data. But droplet diameter measurements may also be conducted in such a fashion as to yield surface- or volume-based results, as noted previously. Cumulative data for either of the latter can obviously be plotted on a log-probability grid, and can be expected to give equally as straight a line as number data for the same emulsion. The 50% point will not be $\overline{D}_{gN}$ in these later instances but will be the geometric mean diameter on a surface or volume basis, designated $\overline{D}_{gA}$ or $\overline{D}_{gV}$, respectively. The geometric standard deviation is a constant for any one emulsion and its value should not depend on the method of measurement.

It is sometimes desirable to relate measurements obtained from different techniques. This can be accomplished by making incremental calculations as outlined previously, but it can also be accomplished using relationships derived from the log-probability function. Surface and number frequency measurements for the same emulsion are related by

$$\log \overline{D}_{gA} = \log \overline{D}_{gN} + 4.6052 \log^2 \sigma_g \tag{15}$$

volume and number frequency data by

$$\log \overline{D}_{gV} = \log \overline{D}_{gN} + 6.9078 \log^2 \sigma_g \tag{16}$$

and volume and surface by

$$\log \overline{D}_{gV} = \log \overline{D}_{gA} + 2.3026 \log^2 \sigma_g \tag{17}$$

To illustrate, Table 6 gives volume- and number-based data as related by incremental calculations. The volume-based data (after percentages are accumulated) are plotted as open circles on Fig. 7. The points are not well represented by a straight line, not even by the criterion of over 80% of the diameter range set forth above. Thus the data cannot fairly be said to correspond to a log-probability distribution. Nevertheless, when a line is drawn through the 25 and 75% point, it suggests a $\overline{D}_{gV}$ value of 6.2 and a $\sigma_g$ value of 12.1/6.2 = 1.95. These numbers inserted into Eq. (16) give a $D_{gN}$ of 1.6, which with the same slope, i.e., value for $\sigma_g$, locate the other line on Fig. 7. The number-based data of Table 6, plotted as closed circles after percentage accumulation, are better represented by this line than might have been expected under the circumstances. With true log-probability data the conversion would, of course, be exact.

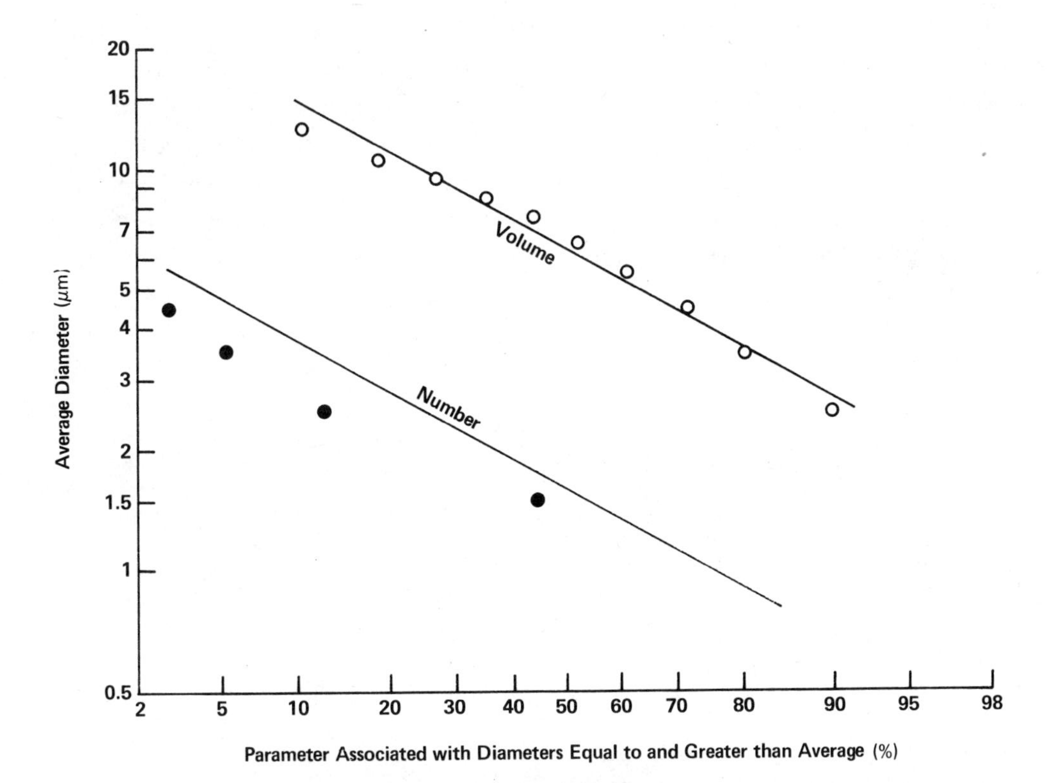


Figure 7. Number distribution from volume distribution.

Recapitulating, a cumulative log-probability plot of sizing data, when a straight line is obtained, permits ready calculation of all mean diameters, simple representation of the spread of diameters about the mean, and comparison of data obtained by different measurement methods. Finally, if droplet size data are obtained by the same measurement technique for two emulsions prepared under different circumstances and if both sets of data plot reasonably straight lines, the results offer a possible means for correlating the relative positions ($\overline{D}_g$) and slopes ($\sigma_g$) of the lines with the parameters of the preparation conditions or with desirable or undesirable properties of the preparations. The latter, of course, is the real objective of droplet size analysis.

B. Modified Log Probability

While the log-probability function frequently describes the general trend of droplet size data, deviations also often occur at the extremes of the size range since probability theory merely requires that both very large and very small droplets occur with extremely low frequencies. Actual emulsion droplets obviously have a limiting maximum size. This has led [23] to an upper limit treatment of data applicable when the deviation occurs primarily at the larger sizes. Here, instead of plotting on log-probability paper the measured diameter versus its cumulative frequency of occurrence, the function $D/(D_{max} - D)$ is plotted versus frequency, where $D_{max}$, the maximum diameter, is determined by trial and error to given the best straight line. The value of $D_{max}$ can usually be found in two or three trials.

When a straight line is attained with a set of diameter versus number frequency data plotted in accordance with this upper limit distribution, mean diameters can be calculated [23] using two terms evaluated from the position and slope of the plotted line. The two terms are

$$p = \frac{1}{[D/(D_{max} - D)]_{\text{at } 50\%}} \tag{18}$$

and

$$q = 0.394 \Big/ \log\left(\frac{[D/(D_{max} - D)]_{\text{at } 90\%}}{[D/(D_{max} - D)]_{\text{at } 50\%}}\right) \tag{19}$$

and the simpler mean diameter expressions are

$$D_{AV} = \frac{D_{max}}{1 + p\,\exp(1/4q^2)} \tag{20}$$

$$D_{LV} = \frac{D_{max}}{[1 + 2p\,\exp(1/4q^2) + p^2\,\exp(1/q^2)]^{1/2}} \tag{21}$$

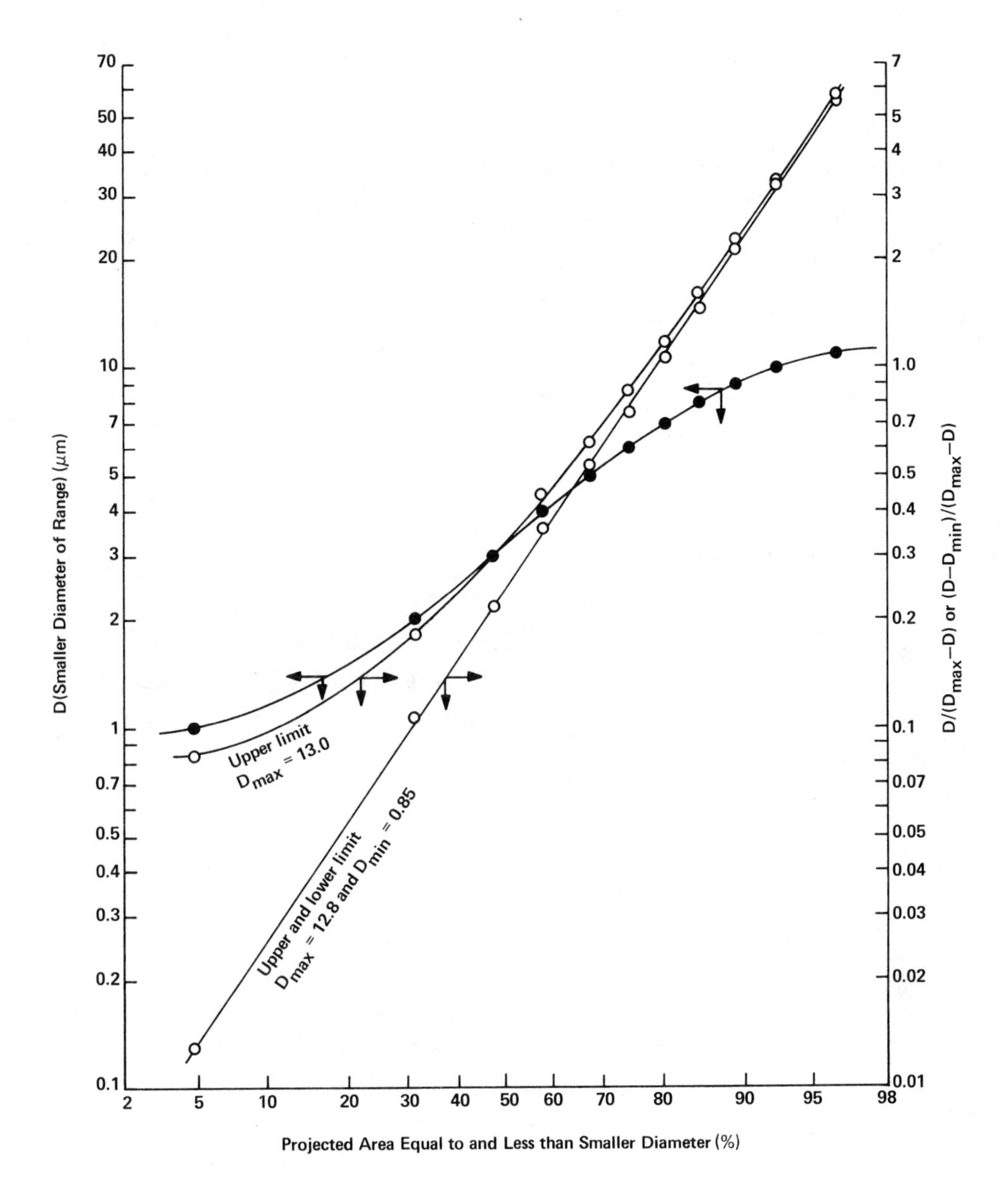


Figure 8. Modified log-probability plots.

and

$$D_{NV} = \frac{D_{max}}{[1 + 3p\,\exp(1/4q^2) + 3p^2\,\exp(1/q^2) + p^3\,\exp(9/4q^2)]^{1/3}} \tag{22}$$

Some data, such as that of Table 5 shown plotted on Fig. 8, reveal deviations at both the upper and lower size regions. Application of the upper limit treatment using $D_{max}$ = 13 μm removes the deviation from a straight line for the larger diameters as also shown on Fig. 8, but curvature remains in the plot for the smaller diameters. A smallest stable diameter of other than zero dimensions is not unlikely with actual emulsions. Assuming both a maximum and a minimum diameter exist, a plot of $(D - D_{min})/(D_{max} - D)$ versus cumulative frequency may be expected to straighten both ends of some distribution plots. This is seen to be the case by the third curve on Fig. 8, obtained using 12.8 and 0.85 μm as the extreme diameters. Mean diameter relationships have not been developed for this last plot, but mean values can be calculated using stepwise procedures as outlined previously. Maximum and minimum size indices add still other parameters to those of mean diameter and slope against which process variables can be examined or product performance compared.

### C. Espenscheid, Kerker, and Matijevic

Another modification [25,30] of the logarithmic function distributes diameter values around the mode, the most frequently occurring size. Intuitively this is satisfying because the modal diameter is an obvious value on a distribution plot, and, practically, it is useful because many emulsion data are described by the expression. It may be written

$$\frac{dP}{dD} = \frac{200}{\sqrt{2\pi}\,\sigma_0 D_m \exp(\sigma_0^2/2)} \exp\left(-\frac{(2.3026)^2[\log(D/2) - \log(D_m/2)]^2}{2\sigma_0^2}\right) \qquad (23)$$

where $D_m$ is the modal diameter and $\sigma_0$, the deviation, controls the spread of the distribution.

The function permits exploration of the effect of changing the spread while holding the mode invariant. The two curves of Fig. 9 are obtained, for example, using the experimental data of Table 2 with 6.0 μm as the modal diameter and deviation values of 0.50 and 0.45. Two values for the deviation are included to show their effect on the shape of the curve, and the histogram of Fig. 1 is repeated (broken line blocks) to show that an approximate fit of these data can be attained. A good fit of experimental data provides two parameters of possible utility in correlating emulsion properties or in preparing emulsion specifications.

### D. Gamma

Another two-parameter expression of potential value in characterizing emulsion droplet distributions is a general gamma-type distribution function [27] written

$$\frac{dN}{dD} = 100\,\frac{D^x \exp(-xD/y)}{\Gamma(x+1)(y/x)^{x+1}} \qquad (24)$$

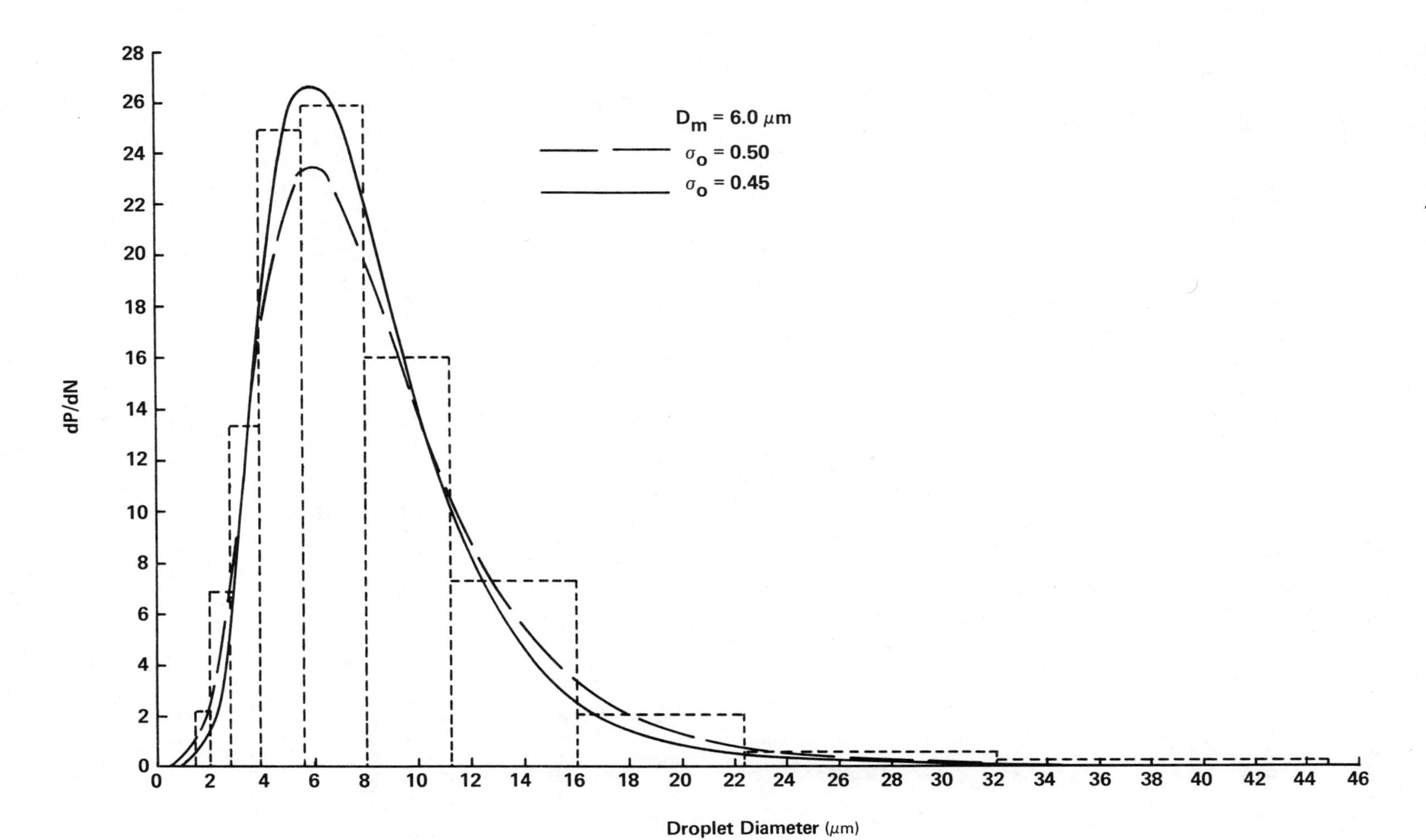


Figure 9. Logarithmic function fit of data from Table 2 to histogram (broken lines) of Fig. 2.

where y is also the most frequently occurring size and the parameter x adjusts the spread and skewing of the distribution. Tables of the gamma function $\Gamma(x + 1)$ are to be found in many handbooks. Increasing values of x, for example, skew the distribution to the larger sizes while narrowing it relative to the mean size. The mean diameters of Table 1 are related to the parameters of Eq. (24) by the following expressions:

$$D_{NL} = y \frac{x + 1}{x} \tag{25}$$

$$D_{NA} = \frac{y}{x} [(x + 2)(x + 1)]^{1/2} \tag{26}$$

$$D_{NV} = \frac{y}{x} [(x + 3)(x + 2)(x + 1)]^{1/3} \tag{27}$$

$$D_{LA} = \frac{y}{x} (x + 2) \tag{28}$$

$$D_{AV} = \frac{y}{x} (x + 3) \tag{29}$$

$$D_{VM} = D_{WM} = \frac{y}{x} (x + 4) \tag{30}$$

The data of Table 8 previously presented as Fig. 4, are most nearly fitted by Eq. (24) when x and y have values of 3.7 and 4.8, respectively. The fit is represented on Fig. 10 with the curve of Fig. 4 reproduced as a broken line for comparison. As is readily seen, the fit is not especially good in this instance. Indeed, a good fit should not be expected since the data have previously been shown to obey a log-probability function with quite acceptable precision. However, when actual distributions conform to Eq. (24), the parameters x and y may be relatable to the emulsion-creating process or to emulsion properties.

## E. Weibull

A statistical distribution function employed widely in diverse fields can be derived from consideration of the probability X(L) of a chain breaking at its weakest link as the load L on the chain increases. This phenomenon has been described [22, 31] by

$$X(L) = 1 - e^{-cL} \tag{31}$$

where the function cL must be positive, must not decrease as L increases, and must vanish at some value of L that may or may not be zero in order to satisfy failure expectations. A simple expression satisfying these criteria is $(L - \gamma)^{\beta}/\alpha$, and using it the distribution function becomes

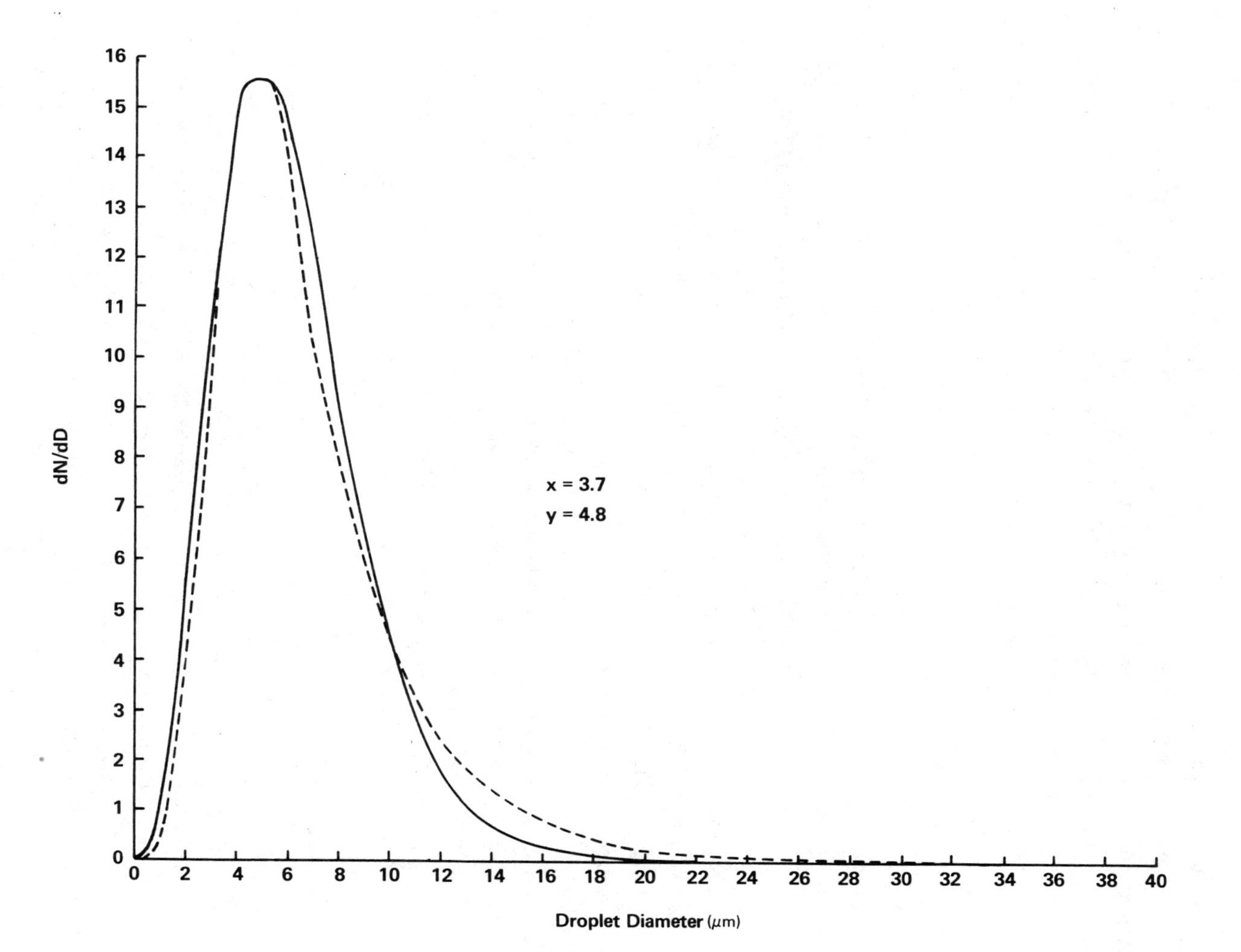


Figure 10. Gamma function fit (solid line) of data from Table 8 to curve of Fig. 4 (broken line).

$$X(L) = 1 - \int_0^L \exp\left(-\frac{(L - \gamma)^\beta}{\alpha}\right) dL \qquad (32)$$

It expresses the probability that a component (link) experiencing a value L (load) equal to or less than a given value is the same as the number of components that meet this condition divided by the total number of components. It is not difficult to suppose that a liquid undergoing a shearing action might break into droplets like a chain failing by the rupture of its weakest link.

A double logarithmic transformation of Eq. (32), after transposing and inverting, permits writing

$$\ln \ln \frac{1}{1 - F(D)} = \beta \ln(D - \gamma) - \ln \alpha \qquad (33)$$

where droplet diameter rather than load is the parameter of interest. Equation (33) is in linear form, thus a plot of the left-hand side of the equation against $\ln(D - \gamma)$ should give a straight line when the proper value of $\gamma$ is included. The slope of the line is $\beta$ and $\alpha$ is a scale parameter.

Weibull probability paper for direct plotting of data is commercially available or it can be prepared rather simply from tables of function values [32,33] using ordinary log-log paper having square coordinates, i.e., paper with cycles of the same dimensions on both axes. Figure 11 shows the data of Table 5 plotted on a Weibull grid. The curved line is the plot of raw data and the straight line the data with $\gamma$ having a value of 0.85. The latter being essentially straight, values for $\beta$ and $\alpha$ are readily determined to be 0.98 and 7.3, respectively.

As with other distribution functions, obtaining values for the several parameters can only be of utility if the values can be related, perhaps through a series of tests, to properties of the system producing the emulsion, such as viscosity, interfacial tension, pressure, or orifice diameter, or to a property of the resulting emulsion such as stability.

### F. Nukiyama-Tanasawa

Liquid atomization into a high-velocity gas frequently produces droplets distributed by diameter in accordance with a purely empirical expression known as the Nukuyama-Tanasawa distribution [24]. Considering the similarities of the atomization process and homogenization techniques for producing emulsions, the equation may be expected to have at least some application to emulsions. It can be written [34]

$$\log \frac{\Delta P}{D^2 \, \Delta D} = \log x - \frac{yD^z}{2.303} \qquad (34)$$

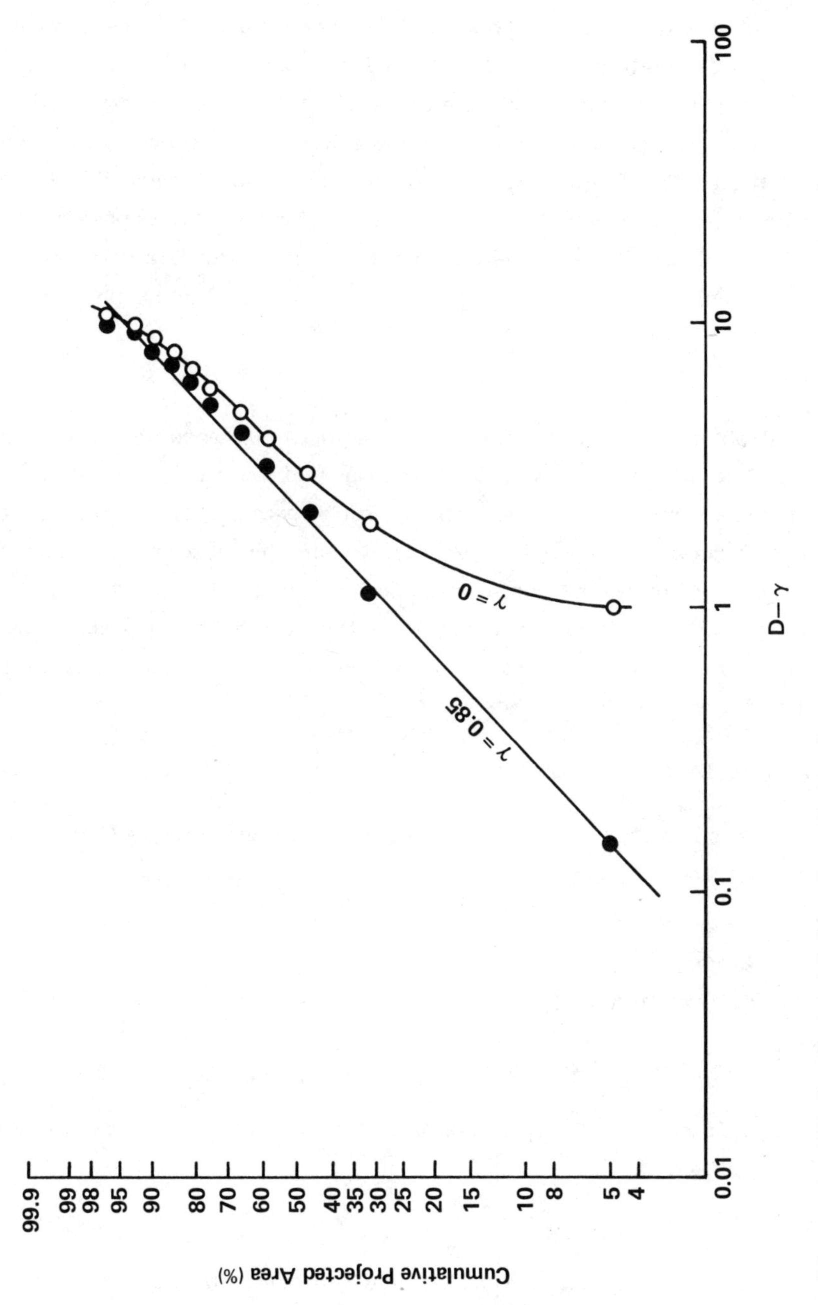


Figure 11. Weibull plot of data from Table 5.

where x, y, and z are constants. Since the equation is of linear form, a plot of $\log [\Delta P/(D^2 \Delta D)]$ against $D^z$ will yield a straight line upon selection of the proper value of z if the relationship applies. The slope of the resulting line is $y/2.303$, and the value of x is determinable from Eq. (34) once y and z are established.

Figure 12 is a cumulative plot of data from Table II in accordance with Eq. (34). A moderately satisfactory line is produced with $z = 1/3$. Again in atomization, the constant z retains a single value for a given nozzle over a wide range of operating conditions, but its value is quite sensitive to nozzle dimensions and design. The same behavior might be expected in pressure emulsification. It is possible to compute the mean diameters of Table I using the evaluated constants of Eq. (34) and a gamma function [23].

## G. Bimodal

Occasionally droplet distribution data will result in two or more peaks when plotted as in Fig. 4. The distribution is said to be bimodal if two peaks are evident and multimodal if more than two are in evidence. Such results typically arise when the output of two or more homogenizers are mixed, when the input material has more than one input level, or when poor process control is maintained.

Bimodal distributions can be represented as the sum of two monomodal functions if the diameters about which measured values are grouped are quite different, but this is rarely the case. The function

$$\frac{dN}{dD} = \exp(-g_4 D^4 + g_3 D^3 + g_2 D^2 + g_1 D + g_0) \tag{35}$$

has been suggested for the representation of bimodal distributions [35]. Equation (35) is of the form

$$-\ln \frac{dN}{dD} = h(D) \tag{36}$$

and h(D) can be rewritten

$$h(D) = g^2(D^4 + g'_3 D^3 + g'_2 D^2 + g'_1 D + g'_0) \tag{37}$$

where $g'_3 = g_3/g^2$; $g'_2 = g_2/g^2$; etc. Then upon replacing D by $D - g'_3/4$, Eq. (36) can be simplified to

$$h(D) = g^2(D^4 + tD^2 + uD + w) \tag{38}$$

where t, u, and w are constants. The bimodal function, Eq. (35), finally can be put in the form

$$\frac{dN}{dD} = k \exp(-g^2(D^4 + tD^2 + uD + w)) \tag{39}$$

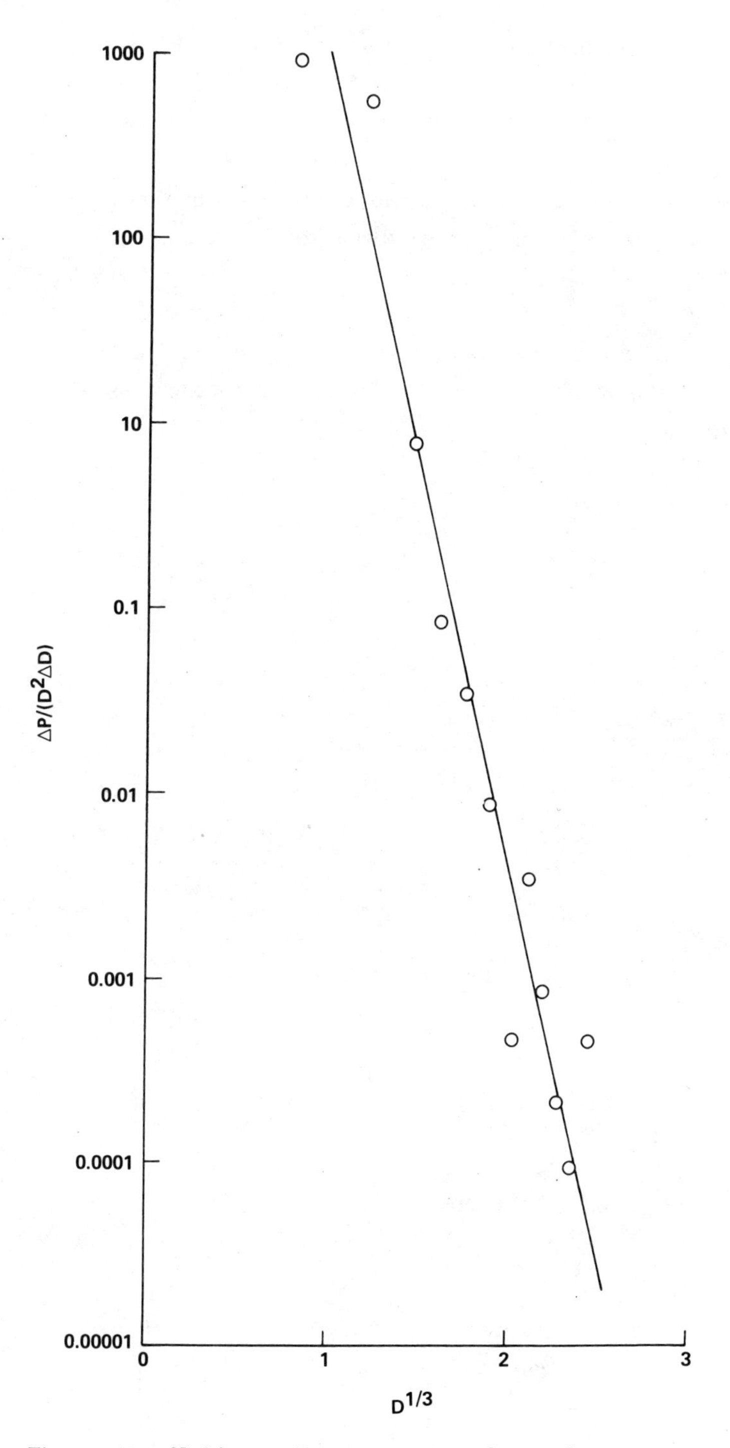


Figure 12. Nukiyama-Tanasawa plot of data from Table 2.

where $k = \exp(-g^2 w)$. Differentiating Eq. (39) and setting the result equal to zero gives

$$4D^3 + 2tD + u = 0 \tag{40}$$

the roots of which correspond to the values of the two modes and the minimum between them when they are real and distinct, the condition for which is

$$-8t^3 > 27\,u^2 \tag{41}$$

There is no simple solution for the area under the curve represented by Eq. (39). Graphical methods are therefore recommended for evaluating the constants for actual distribution data [35].

## Symbols

| | |
|---|---|
| A | surface area |
| c | parameter |
| D | diameter |
| $\overline{D}$ | arithmetic mean diameter |
| F(D) | function of diameter |
| g | group |
| h(D) | function of diameter |
| k | function |
| L | parameter |
| M | mass |
| N | number |
| P | percentage |
| p | special function, Eq. (18) |
| q | special function, Eq. (19) |
| S | projected area |
| t | constant |
| u | constant |
| V | volume |
| w | constant |
| x | spread and skewing parameter, constant |
| y | most frequently occurring size, constant |
| z | constant |
| $\alpha$ | parameter |
| $\beta$ | parameter |
| $\gamma$ | parameter |

$\sigma$ standard deviation

$\sigma_0$ deviation

Subscripts

A area based

AV surface-volume mean

g geometric

LA length-surface mean

LV length-volume mean

M mass based

m model

max maximum

min minimum

N number based

NA number-surface mean

NL number-length mean

NV number-volume mean

V volume based

VM volume-moment mean

WM weight-moment mean

## References

1. J. A. Davidson and H. S. Haller, *J. Colloid Interface Sci.* *47*:459 (1974).
2. E. A. Collins, J. A. Davidson, and C. A. Daniels, *J. Paint Technol.* *47*:35 (1975).
3. J. A. Davidson and E. A. Collins, *J. Colloid Interface Sci.* *40*:437 (1972).
4. C. Holt, A. M. Kimber, B. Brooker, and J. H. Prentice, *J. Colloid Interface Sci.* *65*:555 (1978).
5. S. H. Maron, P. E. Pierce, and I. N. Ulevitch, *J. Colloid Sci.* *18*:470 (1963).
6. P. Walstra, *J. Colloid Interface Sci.* *27*:493 (1968).
7. P. Bagghi and R. D. Vold, *J. Colloid Interface Sci.* *53*:194 (1975).
8. H. R. Carlon, M. E. Milham, and R. H. Frickel, *Appl. Optics* *15*:2454 (1976).
9. A. E. Martens, *Ann. N.Y. Acad. Sci.* *158*:690 (1969).
10. F. Grum, D. M. Paine, and J. L. Simonds, *J. Opt. Soc. Amer.* *61*:70 (1971).
11. A. M. Wims and M. E. Myers, Jr., *J. Colloid Interface Sci.* *39*:447 (1972).
12. R. W. Spinrad, J. R. V. Zaneveld, and H. Pak, *Appl. Optics* *17*:1125 (1978).
13. S. Kinsman, *Ann. N.Y. Acad. Sci.* *158*:703 (1969).
14. P. Walstra, H. Oortwijn, and J. J. de Graaf, *Neth. Milk Dairy J.* *23*:12 (1969).
15. P. Walstra and H. Oortwijn, *J. Colloid Interface Sci.* *29*:424 (1969).
16. W. P. Hendrix and C. Orr, in *Particle Size Analysis* (M. J. Groves and J. L. Wyatt-Sargent, eds.), Society for Analytical Chemistry, London, 1972, pp. 133-44.

17. H. Small, *J. Colloid Interface Sci. 48*:147 (1974).
18. H. Small, F. L. Saunders, and J. Solc, *Adv. Colloid Interface Sci. 6*:237 (1976).
19. H. H. G. Jellinek, *J. Soc. Chem. Ind. 69*:225 (1950).
20. T. Hatch and S. P. Choate, *J. Franklin Inst. 207*:369 (1929).
21. P. S. Roller, *J. Franklin Inst. 223*:609 (1937).
22. W. Weibull, *J. Appl. Mech. 18*:293 (1951).
23. R. A. Mugele and H. D. Evans, *Ind. Eng. Chem. 43*:1317 (1951).
24. S. Nukiyama and Y. Tanasawa, *Trans. Soc. Mech. Engrs. (Jap.)* 5:1 (1939).
25. W. F. Espenscheid, M. Kerker, and E. Matijevic, *J. Phys. Chem. 68*:3093 (1964).
26. R. L. Rowell and A. B. Levit, *J. Colloid Interface Sci. 34*:585 (1970).
27. P. Rosin and E. Rammler, *J. Inst. Fuel* 7:29 (1933).
28. A. D. Randloph and M. A. Larson, *Theory of Particulate Processes*, Academic, New York, 1971.
29. A. N. Kolmogoroff, *C. R. Dokl. Akad Nauk USSR 31*:99 (1941).
30. W. D. Ross, *J. Colloid Interface Sci. 67*:181 (1978).
31. F. H. Steiger, *Chem. Tech. 1*:225 (1971).
32. A. Plait, *Ind. Quality Control 18*:17 (1962).
33. C. A. Moyer, J. J. Bush, and B. T. Ruley, *Mater. Res. Stand. 2*:405 (1962).
34. H. C. Lewis, D. G. Edwards, M. J. Goglia, R. I. Rise, and L. W. Smith, *Ind. Eng. Chem. 40*:67 (1948).
35. J. M. Dalla Valle, C. Orr, Jr., and H. G. Blocker, *Ind. Eng. Chem. 43*:1377 (1951).

# 7

# Rheological Properties of Emulsions

PHILIP SHERMAN / Queen Elizabeth College, University of London, London, England



## I. Introduction

An emulsion contains two liquid immiscible phases one of which is dispersed in the other in the form of drops of microscopie or submicroscopic size. Stability to coalescence is achieved by the presence of an emulsifying agent which is adsorbed around the drop surfaces. The mechanisms of stabilization are reviewed in Chap. 5.

The two liquid phases are usually oil and water into which various additives may be introduced. One of the main reasons why the oil is used in the form of an emulsion rather than in its original state is that a much wider range of flow characteristics and consistencies can be achieved with an emulsion. In order to understand the variations in flow properties which are possible it is necessary to become familiar with the recognized types of flow and their associated theory. A general review of these flow characteristics is presented before considering the flow properties of emulsions.

## II. Rheological Theory

### A. Newtonian Flow

This concept is usually explained by reference to a liquid which is confined between two parallel planes separated by a distance x (Fig. 1). The surface area of each plate is A. A force F' is applied to the upper plate, thus giving rise to a shear stress F'/A, so that it moves with a velocity v in the same direction as that in which the force is applied. The lower plate does not move and the velocity gradient within the liquid is given by dv/dx. If ABCD represents the original location of the liquid between the parallel planes, then EFCD represents its final location since the force F' exerts a progressively decreasing effect within the liquid with increasing distance from the upper plate. The rate of change in velocity is given by dv/dx, which is referred to as the velocity gradient or rate of shear ($\dot{\gamma}$), and the force applied per unit area of liquid is the shear stress ($\tau$). Furthermore,

$$\dot{\gamma} = \frac{1}{\eta}(\tau) \qquad \text{or} \qquad \eta = \frac{\tau}{\dot{\gamma}} \tag{1}$$

where $\eta$ represents the viscosity of the liquid. If one determines the shear stress values resulting from the application of various shear rates to the liquid by using one of the many commercial viscometers which are available (Brookfield, Haake

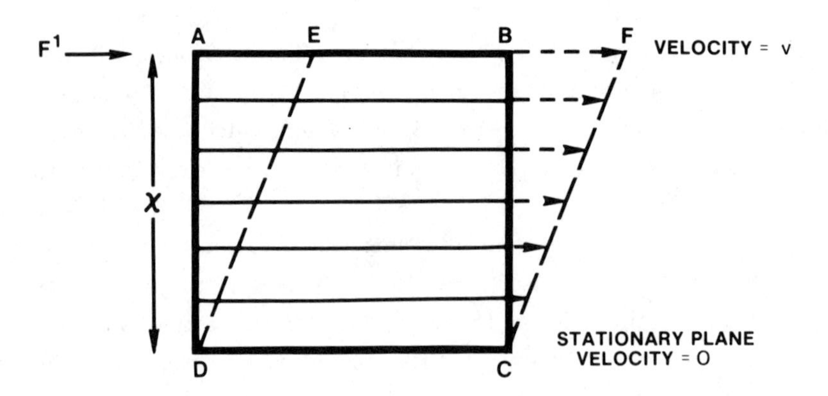


Figure 1. Diagrammatic representation of Newtonian flow.

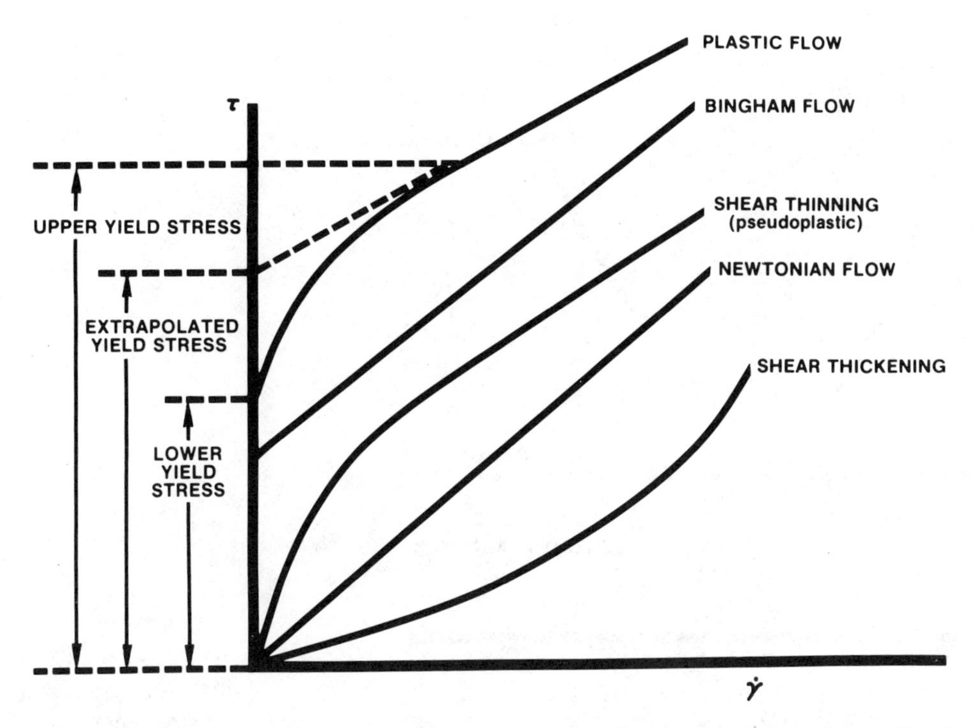


Figure 2. Shear stress-shear rate relationships for Newtonian and non-Newtonian flow.

*Rotovisko*, Contraves, Ferranti, etc.) then the two parameters will prove to be related linearly (Fig. 2) and in accordance with Eq. (1) the ratio of their values represents the viscosity of the liquid. The viscometric data can also be plotted as viscosity against shear rate. When this is done, it is observed (Fig. 3) that the viscosity is not influenced by shear rate, but remains constant at all shear rates.

B. Non-Newtonian Flow

When shear stress and shear rate are not linearly related as in Fig. 1, the flow is described as non-Newtonian. It can arise in four different ways.

i. Shear stress and shear rate are related curvilinearly (Fig. 2), with the relationship eventually becoming linear at high shear rates. The curve passes through the origin of the shear stress-shear rate plot. This behavior characterizes pseudoplastic flow or shear thinning. Viscosity at any selected shear rate is derived by drawing a tangent to the curve in the appropriate region and measuring its gradient. Because of the shape of the curve it is readily apparent that the viscosity decreases with increasing shear rate and that it eventually achieves a constant, minimum value (Fig. 3).

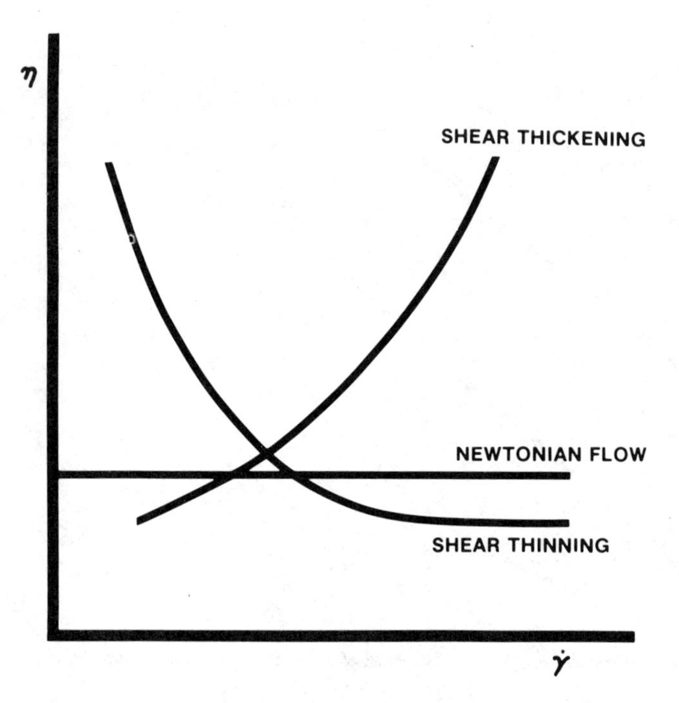


Figure 3. Viscosity-shear rate relationships for Newtonian and non-Newtonian flow.

ii. Shear stress and shear rate are related curvilinearly, as in shear thinning, but now the curve does not pass through the origin of the shear stress-shear rate plot. Instead, a finite shear stress, the lower yield stress, is necessary before flow is initiated (Fig. 2). This constitutes plastic flow. At higher shear rates the curve becomes linear. The shape of the curve is often defined by reference to two additional yield stresses. These are the extrapolated yield stress, which is the value obtained by extrapolating the linear part of the curve back to the shear stress axis, and the upper yield stress, which is the shear stress at which the curve becomes linear. When viscosity is plotted against shear rate the resulting curve has a shape similar to that for shear thinning (Fig. 3). Sometimes an apparent viscosity is reported for shear thinning or plastic systems without quoting a shear rate. In such instances the viscosity refers to the value derived from the linear portion of the shear stress-shear rate plot.

iii. Shear stress and shear rate are related linearly after the development of a finite yield stress (Fig. 2). Such systems are said to exhibit Bingham flow, and the viscosity is not influenced by shear rate. This concept has more theoretical than practical interest. If viscometric studies are extended so as to include sufficiently low shear rates it will be found that few systems, if any, exhibit Bingham flow.

iv. Shear stress and shear rate are related curvilinearly but the curvature increases with increasing shear rate (Fig. 2). This behavior is, therefore, the reverse of shear thinning or plastic flow. It is called *shear thickening*, and viscosity increases with increasing shear rate. A finite yield stress may or may not have to be developed before flow is initiated.

Mathematical representation of the curvilinear relationship between shear stress and shear rate in non-Newtonian flow was first introduced by Ostwald [1] and De Waele [2]. They proposed a relationship in the form of Eq. (1)

$$\dot{\gamma} = \frac{1}{\eta^*}(\tau)^n \tag{2}$$

where $\eta^*$ is a parameter corresponding to viscosity and n is a constant. In the case of plastic flow, the lower yield stress $\tau_0$ is incorporated, and

$$\dot{\gamma} = \frac{1}{\eta^*}(\tau - \tau_0)^n \tag{3}$$

If the flow data conform to Eqs. (2) or (3) then a double logarithmic plot of shear stress against shear rate gives a straight line with the slope equal to n.

Furthermore, when $n < 1$, the equations describe shear thinning or plastic flow, and when $n > 1$, the flow is shear thickening.

The validity of Eqs. (2) and (3) has been questioned on dimensional grounds, and an alternative form has been proposed [3] to overcome this criticism:

$$\tau = k(\dot{\gamma})^n \tag{4}$$

where k is a consistency index.

## C. Viscoelasticity

### 1. *Creep Compliance-Time Behavior*

The non-Newtonian behavior of emulsions is characterized by their flow properties when the shear rate or shear stress is continuously varied. It is also possible to study their response to a constant small shear stress over a period of time [4-7]. This shear stress should be well below the lower yield stress (Fig. 2). In this situation the internal structure undergoes minimal alteration and emulsions exhibiting plastic flow under the influence of a variable shear stress or shear rate now become viscoelastic. That is, they exhibit the characteristics both of solids (elasticity) and of fluids (viscosity).

Figure 4 illustrates the type of curve obtained when the experimental data are plotted as creep compliance against time. Creep compliance is the ratio of the

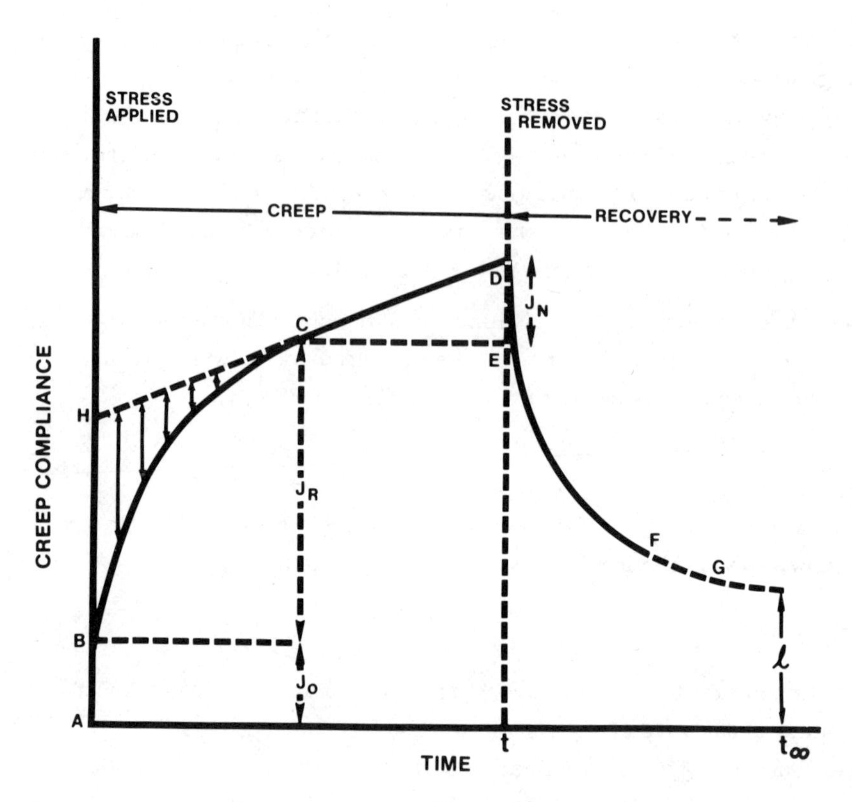


Figure 4. Creep compliance-time behavior of an emulsion exhibiting viscoelasticity.

strain to the applied constant stress. The curve has several distinct regions. Following the application of a stress there is an immediate strain, which is represented by AB. This is referred to as the instantaneous elastic compliance ($J_0$), and if the stress is removed before B, which corresponds to a very short time, the emulsion does not undergo any permanent change. After B the curvature decreases and eventually, beyond C, the creep compliance increases linearly with time. The region BC represents the retarded elastic compliance ($J_R$) and CD is the Newtonian compliance ($J_N$). If the stress is removed at D, strain recovery follows a similar sequence with an instantaneous elastic recovery (DE) followed by a retarded elastic recovery (EF) and an eventual flattening of the curve. The vertical distance $\ell$ from G to the time abscissa represents the nonrecoverable strain per unit stress and it is related to the degree of structural alteration incurred by the sample during the test.

The viscoelastic parameters of the emulsion are derived from an analysis of the creep compliance-time curve ABCD. The total creep compliance J(t) at any time t after the stress has been applied is given by

$$J(t) = J_0 + J_R + J_N \tag{5}$$

and when $t \to 0$,

$$J(t) = J_0 \tag{6}$$

Provided the emulsion exhibits linear viscoelasticity (a linear relationship between stress and strain), which usually applies at low stresses, the data can be analyzed by the following procedure.

$$J_0 = \frac{1}{E_0}$$

where $E_0$ is the instantaneous elastic modulus. Therefore, both parameters are derived by measuring the distance AB in Fig. 4. Furthermore, $J_N = t/\eta_N$, so that $\eta_N$ the Newtonian viscosity, can be calculated.

The retarded elastic compliance region BC incorporates the response of a wide spectrum of bond strengths to the applied stress. The nature of these bonds will be discussed later. Initially, only the weakest bonds are broken, but at longer times somewhat stronger bonds rupture, even though there is no change in the applied stress. The total number of bonds ruptured depends on the magnitude of the stress.

The retarded creep compliance can be represented by the sum of a discrete number i of compliances $J_i$ and the constituent parameters are derived by a graphical procedure [8]

$$J_R = \sum_i J_i \left[1 - \exp\left(-\frac{t}{\tau_i}\right)\right] \tag{7}$$

where $\tau_i$ is the retardation time of the ith component, and $\tau_i = \eta_i J_i = \eta_i/E_i$, where $\eta_i$ is a viscosity and $E_i$ is a retarded elastic modulus.

Now let

$$\sum_i J_i - J_R = Q \tag{8}$$

Then, it follows from Eq. (7) that

$$\sum_i J_i \left[\exp\left(-\frac{t}{\tau_i}\right)\right] = Q \tag{9}$$

and Q represents the distance (Fig. 4) between the extrapolated linear part CH of the creep compliance-time curve and the curved portion CB at any selected time t.

Equation (9) can be written in the form

$$\ln \sum \left(J_i - \frac{t}{\tau_i}\right) = \ln Q \tag{10}$$

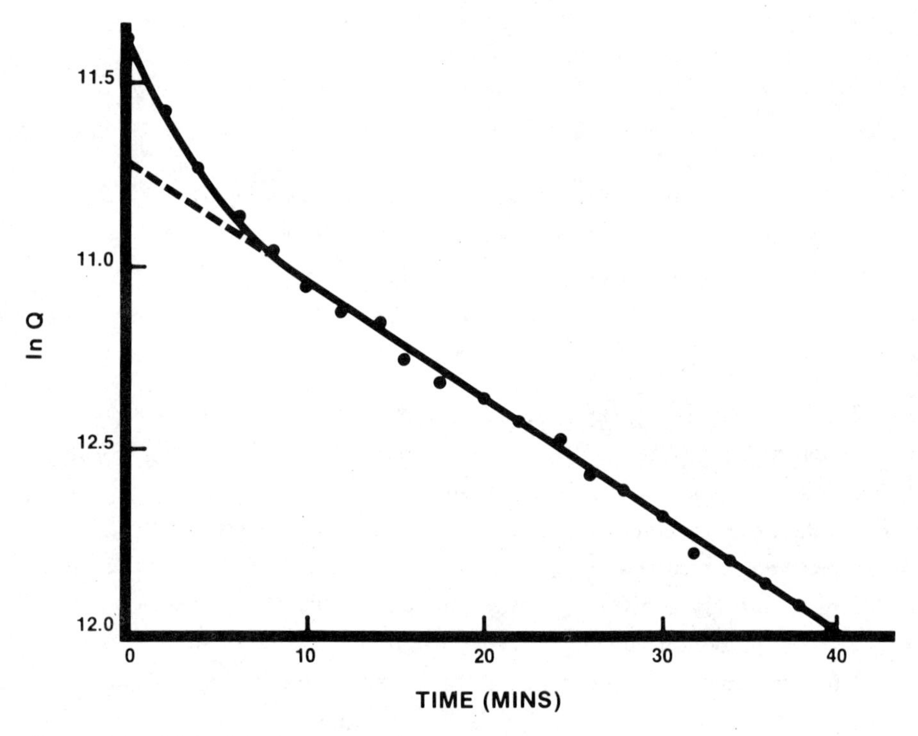


Figure 5. Plot of ln Q versus time.

A plot of ln Q against t gives a straight line at the larger values of t, but there may be some curvature at the low t values, as in Fig. 5. Extrapolation of the linear portion of the plot to the ln Q ordinate axis gives ln $J_1$ and the gradient of the linear portion is $-1/\tau_1$.

When the ln Q-versus-t plot is linear over all values of t the analysis is now complete. However, when curvature is exhibited at low t values the data have to be replotted as $\ln[Q - J_1 \exp(-t/\tau_1)]$ versus t. In the example given in Fig. 6 the plot is linear over all t values. The gradient is $-1/\tau_2$ and the intercept on the ordinate is ln $J_2$. If this second plot does not give a straight line, then it is necessary to plot $\ln[Q - J_1 \exp(-1/\tau_1) - J_2 \exp(-t/\tau_2)]$ versus t, $\ln[Q - J_1 \exp(-t/\tau_1) - J_2 \exp(-t/\tau_2) - J_3 \exp(-t/\tau_3)]$ versus t, etc., until a completely linear plot is achieved.

For the data illustrated in Figs. 4 to 6, Eqs. (5) and (7) become

$$J(t) = J_0 + J_1\left[1 - \exp\left(-\frac{t}{\tau_1}\right)\right] + J_2\left[1 - \exp\left(-\frac{t}{\tau_2}\right)\right] + \frac{t}{\eta_N}$$

$$= J_0 + J_1\left[1 - \exp\left(-\frac{t}{J_1\eta_1}\right)\right] + J_2\left[1 - \exp\left(-\frac{t}{J_2\eta_2}\right)\right] + \frac{t}{\eta_N} \qquad (11)$$

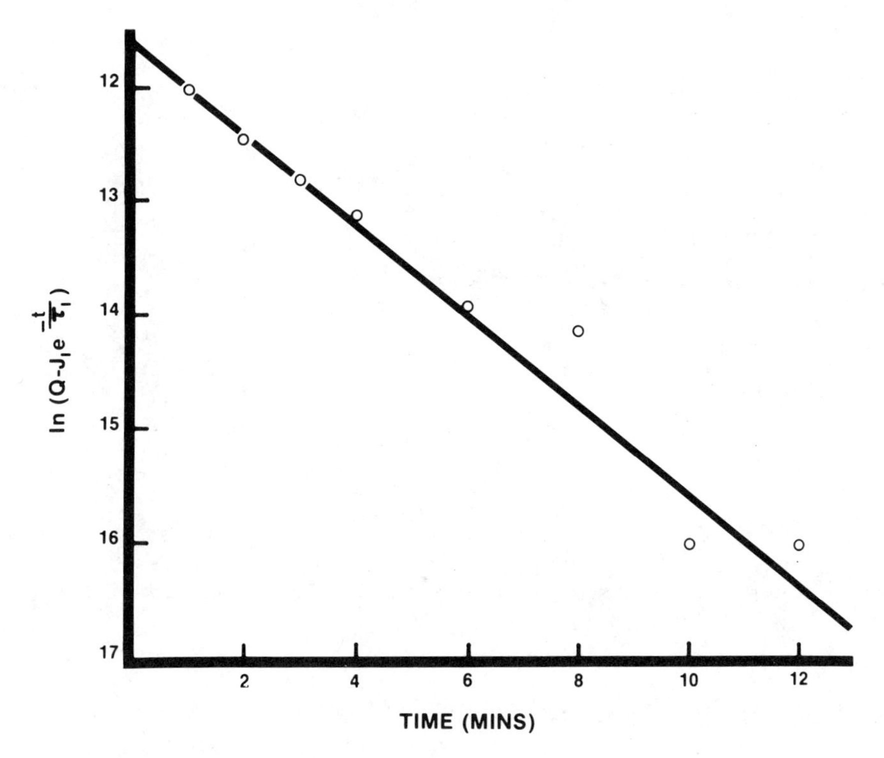


Figure 6. Plot of ln $[Q - J_1 \exp(-t/\tau_1)]$ versus time.

where $J_1 = 1/E_1$ and $J_2 = 1/E_2$, and $E_1$ and $E_2$ are the first and second retarded elastic moduli respectively.

*2. Dynamic Testing*

In creep compliance-time experiments the response of the emulsion with time to a small constant stress is studied. It is also possible to study viscoelastic properties by the response to a stress, which varies periodically with time. When a stress of frequency v (Hz) or $\omega (= 2\pi\nu$ radians $\sec^{-1}$) is applied, the strain amplitude will be proportional to the stress amplitude if the emulsion has linear viscoelastic properties. The frequency $\omega$ is qualitatively equivalent to a transient experiment at time $t = 1/\omega$. Furthermore, the strain alternates sinusoidally, as does the stress, but it will be out of phase with the stress (Fig. 7).

At any frequency the data can be interpreted in terms of a complex shear modulus E*, which is the ratio of stress/strain. It is the sum of two components

$$E^* = E^1 + iE^{11} \tag{12}$$

where $E^1$ is the storage or real shear modulus (component of stress in phase with the strain divided by the strain), $E^{11}$ is the loss or imaginary shear modulus (com-

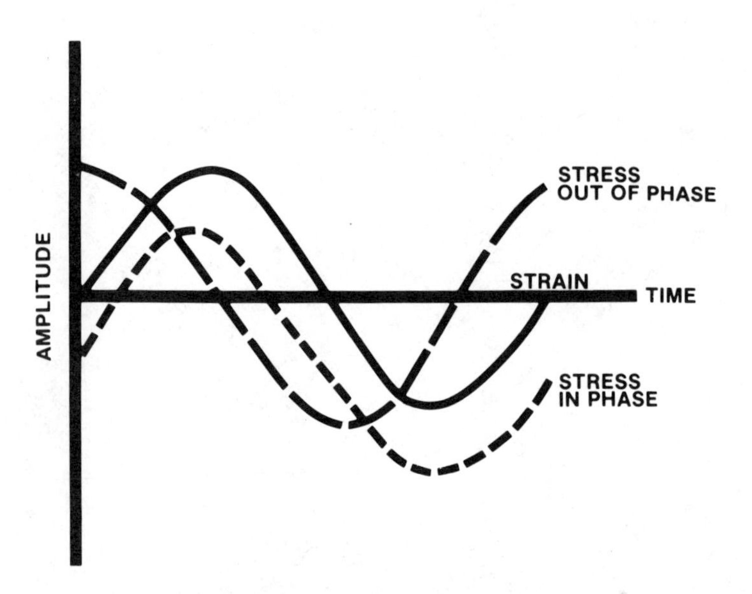


Figure 7. Sinusoidal variation of stress and strain in dynamic rheological studies.

ponent of stress out of phase with the strain divided by the strain) and $i$ is defined by $i^2 = -1$. Similarly, the complex shear compliance (J*) is given by

$$J^* = J^1 - iJ^{11} \tag{13}$$

where $J^1$ and $J^{11}$ are the storage and loss compliances, respectively. The parameters $G^1$ and $J^1$ are associated with the storage and release of energy during the periodic deformation, while $G^{11}$ and $J^{11}$ are associated with the dissipation or loss of energy as heat [9].

In addition,

$$\frac{E^1}{E^{11}} = \frac{J^{11}}{J^1} = \tan\delta \tag{14}$$

where tan δ is the loss tangent, and J* = 1/E*.

Viscoelastic materials also exhibit a complex viscosity η*, representing the dissipative effect of alternating stress, which is the sum of two components as are E* and J*,

$$\eta^* = \eta' - i\eta'' \tag{15}$$

where η' is the stress in phase with the rate of strain divided by the rate of strain, or dynamic viscosity, and η'' is the stress out of phase with the rate of strain divided by the rate of strain or imaginary part of the complex viscosity. Furthermore,

$$\eta^* = \frac{E^*}{i\omega} \tag{16}$$

$$\eta' = \frac{E^{11}}{\omega} \tag{17}$$

and

$$\eta'' = \frac{E^{1}}{\omega} \tag{18}$$

Dynamic studies can be made with the Weissenberg cone-plate rheogoniometer [10-12] and other rheometers of similar design.

## III. Rheological Properties of Dilute Emulsions

There have been relatively few studies of the factors which influence the rheological properties of emulsions, in contrast to the large number of detailed studies on dispersions of solid particles. Consequently there is a tendency to draw upon the theories developed for solid particle dispersions wherever possible.

Very dilute emulsions exhibit a Newtonian viscosity, and this is often defined in terms of the viscosity ($\eta_0$) of the continuous phase and the drop volume fraction ($\phi$) by using the equation proposed by Einstein [13,14],

$$\eta = \eta_0(1 + 2.5\phi)$$

or

$$\frac{\eta}{\eta_0} = \eta_{rel} = 1 + 2.5\phi \tag{19}$$

where $\eta_{rel}$ is the relative viscosity. The factor $\eta_{rel} - 1$ is often quoted; this is the specific increase in viscosity ($\eta_{sp}$). Equation (19) is valid provided: the drops behave as solid rigid spheres; they are large with respect to the size of the continuous phase molecules; there is no hydrodynamic interaction between the drops; and slippage does not occur at the oil/water interface. The increase in viscosity above the value of $\eta_0$ results from energy dissipation when drops of an immiscible liquid are introduced into the continuous phase, and the flow pattern of the latter phase is then modified in the vicinity of the drops.

The limitations imposed by Eq. (19) are often fulfilled by very dilute emulsions, particularly if the drop size does not exceed a few microns and the drops are enveloped by an elastic or viscoelastic film of adsorbed emulsifier molecules. However, when the adsorbed emulsifier film is fluid, as with ionic emulsifiers, Eq. (19) has to be modified to allow for the transmission of normal and tangential components of stress across the interface and into the drops [15]. This produces fluid circulation within the drops and this reduces the distortion of the continuous phase flow pattern around the drops. Equation (19) now becomes

$$\eta_{rel} = 1 + 2.5\left[\frac{\eta_i + (2/5)\eta_0}{\eta_i + \eta_0}\right]\phi \tag{20}$$

where $\eta_i$ is the viscosity of the liquid forming the drops. Therefore, the incremental change in viscosity as compared with the viscosity of a dispersed system of rigid spheres of equal volume concentration is $[\eta_i + (2/5)\eta_0]/(\eta_i + \eta_0)$. When $\eta_i >> \eta_0$ Eq. (20) reduces to Eq. (19), but in all other situations Eq. (20) gives a lower $\eta_{rel}$. Its validity has been confirmed by viscosity studies with a large number of O/W emulsions [16]. Circulation of fluid within the drops was prevented in some of the emulsions by selection of an appropriate emulsifier (Tween 20) and liquid phases (butyl benzoate solution of carbon tetrachloride in 45% aqueous sugar solution or in 33% aqueous glycerine).

When the emulsifier film adsorbed around drops can withstand the increase in the equilibrium interfacial tension resulting from an increase in surface area, the drops behave as solid spheres and the ratio $\eta_i/\eta_0$ does not influence $\eta_{rel}$ [16,17]. The influence of the rehological properties of the adsorbed emulsifier film on $\eta_{rel}$ is given by

$$\eta_{rel} = 1 + \frac{a\{\eta_i + (2/5)\eta_0 + (2/5)[(2\eta_S + 3\eta_\beta)/r]\phi\}}{\eta_i + \eta_0 + (2/5)(2\eta_S + 3\eta_\beta)/r} \tag{21}$$

where a is the shape factor (= 2.5 for rigid spheres), $\eta_S$ is shear viscosity of the emulsifier film, $\eta_\beta$ is its area viscosity, a two-dimensional equivalent of bulk viscosity, and r is drop radius. The quantity $(2\eta_S + 3\eta_\beta)$ ranges from $0.92 \times 10^{-4}$ to $0.014 \times 10^{-4}$ g sec$^{-1}$ for many O/W emulsions [16].

## IV. Flow Rheology of Concentrated Emulsions

In very dilute emulsions $\eta_{rel}$ increases linearly as $\phi$ increases. With more concentrated emulsions $\phi$ exerts a greater influence and the viscosity changes from Newtonian to non-Newtonian. The non-Newtonian character is initially pseudoplastic but in very concentrated systems may become plastic and exhibit viscoelasticity. Very often the influence of $\phi$ on $\eta_{rel}$ is portrayed as in Fig. 8, with $\eta_{rel}$ increasing almost exponentially to a maximum value just prior to emulsion inversion when a critical value of $\phi$ ($\phi_{max}$) is exceeded. This form of presentation is somewhat misleading because it does not show the transition in flow behavior from Newtonian to non-Newtonian and the $\eta_{rel}$ values included for the more concentrated emulsions represent the limiting minimum values obtained at high shear rates. Furhtermore, it is presumed that the drops are non-deformable spheres.

When $\phi$ increases beyond the limit of validity of Eq. (19), the distorted flow patterns around the drops draw close together and eventually they overlap. The

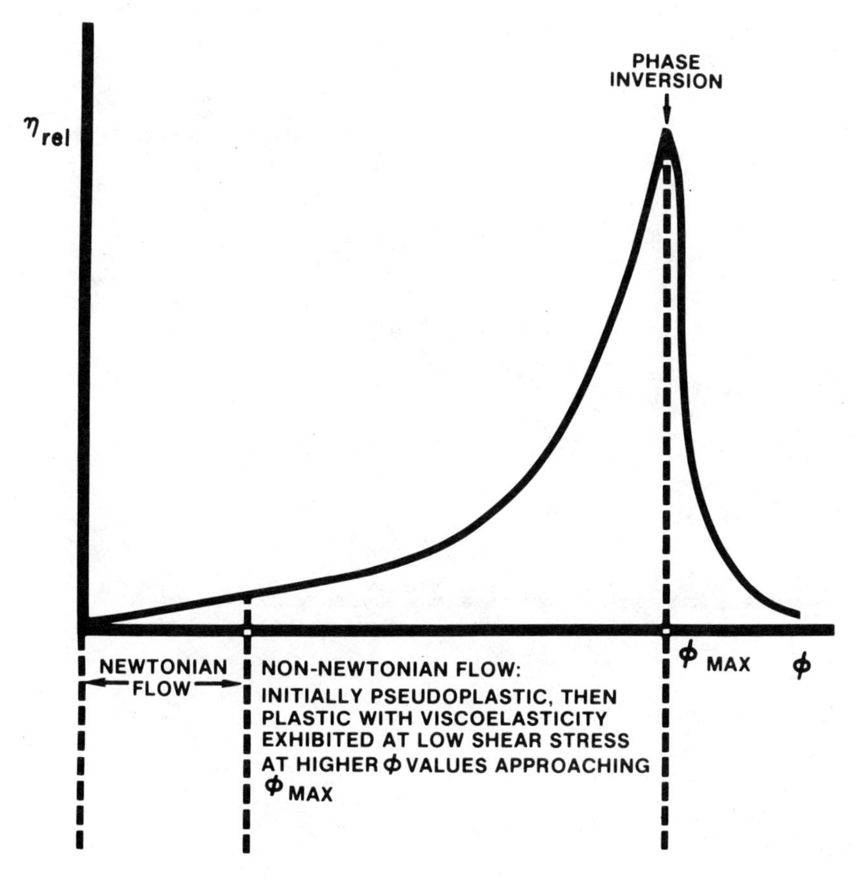


Figure 8. The influence of $\phi$ on $\eta_{rel}$.

resulting hydrodynamic interaction leads to an increased $\eta_{rel}$. This effect has been represented in many different forms but they usually reduce to a power series in $\phi$,

$$\eta_{rel} = 1 + 2.5\phi + b\phi^2 = c\phi^3 + \cdots \tag{22}$$

provided the drops behave as discrete rigid spheres, and where b and c are constants. Many different values of b are quoted in the literature between 0 and 10 for O/W and W/O emulsions, but there are very few values for c [19]. When the emulsions exhibit non-Newtonian flow, $\eta_{rel}$ in Eq. (22) represents the ratio of the limiting viscosity at high shear rates to $\eta_0$.

The hydrodynamic interaction between spherical drops on opposite sides of a hypothetical spherical enclosure and separated by distance f, can be defined by a coefficient $\lambda$, where

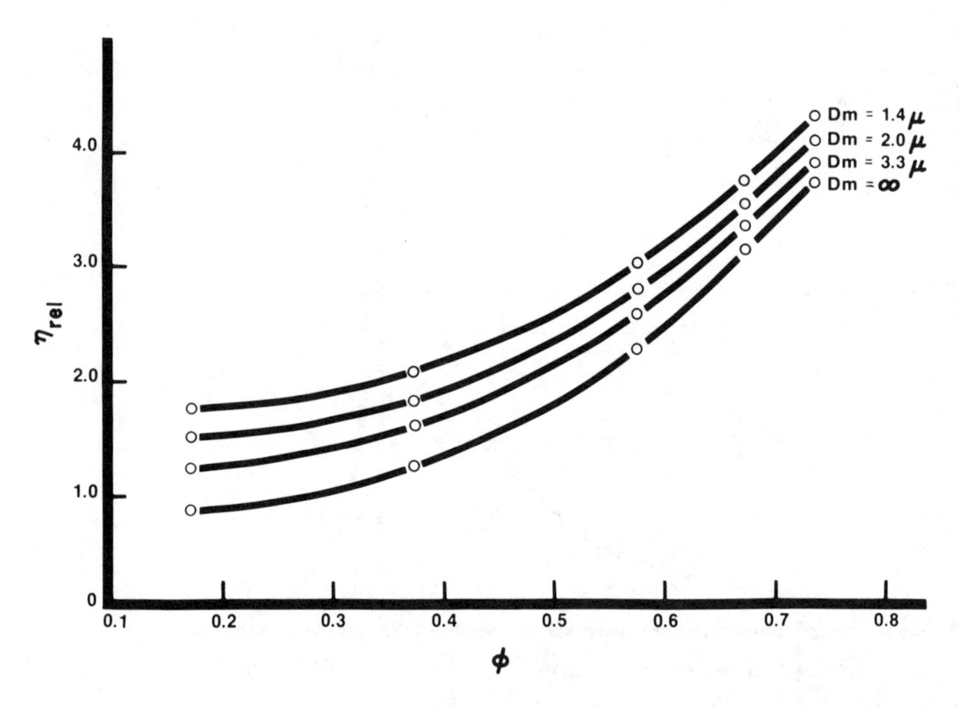


Figure 9. The influence of mean drop size on the $\eta_{rel}$-$\phi$ relationship (W/O emulsion).

$$\lambda = \frac{1 - D/2f}{D/2f} \tag{23}$$

and D is drop diameter. This equation is valid provided D/2f lies between 0.5 and 1 [20,21]. Therefore, the hydrodynamic interaction depends both on the size of drops and on the distance between them. The latter will also be influenced by drop size in that, at constant $\phi$, the value of f will decrease as D decreases. Even if one considers the hydrodynamic interaction between adjacent drops their distance of separation will depend not only on $\phi$ but also on D.

When considering the influence of $\phi$ on emulsion viscosity it is necessary, therefore, to allow for any simultaneous effect due to drop size. This can be illustrated by reference to Fig. 9, which portrays the influence of $\phi$ on $\eta_{rel}$, where $\eta_{rel}$ is the ratio $\eta_\infty/\eta_0$ and $\eta_\infty$ is the emulsion viscosity as measured at very high shear rates, for emulsions with different mean drop sizes [22]. It is readily apparent that the influence of drop size on $\eta_{rel}$ increases as the mean drop size decreases.

The sharp increase in viscosity which is observed in more concentrated emulsions can be explained by applying lubrication theory to calculate the viscous dissipation of energy [23]. When $\phi/\phi_{max} \rightarrow 1$ this gives the relation

$$\eta_{rel} = C \frac{(\phi/\phi_{max})^{1/3}}{1 - (\phi/\phi_{max})^{1/3}} \tag{24}$$

with C = 9/8, if the hypothetical cage model is adopted as in the derivation of Eq. (23). An equation of similar form can be derived by assuming that the thin films of continuous phase between closely packed spherical particles resist the deformation and relative motion of adjacent bounding surfaces and give rise to forces similar to those encountered in hydrodynamic lubrication [24]. This gives an equation of the form

$$\eta = \phi_{max} \mu \theta [1 + O(\varepsilon)] \tag{25}$$

where $\mu$ = Lamé shear modulus

$\theta$ = relaxation time $= \frac{N}{4} \cdot \frac{\eta}{\mu} \frac{R_0}{a_0}$

$\varepsilon = \frac{a_0}{R_0} \to 0$

N = number of neighbors immediately surrounding a particle

$R_0$ = particle radius of curvature

$a_0$ = distance between particles

Further analysis leads to

$$\eta_{rel} = \frac{N\phi_{max}}{8} \frac{(\phi/\phi_{max})^{1/3}}{1 - (\phi/\phi_{max})^{1/3}} \tag{25}$$

When the particles are in a simple cubic array N = 6, and $\phi_{max} = \pi/6$, so that the factor $N\phi_{max}/8 = \pi/8$ instead of 9/8 as in Eq. (24). However, if the particles are in a cubical or hexagonal closest-packed array N = 12, $\phi_{max}$ = 0.75 and $N\phi_{max}/8 \approx$ 9/8.

Equation (24) was found to be valid for some published viscosity data. It was suggested that the rapid increase in viscosity at high $\phi$ values can be interpreted wholly in terms of hydrodynamic interaction between neighboring particles, and that drop collisions, aggregation, and inertial effects exert little effect [23]. The range of viscosity data examined was very limited, however, so that the validity of Eq. (24) was not really substantiated. Furthermore, the shear thinning or viscoelasticity of many concentrated emulsions must involve the disruption of drop aggregates.

The general conclusion to be drawn from a survey of published literature is that, at present, there is no single theory which is universally applicable [25,26]. Nevertheless, the viscosity of concentrated emulsions at high shear rates such that the drops are completely deflocculated can often be satisfactorily described by the relation

$$\frac{\eta_\infty}{\eta_0} = \exp\left(\frac{2.5\phi}{1 - k\phi}\right) \tag{27}$$

where k depends on the hydrodynamic interaction between drops and increases as the drop size decreases [27]. When expanded, this equation gives a power series in $\phi$ similar to Eq. (22) with the values of the constants b and c increasing as drop size decreases. Equation (27) has the same form as that proposed by Mooney [28] with k now being described as a geometric crowding factor such that $1.35 < k < 1.91$.

Practical emulsions are never monodisperse with respect to drop size, and the size distribution characteristics influence the rheological properties. All studies of this effect have, of necessity, been carried out with monodisperse systems of solid spherical particles which were mixed together in varying proportions [29-36]. This procedure can be criticized because it provides a series of discrete sizes rather than a continuous size distribution, but there does not appear to be an alternative approach to the problem at present. Model dispersions containing spheres of two different sizes exhibit a minimum viscosity at a certain ratio of the two sizes. When three or four different sizes of spheres are mixed together, but with the total volume concentration of spheres kept constant, a minimum viscosity is no longer observed. Instead, the viscosity increases curvilinearly as the proportion of the smallest size spheres increases [35].

An emulsion with a bimodal drop size distribution can be regarded as a disperse sytem in which the larger size drops are suspended in a continuous phase represented by the smaller size drops dispersed in the continuous-phase fluid [29, 30]. This model can then be extended to emulsions containing an i-modal size distribution, so that

$$\eta_{rel(\Sigma_i)} = \eta_{rel(1)}\eta_{rel(2)}\eta_{rel(3)} \cdots \eta_{rel(i)} \tag{28}$$

where $\eta_{rel(1)}, \eta_{rel(2)}, \cdots \eta_{rel(i)}$ are the relative viscosities of the different size fractions in the continuous phase at the same volume concentrations that they occupy in the emulsion. In conjunction with Eq. (27) this leads to

$$\eta_{rel(\Sigma_i)} = \exp\left(\frac{2.5\phi_1}{1 - k_1\phi_1}\right) \exp\left(\frac{2.5\phi_2}{1 - k_2\phi_2}\right) \exp\left(\frac{2.5\phi_3}{1 - k_3\phi_3}\right) \cdots \exp\left(\frac{2.5\phi_i}{1 - k_i\phi_i}\right)$$

$$= \Pi \exp\left(\frac{a\phi_i}{1 - k_i\phi_i}\right) \tag{29}$$

An analysis of published experimental data suggests that the relationship between drop size (D) and k can be represented by an empirical relationship of the form

$$k = 1.079 + \exp\left(\frac{0.01008}{D}\right) + \exp\left(\frac{0.00290}{D^2}\right) \quad (30)$$

so that values of $k_1$, $k_2$, $k_3$, . . . , $k_i$, for insertion in Eq. (29), can be calculated if the drop size distribution is known. A comparison of experimental $\eta_{rel}$ data for both O/W and W/O, emulsions with values calculated in accordance with Eqs. (29) and (30) show good agreement, the agreement extending to higher values of $\phi$ for the W/O emulsions (0.5 to 0.6) than for the O/W emulsions ($\sim$ 0.4). The discrepancy arising for the very concentrated O/W emulsions may be due to the drops being larger than in the W/O emulsions, and so the possibility of drop deformation is greater.

The viscosity of an emulsion can also be related to the drop size distribution by an alternate relation

$$\eta = SK - B \quad (31)$$

where K is the rate of change in viscosity with change in the specific surface (S), so that $S = d\eta/dS$, and B is a constant [37]. Analysis of viscosity data for O/W emulsions ($\phi$ = 0.30) indicates that the values of K and B also are influenced by the emulsifier concentration.

Hydration of the emulsifier layer adsorbed around the drops in an emulsion will also influence the rheological properties. The viscosities (Newtonian) of O/W emulsions stabilized with sodium naphtha sulfonates ($\phi$ = 0.0531 – 0.328) increase as the emulsifier concentration is increased at any constant value of $\phi$ [38]. This effect is attributed to an electroviscous effect arising from ionization of the adsorbed sulfonate molecules, which produces a negatively charged layer at the oil/water interface surrounded by a positively charged diffuse layer. The high electric field strength at the particle surface leads to adsorption of water molecules and, consequently, to an apparent increase in drop radius ($\Delta r$). Values of $\Delta r$, which are calculated from the viscosity data using Eq. (19), ranged from 14 to 36 Å. However, these values are suspect in the light of previous discussion of the influence of particle size on viscosity since the drop sizes in the emulsions were extremely small (138 to 1025 Å), and it is extremely doubtful whether the application of Eq. (19) can be justified in these circumstances.

The polyoxyethylene groups of nonionic emulsifiers adsorbed on oil drops in O/W emulsions also bind water molecules and this results in an increase in drop diameter with a consequent increase in viscosity [39,40]. The effective thickness of the adsorbed emulsifier layer and the hydration coefficient increase witb increasing number of oxyethylene groups in the emulsifier molecule. The degree of hydration ($h_a$) is defined by

$$h_a = (n - n_0) \exp\left(-\frac{\Delta G_h}{RT}\right)$$
$$= (n - n_0) \exp\left(-\frac{\Delta H_h}{RT}\right) \exp\left(\frac{\Delta S_h}{RT}\right) \tag{32}$$

where $\Delta G_h$, $\Delta H_h$ and $\Delta S_h$ are, respectively, the free energy, enthalpy, and entropy of hydration, and n and $n_0$ are, respectively, the total number of oxygen atoms in the hydrophilic portion of the emulsifier molecule and the number of oxygen atoms which do not bind water molecules. Addition of electrolytes such as sodium sulfate alters the degree of hydration, the effective thickness of the hydrated adsorbed emulsifier layer decreasing as the electrolyte concentration increases.

Highly concentrated emulsions may exhibit anomalous rheological properties owing to drop deformation. In that case, Eqs. (22) to (31) no longer apply. The original spherical shape of the drops undergoes various changes until at very high $\phi$ values, polyhedra are formed [41].

## V. Rheological Properties of Flocculated Emulsions

Following preparation, the drops in emulsions flocculate and the size of the aggregates so formed increases with storage time. These aggregates immobilize liquid from the continuous phase within the voids between the drops, so that when the emulsions are examined in a viscometer at low shear rates such that the aggregate structure is not seriously damaged, an anomalously high viscosity is exhibited. When the shear rate is increased, the aggregate size is progressively reduced, and so is the volume of continuous phase immobilized. The effect of aggregation on viscosity can be demonstrated by a simple procedure in which an emulsion is first stirred vigorously or subjected to high shear rate and then is examined at a low shear rate [42].

Figure 10 illustrates this procedure as applied to a W/O emulsion ($\phi = 0.47$). Samples are withdrawn up to 405 h after preparing the emulsion and the drop aggregates are destroyed by subjecting each sample to a shear rate of 215.46 $s^{-1}$ for 5 min in a coaxial cylinder viscometer [43]. The shear rate is then quickly reduced to 1.33 $s^{-1}$. Irrespective of age, all samples have the same $\eta_\infty/\eta_0$ value after the 5 min at 215.46 $s^{-1}$. At 1.33 $s^{-1}$ the $\eta/\eta_0$ values of samples increase for about 20 s before a constant maximum value is reached. This stage represents reflocculation and the redevelopment of aggregates. The maximum $\eta/\eta_0$ value achieved depends on sample age and decreases with increasing storage time, because drop coalescence during storage increases the mean drop size.

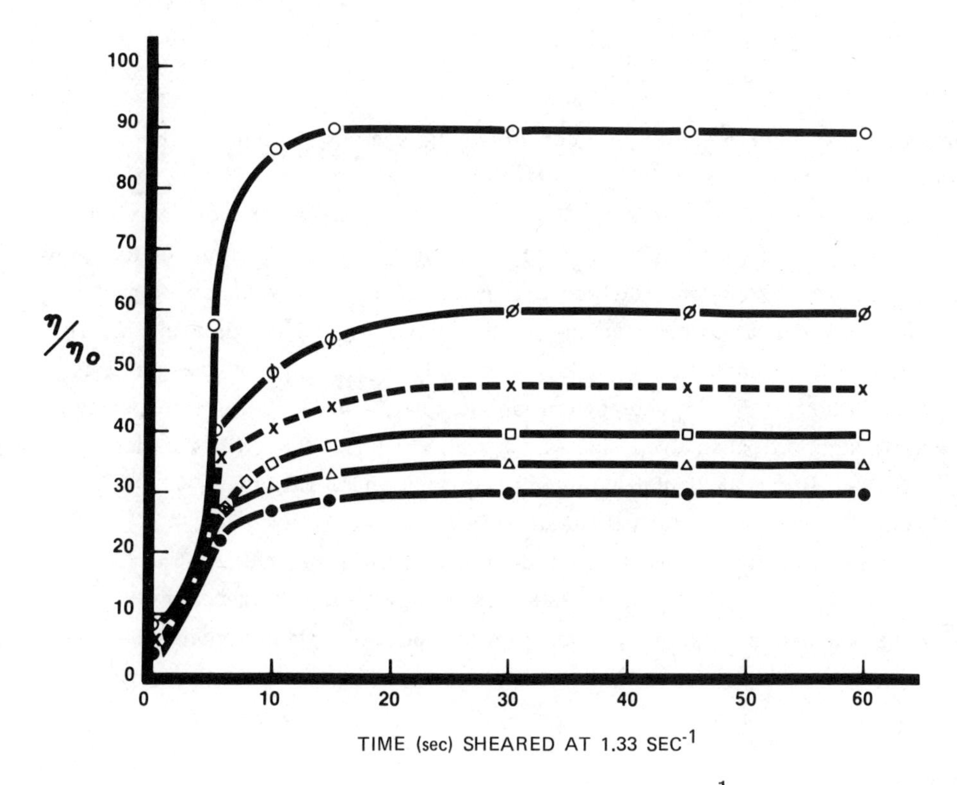


Figure 10. Recovery of $\eta_{rel}$ at a shear rate of 1.33 $sec^{-1}$ after 5 min, at 215.4 $s^{-1}$: (—O—O—) Fresh emulsion, (—ϕ—ϕ—) aged 24 h, (—X— —X—) aged 48 h, (—□—□—) aged 68 h, (—Δ—Δ—) aged 189 h, and (—●—●—) aged 450 h.

In the flocculated state, an emulsion exhibits viscosity comparable to that of a completely deflocculated emulsion with a higher drop concentration. The latter value $\phi_a$ is defined [42] by

$$\phi_a = \frac{\pi(D^3)}{6}[N_1 + f(N_t - N_1)] \tag{33}$$

where f is a *swelling factor* related to the volume of continuous phase immobilized within the drop aggregates, $N_t$ is the total number of drops per milliliter of emulsion at any time t after preparation, and $N_1$ is the number of unassociated drops per milliliter of emulsion. When flocculation proceeds rapidly with respect to the rate of drop coalescence, $N_1 << N_t$ and $f = \phi_a/\phi$. The problem still remains, however, of determining $\phi_a$. Mooney [42] obtained $\phi_a$ from an empirical $\eta_{rel}$-$\phi$ curve [44]

$$\eta_{rel} = \frac{\eta_\infty}{\eta_0} = \frac{\sqrt{1 + 0.5\phi}}{1 - \phi} \exp\left(\frac{2.5\phi}{1 - \phi}\right) \tag{34}$$

by determining the value of $\phi(\phi_a)$ which gives the same $\eta_{rel}$ value as the emulsion exhibited at low shear rate with $\eta_{rel} = \eta/\eta_0$.

In view of the discussion in Sec. IV regarding the complexities of deriving a quantitative $\eta_{rel}$-$\phi$ relationship which accounts for all the influencing factors, the data derived from Eq. (34) are suspect. A more reliable analysis can be made using Eq. (27) and (30) to prepare the $\eta_{rel}$-$\phi$ curve. When the data in Fig. 10 are analyzed with the latter curve it is found that $\eta_{\dot\gamma=215.46}/\eta_0$ corresponds to $\phi = 0.55$, and the $\eta_{\dot\gamma=1.33}/\eta_0$ values correspond to $\phi_a = 0.63 - 0.64$ irrespective of the sample age and mean drop size (1.25 to 3.27 mμ). Using the relation $f = \phi_a/\phi$, f = 1.34 to 1.36, which suggests that at a shear rate of 1.33 $s^{-1}$ the drops occupy ~ 75% of the total volume of each aggregate.

When the same emulsions are subjected to less drastic shear at 21.5 $s^{-1}$ for 5 min and the recovery of $\eta_{rel}$ at 0.133 $s^{-1}$ is followed [43] the recovery is virtually instantaneous (Fig. 11). This implies that much weaker bonds between

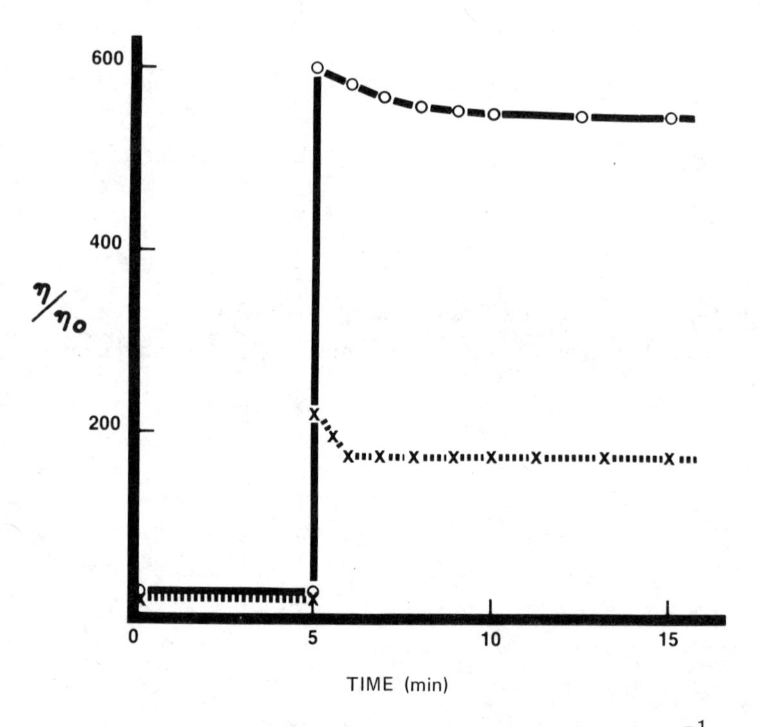


Figure 11. Recovery of $\eta_{rel}$ at a shear rate of 0.133 $s^{-1}$ (—O—O—) or 0.54 $s^{-1}$ (—X— —X—) after shearing at 21.5 $s^{-1}$ or 22.0 $s^{-1}$, respectively (W/O emulsion).

drops are broken at 21.5 $s^{-1}$ than at 215.46 $s^{-1}$ and that the time they require to re-form is much shorter. This type of study provides some indication, therefore, that bonds of different strengths link drops in the aggregates.

In emulsions stabilized by emulsifiers with not too high molecular weights van der Waals attraction forces are primarily responsible for the bonds between drops in the aggregates. This gives rise to viscoelastic properties in the near stationary state. When a small shear stress is applied to the emulsion the resulting time-dependent strain leads to creep compliance-time behavior as illustrated in Fig. 4. For example, both freshly prepared W/O and O/W emulsions ($\phi$ = 0.625) stabilized by Span ® 85 (sorbitan trioleate) and sodium oleate respectively exhibit time-dependent behavior which can be described [45] by Eq. (7). The values of the various parameters decrease as the mean drop size increases with the precise influence of drop size varying from one parameter to another (Figs. 12 and 13). It is noteworthy that in emulsions with relatively small mean drop sizes small changes in mean size can produce substantial changes in the magnitude of the viscoelastic parameters.

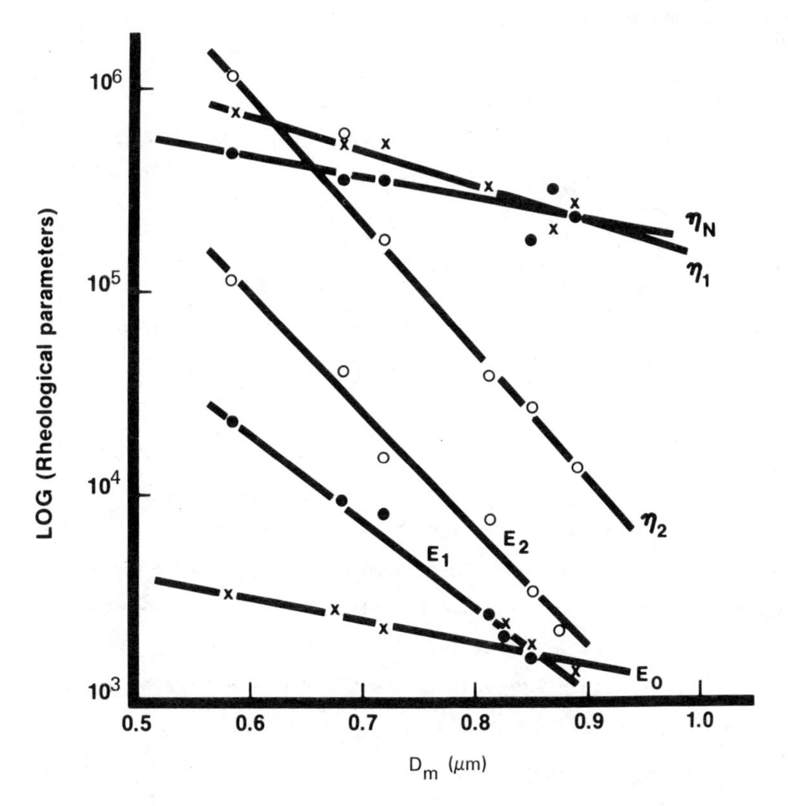


Figure 12. Influence of mean drop size on the creep parameters calculated for W/O emulsions.

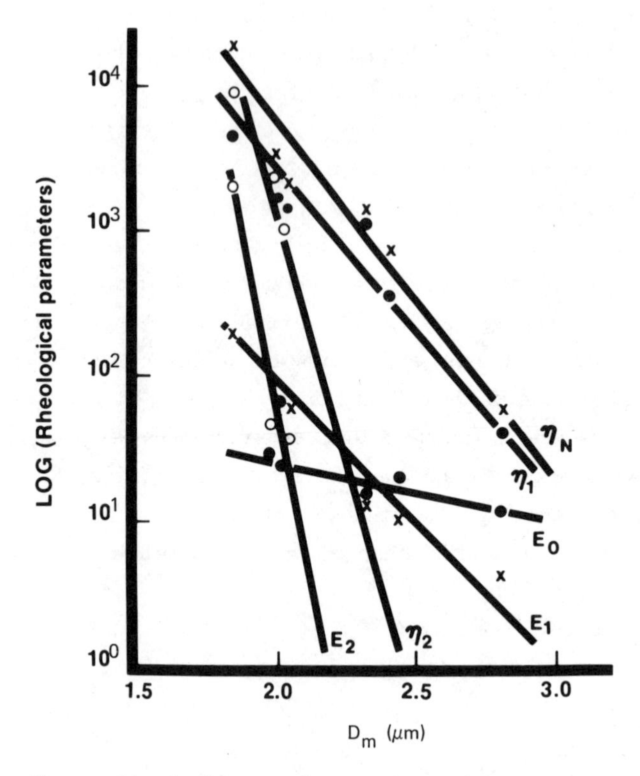


Figure 13. Influence of mean drop size on the creep parameters calculated for O/W emulsions.

An equation for the instantaneous elastic modulus ($E_0$) can be derived in terms of the number of contacts ($N_c$) between a drop and its immediate neighbors and the attraction potential ($\Delta F$) between drops of equal size (D) such that

$$E_0 = \frac{2\phi}{D^2} \frac{N_c}{6} \frac{D}{3} \frac{d^2 \Delta F}{dH_0^2} = \frac{\phi(1 + 1.828v)A}{36\pi D^3 H_0^3} \tag{35}$$

where v is the total volume of continuous phase held in the voids between drops in the aggregated state, $H_0$ is the minimum distance between the drop surfaces, and A is the Hamaker constant (cf. Chap. 4). Thus, at constant $\phi$, $E_0 \propto 1/D^3$ which is reasonably well supported by the experimental data in Figs. 12 and 13.

A somewhat different equation for $E_0$ has been derived for aggregates of rigid particles on the basis of three possible models [46-48]:

i. The particles are arranged in a statistical network of chains.
ii. Some particle aggregates are connected by long chains of particles.
iii. The particles are surrounded by a thick, rigid, continuous-phase layer.

Models i and ii lead to

$$E_0 = \frac{3\phi(D + H_0)}{\pi D^3} \frac{V}{dH_0} \tag{36}$$

where V is the energy binding particles together and $dH_0$ is the small increase in the distance between particles when the shear stress is applied. Model ii differs from model i only in that a smaller fraction of the particles participate in the deformation.

Equation (36) can also be written in the form

$$E_0 = \frac{A\phi(D + H_0)}{8\pi D^2 H_0^2} \tag{37}$$

Experimental studies with dispersions of Aerosil actually suggest a curvilinear relationship between $E_0$ and $\phi$, which is, however, almost linear at low $\phi$ values. A fourth-order polynomial gives a good fit to the experimental data, and this is attributed to increasing interaction between neighboring chains in the aggregates as $\phi$ increases. Equation (36) also indicates a dependence of $E_0$ on D but not to the same degree as indicated by the experimental data. The data suggest that $E_0 \propto 1/D^3$ which reflects a much stronger dependence on particle size than previously proposed equations [49,50] would indicate.

When emulsions are stabilized by polyelectrolyte polymer molecules, their rheological properties, especially after drop flocculation, can be influenced by the conformation of the adsorbed polymer molecules. For example, polymethylmethacrylate (PMA) has a compact hypercoiled (1) conformation at low pH and when the pH exceeds 5.5 it changes to a more extended (2) conformation. The dynamic moduli (E' and E'') and the viscosity (at 7.05 $s^{-1}$) of O/W emulsions ($\phi$ = 0.50) stabilized with PMA (2.2 mg $m^{-2}$), and containing 0.05 M sodium chloride in the aqueous phase, have been examined as a function of $\alpha$, the degree of dissociation [51]. The 1 conformation prevails in PMA at low $\alpha$ and the 2 conformation at high $\alpha$. Figure 14 shows the significant influence of low values of $\alpha$ on all three measured rheological parameters. At high $\alpha$ values the emulsions are very fluid and there is little attraction between the adsorbed PMA layers on adjacent drops. At low $\alpha$ two similar types of attraction prevail:

i. Between PMA loops or tails adsorbed on a single drop
ii. Between extended PMA loops or trains adsorbed on to different drops.

Hydrophobic bonding is believed to play an important role in the prevalence of the 1 conformation at low $\alpha$ [51,52]. This involves the methyl groups in the main chain of the PMA molecule, since there is no hydrophobic interaction when methyl groups are absent [53]. Support for this belief is provided by the increase in E'

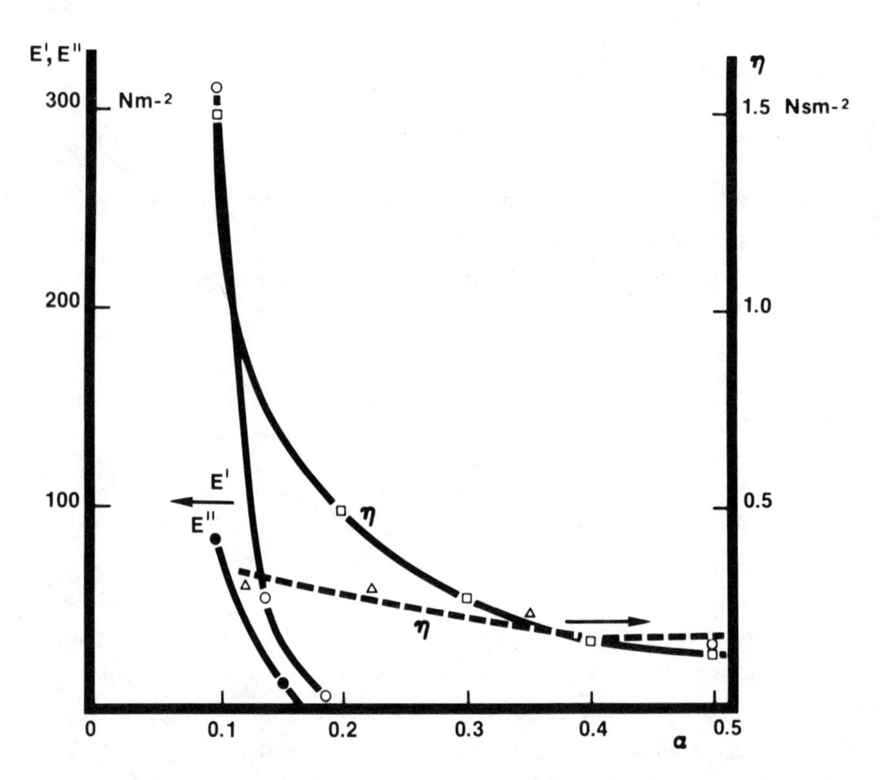


Figure 14. Influence of α on the dynamic viscoelastic parameters for a polyelectrolyte stabilized O/W emulsion. (From Ref. 53.)

and η observed when the temperature is increased, since both the hydrophobic interaction and the proportion of adsorbed PMA molecules in the 1 conformation increases. Both rheological parameters pass through a maximum value as the temperature increases, so it is unlikely that van der Waals attraction is wholly responsible for the 1 conformation.

The O/W emulsions stabilized by PMA molecules can be regarded as having small gel regions linking the oil drops (Fig. 15) when $\alpha = 0.1$, and a plateau region, characteristic of rubberlike elasticity, is exhibited in the E' plot as a function of ω. By considering the extension of a chain of drops and the effective area of interaction within the gel-like regions the contribution to the elastic modulus due to interaction between PMA molecules adsorbed on adjacent drops ($E_{pol}$) is found to be

$$E_{pol} = \frac{B\Gamma N_{av}}{HM} kT \frac{3\phi}{2} \tag{38}$$

where B is a measure of the effectiveness of the cross linking process resulting from intermolecular interactions between loops and tails of adsorbed PMA molecules,

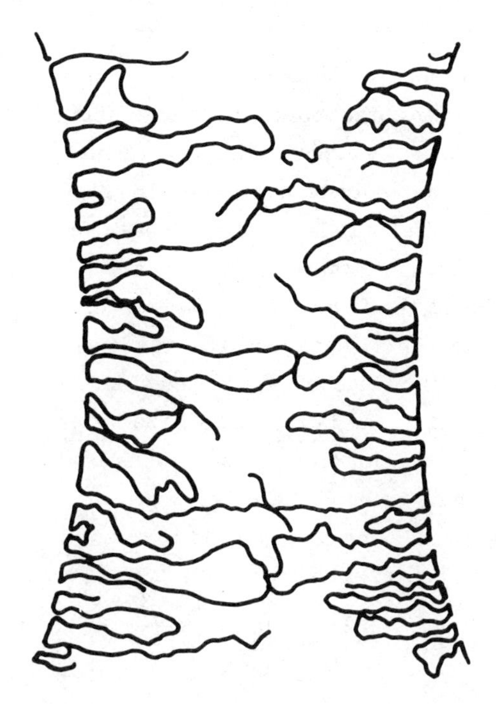

Figure 15. Interaction between polyelectrolyte molecules adsorbed on adjacent oil drops. (From Ref. 54.)

$\Gamma$ is the amount of PMA adsorbed per unit area of drop surface, M is the PMA molecular weight, and H is the distance between the drop surfaces [54]. If one assumes that $E' \approx E_{pol}$, i.e., any contribution by attraction forces can be neglected, which is a justified assumption for this situation, and introducing an experimental value (300 N $m^{-2}$) of E' into Eq. (38) along with $M = 10^5$ and $\Gamma = 1.7$ mg $m^{-2}$

$$\frac{H}{b} = 1.0 \times 10^{-7} \text{ m} \tag{39}$$

If $H \approx 35$ nm, Eq. (39) suggests that only about 33% of the adsorbed PMA molecules contribute to E'.

The theoretical analysis covered by Eqs. (38) and (39) is limited by the assumption that the PMA gel layer responds immediately to an applied tensile or compressive force, whereas a finite time is probably required for the associated water to move in or out of the gel layer when it is deformed.

## VI. Viscoelastic Properties of Emulsions with a Structured Continuous Phase

Oil-in-water emulsions exhibiting viscoelastic properties can be prepared by using a fatty alcohol in conjunction with a commercial grade emulsifier as the stabilizing agents. Detailed studies have been made of isopropyl alcohol or liquid paraffin-in-water emulsions containing sodium dodecylsulfate, potassium laurate, or polyoxyethylene sorbitan monoleate in the aqueous phase and oleyl, lauryl, cetyl, or cetostearyl alcohol in the oil phase [11,12,55,56]. The two liquid phases are heated to a temperature above the melting point of the fatty alcohol, mixed, and emulsified. As the emulsions cool down fatty alcohol diffuses from the oil phase into the aqueous phase, where it is solubilized within micelles of the emulsifier. When the temperature falls below the melting point of the fatty alcohol, the latter can no longer diffuse into the aqueous phase and precipitates instead. This leads to the formation of a viscoelastic gel network which entraps the water drops.

The creep compliance-time behavior of these emulsions is described by an equation similar in form to Eq. (11) but with i = 3 in Eq. (7), so that

$$J(t) = J_0 + J_1\left[1 - \exp\left(-\frac{t}{\tau_1}\right)\right] + J_2\left[1 - \exp\left(-\frac{t}{\tau_2}\right)\right] + J_3\left[1 - \exp\left(-\frac{t}{\tau_3}\right)\right] + \frac{t}{\eta_N} \qquad (40)$$

In a homologous series of $C_{10}$ to $C_{18}$ fatty alcohols the chemical nature of the alcohol has a marked influence on the rheological properties of both freshly prepared and stored emulsions.

The dynamic rheological parameters $E'$ and $\eta'$ increase as the alkyl chain length increases when measurements are made at low frequencies [11,12]. When the frequency is increased $E'$ increases and $\eta'$ decreases by several orders of magnitude.

W/O emulsions containing microcrystalline wax and polyoxyethylene ether as emulsifier also exhibit viscoelastic properties and their creep compliance-time behavior is described by

$$J(t) = J_0 + J_1\left[1 - \exp\left(-\frac{t}{\tau_1}\right)\right] + \frac{t}{\eta_N} \qquad (41)$$

so that fewer parameters are associated with the retarded elastic compliance than for O/W emulsions containing fatty alcohol. As the water content is increased from 0 to 47.6% (w/w), while maintaining the microcrystalline wax contant constant, the values of $E_0$, $E_1$, $\eta_1$, and $\eta_N$ decrease. The effect on $E_0$ is most pronounced. Therefore, the water drops weaken the internal structure. When the water content is kept constant at 47.6% (w/w) and the microcrystalline wax content is increased from 20 to 40% (w/w of oil phase) the values of $E_0$, $E_1$, $\eta_1$ and $\eta_N$ increase, the effect being greatest for $E_1$ and $\eta_1$ [7].

The W/O emulsions are prepared, as were the O/W emulsions containing fatty alcohol, by mixing the preheated oil and water phases. When the agitation is stopped, the water drops begin to flocculate and small aggregates form. However, as the emulsion cools down the viscosity of the continuous phase, in which the microcrystalline wax was previously dissolved, increases continuously and eventually inhibits further flocculation and aggregate growth. The ultimate structure of the cooled W/O emulsions resembles that of O/W emulsions containing fatty alcohol in that small clusters of drops are dispersed in an interlinked solid network of microcrystalline wax. In the W/O emulsions the viscoelastic properties arise primarily from the interlinked wax network, and the influence of the small water-drop clusters is exerted through their effect on the degree of compactness of the network.

The frequency dependence of the dynamic viscoelastic parameters is greatly influenced by the water content and to a much lesser degree by altering the micrcrystalline wax-oil ratio [57]. Microscopic examination of the W/O emulsions indicates a decrease in the number of large anisotropic particles (microcrystalline wax) as the water content increases, and an increase in the number of the much smaller isotropic (water) particles.

## VII. Rheological Properties of Microemulsions

Microemulsions are prepared from water and oil phases with the incorporation of amphipathic molecules in both phases. The drop size is much smaller than in normal emulsions and there has been much discussion as to whether these systems are true emulsions or not. Drop sizes can be as small as 100 Å or less, and consequently the rheological properties of these sytems are of great interest.

In one of the pioneer studies on microemulsions [58,59] it was suggested that microemulsions have a uniform drop size because the viscosity data fitted the equation

$$\eta_{rel} = (1 - 1.135\phi)^{-2.5} \tag{42}$$

which had been proposed for disperse systems with uniform size particles [60,61]. This conclusion is questionable in view of the limited amount of viscosity data available at that time and the observation that considerable deviation occurred when $\phi$ exceeded about 0.35.

Microemulsions of benzene in water containing relatively high concentrations of Tween 20 and Span 20, and with $\phi = 0.03 - 0.406$, exhibit shear thinning at higher $\phi$ values [62]. Because of the high emulsifier concentration employed micelles form in the aqueous phase, and they solubilize a certain volume fraction

($\phi_0$) of benzene. Consequently the volume of dispersed phase ($\phi$) is reduced to $\phi - \phi_0$. Values of $\phi_0$ which can be obtained from turbidity data range from 0.01 to 0.02, the value increasing as the emulsifier concentrations increase. The viscosity data conform to

$$\eta_{rel} = \frac{\eta}{\eta_0} \quad \text{or} \quad \frac{\eta_\infty}{\eta_0} = \exp\left(\frac{a(\phi - \phi_0)}{1 - k(\phi - \phi_0)}\right) \tag{43}$$

where $\eta_0$ is the viscosity of the aqueous phase plus the solubilized benzene. Values of the constant a range from 1.66 to 2.08 depending on the emulsion formulation and it is always lower than the theoretical value of 2.5 for rigid spherical particles. When the drops do not behave as rigid particles and fluid circulation within the drops is not inhibited, then in accordance with Eq. (20),

$$a = 2.5\,\frac{\eta_i/\eta_0 + 2.5}{\eta_i/\eta_0 - 1} \tag{44}$$

Theoretical values of a for these microemulsions should lie between 1.398 and 1.505 so the experimental values fall somewhere between what theory indicates and the values for rigid spheres. The discrepancy may be due to transfer of one of the emulsifiers initially dissolved in the aqueous phase (sorbitan monolaurate) to the benzene.

A modified form of Eq. (43), which allows for solubilization of water, fits the viscosity data obtained for water-in-oil microemulsions prepared from hexadecane, hexanol, potassium oleate, and water with different water-oil ratios [63].

$$\eta_{rel} = [1 - 1.35(\phi - \phi_0)]^{-2.5} \tag{45}$$

provided $\phi - \phi_0$ does not exceed $\sim 0.4$. At very low values of $\phi - \phi_0$, the viscosity data comply satisfactorily with Eq. (19). Experiment indicates that $\phi_0 = 0.028$ for these systems.

With water/oil ratios up to 0.7 viscosity is independent of shear rate. At higher ratios it becomes shear dependent, and this is due to the sperical drops undergoing a transition to water cylinders and ultimately to water lamellae at a water-oil ratio of 1.4.

## VIII. Rheological Properties of Multiple-Phase Emulsions

A multiple-phase emulsion of the W/O/W type can be prepared by a two-stage process which requires a W/O emulsion to be formed initially and this emulsion is then mixed with an aqueous phase containing a hydrophilic emulsifier [64]. When the emulsifier used to prepare the primary W/O emulsion is Span ® 80, and the hydrophilic emulsifier is a polyoxyethylene derivative, the flow is Newtonian and the viscosity data conform to a modified Mooney [28] equation

$$\ln \eta_{rel} = \frac{a(\phi_{wi} + \phi_{oil})}{1 - \lambda_c(\phi_{wi} + \phi_{oil})} \tag{46}$$

where $\lambda_c$ is a crowding factor, $\phi_{wi}$ is the volume fraction (0.165) of inner water phase, and $\phi_{oil}$ is the volume fraction (0.093) of oil. The total volume fraction of disperse phase ($\phi$) in an outer aqueous phase can be regarded as $\phi_{wi} + \phi_{oil}$. Flow remains Newtonian when 1% sodium caseinate is added to the outer aqueous phase at pH 7.30 to 8.90 [65].

Equation (46) can be rearranged so that

$$\frac{\phi_{wi} + \phi_{oil}}{\log \eta_{rel}} = \frac{2.303}{a} - \frac{2.303\lambda_c}{a}(\phi_{wi} + \phi_{oil}) \tag{47}$$

and if the left-hand side of the expression is plotted against $(\phi_{wi} + \phi_{oil})$, a and $\lambda_c$ can be determined.

In the absence of soidum caseinate from the outer aqueous phase a = 3.21 instead of the theoretical value of 2.5 for rigid spheres. When 1% sodium caseinate is added the value of a increases from 2.28 to 2.66 as the pH increases from 7.30 to 8.93. A value of the constant a higher than 2.5 indicates some residual drop flocculation in the emulsions but this is eliminated when sodium caseinate is added, owing to electrostatic repulsion developing between the drops. This effect will also reduce the value of $\lambda_c$.

At very low shear stress (2 to 20 dyn $cm^{-2}$) the W/O/W emulsions exhibit viscoelasticity [66]. With 0.5% glucose present in the outer phase the creep compliance increases with time as the thickness of the oil layer increases from 310 through 4080 Å to infinity.

At a constant oil-layer thickness of 1300 Å the apparent viscosity, measured at any selected shear rate, increases as the glucose concentration in the outer aqueous phase increases from 0.5 to 3.0% (w/w). Increasing the thickness of the oil layer results in an initial decrease in the apparent viscosity and then it increases.

## IX. Rheological Changes in Emulsions During Storage

When freshly prepared emulsions are stored at ambient temperature, the drops flocculate and coalesce for some time before there is visible separation of disperse phase. At the same time, the rheological properties alter significantly provided no other processes are associated with storage [43,67,68].

Viscosity measurements made at high shear rates on W/O and O/W emulsions with medium to high concentrations of dispersed phase indicate a sharp decrease with storage time. This is associated with increasing mean drop size. The kinetics of drop coalescence are defined by

$$N_t = N_0 \exp(-Ct) \tag{48}$$

where $N_0$ and $N_t$ are the number concentrations of drops per milliliter by emulsion initially and at time t, respectively, and C is the rate of drop coalescence, provided the coalescence rate does not depend on the initial value of $N_0$. In terms of drop size Eq. (48) can be written as

$$\ln D_t = \ln D_0 + \frac{Ct}{3} \tag{49}$$

where $D_0$ and $D_t$ are the mean drop sizes initially and at time t, respectively. Therefore, C can be determined by following the rate of change of mean drop size with storage time over a period which depends on whether coalescence proceeds slowly or quickly. The slower the coalescence rate the shorter the period over which measurements have to be made. Two to three weeks are usually sufficient. When the value of C has been established it is possible to predict values of $D_t$ over very long storage periods, and then, by using Eqs. (27) and (30), the associated changes in $\eta_\infty$ can be predicted. For the O/W and W/O emulsions referred to previously there is agreement between experimental and calculated $\eta_\infty$ values with one exception. This is the most concentrated O/W emulsion ($\phi = 0.74$) which exhibited a multiple-phase structure [67].

Similar agreement between experimental and calculated viscosity values at shear rates of 0.133 to 10.77 $s^{-1}$ is observed for W/O emulsions ($\phi = 0.47$). These emulsions flocculate very rapidly and the viscosity, as measured at each shear rate, decreases curvilinearly with time as the mean drop size increases and the flocculate nature alters [43]. It is only because the drops flocculate very rapidly that it is possible to study the influence of mean drop size on viscosity at low shear rates and to use these data to predict changes in viscosity during storage.

O/W emulsions stabilized with sodium oleate ($\phi = 0.532$) and Aerosol OT ($\phi = 0.679$) do not flocculate rapidly because of electrical repulsion between drops. Consequently, the changes in viscosity observed at low shear rates do not resemble the pattern for W/O emulsions. Instead, there is an initial rise over several days to a maximum value, which is maintained for a few days, and then the viscosity decreases sharply to a value which subsequently changes only slowly. Consequently, there is no agreement between experimental viscosity data and values calculated from appropriate viscosity measurements on freshly prepared emulsions with different mean drop sizes [68]. Microphotographs of these emulsions reveal that, after flocculation, the drops are not spherical and their surfaces are flattened in all regions where they face other drops. Their appearance resembles that of a polyhedral foam, so that eventually the volume concentration of drops in the flocculates may well be in excess of 74%.

Creep compliance-time tests indicate that both the W/O and O/W emulsions exhibit viscoelasticity at very low shear stress. The response can be represented by Eq. (11) with the value of each parameter decreasing with storage time [4,68]. For the O/W emulsions the $J_2$ and $\eta_2$ parameters eventually disappear when the mean drop size achieves a certain value. During the first few days of storage time the values of all the parameters increase and then they decrease, so that the general trend is similar to that observed for viscosity at selected low shear rates.

The viscosity of W/O/W emulsions also decreases with storage time [64,55]. For example, with 0.5% glucose in the outer aqueous phase $\eta_{rel}$ decreases from ~ 58 to ~ 4 within 14 days. When 1% BSA (pH 4.58) is also added to the outer aqueous phase, the rate of viscosity change is much slower and in the first few days $\eta_{rel}$ may actually increase before it decreases. At constant shear stresses of 2 to 20 dyn $cm^{-2}$, the time dependence of the creep compliance varies with the thickness of the oil layer. The creep compliance at constant oil-layer thickness increases with storage time as water passes from the inner to the outer aqueous phase.

## REFERENCES

1. W. Ostwald, *Kolloidzeitschrift 36*:99, 157, 248 (1925).
2. A. De Waele, *Kolloidzeitschrift 36*:332 (1925).
3. J. R. Van Wazer, J. W. Lyons, K. Y. Kim, and R. E. Colwell, *Viscosity and Flow Measurement*, Interscience, New York, 1963, p. 202.
4. P. Sherman, *J. Colloid Interface Sci. 24*:107 (1967).
5. P. Sherman, *Proc. 5th Intern. Congr. Rheol. 2*:327 (1970).
6. H. Komatsu and P. Sherman, *J. Texture Studies* 5:97 (1974).
7. H. Komatsu and P. Sherman, in *Colloidal Dispersions and Micellar Behavior* (K. L. Mittal, ed.), ACS Symposium Series No. 9, 1975, p. 126.
8. K. Inokuchi, *Bull Chem. Soc. (Jap.) 28*:453 (1955).
9. J. D. Ferry, *Viscoelastic Properties of Polymers*, 2d ed., Wiley, New York, 1961, pp. 43-44.
10. B. Warburton and S. S. Davis, *Rheol. Acta 8*:205 (1969).
11. B. W. Barry and G. M. Eccleston, *J. Pharm. Pharmacol. 25*:244 (1973).
12. B. W. Barry and G. M. Eccleston, *J. Pharm. Pharmacol. 25*:394, (1973).
13. A. Einstein, *Ann. Phys. 19*:289 (1906).
14. A. Einstein, *Ann. Phys. 24*591 (1911).
15. G. I. Taylor, *Proc. Roy. Soc. (London) A138*:41 (1932).
16. M. A. Nawab and S. G. Mason, *Trans. Faraday Soc. 54*:1712 (1958).
17. J. G. Oldroyd, *Proc. Roy. Soc. (London) A218*:122 (1953).
18. J. G. Oldroyd, *Proc. Roy. Soc. (London) A232*:567 (1955).
19. P. Sherman, in *Emulsion Science* (P. Sherman, ed.), Academic, London, 1968, pp. 287-290.
20. R. Simha, *J. Appl. Phys. 23*:1020 (1952).

21. G. J. Kynch, *Brit. J. Appl. Phys. Suppl. No. 3 55*:(1954).
22. P. Sherman, *Proc. 4th Intern. Congr. Rheol. 3*:605 (1965).
23. N. A. Frankel and A. Acrivos, *Chem. Eng. Sci. 22*:847 (1967).
24. J. D. Goddard, *J. Non-Newtonian Fluid Mechanics 2*:169 (1977).
25. J. Mewis and A. J. B. Spaull, *Adv. Colloid Interface Sci. 6*:173 (1976).
26. D. J. Jeffrey and A. Acrivos, *A.I.Ch.E. J. 22*:417 (1976).
27. F. L. Saunders, *J. Colloid Sci. 16*:13 (1961).
28. M. Mooney, *J. Colloid Sci. 6*:162 (1951).
29. G. F. Eveson, S. G. Ward and R. L. Whitmore, *Disc. Faraday Soc. 11*:11 (1951).
30. G. F. Eveson, in *Rheology of Disperse Systems* (C. C. Mill, ed.), Pergamon, London, 1959, p. 61.
31. J. S. Chong, Doctoral thesis, University of Utah, 1964.
32. F. L. Saunders, *J. Colloid Interface Sci. 23*:230 (1967).
33. R. J. Farris, *Trans. Soc. Rheol. 12*:281 (1968).
34. T. B. Lewis and L. E. Nielsen, *Trans. Soc. Rheol. 12*:421 (1968).
35. C. Parkinson, S. Matsumoto, and P. Sherman, *J. Colloid Interface Sci. 33*:150 (1970).
36. F. Skvára and M. Vancurová, *Silikaty 17*:9 (1973).
37. Lj. Djaković, P. Dokić, P. Radivojević, and V. Kler, *Colloid Polymer. Sci. 254*:907 (1976).
38. M. van der Waarden, *J. Colloid Sci. 9*:215 (1954).
39. A. N. Kutznetsova, V. A. Volkov, and E. M. Aleksandrova, *Colloid J. (USSR) 39*:690 (1978).
40. A. N. Kutznetsova, V. A. Volkov, N. S. Seliverstova, and E. M. Aleksandrova, *Colloid J. (USSR) 39*:694 (1978).
41. K. J. Lissant and K. G. Mayhan, *J. Colloid Interface Sci. 42*:201 (1973).
42. M. Mooney, *J. Colloid Sci. 1*:195 (1946).
43. P. Sherman, *J. Colloid Interface Sci. 24*:97 (1967).
44. H. Eilers, *Kolloid-Z. 97*:313 (1941).
45. P. Sherman, *Proc. 5th Intern. Congr. Rheol. 2*:327 (1970).
46. K. Strenge and H. Sonntag, *Colloid Polymer Sci. 252*:133 (1974).
47. K. Strenge and H. Sonntag, in *Proc. Intern. Conf. Colloid Surface Science* (E. Wolfram, ed.) Elsevier, Amsterdam, 1975, p. 347.
48. H. Sonntag, K. Strenge, and W. N. Schilow, *Colloid Polymer Sci. 255*:292 (1977).
49. E. M. Jager, M. Van den Tempel, and P. De Bruyne, *Proc. Koninkl. Nederl. Akad, Wetensch. Ser. B 66*:17 (1963).
50. J. M. P. Papenhuizen, *Rheol. Acta 11*:73 (1972).
51. T. Van Vliet and J. Lyklema, in *Proc. Int. Conf. Colloid Surface Science* (E. Wolfram, ed.) Elsevier, Amsterdam, 1975, p. 197.
52. J. T. C. Böhm and J. Lyklema, *Theory and Practice of Emulsion Technology* (A. L. Smith, ed.) Academic, London, 1976, p. 23.
53. T. Van Vliet, and J. Lyklema, *J. Colloid Interface Sci. 63*:97 (1978).
54. T. Van Vliet, J. Lyklema and M. van den Tempel, *J. Colloid Interface Sci. 65*:505, (1978).
55. B. W. Barry, *J. Pharm. Pharmac. 21*:533 (1969).
56. B. W. Barry, *J. Colloid Interface Sci. 32*:551 (1970).
57. H. Komatsu, M. Takahanshi, and S. Fukushima, *Trans. Soc. Rheol. 21*:219 (1977).
58. J. E. L. Bowcott, Ph.D. dissertation, Cambridge University, England, 1957.
59. C. E. Cooke Jr. and J. H. Schulman, in *Surface Chem.* (P. Ekwall, K. Groth, and V. Runnstron-Reio, eds.), Academic, New York, 1965, p. 231.

60. R. Roscoe, *Brit. J. Appl. Phys. 3*:267 (1952).
61. R. Brinkman, *J. Chem. Phys. 20*:571 (1952).
62. S. Matsumoto and P. Sherman, *J. Colloid Interface Sci. 30*:525 (1969).
63. J. W. Falco, R. D. Walker, and D. O. Shah, *A.I.Ch.E. J. 20*:510, (1974).
64. S. Matsumoto, Y. Kita, and D. Yonezawa, *J. Colloid Interface Sci. 57*:353 (1976).
65. Y. Kita, S. Matsumoto, and D. Yonezawa, *J. Colloid Interface Sci. 62*:87, (1977).
66. S. Matsumoto and M. Kohda, in *Food Texture and Rheology* (P. Sherman, ed.), Academic, London, 1979, p. 437.
67. P. Sherman, *J. Phys. Chem. 67*:2531 (1963).
68. P. Sherman, *J. Colloid Interface Sci. 27*:282 (1968).

# 8

# Optical Properties of Emulsions

RAYMOND S. FARINATO* and ROBERT L. ROWELL / University of Massachusetts, Amherst, Massachusetts



## I. Introduction

Optical investigations of emulsions may be conveniently grouped according to technique. Most work has been concerned with the spatial distribution of light intensity after a beam has interacted with an emulsion system. Less attention has

---

*Present address: American Cyanamid Company, Stamford, Connecticut

been given to the spectral distribution of this light. Elastic-scattering theories and conventional geometric optics apply in most cases. By far the dominant interest for which optical investigations have been made has been for aspects of emulsion particle size distribution. A variety of physical properties such as rheology, phase stability, interfacial area, appearance, and absorption of energy depend on particle size distribution.

Particle sizes extend from about 10 nm for transparent microemulsions to millimeters for so-called macroemulsions. The lower end of this range precludes direct visual imaging except by some indirect processing such as electron micrographs of freeze-etched samples.

Transmission techniques utilize changes in the energy distribution or state of polarization of the incident light to quantitate system properties. Scattering techniques are used to view the radiation diverted from the incident direction. Microscopy utilizes transmitted or scattered light and will be treated in a separate section. Reflectance techniques are used to probe the emulsion close to the air/emulsion interface. Finally, an assortment of spectroscopic techniques are used to investigate the spectral distribution of scattered or transmitted light. Figure 1 gives a composite view of the techniques to be discussed.

Emulsions are suspensions of one immiscible liquid in another. They are lyophilic (solvent-loving) colloids which are thermodynamically unstable. A dispersed and a continuous phase are recognized. Stability is usually achieved with a surface-active agent operating either sterically or electrostatically to prevent coalescence but not necessarily flocculation. Since the dispersed phase is fluid, the droplets are spherical in the presence of zero shear forces or other external fields, and are usually heterodisperse with a positively skewed unimodal size distribution. Classification of emulsions is by several schemes, e.g., oil in water (O/W) versus water in oil (W/O), internal phase ratio [1], or size range. Microemulsions not only comprise a range of smaller sizes (less than $\lambda/4$) but are thermodynamically stable systems (see Chap. 5).

## II. Optical Microscopy

Optical microscopy offers a reassuring direct measurement of emulsion particle sizes. Emulsions do not usually exhibit the irregularity of shape which introduces some degree of uncertainty and difficulty in pattern recognition. Many of the applications in emulsion science rely on tedious counting and sizing procedures in which the technician is an integral part of the recognition system. Computerization of the recognition and analysis aspect of the data handling has been found extremely useful in biological and clinical applications where the systems are not nearly so geometrically uniform. One can only see economics as the basis for not using state-of-the-art technology in emulsion science.

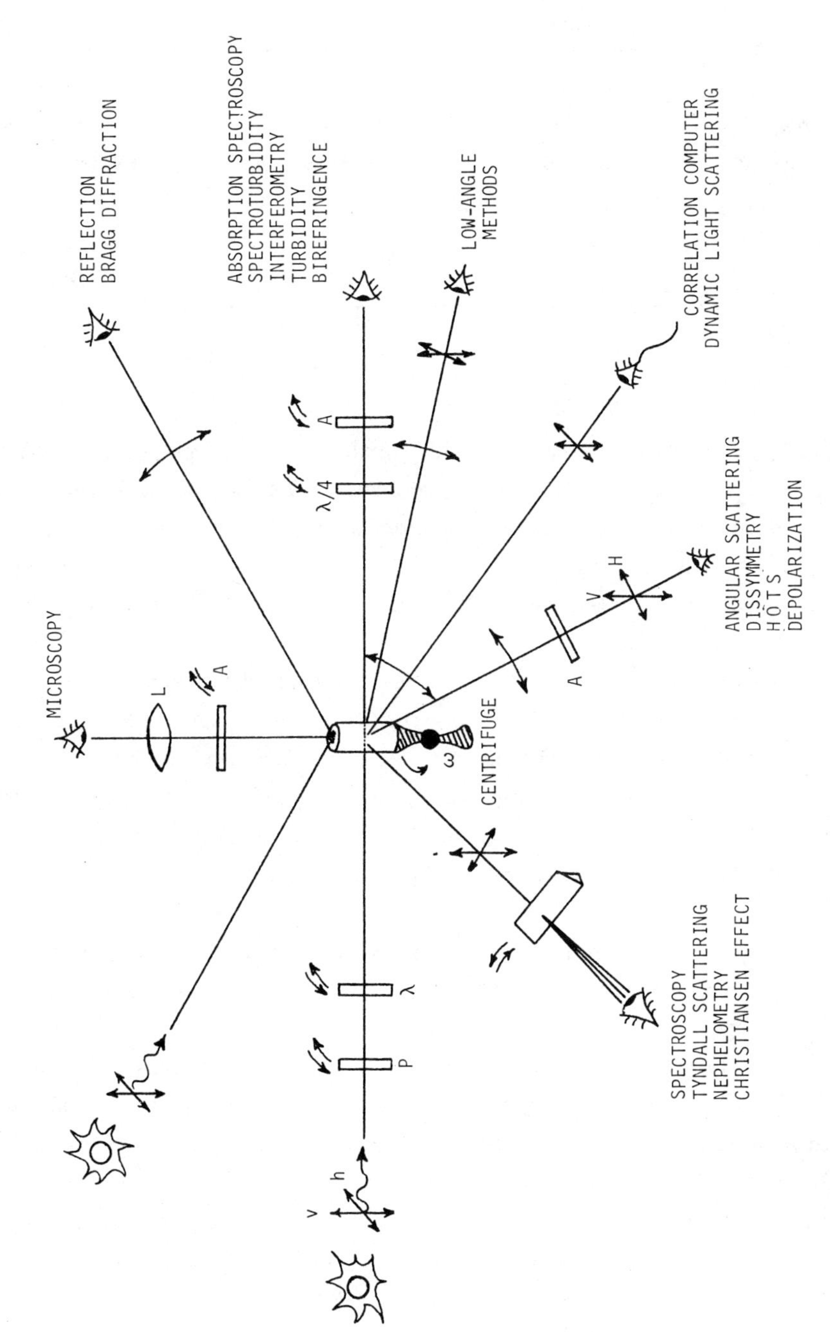


Figure 1. Optical observations of emulsions: P = polarizer, λ = wavelength filter, λ/4 = quarter wave plate, A = analyzer, L = lens system.

**Table 1.** Estimation of Droplet Size of Emulsions by Visual Inspection

| Appearance | Tyndall scattering | Transmitted beam | Diameter (μm) |
|---|---|---|---|
| Two phases | None | No color | Macroglobules |
| Pure white | None | No color | Exceeds 0.5 |
| White to gray | Weak blue | Weak red | 0.3 to 0.1 |
| Gray to translucent | Intense blue | Intense red | 0.14 to 0.01 |
| Transparent | None | No color | Less than 0.01 (microemulsions) |

*Source:* Adapted from Ref. 2.

Table 1, adapted from Prince [2], shows the optical characteristics for the range of particle sizes encountered in emulsions. The generalizations given represent guidelines for simple visual inspection of diluted emulsions. For the size range greater than 0.2 μm, a quantitative characterization can be obtained with the light microscope as discussed in treatises on instrumentation and methodology [3,4].

Below the limit of 0.2 μm, the size of the diffraction pattern in visible light becomes approximately independent of particle size or shape while intensity remains a function of such variables. Accordingly, with techniques of microphotometry and ultramicroscopy the range of measurable particles can be extended down to about 0.01 μm. Microphotometry simply replaces the eye with a more quantitative intensity sensitive device, providing there is enough contrast between object and background.

For small transparent particles in a transparent medium the contrast can be accentuated by using phase contrast techniques which essentially recombine unshifted light directly transmitted through the sample with light whose phase has been slightly shifted because of the refractive index inhomogeneities in the medium due to the presence of emulsion droplets. The interface pattern is modulated by stops to suppress higher than the first order fringes to give a sharp image. A detailed discussion has been given by Pluta [5].

Ultramicroscopy achieves a reduced background intensity by either dark-field illumination, or oblique illumination. In dark-field illumination, the condenser diaphragm is fully opened and a central stop blocks off just the portion of light that would normally enter the objective. Only light scattered from the dispersed particles will enter the objective. In oblique illumination the substage diaphragm is moved to one side, again allowing only diffracted light to reach the objective lens. In either case, for very small particles the light intensity can be measured electronically and these intensities can be calibrated with respect to particle size especially for spherical emulsions. For a description of a scanning microdensitometer see Smith et al. [6a].

Latex particles have been examined by the dark field method by Davidson and Haller [6b]. By adapting a flow system, individual particles may be looked at one at a time and a particle size histogram constructed. Derjaguin and Vlasenko [7] and McFadyen and Smith [8] have described automatic flow ultramicroscopes for submicron particle counting and size analysis.

Whether the emulsion particles are directly visible or require measurement of the diffracted light, the tedium of particle classification and counting still exists. Humphries [9] has given a comprehensive summary of mensuration methods in optical microscopy. The various schemes of classifying irregular particles are not needed for the case of emulsions. Semiautomatic devices, i.e., requiring an operator to determine image class, basically relieve the bookkeeping burden. Two major classes of semiautomatic devices are the intercept measuring devices (including filar or screw micrometer eyepieces and various image shearing devices) and the nominal projection devices (area overlap of a projected image).

Various schemes have been devised for compiling the data. For small numbers of particles, or a cursory view of an emulsion, there are graticular devices relying on either linear scales or image comparison, e.g., globe- and circle-type graticules. The reader is referred to Humphries [9] for a more detailed discussion.

More automated forms of microscopic analysis have resulted from the application of computers and imaging techniques. Such developments have hardly reached routine use in emulsion science, but will be briefly discussed to provide perspective. Mendelsohn et al. [10] give a description of the use of computers in microscopy. They describe the cytophotometric data conversion (CYDAC) system which transforms an optical image into a digital image in a computer memory. This digital transformation not only allows automatic classification and counting of particles, but data processing techniques may be applied for purposes of image sharpening, parameter extraction, and image characterization and discrimination. Preston [11] discusses computer analysis of microscopic images. The CELLSCAN/GLOPR system is even more automated and computer controlled using a special language for communication (GLOL). The images are logged in either digital or quantized form on magnetic tape with a variety of peripheral units for reconstruction and analysis. This has so far found most use in cytological studies.

## III. Transmission Techniques

### A. Turbidity

The extinction of intensity of light passing through a medium results from either absorptive interactions resulting in energy transfer to the medium (nonzero imaginary component of the refractive index), or scattering from optical inhomogeneities in the medium (spatial or temporal variation in the real component of the refractive index). The logarithmic decrease of transmitted light intensity with sample depth is expressed as

$$\frac{I_\ell}{I_0} = \exp(-\tau \ell) \tag{1}$$

where $I_\ell$ is the light intensity at a length $\ell$ along the incident path of a beam of original intensity $I_0$, and $\tau$ is the turbidity.

In general, $\tau$ is composed of two terms, one due to absorption (Beer-Lambert law) and one due to scattering:

$$\tau = N(C_{abs} + C_{sca}) \tag{2}$$

where N is the number of particles per unit volume, and the C are cross sections. The cross sections are conceptualized as areas projected in the plane normal to the direction of propagation of the light, and which remove energy from the beam through either the absorption or scattering process. Cross sections are composed of the projected geometrical area of the particle ($\pi a^2$ for spheres of radius a) multiplied by an efficiency factor, which for scattering depends on the particle size relative to the wavelength of the light, the relative refractive index, the polarization state of the incident beam and the viewing analyzer, and the scattering angle. A more familiar equivalent expression for the absorbance term is $\varepsilon c_m/2.303$, where $\varepsilon$ is the molar extinction coefficient, and $c_m$ is the molar concentration of the dispersed phase. Thus,

$$C_{abs} = \frac{1000\rho V \varepsilon}{2.303M} \tag{3}$$

where $\rho$, V, and M are the density, volume per emulsion particle, and molecular weight of the dispersed phase, respectively. In practical systems V is replaced by the number average particle volume.

For spherical systems such as emulsions $C_{sca}$ can be calculated assuming dilution of the dispersed phase to the extent where no interparticle interactions or multiple scattering effects are important. The general solution for spherical dielectric particles was due to Mie [12], and is applicable to any range of size and relative refraction index. Kerker [13] has described the theoretical and calculational procedures and has compiled sources of numerical computations as well.

Values of the scattering efficiency $Q_{sca} = C_{sca}/\pi a^2$ for spheres, range from near 0 to about 6 for most emulsion sizes. For small spheres $Q_{sca}$ is an oscillatory function of size. However, the turbidity is also proportional to the particle cross-sectional area, which has a mitigating effect. For particles that are very large compared to the wavelength of light but still small compared to the cross section of the beam the limiting value of $Q_{sca} = 2$ is approached. The Rayleigh theory provides an approximate expression for $C_{sca}$ valid in the range of particle radii less than $\sim \lambda/20$ [13]:

$$C_{sca} = \frac{24\pi^3 V^2}{\lambda^4}\left(\frac{m^2 - 1}{m^2 + 2}\right)^2 \tag{4}$$

where $\lambda$ is the wavelength of the light in the medium and m is the relative refractive index of dispersed/continuous phase.

While most emulsions are composed of optically transparent constituents (even though the emulsions themselves may be translucent, colored, or opaque owing to scattering) there are some systems which absorb as well as scatter. The effect of absorption (measured separately in the bulk phases) can be separated out. The generalized Mie theory provides for particles with absorption.

Graphical displays of the efficiency factor for homogeneous dielectric spheres can be found in Kerker [13] and van de Hulst [14]. Lothian and Chappel [15] analyzed the functional dependence of a related factor, the total scattering coefficient, on particle radius relative to the incident wavelength, relative refractive index, and the angular integration of the viewing optics.

Wickramasinghe [16] has tabulated and graphed efficiency factors for homogeneous dielectric spheres of complex refractive index. For $n_2 = n' - in''$ and $\alpha = 2\pi a/\lambda$, Wickramasinghe has computed efficiency factors for the range of parameters: $n' = 1.1(0.1)2.0$, $2.5(0.5)3.5$; $n'' = 0$, $0.05$, $0.1(0.1)0.5$, $1.0(0.5)3.5$; $\alpha = 0.1(0.1)5.0$, $5.2(0.2)15.0$. Thus, for nominal optical wavelengths this would apply to homogeneous emulsions up to a little over 2 μm in diameter. Verner et al. [17] computed Mie coefficients over a larger range in $\alpha(0.05\text{-}100)$ using smaller increments. It has been argued that for nearly monodisperse distributions one must be careful of the increment sizes used for data comparison; much fine structure can be lost [18,19].

The availability of current theories places some restrictions on the interpretation of particle size from turbidimetric measurements. In all cases the systems are assumed dilute enough to preclude multiple scattering. This is accomplished experimentally by plotting the specific turbidity $\tau/c$ versus concentration of the dispersed phase and extrapolating to infinite dilution. The assumption of no interparticle correlations owing to force field interactions, either external or interparticle, is also made. This is critical for emulsions which may have a surface charge. Experimentally, this may be dealt with again by dilution or the addition of salt.

In most cases theoretical computations have been for monodisperse systems. Emulsion droplet size heterogeneity will result in a determination of an averaged quantity from turbidity data (e.g., a volume-average particle volume for small particles). If all the particle sizes are within the Rayleigh criterion ($a < \lambda/20$), then $\tau/c$ is an increasing function of $\alpha$, and may be used to determine the so-called turbidity-average particle radius [20]:

$$a_\tau = \left( \frac{\Sigma_i N_i a_i^6}{\Sigma_i N_i a_i^3} \right)^{1/3} \tag{5}$$

For larger particles typically encountered in emulsions the averaged parameters are more complicated. Meehan and Beattie [21] have discussed the kinds of averages obtained from a variety of light scattering techniques including turbidimetry. The kind of average depends on the particle size range and polydispersity. For a narrow range of particle sizes (about $\alpha = 1/2$ for small sizes) the type of average is determined from the slope of a plot of log ($\tau/c$) versus log $\alpha$. For broad distributions where the largest $\alpha$ is still small (submicron) the average specific turbidity is given to a good approximation by [21]

$$\frac{\tau}{c} = k \frac{\Sigma_i N_i \alpha_i^5}{\Sigma_i N_i \alpha_i^3} \tag{6}$$

where k is a combination of experimental parameters. For continuous distribution functions the summations become integrations. The investigation of Meehan and Beattie [21] utilized a log-normal distribution function.

Dobbins and Jizmagin [22] confirmed the validity of the above equation for $\tau/c$ using a wide variety of monomodal continuous size distributions with particle radii ranging to larger than $\lambda$. Their theoretical curves allowed a determination of a volume-surface mean radius $a_{VS}$ from a measurement of turbidity and concentration.

$$a_{VS} = \frac{\int_0^\infty p(a) a^3 \, da}{\int_0^\infty p(a) a^2 \, da} \tag{7}$$

where p(a) is the particle radius distribution function. For example, $a_{VS}$ for a uniform distribution of particles would ba a/6. For a polydispersion, $a_{VS}$ is 3 times the volume-to-surface area of the dispersed phase and may therefore be used to determine interfacial area.

Langlois, Gullberg, and Vermeulen [23] reported an empirical observation of the transmission properties of unstable emulsions in the particle size range 100 to 1000 μm and relative refractive index close to 1:

$$\frac{I'_0}{I} = 1 + \beta A_{sp} \tag{8}$$

where $I'_0$ is the light intensity transmitted through the pure continuous phase, I the transmitted intensity through the emulsion, $A_{sp}$ the specific surface area per volume of dispersed phase, and β an empirical constant. It is relatively straightforward to show the consistency of this empirical observation with the relations derived by Dobbins and Jizmagin [22]. Starting from the given relations and other standard definitions

$$\frac{\tau}{c} = \frac{3}{4\rho}\frac{\overline{Q}_{sca}}{a_{VS}} \tag{9}$$

where ρ is the density of the dispersed phase and $\overline{Q}_{sca}$ is the mean scattering efficiency ($\bar{k}$ in the notation of Dobbins and Jizmagin) given as [13]

$$\overline{Q}_{sca} = \frac{\int_0^\infty Q_{sca} p(a) a^2 \, da}{\int_0^\infty p(a) a^2 \, da} \tag{10}$$

Since, for very large particles $Q_{sca} = 2$, $\overline{Q}_{sca} = 2$. Also

$$a_{VS} = 3\frac{V}{A} = \frac{3}{A_{sp}} \tag{11}$$

Thus

$$\frac{\tau}{c} = \frac{A_{sp}}{2\rho} \tag{12}$$

Recall

$$\frac{I}{I_0} = \exp(-\tau \ell) \tag{1}$$

where $I_0$ is the incident intensity and ℓ the optical path length. For unit path length (distance through 1 unit volume) and small values of τ (reasonable for emulsions of large transparent particles with $m \simeq 1$), the following expansion applies:

$$\frac{I_0}{I} = 1 + \tau + \cdots$$

$$= 1 + \frac{c}{2\rho} A_{sp} + \cdots \tag{13}$$

For a transparent continuous phase $I_0 \simeq I'_0$, and by truncating Eq. (13) we arrive at the empirical relation (8).

The problem of multiple scattering effects is a difficult one, avoided by most experimentalists except for astronomical or meteorological observers. Multiple scattering is less of a problem in transmission experiments because the scattered and transmitted light can be more easily separated in the forward direction [22]. Schulman and Friend [24] investigated the concentration dependence of the light scattered at 90° from the forward direction (propagation direction of incident beam) for systems of W/O microemulsions whose particle diameters were in the range of several hundred Å. Since these are Rayleigh scatterers the intensity scattered at 90°, I(90), is related to the turbidity by a constant [14]. Their results fit the empirical relation

$$\tau = \tau_0 \exp(-\mu N V^2) \tag{14}$$

where $\tau$ is the apparent turbidity, $\tau_0$ the true value of the turbidity at infinite dilution and $\mu$ a combination of optical constants [24]

$$\mu = \frac{24\pi^3}{\lambda^4}\left(\frac{m^2 - 1}{m^2 + 2}\right)^2 \ell \tag{15}$$

Graciaa et al. [25] also investigated W/O microemulsions in the Rayleigh size range and found empirical evidence for expressing the turbidity in terms of a polynomial expansion in the volume fraction of dispersed phase $\phi$,

$$\tau = \frac{32\pi^3 V}{3\lambda^4}\left(n \frac{dn}{d\phi}\right)^2 \frac{\phi}{1 + A\phi + B\phi^2} \tag{16}$$

where n is the microemulsion refractive index and A and B are constants.

There has been a good deal of theoretical work done on the effects of multiple scattering in both mono- and heterodisperse systems of spherical particles of all sizes [26-28]. There is a noticeable lack of application of these methods in emulsion science because of the computational complexity and machinery involved.

Turbidity measurements per se take the form of a relative intensity transmitted through an emulsion sample. Commercial spectrometers have been used for these purposes, usually with some modification of the optics to increase angular resolution and improve collimation of the source (sometimes by replacement with a laser; see discussion in Sec. III.B. The homemade apparatus is another viable alternative (e.g., Cengel et al. [29]).

The data are analyzed in several ways. For quantitative analysis the specific turbidity, once extrapolated to infinite dilution, is compared with theoretical plots based on either Mie theory calculations or some appropriate approximation depending on the size range. Specific methodologies are outlined in Appendixes 1 and 2.

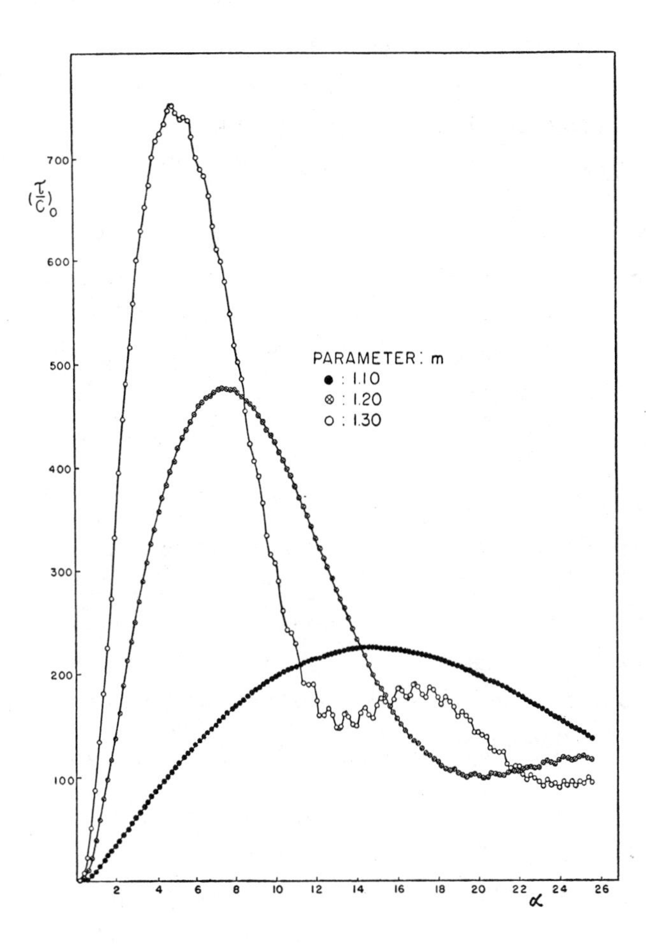


**Figure 2.** Specific turbidity ($\tau/c$) versus size parameter $\alpha$ for different relative refractive indexes m. (From Ref. 123.)

The type of size average being computed for polydisperse samples may also be determined from a plot of log ($\tau/c$) versus log $\alpha$ (see Appendix 1). Figure 2 shows the variation of the specific turbidity as a function of relative particle size.

## B. Spectroturbidimetry

The attractiveness of turbidimetric techniques lies in their simplicity and the high degrees of precision and accuracy obtainable [30]. In most cases the particle size (or average size) is not uniquely determined from one measurement of the specific turbidity (e.g., see graphs in Refs. 13, 31, and 32) unless the concentration of the dispersed phase is also known. In most cases a measurement of $\tau/c$ at two wavelengths will allow a unique size determination.

The extension of the method to measuring the wavelength dependence of turbidity has also been used to determine the average particle size [13,33-36]. Commercial spectrophotometers may be employed for this purpose with sufficient modification to attain the small solid angle of incident beam and viewing optics required [37-39].

The experimental quantity determined is the wavelength exponent of the specific turbidity: $-[d \ln(\tau/g)/d(\ln \lambda_0)]$, where g is the weight fraction of spheres in suspension and $\lambda_0$ is the vacuum wavelength of the incident radiation. Comparison with theoretical plots [34] gives a volume-surface average diameter and total interfacial area [36]. This differential spectrum type approach is more rapid, though less accurate, than a comparison of a normalized turbidity spectrum with theoretical spectra [35,40].

The details of the particle size distribution of emulsion particles may also be investigated by spectrophototurbidimetric techniques. The method of comparing normalized spectra mentioned above is most applicable in this case. Here a distribution function must be assumed. Stevenson et al. [41] used a unimodal positively skewed distribution with a lower limit typical of emulsions. Walstra [36] has explored the use of log-normal, exponential, and upper limit distribution functions. Usually one obtains a modal size, distribution width parameter, and in some cases a size limit parameter.

An effective technique for the measurement of particle size distribution is to combine transmission methods with sedimentation techniques either in a gravitational or centrifugal field. This eliminates the need for prior knowledge of particle size distribution function, a limiting assumption in the other techniques mentioned above. Gumprecht and Sliepcevich [42] used Stokes' law of settling and a corrected form of the light transmission equation (using Mie theory) to determine the size distribution of a kerosene fog in a gravitational settling chamber. Graciaa et al. [43] measured the turbidity at a constant height of a W/O emulsion undergoing settling. The technique, termed differential photosedimentation, employed a He-Ne laser light source, a beam chopper to prevent heating, and a photomultiplier detection system. Turbidity as a function of time is translated into particle size distribution through application of Stokes' law and determination of the extinction coefficient as a function of particle size.

Centrifugal photosedimentation was developed by Kaye [44,45]. The optical attenuation is measured at a given radius in a disk centrifuge. One can start with a single initial loading of the emulsion in the center of the centrifuge, or with a homogeneous emulsion throughout the centrifuge cell. This type of experiment has been used successfully in determining particle size distribution in emulsions [46-49]. Instrumental ingenuity has made the procedure more facile, including data logging [59], use of a laser light source and fiber optics [51,49]. This type of instrument seems usable for particle sizes down to ~ 0.02 μm [49]. The relevant equations are outlined in Groves et al. [51]. Values of the extinction coefficient $Q_{sca} + Q_{abs}$ can be evaluated from the sources listed above in the section on turbidity.

Kerker [13] has mentioned other turbidimetric techniques for determining particle size distributions. These have been used on latexes, but not emulsions per se. Such methods usually consist of less involved versions of the turbidity spectra. For Rayleigh particles the turbidity is independent of distribution breadth, depending only on average size. Gledhill [52] was able to abstract meaningful distribution parameters from submicron samples of spherical particles by measuring $\tau/c$ at one wavelength and the wavelength dependence. Walstra [36] matched turbidity spectra over the entire optical range with theoretical spectra.

### C. Interferometry

The technique of interferometry has seen relatively little practical application in emulsion science. The methodology is well documented [53,54]. O'Brien et al. [55] have used laser light and multiple beam interferometry to study the creaming of pentane/water emulsions. Light rays which were only refracted by the emulsion droplet, and not totally internally reflected, were sufficiently parallel to be collected by a lens. With a small angle wedge between the glass flats of the Fabry-Perot interferometer cell, Fizeau-type fringes resulted. Concentration contours are thus visualized. The fringe patterns sharpen up as the number of droplets in the beam decreases (owing to creaming).

### D. Birefringence

Birefringence in an emulsion results from unequal refractive indices for orthogonal states of polarized light resulting in a phase shift between the two. This is usually the result of molecular ordering within the emulsion system. Shah and Hamlin [56] noted that for a microemulsion system undergoing a phase inversion (e.g., W/O to O/W) the intermediate turbid phase exhibited a birefringence.

Friberg et al. [57] gave arguments for the formation of ordered systems of emulsifier, oil, and water as the cause of enhanced stability to coalescence in some emulsions, especially in regions of the ternary phase diagram corresponding to the presence of a separable mesophase. Further investigations [58,59] confirmed the presence of a stabilizing liquid crystalline layer surrounding emulsion droplets. These were seen under a polarizing light microscope, owing to the birefringence of such ordered layers.

Fukushima et al. [60] observed that whole emulsion droplets of cetyl alcohol in water, stabilized by a nonionic surfactant, were birefringent. These emulsions consisted of a liquid crystal dispersed phase surrounded by a more crystalline material.

## IV. Scattering Techniques

### A. Introduction

Angular scattering methods involve a measurement of the spatial dependence of light scattered from a medium containing optical inhomogeneities, in this case predominantly emulsion droplets. The measurements are made conventionally in either a lateral plane (the plane defined by the incident ray and the scattered ray) or about a small solid angle in the forward direction. Polarization of the incident beam and viewing optics may or may not be considered. The data are usually expressed in terms of an intensity per unit solid angle relative to the incident intensity. Source collimation facilitates quantitative interpretation.

Most theories calculate the scattered field in the far field, quasistatic approximation (i.e., neglecting the time dependence of the incident light wave). Exact solutions applicable to all ranges of the relevant parameters (size, shape, complex refractive index and scattering geometry) have been obtained only for spheroidal particles assuming no interparticle interaction or multiple scattering. Various approximate theories serve to reduce computational difficulties and relax the above restrictions. There are a number of excellent books [13,14] and review articles [31,61-64] on the fundamentals.

The optical inhomogeneities responsible for the scattering are expressed in terms of either structural units of a specific refractive index if the particles are large enough, or fluctuations in the local refractive index owing to fluctuations in concentration, density, anisotropy, etc. (i.e., any physical parameter which would change the local dielectric constant), for molecular species or small aggregates. Most emulsions fall in the range of the former analysis whereas some microemulsions are more readily interpreted using the latter. In any case, one usually measures excess light-scattering quantities (subtracting out contributions due to the continuous phase) for comparison with theory.

Emulsions are usually measured in a high state of dilution to ease interpretation in terms of independent particle scattering models. As mentioned in the section on turbidity, significant progress has been made for situations where multiple scattering cannot be neglected (see sections in Kerker [26] and in Rowell and Stein [28]). The effects of surface charge cannot be neglected except at extreme dilution or in the presence of swamping electrolyte.

The spatial pattern produced in the scattered radiation field is significantly modified by interparticle correlations due to an interactive force field [65,66]. This of course also applies to external fields.

For small particles (Rayleigh region; $a < \lambda/20$) the angular scattering pattern is independent of size and the scattered intensity at any angle may be directly related to the turbidity [see Eq. (20)]. Scattered intensities are usually reported as the Rayleigh ratio $R(\theta)$; the energy scattered by a unit volume in the direction $\theta$ per unit solid angle. If unspecified, $R(\theta)$ is usually for unpolarized incident light. Since most emulsions are inherently optically isotropic and spherical, measurements in the range of small sizes are equivalent to turbidity measurements.

For larger particles the scattering patterns arise mainly from interference of the light scattered from within the particle (assuming no multiple scattering). For particles where the quantity $4\pi a(m - 1)/\lambda << 1$, the phase shift of the light scattered from any two points in the particle depends only on the position of the volume element within the particle. Practically speaking, this corresponds to particle sizes still in the microemulsion range. The analytic scattering equations of Rayleigh-Debye-Gans apply to particles in this size range. Stevenson [67] has formulated analytic expressions which extend this range.

Particles in a still larger size range (i.e., most emulsion droplets) can be treated by the exact theory of scattering from a homogeneous dielectric sphere due to Mie [12], appropriately known as the Lorenz-Mie theory [13]. These expressions involve infinite series of terms composed of Legendre polynomials, Ricatti-Bessel, and Hankel functions. These usually require computer calculation and tables of scattering coefficients which have been compiled (see Kerker [13] for a comprehensive listing, Yajnik et al. [68], Wickramasinghe [16], and Verner et al. [17]).

While this section will be subdivided into different scattering techniques it should be remembered that the particle size range dictates the appropriate theory for analyzing the data. Thus, before considering specific methodologies for determining particle size and distribution, we will give a brief background on the important scattering theories.

The following three sections present such a brief background of the most important scattering theories as a prelude to the last section on scattering measurements. It is important to gain an overall understanding of the theories of light scattering for the most effective utilization. The optical properties are strongly dependent on particle size and it is size range that frequently determines the technique selected.

## B. Rayleigh Theory: Small Isotropic Particles

Incident electromagnetic radiation exerts a periodic force on the bound electrons in a sample of matter. The phase and amplitude of this motion with respect to the driving field depends on the natural frequency of the bound electron oscillator system and the frequency of the incident radiation. The oscillating bound electrons reradiate electromagnetic energy according to the classical laws for an accelerating charge. If the scattering entity is small enough with respect to the wavelength of incident radiation (i.e., at any instant the scattering entity is essentially in a uniform field), it reradiates as would a point dipole in forced oscillation. Frequency modulation of the scattered radiation due to intra- or interparticle processes is neglected, and hence only elastic scattering is considered.

The scattering geometry is depicted in Fig. 3. For the conventional instances of incident polarization being either perpendicular to (v) or parallel to (h) the scattering plane (xy), and the viewing analyzer being either perpendicular to (V) or parallel to (H) the scattering plane, the scattered intensity per unit incident intensity per particle at a distance r from the particle of radius a is given by [13]

$$2V_u = V_v = \frac{16\pi^4 a^6}{r^2\lambda^4}\left(\frac{m^2 - 1}{m^2 + 2}\right)^2 \tag{17}$$

$$2H_u = H_h = \frac{16\pi^4 a^6}{r^2\lambda^4}\left(\frac{m^2 - 1}{m^2 + 2}\right)^2 \cos^2\theta \tag{18}$$

where u stands for unpolarized incident light and m is the refractive index of the scatterers relative to that of the medium. The polarization ratio $\rho(\theta)$ is defined as

$$\rho(\theta) = \frac{H_h}{V_v} = \frac{H_u}{V_u} \tag{19}$$

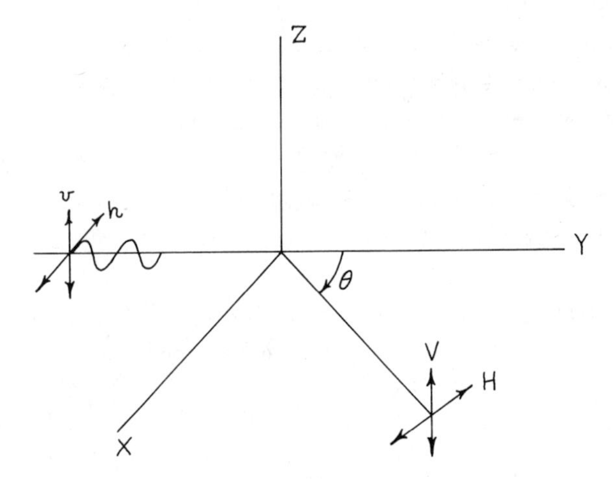


Figure 3. Scattering geometry for incident light along xy direction and scattered-light detector in the xy plane.

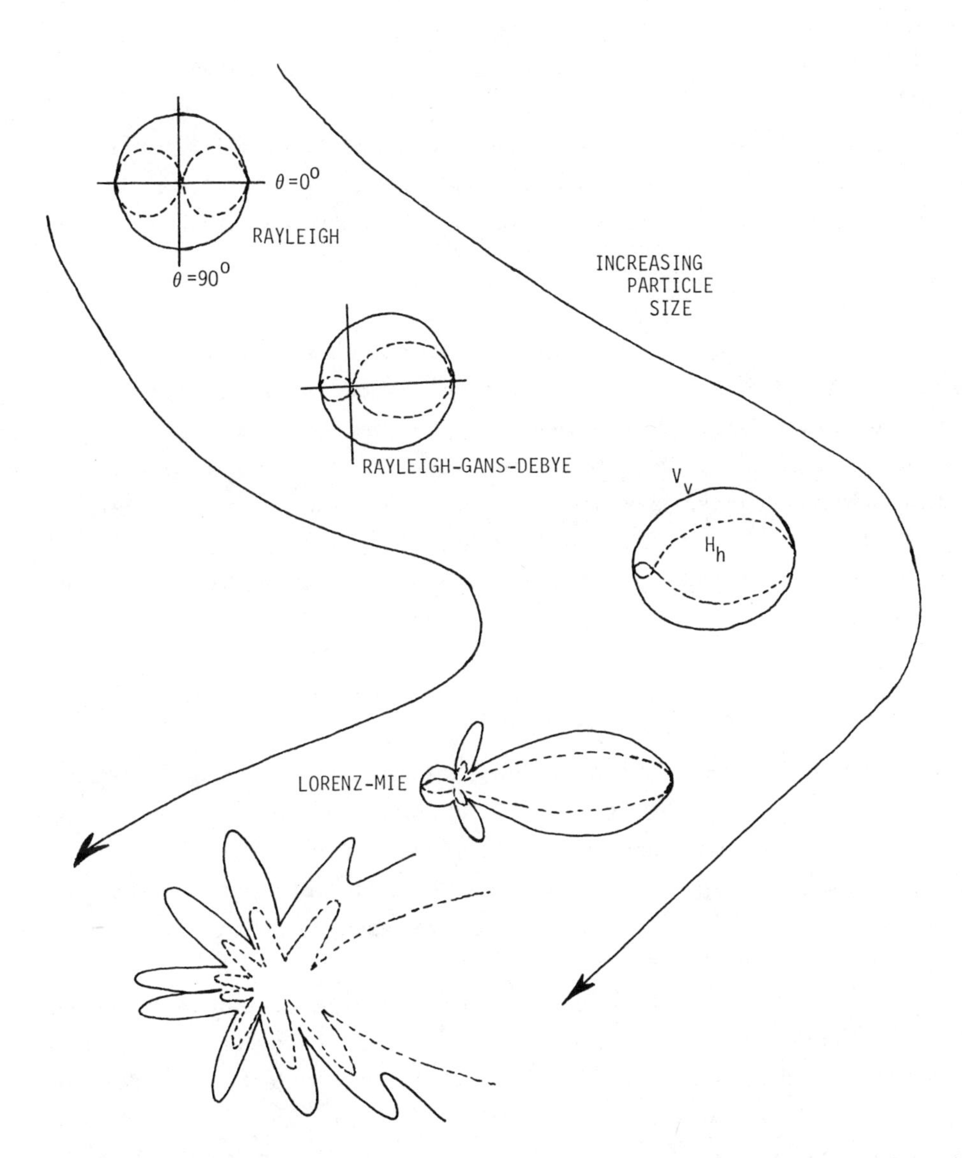


**Figure 4.** Evolution of angular scattering patterns for increasing particle size. Polar plots of intensities for polarized components of the scattered light.

Scattered intensities are usually reported as Rayleigh ratios $R(\theta) = r^2NI(\theta)$, where $I(\theta)$ can be a polarized or an unpolarized intensity (i.e., $V_v$, $H_h$, etc.) for light scattered by N particles per unit volume. Since the size dependent factor $a^6$ and angular factor $(1 + \cos^2 \theta)$ are separable in this range, only the intensity changes while the angular pattern for isotropic particles does not. The basic scattering pattern is seen in Fig. 4 as a polar plot.

Since the scattered intensity at any angle is proportional to $Na^6$ one cannot determine both number concentration and particle radius from a single intensity measurement (or angular scan for that matter), just as is the case with respect to turbidity.

The Rayleigh ratio at one angle is directly proportional to the integrated scattered intensity over all angles (and hence the turbidity) because of the size-independent angular factor. Thus, for unpolarized incident light of unit intensity,

$$R(\theta) = \frac{3}{16\pi} \tau (1 + \cos^2 \theta) \tag{20}$$

Measurement of the Rayleigh ratio offers an advantage over turbidity measurements for only slightly turbid samples; the ratio of signal to background is much greater. This is the principle of nephelometric measurement.

To measure particle size then, a separate measurement of dispersed phase concentration must be made (see Sec. III.A). The theory is well known [13,14], and the result is that for small, isotropic, spherical particles the real part of the emulsion refractive index n of an emulsion which is dilute in the disperse phase would be given by

$$n = \mathrm{Re}(n_1) + 2\pi N \frac{m^2 - 1}{m^2 + 2} a^3 \tag{21}$$

where $\mathrm{Re}(n_1)$ is the real part of the continuous phase refractive index and m is the relative refractive index of the particle. Thus, a measurement of the apparent refractive index of the dispersion in conjunction with $R(\theta)$ can give both N and a.

For a polydisperse sample the average particle radius will be the so-called turbidity-average radius defined in Eq. (5). In most cases the above relations will hold only for microemulsion systems.

The basic Eqs. (17) and (18) account for the blue appearance of the reflected light from the sky or from a cloud of fine smoke particles ($1/\lambda^4$ dependence of the Rayleigh ratio); and the Tyndall effect of the visibility of reflected light from a beam passing through a suspension of particles ($a^6$ term).

The angular terms account for the polarization of sky light or the light scattered from fine suspensions, being nearly completely vertically polarized when viewed at a 90° angle. At high concentrations, multiple scattering accounts for most of the depolarization at this angle.

The range of validity of the Rayleigh equation has been checked against the more general Mie theory and error contours in the $m$-$\alpha$ plane have been determined [13,31,69]. The general rule of thumb of validity (within a couple of percent) of

$a < \lambda/20$ is actually larger than this for relative refractive indices in the range 1.3 to 1.8.

### C. Rayleigh-Debye-Gans Theory: Small Phase Shift

Rayleigh also presented an approximate theory for the scattering by particles of any shape and size but with a small relative refractive index [70]. Kerker [13] has traced the evolution of the theory from the original contributions of Rayleigh in 1881, through additional contributions by Debye in 1915 to the rederivation by Gans in 1925. The approach has been generally known as Rayleigh-Gans scattering although Kerker suggests that it would more appropriately be termed Rayleigh-Debye scattering [13]. We shall use the longer label Rayleigh-Debye-Gans scattering (RDG). The important point is that the RDG approximation is one of small phase shift $4\pi a(m - 1)/\lambda << 1$ so that the theory is applicable to very large particles of sufficiently small refractive index. The extended theory uses a correction factor $P(\theta)$, known as the form factor, which is applicable for small phase shift and has been evaluated for a variety of shapes. For spheres the form factor becomes

$$P(\theta) = \left[\frac{3}{u^3}(\sin u - u \cos u)\right]^2 \tag{22}$$

where $u = (4\pi a/\lambda) \sin(\theta/2)$. For particles in this size range a convenient way to present the results is in terms of the dissymmetry ratio $Z(\theta) = R(\theta)/R(180 - \theta)$. For spheres $Z(\theta)$ is a strong monotonic function of $a/\lambda$ in the appropriate size range [13]. Alfrey et al. [71] have used $Z(45°)$ to determine average particle sizes for poly(vinyltoluene) latexes in the RDG range.

The Zimm method of light-scattering analysis for determining molecular weight M and radius of gyration $R_g$ of particles in solution relies on the approximation

$$\lim_{\theta \to 0} P(\theta) = 1 - \frac{16\pi^2 \sin^2(\theta/2)}{3\lambda^2} R_g^2 \tag{23}$$

and

$$\lim_{c \to 0} \frac{Kc}{R(\theta)} = \frac{1}{M} \tag{24}$$

where K is an optical constant [13].

For a sphere,

$$R_g = \left(\frac{3}{5}\right)^{1/2} a \tag{25}$$

When a sample is polydisperse two kinds of average radii of gyration can be determined from this method: both the number-average $\langle R_g^2 \rangle_n$ and Z-average

$\langle R_g^2 \rangle_Z$, squares of the radii of gyration [72-74]. These different moments of the particle size distribution can be used to assess the degree of polydispersity, and for spheres are defined as follows:

$$\langle R_g^2 \rangle_n = \frac{\int_0^\infty R_g^2(a)p(a)\,da}{\int_0^\infty p(a)\,da} = K_1 \frac{\int_0^\infty a^2 p(a)\,da}{\int_0^\infty p(a)\,da} \tag{26}$$

$$\langle R_g^2 \rangle_Z = \frac{\int_0^\infty R_{g\infty}^2(a)M^2 p(a)\,da}{\int_0^\infty M^2 p(a)\,da} = K_2 \frac{\int_0^\infty M^{8/3} p(a)\,da}{\int_0^\infty M^2 p(a)\,da} = K_2' \frac{\int_0^\infty a^8 p(a)\,da}{\int_0^\infty a^6 p(a)\,da} \tag{27}$$

Stevenson's [67] extension of the Rayleigh equation can also be used for interpreting measurements of $R(\theta)$. Stevenson's theory expands the scattered field in a power series in $a/\lambda$. Numerical results are applicable up to $2\pi a/\lambda \approx 1$. These extensions of the Rayleigh theory then apply up to particle radii in the neighborhood of 0.06 μm for aqueous suspensions using visible light.

Wims [75] has treated the scattering data in a manner similar to Sloan [76,77] and Arrington [78] as discussed in the section on Mie theory. Wims has calculated theoretical curves of log $[I \sin^2(\theta/2)]$ versus log $[\sin(\theta/2)]$ for polydisperse spheres. This allows determination from experimental data of the ratio of the fifth to fourth moment of the size distribution (Z + 2 average). A distribution width parameter is also obtained. This kind of analysis focuses on the forward scattered intensity lobe.

Meehan and Beattie [21] compare the results of several kinds of light scattering particle size analyses in the RDG approximation.

### D. Lorenz-Mie Theory: General Solution

The general solution to the problem of electromagnetic scattering by a homogeneous sphere is widely referred to as the Mie theory [12]. Kerker [13] has pointed out the earlier treatment by Lorenz so that the exact solution might more appropriately be termed the Lorenz-Mie theory.

The mathematical treatment of the problem requires the solution of Maxwell's equations with the appropriate boundary conditions. The electric and magnetic multipole moments of the particle forced into oscillation by the incident field contribute to the scattered field. This is described by a series-expanded solution which may be adequately truncated after a larger number of terms as the particle size increases. Concise expositions of the solution to this problem in addition to Mie's first derivation may be found in Kerker [13], Born and Wolf [79], and van de Hulst [14].

Using Kerker's notation [13], the Rayleigh ratios for polarized light scattered from N noninteracting particles per cubic centimeter are [cf. Eqs. (17) and (18)]

$$V_v(\theta) = \frac{N\lambda^2}{4\pi^2} \left| \sum_{n=1}^{\infty} \frac{2n+1}{n(n+1)} (a_n \pi_n + b_n \tau_n) \right|^2 \tag{28}$$

$$H_h(\theta) = \frac{N\lambda^2}{4\pi^2} \left| \sum_{n=1}^{\infty} \frac{2n+1}{n(n+1)} (a_n \tau_n + b_n \pi_n) \right|^2 \tag{29}$$

The $a_n$ and $b_n$ correspond to the electric and magnetic multipole moments, respectively, and are composed of combinations of Ricatti-Bessel and Hankel functions and their derivatives. The arguments of these functions contain the particle radius, relative refractive index, and incident wavelength. In the far field solution, applicable to most light-scattering studies, these functions simplify somewhat. Also in the far field approximation the $\pi_n$ and $\tau_n$ reduce to functions of first-order Legendre polynomials and are the angular factors, depending only on $\theta$. Computational results are discussed and listed by Kerker [13].

In addition to the references compiled in Kerker [13], Yajnik et al. [68] have calculated and compiled angular scattering functions for heterodisperse systems of spheres using a distribution function found to be applicable to emulsion systems:

$$p(a) = \begin{cases} (a - a_0) \exp\left[-\left(\dfrac{a - a_0}{s}\right)^3\right] & a > a_0 \\ 0 & a \leqslant a_0 \end{cases} \tag{30}$$

Verner et al. [17] have also calculated Mie coefficients over a wide range in $\alpha$ (.05 to 100) and m (1.001 to 1.315).

As a particle increases in size the angular scattering pattern changes from the size-independent Rayleigh pattern, through a pattern characterized by an increased forward lobe in the intermediate Rayleigh-Debye-Gans region, to a structured (though still symmetrical about the 0 to 180° line) pattern with a unique set of maxima and minima (see Fig. 4). The absolute magnitudes of the maxima and minima are extremely sensitive to polydispersity in size.

Since the positions of the maxima and minima are functions of $\alpha = 2\pi a/\lambda$, the observational angles where constructive interference occurs will depend on wavelength for a given sample. If the scattering pattern is not too complex and these maxima are resolvable for different colors one observes a higher order Tyndall spectrum (HOTS) for monodisperse samples in the range ~0.1 to 0.8 μm.

Sinclair and La Mer [80] stated as a rule of thumb: the particle size in μm is approximately equal to 1/10 the number of red or green orders seen between $\theta$ = 0 to 180°. This phenomenon has been taken as an indication of monodispersity; however, Kerker et al. [81] have shown that in some cases broad distribution of small particles may also exhibit the HOTS.

## E. Scattering Measurements

Two classes of experimental observations are usually made: either the variation of a scattering function [e.g., $V_v(\theta)$, $H_h(\theta)$, $\rho(\theta)$] with angle, using a monochromatic light source, or the wavelength dependence of a scattering function at a fixed angle (usually 90°).

Machine computations and fitting routines allow for multiparameter fits of the data to polydisperse models of general size once a distribution function is assumed.

Light-scattering measurements of the particle size distribution of emulsions without assuming an a priori analytical form have been mostly made using the centrifugal photosedimentation techniques mentioned in the Turbidity section to prepare monodisperse fractions. Wallace et al. [82] have used a combination zonal centrifugation, angular light-scattering technique to measure the size distribution of latex particles directly.

### 1. *Angular Methods*

One of the most widely used methods for determining the parameters of a size distribution of emulsions is the polarization method [83]. The quantity $\rho(\theta) = H_h(\theta)/V_v(\theta)$ is measured as a function of $\theta$ and compared with theoretical plots. This quantity, the polarization ratio, has the advantage that one need not determine instrument constants, the maxima and minima for monodisperse emulsions are very well defined, for mildly disperse emulsions the positions of the extrema remain fairly constant, and for heterodisperse emulsions $\rho(\theta)$ retains a simple and strong limiting dependence on average particle size.

Graessley and Zufall [84] have shown that for particles in the size range a = 0.2 to 1.5 μm; if the distribution breadth is at least 0.1 to 0.2 μm, then the average value of $\rho(90°)$ is a linear function of $\alpha$ and m:

$$\overline{\rho(90°)} = 1.89(m - 1)^2 \bar{\alpha} \tag{31}$$

The average radius obtained is

$$\bar{a} = \frac{\int_0^\infty a^{7/2} p(a)\, da}{\int_0^\infty a^{5/2} p(a)\, da} \tag{32}$$

Measurements on O/W emulsions have confirmed these results. Note that their quantity $\rho$, the depolarization ratio, is the same as the polarization ratio of Kerker et al. [83] since $H_h(\theta)/V_v(\theta) = H_u(\theta)/V_u(\theta)$ for isotropic spheres.

Using the positions of the maxima and minima of the polarization ratio, or any light-scattering function for that matter, to determine particle size is useful only with fairly monodisperse samples. Kerker et al. [83] showed that for a log-normal distribution of particles with a modal $\alpha_m = 5.0$, the structure in $\rho(\theta)$ was nearly washed out for a log standard deviation of the particle distribution of $\sigma_0 = 0.3$. The two other commonly used methods of light-scattering particle size analysis by determination of extrema in the angular scan, [i.e., in the HOTS and in $V_v(\theta)$] are also subject to similar limitations. Monk et al. [85] showed that for a broad emulsion distribution (in the region $\alpha_m = 2.8$, $\sigma_0 = 0.2$) a large range of size-distribution parameters were consistent with the experimental results. They measured a slightly different quantity termed the polarization

$$P(\theta) = \frac{H_h - V_v}{H_h + V_v} \tag{33}$$

$$P(\theta) = \frac{\rho(\theta) - 1}{\rho(\theta) + 1} \tag{34}$$

Note that this is not to be confused with the particle-scattering function.

A critical test of a size-distribution result is to determine how well it fits the data at different wavelengths. Error contour maps of a statistical-fit parameter in $\alpha_m - \sigma_0$ space greatly aid in assessing realistic confidence in the results [86,13].

Similar analysis by means of the polarized Rayleigh ratios $[V_v(\theta), H_h(\theta)]$ has been carried out mainly on nonemulsion systems such as sulfur sols, aerosols, and polystyrene latexes [87-92]. Again, reasonably single-valued fits of the data are possible for fairly monodisperse samples. This method also has the advantage that number concentrations and interfacial areas [90-92] may be determined.

Sloan [76,77], Arrington [78], and Wortz [93] used a different data handling procedure for angular-dependent intensity measurements to deduce the particle size distribution of polydisperse systems. Their data was plotted in the form log $I\theta^2$ versus log $\theta$.

Kaye [94,95] has developed a low angle laser light scattering device which accurately measures absolute intensities in the forward direction to very low angles. It is desirable to work with this portion of the scattering diagram since $I(\theta)$ is monotonic in particle diameter up to about 10 μm, but only for very small angles (several degrees or less).

Other related methods for the determination of average particle size are collected and discussed in Kerker [13]. These include using the extrema in various functions listed above and the phase difference of H- and V-analyzed light when the incident light is polarized obliquely to the scattering plane.

### 2. *Spectral Methods*

The second class of light-scattering methods for determining either average sizes or size distributions consists of determining the spectral variations in the light-scattering functions already discussed. $R(\theta)$ is monotonic up to $\alpha \approx 2$, whereupon this function begins to oscillate. Thus, for larger values of $\alpha$, single measurements of $R(\theta)$ at one angle will not uniquely determine size. However, measurements at two wavelengths can pinpoint a unique size. This is also true of the various light-scattering functions as discussed in Sec. III.A.

Stevenson et al. [41] developed the method of determining particle size distribution parameters from the spectra of the scattering ratio (depolarization). In the notational scheme developed in this chapter their scattering ratio $\sigma$ would be defined as

$$\sigma = \frac{U_h}{U_v} \tag{35}$$

that is the observing channel has no analyzer but the incident light is polarized either parallel (h) or perpendicular (v) to the scattering plane. For homogeneous isotropic spheres $\sigma(\theta) = \rho(\theta)$. They assumed a unimodal, two-parameter, positively skewed distribution function, such as is commonly found in emulsions. This method is sensitive enough to compete with electron microscopic histograms [96] and if measured over a range of 1000 Å in $\lambda$ can give reasonable assurance as to the form of the distribution function as well as determining its parameters [41].

### 3. *Comparison of Methods*

The various light-scattering methods have been compared in a number of studies [97-101].

The comment has been often made that sizes determined from light scattering are smaller than those from microscopic (optical or electron) investigations [102]. In most cases light-scattering measurements sample a higher moment of the particle size distribution function and would in principle be expected to yield a *larger* average size. Practically speaking, however, the effects of polydispersity on the form of the angular scattering pattern, with the tendency to flatten extrema, could lead to interpretations of a smaller average size by some techniques.

It has been recently shown, however, that particle size determination of a polystyrene latex by a variety of light scattering techniques agrees within experimental error with the electron-microscopic average size [101]. Initial discrepancies have been reconciled by a recalibration of the EM system.

A condensed summary of scattering techniques is presented in Appendix 3.

Most emulsion systems provide a fairly idealized system of homogeneous isotropic spheres for which the theory is well developed. Friberg and Mandell [57, 58], using microscopic investigation in polarized light, have shown that some emulsion systems are stabilized by a liquid crystalline outer shell. The exact solution of the light scattering problem for stratified spheres, analogous to that of Mie, was first derived by Aden and Kerker [103], then by Güttler [104]. Approximate solutions in the Rayleigh-Debye-Gans region have also been formulated [105]. These liquid crystalline-encapsulated emulsions would in principle be amenable to analysis with a concentric-sphere model.

## V. Reflectance Techniques

Lloyd [106] demonstrated an empirical relationship between the reflectance R of an O/W emulsion whose dispersed phase had been dyed, and the surface-average particle diameter $D_s$ or, equivalently, the specific interfacial area $A_{sp}$

$$\log R = -k \log D_s + \log k' \tag{36}$$

$$\log R = k \log A_{sp} + \log k'' \tag{37}$$

where the k are constants dependent upon dispersed phase concentration, absorption coefficient of the droplets, relative refractive index, and incident wavelength. The above relation was seen to hold at wavelengths partially absorbed by the colored disperse phase (in this case 450-nm light was used). This method was used as a diagnostic of emulsion stability and coalescence kinetics.

The Bragg reflection of light from ordered arrays of colloidal particles has been used to determine particle size, lattice parameters, and phase transitions in the lattice structure [71,107,108]. The mother-of-pearl-like iridescence seen in high concentrations of latex particles is a manifestation of this diffraction effect.

In most cases emulsions will undergo a phase inversion when the dispersed-phase concentration becomes too high; however, there are a number of emulsion systems stable enough to approach close-packed structures. While polymer latices form excellent hard sphere model systems with which to study order-disorder transitions, emulsion systems would include the additional variable of particle flexibility. Certainly cluster structures in emulsion systems have been seen by microscopic investigation. The Bragg diffraction technique can be used to study the interparticle forces involved.

## VI. Spectroscopic Techniques

Traditional spectroscopy finds practical use in monitoring chemical composition which affects emulsion stability. For example, Penney et al. [109] used infrared absorbance analysis to control and correct emulsion composition in an industrial process. Since, in emulsions, or actually in any colloidal system, a substantial amount of material exists in an interfacial situation, spectroscopic studies may be used for structural studies at interfaces.

Larsson [110] studied the structure at triolein/water/monostearin interfaces with Raman spectroscopy.

Many oils have fluorescent chromophore groups. This property has been used to identify emulsion type by microscopic investigation [102]. Depolarization of fluorescence measurements have been used to determine micellar properties, such as critical micelle concentration and local internal viscosity [111].

Eicke and Zinsli [112] have used nonosecond spectroscopy to investigate the structural details of W/O microemulsion interfaces. Rotational correlation times were determined by monitoring the polarization anisotropy as a function of time.

Light-beating (correlation) spectroscopy, also termed *dynamic light scattering*, has been used by Graciaa et al. [113] to measure the translational diffusion coefficient of monodisperse charged microemulsions. In this type of spectroscopy monochromatic coherent light which is inelastically scattered from a system is recombined either with itself, of with unshifted (in frequency) light. Beat frequencies resulting from this superposition on a square-law detector (i.e., photomultiplier) can be analyzed by conventional audio and radio-frequency techniques. Beat frequencies in this range derive from Doppler modulation of the scattered beam, which for the case of isotropic emulsions is due to thermally initiated translational motion. It is convenient to analyze the autocorrelation function of the photomultiplier signal and relate this to diffusion coefficients [114,115]. Emulsion charge affects interparticle correlations and cannot be neglected. A fiber optic Doppler anemometer (FODA) developed by Dyott [116] has been used on silica dispersions [117] and should be applicable to emulsion systems.

An intriguing phenomenon, though not used in emulsion analysis, is the structural colors present in some emulsions. These were first observed in emulsions by Bodroux [118] and are of the same sort discovered by Christiansen [119] in quartz dispersions. For the case of emulsions the disperse and continuous phases have the same refractive index at one wavelength, though their dispersions ($dn/d\lambda$) are different. When a beam of white light is passed through the emulsion, the transmitted and scattered light appear colored. Diffuse incident light tends to wash out the effect. Francis [120] has made a study of a variety of these kinds of emulsions. The effect is not simply prismatic since only complementary colors are seen. The disperse/continuous phase optical interface is transparent for one color.

This color is freely transmitted through the emulsion. However, because of the dispersion differences and hence refractive index differences between materials, the other colors will be reflected in varying degrees by the interface. The appearance of the emulsion when viewed from the side of the beam then is the full spectrum minus the color for which the emulsion is optically homogeneous. The colors seen will change with composition and temperature. Francis [120] gives a variety of formulations.

## VII. Summary

In this section we present some guidelines for choosing the optimal optical techniques for investigating emulsions. There are some obvious initial investigations to be made with the naked eye. These include the overall appearance as viewed by the scattered light and the transmitted light, absorbancies of the individual components, and the appearance of the emulsion as it is diluted in the continuous phase.

Table 1 lists some interpretations of particle size range in terms of appearance. Caution is suggested. Highly turbid samples can wash out a distinguishing appearance, owing to multiple scattering. This can also be true of very polydisperse samples. A collimated light source is best for observations, since diffuse light will tend to wash out characteristic appearances. If white light is used and the colored fringes of the HOTS are visible then the average particle size in micrometers is likely to be in the range of 1/10 the number of red or green orders seen between $\theta = 0$ and 180°. This generalization applies in the range 0.1 to 0.8 $\mu$m.

If the 90° scattered light is polarized almost completely perpendicular to the scattering plane, then the particles fall in the Rayleigh range ($a < \lambda/20$).

For the unusual case of a chromatic emulsion which appears the same color at all angles and whose transmitted color is the complement of the scattered color, the Christiansen effect is implied. This occurs when the refractive indexes of dispersed and continuous phase are matched at one wavelength.

Microemulsions, while transparent, will scatter more light than a binary or ternary liquid. Even qualitative differences may well need a photometer to be measured.

There are several physical methods for distinguishing between the emulsion types O/W and W/O [102]. Optical methods for this purpose include observing the diffusion of an oil- or water-soluble dye placed gently on the surface of the emulsion. For example, an oil-soluble dye will diffuse into a W/O emulsion, but will remain on the surface for an O/W emulsion. Another technique is to observe a parallel beam going through an emulsion of low turbidity; convergence usually means O/W and divergence W/O (owing to the relative refractive index of most oils and water [121]). The interpretation of all visual information also strongly depends on inherent absorbancies of the individual phases in the bulk.

Microscopic investigation allows a more quantitative assessment of particle size distribution. Even in a cursory investigation one can often tell something about the average particle size and whether the distribution is uni- or multimodal, and something about the breadth of polydispersity. Quantitative determination requires as much effort as constructing the entire size histogram. The various semiautomatic and computerized methods for reducing the tedium of particle classification and counting have been discussed above. The simplification of usually having only spherical particles in emulsions reduces the classification problem, though computerized techniques are not in wide use in emulsion science. Static or only slowly varying systems can be observed by microscopy. This proves adequate, however, for even many unstable emulsions. The various forms of microscopy (transmission, phase contrast, ultramicroscopy) need to be considered for optimal viewing. Particles smaller than about 0.25 μm will require photometric measurement for quantitative determinations. Above this size limit, size-dependent images are resolvable. The observation of emulsions under polarized light can reveal liquid crystalline regions in a sample, which is of special interest in some systems stabilized by such layers around the emulsion particles.

Photometric intensity measurements are divided into transmission and scattering methods. For particles in the Rayleigh size range, distinguished by a $1 + \cos^2 \theta$ dependence of unpolarized scattered light and complete polarization of the light scattered at 90°, turbidimetric and scattering measurements are equivalent. Scattering measurements are better for less turbid samples (say if $< 2\%$ of the light is scattered) since the ratio of signal to background is better.

Larger particles, evidenced by a dissymmetry substantially different from unity, can be analyzed by a variety of techniques. The effects of multiple scattering, determined by concentration dependence of the measured quantity, must be assessed in any case. It is usually preferred to work in the independent particle realm, though empirical and computational methods for dealing with multiple-scattering situations are available. Many times these effects would be indistinguishable from those due to sample polydispersity. Interparticle interaction, especially in the case of charged emulsions must also be assessed, either by dilution or by changing the continuous phase dielectric constant. This latter approach often leads to other consequences in the chemistry of the situation.

Most turbidity and scattering parameters are monotonic functions of particle size up to $\alpha = 2\pi a/\lambda \approx 2$ (or greater for lower relative refractive index). To determine a unique size above this range, the quantity is usually measured at several wavelengths.

The specific turbidity and depolarization (also termed *polarization ratio* or *scattering ratio*, and related to the polarization) and their spectral dependence have been successfully used in the submicrometer-particle-size range and larger. The angular variation of various quantities [$R(\theta)$, $V_v(\theta)$, $H_h(\theta)$, $\rho(\theta)$] give characteristic fingerprint patterns in the Mie region. Complete analysis often requires machine computation and fitting, though simpler methods based on the position of extrema and the limiting slope of the scattering pattern at very low angles have been devised.

Most of the scattering methods mentioned above have been used to assess polydispersity by first assuming a distribution function, then obtaining two-parameter fits (e.g., a modal size and a measure of breadth) to the data. For larger particles, angular variation data usually require machine computations for fitting, while spectral variation data can be graphically analyzed. For smaller particles (Rayleigh-Debye-Gans region and below) the extraction of particle size distribution parameters using these procedures is more difficult.

The application of centrifugal fields to the emulsion can provide a more direct determination of particle size distribution without assuming a functional form. Most of these devices at the present time use turbidimetric types of analysis.

The more conventional technique of Schlieren optics in standard centrifugation experiments has also been used for microemulsion systems [122].

For less involved data analysis one obtains averaged quantities which are various moments of the size distribution. The most definitive work has been done with unimodal distributions. The type of average depends on the quantity measured and in some cases the size range of the dispersed phase, detailed in the above sections.

Perhaps the greatest advantage of photometric over microscopic techniques is the rapidity of data analysis, given the currently available instrumentation. Turbidimetric, scattering, and reflectance measurements can be used in a real-time analysis of kinetic systems which are rapidly changing.

Most conventional spectroscopic investigations can be applicable insofar as the chemical factors affecting emulsion stability are known, and are also useful in determining molecular properties at interfaces.

Correlation spectroscopy (dynamic light scattering) requires more instrumentation then conventional light scattering, but yields a measure of the translational diffusion coefficient, which in the case of spherical emulsions is readily interpretable.

The major thrust of optical measurements of emulsions then is the determination of particle size distribution or average size. This information for a static emulsion relates to rheological, interfacial, and energy absorption properties, and appearance. For a dynamic system the factors affecting emulsion stability and the kinetics of flocculation and coalescence may be determined.

**Table 2.** Optical Measurements and Major Physical Effects

| | Amplitude | Phase | Polarization | Frequency |
|---|---|---|---|---|
| Transmission | Turbidity and Spectrophototurbidity | Phase contrast microscopy | Birefringence | Absorption at specific $\lambda$ |
| | Microphotometry | | | |
| | Conventional microscopy | Ultramicroscopy | | |
| | Rayleigh-Debye-Gans | | | |
| Scattering | Rayleigh Mie | | Depolarization | HOTS |
| | | | | Fluorescence |
| | | | | Raman |
| | | | | Correlation spectroscopy |
| | | | | Christiansen effect |
| Reflected | Reflection intensity | Bragg reflection | | |

Table 2 shows the various types of optical measurements on emulsions and groups them according to the major physical effect on the properties of the incident light wave.

In the Appendix we have compiled a condensed outline of the measured quantities, calculated parameters, general equations, methods of data analysis, and reference sources for the major kinds of photometric measurements.

Figure 1 depicts the array of optical investigations possible.

## Appendix

### 1. Turbidity

*Measure*

i. $\tau/c$ versus c, extrapolate to c = 0;

$$\frac{\tau}{c} = \frac{cm^{-1}}{g\ cm^{-3}} = \frac{cm^2}{g} \text{ specific turbidity}$$

ii. Relative refractive index or dn/dc

*Calculate Average Particle Size*

i. Volume-surface mean radius

$$a_{VS} = \frac{\int_0^\infty p(a)a^3\, da}{\int_0^\infty p(a)a^2\, da}$$

ii. Turbidity average (Rayleigh particles)

$$a_\tau = \left[\frac{\int_0^\infty p(a)a^6\, da}{\int_0^\infty p(a)a^3\, da}\right]^{1/3}$$

*General Equations (Nonabsorbing Spheres)*

$$\frac{\tau}{c} = \frac{3}{4\rho_2} \frac{\overline{Q}_{sca}}{a_{VS}} \tag{A.1}$$

$$\overline{Q}_{sca} = \frac{\int_0^\infty Q_{sca} p(a)a^2\, da}{\int_0^\infty p(a)a^2\, da} \tag{A.2}$$

$$Q_{sca} \begin{cases} \dfrac{32\pi^3 a}{\lambda^4} V \left(\dfrac{m^2-1}{m^2+2}\right)^2 = \dfrac{8}{3}\alpha^4\left(\dfrac{m^2-1}{m^2+2}\right)^2 & \text{(Rayleigh)} \\ \dfrac{2}{\alpha^2}\sum_{n=1}^{\infty}(2n+1)(|a_n|^2+|b_n|^2) & \text{(Mie)} \end{cases} \tag{A.3}$$

$$\frac{\tau_E}{c} = \frac{32\pi^3 n^2}{3N_A\lambda_0^4}\left(\frac{dn}{dc}\right)^2 M \qquad \text{(Rayleigh)} \tag{A.4}$$

$$p(a) = c(a - a_0)\exp - \left(\frac{a-a_0}{s}\right)^3 \qquad \text{(Stevenson et al. [41])}$$

$$\left.\begin{aligned} p(a) &= \frac{1}{\sqrt{2\pi}\,\sigma a}\exp\left[\frac{-(\log a - \log a_m)^2}{2\sigma^2}\right] && \text{log-normal} \\ p(a) &= \frac{1}{\sqrt{2\pi}\,\sigma a_m}\exp\left(\frac{\sigma^2}{2}\right)\exp\left[\frac{-(\log a - \log a_m)^2}{2\sigma^2}\right] && \text{ZOLD} \end{aligned}\right\} \text{equivalent forms} \tag{A.5}$$

*Methods*

i. Compare experimental quantity $(4\rho_2/3)\tau/c$ with curve of $\overline{Q}_{sca}/a_{VS}$ versus $a_{VS}$.

ii. Experimental plot of log $[(4\rho_2/3)\tau/c]$ versus log $[4\pi(m-1)/\lambda]$ for at least two wavelengths is overlaid on theoretical plot of log $Q_{sca}$ versus log $2\alpha(m-1)$. The displacement of either axis gives log a.

iii. To determine type of average, plot log $(\tau/c)$ versus log $\alpha$. If value of slope = m, then

$$\langle a \rangle = \frac{\int_0^\infty p(a)a^{m+1}\,da}{\int_0^\infty p(a)a^m\,da}$$

$$\frac{\tau}{c} = k_1\alpha^{\overline{m}}$$

m is usually 2 to 3 for most emulsions.

*Sources of Numerical Computations and Graphical displays*

i. Kerker [13]

ii. Van de Hulst [14]

iii. Lothian and Chappel [15]

iv. Wickramasinghe [16]

2. Turbidity Spectra

*Measure*

i. $\tau/c$ versus $\lambda$ in high dilution
ii. Maximum in $\tau/c$ versus $\lambda$
iii. Relative refractive index
iv. Dispersion of refractive indices

*Calculate*

i. Average particle size
ii. Size-distribution parameters

*General Equations (See Sec. III.A)*

$$-\frac{d[\ln(\tau/w)]}{d(\ln\lambda_0)} = g_0$$

$$g_0 = 4\left(1 - \frac{\lambda_0}{n_2}\frac{dn_2}{d\lambda_0}\right) - \frac{12m}{(m^2-1)(m^2+2)}\frac{\lambda}{n_2}\left(\frac{dn_1}{d\lambda_0} - m\frac{dn_2}{d\lambda_0}\right) \quad \text{(Rayleigh)}$$

$$= m\left[1 - \frac{d(\ln n_2)}{d(\ln\lambda_0)}\right] - \frac{dm}{d(\ln\lambda_0)}\left[\frac{12m}{(m^2-1)(m^2+2)} + \frac{48\alpha^2 m}{5(m^2+2) - 6(m^2-2)\alpha^2}\right] \quad \text{(Stevenson)}$$

$$= \left\{\alpha\frac{\partial[\ln\sum_{n=1}^{\infty}(2n+1)(|a_n|^2+|b_n|^2)]}{\partial\alpha} - 2\right\}\left[1 - \frac{d(\ln n_2)}{d(\ln\lambda_0)}\right] - \left\{\frac{d[\ln\sum_{n=1}^{\infty}(2n+1)(|a_n|^2+|b_n|^2)]}{dm}\right\}\frac{dm}{d(\ln\lambda_0)} \quad \text{(Mie)} \tag{A.7}$$

$$\tau = \frac{\lambda^2}{2\pi}\int_{a_0}^{\infty}\left[\sum_{n=1}^{\infty}(2n+1)(|a_n|^2+|b_n|^2)\right]p(a)\,da \tag{A.8}$$

$$\frac{\lambda\tau}{\phi} = \frac{\lambda\tau}{\int_{a_0}^{\infty}(4\pi a^3/3)p(a)\,da} \tag{A.9}$$

*Methods*

i. Plot experiment values of $(\lambda\tau/\phi)_\lambda/(\lambda\tau/\phi)_{\lambda_R}$ versus $\lambda$ and compare with theoretical plots of same (size-distribution parameters being varied).

ii. Compute wavelength exponent of the specific turbidity $g_0$ and compare with theoretical values. Good for monodisperse emulsions or will give averaged size (see Sec. III.A).

iii. Use $\tau/c$ at two wavelengths to determine unique averaged size from theoretical plot of $\tau/c$ versus $\alpha$.

iv. Compare first maximum in theoretical curve of $Q_{sca}$ versus $\alpha$ with corresponding maximum in experimental curve of OD versus $\lambda$. Neglecting dispersion,

$$a = \lambda_{max} \frac{\alpha_{theor}}{2\pi}$$

$$N = \frac{2.3OD}{Q_{sca} \pi \ell a^2}$$

v. Compare dispersion quotient $DQ = \tau_1 \lambda_2^2 / \tau_2 \lambda_1^2$ with theoretical plot of DQ versus $\alpha$ in Rayleigh range.

*Sources of Computations*

i. Kerker [13]
ii. Wallach et al. [35]
iii. Heller et al. [34]
iv. Heller [33]
v. Meehan and Beattie [21]

3. Scattering Techniques

*Measure*

i. Angular variation of
   a. Rayleigh ratio, $R(\theta)$
   b. Polarized Rayleigh ratios, $V_v(\theta)$, $H_h(\theta)$
   c. Depolarization ratio, $\rho(\theta) = H_h(\theta)/V_v(\theta) = \sigma(\theta)$, the scattering ratio
   d. Polarization, $P(\theta) = (H_h - V_v)/(H_h + V_v) = [\rho(\theta) - 1]/[\rho(\theta) + 1]$
ii. Spectral variation of above quantities
iii. Dissymmetry, $Z(\theta) = R(\theta)/R(180 - \theta)$
vi. Angular positions of maxima and minima
v. HOTS
vi. Relative refractive index or dn/dc

*Calculate*

i. Average particle size
ii. Radius of gyration (number and Z averages), $R_g$

iii. Particle-size-distribution parameters
iv. Particle interference factor, $P(\theta)$
v. Sloan-Arrington method: $\log I\theta^2$ versus $\log \theta$
vi. Interfacial area
vii. Particle concentration

*General Equations*

$$R(\theta) = I(\theta)\, Nr^2 \tag{A.10}$$

$$V_v = 2V_u \qquad H_h = 2H_u \tag{A.11}$$

$$I_u(\theta) = V_u(\theta) + H_u(\theta) \tag{A.12}$$

$$\langle R_g^2 \rangle_n = \frac{\int_0^\infty R_g^2(a) p(a)\, da}{\int_0^\infty p(a)\, da} \tag{A.13}$$

$$\langle R_g^2 \rangle_Z = \frac{\int_0^\infty R_g^2(a) M^2 p(a)\, da}{\int_0^\infty M^2 p(a)\, da} \tag{A.14}$$

$$\frac{H_h}{V_v} = \frac{H_u}{V_u} = \frac{U_h}{U_v} \text{ (isotropic spheres)} \tag{A.15}$$

Rayleigh region (Rayleigh ratios):

$$V_v + H_h = \frac{N\, 16\pi^4 a^6}{\lambda^4} \left(\frac{m^2 - 1}{m^2 + 2}\right)^2 (1 + \cos^2 \theta) \tag{A.16a}$$

$$R(\theta) = \frac{3}{16\pi} \tau\, (1 + \cos^2 \theta) \tag{A.16b}$$

Rayleigh-Debye-Gans region $[4\pi a(m - 1)/\lambda \ll 1]$:

$$R_{RDG}(\theta) = R_{Rayleigh}(\theta)\, P(\theta) \tag{A.17a}$$

For spheres,

$$P(\theta) = \left[\frac{3}{u^3} (\sin u - u \cos u)\right]^2 \qquad u = \frac{4\pi a}{\lambda} \sin \frac{\theta}{2} \tag{A.17b}$$

$$R_g^2 = -\left[\frac{dP(\theta)}{d[\sin^2(\theta/2)]}\right]_{\theta \to 0} \frac{3\lambda^2}{16\pi^2} \tag{A.17c}$$

For spheres,

$$R_g = \left(\frac{3}{5}\right)^{1/2} a \tag{A.17d}$$

Mie region (N particles):

$$V(\theta) = \frac{N\lambda^2}{4\pi^2} \left| \sum_{n=1}^{\infty} \frac{(2n+1)}{n(n+1)} (a_n \pi_n + b_n \tau_n) \right|^2$$

$$H_h(\theta) = \frac{N\lambda^2}{4\pi^2} \left| \sum_{n=1}^{\infty} \frac{(2n+1)}{n(n+1)} (a_n \tau_n + b_n \pi_n) \right|^2$$

*Methods*

i. Position of extrema in $R(\theta)$, $V_v(\theta)$, $H_h(\theta)$, $P(\theta)$, $\rho(\theta)$.
ii. Position of colors with incident white light (HOTS method)
iii. Zimm plot, $Kc/R(\theta)$ versus $\sin^2 \theta/2 + gc$:

$$K = \frac{2\pi^2 n^2}{\lambda_0^4 N_A} \left(\frac{\partial n}{\partial c}\right)_{T,P}^2 (1 + \cos^2 \theta)$$

$g$ = arbitrary constant

iv. Plot $\log I\theta^2$ versus $\log \theta$ (Sloan-Arrington).
v. Match theoretical curves with $R(\theta)$, $V_v(\theta)$, $H_h(\theta)$, $\rho(\theta)$, and $P(\theta)$. Usually requires a computer to determine best fit of a distribution of sizes in the Mie range.
vi. For monodispersions, Rayleigh ratios or $\rho(\theta)$ at two wavelengths can give a unique size (N must be known if Rayleigh ratios are used).

*Sources of Computations (Mie Functions)*

i. See Kerker [13], p. 78.
ii. Yanjik et al. [68].

## Symbols

| | |
|---|---|
| $A$ | surface area of particle; empirical constant |
| $A_{sp}$ | specific surface area per volume of dispersed phase |
| $a_{VS}$ | volume-surface mean radius |
| $a$ | radius of a sphere |
| $\alpha$ | $2\pi a/\lambda$, size parameter |
| $a_\tau$ | turbidity average radius |
| $\beta$ | empirical constant |
| $C_{abs}$, $C_{sca}$ | absorption and scattering cross sections |
| $c_m$ | molar concentration |
| $c$ | concentration g $cm^{-3}$ |

| | |
|---|---|
| $\varepsilon$ | molar extinction coefficient |
| g | concentration in g $g^{-1}$ |
| h | incident light polarized parallel to scattering plane |
| H | viewing optics polarized parallel to scattering plane |
| $I(\theta)$ | intensity scattered at angle $\theta$ |
| $i$ | $\sqrt{-1}$; as subscript implies ith species |
| $I_\ell$ | intensity at sample depth $\ell$ |
| $I_0$ | incident intensity |
| $I'_0$ | intensity transmitted through continuous phase |
| I | intensity transmitted through emulsion |
| $k, k_1, k_2, k'_2, k', k''$ | experimental constants |
| $\lambda_0$ | vacuum wavelength |
| $\lambda$ | wavelength of light in medium |
| $\ell$ | sample depth; distance |
| m | relative refractive index (dispersed/continuous phase) |
| M | molecular weight of scattering entity or dispersed-phase constituent |
| N | number of particles per cubic centimeter |
| $n_2$ | refractive index of dispersed droplet |
| n' | real part of refractive index |
| n'' | imaginary part refractive index |
| n | refractive index of emulsion; indexing parameter |
| $n_1$ | refractive index of continuous phase |
| p(a) | particle radius distribution function |
| $Q_{abs}$, $Q_{sca}$ | absorption and scattering efficiency factors |
| $\pi_n$ | Mie factor |
| $\phi$ | volume fraction of dispersed phase |
| $\rho$ | density |
| $R(\theta)$ | Rayleigh ratio |
| r | detector distance from scattering volume |
| $\tau$ | turbidity |
| $\tau_0$ | infinite dilution value of turbidity |
| $\theta$ | scatterning angle |
| U | unpolarized viewing optics |
| u | unpolarized incident light; $(4\pi a/\lambda)\sin(\theta/2)$ |
| V | volume of emulsion particle; viewing optics polarized perpendicular to scattering plane |
| v | incident light polarized perpendicular to scattering plane |
| $Z(\theta)$ | dissymmetry ratio |
| Z | subscript, Z average |

| | |
|---|---|
| $\mu$ | optical constant |
| B | empirical constant |
| $\rho(\theta)$ | polarization ratio; depolarization |
| $P(\theta)$ | interference factor |
| $R_g$ | radius of gyration |
| n | subscript, number average |
| $a_n$ | Mie coefficient |
| $b_n$ | Mie coefficient |
| $\tau_n$ | Mie factor |
| $a_0$ | lower limit particle radius |
| s | distribution width parameter |
| $\alpha_m$ | modal value of $\alpha$ |
| $\sigma_0$ | log standard deviation of particle size |
| $P(\theta)$ | polarization |
| $\sigma$ | scattering ratio |
| R | reflectance |
| $D_s$ | surface-average particle diameter |

## References


1. K. J. Lissant, in *Emulsions and Emulsion Technology* (K. J. Lissant, ed.), Marcel Dekker, New York, 1974.
2. L. M. Prince, in *Emulsions and Emulsion Technology* (K. J. Lissant, ed.), Marcel Dekker, New York, 1974.
3. W. C. McCrone and R. I. Johnson (eds.), *Techniques, Instruments and Accessories for Microanalysts: a User's Manual,* W. C. McCrone, Chicago, Ill., 1974.
4. G. Clark (ed.), *The Encyclopedia of Microscopy,* Reinhold, New York, 1961.
5. M. Pluta, in *Advances in Optical and Electron Microscopy,* Vol. 6 (R. Barer and V. E. Cosslett, eds.) Academic, New York 1975.
6. (a) F. H. Smith, D. S. Moore, and D. J. Goldstein, in *Advances in Optical and Electron Microscopy,* Vol. 6 (R. Barer and V. E. Cosslett, eds.), Academic, New York 1975; (b) J. A. Davidson and H. S. Haller, *J. Colloid Interface Sci. 55*:170 (1976).
7. B. V. Derjaguin and G. Ja. Vlasenko, *J. Colloid Interface Sci. 17*:605 (1962).
8. P. McFayden and A. L. Smith, *J. Colloid Interface Sci. 45*:573 (1973).
9. D. W. Humphries, in *Advances in Optical and Electron Microscopy,* Vol. 3 (R. Barer and V. E. Cosslett, eds.), Academic, New York, 1969.
10. M. L. Mendelsohn, B. H. Mayall, J. M. S. Prewitt, R. C. Bostrom and W. G. Holcomb, in *Advances in Optical and Electron Microscopy,* Vol. 2 (R. Barer and V. E. Cosslett, eds.), Academic, New York, 1968.
11. K. J. Preston, Jrs., in *Advances in Optical and Electron Microscopy,* Vol. 5 (R. Barer and V. E. Cosslett, eds.), Academic, New York, 1973.
12. G. Mie, *Ann. Physik 25*:377 (1908).
13. M. Kerker, *The Scattering of Light and Other Electromagnetic Radiation,* Academic, New York, 1969.

14. H. C. van de Hulst, *Light Scattering by Small Particles*, Wiley, New York, 1957.
15. G. F. Lothian and F. D. Chappel, *J. Appl. Chem. 1*:475 (1951).
16. N. C. Wickramasinghe, *Light Scattering Functions for Small Particles With Applications in Astronomy*, Wiley, New York, 1973.
17. B. Verner, M. Bârta, and B. Sedlácek, *Tables of Scattering Functions for Spherical Particles*, Macro, Prague, 1976.
18. B. Verner, M. Bârta, and B. Sedlácek, *J. Colloid Interface Sci. 62*:348 (1977).
19. P. Chýlek, J. T. Kiehl, and M. K. W. Ko, *J. Colloid Interface Sci. 64*:595 (1978).
20. G. L. Bayer in *Techniques of Organic Chemistry* (A. Weissberger, ed.) Vol. 1, Part 1, Wiley-Interscience, New York, 1959, p. 191.
21. E. J. Meehan and W. H. Beattie, *J. Phys. Chem. 64*:1006 (1960).
22. R. A. Dobbins and G. S. Jizmagin, *J. Opt. Soc. Amer. 56*:1345, 1351 (1966).
23. G. E. Langlois, J. E. Gullberg, and T. Vermeulen, *Rev. Sci. Inst. 25*:360 (1954).
24. J. H. Schulman and J. A. Friend, *J. Colloid Sci. 4*:497 (1949).
25. A. Graciaa, J. Lachaise, A. Martinez, M. Bourrel, and C. Chambu, *Comptes Rend. Acad. Sci. B282*:547 (1976).
26. M. Kerker (ed.), *Electromagnetic Scattering*, Pergamon, Oxford, 1963.
27. C. Smart, R. Jacobsen, M. Kerker, J. P. Kratohvil, and E. Matijević, *J. Opt. Soc. Amer. 55*:947 (1965).
28. R. L. Rowell and R. S. Stein (eds.), *Electromagnetic Scattering*, Gordon and Breach, New York, 1967.
29. J. A. Cengel, J. G. Knudsen, A. Landsberg, and A. A. Faruqui, *Can. J. Chem. Eng. 39*:189 (1961).
30. R. Tabibian, W. Heller, and J. N. Epel, *J. Coll. Sci. 11*:195 (1956).
31. M. Kerker, *Amer. Perf. Cosm. 85*:43 (1970).
32. W. Heller in *Electromagnetic Scattering* (M. Kerker, ed.) Pergamon, Oxford, 1963.
33. W. Heller, *J. Chem. Phys. 40*:2700 (1964).
34. W. Heller, H. L. Bhatnagar, and M. Nakagaki, *J. Chem. Phys. 36*:1163 (1962).
35. M. L. Wallach, W. Heller, and A. F. Stevenson, *J. Chem. Phys. 34*:1796 (1961).
36. P. Walstra, *J. Colloid Interface Sci. 27*:493 (1968).
37. F. W. Billmeyer, *J. Amer. Chem. Soc. 76*:4636 (1954).
38. G. Deželić, *Croatica Chem. Acta 33*:51 (1961).
39. W. Heller and R. Tabibian, *J. Coll. Sci. 12*:25 (1957).
40. M. L. Wallach and W. Heller, *J. Phys. Chem. 68*:924 (1964).
41. A. F. Stevenson, W. Heller, and M. L. Wallach, *J. Chem. Phys. 34*:1789 (1961).
42. R. O. Gumprecht and C. M. Sliepcevich, *J. Phys. Chem. 57*:95 (1953).
43. A. Graciaa, J. Lachaise, A. Martinez, and E. Kuten, *Comptes Rend. Acad. Sci. B281*:595 (1975).
44. B. H. Kaye, Brit. Pat. 895,222 (1962).
45. B. H. Kaye and M. R. Jackson, *Powder Tech. 1*:81 (1967).
46. M. J. Groves and D. C. Freshwater, *J. Pharm. Sci. 57*:1273 (1968).
47. C. Parkinson, S. Matsumoto, and P. Sherman, *J. Colloid Interface Sci. 33*:150 (1970).
48. N. Fischer, *Polym. Sci. Eng. 14*:332 (1974).
49. M. J. Groves and H. S. Yalabik, *Powder Tech. 17*:213 (1977).

50. J. A. Tempel and M. J. Groves, *Powder Tech.* *9*:147 (1974).
51. M. J. Groves, H. S. Yalabik, and J. A. Tempel, *Powder Tech.* *11*:245 (1975); M. J. Groves, *J. Dispersion Sci. Technol.* *1*:97 (1980).
52. R. J. Gledhill, *J. Phys. Chem.* *66*:458 (1962).
53. R. N. O'Brien in *Techniques in Chemistry, Physical Methods, Pt. III* (A. Weissberger and B. Rassiter, eds.) Interscience, New York, 1971.
54. M. Francon, *Optical Interferometry* (J. Wilmann, transl.) Academic, New York, 1966.
55. R. N. O'Brien, A. I. Feher, and J. Leja, *J. Colloid Interface Sci.* *42*:218 (1973).
56. D. O. Shah and R. M. Hamlin, Jr., *Science*, *171*:483 (1971).
57. S. Friberg, L. Mandell, and M. Larsson, *J. Colloid Interface Sci.* *29*:155 (1969).
58. S. Friberg and L. Mandell, *J. Pharm. Sci.* *59*:1001 (1970).
59. S. Friberg and L. Tydhag, *Kolloid-Z. Z. Polym.* *244*:233 (1971).
60. S. Fukushima, M. Yamaguchi, and F. Harusawa, *J. Colloid Interface Sci.* *59*:159 (1977).
61. R. S. Stein, R. L. Rowell, and H. Brumberger, *Intern. Sci. Tech.* *83*:34-42 (1968).
62. J. P. Kratohvil, *Anal. Chem.* *26*:458R (1964).
63. J. P. Kratohvil, *Anal. Chem.* *38*:517R (1966).
64. H. Utiyama in *Light Scattering from Polymer Solutions* (M. B. Huglin, ed.) Academic, London, 1972.
65. H. Benoit and W. H. Stockmayer, *J. Phys. Radium* *17*:21 (1956).
66. J. C. Brown, P. N. Pusey, J. W. Goodwin, and R. H. Ottewill, *J. Phys.* *A8*:664 (1975).
67. A. F. Stevenson, *J. Appl. Phys.* *24*:1134, 1143 (1953).
68. M. Yajnik, W. Heller and J. Witeczek, *Tables of Angular Scattering Functions for Heterodisperse Systems of Spheres*, Wayne State University, Detroit, Mich., 1969.
69. W. Heller, *J. Chem. Phys.* *42*:1609 (1965).
70. Lord Rayleigh, *Phil. Mag.* *12*:81 (1881).
71. T. Alfrey, Jr., E. B. Bradford, J. W. Vanderhoff, and G. Oster, *J. Opt. Soc. Amer.* *44*:603 (1954).
72. C. Sadron, H. Benoit, and M. Daune, *Int. Symp. Macromolecules*, Milan, Turin, 1954.
73. K. A. Stacey, *Light Scattering in Physical Chemistry*, Butterworths, London, 1956.
74. H. Benoit, A. M. Holtzer, and P. Doty, *J. Phys. Chem.* *58*:635 (1954).
75. A. M. Wims, *J. Colloid Interface Sci.* *44*:361 (1973).
76. C. K. Sloan, Abstracts 125th ACS Meeting, Kansas City, April, 1954.
77. C. K. Sloan, *J. Phys. Chem.* *59*:834 (1955).
78. C. H. Arrington, Abstracts 125th ACS Meeting, Kansas City, April, 1954.
79. M. Born and E. Wolf, *Principles of Optics*, Pergamon, Oxford, 1959.
80. D. Sinclair and V. K. La Mer, *Chem. Rev.* *44*:245 (1949).
81. M. Kerker, W. Farone, and W. Espenscheid, *J. Colloid Interface Sci.* *21*:459 (1966).
82. T. P. Wallace, R. J. Cembrola, A. J. Miglione, and D. E. De Cann, *J. Colloid Interface Sci.* *51*:283 (1975).
83. M. Kerker, E. Matijević, W. Espenscheid, W. Farone, and S. Kitani, *J. Colloid Sci.* *19*:213 (1964).
84. W. W. Graessley and J. H. Zufall, *J. Colloid Interface Sci.* *19*:516 (1964).
85. D. Monk, E. Matijević and M. Kerker, *J. Colloid Interface Sci.* *29*:164 (1969).

86. T. P. Wallace and J. P. Kratohvil, *Polymer Lett.* *5*:1139 (1967).
87. M. Kerker, E. Daby, G. L. Cohen, J. P. Kratohvil, and E. Matijević, *J. Phys. Chem.* *67*:2105 (1963).
88. J. P. Kratohvil, and C. Smart, *J. Colloid Sci.* *20*:875 (1965).
89. E. Matijević, S. Kitani, and M. Kerker, *J. Colloid Sci.* *19*:223 (1964).
90. R. L. Rowell, T. P. Wallace, and J. P. Kratohvil, *J. Colloid Interface Sci.* *26*:494 (1968).
91. A. B. Levit and R. L. Rowell, *J. Colloid Interface Sci.* *50*:162 (1975).
92. R. L. Rowell and R. S. Farinato in *Characterization of Metal and Polymer Surfaces* (L.-H. Lee, ed.), Vol. 2, Academic, New York, 1977.
93. C. G. Wortz, Abstracts 125th ACS Meeting, Kansas City, April, 1954.
94. W. Kaye, *Anal. Chem.* *45*:221A (1973).
95. W. Kaye, *J. Colloid Interface Sci.* *44*:384 (1973).
96. W. Heller and M. L. Wallach, *J. Phys. Chem.* *67*:2577 (1963).
97. Gj. Dežel𝑖ć and J. P. Kratohvil, *Kolloid Z.* *173*:38 (1960).
98. Gj. Deželić and J. P. Kratohvil, *J. Colloid Sci.* *16*:561 (1961).
99. S. H. Maron and M. E. Elder, *J. Colloid Sci.* *18*:107, 199 (1963).
100. S. H. Maron, P. E. Pierce, and M. E. Elder, *J. Colloid Interface Sci.* *18*:391, 733 (1963).
101. R. L. Rowell, R. S. Farinato, J. W. Parsons, J. R. Ford, K. H. Langley, J. R. Stone, T. R. Marshall, C. S. Parmenter, M. Seaver, and E. B. Bradford, *J. Colloid Interface Sci.* *69*:590 (1979).
102. P. Becher, *Emulsions: Theory and Practice,* 2d ed., ACS Monograph No. 162, Reinhold, New York, 1965 (reprinted by Krieger Publishing Co., Huntington, New York, 1977).
103. A. L. Aden and M. Kerker, *J. Appl. Phys.* *22*:1242 (1951).
104. A. Güttler, *Ann. Physik* [6] *11*:65 (1952).
105. M. Kerker, J. P. Kratohvil, and E. Matijević, *J. Opt. Soc. Amer.* *52*:551 (1962).
106. N. E. Lloyd, *J. Colloid Interface Sci.* *14*:441 (1959).
107. I. M. Krieger and F. O'Neill, *J. Amer. Chem. Soc.* *90*:3114 (1968).
108. I. M. Krieger and P. A. Hiltner, in *Polymer Colloids* (R. M. Fitch, ed.) Plenum, New York, 1971.
109. F. R. Penney, I. S. Kolarik, and I. P. Hammer, *Lub. Eng.* *32*:238 (1975).
110. K. Larsson, in *Lipids* (R. Paoletti, G. Porcellati, and G. Jacini, eds.) Vol. 2, Raven, New York, 1976.
111. M. Shinitzky, A. C. Dianoux, G. Gitler, and G. Weber, *Biochemistry* *10*:2106 (1971).
112. H.-F. Eicke and P. E. Zinsli, *J. Colloid Interface Sci.* *65*:131 (1978).
113. A. Graciaa, J. Lachaise, P. Chabrat, L. Letamendia, J. Rouch, C. Vancamps, M. Bourrel, and C. Chambu, *J. Phys. (Paris) Lett.* *38*:253 (1977).
114. B. Chu, *Laser Light Scattering,* Academic, New York, 1974.
115. B. J. Berne and R. Pecora, *Dynamic Light Scattering,* Wiley, New York, 1976.
116. R. B. Dyott, *IEE J. Microwaves Optics Acoust.* *2*:13 (1978).
117. D. A. Ross, H. S. Dhadwal and R. B. Dyott, *J. Colloid Interface Sci.* *64*:533 (1978).
118. F. Bodroux, *Comp. Rend.* *156*:772 (1913).
119. R. W. Wood, *Physical Optics,* Macmillan, New York, 1921.
120. A. W. Francis, *J. Phys. Chem.* *56*:510 (1952).
121. H. Bennett, J. L. Bishop, Jr., and M. F. Wulfringhoff, *Practical Emulsions,* Chemical Publishing Co., Brooklyn, New York, 1968.
122. G. D. Smith, C. E. Donalan, and R. E. Barden, *J. Colloid Interface Sci.* *60*:488 (1977).
123. W. Heller and J. H. McCarty, *J. Chem. Phys.* *29*:78 (1958).

# 9

# Dielectric Properties of Emulsions and Related Systems

MARC CLAUSSE[†] / Université de Pau et des Pays de l'Adour, Pau, France



[†] Present address: Université de Technologie de Compiègne, Compiègne, France

During the past 20 years, the conductive and dielectric properties of emulsions have been reviewed by several authors [1-4]. Reports of data concerning this subject can also be found in books or extensive papers dealing with the dielectric behavior of heterogeneous systems or dielectric relaxation processes in condensed media [5-10].

To avoid tedious unnecessary comments, well-disseminated results will be reported and analyzed briefly, emphasis being put rather on recent theoretical and experimental developments relating to the conductive and dielectric properties of emulsions and closely related systems, such as emulsions undergoing freezing, dispersions of ice microcrystals obtained from water-in-oil-type emulsions, and microemulsions. Consequently, no mention, but incidental necessary quotations, will be made of data concerning colloidal suspensions, gels, sols, solutions of macromolecules, biological systems, liquid crystals, and mixtures of mutually miscible liquids which are not directly relevant to the subject. Analyses of the dielectric properties of systems of these types are available in the current literature, and numerous references can be found in Refs. 3 and 7 to 28.

Moreover, apart from essential notions, no thorough development of the general electromagnetic theory, the description of dielectric mechanisms, and their relation to molecular behavior and structure, and dielectric instrumentation methods and techniques will be made. All these subjects have been treated extensively in classical books and surveys [29-49].

## I. Introduction

In Emulsions: Theory and Practice [1], Becher expressed some surprise about the scarcity of research devoted to the dielectric properties of emulsions, as investigations in this field could bring valuable information concerning emulsion properties. He reported some of the results available at that time, dealing with permittivity and conductivity studies, such as those of Fradkina and others [50-54], Kubo and Nakamura [55], Ben'kovskii [56], Bathnagar [57], Lifshits and Teodorovich [58], and Meredith and Tobias [59]. He quoted as well earlier papers published by Hanai on the subject [60-62].

Since that time, approximately during the past 20 years, the problem has been given much more attention and many papers have been published, giving comprehensive descriptions of the dielectric behavior of emulsions. It is worth pointing out that, in the author's opinion, this progress may be ascribed to the fact that both experimental and theoretical aspects were examined simultaneously and that the investigations were developed in terms of complex permittivity, not in terms of dielectric constant, on the one hand, and of electrical conductivity, on the other hand, as had been done previously. Except for isolated studies [59,63-68], most of the papers available in the recent current literature and dealing specifically with the dielectric properties of emulsions have issued from two institutions, namely, Kyoto University, Japan, and Pau University, France.

Hanai and co-workers were the first to approach the subject extensively. They studied the dielectric and conductive properties of emulsions, either of the oil-in-water (O/W) type or of the water-in-oil (W/O) type, and compared their experimental data [60-62,69-72] to a theoretical formula that had been proposed by Hanai [62,73,74] to represent the complex permittivity of emulsions. Although these authors found that emulsions exhibit dielectric relaxations in the kilohertz frequency range, they concentrated their investigations mainly on emulsion conductive and dielectric properties at both ends of the relaxation frequency domain. The results obtained by Hanai and his co-workers have been published extensively in two review papers, [2,3].

The study of the dielectric relaxation exhibited by emulsions was carried out by Clausse and others [4,75-80]. They analyzed their experimental results by using a computerized procedure designed by Clausse and Royer [4,81,82] to obtain from Hanai's formula, theoretical models for the dielectric behavior of emulsion-type systems [83]. This procedure also was later used by Hanai and co-workers [84, 85] to obtain models for the dielectric behavior of disperse systems.

The good agreement obtained between experiments and theoretical predictions provided a sound basis for applying the same investigation method to the dielectric behavior of W/O-type emulsions undergoing progressive freezing [86-92] and of frozen W/O-type emulsions [93-98], to which earlier studies had been devoted within the framework of research on the dielectric properties of polycrystalline ice [99-114].

Emulsion physical properties, especially the conductive and dielectric ones, can be approached along two different, although complementary, main lines. The first one is based on the fact that emulsions may be regarded as belonging to the field of colloids which covers a multitude of physical systems such as solutions of polymers, micellar solutions, gels, dispersions of solid particles in electrolytic solutions, biological suspensions, and emulsions. All these systems occupy, to different extents, an intermediate position between molecular solutions and coarse heterogeneous materials. This *colloidal system* approach lies rather on the fundamental side as its main concern is the investigation of the structural properties of colloidal systems through the phenomena arising from or associated with the existence of interfaces between the disperse particles and the immersion medium. In their comprehensive review of the subject, Dukhin and Shilov [8] have thoroughly examined different aspects of the dielectric properties of disperse systems, particularly the influence of electrodiffusion processes in the vicinity of interfaces and of electrical double-layer polarization phenomena. They stressed the usefulness of the dielectric method and pointed out as well some difficulties attached to it. Dukhin and Shilov gave a classification of the available theories of the dielectric properties of disperse systems whose components do not exhibit intrinsic dielectric relaxations, with respect to the disperse particle electrisation state and to the product $\kappa\ell$, where $\ell$ is the particle linear dimension and $\kappa$ the inverse of the Debye screening radius, which is a measure of the thickness of the diffuse double layer surrounding the particles. In the case of systems whose disperse particles are uncharged and large enough to assume that $\kappa^{-1}$ is negligible compared with $\ell$ ($\kappa\ell \gg 1$), the Maxwell-Wagner-Sillars (MWS) model is applicable, as it postulates that polarization-free charges are generated at the interfaces in a region which is infinitesimally thin [115-118]. If this is not the case, the MWS model must be replaced by an amended version designed by Trukhan [119] who took into consideration the case of arbitrary values of $\kappa$. For systems containing particles bearing a net electrical charge, more sophisticated models should be used, such as those designed by O'Konski ($\kappa\ell \gg 1$, no diffusion) [14,120], Shilov and Dukhin ($\kappa\ell \gg 1$, diffusion) [121,122,123], Debye and Falkenhagen ($\kappa\ell \ll 1$, diffusion) [124, 125].

As pointed out by Dukhin [7], common emulsion type systems fall within the limits of applicability of the MWS model, since the disperse particles are generally a few micrometers in diameter, which allows us to assume that $\kappa\ell \gg 1$ when either the disperse phase or the continuous phase, depending on the case, is fairly conducting. Consequently, if surface specific phenomena can be disregarded at first approximation, the study of emulsion dielectric properties can be approached from the "composite material" standpoint. This is particularly relevant because of the multiple possible applications of emulsions in various fields of industry which concern, to give but a few examples, the manufacturing of foods, beauty products and drugs, paints and coatings, agricultural sprays, drilling fluids and polymeric materials, the transport of dangerous substances, the efficiency of detergents, and even the storage and restitution of heat, as it has been recently suggested by Babin and co-workers [126]. This composite material approach toward the understanding of emulsion physical properties lies rather on the technological side, as part of the ever-growing concern in both theoretical and experimental studies of heterogeneous substances which are of great importance in many fields of technology and industry. However, it has been followed as well by fundamental investigators who used emulsions as a convenient technique to investigate the thermodynamic metastability of liquids and their breakdown [68,127-179]. As concerns systems involving aqueous phases, a review paper on the subject of metastability, contributed by Angell, is to appear in a forthcoming volume of *Water: A Comprehensive Treatise,* edited by Franks [180].

The colloidal side of disperse systems, including emulsions, having been treated recently in a comprehensive way by Dukhin and Shilov [8], the present contribution will be concentrated on the composite material aspect of emulsions. In that respect, dielectric studies appear to be a powerful method of investigation which has attracted the attention of many scientists and technologists.

In his book on aqueous dielectrics, Hasted [9] indicates that, in competition with other techniques such as neutron scattering, infrared spectroscopy and ultrasonic spectronomy, dielectric studies have an important role to play in earth sciences. For instance, electrical conductometry and dielectrometry can deliver valuable information concerning the structure, composition, and moisture content of the ground and of soils, thus helping us to find correct answers to technological problems in several branches of engineering, such as the grounding of aerials and electrical network poles, the propagation of radio waves through the ground and the adjacent atmosphere, prospecting for underground mineral veins and water reservoirs, and the packing of snow and age sampling of ice in glaciers and polar regions [181-194].

Since the determination of moisture content of materials is a major concern in economically important branches of industry, many dielectric investigations have been devoted to this problem, and technical procedures and apparatus have been designed which enable one to get quick and reliable indications of moisture content through complex permittivity measurements. Reports of data in this field, concerning building materials, textiles, casting resins, detergents, fertilizers, and foods are to be found in Refs. 195-220. In connection with this aspect of dielectric studies, it is worth mentioning that dielectrometry has been used to investigate, from a fundamental point of view, the structural properties of adsorbed or crystallization water in solids, either of inorganic or of organic and even biological nature [221-236].

Of particular relevance to the subject of emulsion dielectric behavior is the dielectric determination of water content of oil and oil derivatives [237-246], the resolution of W/O-type emulsions by electrical treatment [247], the evaluation of emulsion stability by permittivity measurements [248], and the study of the dielectric properties of dispersions, in connection with that of their rheological properties, with the view of investigating particle agglomeration and reticulation in thixotropic and dilatant systems [249-272].

To add a complementary touch to this overview of the dielectric properties of heterogeneous systems, as applicable to practical problems, it may be indicated that electrical conductometry and dielectrometry have been used in the field of chemical engineering, for instance to determine liquid solidification pressures [273], to investigate porosity and its local fluctuations in fluidized beds, and to study the soaking and trickling of liquids through porous media and packed columns [274-278].

## II. Fundamental Concepts of Electromagnetism and Dielectric Phenomena

### A. The General Laws of Electromagnetism

According to Maxwell's phenomenological theory, the macroscopic electromagnetic phenomena can be described by means of five vectors, $\mathbf{E}$, $\mathbf{H}$, $\mathbf{D}$, $\mathbf{B}$, and $\mathbf{J}$. At every ordinary space point, these vectors are linked by two differential equations,

$$\frac{\partial \mathbf{D}}{\partial t} = \text{curl } \mathbf{H} - \mathbf{J} \tag{1}$$

$$\frac{\partial \mathbf{B}}{\partial t} = \text{curl } \mathbf{E} \tag{2}$$

and three constitutive equations connecting $\mathbf{D}$ with $\mathbf{E}$, $\mathbf{B}$ with $\mathbf{H}$, and $\mathbf{J}$ with $\mathbf{E}$, respectively

$$\mathbf{D} = ||\hat{\varepsilon}^*||\mathbf{E} \tag{3}$$

$$\mathbf{B} = ||\hat{\mu}^*||\mathbf{H} \tag{4}$$

$$\mathbf{J} = ||\sigma^*||\mathbf{E} \tag{5}$$

$\mathbf{E}$ is the electric field, $\mathbf{D}$ the electric induction (or electric displacement), $\mathbf{H}$ the magnetic field, $\mathbf{B}$ the magnetic induction, and $\mathbf{J}$ the current density. The factors appearing in Eqs. (3) to (5) are respectively the absolute electric permittivity, the absolute magnetic permeability, and the conductivity of the medium in which the electromagnetic phenomena take place. These quantities characterize the electric and magnetic properties of the medium.

The preceding set of equations must be completed by the equation of continuity which formulates the principle of conservation of electricity,

$$\text{div } \mathbf{J} + \frac{\partial \rho}{\partial t} = 0 \tag{6}$$

$\rho$ being the volume density of electrical charge. By combining Eqs. (1) and (6), a relation is obtained which connects $\rho$ with the electric displacement vector $\mathbf{D}$:

$$\text{div } \mathbf{D} = \rho \tag{7}$$

Thus the divergence of $\mathbf{D}$ appears to represent the volumic distribution of electrical charges throughout space. Similarly, by taking the divergence of both sides of Eq. (2), a relation is obtained concerning the magnetic induction $\mathbf{B}$:

$$\text{div } \mathbf{B} = 0 \tag{8}$$

This equation indicates that there is no magnetic equivalent to the net volume density of electrical charges.

If space is divided into two regions by a discontinuity surface S corresponding to abrupt changes in the quantities $||\varepsilon^*||$, $||\hat{\mu}^*||$, and $||\sigma^*||$, as shown in Fig. 1, an additional set of equations is necessary to describe fully the electromagnetic phenomena. The surface equations analogous to Eqs. (6) to (8) are, respectively,

$$J_{n_1} + J_{n_2} = \frac{\partial \sigma}{\partial t} \tag{9}$$

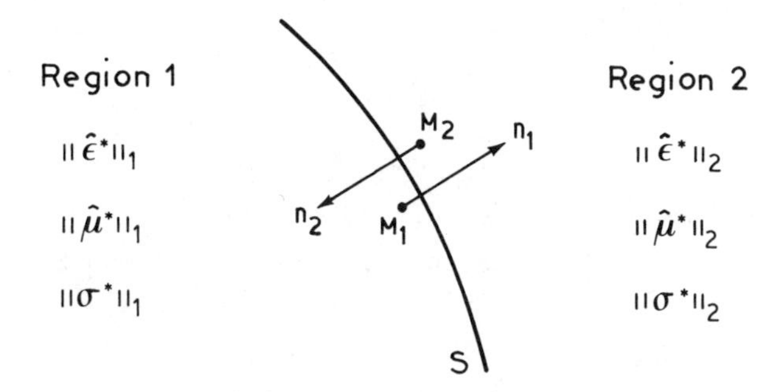


**Figure 1.** Schematic representation of a discontinuity surface S dividing space into two regions characterized by different electrical and magnetic properties. Each of the normal unit vectors $\mathbf{n}_1$ and $\mathbf{n}_2$ is directed outside the corresponding region.

$$D_{n_1} + D_{n_2} = -\underline{\sigma} \quad (10)$$

$$B_{n_1} + B_{n_2} = 0 \quad (11)$$

$\underline{\sigma}$ represents the surface density of electrical charges on S. Eq. (9) is the surface formulation of the principle of conservation of electricity. The discontinuity of the normal component of D across S appears to characterize the distribution of electrical charges all over S. As concerns B, Eq. (11) implies that there is no magnetic equivalent to the net surface density of electrical charges.

Two more equations concern the continuity of the tangential components of the electrical field vector E and magnetic field vector H across S:

$$E_{T_1} = E_{T_2} \quad (12)$$

$$H_{T_1} = H_{T_2} \quad (13)$$

Equation (12) implies that the electrical potential can undergo a discontinuity when the surface S is passed through

$$(\phi)_{M_2} = (\phi)_{M_1} + e_{12} \quad (14)$$

This discontinuity $e_{12}$, termed the *contact electrical potential,* arises from the difference in the chemical nature and thermodynamic state of the two media whose interface is S. If adequate physical conditions are fulfilled, contact electrical potentials are sources of nonconservative electrical fields giving rise to electrical currents. In the following, it will be considered that, in any space region that does not contain electrical current primary sources, the electrical potential $\phi$ is continuous when crossing the interface between two media, i.e.,

$$(\phi)_{M_2} = (\phi)_{M_1} \quad (15)$$

## B. The Constitutive Equations

From the most general point of view, the absolute electric permittivity $||\hat{\varepsilon}^*||$, the absolute magnetic permeability $||\hat{\mu}^*||$, and the conductivity $||\sigma^*||$ are tensor-type quantities whose coefficients depend on the coordinates and on the strength and frequency spectrum of the corresponding field vector. In the case of monochromatic electromagnetic phenomena taking place in linear, homogeneous, and isotropic media (LHI) media), they are functions only of the frequency f (or the angular frequency $\omega$, with $\omega = 2\pi f$) of the applied electromagnetic field. Then, the constitutive equations become linear relations linking D to E, B to H, and J to E:

$$D = \hat{\varepsilon}^*(f)E \tag{16}$$

$$B = \hat{\mu}^*(f)H \tag{17}$$

$$J = \sigma^*(f)E \tag{18}$$

In the case of free space, whatever the value of f, $\hat{\varepsilon}^*(f)$ is equal to $\varepsilon_0$, the absolute permittivity or dielectric constant of vacuum [$\varepsilon_0 = (1/36\pi) \times 10^{-9}$ F m$^{-1}$], and $\hat{\mu}^*(f)$ to $\mu_0$, the absolute magnetic permeability of vacuum ($\mu_0 = 4\pi \times 10^{-7}$ H m$^{-1}$). Consequently, it is possible to characterize the dielectric and magnetic properties of a substance by its relative permittivity $\varepsilon^*(f)$ and relative permeability $\mu^*(f)$, which are defined by

$$\varepsilon^*(f) = \frac{\hat{\varepsilon}^*(f)}{\varepsilon_0} \tag{19}$$

$$\mu^*(f) = \frac{\hat{\mu}^*(f)}{\mu_0} \tag{20}$$

In dielectrics common language, $\varepsilon^*(f)$ and $\mu^*(f)$ are referred to simply as the complex permittivity and the complex permeability.

Generally, $\varepsilon^*(f)$, $\mu^*(f)$, and $\sigma^*(f)$ are complex functions of the frequency f, giving at the same time the intensity ratio and the phase angle of D and E, B and H, and J and E, respectively. They can be split into their real and imaginary parts and written as indicated below, j being defined by $j^2 = -1$:

$$\varepsilon^*(f) = \varepsilon'(f) - j\varepsilon''(f) \tag{21}$$

$$\mu^*(f) = \mu'(f) - j\mu''(f) \tag{22}$$

$$\sigma^*(f) = \sigma'(f) - j\sigma''(f) \tag{23}$$

The function $\varepsilon'(f)$ describes dielectric dispersion phenomena and $\varepsilon''(f)$ dielectric absorption phenomena, both functions being generally interconnected. When $f \to 0$, $\varepsilon''(f)$ vanishes while $\varepsilon'(f)$ reaches its static value $\varepsilon_s$, the relative static permittivity (or, more simply, the static permittivity) of the medium. According to the classical dissipation theories, when $f \to \infty$, $\varepsilon''(f)$ vanishes while $\varepsilon'(f)$ reaches asymptotically a value equal to 1. In between, $\varepsilon'(f)$ and $\varepsilon''(f)$ undergo generally nonmonotonic changes as f increases and takes on values corresponding to molecular dipole relaxations and to atomic and electronic oscillator resonances.

A good example of this general behavior is the case of water, which is illustrated by Fig. 2., taken from Hasted's book [9].

Over the 1- to 300-GHz frequency range, water exhibits a dielectric relaxation of dipolar origin. As f increases, the dispersion curve $\varepsilon'(f)$ starts from a plateau value which is the static permittivity of water $\varepsilon_s$ (equal to 78.3 at 298 K), and falls smoothly down to a value which is the high-frequency limiting permittivity of water $\varepsilon_d$ (almost temperature independent and equal to 4.25). Correlatively, $\varepsilon''(f)$

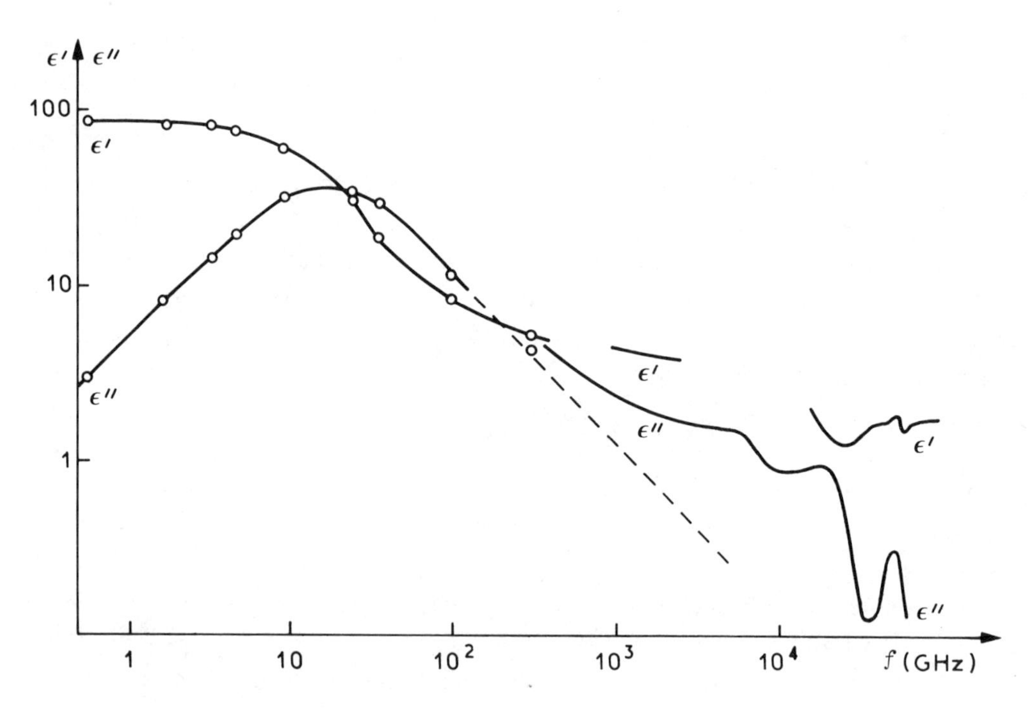


Figure 2. Frequency variations of ε' and ε'' for liquid $H_2O$ at 298.2 K. (From Ref. 9, figure designed by R. Abrahams.)

increases, culminates at a value sensibly equal to $(\varepsilon_s - \varepsilon_d)/2$ and then decreases. At higher frequencies, corresponding to submillimeter wavelengths, the features of the ε'(f) and ε''(f) curves are more complicated because of the existence of several dispersion and absorption phenomena.

A general conventional rule divides the studies of the complex permittivity into two fields of interest. Phenomena located in the submillimeter wavelength range belong to the field of optics, while phenomena taking place in the radio-frequency range (0 to 300 GHz) belong to the field of dielectric studies. An ideal dividing criterion would be the frequency $f_d$ at which dipolar relaxation mechanisms contribute nothing to the permittivity while atomic and electronic resonances give their full contribution. Then, the value of $\varepsilon_d$, measured at $f_d$ through dielectric methods would be equal to $n^2$, n being the refractive index determined at $f_d$ by optical methods.

The actual situation is not so simple because of the overlapping of dielectric and optical phenomena in the transition region between radio frequencies and infrared radiation. The most notable example is that of water, which is thought to exhibit an additional very small dipolar relaxation bridging the gap between $\varepsilon_d$ (4.25) and $n_D^2$ (1.78). Similar differences between the values of $\varepsilon_d$ and $n_D^2$ occur in some alcohols, in nitrobenzene, and even in relatively simple liquids such as chlorobenzene. Comments about this problem can be found in Refs. 9 and 10.

In spite of its great interest, this question will be neglected when the microwave properties of emulsions involving water is analyzed, and no incursion will be made in the field of emulsion optical properties, which are treated in Chap. 8.

As concerns conductivity phenomena, if frequency effects such as those observed in electrolytic solutions in the microwave region [9] are neglected, $\sigma^*(f)$ can be assimilated to its real part $\sigma'$ whose value is considered to be equal to that of the steady conductivity $\sigma$, whatever the frequency of the applied field. Consequently, in most cases, the conductivity contribution to the global complex permittivity takes the form of an imaginary quantity, $\sigma/(2\pi j\varepsilon_0 f)$, which constitutes a term additional to the imaginary part of the dipolar-type relaxation permittivity, as will be shown later. Then, the total complex permittivity of a substance can be written as

$$\begin{aligned} \varepsilon^*(f) &= \varepsilon'(f) - j\varepsilon''_T(f) = \varepsilon'(f) - j[\varepsilon''_R(f) + \varepsilon''_C(f)] \\ &= \varepsilon'(f) - j\varepsilon''_R(f) - j\frac{\sigma}{2\pi\varepsilon_0 f} \end{aligned} \tag{24}$$

$\varepsilon''_T$ denotes the total absorption coefficient, $\varepsilon''_R$ being the contribution of dipolar type relaxation processes, and $\varepsilon''_C$ the contribution of conduction phenomena. It is readily seen from Eq. (24) that the conduction absorption term is preponderant at low frequencies while its contribution can become negligible at high frequencies. If the steady conductivity $\sigma$ is great enough, a wide overlapping of relaxation and conduction phenomena can occur, which yields serious complications in the carrying out of complex permittivity measurements and in their interpretation.

The complex permittivity of a material is almost always measured by introducing a sample into a cell which is part of an electrical circuit, either an impedance bridge, a resonant circuit, a coaxial line, a waveguide apparatus, or a resonant cavity. The comparison of the response of the circuit to an electrical excitation, with and without the sample, allows the determination of $\varepsilon'(f)$ and $\varepsilon''_T(f)$. The knowledge of $\varepsilon''_R(f)$ is obtained by subtracting from $\varepsilon''_T(f)$ the conduction contribution at the same frequency $\varepsilon''_C(f)$, which can be calculated from conductivity measurements at low frequencies. Complex permittivity measurements in the millimeter wavelength range can also be carried out through free-space techniques. As a general rule, it can be said that the precision is the poorer as the sample losses are the higher. Moreover, one of the major sources of errors in dielectric measurements is the existence of parasitic and stray phenomena such as, for instance, inductance effects occurring above 1 MHz in impedance bridges, and the build-up of space charges at the measuring cell electrodes, which is particularly inconvenient in the lower frequency range. Details concerning dielectric measuring apparatus and methods can be found in Refs. 10, 39, 42-44, 46-49.

C. Time-Independent Electrical Phenomena

When electromagnetic phenomena do not depend upon time, Eqs. (1) and (2) become

$$\text{curl } \mathbf{H} = \mathbf{J} \tag{25}$$

$$\text{curl } \mathbf{E} = 0 \tag{26}$$

The continuity Eqs. (6) and (9) are replaced by

$$\text{div } \mathbf{J} = 0 \tag{27}$$

$$J_{n_1} + J_{n_2} = 0 \tag{28}$$

Two subcases must be considered. If there is no net flow of electricity, the density current vector J is equal to zero everywhere in space, either because $\sigma$ is equal to zero (insulating media) or because the electrical field E does not exist (conducting media). Electrical and magnetic phenomena are decoupled and can be studied independently. If electrical-charge steady flows exist somewhere in space, J is not identically null and is the source of magnetic phenomena that do not depend upon time.

The first case corresponds to electrostatic phenomena, and the second to steady (or stationary) electrokinetic phenomena.

*1. Static Electrical Phenomena*

a. General Considerations. The set of equations describing static electrical phenomena (electrostatic phenomena) is the following:

$$\mathbf{J} = 0 \tag{29}$$

$$\text{curl } \mathbf{E} = 0 \tag{30}$$

$$E_{T_1} = E_{T_2} \tag{31}$$

$$\text{div } \mathbf{D} = \rho \tag{32}$$

$$D_{n_1} + D_{n_2} = -\underline{\sigma} \tag{33}$$

If space is filled with a LHI medium, the constitutive equation is written

$$\mathbf{D} = \varepsilon_0 \varepsilon_s \mathbf{E} \tag{34}$$

$\varepsilon_s$ being the (relative) static permittivity which is characteristic for the electrostatic properties of the medium. If space is filled with different LHI media separated by boundary surfaces, Eq. (34) is valid everywhere, provided that $\varepsilon_s$ be given the value corresponding to the medium over which the phenomena are considered.

In the case of free space, the electrical field vector E is linked to the electrical displacement vector D simply by

$$D = \varepsilon_0 E \tag{35}$$

The volume and surface divergence of E are given by

$$\text{div } E = \frac{\rho}{\varepsilon_0} \tag{36}$$

$$E_{n_1} + E_{n_2} = -\frac{\sigma}{\varepsilon_0} \tag{37}$$

$E_{n_1}$ and $E_{n_2}$ are the components of E taken along the normal to the surface S bearing $\sigma$, the orientations of $n_1$ and $n_2$ on both sides of S being similar to those chosen in the case of a surface separating two different media (see Fig. 1).

Consequently, the potential function $\phi$ obeys Poisson's equation,

$$\nabla^2\phi + \frac{\rho}{\varepsilon_0} = 0 \tag{38}$$

and is continuous when S is passed through

$$(\phi)_{M_1} = (\phi)_{M_2} \tag{39}$$

while its derivatives taken along the normal to S (oriented positively from side 1 to side 2) at two points separated by S, but infinitely close to each other, are linked by

$$\left(\frac{\partial \phi}{\partial n}\right)_{M_2} - \left(\frac{\partial \phi}{\partial n}\right)_{M_1} = -\frac{\sigma}{\varepsilon_0} \tag{40}$$

If space is filled with a LHI medium characterized by its static permittivity $\varepsilon_s$, Eqs. (36) to (40) become, respectively

$$\text{div } E = \frac{\rho}{\varepsilon_0 \varepsilon_s} \tag{41}$$

$$E_{n_1} + E_{n_2} = -\frac{\sigma}{\varepsilon_0 \varepsilon_s} \tag{42}$$

$$\nabla^2\phi + \frac{\rho}{\varepsilon_0 \varepsilon_s} = 0 \tag{43}$$

$$(\phi)_{M_1} = (\phi)_{M_2} \tag{44}$$

$$\left(\frac{\partial \phi}{\partial n}\right)_{M_2} - \left(\frac{\partial \phi}{\partial n}\right)_{M_1} = -\frac{\sigma}{\varepsilon_0 \varepsilon_s} \tag{45}$$

Thus, it appears that, if a set of source densities $\rho$ and $\underline{\sigma}$ is given, the electrical field and electrical potential describing the electrostatic phenomena in a LHI medium filling space are proportional to the electrical field and the electrical potential describing the electrostatic phenomena in free space, the multiplying factor being equal to $1/\varepsilon_s$.

Equations (41), (43), and (44) remain valid when space is filled by different LHI media, provided that $\varepsilon_s$ is given the value corresponding to the medium over which the phenomena are considered, but Eqs. (42) and (45) must be replaced by the more general ones

$$\varepsilon_{1s} E_{n_1} + \varepsilon_{2s} E_{n_2} = -\frac{\underline{\sigma}}{\varepsilon_0} \tag{46}$$

$$\varepsilon_{2s}\left(\frac{\partial \phi}{\partial n}\right)_{M_2} - \varepsilon_{1s}\left(\frac{\partial \phi}{\partial n}\right)_{M_1} = -\frac{\underline{\sigma}}{\varepsilon_0} \tag{47}$$

In the following, it will be assumed that the system being studied forms a region of space that does not contain source densities. Consequently, in this case, the set of equations describing electrostatic phenomena throughout the system is written as $\rho = 0$ and $\underline{\sigma} = 0$.

Field equations:

$$\text{curl } E = 0 \tag{48}$$

$$E_{T_1} = E_{T_2} \tag{49}$$

$$\text{div } E = 0 \tag{50}$$

$$\varepsilon_{1s} E_{n_1} + \varepsilon_{2s} E_{n_2} = 0 \tag{51}$$

Potential equations:

$$(\phi)_{M_1} = (\phi)_{M_2} \tag{52}$$

$$\nabla^2 \phi = 0 \tag{53}$$

$$\varepsilon_{2s}\left(\frac{\partial \phi}{\partial n}\right)_{M_2} - \varepsilon_{1s}\left(\frac{\partial \phi}{\partial n}\right)_{M_1} = 0 \tag{54}$$

b. Polarization of Material Media

*(1) The static polarization vector* $P_s$. It has been shown above that, if a set of source densities is given, the electrical field and potential describing the electrostatic phenomena in a LHI medium filling space can be obtained by multiplying the electrical field and potential describing the electrostatic phenomena in free space by $1/\varepsilon_s$. It is possible to solve the problem as well by considering

the deviations which occur in electrostatic phenomena when free space is replaced by a LHI medium filling it completely, the source densities being unchanged. If $\rho$ and $\underline{\sigma}$ are given, the displacement vector **D** is determined entirely by Eqs. (32) and (33), and the electrical field can be derived by means of the constitutive Eq. (34). Equation (34) may be written

$$\mathbf{D} = \varepsilon_0\mathbf{E} + \mathbf{P}_s \tag{55}$$

Comparison of Eqs. (35) and (55) shows that the vector $\mathbf{P}_s$, called the *static polarization vector*, represents the discrepancies existing between electrostatic phenomena in free space and in a LHI medium filling it, for a given set of source densities:

$$\mathbf{P}_s = \varepsilon_0(\mathbf{E}_0 - \mathbf{E}) \tag{56}$$

$\mathbf{E}_0$ is the electrical field appearing in Eq. (35) and relative to free space and **E** the electrical field in the LHI medium. By combining Eqs. (34) and (55), $\mathbf{P}_s$ can be expressed in terms of **E** only:

$$\mathbf{P}_s = \varepsilon_0(\varepsilon_s - 1)\mathbf{E} \tag{57}$$

From Eq. (57), it is readily seen that $\mathbf{P}_s$ vanishes if $\varepsilon_s \to 1$.

By taking the divergence of both sides of Eq. (55), it follows

$$\operatorname{div} \mathbf{E} = \frac{\rho + \rho'}{\varepsilon_0} \tag{58}$$

with

$$\rho' = -\operatorname{div} \mathbf{P}_s \tag{59}$$

Similarly, the surface divergence of **E** is given by

$$\mathbf{E}_{n_1} + \mathbf{E}_{n_2} = -\frac{\underline{\sigma} + \underline{\sigma}}{\varepsilon_0} \tag{60}$$

with

$$\underline{\sigma}' = \mathbf{P}_{sn_1} + \mathbf{P}_{sn_2} \tag{61}$$

Thus, by introducing the static polarization vector $\mathbf{P}_s$, it is possible to define fictitious volume and surface source densities $\rho'$ and $\sigma'$, respectively, which allow the description of electrostatic phenomena in LHI media by means of a set of equations identical to that relative to free space, Eqs. (36) to (40), except for the substitution of $\rho_T = \rho + \rho'$ for $\rho$ and $\sigma_T = \underline{\sigma} + \underline{\sigma}'$ for $\underline{\sigma}$,

$$\operatorname{div} \mathbf{E} = \frac{\rho_T}{\varepsilon_0} \tag{62}$$

$$\mathbf{E}_{n_1} + \mathbf{E}_{n_2} = -\frac{\underline{\sigma}_T}{\varepsilon_0} \tag{63}$$

$$\nabla^2 \phi + \frac{\rho_T}{\varepsilon_0} = 0 \tag{64}$$

$$(\phi)_{M_1} = (\phi)_{M_2} \tag{65}$$

$$\left(\frac{\partial \phi}{\partial n}\right)_{M_2} - \left(\frac{\partial \phi}{\partial n}\right)_{M_1} = -\frac{\underline{\sigma}_T}{\varepsilon_0} \tag{66}$$

It must be stressed that this set of equations is more general than Eqs. (41) to (45) because Eqs. (63) and (66) are substitutes for Eqs. (46) and (47), as well as for Eqs. (42) and (45). Consequently, it may be applied when space is filled either with a unique LHI medium or by several LHI media separated by boundary surfaces.

In a space region that does not contain any volume source density, $\rho$ is equal to zero, which implies $\rho' = 0$ and $\rho_T = 0$. Similarly, if the region considered is cut out of a LHI medium, the fact that $\underline{\sigma} = 0$ yields $\underline{\sigma}' = 0$ and $\underline{\sigma}_T = 0$. But if there are discontinuity surfaces at the boundary of or inside the region, $\underline{\sigma}'$ has a nonzero value even for $\underline{\sigma} = 0$ and contributes to the electrical field and electrical potential. In that case, the set of equations describing the electrostatic phenomena in the region, and equivalent to Eqs. (48) through (54), is the following:

Field equations:

$$\text{curl } \mathbf{E} = 0 \tag{67}$$

$$\mathbf{E}_{T_1} = \mathbf{E}_{T_2} \tag{68}$$

$$\text{div } \mathbf{E} = 0 \tag{69}$$

$$\mathbf{E}_{n_1} + \mathbf{E}_{n_2} = -\frac{\underline{\sigma}'}{\varepsilon_0} \tag{70}$$

$$(\phi)_{M_1} = (\phi)_{M_2} \tag{71}$$

Potential equations:

$$\nabla^2 \phi = 0 \tag{72}$$

$$\left(\frac{\partial \phi}{\partial n}\right)_{M_2} - \left(\frac{\partial \phi}{\partial n}\right)_{M_1} = -\frac{\underline{\sigma}'}{\varepsilon_0} \tag{73}$$

$\rho_T$, the total volume source density, is the sum of $\rho$, the true charge volume density, and of $\rho'$, the fictitious charge volume density, often called *bound charge volume density*. Similarly, $\underline{\sigma}_T$, the total surface source density, is the sum of $\underline{\sigma}$, the true charge surface density, and of $\underline{\sigma}'$, the fictitious charge surface density, often called *bound charge surface density*. $\rho$ and $\underline{\sigma}$ are the primary sources of the electrical field, $\rho'$ and $\underline{\sigma}'$ being secondary sources resulting from what is defined as the *polarization of matter.* By evaluating the electrical potential function, it is possible to show that the contribution of $\rho'$ and $\underline{\sigma}'$ can be considered as the potential arising from a volumic distribution of dipoles, the dipolar moment per volume unit being equal to $P_s$,

$$P_s = \frac{dp_s}{dV} \tag{74}$$

where $dp_s$ represents the dipolar moment of an elementary volume dV of the material medium.

Thus, $P_s$ can be given a significance based upon the interaction of matter with the electrical field arising from the existence of the primary source densities $\rho$ and $\underline{\sigma}$. Under the action of the initial field created by $\rho$ and $\underline{\sigma}$, any elementary volume of the material acquires a dipolar moment because of the creation of molecular dipoles due to the shift of electrons, atoms, or groups of atoms from their initial statistical equilibrium positions and/or the orientation of polar molecules along the field. These elementary dipoles add a contribution to the field and a repetitive feedback-type process develops until the electrostatic equilibrium is reached. Then, $P_s$ is linked to the total actual electrical field by Eq. (57). The electrostatic polarization vector $P_s$ can be written as the sum of two polarization vectors $P_{os}$ and $P_d$:

$$P_s = P_{os} + P_d \tag{75}$$

$P_d$, sometimes written $P_\infty$, is called the *distortion polarization vector.* It represents the contribution to the total polarization of molecular dipoles created by the shift of electrons, atoms, or groups of atoms from their equilibrium positions. Because of the small inertia of electrons and atoms, the distortion polarization vector reaches its final value almost instantaneously. By introducing a coefficient called the *polarizability of the molecules* $\alpha_d$, $P_d$ can be written

$$P_d = N\alpha_d E_i \tag{76}$$

where $E_i$ represents the effective field, or internal electrical field, acting on molecules and N is the number of molecules per volume unit. As $\alpha_d$ is a characteristic of interactions existing within molecules, it is almost independent of the external thermodynamic conditions.

$P_{os}$, called the *static orientation polarization vector,* represents the contribution brought by the statistically preferred orientation of molecules along the direction of the electrical field, as opposed to the random orientation caused by thermal agitation. As in the case of $P_d$, $P_{os}$ can be written

$$P_{os} = N\alpha_o E_r \tag{77}$$

where $\alpha_o$ denotes the molecular orientation polarizability and $E_r$ denotes the effective directing electrical field. Because of the competition between the ordering action of $E_r$ and the thermal agitation, $\alpha_o$ is strongly dependent upon the thermodynamic conditions. Moreover, the orientation polarization gradually will reach its static equilibrium value $P_{os}$ after the primary electrical field has been applied to the system.

The total polarization being expressed by means of $\alpha_o$, $\alpha_d$, $E_r$, and $E_i$,

$$P_s = N(\alpha_o E_r + \alpha_d E_i) \tag{78}$$

the static permittivity is given by

$$\varepsilon_0(\varepsilon_s - 1)E = N(\alpha_o E_r + \alpha_d E_i) \tag{79}$$

or by

$$\varepsilon_0(\varepsilon_s - 1)E = \frac{\mathcal{N}\bar{\omega}}{M}(\alpha_o E_r + \alpha_d E_i) \tag{80}$$

$\mathcal{N}$ being Avogadro's constant, $\bar{\omega}$ the specific mass of the substance, and M its molar weight.

From Eq. (80), it appears that the static permittivity can be related to the specific mass which is a quantity reflecting the state of condensation of matter and to molecular characteristics such as the polarizability $\alpha_d$ or the orientation polarizability $\alpha_o$ which is linked to the permanent dipole moment of the molecule. To obtain usable formulas, the problem to be solved is the evaluation of the internal field $E_i$ and/or the directing field $E_r$. For that purpose, several theoretical models have been designed by Lorentz, Onsager, and others; see Refs. 36 to 41. For instance, by assuming that both $E_r$ and $E_i$ could be expressed by means of the formula derived by Lorentz for the internal field,

$$E_i = \frac{\varepsilon_s + 2}{3} E \tag{81}$$

Debye [279] proposed the equation

$$\frac{M}{\bar{\omega}} \frac{\varepsilon_s - 1}{\varepsilon_s + 2} = \frac{\mathcal{N}}{3\varepsilon_0}(\alpha_o + \alpha_d) \tag{82}$$

As the orientation polarizability of a polar molecule in thermal equilibrium with a large number of neighbors is given by

$$\alpha_o = \frac{\underline{\mu}^2}{3kT} \tag{83}$$

$\underline{\mu}$ being the molecule dipole moment, k Boltzmann's constant, and T the kelvin temperature, Eq. (82) becomes

$$\frac{M}{\omega}\frac{\varepsilon_s - 1}{\varepsilon_s + 2} = \frac{N}{3\varepsilon_0}\left(\alpha_d + \frac{\mu^2}{3kT}\right) \tag{84}$$

Although Eq. (84) is an approximation of a more general and complicated one, it correctly represents the experimental dielectric behavior of gases at moderate pressures. In the case of nonpolar substances, $\mu$ is equal to zero and Eq. (84) is simplified to

$$\frac{M}{\omega}\frac{\varepsilon_s - 1}{\varepsilon_s + 2} = \frac{N}{3\varepsilon_0}\alpha_d \tag{85}$$

which is the well-known Clausius-Mosotti formula. If it is permissible to substitute the square of the refractive index for the permittivity, Eq. (85) becomes the Lorentz-Lorenz formula,

$$\frac{M}{\omega}\frac{n^2 - 1}{n^2 + 2}\frac{N}{3\varepsilon_0}\alpha_d \tag{86}$$

which is applicable, as a first approximation, to the optical properties of nonpolar substances, or of polar substances beyond the high-frequency end of their orientation polarization domain.

In spite of the crudeness of the hypotheses sustaining them, the Debye and Clausius-Mosotti formulas are good illustrations of the usefulness of dielectric studies among the numerous methods developed in order to investigate the physicochemical properties of matter.

*(2) Classical macroscopic polarization problems with spherical geometry.* Polarization problems with spherical geometry are of interest for the dielectric study of emulsions because these systems are made generally of spherical droplets imbedded in a continuous medium. The basic problem is the following. A uniform electrical field $\mathbf{E}_0$ is produced by a convenient set of fixed source densities located in a limited region of space, while an insulating (or dielectric) spherical body of radius a, characterized by its static permittivity $\varepsilon_{1s}$, is introduced elsewhere in space. The electrostatic phenomena can be described by means of the electrical potential function $\phi$ which must obey Eqs. (52) to (54) and be equal to the potential function $\phi_0$ of $\mathbf{E}_0$ at great distances from the spherical body. The sphere center is chosen as the coordinate origin, the z axis coinciding with the direction of $\mathbf{E}_0$, as indicated in Fig. 3. The resolution of the Laplace equation by the method of zonal harmonics shows that $\phi$ within the spherical body is given by

$$\phi_1 = (1 - A)\phi_0 \tag{87}$$

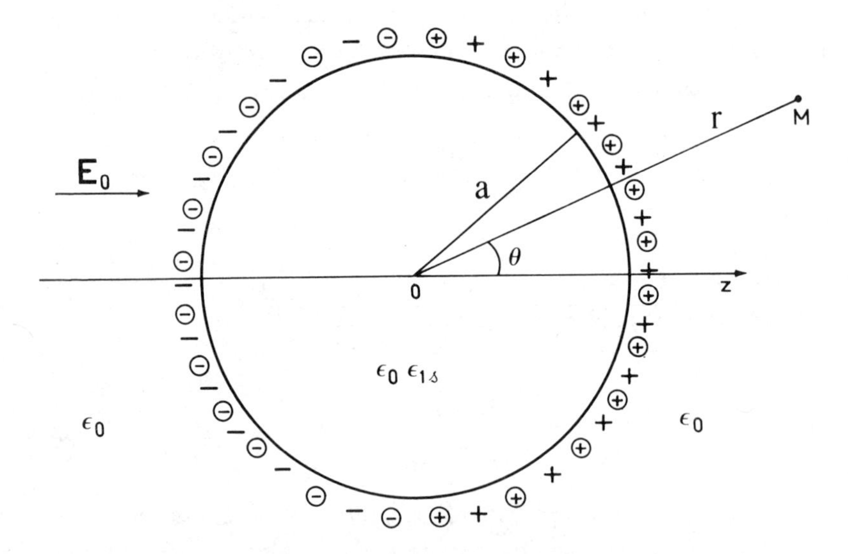


**Figure 3.** Polarization of an imperfect insulating spherical body. ⊕ and ⊖ represent bound charges, and + and − free charges.

and outside the spherical body by

$$\phi_2 = \left(1 - A \frac{a^3}{r^3}\right) \phi_0 \tag{88}$$

$\phi_0$ being related to $E_0$ by

$$\phi_0 = - E_0 z = -E_0 r \cos \theta \tag{89}$$

By using Eq. (54), where $\varepsilon_{2s}$ is to be replaced by 1, the medium surrounding the sphere being free space, it follows that

$$A = \frac{\varepsilon_{1s} - 1}{\varepsilon_{1s} + 2} \tag{90}$$

From Eq. (88), it is readily seen that the modifications brought into the initial electrical situation by the presence of the dielectric sphere can be regarded as arising from a dipole located at its center and whose moment $p_s$ is given by

$$p_s = \left(\frac{4}{3} \pi a^3\right) 3\varepsilon_0 \frac{\varepsilon_{1s} - 1}{\varepsilon_{1s} + 2} E_0 \tag{91}$$

Thus as far as electrostatic phenomena are concerned, the spherical body is equivalent to a dipole located at its center. The polarization vector being defined as the dipole moment per volume unit, $P_{1s}$ is given by

$$P_{1s} = 3\varepsilon_0 \frac{\varepsilon_{1s} - 1}{\varepsilon_{1s} + 2} E_0 \tag{92}$$

This formula can be derived as well by applying Eq. (57) to the present case, $\varepsilon_s$ being replaced by $\varepsilon_{1s}$ and E by $E_1$, the electrical field corresponding to the potential function $\phi_1$ given by Eq. (87). It can be checked that $P_{1s}$ has a zero divergence, which is consistent with the fact that, within the spherical body, there is no true charge volume density and consequently no bound charge volume density. On the boundary surface of the spherical body, the normal component of $P_s$ undergoes a discontinuity, implying the existence of a bound charge surface density $\underline{\sigma}'$ which is given by

$$\underline{\sigma}' = 3\varepsilon_0 \frac{\varepsilon_{1s} - 1}{\varepsilon_{1s} + 2} |E_0| \cos\theta \tag{93}$$

A similar approach holds in the case of a perfectly conducting spherical body. Then A is equal to 1 and the moment of the dipole equivalent to the conducting sphere $P_{cs}$ and the charge surface density $\underline{\sigma}$ on the boundary surface are, respectively, given by

$$p_{cs} = \left(\frac{4}{3}\pi a^3\right) 3\varepsilon_0 E_0 \tag{94}$$

$$\underline{\sigma} = 3\varepsilon_0 |E_0| \cos\theta \tag{95}$$

It must be pointed out that, in this case, $\underline{\sigma}$ does not represent bound electrical charges but free electrical charges in equilibrium on the boundary surface under the influence of the applied electrical field. This can be checked by calculating the surface divergence of D, which is found to be nonzero and equal to $\underline{\sigma}$. The free charges whose surface distribution is represented by $\sigma$ can be either electrons and positive vacancies in the case of metal-type substances, or anions and cations in the case, for instance, of electrolytic solutions.

In spite of the artificial character of this procedure, it is worth mentioning that Eqs. (94) and (95) can be obtained simply by taking $\varepsilon_{1s} = \infty$ in Eqs. (91) and (93).

If the spherical body is made of an imperfectly insulating, or imperfectly conducting, substance, its contribution to the electrostatic phenomena is at the same time that of a dielectric sphere, as described by Eqs. (91) and (93), and that of a conducting sphere, as described by Eqs. (94) and (95).

All the preceding results can be given a more general significance by considering that the medium surrounding the spherical body is not vacuum but a material characterized by its static permittivity $\varepsilon_{2s}$. For instance, Eqs. (91) and (93) become, respectively,

$$p_s = \left(\frac{4}{3}\pi a^3\right) 3\varepsilon_0 \frac{\varepsilon_{1s} - \varepsilon_{2s}}{\varepsilon_{1s} + 2\varepsilon_{2s}} E_0 \qquad (96)$$

$$\underline{\sigma}' = 3\varepsilon_0 \frac{\varepsilon_{1s} - \varepsilon_{2s}}{\varepsilon_{1s} + 2\varepsilon_{2s}} |E_0| \cos\theta \qquad (97)$$

2. *Stationary Electrical Phenomena*

The set of equations describing stationary electrical phenomena is

$$J \neq 0 \qquad (98)$$

$$\text{curl } E = 0 \qquad (99)$$

$$E_{T_1} = E_{T_2} \qquad (100)$$

$$\text{div } J = 0 \qquad (101)$$

$$J_{n_1} + J_{n_2} = 0 \qquad (102)$$

If space is filled with a LHI conducting medium, the constitutive equation is written

$$J = \sigma E \qquad (103)$$

$\sigma$ being the stationary conductivity which is characteristic of the conductive properties of the medium. If space is filled with different LHI conducting media separated by boundary surfaces, Eq. (103) is valid everywhere provided that $\sigma$ is given the value corresponding to the medium over which the phenomena are considered.

The existence of flows of electrical charges induces magnetic phenomena. Once the steady equilibrium is established, the magnetic phenomena are independent of time and do not influence the electrical ones. The relationship between magnetic and electric phenomena is given by

$$\text{curl } H = J \qquad (104)$$

or

$$\text{curl } H = \sigma E \qquad (105)$$

In any homogeneous space region, characterized by its stationary conductivity $\sigma$, Eq. (101) yields

$$\text{div } E = 0 \qquad (106)$$

Consequently, the electrical potential function $\phi$ obeys the Laplace equation

$$\nabla^2 \phi = 0 \tag{107}$$

If the space region considered is filled by different LHI media, Eq. (107) holds in every homogeneous subregion. When the interface of two subregions is passed through, the discontinuity of the normal component of E is given by Eq. (102), which can be written

$$\sigma_1 E_{n_1} + \sigma_2 E_{n_2} = 0 \tag{108}$$

By introducing the derivatives of $\phi$ along the normal to the interface S oriented from medium 1 to medium 2, taken at two points infinitely close to each other but separated by S, Eq. (108) is transformed into

$$\sigma_2 \left(\frac{\partial \phi}{\partial n}\right)_{M_2} - \sigma_1 \left(\frac{\partial \phi}{\partial n}\right)_{M_1} = 0 \tag{109}$$

Moreover, if the region does not contain any current source characterized by a contact electrical potential, the potential function $\phi$ is continuous when crossing the surface S, i.e.,

$$(\phi)_{M_1} = (\phi)_{M_2} \tag{110}$$

It appears that the equations describing stationary electrical phenomena have the same formulation as those describing electrostatic phenomena, except for the substitution of $\sigma$ for $\varepsilon_0 \varepsilon_s$.

Field equations:

$$\text{curl } E = 0 \tag{111}$$

$$E_{T_1} = E_{T_2} \tag{112}$$

$$\text{div } E = 0 \tag{113}$$

$$\sigma_1 E_{n_1} + \sigma_2 E_{n_2} = 0 \tag{114}$$

Potential equations:

$$(\phi)_{M_1} = (\phi)_{M_2} \tag{115}$$

$$\nabla^2 \phi = 0 \tag{116}$$

$$\sigma_2 \left(\frac{\partial \phi}{\partial n}\right)_{M_2} - \sigma_1 \left(\frac{\partial \phi}{\partial n}\right)_{M_1} = 0 \tag{117}$$

Consequently, any general solution of the Laplace equation describing electrostatic phenomena is applicable also to stationary electrical phenomena, provided that the

LHI dielectric media filling space are replaced by LHI conducting media of identical geometry. This result is an illustration of a general principle, known as the *principle of generalized conductivity* [7,8,280-282], which states the similarity of the general solutions of stationary flow problems involving transport coefficients such as permittivity, conductivity, magnetic permeability, thermal conductivity, and diffusivity. This principle is of importance for the study of the dielectric properties of heterogeneous media such as emulsions because it permits the generalization of the results obtained for static permittivity or stationary conductivity to the case of complex permittivity (or complex conductivity) [4,7,8]. If the LHI media filling space are imperfect conductors (or imperfect insulators) the results stated above are still valid, the global phenomena resulting from the superposition of electrostatic phenomena and of stationary electrical phenomena which are ruled by equations of the same type. It should be noted that in a region that does not contain source densities, the interface between two adjacent media will bear at the same time a bound charge surface density arising from polarization phenomena and a free charge surface density whose value $\underline{\sigma}$ can be evaluated by combining Eqs. (46) and (114).

## D. Time-Dependent Electrical Phenomena

### 1. *General Considerations*

In the case of time-dependent electromagnetic phenomena, electrical and magnetic phenomena are connected and cannot be studied separately, as in the static and stationary cases. In a homogeneous space region where there is no true charge density and no current source, the electrical field obeys the following equation, obtained by transforming and combining Eqs. (1) and (2):

$$\frac{\varepsilon\mu}{c^2}\frac{\partial^2\mathbf{E}}{\partial t^2} + \sigma\mu\mu_0\frac{\partial\mathbf{E}}{\partial t} = \nabla^2\mathbf{E} \tag{118}$$

$\varepsilon$, $\mu$, and $\sigma$ are the relative permittivity, the relative magnetic permeability, and the conductivity of the medium, c being the velocity of electromagnetic signals traveling in free space ($\varepsilon_0\mu_0c^2 = 1$). The magnetic field obeys an equation of the same type.

If the medium is nondispersive, $\varepsilon$ is equal to $\varepsilon_s$, and $\sigma$ represents the stationary conductivity. The quantity defined by

$$T = \frac{\varepsilon_0\varepsilon_s}{\sigma} \tag{119}$$

has the physical dimension of a time and is termed the *time constant* of the medium. $T$ is a characteristic for the medium in that it is an indication of the time rate at which stationary equilibrium will be reached after the initiation of a time-indepenent electrical field. Moreover, if the case of ferromagnetic media is excluded, the relative magnetic permeability may be taken equal to unity.

Equation (118) becomes

$$\frac{\partial^2 E}{\partial t^2} + \frac{1}{T}\frac{\partial E}{\partial t} = v^2 \nabla^2 E \tag{120}$$

where v is the phase velocity of the electromagnetic signal traveling in the medium.

In a nonconducting medium $1/T = 0$. The second term on the left-hand side of Eq. (120) vanishes, leaving a propagation equation. In an essentially conducting medium $1/T$ is generally large and the first term on the left-hand side of Eq. (120) is usually negligible compared to the second one; then Eq. (120) gives place to a diffusion equation. In the case of an imperfectly conducting, or imperfectly insulating, medium, both terms must be retained. Owing to the existence of a conductivity, the propagation in the medium will be affected by dispersion phenomena if the electromagnetic signal is nonmonochromatic.

In a dispersive medium, the permittivity and magnetic permeability depend upon the frequency of the electromagnetic signal. If the signal is monochromatic, the electromagnetic phenomena are similar to those occurring in a nondispersive medium, but the relationship between D and E, or P and E, depends upon the frequency. If the electromagnetic signal is nonmonochromatic, being for instance of the pulse type, the propagation is affected by dispersion and absorption phenomena. The dispersive and absorptive properties of the medium, represented by the variations of the permittivity and magnetic permeability with the frequency, may be investigated globally by means of Fourier analysis (TDS experiments), if the signal is a pulse, or through step-by-step studies of polarization phenomena at numerous discrete frequencies.

If space is filled with different LHI media, the situation is more complicated because of the existence of refraction and reflection phenomena at the interfaces.

### 2. *The Quasi-Stationary Approximation*

The quasi-stationary approximation allows the application of the laws ruling steady phenomena to the case of time-dependent phenomena, provided that the steady quantities be replaced by instantaneous ones. A good criterion for the application of this approximation can be found by comparing the wavelength of the electromagnetic signal to the linear dimensions of the systems under consideration. If, for instance, the wavelength is 10 times greater than the dimension of a system, it can be considered that, at a given instant, the electrical situation is the same throughout the system, as in the stationary case. The main implication of the quasi-stationary approximation being the disregard of propagation phenomena, it must be handled carefully, especially in the higher radio-frequency range for which wavelengths approach sample dimensions. However, this approximation is a very helpful tool for the study of time-dependent dielectric phenomena because of the simplifications it permits in the calculations.

### 3. *Field and Potential Equations in the Quasi-Stationary Approximation*

Electromagnetic phenomena linked to monochromatic sinusoidal signals may be easily dealt with by using the complex representation for scalar and vector functions [283]. If $\mathbf{E}$, $\mathbf{D}$, $\mathbf{H}$, $\mathbf{B}$, and $\mathbf{J}$ are assumed to be complex vector functions of time of the following type:

$$X = \overline{X}\, e^{j\omega t} \tag{121}$$

$\omega$ being the angular frequency ($\omega = 2\pi f$), $\overline{X}$ the vector complex amplitude, and j defined by $j^2 = -1$, Eqs. (1) and (2) are transformed into

$$\text{curl } \mathbf{H} = \mathbf{J} + j\omega \mathbf{D} \tag{122}$$

$$\text{curl } \mathbf{E} = -j\omega \mathbf{B} \tag{123}$$

$\mathbf{B}$ having a zero divergence, it can be regarded as a curl vector

$$\mathbf{B} = \text{curl } \mathbf{A} \tag{124}$$

$\mathbf{A}$ is the vector potential which can be defined completely by what is known as the Lorentz condition. Introducing $\mathbf{A}$ in Eqs. (123) leads to

$$\text{curl } (\mathbf{E} + j\omega \mathbf{A}) = 0 \tag{125}$$

As $\mathbf{D}$ and $\mathbf{J}$ can be expressed in terms of $\mathbf{E}$ by means of the relevant constitutive equations, Eq. (122) can be written

$$\text{curl } \mathbf{H} = [\sigma + j\omega \varepsilon_0 \varepsilon^*(\omega)]\mathbf{E} \tag{126}$$

If the factor premultiplying $\mathbf{E}$ is defined as the effective complex conductivity of the medium,

$$\overline{\sigma}^*(\omega) = \sigma + j\omega \varepsilon_0 \varepsilon^*(\omega) \tag{127}$$

Eq. (126) takes the form

$$\text{curl } \mathbf{H} = \overline{\sigma}^*(\omega)\, \mathbf{E} \tag{128}$$

It is possible to define alternatively an effective complex permittivity by

$$\overline{\varepsilon}^*(\omega) = \varepsilon^*(\omega) + \frac{\sigma}{j\omega \varepsilon_0} \tag{129}$$

and Eq. (128) becomes

$$\text{curl } \mathbf{H} = j\omega\,\varepsilon_0 \bar{\varepsilon}^*(\omega)\,\mathbf{E} \tag{130}$$

Equations (128) and (130) are strictly equivalent since

$$\bar{\sigma}^*(\omega) = j\omega\,\varepsilon_0 \bar{\varepsilon}^*(\omega) \tag{131}$$

If the space region considered is filled with different LHI media which are, in the general case, imperfect insulators, the surfaces separating two different media bear, in addition to bound charge densities, free charge densities, although it is assumed that the region does not contain primary source densities. The free charge density $\underline{\sigma}$ is related to the surface divergence of $\mathbf{D}$ by

$$D_{n_1} + D_{n_2} = -\underline{\sigma} \tag{132}$$

and its time derivative is related to the surface divergence of $\mathbf{J}$ by

$$J_{n_1} + J_{n_2} = \frac{\partial \underline{\sigma}}{\partial t} \tag{133}$$

As $\mathbf{D}$ and $\mathbf{J}$ are of the type described by Eq. (121), Eqs. (132) and (133) can be combined into a unique equation,

$$(J_{n_1} + j\omega D_{n_1}) + (J_{n_2} + j\omega D_{n_2}) = 0 \tag{134}$$

or, alternatively,

$$(\mathbf{J} + j\omega \mathbf{D})_{n_1} + (\mathbf{J} + j\omega \mathbf{D})_{n_2} = 0 \tag{135}$$

which expresses the fact that the complex vector function $\mathbf{J} + j\omega\mathbf{D}$ has a zero surface divergence. By introducing the effective complex conductivity defined by Eq. (127), Eq. (135) yields

$$\bar{\sigma}_1^*(\omega) E_{n_1} + \bar{\sigma}_2^*(\omega) E_{n_2} = 0 \tag{136}$$

$\bar{\sigma}_1^*(\omega)$ is the effective complex conductivity of medium 1 and $\bar{\sigma}_2^*(\omega)$ that of medium 2. Equation (136) can be written in an alternative form by using the effective complex permittivity instead of the effective complex conductivity:

$$\bar{\varepsilon}_1^*(\omega) E_{n_1} + \bar{\varepsilon}_1^*(\omega) E_{n_2} = 0 \tag{137}$$

If, in application of the quasi-stationary approximation, it is permissible to consider in Eq. (125) $j\omega\mathbf{A}$ as negligible compared to $\mathbf{E}$, the electric field appears to be an irrotational vector. Consequently, if the space region considered does not contain source densities, the electrical field obeys the following set of equations:

$$\text{curl } E = 0 \tag{138}$$

$$E_{T_1} = E_{T_2} \tag{139}$$

$$\text{div } E = 0 \tag{140}$$

$$\bar{\sigma}_1^* E_{n_1} + \bar{\sigma}_2^* E_{n_2} = 0 \tag{141}$$

which have the same formulation as Eq. (111) to (114) that rule steady electrical phenomena, except for the substitution of $\bar{\sigma}^*$ for $\sigma$. By using $\bar{\varepsilon}^*$ instead of $\bar{\sigma}^*$, Eqs. (138) to (141) may be written, alternatively,

$$\text{curl } E = 0 \tag{142}$$

$$E_{T_1} = E_{T_2} \tag{143}$$

$$\text{div } E = 0 \tag{144}$$

$$\bar{\varepsilon}_1^* E_{n_1} + \bar{\varepsilon}_2^* E_{n_2} = 0 \tag{145}$$

These equations have the same formulation as Eqs. (48) to (51) describing electrostatic phenomena, except for the substitution of $\bar{\varepsilon}^*$ for $\varepsilon_s$.

The complex electrical potential function obeys a set of equations which are similar to Eqs. (52) to (54) or (115) to (117):

$$(\phi)_{M_1} = (\phi)_{M_2} \tag{146}$$

$$\nabla^2 \phi = 0 \tag{147}$$

$$\bar{\varepsilon}_2^* \left(\frac{\partial \phi}{\partial n}\right)_{M_2} - \bar{\varepsilon}_1^* \left(\frac{\partial \phi}{\partial n}\right)_{M_1} = 0 \tag{148}$$

or, alternatively,

$$\bar{\sigma}_2^* \left(\frac{\partial \phi}{\partial n}\right)_{M_2} - \bar{\sigma}_1^* \left(\frac{\partial \phi}{\partial n}\right)_{M_1} = 0 \tag{149}$$

These results are of great importance in dielectric studies since they show that, when the quasi-stationary approximation holds, all formulas derived in the case of static permittivity or steady conductivity can be transposed to the case of effective complex permittivity or effective complex conductivity, as an application of the generalized conductivity principle. This theorem has been stated clearly only in the past few years [4,7,8].

In the following, all formulas relevant to time-dependent dielectric phenomena will not be written in terms of $\bar{\sigma}^*$ but in terms of $\bar{\varepsilon}^*$. For the sake of simplicity, $\bar{\varepsilon}^*$ will be noted $\varepsilon^*$ as has already been done in Eq. (24).

*4. Macroscopic Description of Time-Dependent Polarization Phenomena*

It has been mentioned above that the orientation polarization $P_o$ reaches its static equilibrium value $P_{os}$ after the application of the primary electrical field to the system. The simplest assumption concerning the description of the transient variation of $P_o$ is to take the speed of approach to equilibrium proportional to the distance from equilibrium, so that

$$\frac{dP_o}{dt} = \frac{1}{\tau}(P_{os} - P_o) \tag{150}$$

$\tau$ being a constant called the *macroscopic relaxation time* of the orientation process.

By expressing $P_{os}$ in terms of $P_s$ and $P_d$ by means of Eq. (75), it is possible to transform Eq. (150) to

$$\frac{dP_o}{dt} + \frac{1}{\tau}P_o = \frac{1}{\tau}(P_s - P_d) \tag{151}$$

The solution of this differential equation is given by

$$P_o = (P_s - P_d)(1 - e^{-t/\tau}) \tag{152}$$

$P_o$ varies from zero to $P_{os}$ (equal to $P_s - P_d$) according to an exponential law. As for the global polarization P, it varies from $P_d$ to $P_s$ since

$$P = P_o + P_d \tag{153}$$

Equation (151) holds when the acting electrical field is a function of time, $E = E(t)$, provided that $P_{os}$ be given an instantaneous value, that is, the value $P_o$ would reach if a static electrical field equal to E(t) were applied to the system during an infinite lapse of time:

$$P_{os}(t) = P_s(t) - P_d(t) \tag{154}$$

As $P_s(t)$ and $P_d(t)$ are given, respectively, by

$$P_s(t) = \varepsilon_0(\varepsilon_s - 1)E(t) \tag{155}$$

$$P_d(t) = \varepsilon_0(\varepsilon_d - 1)E(t) \tag{156}$$

Eq. (154) becomes

$$P_{os}(t) = \varepsilon_0(\varepsilon_s - \varepsilon_d)E(t) \tag{157}$$

Introducing this expression for $P_{os}(t)$ into Eq. 151 yields

$$\frac{dP_o}{dt} + \frac{P_o}{\tau} = \frac{\varepsilon_o}{\tau}(\varepsilon_s - \varepsilon_d)E(t) \tag{158}$$

If E(t) is a sinusoidal function of time as indicated by Eq. (121), the general solution of Eq. (158) is the sum of a decaying exponential function of time, which is the transient solution, and of a sinusoidal function of time which is the permanent solution. When the transient solution has faded, $P_o$ is linked to E by

$$P_o(t) = \varepsilon_0 \frac{\varepsilon_s - \varepsilon_d}{1 + j\omega\tau} E(t) \tag{159}$$

and the total polarization is given by

$$P(t) = \varepsilon_0 \left( \varepsilon_d + \frac{\varepsilon_s - \varepsilon_d}{1 + j\omega\tau} - 1 \right) E(t) \tag{160}$$

It is readily seen from Eq. (160) that the complex permittivity of the system is expressed by

$$\varepsilon^* = \varepsilon_d + \frac{\varepsilon_s - \varepsilon_d}{1 + j\omega\tau} \tag{161}$$

Equation (161) is known as the Debye formula, derived by him on a molecular basis. In fact, it may be considered as a phenomenological equation describing the dependency upon frequency of the complex permittivity associated with any polarization process that obeys Eq. (151). It will be shown later on that Eq. (161) can represent as well the features of dielectric relaxations arising from interfacial polarization phenomena in heterogeneous systems.

The real and imaginary parts of $\varepsilon^* = \varepsilon' - j\varepsilon''$ are given by

$$\varepsilon' = \varepsilon_d + \frac{\varepsilon_s - \varepsilon_d}{1 + \omega^2\tau^2} \tag{162}$$

$$\varepsilon'' = \frac{(\varepsilon_s - \varepsilon_d)\omega\tau}{1 + \omega^2\tau^2} \tag{163}$$

When $\omega$ increases, $\varepsilon'$ decreases from $\varepsilon_s$ to $\varepsilon_d$, while $\varepsilon''$ increases from zero up to a maximum value and then decreases to zero. The maximum value of $\varepsilon''$, given by

$$\varepsilon''_m = \frac{\varepsilon_s - \varepsilon_d}{2} \tag{164}$$

is reached when

$$\omega = \frac{1}{\tau} = \omega_c = 2\pi f_c \tag{165}$$

Then $\varepsilon'$ is equal to

$$\varepsilon'_m = \frac{\varepsilon_s + \varepsilon_d}{2} \tag{166}$$

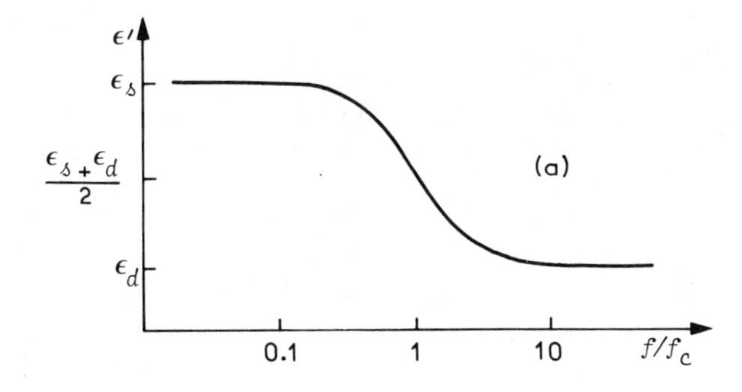


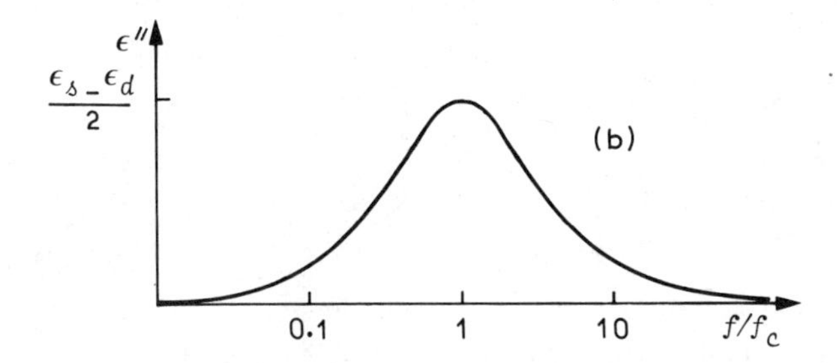


**Figure 4.** Debye-type dielectric relaxation: (a) variations of $\varepsilon'$ versus reduced frequency $f/f_c$ and (b) variations of $\varepsilon''$ versus reduced frequency $f/f_c$.

Figures 4a and b show typical variations of $\varepsilon'$ and $\varepsilon''$ with frequency. By combining Eqs. (162) and (163) it is possible to express $\varepsilon''$ as an implicit function of $\varepsilon'$:

$$\left(\varepsilon' - \frac{\varepsilon_s + \varepsilon_d}{2}\right)^2 + \varepsilon''^2 = \frac{(\varepsilon_s - \varepsilon_d)^2}{4} \tag{167}$$

An elegant geometrical method has been designed by Cole and Cole [284] to depict dielectric relaxation features. It consists of plotting $\varepsilon''$ versus $\varepsilon'$ in cartesian coordinates. An analysis of the diagram thus obtained, which is known as the Cole-Cole plot, allows the characterization of dielectric relaxation phenomena. If the relationship between $\varepsilon'$ and $\varepsilon''$ is given by Eq. (167), the dielectric relaxation being of the Debye type, the Cole-Cole plot is a semicircle whose center lies on the $\varepsilon'$ axis, as shown in Fig. 5. A given point of the semicircle corresponds to a given $\omega$, the apex corresponding to $\omega = \omega_c = 1/\tau$.

In practice, dielectric relaxation phenomena are dealt with in terms of frequency f rather than of angular frequency $\omega$. As the critical frequency $f_c$, corresponding to the maximum value of $\varepsilon''$, is defined by

$$2\pi \tau f_c = 1 \tag{168}$$

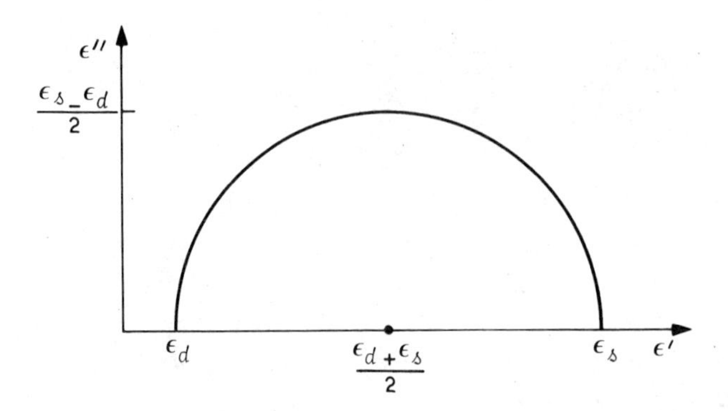


Figure 5. Cole-Cole plot of a Debye-type dielectric relaxation.

Eq. (161) can be written, alternatively,

$$\varepsilon^* = \varepsilon_d + \frac{\varepsilon_s - \varepsilon_d}{1 + jf/f_c} \tag{169}$$

Several alternative formulas have been proposed for the description of dielectric behavior. For instance, Cole and Cole [284] generalized Eq. (169) so that it can be applied to the case of orientation polarization processes involving a symmetrical distribution of relaxation times

$$\varepsilon^* = \varepsilon_d + \frac{\varepsilon_s - \varepsilon_d}{1 + (jf/f_c)^{(1-h)}} \tag{170}$$

$\varepsilon_d$ and $\varepsilon_s$ have the same significance as in Eq. (169), h is a quantity called the *frequency spread parameter* and related to the geometry of the Cole-Cole plot, which is a circular arc whose center lies below the ε' axis, as shown in Fig. 6. In this case, the dielectric behavior is considered as arising from the existence of a continuous distribution of relaxation times, each of them corresponding to a

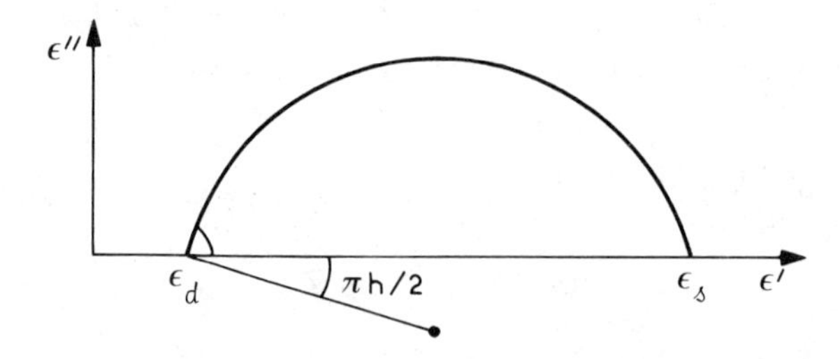


Figure 6. Cole-Cole plot of a Cole-Cole-type dielectric relaxation.

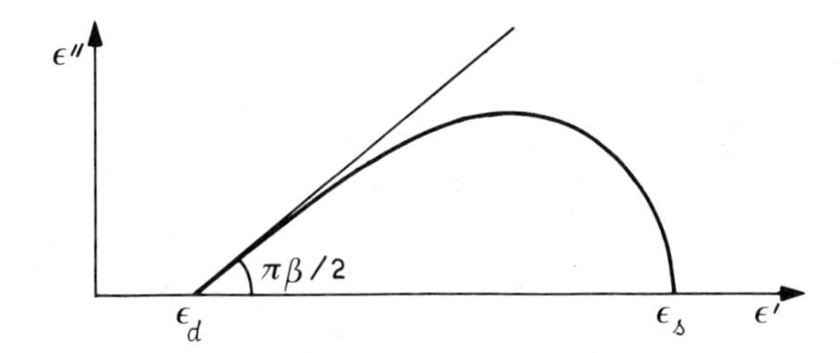


Figure 7. Cole-Cole plot of a Davidson-Cole-type dielectric relaxation.

Debye-type dielectric relaxation. $f_c$ is the critical frequency associated with the most probable relaxation time of the distribution, assumed to be symmetrical and nearly of the Gaussian type. Another formula has been proposed by Davidson and Cole [285] and also by Denney [286]:

$$\varepsilon^* = \varepsilon_d + \frac{\varepsilon_s - \varepsilon_d}{(1 + jf/f_c)^{\beta}} \tag{171}$$

$\beta$ is again a quantity related to the geometry of the Cole-Cole plot which is a skewed arc, as shown in Fig. 7. The corresponding dielectric behavior can be understood as resulting from a series of Debye-type relaxation mechanisms whose intensities are less as their relaxation times are shorter. $f_c$ is the critical frequency associated with the cutoff relaxation time of the distribution which is, in this case, an asymmetrical one.

Equations (170) and (171) should be considered, like Eq. (169), as phenomenological equations describing different dielectric relaxation types. Their relationships with molecular behavior and structure cannot be expressed in a straightforward manner. Comprehensive discussions of this problem can be found in Refs. 6, 10, and 36 to 42.

Many materials show dielectric relaxations of the Cole-Cole type while examples of dielectric relaxations of the pure Debye type are very scarce in the literature. A comparative study made by Davidson and Cole [285] showed that the dielectric behavior of 1-propanol could be well fitted by a Debye-type equation, while glycerol and propylene glycol exhibit dielectric relaxations that conform to Eq. (171).

If the medium is a conductive one, Eqs. (169) to (171) should be corrected by an ohmic term representing the contribution of conduction phenomena to the total loss factor of the substance:

$$\varepsilon^* = \varepsilon_d + \frac{\varepsilon_s - \varepsilon_d}{1 + jf/f_c} - j\,\frac{\sigma}{2\pi\,\varepsilon_0 f} \tag{172}$$

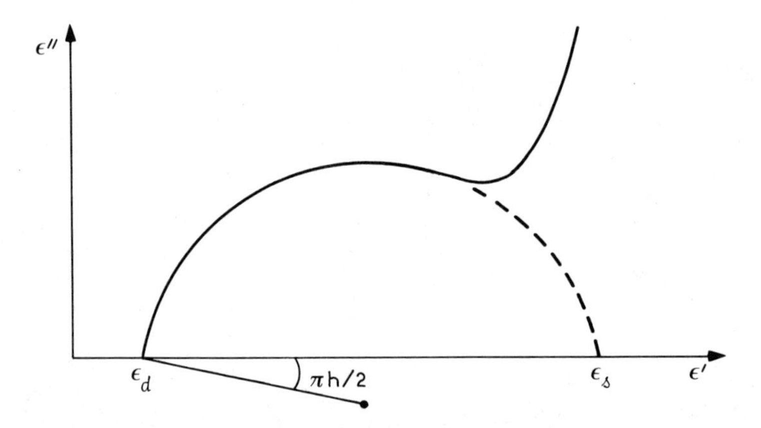


Figure 8. Cole-Cole plot of a Cole-Cole-type dielectric relaxation as modified by the existence of a conduction absorption.

$$\varepsilon^* = \varepsilon_d + \frac{\varepsilon_s - \varepsilon_d}{1 + (jf/f_c)^{(1-h)}} - j\,\frac{\sigma}{2\pi\,\varepsilon_0 f} \tag{173}$$

$$\varepsilon^* = \varepsilon_d + \frac{\varepsilon_s - \varepsilon_d}{(1 + jf/f_c)^{\beta}} - j\,\frac{\sigma}{2\pi\,\varepsilon_0 f} \tag{174}$$

The existence of a conduction absorption term occasions distortions in the dielectric relaxation Cole-Cole plots, as shown in Fig. 8. The distortion in Cole-Cole plots can be very drastic if $\sigma$ is so great that the conduction contribution remains significant even at the higher end of the dielectric relaxation frequency range. The Cole-Cole plot method is a tool which is very helpful in characterizing the type of dielectric relaxation exhibited by a material and in selecting the appropriate phenomenological formula. With the development of computerized numerical analysis techniques, curve-fitting procedures have been designed that allow the determination of dielectric relaxation parameters from the $\varepsilon'(f)$ experimental curve and from knowledge of the steady conductivity [287-289].

The typical temperature dependence of the critical frequency $f_c$ of a dielectric relaxation is given by

$$f_c = Ae^{-u/kT} \tag{175}$$

A is the frequency factor and u the activation enthalpy (almost always referred to as the activation energy), k being Boltzmann's constant, and T the kelvin temperature. If, according to the empirical Arrhenius law, A is a constant, it is possible to evaluate u by plotting the logarithm of $f_c$ against the inverse temperature, the diagram being a straight line whose slope is proportional to u. However, this procedure is not a generally acceptable one because experiments performed over large temperature ranges show that the diagram is very often a smooth curve. This behavior can be ascribed to variations of A with temperature, and different theories have been built thereon [10]. The curvature of the diagram may be explained as well by assuming changes in the activation energy. In any case, within limited temperature variation ranges, it can be considered in practice that the Arrhenius law is applicable to dielectric relaxations. A similar approach can be followed as concerns the conductivity activation energy.

The dielectric dispersion and absorption curves of water have been presented in Fig. 2. From earlier analyses of the data it was established that water exhibited a Debye-type dielectric relaxation in the gigahertz frequency range. Later, it was found [290,291] that a Cole-Cole-type formula with a small relaxation time spread ($h \simeq 0.015$) fitted the phenomenon better. More recently, Sheppard and Grant [292] pointed out that, for dielectric relaxations whose increment is great, as in the water case, it is difficult to distinguish between a Cole-Cole-type relaxation with a small relaxation time spread and the superposition of two Debye-type relaxations whose relaxation times are in the ratio 1/2 or less. Although these analyses are essential within the framework of water molecular structure investigations, they are not of paramount importance for emulsion dielectric studies as the discrepancies they could induce in the complex permittivity of emulsions would be scarcely noticeable compared to experimental uncertainties. Consequently, to avoid unnecessary complications in formulas and calculations, it will be assumed in the following that water and dilute saline aqueous solutions exhibit dielectric relaxations of the Debye type, characterized by a single relaxation time. The data that will be used concerning $\varepsilon_s$, $\varepsilon_d$, and $\tau$, and their variations with temperature, are those reported by Hasted [9,291]. A similar approach has been followed as concerns ice, to which a considerable amount of work has been devoted [293-295].

## III. Permittivity of Heterogeneous Systems

### A. Historical Background and General Considerations

The prediction of physical parameters of heterogeneous systems is a long-standing and classical problem that has attracted the attention of scientists since the middle of the past century. Beyond the evident fundamental interest lying in the adaptation of theoretical laws and experimental results valid for homogeneous media to the case of heterogeneous ones, the main concern in this field can be ascribed to several factors. Heterogeneous substances are omnipresent naturally in everyday life as, for example, foods, pharmaceuticals, textiles, and building materials, and for this simple reason, deserve technological attention. The development of scientific investigation techniques and methods and that of computerized numerical procedures have given scientists and engineers powerful tools allowing them to tackle, among others, the problem of composite materials which are of growing importance because of their multiple potential applications in many branches of industry. Recently, the problem has been rejuvenated and given a new dimension by physicists interested in disordered materials presenting microstructures and so-called microscopic inhomogeneities associated with short-range fluctuations in bonding configurations, composition, or density. Thus, it appears that the studies of the physical properties of heterogeneous systems are involved in a wide range of subjects and activities spanning from crude technological applications to sophisticated problems in solid-state physics, supporting the statement made by Krumhansl that "it's a random world" [296]. Some examples are given below to illustrate the plurality of the subject.

As an application of the theory developed by Fricke [297,298], concerning the conductive and dielectric properties of suspensions of spherical particles, Fricke and Morse [299] proposed a method of determining fat contents in milk and cream. Fricke extended his calculations to the case of membrane-covered ellipsoids [300, 301], and considered MWS effects in suspensions of ellipsoidal particles [302, 303], with the view of studying electrical properties of blood and biological suspensions [304-309]. Woodside and Messmer [310] pointed out the necessity of determining the thermal conductivity of heterogeneous media and gave a nonexhaustive list of situations requiring such information in the field of geophysics: for instance, thermal enhancement of oil production, evaluation of heat dissipation from underground nuclear explosions, and calculations of heat losses from buried steam and hot-water pipes. Earlier, Russel [311] remarked that all insulating structures are porous and he gave some principles of heat flow through materials such as bricks, containing voids commonly filled with air. Wang and Knudsen [312] examined the problem of emulsion thermal conductivity in order to predict heat transfer rates for flows in pipelines. Nielsen [313] has devoted two chapters in his book on polymer rheology to the rheological behavior of suspensions, latices, plastisols, powders, and granular media and described the relevance of such studies to practical problems occurring in the polymer industry, especially as concerns polymeric materials loaded with rigid fillers. The elastic properties of multilayered and fiber or particulate reinforced materials were investigated by Kerner [314], Hashin and Shtrikman [315], Hill [316], Halpin [317], Tsai [318], Miller [319], Korringa [320], Lewis and Nielsen [321], and Nielsen [322,323], who also gave attention to the thermal and electrical properties of heterogeneous structures [324,325]. Mixtures of electrically conducting and isolating substances have many practical applications such as the design of high-quality inductance coils [326], semiconducting screens for power cables [327,328,329], temperature- and frequency-dependent switching materials [330], and devices for microwave detection [331,332]. Studies of electrical conduction in heterogeneous media are also of interest for the evaluation of bulk foam density, [333,334], the design of fixed and fluidized beds of metal particles as extended surface electrodes in electrochemical reactors [335,336,337], and the evaluation of conductivity and Hall constant in disordered polycrystalline solids [338,339]. The principles and methods of investigation of composite material properties, relabeled *effective medium theory* (EMT), or *self-consistent scheme* (SCS), have been applied in recent years, along with the percolation [340-342] to problems in condensed matter, such as the conductive, dielectric, and optical properties of metal-ammonia solutions, alkali-tungsten bronzes, and amorphous germanium and granular metal films [343-346].

Because of and, unfortunately, in spite of the great interest attached to heterogeneous material physical properties, the literature on the subject is abundant and confusing. Among all physical properties of matter, transport coefficients such as permittivity, magnetic permeability, electrical conductivity, thermal conductivity, and diffusivity can be treated on the same footing by virtue of the generalized conductivity principle to be explained later. Any solution derived for any of these coefficients can be transposed to the remaining ones, owing to the formal coincidence of the constitutive relations and differential equations describing the different phenomena. On the basis of the theoretical studies made by Maxwell [115], Rayleigh [347], Garnett [348,349], Wiener [350], Wagner [116, 117], Lichtenecker [351-355], Lichtenecker and Rother [356], Bruggemann [357], Bottcher [358], Landauer [359], and other authors, a multitude of papers have been published mainly concerning the static permittivity of composites. Some of the reasons for this special interest in permittivity problems were the development of artificial dielectrics for technological purposes [360], the growing use of electrical techniques to evaluate moisture contents in materials [195-220], and the necessity of determining the dielectric constants of pulverulent substances by the immersion method [361-381]. The reader can find in the literature on mixture permittivities many errors and misinterpretations that are currently overshadowed by experimental uncertainties, especially when the two component permittivities are not very different. For instance, the MWS effect, which is of paramount importance when dealing with mixtures of conducting components, was ignored by several authors who did not concern themselves with the dependency of the permittivity of such systems upon either the frequency of the applied electrical field or the temperature. When the MWS effect was taken into account, its possible interference with intrinsic dielectric relaxations exhibited by either component of the systems was generally disregarded. On the theoretical side, approximate equations, derived from basic formulas and applicable to special cases, were considered as rigorous ones and utilized with no discernment. The same remark applies as to equations containing empirical coefficients introduced by some authors to fit their experimental results. The geometrical configuration of the system under study was sometimes disregarded and formulas valid for particular geometries were applied to entirely different ones, although several authors had pointed out that the overall permittivity of a heterogeneous system is strongly dependent upon its geometry [357,382,383].

One of the most striking errors arises from a wrong interpretation of a would-be principle of symmetry, considered by many authors as compulsory for any mixture law in any situation. As far as statistical mixtures are concerned (for instance, mixtures of two different pulverulent substances), it may be assumed reasonably, in a first rough analysis, that the system total transport coefficient $\Theta$ will be given by a formula of the type

$$\mathfrak{G}(\Theta) = \Phi\mathfrak{G}(\Theta_1) + (1 - \Phi)\mathfrak{G}(\Theta_2) \tag{176}$$

where $\Phi$ and $1 - \Phi$ are the volume fractions of phases 1 and 2 whose transport coefficients are $\Theta_1$ and $\Theta_2$. To describe the electrical properties of mixtures, Lichtenecker [351-356] proposed the family of formulas

$$\Theta^k = \Phi\Theta_1^k + (1 - \Phi)\Theta_2^k \tag{177}$$

to which must be added a particular one, known as the *logarithmic law*,

$$\log \Theta = \Phi \log \Theta_1 + (1 - \Phi) \log \Theta_2 \tag{178}$$

With k equal to $\pm 1$, the general Eq. (177) yields two special equations, written

$$\Theta = \Phi\Theta_1 + (1 - \Phi)\Theta_2 \tag{179}$$

$$\frac{1}{\Theta} = \frac{\Phi}{\Theta_1} + \frac{1 - \Phi}{\Theta_2} \tag{180}$$

It will be shown below that the solutions of Eqs. (179) and (180), which are exact ones in the case of stratified structures whose stratification axis is either perpendicular or parallel to the flux direction, are upper and lower bounds for the effective value of the total transport coefficient of any heterogeneous system, whatever its geometry. With $k = 1/3$, Eq. (177) gives the following

$$\Theta^{1/3} = \Phi\Theta_1^{1/3} + (1 - \Phi)\Theta_2^{1/3} \tag{181}$$

which has been derived by Landau and Lifshitz [35], and later by Looyenga [384], in the case of permittivity. With $k = 1/2$, Eq. (177) yields

$$\Theta^{1/2} = \Phi\Theta_1^{1/2} + (1 - \phi)\Theta_2^{1/2} \tag{182}$$

This formula, that was proposed by Beer [385], has been used by Kraszewski and others [386,387]. Upon exchanging $\Phi$ and $1 - \Phi$ and $\Theta_1$ and $\Theta_2$, the value of $\Theta$, as given by the general formula (176) or by any particular formula of the same type, remains the same. This result can be admitted, under certain circumstances, for statistical mixtures, but it does not hold when dispersions are concerned, as one of the two phases is continuous, while the other one consists of separate particles. This fact, which was clearly expressed by Poley more than twenty years ago [388], can be illustrated clearly from Lichtenecker's logarithmic law which, curiously enough, has been much in favor in the past. The transport coefficient considered being the electrical conductivity, the logarithmic law implies that a continuous conductor will act as a perfectly insulating medium if it contains at most one small inclusion made of a perfect insulator, which is an absurd result. Similar absurdities may be derived from all symmetrical equations which, consequently, cannot describe the transport properties of systems consisting of separate particles suspended in a continuous matrix. Even in the case of statistical mixtures, symmetrical equations as expressed by general Eq. (176) cannot be applied in every situation because of the existence of percolation thresholds [342]. This restriction can be understood by considering, for instance, the case of conduction in a heterogeneous disordered medium that will not act as a conductor if the conducting phase content does not reach a critical value that allows the formation of a conducting path stretched throughout the sample ("infinite cluster").

From the most general standpoint, the total transport coefficient value of a binary heterogeneous system $\Theta$ is expressed by a general relation involving $\Theta_1$, $\Theta_2$, the values of the transport coefficient for constituents 1 and 2, and a parameter characteristic for the constituent proportions, for instance $\Phi$, the volume fraction of constituent 1, i.e.,

$$\Theta = \mathfrak{G}(\Theta_1, \Theta_2, \Phi) \tag{183}$$

The function $\mathfrak{G}$ must fulfill adequate analytical conditions and must contain all the information about the geometry of the system and the physical interactions taking place within the system.

The evaluation of the permittivity of a composite system from the known properties of the components must be performed by solving the general equations of electromagnetism, with respect to the system characteristics. In the electrostatic case, if the source densities are assumed to be localized outside the composite, the electrical field inside the system $\mathbf{E}$ is a curl-free vector and the electrical displacement $\mathbf{D}$ is a divergence-free vector. As $\mathbf{D}$ and $\mathbf{E}$ are linked by

$$\mathbf{D} = \varepsilon_0 \varepsilon_l \mathbf{E} \tag{184}$$

where $\varepsilon_l$ is the local static permittivity which is a function of the coordinates, $\mathbf{E}$ satisfies the equations

$$\text{curl } E = 0 \tag{185}$$

$$\text{div}(\varepsilon_l E) = 0 \tag{186}$$

A definition of the total static permittivity of the heterogeneous system (often called the effective static permittivity) can be obtained by evaluating the electrostatic energy stored by a parallel-plate capacitor filled with the composite [389, 390],

$$W = \frac{\varepsilon_0}{2}\int_V \varepsilon_l E^2 \, dV \tag{187}$$

V being the internal volume of the capacitor. If the composite is replaced by a homogeneous filler characterized by a static permittivity $\varepsilon_s$ so that W is not changed, then

$$W = \frac{1}{2} \varepsilon_0 \varepsilon_s \bar{E}^2 V \tag{188}$$

$\bar{E}$ being the averaged electrical field, given by

$$\bar{E} = \frac{1}{V}\int_V E \, dV \tag{189}$$

Thus, the effective static permittivity of the composite material is expressed by

$$\varepsilon_s = \frac{\int_V \varepsilon_l E^2 \, dV}{\bar{E}^2 V} \tag{190}$$

Owing to the properties of E and D [391], this definition is equivalent to

$$\bar{D} = \varepsilon_s \bar{E} \tag{191}$$

where $\bar{D}$ and $\bar{E}$ are the averaged values of D and E over V [392-394]. As pointed out by Bergman [389], the problem of solving Eq. (190) so as to obtain a formula of the general type of Eq. (183) is usually a formidable one, even if the geometry of the system is completely known and is fairly regular.

In a limited number of simple cases, exact solutions or solutions with good approximations can be easily computed by using generally the properties of the electrical potential function as described by Eqs. (52) to (54). For instance, in binary-layered systems where $\varepsilon_l$ is constant over constituent 1 and equal to $\varepsilon_{1s}$ and constant over constituent 2 and equal to $\varepsilon_{2s}$, the effective static permittivity of the system is given by

$$\varepsilon_s = \Phi \varepsilon_{1s} + (1 - \Phi)\varepsilon_{2s} \tag{192}$$

when the electrical field is perpendicular to the stratification axis or by

$$\frac{1}{\varepsilon_s} = \frac{\Phi}{\varepsilon_{1s}} + \frac{1 - \Phi}{\varepsilon_{2s}} \tag{193}$$

when the field is parallel to the stratification axis. Equations (192) and (193) are valid only if it is assumed that no special phenomena occur at the interfaces. Similarly, the effective static permittivity of a system consisting of spherical particles characterized by $\varepsilon_{1s}$ and sparsely dispersed within a continuous medium characterized by $\varepsilon_{2s}$ is expressed as follows,

$$\frac{\varepsilon_s - \varepsilon_{2s}}{\varepsilon_s + 2\varepsilon_{2s}} = \frac{\varepsilon_{1s} - \varepsilon_{2s}}{\varepsilon_{1s} + 2\varepsilon_{2s}} \Phi \tag{194}$$

$\Phi$ being the volume fraction of disperse globules. This equation, known as the *Wiener formula* [350], is valid for small values of $\Phi$ because particle electrostatic interactions are neglected in its derivation. If special phenomena arise at the interfaces between the particles and the continuous matrix, amendments such as those suggested by Trukhan [119] must be introduced in Eq. (194). In fact, Eq. (194) is the spherical formulation of a more general equation valid for dilute dispersions of ellipsoidal particles in a continuous medium, [297,382,394-396].

Details concerning the effective static permittivity of mixtures and composites can be found in Refs. 3 to 5, 7, 8, 63, 201, 359, 372, 374, and 394 to 403. Also relevant to the subject are two papers by Meredith and Tobias who consider the problem of flow through heterogeneous systems [404,405], an article by Tinga and coauthors [406], a recent book by Nielsen who made an attempt to clarify the subject of the physical properties of mixtures [407], and a paper by Landauer which is a review of electrical conductivity in inhomogeneous media [408].

Since no explicit solution of Eq. (190) can be easily derived in the general case, when the system geometry is too complicated or not accurately known and/or when $\varepsilon_l$ varies from point to point, one of the modern trends consists of establishing upper and lower bounds for $\varepsilon_s$ by using the available information on constituent volume fractions and on composite macroscopic isotropy [389-393, 409-420].

In the following, after a presentation of the generalized conductivity principle, the problem of upper and lower bounds on $\varepsilon_s$ will be reported briefly, along with the mathematical *theory of homogenization* which has been reported recently [421-425]. Following this, an analysis will be given concerning formulas for the static permittivity of systems of the spherical dispersion type, these formulas being relevant directly to the dielectric study of emulsions.

## B. The Generalized Conductivity Principle

Among the physical properties, a special class is formed by the so-called transport coefficients in motionless media. As pointed out by Dukhin and Shilov [7,8], this group of properties may be combined under the common term *generalized conductivity* because of the formal coincidence of the differential equations describing steady fluxes in different situations. This approach applies to the static permittivity, the steady conductivity, the magnetic permeability, the thermal conductivity, and the diffusivity. The cases of viscosity and elastic modulus, although closely related to the preceding ones, are somewhat different.

Using the vocabulary of the thermodynamics of irreversible processes, it can be said that the generalized conductivity $\Theta$ is a phenomenological or kinetic coefficient appearing in the relation linking the flux vector $\mathcal{J}$ to the thermodynamic force $\mathfrak{X}$:

$$\mathcal{J} = \Theta \mathfrak{X} \tag{195}$$

In any region not containing source densities, the steady flux density vector is divergence free,

$$\text{div } \mathcal{J} = 0 \tag{196}$$

which implies, if the region is filled with a LHI medium,

$$\text{div } \mathfrak{X} = 0 \tag{197}$$

In the case of adjacent LHI media, Eq. (195) is valid throughout the bulk of each medium provided that $\Theta$ is given the adequate values. But Eq. (196) must be completed by an analogous surface equation stating the conservation of flux at the interfaces

$$\mathcal{J}_{n_1} + \mathcal{J}_{n_2} = 0 \tag{198}$$

implying

$$\Theta_1 \mathfrak{X}_{n_1} + \Theta_2 \mathfrak{X}_{n_2} = 0 \tag{199}$$

An additional equation concerns the continuity of the tangential component of the thermodynamic force when crossing an interface,

$$\mathfrak{X}_{T_1} = \mathfrak{X}_{T_2} \tag{200}$$

Equations (195), (197), (199), and (200) constitute a general formulation for particular equations relative to steady electrical, magnetic, diffusion, and heat-conduction phenomena. This result is of particular interest for the determination of total transport coefficients of composite systems. Any solution derived for a particular coefficient is applicable to any other one, provided that the system characteristics are identical in both cases. For instance, the total static permittivity of heterogeneous systems is expressed by the same equation as that relative to the total steady conductivity. It has been shown previously that equations similar to (195), (197), (199) and (200) hold for time-dependent sinusoidal electrical fields, provided that the generalized conductivity coefficient represents the complex permittivity. If the quasi-stationary approximation is assumed to be applicable, the equation expressing the complex permittivity of a heterogeneous system with given geometry has a formulation identical to that of the equation giving the static permittivity, and consequently, any formula valid for the static permittivity of composite media can be transposed to the complex permittivity case [4,7,8].

## C. Modern Approaches to the Problem of Heterogeneous System Physical Properties

### 1. *The Theory of Homogenization*†

The problem of heterogeneous system physical properties has been investigated recently by mathematicians who gave a rigorous formulation involving variational methods in classical Sobolev spaces and G-convergence principles. A book on the subject has been written recently by Bensoussan et al. [425], in which are quoted papers by other scientists interested in the same topic or in closely related ones. The spirit of the theory of homogenization can be illustrated by a simple example, as follows.

A binary composite material with two-dimensional periodic structure can be represented, in $\mathbb{R}^2$, by a domain $\Omega$ divided into a great number of adjacent cells (Fig. 9a). Each cell, assumed to be a square for the sake of simplicity, is the homothetic by ratio $\eta$ of a basic unit square cell containing all the information about the geometrical distribution of the system components whose physical properties are different (Fig. 9b).

If the physical property considered is a phenomenological coefficient $\Theta$, relating a flux density to a thermodynamic force deriving from a potential, in the most general case the potential function satisfies a generalized Poisson-type equation,

$$\sum_{i=1}^{2} \sum_{j=1}^{2} \frac{\partial}{\partial x_j} \left[ \Theta_{ij}\left(\frac{x_1}{\eta}, \frac{x_2}{\eta}\right) \frac{\partial \phi_\eta}{\partial x_i} \right] = f \tag{201}$$

with boundary conditions for $\phi_\eta$. The theory of homogenization states that when $\eta \to 0$, $\phi_\eta$ converges, in a certain sense, toward a function $\phi$ which satisfies an

---

†The author is indebted to Professor J. Genet (Department of Mathematics, University of Pau, France) for fruitful discussions on the subject of homogenization.

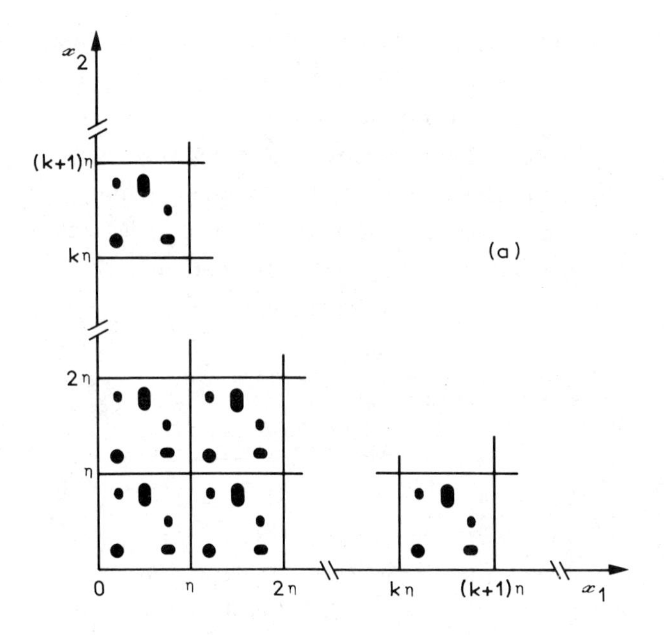


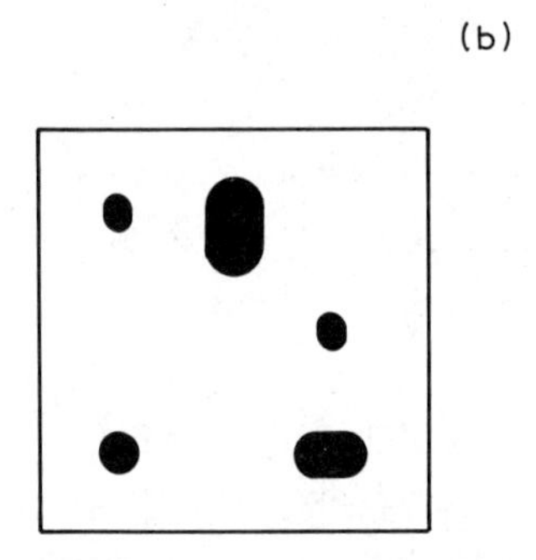


**Figure 9.** Schematic representation of a binary composite material with periodic structure, in $\mathbb{R}^2$: (a) regular arrangement of identical cells and (b) basic cell showing the geometrical distribution of the two components.

equation similar to Eq. (201) except for the replacement of the $\Theta_{ij}$ with *homogenized* coefficients $\tilde{\Theta}_{ij}$

$$\sum_{i=1}^{2} \sum_{j=1}^{2} \tilde{\Theta}_{ij} \frac{\partial^2 \phi}{\partial x_i \, \partial x_j} = f \tag{202}$$

The solution of Eq. (202) gives an accurate representation of the total behavior of the composite material toward the phenomenon considered. In addition, it is an answer to the problem of computing $\phi_\eta$, which is a difficult one since, generally, the phenomenological coefficient $\Theta$ is discontinuous at the interfaces and may undergo drastic variations. It must be remarked that the subscripts i and j in $\tilde{\Theta}_{ij}$ indicate that the composite material may be macroscopically anisotropic.

The result of homogenization can be simply described for a model problem as shown hereafter. Assuming that the model system has a one-dimensional periodic structure as shown in Fig. 10, the homogenized coefficient $\tilde{\Theta}$ is given by

$$\tilde{\Theta} = \frac{1}{m(1/\Theta)} \tag{203}$$

with

$$m\left(\frac{1}{\Theta}\right) = \int_0^1 \frac{du}{\Theta(u)} \tag{204}$$

u being the normalized space coordinate ($u = x/L$). If, in each $\eta$ cell, $\Theta$ takes a constant value $\Theta_1$ over the length fraction $\eta\Phi$ ($0 \leqslant \Phi \leqslant 1$), and a constant value $\Theta_2$ over the complementary length fraction $\eta(1 - \Phi)$, as represented on Fig. 11, then $\tilde{\Theta}$ is given by

$$\frac{1}{\tilde{\Theta}} = \frac{\Phi}{\Theta_1} + \frac{1 - \Phi}{\Theta_2} \tag{205}$$

If the physical coefficient considered is the static permittivity, Eq. (205) yields Eq. (193) which gives the total static permittivity of a binary layered medium whose stratification axis is parallel to the static electrical field direction.

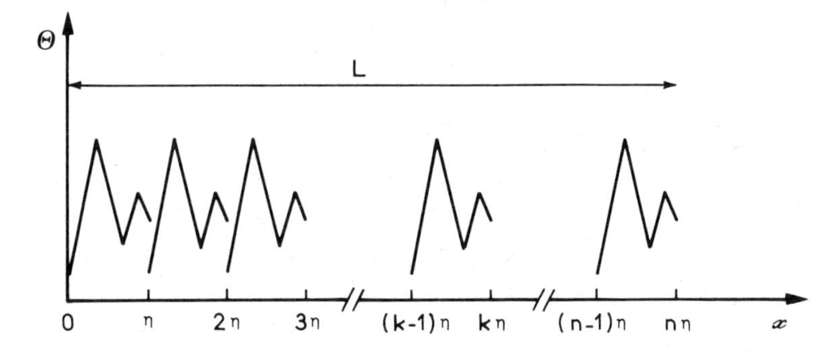


**Figure 10.** One-dimensional representation of a binary composite cellular material with periodic structure, showing the variation of $\Theta$ with x.

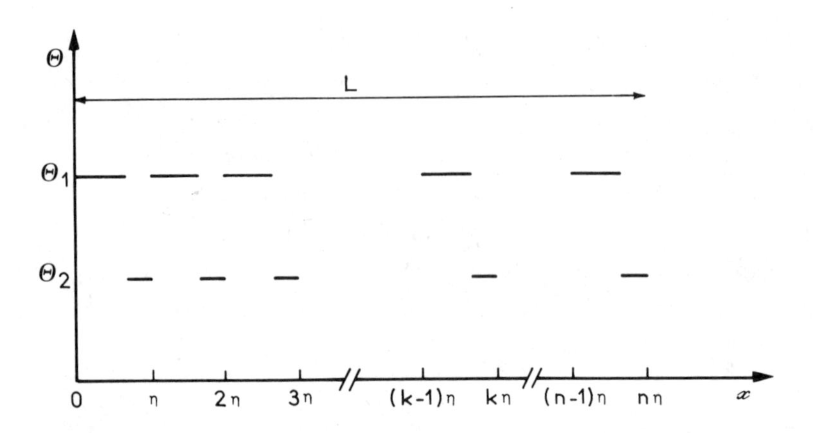


Figure 11. One-dimensional representation of a binary composite cellular material with periodic structure, $\Theta$ periodically taking two constant values, $\Theta_1$ and $\Theta_2$.

## 2. *The Upper and Lower Bounds Method*

As has been previously pointed out, the computation of the total physical coefficient of a composite material from general formulas is usually a formidable task because the result cannot be defined completely from the physical properties and volume fractions of the constituents. This aspect of the problem has been expressed clearly by Brown [392], who stressed that statistical information about the geometrical distribution of the constituents is needed to approach the actual result accurately. Brown derived different alternative formulas in the form of series expansions showing that equations of the types (178), (179), (180), and (194) yield comparable values, whatever the geometry, provided that $\Phi$ and $|\varepsilon_{1s} - \varepsilon_{2s}|$ be small. For better approximations, terms of order equal to or higher than the third and involving geometrical characteristics must be retained in the series expansions.

Following Brown's analysis, several authors have calculated more and more stringent upper and lower bounds for the physical properties of heterogeneous materials, in particular for the so-called transport coefficients. By virtue of the generalized conductivity principle, bounds derived for a given transport coefficient are applicable in a straightforward manner to others.

Speaking in terms of static permittivity, the less restrictive upper and lower bounds that can be calculated for the effective static permittivity of a heterogeneous medium, without any information on its geometry, are the arithmetic average and the harmonic average taken over the system [391,411] as indicated in the following formula:

$$\frac{1}{\overline{1/\varepsilon_l}} \leqslant \varepsilon_s \leqslant \tilde{\varepsilon}_l \tag{206}$$

$\varepsilon_l$ representing the local static permittivity. If the heterogeneous medium considered is a binary structure, Eq. (206) means that the effective static permittivity has a value included between an upper bound given by Eq. (192) and a lower bound given by Eq. (193). This result was suggested a long time ago by Wiener [350], and Eqs. (192) and (193) are known as *Wiener's limiting formulas.*

After a study of the statistical properties of the electrical field in a medium with small random variations in static permittivity, Beran [410] used a variational approach to derive upper and lower bounds which are expressed by intricate formulae involving third order correlation functions [411]. In the limit of small perturbations in permittivity, each bound converges toward the other, which yields, for the effective permittivity, the following equation:

$$\varepsilon_s = \bar{\varepsilon}_l \left( 1 - \frac{1}{3} \frac{\overline{\delta}^2}{\bar{\varepsilon}_l^2} \right) \tag{207}$$

where $\bar{\varepsilon}_l$ is the average value of $\bar{\varepsilon}_l$ in an ensemble sense and $\delta$ represents the fluctuating part of $\varepsilon_l$, $\overline{\delta} = 0$. In the case of statistically homogeneous and isotropic binary composite media, Eq. (207) is transformed into

$$\varepsilon_s = \tilde{\varepsilon}_l \left[ 1 - \frac{\Phi(1 - \Phi)(\varepsilon_{1s} - \varepsilon_{2s})^2}{3\tilde{\varepsilon}_l^2} \right] \tag{208}$$

$\tilde{\varepsilon}_l$ is Wiener's upper bound, $\varepsilon_{1s}$ and $\varepsilon_{2s}$ the static permittivities of constituents 1 and 2, $\Phi$ being the volume fraction of constituent 1. Equation (208) is a truncated form, to order 2, of the general series expansion formula previously derived by Brown [392].

Equation (207), which has been suggested as well by Landau and Lifshitz [35], being valid in the case of small perturbations, Eq. (208) applies when $\varepsilon_{1s}$ and $\varepsilon_{2s}$ are not too different one from the other. More stringent bounds than Wiener's have been calculated by Hashin and Shtrikman [391, 409], who applied variational methods to maximize or minimize the Gibbs free energy of a dielectric body. In the case of binary composite media which are statistically isotropic and homogeneous, the upper bound $\varepsilon_s^+$ and the lower bound $\varepsilon_s^-$ are expressed as follows, when, for instance, $\varepsilon_{1s} > \varepsilon_{2s}$:

$$\frac{\varepsilon_s^+ - \varepsilon_{1s}}{\varepsilon_s^+ + 2\varepsilon_{1s}} = \frac{\varepsilon_{2s} - \varepsilon_{1s}}{\varepsilon_{2s} + 2\varepsilon_{1s}} (1 - \Phi) \tag{209}$$

$$\frac{\varepsilon_s^- - \varepsilon_{2s}}{\varepsilon_s^- + 2\varepsilon_{2s}} = \frac{\varepsilon_{1s} - \varepsilon_{2s}}{\varepsilon_{1s} + 2\varepsilon_{2s}}\phi \tag{210}$$

If $\varepsilon_{1s} < \varepsilon_{2s}$, Eq. (209) gives the lower bound and Eq. (210) the upper bound. As $\varepsilon_s^+$ and $\varepsilon_s^-$ are given by equations similar to Eq. (194), it appears that the effective static permittivity of a statistically isotropic and homogeneous binary composite system is bounded by the static permittivity of a dispersion of spherical particles with $\varepsilon_{1s}$ in a continuous matrix with $\varepsilon_{2s}$ and by the static permittivity of a dispersion of spherical particles with $\varepsilon_{2s}$ in a continuous matrix with $\varepsilon_{1s}$, the constituent contents being the same in both situations. Consequently, $\varepsilon_s^+$ or $\varepsilon_s^-$ can be reached in the case of systems consisting of spherical particles of one constituent sparsely dispersed within a continuous matrix made of the other constituent. In a later paper [412], Hashin showed that, for binary materials of arbitrary phase geometry, the EMT theory does not yield more information about the effective static permittivity than the upper and lower bounds as expressed by Eqs. (209) and (210). Improvements on Hashin and Shtrikman's bounds have been proposed by Brown [393] on the basis of variational theorems by Miller [414], who used the numerical values of three-point correlation functions in cell materials, by Prager [415] and Bergman [389], who suggested the introduction of experimental information. Comprehensive comments on the subject of upper and lower bounds can be found in Refs. 390, 393, and 419.

### D. Formulas for the Static Permittivity of Dispersions of Spheres in a Continuous Matrix

From the preceding developments, it can be inferred that it is possible to make theoretical quantitative predictions as concerns the permittivity of binary systems of the spherical dispersion type. On the basis of the theory of homogenization, it is ascertained that the problem is a soluble one, and the method of upper and lower bounds permits us to define the domain where the solution must be sought. At low disperse particle contents, the exact solution is given by either $\varepsilon_s^+$, Eq. (209), or $\varepsilon_s^-$, Eq. (210), depending on the case. For higher disperse phase contents, the exact solution cannot be formulated simply and approximations must be made so as to obtain practical working equations. In the following, a critical review will be made for equations which deserve attention when studying the dielectric behavior of emulsions, which generally consist of spherical droplets dispersed within a continuous matrix. These equations have a common basis, that is, the electrical polarization of a spherical body under the influence of an applied uniform electrical field.

### 1. *Basic Formulas*

a. Wiener's Equation for Dilute Dispersions of Spherical Particles. This equation, which has been already mentioned, Eq. (194), can be demonstrated quite simply by applying Eq. (96) to the evaluation of the equivalent dipole moment of a system of spherical particles with static permittivity $\varepsilon_{1s}$ imbedded in a continuous matrix with static permittivity $\varepsilon_{2s}$, the system being submitted to the influence of an initially uniform applied electrical field designated by $E_0$ (Fig. 12a).

Should any of the particles be isolated, its equivalent dipole moment would be

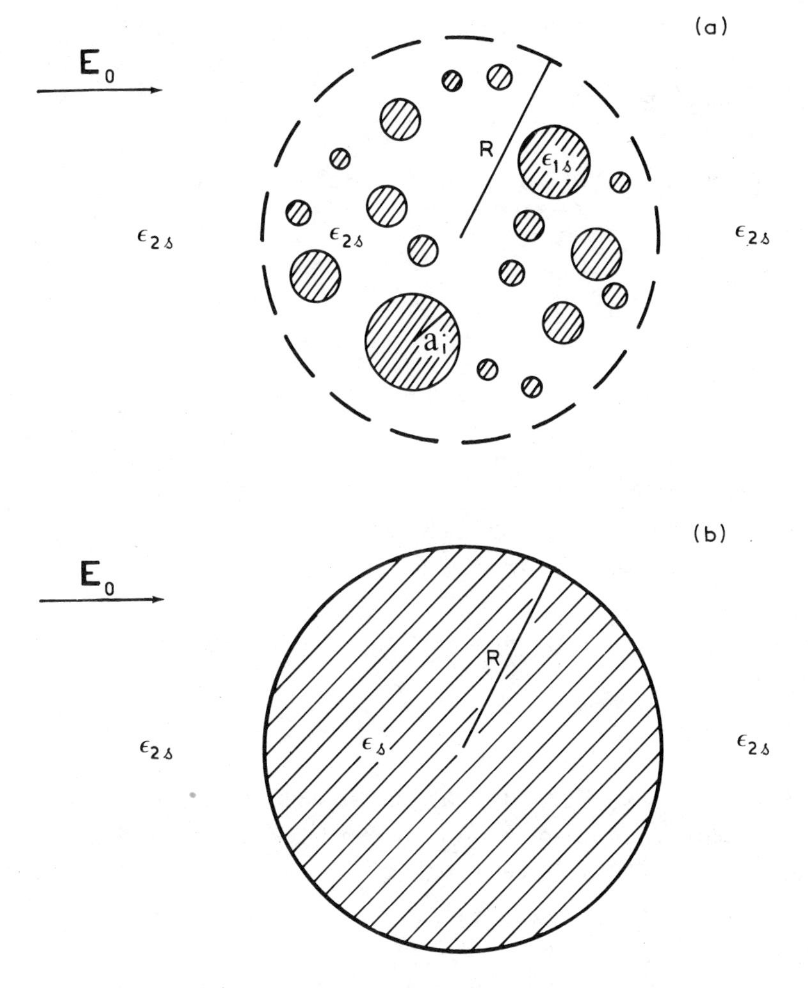


Figure 12. Wagner's model for spherical dispersion type systems with low-disperse-phase contents: (a) schematic representation of the actual situation and (b) equivalent spherical body.

$$p_i = \left(\frac{4}{3}\pi a_i^3\right) 3\varepsilon_0 \frac{\varepsilon_{1s} - \varepsilon_{2s}}{\varepsilon_{1s} + 2\varepsilon_{2s}} E_0 \tag{211}$$

If the interparticle distances are large enough so that dipolar and multipolar interactions can be neglected, $p_i$ as given by Eq. (211) represents the contribution of an individual particle to the equivalent total dipole moment. This then can be expressed as follows, $N_i$ representing the number of particles with radius $a_i$:

$$p_T = \sum_i N_i p_i \tag{212}$$

Assuming that the spherical region of radius R in which all the particles lie can be considered as a macroscopically homogeneous spherical body characterized by a static permittivity $\varepsilon_s$ and imbedded in the medium with static permittivity $\varepsilon_{2s}$ (Fig. 12b), the equivalent dipole moment of this spherical body is given by

$$p' = \left(\frac{4}{3}\pi R^3\right) 3\varepsilon_0 \frac{\varepsilon_s - \varepsilon_{2s}}{\varepsilon_s + 2\varepsilon_{2s}} E_0 \tag{213}$$

The identification of p' with $p_T$ yields the equation which must be satisfied by $\varepsilon_s$,

$$\frac{\varepsilon_s - \varepsilon_{2s}}{\varepsilon_s + 2\varepsilon_{2s}} = \frac{\varepsilon_{1s} - \varepsilon_{2s}}{\varepsilon_{1s} + 2\varepsilon_{2s}} \Phi \tag{214}$$

As $\Phi$ is given by

$$\Phi = \frac{\Sigma_i N_i a_i^3}{R^3} \tag{215}$$

it represents the volume fraction corresponding to the disperse particles. It must be stressed that Eq. (214) is applicable only to dilute systems, since the mutual electrical interactions between disperse particles have been neglected during its derivation.

For the sake of simplicity, Eq. (214) will be written in some later developments in a condensed form, as

$$\varepsilon_s = \mathfrak{M}(\varepsilon_{1s}, \varepsilon_{2s}, \Phi) \tag{216}$$

Equation (214) was proposed by Maxwell [115], and later by Cole [426], to express the equivalent permittivity of a spherical body consisting of a spherical core with $\varepsilon_{1s}^c$ surrounded by a concentric shell with $\varepsilon_{1s}^s$, as shown in Fig. 13.

Thus, the static permittivity $\varepsilon_{1s}$ of such a shell-covered spherical particle is expressed by

$$\frac{\varepsilon_{1s} - \varepsilon_{1s}^s}{\varepsilon_{1s} + 2\varepsilon_{1s}^s} = \frac{\varepsilon_{1s}^c - \varepsilon_{1s}^s}{\varepsilon_{1s}^c + 2\varepsilon_{1s}^s} \underline{\phi} \tag{217}$$

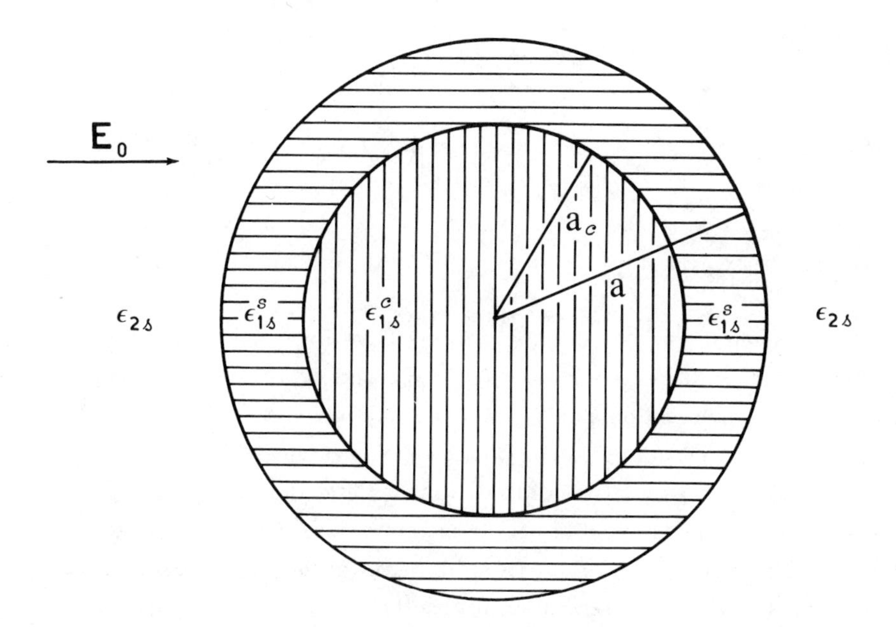


**Equation 13.** Schematic representation of a spherical particle covered with a shell: a, external radius; $a_c$ core radius; $\varepsilon_{1s}^{s}$, shell static permittivity; $\varepsilon_{1s}^{c}$, core static permittivity.

with $\underline{\phi}$, the volume fraction corresponding to the core, given by

$$\underline{\phi} = \left(\frac{a_c}{a}\right)^3 = \frac{a_c^3}{(a_c + d)^3} \tag{218}$$

where $a_c$ is the core radius, a the shell external radius and d the shell thickness. In that case, Eq. (216) remains valid for any value of $\underline{\phi}$.

As has been shown for (214), Eq. (217) can be formulated as follows:

$$\varepsilon_s = \mathfrak{W}(\varepsilon_{1s}^{c}, \varepsilon_{1s}^{s}, \underline{\phi}) \tag{219}$$

b. Amended Equations Taking into Account Particle Interactions. When the disperse-phase content increases, the interparticle distances decrease and the mutual electrical interactions between particles must be taken into account. The first theoretical investigation in this direction was performed by Rayleigh [347], who considered rectangular arrays of either cylindrical or spherical obstacles. For identical spherical obstacles arranged in rectangular order in a continuous medium, Rayleigh arrived at the formula

$$\varepsilon_s = \varepsilon_{2s}\left\{1 + 3\Phi\left[\frac{\varepsilon_{1s} + 2\varepsilon_{2s}}{\varepsilon_{1s} - \varepsilon_{2s}} - \Phi - 1.65\,\frac{\varepsilon_{1s} - \varepsilon_{2s}}{\varepsilon_{1s} + (4/3)\varepsilon_{2s}}\,\Phi^{10/3}\right]^{-1}\right\} \tag{220}$$

which is, in fact, the beginning of a series expansion. This equation has been later corrected by Runge [427], who substituted the coefficient 0.525 for 1.65, which appears in Eq. (220), because of the omission of a factor of $1/\pi$ by Rayleigh. When $\Phi$ is small, the term containing $\Phi^{10/3}$ is negligible, and Eq. (220) reduces to Wiener's Eq. (214).

Other attempts to account for particle interactions have been made by Kharadly and Jackson [428], by Meredith and Tobias [404,405], who improved on Rayleigh's derivation, and by Gunther and Heinrich [429], who performed the evaluation of higher order multipolar contributions to the particle polarization and to the effective permittivity of the heterogeneous medium. Gunther and Heinrich's calculations are an illustration of the formidable problems that arise when attempting the exact computation of Eqs. (190) or (191). In spite of its great theoretical interest, Gunther and Heinrich's method is intractable in practice.

c. Approximate Equations for Concentrated Dispersions of Spherical Particles (EMT). Several formulas have been proposed to represent the static permittivity of concentrated systems without resorting to complicated developments such as those made by Gunther and Heinrich. Although it is somewhat difficult to evaluate clearly the approximations they contain, these formulas are, from the practical point of view, very useful ones because they can be handled without too many difficulties. In any case, the values of the static permittivity that can be derived from them fall between the upper and lower bounds that have been calculated by different authors.

*(1) Bottcher-Landauer equation.* This equation has been derived by Landauer [359], on the basis of Eq. (96), but with the assumption that both constituents of the composite material exist as adjacent particles and that the medium surrounding each particle can be considered as a homogeneous one, characterized by $\varepsilon_s$, the effective static permittivity sought after. Landauer's equation is written

$$\frac{\varepsilon_{1s} - \varepsilon_s}{\varepsilon_{1s} + 2\varepsilon_s}\,\Phi + \frac{\varepsilon_{2s} - \varepsilon_s}{\varepsilon_{2s} + 2\varepsilon_s}\,(1 - \Phi) = 0 \tag{221}$$

This equation has a symmetric form, as could have been predicted from the assumption made as concerns the constituent geometrical distribution. Consequently, it is evident that Eq. (221) cannot be utilized when dissymetric systems such as emulsions are considered. However, recent experimental data tend to show that Eq. (221) could be used to account for the conductive behavior of certain microemulsion systems.

Landauer's formula was given previously by Bruggeman [357] in the same form, and also by Bottcher [358], with the following alternative formulation:

$$\frac{\varepsilon_s - \varepsilon_{2s}}{3\varepsilon_s} = \frac{\varepsilon_{1s} - \varepsilon_{2s}}{\varepsilon_{1s} + 2\varepsilon_s}\,\Phi \tag{222}$$

*(2) Bruggeman equation.* Bruggeman [357] used a very peculiar and ingenious procedure by considering that a spherical dispersion system can be built up by repetitive additions of disperse phase in infinitesimal amounts. At a given state, the static permittivity of the system is equal to $\varepsilon_s$ and the disperse-phase volume fraction to $\Phi'$. A further addition of a small disperse-phase amount induces an increase in $\Phi'$ equal to $d\Phi'$ and a variation of $\varepsilon_s$ equal to $d\varepsilon_s$. Bruggeman considers that the new value of the system static permittivity $\varepsilon_s + d\varepsilon_s$ can be expressed by Wiener's equation where $\varepsilon_s$ is replaced by $\varepsilon_s + d\varepsilon_s$, $\varepsilon_{2s}$ by $\varepsilon_s$, and $\Phi$ by $d\Phi'/(1 - \Phi')$, which represents the volume fraction of the added small amount of disperse phase. Thus, a differential equation linking $\varepsilon_s$ to $\Phi'$ is obtained:

$$\frac{2\varepsilon_s + \varepsilon_{1s}}{3\varepsilon_s(\varepsilon_s - \varepsilon_{1s})}\,d\varepsilon_s = -\frac{d\Phi'}{1 - \Phi'} \tag{223}$$

This equation can be integrated without any difficulty for $0 \leqslant \Phi < 1$, since it is granted that $\varepsilon_s$ is different from zero and, according to the method of upper and lower bounds, different from $\varepsilon_{1s}$ as well. So, when $\Phi'$ varies from zero to its final value $\Phi$, $\varepsilon_s$ is given by the integral equation

$$\int_{\varepsilon_{2s}}^{\varepsilon_s} \frac{2x + \varepsilon_{1s}}{3x(x - \varepsilon_{1s})}\,dx = \int_0^{\Phi} \frac{-d\Phi'}{1 - \Phi'} \tag{224}$$

which yields

$$\left(\frac{\varepsilon_{1s} - \varepsilon_{2s}}{\varepsilon_{1s} - \varepsilon_s}\right)^3 \frac{\varepsilon_s}{\varepsilon_{2s}} = \frac{1}{(1 - \Phi)^3} \tag{225}$$

It is somewhat difficult to point out exactly what are the implicit approximations inherent to Bruggeman's method. As will be shown later, Bruggeman's equation has been checked by different authors for different situations. Moreover, numerical calculations performed by Gunther and Heinrich [429] show that their own equation and Eq. (225) gives quite comparable values for $\varepsilon_s$, up to about $\Phi = 0.40$ when $\varepsilon_{1s} \gg \varepsilon_{2s}$ and up to about $\Phi = 0.60$ when $\varepsilon_{1s} \ll \varepsilon_{2s}$. One of the main arguments developed against Bruggeman's formula is that it should not hold for highly concentrated systems because of the existence of the maximum packing limit which is equal to $\pi\sqrt{2}/6$ for spheres of identical size. This argument is not valid since Bruggeman's derivation does not contain any hypothesis stating monodispersity.

As for the Wiener equation, Bruggeman's can be given a condensed formulation:

$$\varepsilon_s = \mathbf{B}(\varepsilon_{1s}, \varepsilon_{2s}, \Phi) \tag{226}$$

2. *Iterated and Combined Formulas*

If the system considered is a ternary one, it is possible to express its total static permittivity by iterating or combining formulas valid for binary systems.

For instance, a double spherical dispersion-type system having two different disperse phases 1a and 1b, characterized respectively by $\varepsilon_{1s}^{a}$ and $\varepsilon_{1s}^{b}$, can be considered either as a dispersion of 1a spherical particles in a continuous matrix consisting of a dispersion of 1b spherical particles embedded in a continuous medium characterized by $\varepsilon_{2s}$, or alternatively, as a dispersion of 1b spherical particles in a continuous matrix consisting of a dispersion of 1a spherical particles embedded in a continuous medium characterized by $\varepsilon_{2s}$. In the first situation, the effective static permittivity of the system $\varepsilon_s$ will be expressed as follows, in the case of dilute systems:

$$(\varepsilon_s)^{ba} = \mathbf{W}[\varepsilon_{1s}^{a}, \mathbf{W}(\varepsilon_{1s}^{b}, \varepsilon_{2s}, \Psi_b), \Phi_a] \tag{227}$$

$\Phi_a$ is the volume fraction of phase 1a in the total system and $\Psi_b$ the volume fraction of phase 1b in the initial dispersion of phase 1b particles within phase 2. In the second situation, $\varepsilon_s$ will be given by a formula similar to Eq. (227) and deduced from it by exchanging subscripts and superscripts a and b:

$$(\varepsilon_s)^{ab} = \mathbf{W}[\varepsilon_{1s}^{b}, \mathbf{W}(\varepsilon_{1s}^{a}, \varepsilon_{2s}, \Psi_a), \Phi_b] \tag{228}$$

When concentrated systems are concerned, the same procedure can be applied, provided that the Bruggeman equation is substituted for Wiener's in Eq. (227) and (228). In that case, $(\varepsilon_s)^{ba}$ and $(\varepsilon_s)^{ab}$ are expressed, respectively, by

$$(\varepsilon_s)^{ba} = \mathbf{B}[\varepsilon_{1s}^{a}, \mathbf{B}(\varepsilon_{1s}^{b}, \varepsilon_{2s}, \Psi_b), \Phi_a] \tag{229}$$

and

$$(\varepsilon_s)^{ab} = \mathbf{B}[\varepsilon_{1s}^{b}, \mathbf{B}(\varepsilon_{1s}^{a}, \varepsilon_{2s}, \Psi_a), \Phi_b] \tag{230}$$

The question must be raised whether Eqs. (227) and (228), or (229) and (230), yield identical values for $(\varepsilon_s)^{ba}$ and $(\varepsilon_s)^{ab}$. The general answer is probably negative [87,88], but it is ascertained that $(\varepsilon_s)^{ba}$ and $(\varepsilon_s)^{ab}$ are close to each other since both of them lie between the upper and lower bounds. It will be shown further that, in special cases, $(\varepsilon_s)^{ba}$ and $(\varepsilon_s)^{ab}$ can be considered as being equal, owing to the usual uncertainties attached to dielectric experiments.

A similar approach can be followed in the case of dispersions of spherical particles covered with shells. For dilute systems, Eqs. (214) and (217) must be combined, as indicated:

$$\varepsilon_s = \mathfrak{W}[\mathfrak{W}(\varepsilon_{1s}^c, \varepsilon_{1s}^s, \underline{\phi}), \varepsilon_{2s}, \Phi] \tag{231}$$

For concentrated systems, $\varepsilon_s$ is expressed by

$$\varepsilon_s = \mathfrak{B}[\mathfrak{W}(\varepsilon_{1s}^c, \varepsilon_{1s}^s, \underline{\phi}), \varepsilon_{2s}, \Phi] \tag{232}$$

In contrast to the preceding situation, Eqs. (231) and (232) have a unique possible physical solution.

## E. Extension of the Wiener and Bruggeman Equations to the Case of Complex Permittivity

By virtue of the prinicple of generalized conductivity, Wiener's equation can be extended to the case of complex permittivities provided that the quasi-stationary approximation holds, yielding the following equation, known as the *Wagner formula* [116,117]:

$$\frac{\varepsilon^* - \varepsilon_2^*}{\varepsilon^* + 2\varepsilon_2^*} = \frac{\varepsilon_1^* - \varepsilon_2^*}{\varepsilon_1^* + 2\varepsilon_2^*}\,\Phi \tag{233}$$

Lewin [430] has shown that Eq. (233) holds at high frequencies (for which the applicability of the quasi-stationary approximation would be doubtful), and his theoretical predictions have been supported by experiments performed by Corkum [431] on artificial dielectrics consisting of cubical arrays of metallic or dielectric spheres embedded in a matrix made of Styrofoam (Dow Chemical Co.).

Similarly, in the case of concentrated systems, the extension of the Bruggeman equation to complex permittivities yields the following formula, first suggested by Hanai [62,73]:

$$\left(\frac{\varepsilon_1^* - \varepsilon_2^*}{\varepsilon_1^* - \varepsilon^*}\right)^3 \frac{\varepsilon^*}{\varepsilon_2^*} = \frac{1}{(1 - \Phi)^3} \tag{234}$$

As in the case of the Wiener and Bruggeman equations, Eqs. (233) and (234) can be given condensed formulations, as follows:

$$\varepsilon^* = \mathfrak{W}(\varepsilon_1^*, \varepsilon_2^*, \Phi) \tag{235}$$

for the Wagner formula, and

$$\varepsilon^* = \mathfrak{C}(\varepsilon_1^*, \varepsilon_2^*, \Phi) \tag{236}$$

for the Hanai formula.

When systems of the double spherical dispersion type or systems of shell-covered spherical particles embedded in a continuous medium are considered, their dielectric behavior can be described tentatively by iterated or combined formulas, analogous to Eqs. (227) to (232), namely, for dilute double spherical dispersion systems

$$(\varepsilon^*)^{ba} = \mathfrak{W}[\varepsilon_1^{*a}, \mathfrak{W}(\varepsilon_1^{*b}, \varepsilon_2^*, \Psi_b), \Phi_a] \tag{237}$$

$$(\varepsilon^*)^{ab} = \mathfrak{W}[\varepsilon_1^{*b}, \mathfrak{W}(\varepsilon_1^{*a}, \varepsilon_2^*, \Psi_a), \Phi_b] \tag{238}$$

for concentrated double spherical dispersion systems,

$$(\varepsilon^*)^{ba} = \mathfrak{T}[\varepsilon_1^{*a}, \mathfrak{T}(\varepsilon_1^{*b}, \varepsilon_2^*, \Psi_b), \Phi_a] \tag{239}$$

$$(\varepsilon^*)^{ab} = \mathfrak{T}[\varepsilon_1^{*b}, \mathfrak{T}(\varepsilon_1^{*a}, \varepsilon_2^*, \Psi_a), \Phi_b] \tag{240}$$

for dilute dispersions of shell covered spherical particles,

$$\varepsilon^* = \mathfrak{W}[\mathfrak{W}(\varepsilon_1^{*c}, \varepsilon_1^{*s}, \underline{\phi}), \varepsilon_2^*, \Phi] \tag{241}$$

and for concentrated dispersions of shell-covered spherical particles,

$$\varepsilon^* = \mathfrak{T}[\mathfrak{W}(\varepsilon_1^{*c}, \varepsilon_1^{*s}, \underline{\phi}), \varepsilon_2^*, \Phi] \tag{242}$$

## F. Comparison of the Wiener and Bruggemann Formulas with Experimental Data

Comparative reviews of theoretical equations and experimental data available in the literature as concerns systems involving spherical particles have been made by Van Beek [5] and Hanai [3]. From these reviews, it appears that Wiener's formula, for dilute systems only, and Bruggeman's formula, for dilute and concentrated ones, can represent convincingly the static permittivity and steady conductivity of dispersions of spherical particles in a continuous matrix, while the Bottcher-Landauer formula would be more adequate for statistical mixtures of spherical particles.

Guillien [361] investigated at room temperature, between 70 kHz and 3.5 MHz, the permittivity of suspensions of spherical particles as a function of the disperse-phase content. He found, for low disperse-phase contents, a reasonable agreement between his experimental data and theoretical predictions. For mercury-in-oil emulsions, his results are fitted quite correctly by Eq. (360), which is the special form of Bruggeman's formula applicable to dispersions of highly conducting spherical particles. Guillien noted that deviations from Eq. (360) are observable if the suspended particles are not spherical. The validity of Eq. (360) as applied to the case of dispersions of highly conducting spherical particles has been checked as well by Naiki et al. [432], and by Clausse and Boned [4,77,104] on mercury-in-oil systems.

It has been already reported that, according to Lewin's theoretical analysis [430], Wiener's formula is applicable at high radio frequencies, provided that the wavelength of the monochromatic electromagnetic signal is large compared to the spherical particle radius. Lewin's predictions were checked quite satisfactorily by Corkum [431], who measured, at 5 GHz, the dielectric properties of artificial dielectrics consisting of regular arrays of spheres imbedded in a continuous matrix. Pearce [398] gathered a certain number of reliable experimental results and found a good agreement between them and Bruggeman's formulas for porphyritic and spherical dispersion systems, depending on the particle shapes. Voet [250] introduced into Eq. (360) a form factor to take into account the effect of particle shape and found that the modified equation depicted satisfactorily the dielectric behavior at 5 kHz of nonaqueous dispersions. Good agreement of Bruggeman's formula with experimental data was also found by Eichbaum [370], with respect to the permittivity of suspensions of window glass powder in polystyrene and by De La Rue and Tobias [433] for the conductivity of dispersions of glass beads in nearly saturated $ZnBr_2$ aqueous solutions. De La Rue and Tobias made remarks similar to those of Guillien [361] and Voet [250] concerning the influence of particle shape.

## IV. The Maxwell-Wagner-Sillars Effect

### A. General Remarks

The Maxwell-Wagner-Sillars (MWS) effect, alternatively called *interfacial polarization* or *migration polarization*, is a dielectric phenomenon typical of heterogeneous dielectrics with at least one conducting component. As Van Beek pointed out [5], it is quite surprising that many people interested in the dielectric properties of mixtures have disregarded the MWS effect when comparing experimental data and theoretical predictions. Even in fairly recent papers, erroneous interpretations can be found because the authors were not aware that, owing to migration polarization, the permittivity of a heterogeneous material could depend upon the frequency, even if none of its individual components exhibited an intrinsic dielectric relaxation of molecular or ionic origin. Besides, once the MWS effect may give rise to dielectric relaxations of the Debye or Cole-Cole type, confusion can exist as to the origin of dielectric relaxations observed in materials if heterogeneities of a chemical nature (foreign particles) or of a physical nature (cracks and voids) are present in the bulk or at the surface of the samples. Volger [434] has discussed some aspects of this problem in the case of semiconductors. Even if great care is devoted to the preparation of the samples, a stray dielectric relaxation due to the MWS effect can arise because of a deficient electrical contact, whatever its origin, between the sample and the electrodes of the measuring cell. Illustrations of this situation have been given by Cross and Hart [435], and by Wimmer and Tallan [436] in the case of dielectric studies of 1-tetradecanol and of NaCl single crystals, respectively.

In spite of some negative consequences such as those reported above, the MWS effect is a phenomenon that deserves attention since it can be, when carefully identified and controlled, a useful technique in the investigation of the physical properties of materials or of physicochemical processes developing or existing within them. Dansas and Dixou [437] suggested the use of the MWS effect to eliminate electrode polarization during the study of the conductive and dielectric behavior of conducting liquids. This can be achieved by sandwiching a sample of the substance under investigation between two sheets of a highly isolating material such as mica. Thus, the accumulation of electrical carriers at the electrodes is inhibited, eliminating inconvenient electrode polarization phenomena, and the conductive properties of the substance can easily be studied through the dielectric relaxation arising from migration polarization. This blocking-electrode method was applied later by Mounier and Sixou [438] to the study of the dielectric behavior of pure and doped ice samples. Other applications of the MWS effect concern, for instance, the evaluation of the agglomeration of elongated magnetic oxide particles in polymer substrates [439], and the investigation of the temperature dependence of the static dielectric constant and conductivity of water at subzero temperatures [68]. The MWS effect is also at the basis of the dielectric studies that have been performed on water-in-oil-type emulsions undergoing a monothermal progressive freezing, [86-92]. Not to be forgotten are the strong connections existing between interfacial polarization phenomena and the determination of moisture content in materials, a subject which has been sketched rapidly in the introductory part of this chapter. An analysis of the origins and features of different contributions to the total complex permittivity of moist materials has been made by Lebrun [440].

The MWS effect was introduced first by Maxwell [115] in the case of layered structures, then by Wagner [116,117] in the case of dispersions of spherical particles. Later, Sillars [118] discussed Wagner's model and extended it to the case of dispersions of spheroidal particles, which were also studied by Mandel [441]. Trukhan [119] proposed an amended formulation of the problem by taking into account the finite thickness of the region of polarization-free charge localization which had been neglected in earlier studies. Goffaux [442-444] added a contribution to the understanding of the MWS effect by analyzing the mechanism of electrical charge transfer across the interfaces. Clausse [445] discussed the theoretical conditions for the existence of a dielectric relaxation due to interfacial polarization in heterogeneous systems.

Many experimental data concerning the MWS effect can be found in the current literature. Sillars [118] reported experiments on water droplets suspended in paraffin wax. As expected, he found a dielectric dispersion and absorption located around 1 MHz, and he explained the quantitative discrepancies observed between experimental results and theoretical predictions by the formation of fissures filled with water. Hamon [446] investigated the dielectric behavior of heterogeneous substances consisting of 1-decanol or copper phthalocyanine particles dispersed within a base material, either paraffin wax, 1-docosane or poly(tetrafluoroethylene), and found a satisfactory agreement with Wagner's theory. Kharadly and Jackson [428], after a thorough analysis of the effective permittivity of arrays of perfectly conducting elements of simple geometrical shapes, studied over the frequency range 400 Hz to 10 kHz the complex permittivity of a cubical array of spherical elements made of an imperfectly conducting material. They reported dispersion and absorption curves which were in better agreement with a formula taking into account dipolar interactions than with Wagner's formula. In order to check the validity of the MWS model, artificial dielectric materials were designed by Van Beek and others [447,448], who used spheres and cylinders of polyethylene filled with distilled water and embedded in an insulating medium made of a mixture of paraffin wax and vaseline; these authors found a good agreement between theory and experiments, especially in the case of spheres. Interfacial polarization phenomena were also found by Hanai et al. [449], who studied the complex capacitance of a model system analogous to a layered structure and involving an egg-lecithin film.

A comprehensive study of the two-layer system has been made by Sixou [403], who investigated theoretically and experimentally the total dielectric response of samples whose components exhibited dipolar-type absorption along with conduction. This kind of study, which is of importance for a complete understanding of the dielectric behavior of heterogeneous materials, especially of those involving water, has been performed also by De Loor [450,451] and by Clausse who gave theoretical models for the complex permittivity of spherical dispersion systems whose disperse phase exhibits a dielectric relaxation which may or may not be overlapped by a conduction absorption, [4,78]. Partial analyses of this problem can also be found in Refs. 452 to 454.

Interfacial polarization phenomena also have attracted the attention of biophysicists interested in the electrical properties of biological systems. Numerous references to the subject can be found in Ref. 19. Most of the theoretical developments in this field rely on earlier calculations made by Maxwell [115], and later by Cole [426] to express the resistivity of a spherical body consisting of a spherical core surrounded by a shell, the resistivities of the core and of the shell being different. When the problem is stated in terms of conductivity or permittivity, the total conductivity or permittivity of a spherical body, considered as being homogeneous, is given simply by a Wiener-type equation, as has been previously shown. In the case of alternating sinusoidal electrical fields, the dielectric properties of the spherical body will be expressed by its complex permittivity, which is then given by the Wagner formula, implying a dielectric relaxation of the MWS type originating from the existence of an interface between the core and the shell. Pauly and Schwan [455] gave a theoretical analysis of the dielectric behavior of a suspension consisting of shell-covered spherical particles dispersed within a conducting medium. They showed that such a system is expected to exhibit two dielectric relaxations, the first arising from the heterogeneous nature of the disperse globules, and the other from the heterogeneous nature of the dispersion itself. Their analysis has been reviewed in detail by Hanai and others [3, 456], who brought to it some slight amendments. A computerized numerical estimation of the Pauly and Schwan model has been performed by Redwood and co-workers [457] in the case of phosphatidylcholine vesicles. The shell-covered sphere model is also of interest in the investigation of the dielectric and electrophoretic properties of disperse systems in which the electrical double layer plays an important role. Miles and Robertson [11] gave a treatment of the dielectric behavior of colloidal particle systems by assuming that the particles were spheres surrounded by an electrical double layer represented by a concentric conducting shell. More recent contributions to this problem are those of O'Konski [14], Schwarz [458], Schuur [459], and Shilov and Dukhin [8,123], which are quoted in a paper by De Backer and Watillon [460] devoted to the theoretical interpretation of experimental data obtained on porous Pyrex plugs in the presence of very dilute solutions of KCl [461]. An interesting review of the polarization of shell-covered globules, in connection with electrophoresis, has been given by Briant and others [462], who quotes the earlier contributions of Henry [463,464], Booth [465,466], and Overbeek [467,468].

### B. Theoretical Evidence for Dielectric Relaxation Due to Interfacial Polarization

The existence of dielectric relaxation due to interfacial polarization can be deduced from any theoretical formula giving the complex permittivity of a heterogeneous system. The qualitative features of the dielectric relaxation are insensitive to the system geometry which influences only the values of the dielectric relaxation parameters.

In the following, the analysis will be restricted to two binary models of simple geometry, namely, the two-layer structure, because of its pedagogical interest, and the dispersion of spherical particles within a continuous matrix, because of its evident relevancy to the emulsion case. For spherical dispersion systems, the amendments introduced by Trukhan [119], who took into account the finite thickness of the region of localization of polarization-free charges, will be briefly indicated. Although these systems are in fact ternary ones, dispersions of shell-covered spherical particles also will be considered, in order to show how the phenomena observed for binary spherical dispersions are modified when the transition region between the disperse particles and the continuous matrix is depicted by well-defined conductive and dielectric parameters.

The complex permittivities of the system components will be taken as given by the following general formula:

$$\varepsilon^* = \varepsilon_s + \frac{\sigma}{j\omega\varepsilon_0} \tag{243}$$

where $\varepsilon_s$ represents the static permittivity, $\sigma$ the steady conductivity, and $\varepsilon_0$ the dielectric constant of free space, $\omega$ the angular frequency of the applied electrical field, j is defined by $j^2 = -1$.

By introducing the time constant $T$ defined by

$$T = \frac{\varepsilon_0 \varepsilon_s}{\sigma}$$

Eq. (243) can be written, alternatively,

$$\varepsilon^* = \varepsilon_s \left(1 + \frac{1}{j\omega T}\right) \tag{244}$$

It is assumed that the interfaces do not bear true electrical charges.

Except for their conduction losses, the components of the system will be considered to be loss free, as expressed by Eqs. (243) and (244). The problem of the interference of interfacial polarization phenomena with intrinsic dielectric relaxations exhibited by one or both components will be examined later in the specific case of Hanai's formula as applied to emulsion dielectric properties.

### 1. *Maxwell-Wagner-Sillars Effect in Two-Layer Heterogeneous Structures*

As has been shown earlier, when the electrical field direction is parallel to the stratification axis, the static dielectric constant $\varepsilon_s$ of a binary multilayered structure is given by

$$\frac{1}{\varepsilon_s} = \frac{\Phi}{\varepsilon_{1s}} + \frac{1 - \Phi}{\varepsilon_{2s}} \tag{245}$$

where $\varepsilon_{1s}$ and $\Phi$ are the static permittivity and volume fraction of the first component and $\varepsilon_{2s}$ and $1 - \Phi$ those of the second. A similar formula holds for the steady conductivity $\sigma$,

$$\frac{1}{\sigma} = \frac{\Phi}{\sigma_1} + \frac{1 - \Phi}{\sigma_2} \tag{246}$$

If the system is submitted to an alternating sinusoidal electrical field, its dielectric properties are depicted by its complex permittivity $\varepsilon^*$ given by the following formula, known as the *Maxwell formula*:

$$\frac{1}{\varepsilon^*} = \frac{\Phi}{\varepsilon_1^*} + \frac{1 - \Phi}{\varepsilon_2^*} \tag{247}$$

According to Eq. (243), $\varepsilon_1^*$ and $\varepsilon_2^*$ are written, respectively,

$$\varepsilon_1^* = \varepsilon_{1s} + \frac{\sigma_1}{j\omega\varepsilon_0} \tag{248}$$

$$\varepsilon_2^* = \varepsilon_{2s} + \frac{\sigma_2}{j\omega\varepsilon_0} \tag{249}$$

or alternatively,

$$\varepsilon_1^* = \varepsilon_{1s}\left(1 + \frac{1}{j\omega T_1}\right) \tag{250}$$

$$\varepsilon_2^* = \varepsilon_{2s}\left(1 + \frac{1}{j\omega T_2}\right) \tag{251}$$

$T_1$ and $T_2$ being the time constants of components 1 and 2.

Equation (247) can be written as follows:

$$\varepsilon^* = \frac{\varepsilon_1^*\varepsilon_2^*}{\varepsilon_1^* + \Phi(\varepsilon_2^* - \varepsilon_1^*)} \tag{252}$$

By expressing $\varepsilon_1^*$ and $\varepsilon_2^*$ with the aid of Eqs. (248) and (249), Eq. (252) is transformed into

$$\varepsilon^* = \varepsilon_h + \frac{\varepsilon_\ell - \varepsilon_h}{1 + j\omega\bar{\tau}} + \frac{\sigma_\ell}{j\omega\varepsilon_0} \tag{253}$$

which is an equation of the Debye type with an additional ohmic term.

Thus, provision being made for certain special conditions to be discussed later, a two-layer heterogeneous system will exhibit, theoretically, a dielectric relaxation of the Debye type, along with a conduction absorption.

The high-frequency limiting permittivity $\varepsilon_h$ is expressed by the equation

$$\varepsilon_h = \frac{\varepsilon_{1s}\varepsilon_{2s}}{\varepsilon_{1s} + \Phi(\varepsilon_{2s} - \varepsilon_{1s})} \tag{254}$$

which is an alternative formulation of Eq. (245), $\varepsilon_h$ representing the static permittivity which would characterize the heterogeneous system if neither of its constituents were conducting. This is the reason why $\varepsilon_h$ is the permittivity at the high-frequency end of the relaxation domain, where the conduction contribution to the dielectric properties vanishes.

Similarly, $\sigma_\ell$ the steady conductivity, is expressed by

$$\sigma_\ell = \frac{\sigma_1 \sigma_2}{\sigma_1 + \Phi(\sigma_2 - \sigma_1)} \tag{255}$$

which is an alternative formulation of Eq. (246). It is readily seen from Eq. (255) that $\sigma_\ell$ is equal to zero if either $\sigma_1$ or $\sigma_2$ is equal to zero. In that case, i.e., if either constituent is a perfect insulator, the ohmic term in Eq. (253) disappears, implying that the heterogeneous system will exhibit a dielectric relaxation only.

While $\varepsilon_h$ is a function of $\varepsilon_{1s}$, $\varepsilon_{2s}$, and $\Phi$ only, and $\sigma_\ell$ a function of $\sigma_1$, $\sigma_2$, and $\Phi$ only, the low-frequency limiting permittivity $\varepsilon_\ell$ does depend simultaneously upon the static permittivities, the steady conductivities, and the volume ratio of both components

$$\varepsilon_\ell = \frac{\varepsilon_{2s}\sigma_1^2 + \Phi(\varepsilon_{1s}\sigma_2^2 - \varepsilon_{2s}\sigma_1^2)}{[\sigma_1 + \Phi(\sigma_2 - \sigma_1)]^2} \tag{256}$$

The relaxation increment $\varepsilon_\ell - \varepsilon_h$, which is a measure of the dielectric relaxation intensity, is given by

$$\varepsilon_\ell - \varepsilon_h = \frac{\Phi(1 - \Phi)(\varepsilon_{1s}\sigma_2 - \varepsilon_{2s}\sigma_1)^2}{[\varepsilon_{1s} + \Phi(\varepsilon_{2s} - \varepsilon_{1s})][\sigma_1 + \Phi(\sigma_2 - \sigma_1)]^2} \tag{257}$$

and the relaxation time $\bar{\tau}$ by

$$\bar{\tau} = \varepsilon_0 \frac{\varepsilon_{1s} + \Phi(\varepsilon_{2s} - \varepsilon_{1s})}{\sigma_1 + \Phi(\sigma_2 - \sigma_1)} \tag{258}$$

If, to avoid possible confusion with molecular relaxation times, it is thought preferable to speak in terms of critical frequency, Eq. (253) is transformed into

$$\varepsilon^* = \varepsilon_h + \frac{\varepsilon_\ell - \varepsilon_h}{1 + jf/\bar{f}_c} + \frac{\sigma_\ell}{2\pi j \varepsilon_0 f} \tag{259}$$

with

$$\bar{f}_c = \frac{1}{2\pi\varepsilon_0} \frac{\sigma_1 + \Phi(\sigma_2 - \sigma_1)}{\varepsilon_{1s} + \Phi(\varepsilon_{2s} - \varepsilon_{1s})} \tag{260}$$

### 2. *Maxwell-Wagner-Sillars Effect in Suspensions of Spherical Particles*

The static dielectric constant $\varepsilon_s$ of a system consisting of spherical particles suspended sparsely within a continuous matrix is given by

$$\frac{\varepsilon_s - \varepsilon_{2s}}{\varepsilon_s + 2\varepsilon_{2s}} = \frac{\varepsilon_{1s} - \varepsilon_{2s}}{\varepsilon_{1s} + 2\varepsilon_{2s}} \Phi \tag{261}$$

where $\varepsilon_{1s}$ and $\varepsilon_{2s}$ are the static permittivities of the disperse and of the continuous phases, respectively, and $\Phi$ the volume fraction of the disperse phase. A similar formula applies to the steady conductivity

$$\frac{\sigma - \sigma_2}{\sigma + 2\sigma_2} = \frac{\sigma_1 - \sigma_2}{\sigma_1 + 2\sigma_2} \Phi \tag{262}$$

In the case of alternating sinusoidal electrical fields, the dielectric behavior of the system is depicted by its complex permittivity $\varepsilon^*$, which is given by the following equation, known as the *Wagner formula*:

$$\frac{\varepsilon^* - \varepsilon_2^*}{\varepsilon^* + 2\varepsilon_2^*} = \frac{\varepsilon_1^* - \varepsilon_2^*}{\varepsilon_1^* + 2\varepsilon_2^*} \Phi \tag{263}$$

a. Suspensions of Spherical Particles Without Shells. *(1) General equations in the case of infinitesimally thin diffusion layers.* Equation (263) can be rewritten as follows:

$$\varepsilon^* = \varepsilon_2^* \frac{\varepsilon_1^* + 2\varepsilon_2^* + 2\Phi(\varepsilon_1^* - \varepsilon_2^*)}{\varepsilon_1^* + 2\varepsilon_2^* - \Phi(\varepsilon_1^* - \varepsilon_2^*)} \tag{264}$$

By expressing, as before, $\varepsilon_1^*$ and $\varepsilon_2^*$ by means of Eqs. (248) and (249), Eq. (264) can be transformed into a Debye-type equation with an additional ohmic term,

$$\varepsilon^* = \varepsilon_h + \frac{\varepsilon_\ell - \varepsilon_h}{1 + j\omega\bar{\tau}} + \frac{\sigma_\ell}{j\omega\varepsilon_0} \tag{265}$$

Thus, as in a two-layer structure, a suspension of spherical particles will exhibit, in the general case, a dielectric relaxation of the Debye type, along with a conduction absorption.

The high-frequency limiting permittivity $\varepsilon_h$ is given by

$$\varepsilon_h = \varepsilon_{2s} \frac{\varepsilon_{1s} + 2\varepsilon_{2s} + 2\Phi(\varepsilon_{1s} - \varepsilon_{2s})}{\varepsilon_{1s} + 2\varepsilon_{2s} - \Phi(\varepsilon_{1s} - \varepsilon_{2s})} \tag{266}$$

which is an alternative form of Eq. (261). $\varepsilon_h$ represents the static permittivity that would characterize the system if none of its constituents were conducting. It is the reason why $\varepsilon_h$ is the permittivity at the high-frequency end of the relaxation domain where the conduction contribution to the dielectric properties vanishes.

Similarly, $\sigma_\ell$ the steady conductivity, is expressed by

$$\sigma_\ell = \sigma_2 \frac{\sigma_1 + 2\sigma_2 + 2\Phi(\sigma_1 - \sigma_2)}{\sigma_1 + 2\sigma_2 - \Phi(\sigma_1 - \sigma_2)} \tag{267}$$

which is an alternative form of Eq. (262). The symmetry observed for the two-layer structure as to the role of $\sigma_1$ and $\sigma_2$ does not hold in the present case. If the continuous matrix is made of a perfectly insulating material ($\sigma_2 = 0$), $\sigma_\ell$ is equal to zero and Eq. (265) is reduced to a simple Debye-type formula. On the contrary, when $\sigma_1$ is equal to zero, the ohmic term still exists and the system exhibits a bulk conduction.

The low-frequency limiting permittivity $\varepsilon_\ell$ depends upon the disperse-phase volume fraction $\Phi$, and upon the static permittivities $\varepsilon_{1s}$ and $\varepsilon_{2s}$, and the conductivities $\sigma_1$ and $\sigma_2$ of each component:

$$\varepsilon_\ell = \varepsilon_{2s} \frac{\sigma_\ell}{\sigma_2} + \frac{9\Phi\sigma_2(\varepsilon_{1s}\sigma_2 - \varepsilon_{2s}\sigma_1)}{[\sigma_1 + 2\sigma_2 - \Phi(\sigma_1 - \sigma_2)]^2} \tag{268}$$

The relaxation increment $(\varepsilon_\ell - \varepsilon_h)$ and the relaxation time $\bar{\tau}$ are expressed by

$$\varepsilon_\ell - \varepsilon_h = \frac{9\Phi(1 - \Phi)(\varepsilon_{1s}\sigma_2 - \varepsilon_{2s}\sigma_1)^2}{[\varepsilon_{1s} + 2\varepsilon_{2s} - \Phi(\varepsilon_{1s} - \varepsilon_{2s})][\sigma_1 + 2\sigma_2 - \Phi(\sigma_1 - \sigma_2)]^2} \tag{269}$$

$$\bar{\tau} = \varepsilon_0 \frac{\varepsilon_{1s} + 2\varepsilon_{2s} - \Phi(\varepsilon_{1s} - \varepsilon_{2s})}{\sigma_1 + 2\sigma_2 - \Phi(\sigma_1 - \sigma_2)} \tag{270}$$

If the critical frequency $\bar{f}_c$ is introduced, Eq. (265) becomes

$$\varepsilon^* = \varepsilon_h + \frac{\varepsilon_\ell - \varepsilon_h}{1 + jf/\bar{f}_c} + \frac{\sigma_\ell}{2\pi j \varepsilon_0 f} \tag{271}$$

with

$$\bar{f}_c = \frac{1}{2\pi\varepsilon_0} \frac{\sigma_1 + 2\sigma_2 - \Phi(\sigma_1 - \sigma_2)}{\varepsilon_{1s} + 2\varepsilon_{2s} - \Phi(\varepsilon_{1s} - \varepsilon_{2s})} \tag{272}$$

*(2) Amended formulation in the case of arbitrary values of the Debye screening radius.* Trukhan [119] corrected the MWS theory by taking into account the non-uniformity of the concentration of charge carriers. He introduced in the expression for the current density J an ionic diffusion term, which leads, for the electrical potential, to the substitution of the Poisson equation for the Laplace equation and, consequently, to a modification of the formulas describing the complex permittivity of heterogeneous systems.

For suspensions of spherical particles, Trukhan obtained the following expression:

$$\varepsilon^* = \varepsilon_2^* \frac{\varepsilon_1^* + 2\varepsilon_2^* - 2\beta^* + 2\Phi(\varepsilon_1^* - \varepsilon_2^* + \beta^*)}{\varepsilon_1^* + 2\varepsilon_2^* - 2\beta^* - \Phi(\varepsilon_1^* - \varepsilon_2^* + \beta^*)} \tag{273}$$

where $\beta^*$ is an intricate complex function of the particle radius a, of the inverse of the Debye screening distance $\kappa$, of the ionic diffusion coefficient D, and of the frequency of the applied electrical field f,

$$\beta^* = \varepsilon_2^* \frac{\kappa^2[3 + (\gamma a)^2] \tanh(\gamma a) - 3\gamma a}{\gamma^2[2 + (\gamma a)^2] \tanh(\gamma a) - 2\gamma a} \tag{274}$$

$\gamma$ being given by

$$\gamma^2 = \kappa^2 + \frac{j\omega}{D} \tag{275}$$

As $\beta^*$ vanishes at high frequencies, $\varepsilon_h$ is still expressed by Eq. (266). But now $\varepsilon_\ell$ and $\bar{\tau}$ (or $\bar{f}_c$) depend upon $\gamma$ and upon a, the particle radius. Thus, according to Trukhan's calculations, the features of the dielectric relaxation should depend upon the size of the disperse particles. In the case of polydisperse systems, a distortion of the Cole-Cole diagram should be observed arising from the critical frequency scattering.

b. Suspensions of Spherical Particles Covered with Shells. Maxwell's Eq. (217), giving the equivalent static permittivity of a spherical body consisting of a spherical core covered with a concentric shell can be generalized to the complex permittivity case, in application of the generalized conductivity principle. If $\varepsilon_1^{*c}$ is the complex permittivity of the core and $\varepsilon_1^{*s}$ that of the shell, the complex permittivity $\varepsilon_1^*$ of the spherical body, considered as being homogeneous, is expressed by an equation which has the same formulation as the Wagner equation for a dispersion of spherical particles in a continuous matrix

$$\frac{\varepsilon_1^* - \varepsilon_1^{*s}}{\varepsilon_1^* + 2\varepsilon_1^{*s}} = \frac{\varepsilon_1^{*c} - \varepsilon_1^{*s}}{\varepsilon_1^{*c} + 2\varepsilon_1^{*s}} \underline{\phi} \tag{276}$$

with $\underline{\phi}$, the fraction of volume occupied by the core in the spherical body, given by

$$\underline{\phi} = \left(\frac{a_c}{a}\right)^3 = \frac{a_c^3}{(a_c + d)^3} \tag{277}$$

where a is the external radius of the spherical body, $a_c$ that of the core and d the shell thickness.

Since, according to the general Eq. (243), $\varepsilon_1^{*c}$ and $\varepsilon_1^{*s}$ are written

$$\varepsilon_1^{*c} = \varepsilon_{1s}^{c} + \frac{\sigma_1^c}{j\omega\varepsilon_0} \tag{278}$$

$$\varepsilon_1^{*s} = \varepsilon_{1s}^{c} + \frac{\sigma_1^c}{j\omega\varepsilon_0} \tag{279}$$

Eq. (276) can be transformed into a Debye-type formula with an additional ohmic term. Thus, it appears that a heterogeneous spherical particle consisting of a spherical core covered with a concentric shell displays generally a dielectric relaxation of the Debye type, along with a conduction. If the shell is made of a perfectly insulating material, no bulk conduction phenomena occur and the particle exhibits a dielectric relaxation only.

The complex permittivity $\varepsilon^*$ of a suspension of shell-covered spherical particles dispersed sparsely within a continuous medium is given by the Wagner formula,

$$\frac{\varepsilon^* - \varepsilon_2^*}{\varepsilon^* + 2\varepsilon_2^*} = \frac{\varepsilon_1^* - \varepsilon_2^*}{\varepsilon_1^* + 2\varepsilon_2^*}\Phi \tag{280}$$

where $\varepsilon_2^*$ is the complex permittivity of the suspending medium, $\varepsilon_1^*$ the disperse globule complex permittivity as given by Eq. (276) and $\Phi$ the disperse phase volume fraction. Combining Eqs. (276) and (280) yields

$$\varepsilon^* = \varepsilon_2^* \frac{1 + 2\Phi + 2(1 - \Phi)K}{1 - \Phi + (2 + \Phi)K} \tag{281}$$

with

$$K = \frac{\varepsilon_2^*}{\varepsilon_1^{*s}} \frac{(1 - \underline{\phi})\varepsilon_1^{*c} + (2 + \underline{\phi})\varepsilon_1^{*s}}{(1 + 2\underline{\phi})\varepsilon_1^{*c} + 2(1 - \underline{\phi})\varepsilon_1^{*s}} \tag{282}$$

Since $\varepsilon_1^{*c}$, $\varepsilon_1^{*s}$, and $\varepsilon_2^*$ are expressed by the general Eq. (243), Eq. (281) can be transformed into

$$\varepsilon^* = \varepsilon_h + \frac{\varepsilon_i - \varepsilon_h}{1 + j\omega\overline{\tau}_Q} + \frac{\varepsilon_\ell - \varepsilon_i}{1 + j\omega\overline{\tau}_P} + \frac{\sigma_\ell}{j\omega\,\varepsilon_0} \tag{283}$$

Equation (283) shows that the system displays, along with a bulk conduction, two distinct dielectric relaxations of the Debye type. The features of the total dielectric behavior of the system will depend strongly upon the ratio $a_c/a$ and upon the relative magnitudes of the permittivities and conductivities of the core, the shell, and the suspending medium, since different situations may occur as to the overlapping of the two relaxations and of the bulk conduction with the relaxations.

### C. Qualitative Theoretical Approach of the MWS Effect and Theoretical Conditions for the Existence of a Dielectric Relaxation

The fact that Eqs. (247) and (263) can be transformed into Debye-type equations with additional ohmic terms proves that a polarization mechanism describable by a phenomenological equation similar to Eq. (150) takes place within binary heterogeneous materials when at least one of their components is conducting. Thenceforth, an analysis of the MWS effect can be built by following the same lines as in the case of orientation polarization phenomena [4].

It has been shown earlier that a dielectric spherical particle submitted to the action of an initially uniform electrical field acquires a dipole moment $p_o$, owing, in a simple model, to the statistical orientation of molecular dipoles along the field lines. The moment $p_o$ is related to the surface density $\underline{\sigma}'$ of bound electrical charges appearing at the interface between the particle and the surrounding medium. If, in addition, the particle is conducting, it acquires an additional dipole moment $p_c$ arising from the shift of free charge carriers, opposed to the field direction for negative ones and along the field direction for the positive ones. $p_c$ is related to the surface density $\underline{\sigma}$ of free electrical charges spread over the interface between the particle and the surrounding medium. $\underline{\sigma}$ results from an accumulation of negative and positive charge carriers at the interface, because of the difference in the values of the static permittivity and the steady conductivity on both sides of the interface. Thus, the total dipole moment p of the particle is given by

$$p = p_c + p_o + p_d \tag{284}$$

and its total polarization vector P by

$$P = P_c + P_o + P_d \tag{285}$$

where $P_d$ represents the contribution of the distortion polarization, $P_o$ the contribution of the orientation polarization, and $P_c$ the contribution of the migration polarization. Like $P_o$, $P_c$ will reach its equilibrium value $P_{cs}$, as expressed by the phenomenological Eq. (150).

When an alternative sinusoidal field is acting upon the system, the resolution of Eq. (150) applied to $P_c$ yields a solution similar to that obtained for $P_o$ as expressed by Eq. (160). The particle can be considered as electrically equivalent to two pulsating dipoles located at its center. The individual contributions of both dipoles may overlap or not, according to the relative values of $\bar{\tau}$ and $\tau$, $\bar{\tau}$ being the migration polarization relaxation time and $\tau$ the orientation polarization relaxation time.

$P_c$ being proportional to $\sigma$, the surface density of free charges spread over the interface, it is possible to determine, through a study of $\underline{\sigma}$, the conditions for the existence of a dielectric relaxation arising from migration polarization. According to Eq. (132), $\underline{\sigma}$ is given by

$$\varepsilon_{1s} E_{n_1} + \varepsilon_{2s} E_{n_2} = -\frac{\underline{\sigma}}{\varepsilon_0} \tag{286}$$

$E_{n_1}$ and $E_{n_2}$, the normal components of the electrical field on both sides of the interface are linked by Eq. (137) which can be transformed into

$$\frac{E_{n_1}}{E_{n_2}} = \frac{\varepsilon_{2s}}{\varepsilon_{1s}} \frac{1 + 1/j\omega T_2}{1 + 1/j\omega T_1} \tag{287}$$

by introducing the time constants $T_1$ and $T_2$ of the particle material and of the surrounding medium.

Combining Eqs. (286) and (287) leads to

$$\underline{\sigma} = -\varepsilon_0 \varepsilon_{2s} \left(1 - \frac{1 + 1/j\omega T_2}{1 + 1/j\omega T_1}\right) E_{n_2} \tag{288}$$

or, alternatively, to

$$\underline{\sigma} = -\varepsilon_0 \varepsilon_{1s} \left(1 - \frac{1 + 1/j\omega T_1}{1 + 1/j\omega T_2}\right) E_{n_1} \tag{289}$$

It is readily seen from Eqs. (288) and (289) that $\underline{\sigma}$ is equal to zero when

$$T_1 = T_2 \tag{290}$$

or, alternatively, when

$$\varepsilon_{1s}\sigma_2 - \varepsilon_{2s}\sigma_1 = 0 \tag{291}$$

Thus, no dielectric relaxation arising from interfacial polarization can occur if the time constants of the two components of the heterogeneous system are identical. Qualitatively, this means that a dielectric relaxation cannot exist if the rate of flow of electrical carriers is the same in both constituents, implying that there is no accumulation of electrical charges at the interface [4].

Moreover, if either $T_1$ or $T_2$ is equal to zero, the electrical field in the corresponding phase, which is then a perfect conductor, vanishes. The electrical phenomena are located entirely in the other phase and the heterogeneous system does not exhibit any dielectric relaxation arising from migration polarization. (In fact, a dielectric relaxation could be observed at infinite values of the frequency of the applied electrical field.) This theoretical result has been checked by Clausse [4,77] on mercury-in-oil dispersions. The condition describing these cases of nonrelaxation is

$$T_1 T_2 = 0 \tag{292}$$

Finally, no relaxation is displayed if the system is not heterogeneous; therefore when $\Phi$ is equal either to 0 or to 1 (trivial conditions),

$$\Phi(1 - \Phi) = 0 \tag{293}$$

Consequently, a binary heterogeneous system will exhibit a dielectric relaxation arising from migration polarization if, and only if, Eqs. (290), (292), and (293) do not hold simultaneously, as formulated [4,445]:

$$T_1 T_2 (T_1 - T_2) \Phi(1 - \Phi) \neq 0 \tag{294}$$

It can be shown that conditions (290), (292), and (293), and, consequently, the above inequality, can be derived from Eqs. (257) and (258), relative to the stratified model, and Eqs. (269) and (270), relative to the spherical dispersion model. In both cases, the relaxation term in the Debye-type formula representing the complex permittivity of the system cancels if either $\varepsilon_\ell - \varepsilon_h$ or $\overline{\tau}$ or $1/\overline{\tau}$ is equal to zero, leading to conditions (290), (292), and (293). It must be noted that, when $T_1$ is equal to $T_2$, the heterogeneous medium behaves as a homogeneous one whose time constant $T$ is equal to $T_1$ and $T_2$. This indicates a decoupling of conductive and capacitive effects within the system.

### D. Experimental Evidence for the Existence of Interfacial Polarization Phenomena in Emulsions

As was reported earlier, Sillars [118] found that suspensions of water droplets in paraffin wax exhibited dielectric relaxations located around 1 MHz. Later, Dryden and Meakins [64] studied the dielectric behavior of emulsions of water in wool wax or in petroleum jelly with a few percent of added wool wax and reported that in the 1-kHz to 100-MHz frequency range, there occurred dielectric relaxation whose intensity increased with the water content. Chapman [65] investigated the influence of the emulsifying agent upon the dielectric relaxation displayed in the 1-MHz region by emulsions of water in paraffin oil. Hanai and co-workers [60-62,69-74] found that water-in-oil emulsions, with Nujol as the dispersing medium, exhibited dielectric relaxations between 100 kHz and 100 MHz, the features of which were dependent upon the shearing stress applied to the emulsions. They explained this effect by assuming that the application of the shearing stress counterbalanced disperse droplet agglomeration. These authors showed that dielectric relaxations exist as well in emulsions of nitrobenzene in water. The dielectric behavior of water-in-oil emulsions was studied also by Clausse [4,75-77], who reported relaxations of the Cole-Cole type located around 1 MHz, as shown in Fig. 14.

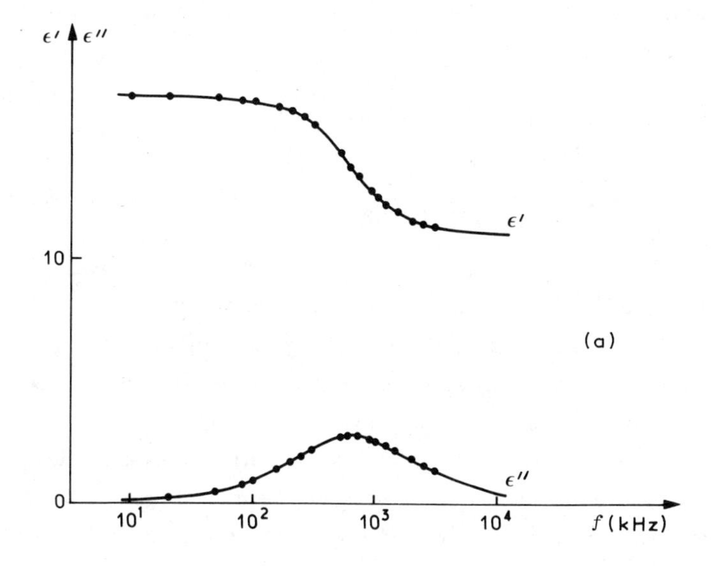


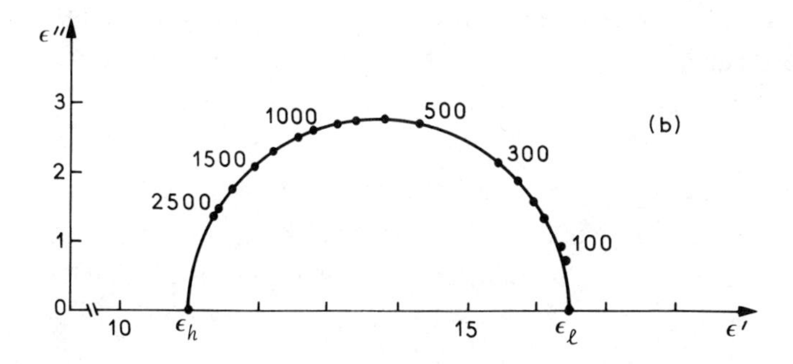


**Figure 14.** Features of the dielectric relaxation exhibited by an emulsion of water in vaseline oil, at 296 K: $\Phi = 0.47$; (a) dispersion and absorption curves and (b) Cole-Cole plot. (From Ref. 75, Gauthier-Villars, Paris.)

From this brief review of data available in the current literature, it appears that the dielectric relaxations exhibited around 1 MHz by emulsion-type systems must be ascribed to migration polarization since, at these frequencies, the intrinsic dipolar dielectric relaxation displayed by water in the 10 GHz region cannot contribute anything to the observed phenomena. On this basis, Hodge and Angell [68] investigated, through interfacial polarization studies of emulsions of water in a mixture of heptane and carbon tetrachloride, the temperature dependence of the dielectric constant and conductivity of supercooled water between 0 and −35°C. Their method is somewhat similar to that used by Hasted and Shahidi [469] to determine the static dielectric constant of water between +20 and

$-36°C$ from measurements of the real part of the complex permittivity of emulsions containing about 1% water dispersed in hexane.

## V. Comparative Analyses of the Wagner and Hanai Formulas

### A. Introductory Remarks

Because of its intricate implicit form, the general Eq. (234), proposed by Hanai [62,73] as an extension of Bruggeman's formula [357] to the case of complex permittivities, has discouraged many authors who have disregarded or even rejected it without serious reasons. Even Hanai and his co-workers restricted their analysis to the comparison of their conductivity and low-frequency and high-frequency limiting permittivity experimental data on emulsions with some particular equations derived from the general formula. This is somewhat surprising, since the theoretical study of the Hanai formula can be performed through numerical analysis, by using computerized calculation methods. The theoretical aspects of Clausse's investigations of the dielectric behavior of emulsions [4,81,83] were approached on the basis of such a computerized numerical procedure, which has been reported with full details elsewhere [82]. Later, Hanai and co-workers [84, 85] also applied systematic numerical analyses to determine dielectric parameters from experimental data on dielectric relaxations in disperse systems.

Owing to the generalized conductivity principle, the limitations imposed on Hanai's formula as to its applicability are the same as those existing for Bruggeman's equation. On the basis of the calculations made by Gunther and Heinrich [429] in order to compare their theory to existing formulas for the dielectric constants of spherical dispersion systems, it can be taken for granted that Hanai's formula can be used without restriction up to values of $\Phi$, the disperse phase volume fraction, equal to 0.40 or so when the conductivity and static permittivity of the continuous phase are small compared to those of the disperse phase, and up to about 0.60 in the reverse case. Experimental data obtained on emulsion-type systems tend to prove that these limiting values of $\Phi$ might be too strict and that the Hanai formula could be used for higher values of the disperse phase content. An explanation for such an extension of the limits of applicability of Hanai's formula could be based on the fact that, as opposed to Bruggeman's equation, the Gunther and Heinrich equation is restricted to the case of suspensions of spheres of identical size, the maximum packing being then equal to $\pi\sqrt{2}/6$ (10.741).

Whatever the limitations existing as for the application of Hanai's formula to interpret actual experimental data, an analysis covering all the range of volume fractions ($0 \leqslant \Phi \leqslant 1$) is interesting enough in itself to be performed. This is the case as well for Wagner's formula although it is valid only when $\Phi$ is small compared to 1. In the following, parallel theoretical predictions will be presented for both Wagner's and Hanai's formulas, the emphasis being put on the two situations described below.

Both of the phases are nonpolar but exhibit conduction absorptions. One phase may be nonconducting.

The disperse phase exhibits a Debye-type dipolar relaxation along with a conduction absorption, the continuous phase being nonpolar and nonconductive. The disperse phase may be nonconducting and exhibit a Debye-type relaxation only.

These two situations have been selected among all the possible ones because of their particular relevance to the dielectric studies of emulsions involving nonpolar oils and water, which are the most common ones.

### B. General Principles of the Numerical Solution of the Hanai Formula

To obtain models for the dielectric behavior of binary disperse systems, Clausse [4,81] proposed considering Hanai's equation as a formula generating a family of geometric transformations $\mathfrak{T}$ in an Argand diagram, with $\Phi$, the volume fraction of disperse phase, taken as a parameter. With $\Phi$ being given, the point representative of $\varepsilon^*$, the complex permittivity of the system, can then be considered as the image through $\mathfrak{T}$ of the points representative of $\varepsilon_1^*$ and $\varepsilon_2^*$, the complex permittivities of the disperse and the continuous phases, as shown in Fig. 15.

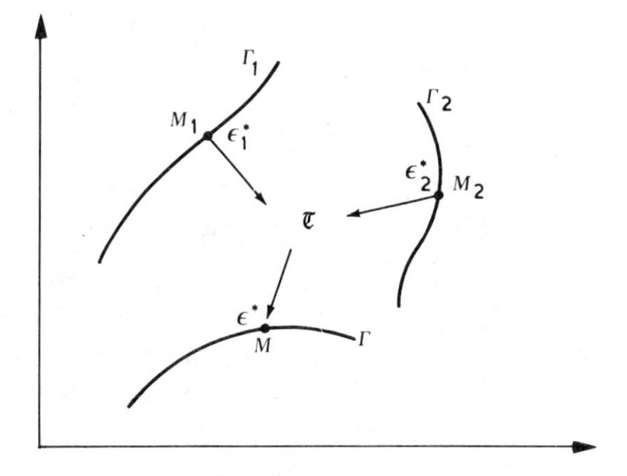


**Figure 15.** Schematic representation of the relationship among $\varepsilon_1^*(f)$, $\varepsilon_2^*(f)$, and $\varepsilon^*(f)$, through the transformation $\mathfrak{T}$, in an Argand diagram.

As $\varepsilon_1^*$ and $\varepsilon_2^*$ are functions of the frequency f of the applied field, their representative points, $M_1$ and $M_2$, circulate respectively on a curve $\Gamma_1$ and on a curve $\Gamma_2$ when f varies. Their image through $\mathfrak{T}$, which is the point representative of $\varepsilon^*$, circulates on a curve $\Gamma$ which depicts geometrically the dielectric properties of the system. From a mathematical analysis of $\Gamma$, it is possible to obtain its analytical representation and therefore the function $\varepsilon^*(f)$ characteristic for the dielectric behavior of the system at a given value of $\Phi$. Repeating this process for other values of $\Phi$ enables us to determine the dependency of $\varepsilon^*(f)$ upon $\Phi$.

Thus, Hanai's formula, which is written

$$\left(\frac{\varepsilon_1^* - \varepsilon_2^*}{\varepsilon_1^* - \varepsilon^*}\right)\frac{\varepsilon^*}{\varepsilon_2^*} = \frac{1}{(1 - \Phi)^3} \tag{295}$$

can be put in the following form

$$\varepsilon^* = \mathfrak{T}(\varepsilon_1^*, \varepsilon_2^*, \Phi) \tag{296}$$

By solving Eq. (296) numerically for different values of $\Phi$ ranging from 0 to 1 and for standard functions $\varepsilon_1^*(f)$ and $\varepsilon^*(f)$ representing conduction absorptions and/or dielectric relaxations, it is possible to determine models for the dielectric behavior of binary disperse systems, especially of emulsions, and, so to speak, to build abaci in order to compare theoretical predictions with experimental data.

This procedure can be extended to the study of more complicated systems. For instance, it was applied by Clausse and Lachaise [87,88] to the investigation of the monothermal progressive freezing of water-in-oil type emulsions which, at a given instant, can be considered as consisting either of ice globules suspended in a water-in-oil emulsion or of water droplets suspended within a dispersion of ice globules in oil. The complex permittivity of the system can be obtained through an iterated application of Hanai's formula. Similarly, a theoretical model for the dielectric behavior of frozen water-in-oil emulsions has been built by Lagourette and co-workers [93,94,95], who used an intricate iterated geometrical transformation combining Hanai's and Wagner's formulas.

Owing to the cubic form of Hanai's equation, a problem arises as to the choice of the proper solution, which must be determined from physical considerations such as the sign of $\varepsilon'$ and $\varepsilon''$ (which must be positive when the complex permittivity is written $\varepsilon^* = \varepsilon' - j\varepsilon''$, as usual), the relative magnitudes of $\varepsilon'$ and the convergence of $\varepsilon^*$ toward the real solution of Bruggeman's equation when f increases. Other computing difficulties arise from the drastic variations undergone by the imaginary parts of $\varepsilon_1^*$ and $\varepsilon_2^*$ when the frequency f varies, and adequate precautions must be taken to avoid overflows and catastrophic losses of precision. Details on the mathematical procedure and a routine written in FORTRAN IV language can be found in a paper published by Clausse and Royer [82].

C. Models for the Dielectric Behavior of Emulsions

*1. Both Phases are Conducting but do not Exhibit Intrinsic Dielectric Relaxations*

In this case, the complex permittivity of the disperse phase, volume fraction equal to $\Phi$, is given by

$$\varepsilon_1^* = \varepsilon_{1s} + \frac{\sigma_1}{j\omega\varepsilon_0} \tag{297}$$

and that of the continuous phase, volume fraction equal to $1 - \Phi$, by

$$\varepsilon_2^* = \varepsilon_{2s} + \frac{\sigma_2}{j\omega\varepsilon_0} \tag{298}$$

a. Dilute Systems (Wagner's Formula). *(1) General equations.* As has been previously shown, Wagner's formula, which is written,

$$\frac{\varepsilon^* - \varepsilon_2^*}{\varepsilon^* + 2\varepsilon_2^*} = \frac{\varepsilon_1^* - \varepsilon_2^*}{\varepsilon_1^* - \varepsilon_2^*}\Phi \tag{299}$$

can be transformed into

$$\varepsilon^* = \varepsilon_h + \frac{\varepsilon_\ell - \varepsilon_h}{1 + j\omega\bar{\tau}} + \frac{\sigma_\ell}{j\omega\,\varepsilon_0} \tag{300}$$

with $\sigma_\ell$, $\varepsilon_h$, $\varepsilon_\ell$, and $\varepsilon_\ell - \varepsilon_h$ given by

$$\sigma_\ell = \sigma_2 \frac{\sigma_1 + 2\sigma_2 + 2\Phi(\sigma_1 - \sigma_2)}{\sigma_1 + 2\sigma_2 - \Phi(\sigma_1 - \sigma_2)} \tag{301}$$

$$\varepsilon_h = \varepsilon_{2s} \frac{\varepsilon_{1s} + 2\varepsilon_{2s} + 2\Phi(\varepsilon_{1s} - \varepsilon_{2s})}{\varepsilon_{1s} + 2\varepsilon_{2s} - \Phi(\varepsilon_{1s} - \varepsilon_{2s})} \tag{302}$$

$$\varepsilon_\ell = \varepsilon_{2s} \frac{\sigma_\ell}{\sigma_2} + \frac{9\Phi\,\sigma_2(\varepsilon_{1s}\sigma_2 - \varepsilon_{2s}\sigma_1)}{[\sigma_1 + \sigma_2 - \Phi(\sigma_1 - \sigma_2)]^2} \tag{303}$$

$$\varepsilon_\ell - \varepsilon_h = \frac{9\Phi(1 - \Phi)(\varepsilon_{1s}\sigma_2 - \varepsilon_{2s}\sigma_1)^2}{[\varepsilon_{1s} + 2\varepsilon_{2s} - \Phi(\varepsilon_{1s} - \varepsilon_{2s})][\sigma_1 + 2\sigma_2 - \Phi(\sigma_1 - \sigma_2)]^2} \tag{304}$$

The macroscopic relaxation time $\bar{\tau}$ is expressed by

$$\bar{\tau} = \varepsilon_0 \frac{\varepsilon_{1s} + 2\varepsilon_{2s} - \Phi(\varepsilon_{1s} - \varepsilon_{2s})}{\sigma_1 + 2\sigma_2 - \Phi(\sigma_1 - \sigma_2)} \tag{305}$$

and consequently the critical frequency $\bar{f}_c$ by

$$\bar{f}_c = \frac{1}{2\pi\varepsilon_0} \frac{\sigma_1 + 2\sigma_2 - \Phi(\sigma_1 - \sigma_2)}{\varepsilon_{1s} + 2\varepsilon_{2s} - \Phi(\varepsilon_{1s} - \varepsilon_{2s})} \tag{306}$$

Thus, in the general case, a dilute spherical dispersion system exhibits a dielectric relaxation of the Debye type, along with a bulk conduction. From Eqs. (304) and (305), it is readily seen that the condition for the existence of a dielectric relaxation is expressed by the inequality (294), as expected. If the continuous phase is nonconducting, the bulk conductivity $\sigma_\ell$ as given by Eq. (301) is equal to zero and the system exhibits a dielectric relaxation only. The limiting values of $\bar{f}_c$ when $\Phi$ approaches either 0 or 1 are

$$\bar{f}_c(0) = \frac{1}{2\pi\varepsilon_0} \frac{\sigma_1 + 2\sigma_2}{\varepsilon_{1s} + 2\varepsilon_{2s}} \tag{307}$$

$$\bar{f}_c(1) = \frac{1}{2\pi\varepsilon_0} \frac{\sigma_2}{\varepsilon_{2s}} \tag{308}$$

*(2) Equations applying to special cases.*

$\sigma_1 = \sigma_2$

If the conductivities of the disperse and continuous phases are equal, then $\sigma_\ell$ is equal to $\sigma_1$ and $\sigma_2$,

$$\sigma_\ell = \sigma_1 = \sigma_2 = \sigma \tag{309}$$

and $\varepsilon_\ell$ is given simply by

$$\varepsilon_\ell = \varepsilon_{1s}\Phi + \varepsilon_{2s}(1 - \Phi) \tag{310}$$

As $\varepsilon_h$ is still expressed by Eq. (302), the dielectric increment is given by

$$\varepsilon_\ell - \varepsilon_h = \frac{\Phi(1 - \Phi)(\varepsilon_{1s} - \varepsilon_{2s})^2}{\varepsilon_{1s} + 2\varepsilon_{2s} - \Phi(\varepsilon_{1s} - \varepsilon_{2s})} \tag{311}$$

The formula giving $\bar{f}_c$ is transformed into

$$\bar{f}_c = \frac{1}{2\pi\varepsilon_0} \frac{3\sigma}{\varepsilon_{1s} + 2\varepsilon_{2s} - \Phi(\varepsilon_{1s} - \varepsilon_{2s})} \tag{312}$$

$\sigma_1 << \sigma_2$

If the conductivity of the disperse phase is small compared to that of the continuous one, then $\varepsilon_h$ is still given by Eq. (302) but $\sigma_\ell$, $\varepsilon_\ell$, and $\bar{f}_c$ are then expressed by

$$\sigma_\ell = \sigma_2 \frac{2(1 - \Phi)}{2 + \Phi} \tag{313}$$

$$\varepsilon_\ell = \varepsilon_{2s} \frac{2(1 - \Phi)}{2 + \Phi} + \frac{9\Phi(\varepsilon_{1s}\sigma_2 - \varepsilon_{2s}\sigma_1)}{(2 + \Phi)^2 \sigma_2} \tag{314}$$

$$\bar{f}_c = \frac{1}{2\pi\varepsilon_0} \frac{(2 + \Phi)\sigma_2}{\varepsilon_{1s} + 2\varepsilon_{2s} - \Phi(\varepsilon_{1s} - \varepsilon_{2s})} \tag{315}$$

If it is permissible to neglect $\varepsilon_{2s}\sigma_1$, Eq. (314) reduces to

$$\varepsilon_\ell = \varepsilon_{2s} \frac{2(1 - \Phi)}{2 + \Phi} + \varepsilon_{1s} \frac{9\Phi}{(2 + \Phi)^2} \tag{316}$$

It must be noted that Eqs. (313), (315), and (316) are exact when $\sigma_1$ is strictly equal to zero.

Moreover, if $\varepsilon_{1s}$ is small, compared to $\varepsilon_{2s}$, as in the case of emulsions of nonpolar oils in water, $\varepsilon_h$, $\varepsilon_\ell$, and $\bar{f}_c$ are given to a good approximation by

$$\varepsilon_h = \varepsilon_{2s} \frac{2(1 - \Phi)}{2 + \Phi} \tag{317}$$

$$\varepsilon_\ell = \varepsilon_{2s} \frac{2(1 - \Phi)}{2 + \Phi} + \varepsilon_{1s} \frac{9\Phi}{(2 + \Phi)^2} \tag{318}$$

$$\bar{f}_c = \frac{1}{2\pi\varepsilon_0} \frac{\sigma_2}{\varepsilon_{2s}} \tag{319}$$

As Wagner's equation applies to dilute systems, $\Phi$ is normally small. The second term on the right-hand side of Eq. (316) can then be neglected compared to the first. The dielectric increment being thus negligible, $\varepsilon_\ell$ and $\varepsilon_h$ have nearly the same value. This result shows that emulsions of nonpolar oils in water with low oil contents will display only very small dielectric relaxations, which are expected to be barely observable, and that their total behavior will be essentially of a conductive kind. Such systems can be considered as being characterized by a frequency-independent real permittivity and a steady conductivity, respectively given by

$$\varepsilon' = \varepsilon_{2s} \frac{2(1 - \Phi)}{2 + \Phi} \tag{320}$$

$$\sigma = \sigma_2 \frac{2(1 - \Phi)}{2 + \Phi} \tag{321}$$

$\sigma_1 >> \sigma_2$

If the conductivity of the continuous phase is small compared to that of the disperse one, then $\varepsilon_h$ is still given by Eq. (302), but $\sigma_\ell$, $\varepsilon_\ell$, and $\bar{f}_c$ are then expressed by

$$\sigma_\ell = \sigma_2 \frac{1 + 2\Phi}{1 - \Phi} \tag{322}$$

$$\varepsilon_\ell = \varepsilon_{2s} \frac{1 + 2\Phi}{1 - \Phi} + \frac{9\Phi\,\sigma_2(\varepsilon_{1s}\sigma_2 - \varepsilon_{2s}\sigma_1)}{\sigma_1^2(1 - \Phi)^2} \tag{323}$$

$$\overline{f}_c = \frac{1}{2\pi\varepsilon_0} \frac{(1 - \Phi)\sigma_1}{\varepsilon_{1s} + 2\varepsilon_{2s} - \Phi(\varepsilon_{1s} - \varepsilon_{2s})} \tag{324}$$

By a further approximation, Eq. (323) yields

$$\varepsilon_\ell = \varepsilon_{2s} \frac{1 + 2\Phi}{1 - \Phi} \tag{325}$$

If $\sigma_2$ is strictly equal to zero, $\sigma_\ell$ is equal to zero as well, $\varepsilon_\ell$ and $\overline{f}_c$ being given by Eqs. (325) and (324), which are then exact. It must be noted that, as opposed to the preceding situation, $\sigma_\ell$ and $\varepsilon_\ell$ undergo a discontinuity for $\Phi = 1$. This is due to the asymmetric configuration of the system.

If $\sigma_1$ obeys an Arrhenius-type law,

$$\sigma_1 = \sigma_{1\infty}\, e^{-u_1/kT} \tag{326}$$

the critical frequency $\overline{f}_c$ as given by Eq. (324) obeys a similar law, provided that $\varepsilon_{1s}$, $\varepsilon_{2s}$, and $\Phi$ are not dependent, or depend only slightly, upon the temperature:

$$\overline{f}_c = \overline{A} e^{-\overline{u}/kT} \tag{327}$$

The activation energy $\overline{u}$ of the dielectric relaxation of the system is equal to the activation energy of the disperse-phase conductivity. This result proves that the temperature dependence of the conductivity of a substance dispersed within a nonconducting matrix can be obtained from temperature studies of the MWS dielectric relaxation,

$$\sigma_1 \to \infty, \qquad \sigma_2 = 0$$

If the conductivity of the disperse phase is very high, the continuous phase being nonconducting, $\sigma_\ell$ is equal to zero, and $\varepsilon_\ell$ is given by

$$\varepsilon_\ell = \varepsilon_{2s} \frac{1 + 2\Phi}{1 - \Phi} \tag{328}$$

This is one of the situations where the dielectric relaxation cannot be observed, which implies that Eq. (328) is valid at any radio frequency, since $\overline{f}_c \to \infty$.

b. Concentrated Systems (Hanai's Formula). *(1) General equations.* From Hanai's formula, which is written

$$\left(\frac{\varepsilon_1^* - \varepsilon_2^*}{\varepsilon_1^* - \varepsilon^*}\right)^3 \frac{\varepsilon^*}{\varepsilon_2^*} = \frac{1}{(1 - \Phi)^3} \tag{329}$$

limiting equations at high and low frequencies can be derived without resorting to a numerical analysis. At high frequencies, where the contribution of conduction phenomena vanishes, $\varepsilon^*$ reaches a limit which is simply the real solution of Bruggeman's equation. Hence, $\varepsilon_h$, the high-frequency limiting permittivity, is given by

$$\left(\frac{\varepsilon_{1s} - \varepsilon_{2s}}{\varepsilon_{1s} - \varepsilon_h}\right)^3 \frac{\varepsilon_h}{\varepsilon_{2s}} = \frac{1}{(1 - \Phi)^3} \tag{330}$$

Symmetrically, at low frequencies, where the contribution of conduction phenomena is preponderant, the imaginary part of $\varepsilon^*$ becomes identified with the contribution of $\sigma_\ell$, the bulk conductivity of the system, which is given by

$$\left(\frac{\sigma_1 - \sigma_2}{\sigma_1 - \sigma_\ell}\right)^3 \frac{\sigma_\ell}{\sigma_2} = \frac{1}{(1 - \Phi)^3} \tag{331}$$

As for $\varepsilon_\ell$, the low-frequency limiting permittivity, it is expressed by the following equation:

$$\varepsilon_\ell\left(\frac{3}{\sigma_\ell - \sigma_1} - \frac{1}{\sigma_\ell}\right) = 3\left(\frac{\varepsilon_{2s} - \varepsilon_{1s}}{\sigma_2 - \sigma_1} + \frac{\varepsilon_{1s}}{\sigma_\ell - \sigma_1}\right) - \frac{\varepsilon_{2s}}{\sigma_2} \tag{332}$$

The fact that $\varepsilon_h$ and $\varepsilon_\ell$, as given by Eqs. (330) and (332), do not have the same value implies that Hanai's general formula is representative of dielectric relaxations whose features must be determined through numerical calculations.

Figures 16, 17, 18, and 20 illustrate the dielectric relaxations which can be obtained from numerical analyses of Hanai's formula in different situations. As a general rule, it can be said that, up to $\Phi = 0.40$ or so, the Cole-Cole plots representative of the relaxations are circular arcs. Therefore, the complex permittivity of the system can be expressed by means of a Cole-Cole-type formula with an additional ohmic term,

$$\varepsilon^* = \varepsilon_h + \frac{\varepsilon_\ell - \varepsilon_h}{1 + (jf/\bar{f}_c)^{1-h}} + \frac{\sigma_\ell}{2\pi j \varepsilon_0 f} \tag{333}$$

$\sigma_\ell$, $\varepsilon_h$, and $\varepsilon_\ell$ being given by Eqs. (331), (330), and (332), respectively. For higher values of $\Phi$, the Cole-Cole plots become more and more distorted as $\Phi$ increases and Eq. (333) no longer holds, but $\sigma_\ell$, $\varepsilon_h$, and $\varepsilon_\ell$ still conform to Eqs. (331), (330), and (332), respectively.

The dependence of $\sigma_\ell$, $\varepsilon_h$, $\varepsilon_\ell$ upon $\Phi$ can be studied in the general case by using reduced parameters and unknowns, as indicated:

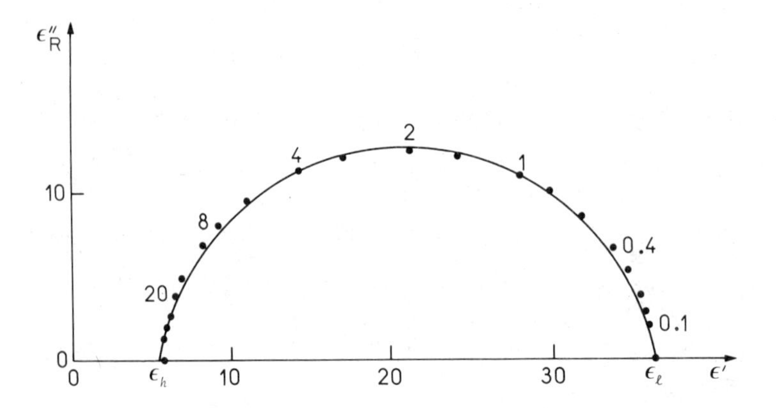


**Figure 16.** Cole-Cole plot of the dielectric relaxation of a system characterized by $\varepsilon_{1s} = 78.3$; $\sigma_1 = 10^{-6}$ S $m^{-1}$; $\varepsilon_{2s} = 2.25$; $\sigma_2 = 3.10^{-6}$ S $m^{-1}$; $\Phi = 0.30$. The frequencies are indicated in kilohertz. (From Ref. 83, Dr. Dietrich Steinkopff Verlag, Darmstadt, Germany.)

$$m = \frac{\sigma_2}{\sigma_1} \qquad x = \frac{\sigma_\ell}{\sigma_1} \qquad n = \frac{\varepsilon_{2s}}{\varepsilon_{1s}} \qquad y = \frac{\varepsilon_h}{\varepsilon_{1s}} \qquad z = \frac{\varepsilon_\ell}{\varepsilon_{1s}}$$

Introducing these quantities into Eqs. (330) to (332) yields

$$\left(\frac{1-m}{1-x}\right)^3 \frac{x}{m} = \frac{1}{(1-\Phi)^3} \tag{334}$$

$$\left(\frac{1-n}{1-y}\right)^3 \frac{y}{n} = \frac{1}{(1-\Phi)^3} \tag{335}$$

$$z = x + (1-\Phi)\left(\frac{x}{m}\right)^{4/3} \frac{2m+1}{2x+1}(n-m) \tag{336}$$

Since Eqs. (334) and (335) have the same form, the variations of x with m and $\Phi$ and those of y with n and $\Phi$ are ruled by the same law. Numerical values of x (or y) have been reported versus $\Phi$ by Boned et al. [83] for different values of m (or n). It is readily seen from Eq. (336) that, for a given value of n, z depends only upon m and $\Phi$.

At any value of $\Phi$, n being given, the curve representing the variation of z with m shows a minimum which corresponds to m = n, namely,

$$\varepsilon_{1s}\sigma_2 - \varepsilon_{2s}\sigma_1 = 0 \tag{337}$$

or, alternatively,

$$T_1 = T_2 \tag{338}$$

The value of $\varepsilon_\ell$ corresponding to this minimum is equal to that of $\varepsilon_h$ at the same $\Phi$, indicating that the system does not exhibit any dielectric relaxation, in conformity with the general analysis made previously as to the existence of a dielectric relaxation owing to MWS effects in heterogeneous dielectrics. This result could have been derived directly from the general Hanai formula, as well as the other two conditions given by Eq. (292) and (293). Thus, the general condition expressed by the inequality (294) for the existence of a dielectric relaxation due to interfacial polarization can be checked in the case of Hanai's formula as it can in the case of Maxwell's and Wagner's. As concerns the critical frequency $\bar{f}_c$, its variation versus $\varepsilon_{1s}$, $\varepsilon_{2s}$, $\sigma_1$, $\sigma_2$, and $\Phi$ are quite complicated and no general law can be simply formulated. Examples of the dependence of $\bar{f}_c$ upon $\Phi$ will be given below, for special cases.

*(2) Equations applying to special cases.*

$\sigma_1 = \sigma_2$

If the conductivities of the disperse and continuous phases are equal, then for Wagner's equation, $\sigma_\ell$ is equal to $\sigma_1$ and $\sigma_2$, and $\varepsilon_\ell$ is expressed simply by

$$\varepsilon_\ell = \varepsilon_{1s}\Phi + \varepsilon_{2s}(1 - \Phi) \tag{339}$$

$\sigma_1 << \sigma_2$

If the conductivity of the disperse phase is small compared to that of the continuous one, $\varepsilon_h$ is still given by Eq. (330), but $\sigma_\ell$ and $\varepsilon_\ell$ are then expressed by

$$\frac{\sigma_2^2\sigma_\ell}{(\sigma_\ell - \sigma_1)^3} = \frac{1}{(1 - \Phi)^3} \tag{340}$$

$$\varepsilon_\ell = \frac{\sigma_\ell[3\varepsilon_{1s}\sigma_2 + (2\varepsilon_{2s} - 3\varepsilon_{1s})(\sigma_\ell - \sigma_1)]}{\sigma_2(2\sigma_\ell + \sigma_1)} \tag{341}$$

If it is permissible to take $\sigma_1$ as negligible compared to $\sigma_\ell$, which is the case for moderate values of $\Phi$, Eqs. (340) and (341) become, respectively,

$$\sigma_\ell = \sigma_2(1 - \Phi)^{3/2} \tag{342}$$

$$\varepsilon_\ell = \frac{3}{2}\varepsilon_{1s} + \left(\varepsilon_{2s} - \frac{3}{2}\varepsilon_{1s}\right)(1 - \Phi)^{3/2} \tag{343}$$

Equation (343) can be written, alternatively

$$\frac{2\varepsilon_\ell - 3\varepsilon_{1s}}{2\varepsilon_{2s} - 3\varepsilon_{1s}} = (1 - \Phi)^{3/2} \tag{344}$$

It must be noted that Eqs. (343) or (344) and (342) are exact when $\sigma_1$ is strictly equal to zero.

Moreover, if $\varepsilon_{1s}$ is small compared to $\varepsilon_{2s}$, as in the case of emulsions of nonpolar oils in water, Eqs. (330) and (343) can be approximated by the same expression, so that

$$\varepsilon_\ell \simeq \varepsilon_h \simeq \varepsilon_{2s}(1 - \Phi)^{3/2} \tag{345}$$

Thus, as for Wagner's formula, it can be inferred from Hanai's formula that emulsions of poorly conducting nonpolar oils in water will exhibit very small dielectric relaxations which are expected to be hardly observable. Therefore, such systems can be characterized by a frequency-independent real permittivity and a steady conductivity respectively, given by

$$\varepsilon' = \frac{3}{2}\varepsilon_{1s} + \left(\varepsilon_{2s} - \frac{3}{2}\varepsilon_{1s}\right)(1 - \Phi)^{3/2} \tag{346}$$

$$\sigma = \sigma_2(1 - \Phi)^{3/2} \tag{347}$$

$\sigma_1 >> \sigma_2$

If the conductivity of the continuous phase is small compared to that of the disperse one, the equation giving $\varepsilon_h$ is unchanged, but those expressing $\sigma_\ell$ and $\varepsilon_\ell$ are replaced by the following approximations:

$$\frac{\sigma_1^3}{(\sigma_1 - \sigma_\ell)^3}\frac{\sigma_\ell}{\sigma_2} = \frac{1}{(1 - \Phi)^3} \tag{348}$$

$$\varepsilon_\ell = \varepsilon_{2s}\frac{\sigma_\ell(\sigma_1 - \sigma_\ell)}{\sigma_2(\sigma_1 + 2\sigma_\ell)} \tag{349}$$

If it is permissible to take $\sigma_\ell$ as negligible compared to $\sigma_1$ (the case for moderate values of $\Phi$), Eqs. (348) and (349) become, respectively,

$$\sigma_\ell = \sigma_2\frac{1}{(1 - \Phi)^3} \tag{350}$$

$$\varepsilon_\ell = \varepsilon_{2s}\frac{1}{(1 - \Phi)^3} \tag{351}$$

When $\sigma_2$ is strictly equal to zero, $\sigma_\ell$ is equal to zero as well and $\varepsilon_\ell$ is given exactly by Eq. (351), $\varepsilon_h$ being still expressed by Eq. (330). This situation corresponds to the case of emulsions of water in nonconducting oil phases.

When the continuous medium is nonconducting and is characterized by a static permittivity which is small compared to that of the disperse phase ($\varepsilon_{2s} << \varepsilon_{1s}$), it is possible to show that, for low values of $\Phi$, Hanai's general formula can be transformed into a Debye-type equation. With $\varepsilon_{2s} << \varepsilon_{1s}$ and $\sigma_2 = 0$, by assuming that $|\varepsilon^*| << |\varepsilon_1^*|$ when $\Phi$ is small enough, Eq. (329) yields the following approximate equation:

$$\varepsilon^* = \varepsilon_{2s} \frac{\varepsilon_1^*}{3\varepsilon_{2s} + (\varepsilon_1^* - 3\varepsilon_{2s})(1 - \Phi)^3} \tag{352}$$

which can be transformed into

$$\varepsilon^* = \varepsilon_h + \frac{\varepsilon_\ell - \varepsilon_h}{1 + jf/\bar{f}_c} \tag{353}$$

by taking

$$\varepsilon_h = \varepsilon_{2s} \frac{\varepsilon_{1s}}{3\varepsilon_{2s} + (\varepsilon_{1s} - 3\varepsilon_{2s})(1 - \Phi)^3} \tag{354}$$

$$\varepsilon_\ell = \varepsilon_{2s} \frac{1}{(1 - \Phi)^3} \tag{355}$$

$$\bar{f}_c = \frac{1}{2\pi\varepsilon_0} \frac{\sigma_1}{\varepsilon_{1s} + 3\varepsilon_{2s}[1/(1 - \Phi)^3 - 1]} \tag{356}$$

Equation (354) is only an approximate form of Eq. (330), valid for $\varepsilon_{2s} << \varepsilon_{1s}$. Equation (356) shows that $\bar{f}_c$ is an increasing function of $\sigma_1$ and a decreasing function of $\varepsilon_{1s}$, $\varepsilon_{2s}$, and $\Phi$. This is consistent with the predictions derivable from Eq. (342). Moreover, the comparison of Eqs. (356) and (307) shows that, when $\Phi$ vanishes, $\bar{f}_c$ as given by Eq. (356) approaches $\bar{f}_c(0)$, taken for $\sigma_2 = 0$, which is the limiting critical frequency at $\Phi = 0$ for Wagner's formula. As in the case of dilute spherical dispersions, if $\sigma_1$ obeys an Arrhenius law, $\bar{f}_c$ follows a similar law, provided that $\varepsilon_{1s}$, $\varepsilon_{2s}$, and $\Phi$ do not depend, or depend slightly only, upon the temperature.

The extension of all the preceding results to higher values of $\Phi$ can be obtained easily from numerical calculations. As an example, in Figs. 17a and b are plotted the variations of $\varepsilon'$ and $\varepsilon''$ with the frequency f, for a system characterized by $\varepsilon_{1s} = 80$, $\sigma_1 = 10^{-3}$ S m$^{-1}$, $\varepsilon_{2s} = 2.5$, $\sigma_2 = 0$, and $\Phi = 0.4$, while Fig. 17c gives the corresponding Cole-Cole diagram. Figures 18a to c illustrate the strong dependence of the dielectric relaxation upon the disperse-phase volume fraction and show how the Cole-Cole plots become distorted at higher values of $\Phi$. The Cole-Cole plots can be considered as being circular arcs up to $\Phi = 0.6$ or so, a higher limit than $\Phi = 0.4$ which was found in the general case, and consequently, the complex permittivity is expressed, for $\Phi \leqslant 0.7$ or so, by a Cole-Cole type equation

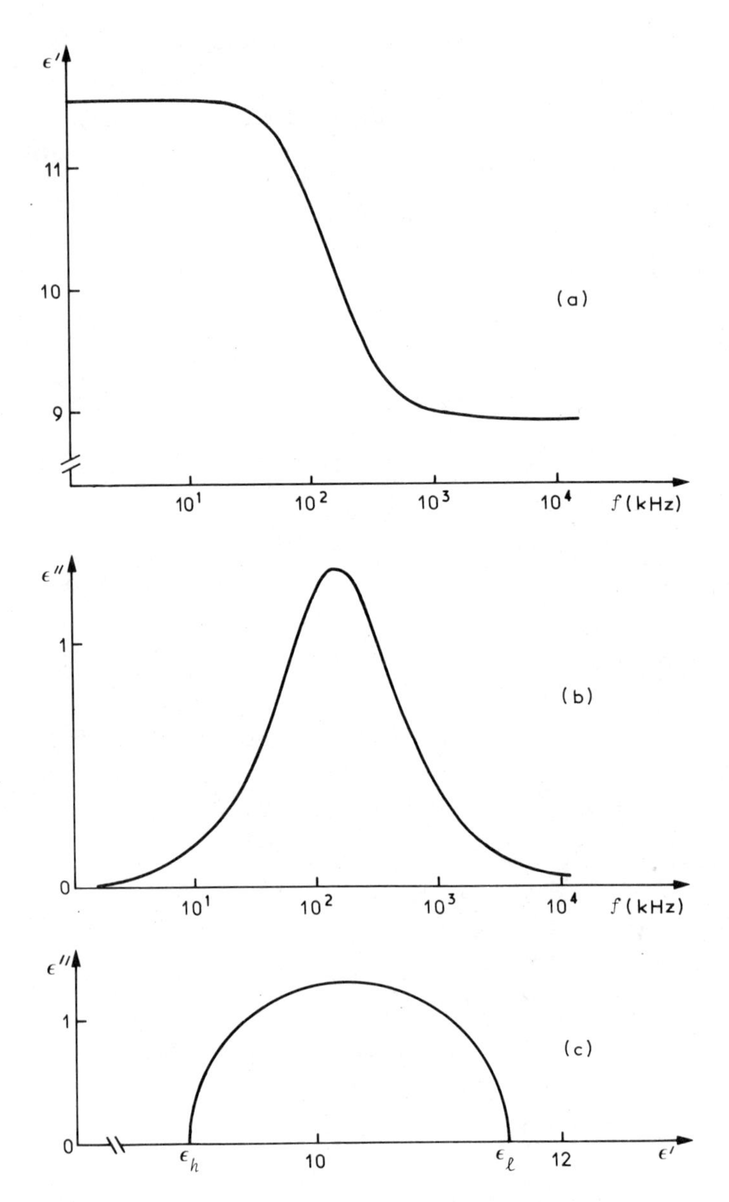


**Figure 17.** Theoretical features of the dielectric relaxation exhibited by a system characterized by $\varepsilon_{1s} = 80$; $\sigma_1 = 10^{-3}$ S m$^{-1}$; $\varepsilon_{2s} = 2.5$; $\sigma_2 = 0$; $\Phi = 0.40$: (a) dispersion curve, (b) absorption curve, and (c) Cole-Cole plot. (Modified from Ref. 76, Gauthier-Villars, Paris.)

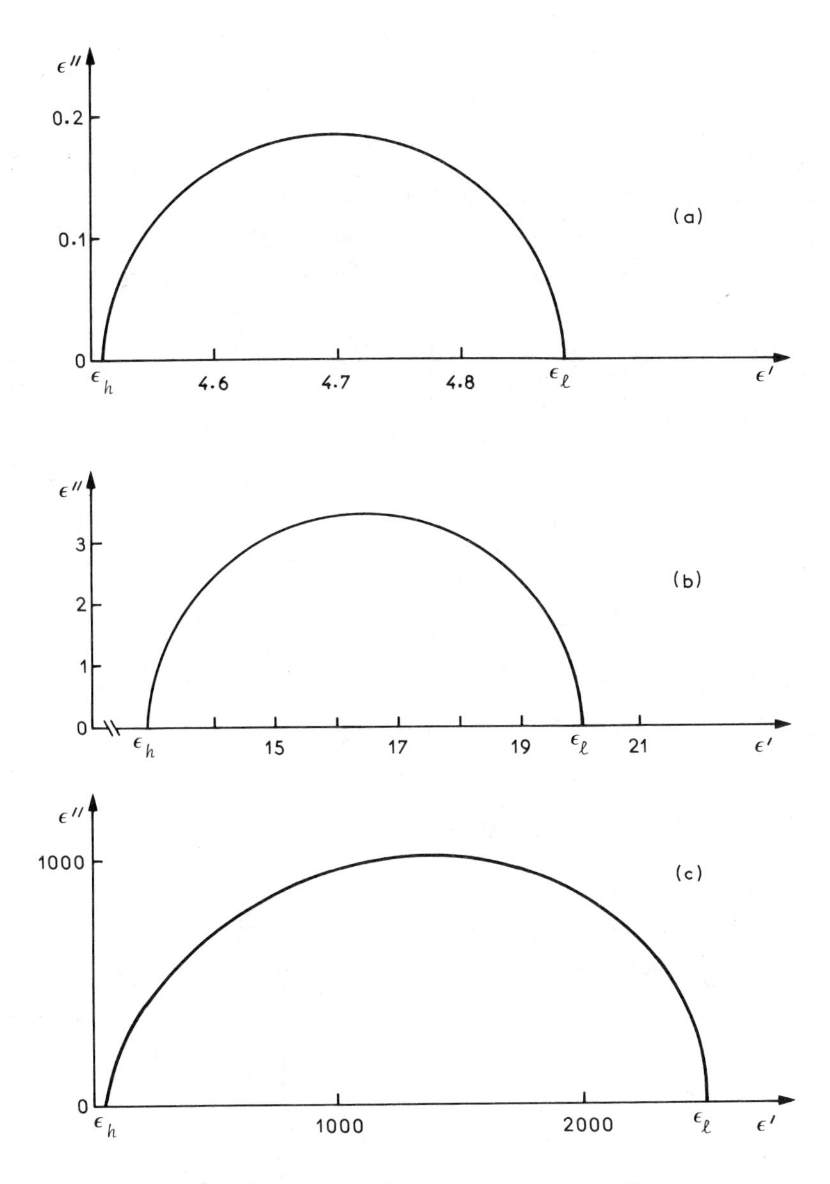


**Figure 18.** Cole-Cole plots of the relaxations exhibited by systems characterized by $\varepsilon_{1s} = 80$; $\sigma_1 = 10^{-3}$ S m$^{-1}$; $\varepsilon_{2s} = 2.5$; $\sigma_2 = 0$: (a) $\Phi = 0.20$, (b) $\Phi = 0.50$, and (c) $\Phi = 0.90$. (a and b from Ref. 4; c modified from Ref. 76, Gauthier-Villars, Paris.)

$$\varepsilon^* = \varepsilon_\ell + \frac{\varepsilon_\ell - \varepsilon_h}{1 + (jf/\bar{f}_c)^{1-h}} \tag{357}$$

where $\varepsilon_h$ and $\varepsilon_\ell$ are given by Eqs. (330) and (351). The parameter h is always less than 0.1. For values of $\Phi$ higher than about 0.7 the Cole-Cole plots become more and more distorted and depressed as $\Phi$ increases and they present an umbilical singularity for f = 0. However, $\varepsilon_h$ and $\varepsilon_\ell$ are still given by Eqs. (330) and (351) and a critical frequency $\bar{f}_c$ can be defined from the absorption curves. Figure 19 shows the variation of $\bar{f}_c$ with $\Phi$, comparing Hanai's and Wagner's formulas. For vanishing values of $\Phi$, $\bar{f}_c$ is the same for both laws, as expected, but beyond $\Phi$ = 0.10 the discrepancy between the two curves increases as $\Phi$ increases,

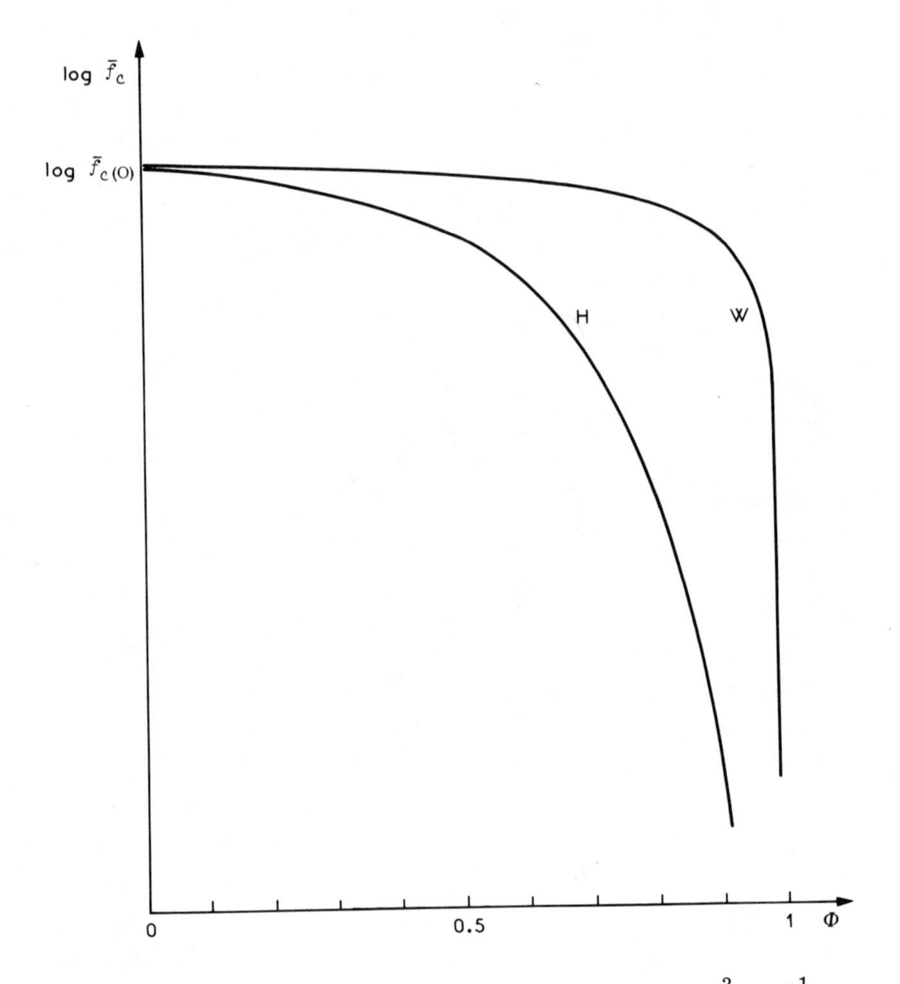


**Figure 19.** Variation of $\bar{f}_c$ versus $\Phi$. $\varepsilon_{1s} = 80$; $\sigma_1 = 10^{-3}$ S m$^{-1}$; $\varepsilon_{2s} = 2.5$; $\sigma_2 = 0$; W, Wagner formula; H, Hanai formula. (From Ref. 4.)

the curve corresponding to Hanai's formula being the lower. It can be checked easily that $\bar{f}_c$ is given by

$$\bar{f}_c = \bar{A}e^{-\bar{u}/kT} \tag{358}$$

if $\sigma_1$ follows an Arrhenius-type law. The dielectric relaxation activation energy $\bar{u}$ is equal, whatever the value of $\Phi$, to the conductivity activation energy of the disperse phase. This theoretical result justifies the use of emulsions for dielectric studies of liquids in metastable thermodynamic states. For instance, the conductivity activation energy of water and aqueous saline solutions can be investigated from a study, at subzero temperatures, of the interfacial polarization of water-in-oil type emulsions.

Figure 20a and b reports theoretical Cole-Cole plots for a system in which $\varepsilon_{1s} << \varepsilon_{2s}$ and for a system in which $\varepsilon_{1s} >> \varepsilon_{2s}$, $\varepsilon_{1s}$ and $\varepsilon_{2s}$ having the same values and the constituent proportions being the same in both cases. These two figures illustrate the discrepancies existing between the dielectric behavior of an emulsion of a nonconducting and nonpolar oil phase in water and an emulsion of water in a nonconducting and nonpolar oil phase, at a given content of oil (or water). As has been demonstrated previously, it can be expected that such O/W systems will exhibit only very small dielectric relaxations (hardly measurable be-

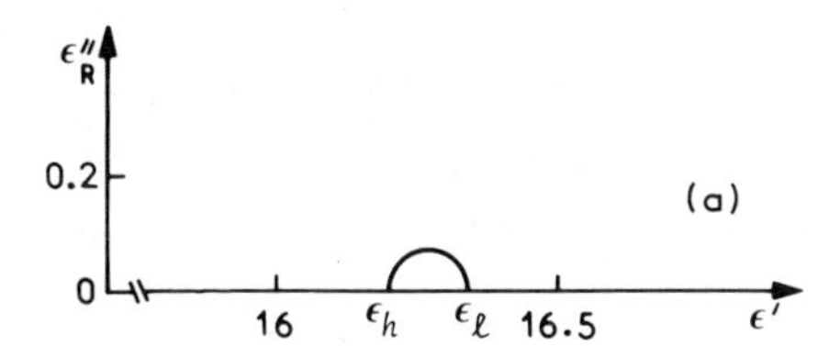


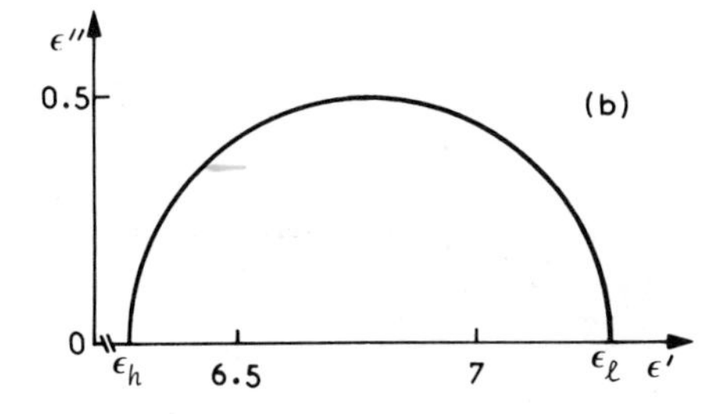


**Figure 20.** Cole-Cole plots of the relaxations exhibited by two symmetrical systems characterized by equal oil contents (or water contents): (a) O/W system, $\Phi_{oil} = \Phi = 0.70$; (b) W/O system, $\Phi_{oil} = 1 - \Phi = 0.70$. (From Ref. 79, Dr. Dietrich Steinkopff Verlag, Darmstadt, Germany.)

cause of experimental uncertainties), and consequently that their total behavior will be essentially conductive. On the other hand, W/O systems of the symmetric type will be characterized by a zero bulk conductivity and will display striking dielectric relaxations which will be, in any case, much greater than those of the corresponding O/W emulsions.

$\sigma_1 \to \infty$, $\sigma_2 = 0$

If the conductivity of the disperse phase is very high, the continuous phase being nonconducting, then when $\sigma_1 \to \infty$, Hanai's formula reduces to

$$\varepsilon^* = \varepsilon_2^* \frac{1}{(1 - \Phi)^3} \tag{359}$$

If $\sigma_2$ is equal to zero, $\varepsilon_2^*$ is equal to $\varepsilon_{2s}$ and $\varepsilon^*$ to $\varepsilon_\ell$, given by

$$\varepsilon_\ell = \varepsilon_{2s} \frac{1}{(1 - \Phi)^3} \tag{360}$$

This is one of the situations where the dielectric relaxation cannot be observed, Eq. (360) being valid at any radio frequency since $\bar{f}_c \to \infty$.

### 2. *The Disperse Phase Exhibits a Debye-Type Dipolar Relaxation, the Continuous Phase Being Nonpolar and Nonconducting*

In this case, the complex permittivity of the continuous phase, volume fraction equal to $1 - \Phi$, is given simply by

$$\varepsilon_2^* = \varepsilon_{2s} = \varepsilon_2 \tag{361}$$

while the complex permittivity of the disperse phase is expressed by

$$\varepsilon_1^* = \varepsilon_{1d} + \frac{\varepsilon_{1s} - \varepsilon_{1d}}{1 + j\omega\tau_1} + \frac{\sigma_1}{j\omega\varepsilon_0} \tag{362}$$

or, alternatively, introducing the time constant $T_1$

$$\varepsilon_1^* = \varepsilon_{1d} + \frac{\varepsilon_{1s} - \varepsilon_{1d}}{1 + j\omega\tau_1} + \frac{\varepsilon_{1s}}{j\omega T_1} \tag{363}$$

a. Dilute Systems (Wagner's Formula). *(1) The disperse phase is nonconducting* ($\sigma_1 = 0$). Introducing in Wagner's equation $\varepsilon_1^*$, given by

$$\varepsilon_1^* = \varepsilon_{1d} + \frac{\varepsilon_{1s} - \varepsilon_{1d}}{1 + j\omega\tau_1} \tag{364}$$

yields, for $\varepsilon^*$ the complex permittivity of the system,

$$\varepsilon^* = \varepsilon_d + \frac{\varepsilon_h - \varepsilon_d}{1 + j\omega\tau} \tag{365}$$

$\varepsilon_d$ the high frequency limiting permittivity, $\varepsilon_h$ the low frequency limiting permittivity, and $\tau$ the macroscopic relaxation time, are given respectively by

$$\varepsilon_d = \varepsilon_2 \frac{\varepsilon_{1d} + 2\varepsilon_2 + 2\Phi(\varepsilon_{1d} - \varepsilon_2)}{\varepsilon_{1d} + 2\varepsilon_2 - \Phi(\varepsilon_{1d} - \varepsilon_2)} \tag{366}$$

$$\varepsilon_h = \varepsilon_2 \frac{\varepsilon_{1s} + 2\varepsilon_2 + 2\Phi(\varepsilon_{1s} - \varepsilon_2)}{\varepsilon_{1s} + 2\varepsilon_2 - \Phi(\varepsilon_{1s} - \varepsilon_2)} \tag{367}$$

$$\tau = \tau_1 \frac{(\varepsilon_{1d} + 2\varepsilon_2)(\varepsilon_s + 2\varepsilon_2)}{(\varepsilon_d + 2\varepsilon_2)(\varepsilon_{1s} + 2\varepsilon_2)} \tag{368}$$

Equation (368) can be written alternatively, in terms of the critical frequency,

$$f_c = f_{c_1} \frac{(\varepsilon_d + 2\varepsilon_2)(\varepsilon_{1s} + 2\varepsilon_2)}{(\varepsilon_{1d} + 2\varepsilon_2)(\varepsilon_h + 2\varepsilon_2)} \tag{369}$$

The limiting values of $f_c$ as $\Phi \to 0$ and $\Phi \to 1$ are, respectively,

$$f_c(0) = f_{c_1} \frac{\varepsilon_{1s} + 2\varepsilon_2}{\varepsilon_{1d} + 2\varepsilon_2} \tag{370}$$

$$f_c(1) = f_{c_1} \tag{371}$$

As in any case, $\varepsilon_{1s}$ is greater than $\varepsilon_{1d}$, $f_c(0)$ is greater than $f_c(1)$, and $f_c$ is a decreasing function of $\Phi$.

Equation (365) expresses the fact that the system displays a Debye-type dielectric relaxation which is the image of the disperse-phase Debye-type dielectric relaxation through Wagner's formula. The low-frequency limiting permittivity $\varepsilon_h$ is given by the same equation as (302), satisfied by the high-frequency limiting permittivity of the dielectric relaxation arising from interfacial polarization. If $f_{c_1}$ obeys an Arrhenius-type law, $f_c$ follows a similar law, provided that $\varepsilon_{1s}$ and $\Phi$ do not depend, or depend slightly, upon the temperature. The activation energy of the system dielectric relaxation is equal to the activation energy of the Debye relaxation exhibited by the disperse phase.

*(2) The disperse phase is conducting* ($\sigma_1 \neq 0$). The complex permittivity of the system is then given by

$$\varepsilon^* = \varepsilon_d + \frac{\varepsilon_h - \varepsilon_d}{1 + j\omega\tau} + \frac{\varepsilon_\ell - \varepsilon_h}{1 + j\omega\bar{\tau}} \tag{372}$$

with $\varepsilon_d$, $\varepsilon_h$, $\varepsilon_\ell$, $\tau$, and $\bar{\tau}$ given respectively by Eqs. (366), (367), (325), (368), and (305).

The dielectric relaxation characterized by $\tau$ is connected with the Debye dielectric relaxation of the disperse phase while the dielectric relaxation characterized by $\bar{\tau}$ arises from interfacial polarization phenomena. Depending on the respective values of $\tau_1$ and $T_1$, the two relaxations may or may not overlap each other. If $\tau_1$ and $T_1$ are very different, the two relaxations exhibited by the system are decoupled and the total Cole-Cole plot consists of two adjacent semicircles whose unique contact point lies on the $\varepsilon'$ axis at $\varepsilon' = \varepsilon_h$.

Equation (372) shows that by studying the total dielectric behavior of a heterogeneous system of the spherical dispersion type, it is possible to investigate easily the dielectric and conductive properties of the disperse phase. This is of particular interest in the case of systems involving as the disperse phase water or an aqueous saline solution in a metastable thermodynamic state.

It must be noted that, if the continuous phase is conducting, the system will exhibit a steady conductivity $\sigma_\ell$, given by Eq. (301), and an additional ohmic term will appear in Eq. (372).

b. Concentrated Systems (Hanai's Formula). *(1) The disperse phase is nonconducting* ($\sigma_1 = 0$). From Hanai's general formula, it is possible to obtain directly equations for $\varepsilon_d$, the high-frequency limiting permittivity of the system, and for $\varepsilon_h$, the low-frequency limiting permittivity of the system. When $f \to \infty$, $\varepsilon^*$ approaches a real value $\varepsilon_d$, which is given by

$$\left(\frac{\varepsilon_{1d} - \varepsilon_2}{\varepsilon_{1d} - \varepsilon_d}\right)^3 \frac{\varepsilon_d}{\varepsilon_2} = \frac{1}{(1 - \Phi)^3} \tag{373}$$

Similarly, when $f \to 0$, $\varepsilon^*$ approaches a real value $\varepsilon_h$, which is given by

$$\left(\frac{\varepsilon_{1s} - \varepsilon_2}{\varepsilon_{1s} - \varepsilon_h}\right)^3 \frac{\varepsilon_h}{\varepsilon_2} = \frac{1}{(1 - \Phi)^3} \tag{374}$$

As $\varepsilon_d$ and $\varepsilon_h$, given by Eqs. (373) and (374), do not have the same value, it can be expected that the system will exhibit a dielectric relaxation. Further information can be gained by resorting to a numerical analysis of Hanai's general formula in which are inserted $\varepsilon_1^*$ and $\varepsilon_2^*$ as expressed by Eqs. (364) and (361).

Systematic calculations made by Clausse [4] proved that when $\varepsilon_{1s} >> \varepsilon_2$, as in water-in-oil-type emulsions, Hanai's formula can be transformed into a Cole-Cole-type equation, for any value of $\Phi$,

$$\varepsilon^* = \varepsilon_d + \frac{\varepsilon_h - \varepsilon_d}{1 + (jf/f_c)^{1-h}} \tag{375}$$

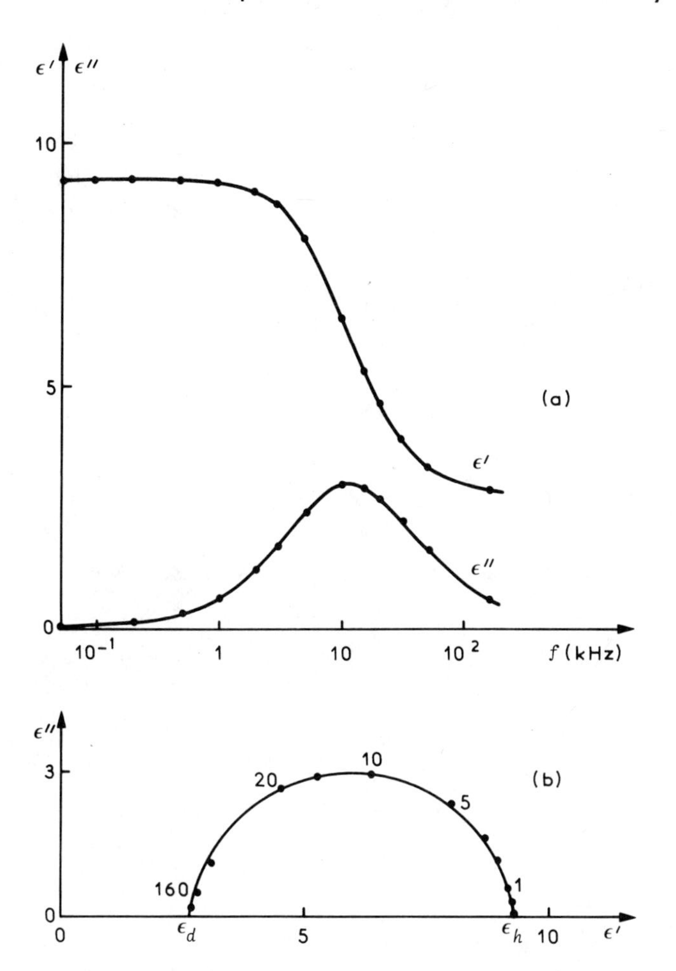


Figure 21. Theoretical features of the dielectric relaxation exhibited by a system characterized by $\varepsilon_{1s} = 95$; $\varepsilon_{1d} = 3.08$; $f_{c1} = 2.65$ kHz; $\sigma_1 = 0$; $\varepsilon_{2s} = 2.5$; $\sigma_2 = 0$; $\Phi = 0.40$: (a) dispersion and absorption curves and (b) Cole-Cole plot. (From Ref. 78, Gauthier-Villars, Paris.)

$\varepsilon_d$ and $\varepsilon_h$ being given by Eqs. (373) and (374). The parameter h is always lower than 0.1 over the entire volume fraction range.

In Fig. 21a are reported the dispersion and absorption curves for a system characterized by $\Phi = 0.4$, and in Fig. 21b the corresponding Cole-Cole plot. Figure 22 compares the variation of $f_c/f_{c1}$ for Hanai's and Wagner's formulas. $f_c$ is a decreasing function of $\Phi$ for both formulas, and, as expected, $f_c$ is the same for both formulas at vanishing values of $\Phi$ and $1 - \Phi$. In between, the discrepancy between the curves varies with $\Phi$, the curve corresponding to Hanai's formula lying below the other. It can be easily checked that, as in the case of

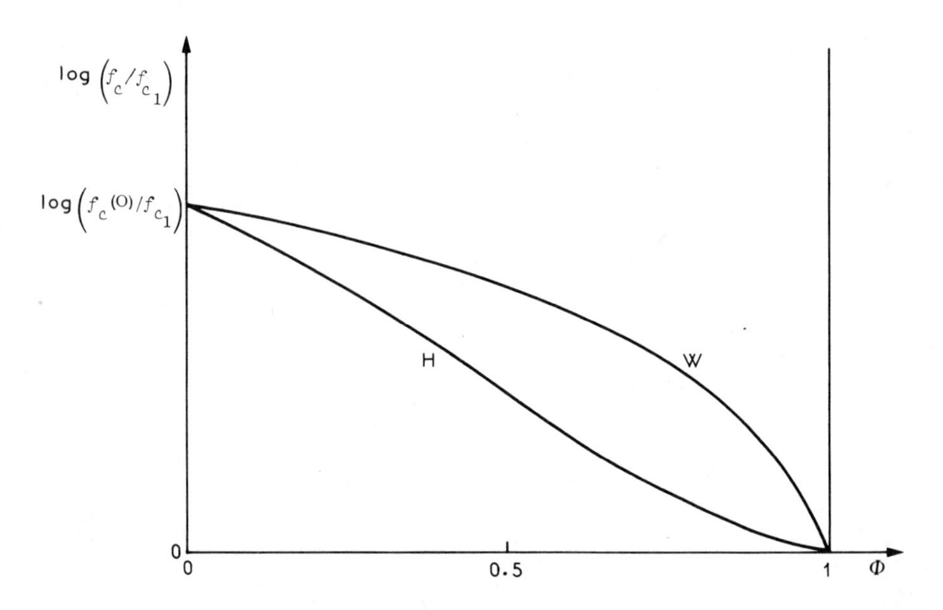


Figure 22. Variation of $(f_c/f_{c_1})$ versus $\Phi$ for Wagner's formula (W), and for Hanai's formula (H). (From Ref. 4.)

Wagner's formula, $f_c$ follows an Arrhenius-type law if $f_{c_1}$ does, the activation energy of the system dielectric relaxation being equal to the activation energy of the relaxation exhibited by the disperse phase.

*(2) The disperse phase is conducting* ($\sigma_1 \neq 0$). From the results reported previously, it can be expected that the system will exhibit two dielectric relaxations, the first being connected with the disperse-phase Debye relaxation and characterized by the dielectric increment $\varepsilon_h - \varepsilon_d$, and a second one arising from interfacial polarization phenomena and characterized by the dielectric increment $(\varepsilon_\ell - \varepsilon_h)\varepsilon_d$. $\varepsilon_h$ and $\varepsilon_\ell$ are given by Eqs. (373), (374), and (351), respectively. This theoretical behavior was demonstrated by Clausse [4] from numerical analyses of Hanai's general formula, in the case of systems of the water-in-oil type.

Figure 23 shows the features of the dielectric behavior of a system with volume fraction equal to 0.7, and Fig. 24 gives the corresponding Cole-Cole plot. In that case, the two relaxations are decoupled and do not overlap. This situation corresponds to the case of emulsions of water and aqueous saline solutions for which the conduction absorption does not interfere with the dipolar relaxation of water. For values of $\Phi$ ranging from zero to 0.7 or so, the complex permittivity can be written

$$\varepsilon^* = \varepsilon_d + \frac{\varepsilon_h - \varepsilon_d}{1 + j\omega\tau} + \frac{\varepsilon_\ell - \varepsilon_h}{1 + j\omega\bar{\tau}} \qquad (376)$$

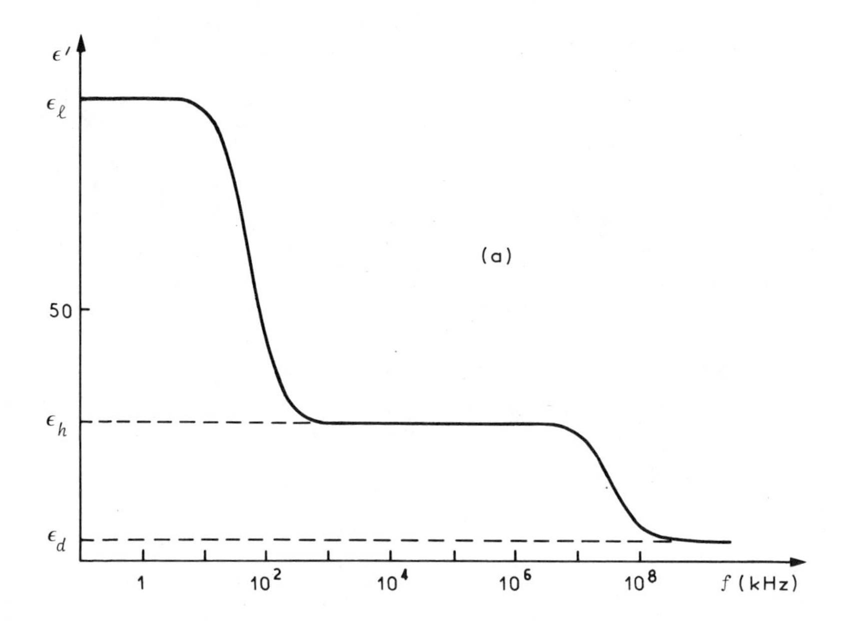


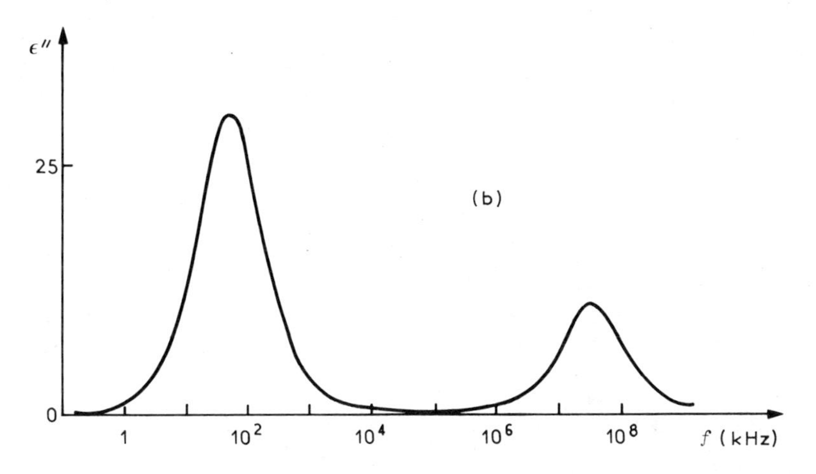


Figure 23. Theoretical features of the dielectric behavior of a water-in-oil emulsion-type system over the entire radio-frequency range: $\Phi = 0.70$; (a) dispersion curve and (b) absorption curve. (From Ref. 4.)

For higher values of $\Phi$, this formula no longer holds because of the distortions appearing in the MWS contribution. As $\Phi$ increases from 0 to 1, the magnitudes of the two relaxations vary as indicated in Fig. 25, which shows the dependence of $\varepsilon''_m$, the maximum value of $\varepsilon'$, upon $\Phi$ for both relaxations. When $\Phi$ is lower than about 0.55, the MWS-type relaxation is smaller than the other, the reverse being true for $\Phi > 0.55$.

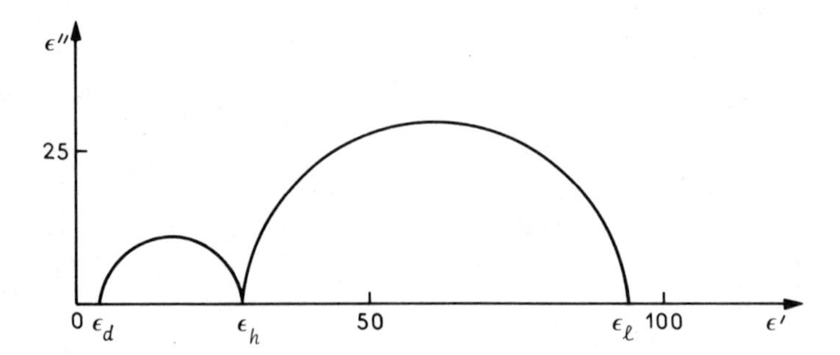


Figure 24. Theoretical Cole-Cole plot of the dielectric behavior of a water-in-oil emulsion system over the entire radio-frequency range. $\Phi = 0.70$. (From Ref. 4.)

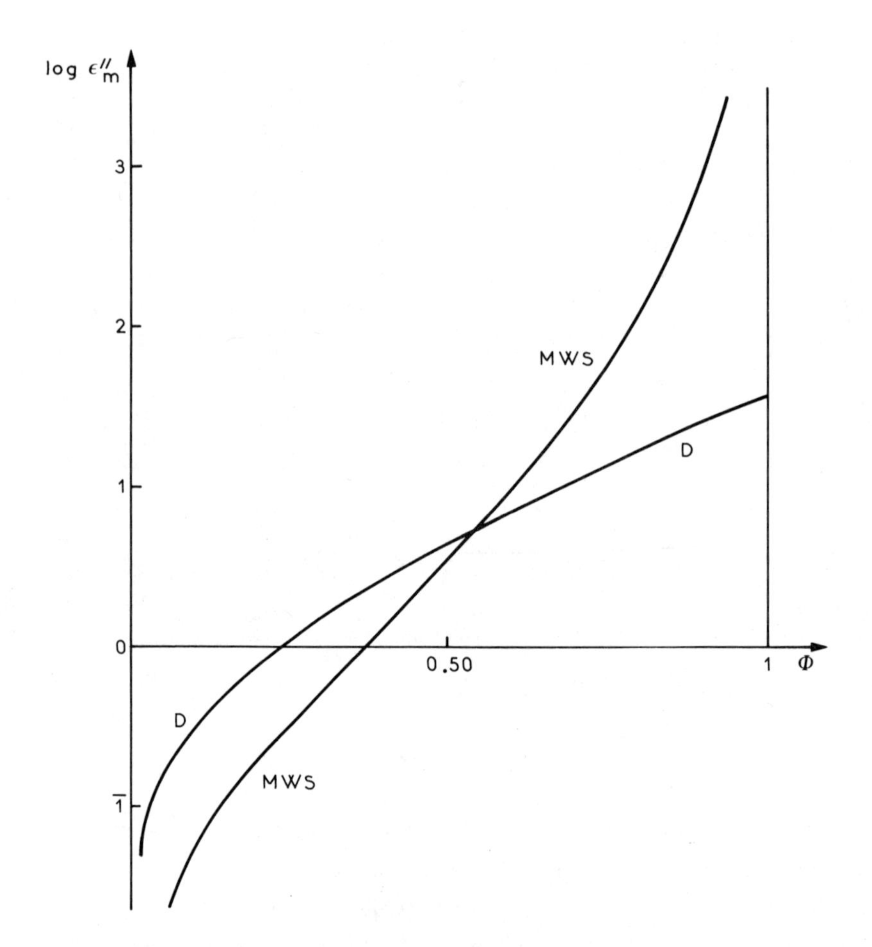


Figure 25. Comparative variation versus $\Phi$ of the absorption maximum $\varepsilon''_m$ of the two dielectric relaxations exhibited by water-in-oil emulsion systems: D, relaxation connected with the intrinsic dipolar relaxation of water; MWS, relaxation arising from interfacial polarization phenomena. (From Ref. 4.)

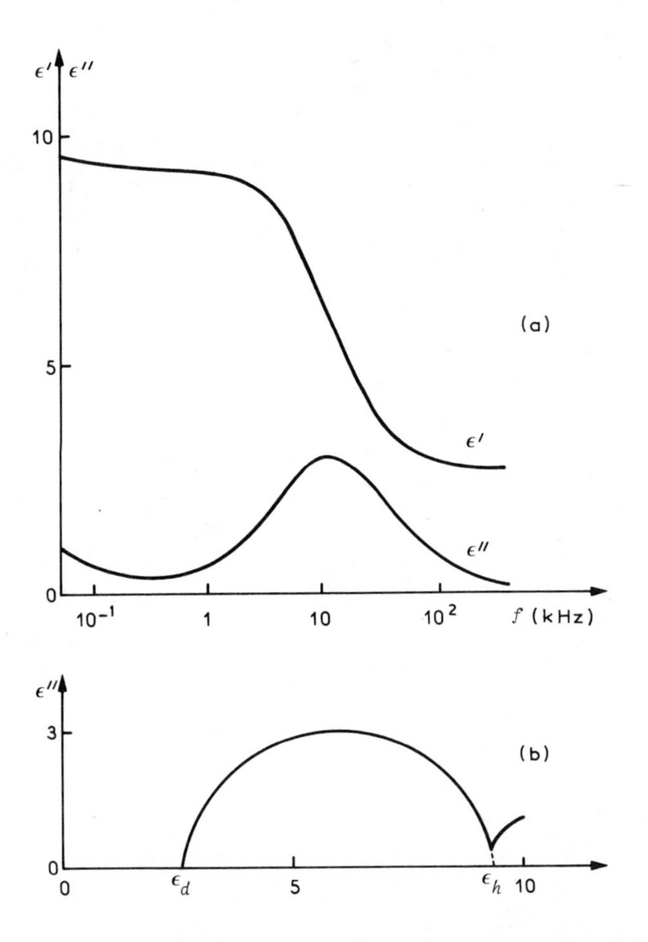


Figure 26. Theoretical features of the dielectric behavior of a water-in-oil emulsion system in which the MWS effect interferes slightly with the dielectric relaxation connected with the disperse-phase intrinsic dielectric relaxation: $\Phi = 0.40$; (a) dispersion and absorption curves and (b) Cole-Cole plot. (From Ref. 4.)

If the conductivity of the disperse phase is high enough, conduction phenomena interfere with dipolar relaxation phenomena and the total dielectric behavior of the system will result from an overlapping of the dielectric relaxation of dipolar origin and of the MWS dielectric relaxation, which is shifted toward higher frequencies. This situation, which is encountered in emulsions of concentrated saline solutions in oil and in dispersions of pure or doped ice particles in oil, is illustrated by Figs. 26 to 28 which correspond to a slight, a medium, and a large overlapping. It is readily seen from these figures that, especially in the last two cases, curve analysis methods are necessary to distinguish between the contribution of the MWS mechanism and the contribution of the dipolar mechanism.

**Table 1.** Values of $T_1$ and $\tau_1$ for Some Submolar Electrolytic Solutions

| Substance | $T_1$ (s) | $\tau_1$ (s) |
|---|---|---|
| Nonelectrolytic water | $1.8 \times 10^{-4}$ | $0.9 \times 10^{-11}$ |
| 0.01 M acetic acid | $1.8 \times 10^{-7}$ | $0.9 \times 10^{-11}$ |
| 0.01 M KCl | $5.1 \times 10^{-8}$ | $0.9 \times 10^{-11}$ |
| 0.10 M KCl | $5.5 \times 10^{-10}$ | $0.9 \times 10^{-11}$ |

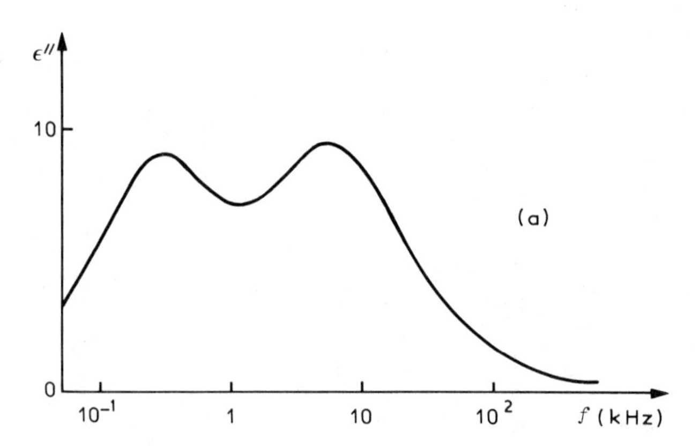


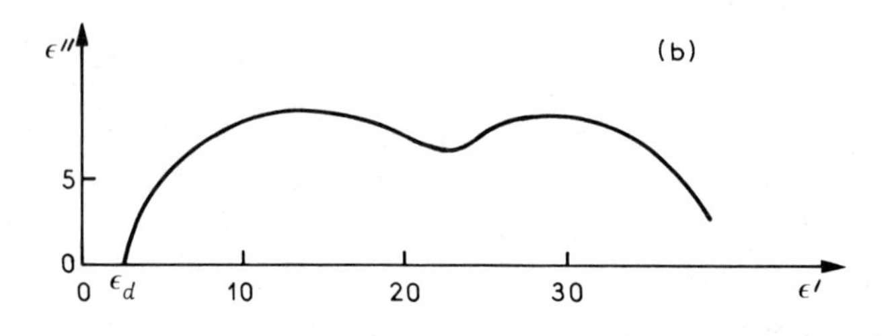


**Figure 27.** Theoretical features of the dielectric behavior of a water-in-oil emulsion system with medium overlapping of the MWS effect and of the dielectric relaxation connected with the disperse-phase intrinsic dielectric relaxation: $\Phi$ = 0.60; (a) absorption curve and (b) Cole-Cole plot. (From Ref. 4.)

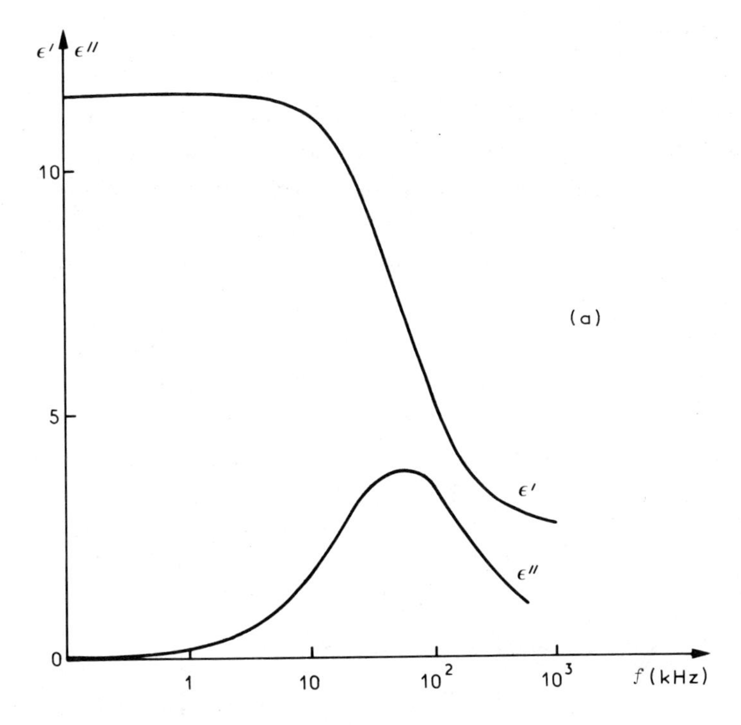


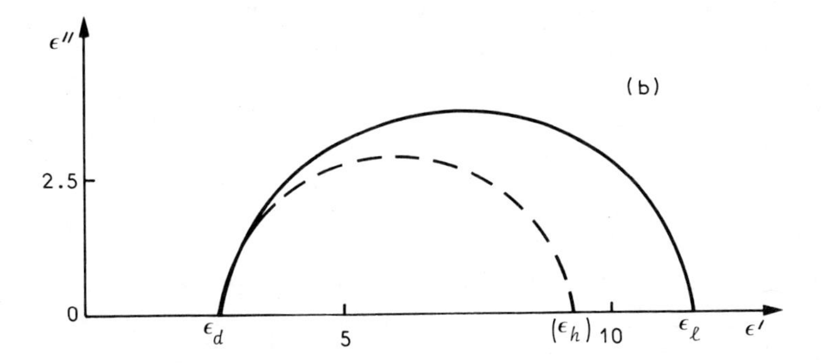


Figure 28. Theoretical features of the dielectric behavior of a water-in-oil emulsion system in which the MWS effect interferes strongly with the dielectric relaxation connected with the disperse-phase intrinsic dielectric relaxation: $\Phi = 0.40$; (a) dispersion and absorption curves and (b) Cole-Cole plot. (The dashed circular arc represents the Cole-Cole plot corresponding to $\sigma_1 = 0$.) (From Ref. 4.)

The overlapping of the MWS dielectric relaxation and of the dielectric relaxation of dipolar origin can be predicted from the comparison of $T_1$, the time constant of the disperse material, with $\tau_1$, the relaxation time of the dipolar process. For instance, if the dipolar relaxation time and static permittivity of submolar electrolytic solutions are assumed to be roughly equal to those of pure water, the data reported in Table 1 are obtained for temperatures around 295 K.

From the figures reported in Table 1, it can be concluded that, in an emulsion containing nonelectrolytic water as the disperse phase, the dielectric relaxation arising from interfacial polarization will not interfere with the relaxation of dipolar origin, as shown in Figs. 23 and 24, while, for emulsions of electrolytic aqueous solutions, both relaxations will overlap, the overlapping being the larger as $T_1$ approaches to $\tau_1$. When the conductivity of the disperse phase is sufficiently high, as in the case of 0.1 molar KCl, $T_1$ and $\tau_1$ are of the same magnitude, implying strong interference of the two dielectric relaxations.

## VI. Experimental Studies of Emulsions and Closely Related Systems

Typical experimental results will be reported below, showing that the theoretical models derived from Bruggeman's and Hanai's formulas can describe correctly the dielectric behavior of emulsions and closely related systems, such as frozen water-in-oil emulsions and water-in-oil emulsions undergoing progressive monothermal freezing. Mention will also be made of the relations existing between the dielectric and rheological properties of some types of emulsions and dispersions.

### A. Dielectric Studies of Emulsions

#### 1. *Emulsions of Water in Nonpolar Oils*

a. Typical Dielectric Behavior. *(1) Low-frequency studies.* Earlier works dealing with emulsion systems have revealed the existence of dielectric relaxations located in the 1-MHz region and arising from interfacial polarization phenomena [60-62,64,65,69-74,118]. These findings are consistent with the results of the theoretical analysis performed on Eq. (329), from which it can be inferred that emulsions of water phases in nonpolar and nonconducting oils should not be conducting and should display striking Cole-Cole-type dielectric relaxations arising from migration polarization phenomena.

These theoretical predictions were checked quite accurately by Clausse and coworkers, from a systematic study of the low-frequency dielectric behavior of W/O-type emulsions. The influence of $\Phi$, the disperse phase volume fraction, of $\sigma_1$, the conductivity of the disperse phase, and of the temperature T was investigated [4,

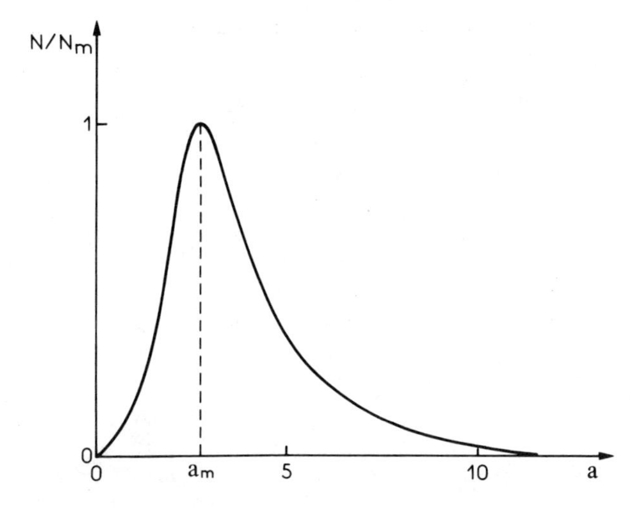


Figure 29. Droplet radius distribution in an emulsion of water in paraffin oil chacterized by $\Phi = 0.30$, the surfactant concentration in the oil phase being equal to 4.5% (W/W). a is indicated in micrometers.

75-77]. The experiments were performed on systems using resin-exchanged water or dilute aqueous solutions of alkali metal chlorides as the disperse phase, and nonpolar and nonconducting liquids such as vaseline or paraffin oil (Prolabo R.P., France) or silicone oils (Société Industrielle des Silicones, France) as the continuous phase, the surfactant generally being purified lanolin (Prolabo R.P., France). The emulsification was carried out, at room temperature, by means of a high-speed propeller-type homogenizer. Microscopic observations proved that the emulsions thus obtained consisted of dispersions of minute spherical aqueous droplets whose radius distribution was fairly selective, as shown in Fig. 29. It was found that the mean droplet radius is smaller as the proportion of surfactant is larger.

It was observed that these emulsions were mechanically stable over long periods of time (several weeks) when stored at room temperature, and repetitive measurements at regular time intervals showed no drift in the dielectric parameters.

An example of the dielectric relaxations observed between 30 and $-10°C$, in the 10-kHz to 10-MHz frequency range, for disperse-phase volume fractions ranging from 0 to 0.60 or so, is given in Fig. 14a and b. The continuous phase being nonconducting, the dielectric relaxations are not encumbered by bulk conduction phenomena, and nondistorted Cole-Cole plots are obtained, as shown in Fig. 14b. The analysis of the distribution of the experimental points on the Cole-Cole plots by means of the Smyth method [37] proves that the emulsion complex

Table 2. Values of $\varepsilon_h$ and $\varepsilon_\ell$ versus $\Phi$ for Emulsions of Water in Vaseline Oil

| $\Phi$ | 0 | 0.235 | 0.280 | 0.330 | 0.360 | 0.410 | 0.470 | 0.525 |
|---|---|---|---|---|---|---|---|---|
| $(1 - \Phi)^{-3}$ | 1 | 2.23 | 2.68 | 3.32 | 3.81 | 4.87 | 6.72 | 9.33 |
| $\varepsilon_h$ | 2.30 | 4.61 | 5.35 | 6.42 | 7.25 | 8.63 | 10.98 | 13.52 |
| $\left(\frac{\varepsilon_{1s} - \varepsilon_{2s}}{\varepsilon_{1s} - \varepsilon_h}\right)^3 \frac{\varepsilon_h}{\varepsilon_{2s}}$ | 1 | 2.19 | 2.62 | 3.29 | 3.84 | 4.84 | 6.81 | 9.38 |
| $\varepsilon_\ell$ | 2.50 | 5.66 | 6.70 | 8.28 | 9.50 | 12.00 | 16.70 | 22.87 |
| $\frac{\varepsilon_\ell}{\varepsilon_{2s}}$ | 1 | 2.26 | 2.68 | 3.31 | 3.80 | 4.80 | 6.67 | 9.15 |

*Source:* Modified from Refs. 4 and 77, Gauthier-Villars, Paris.

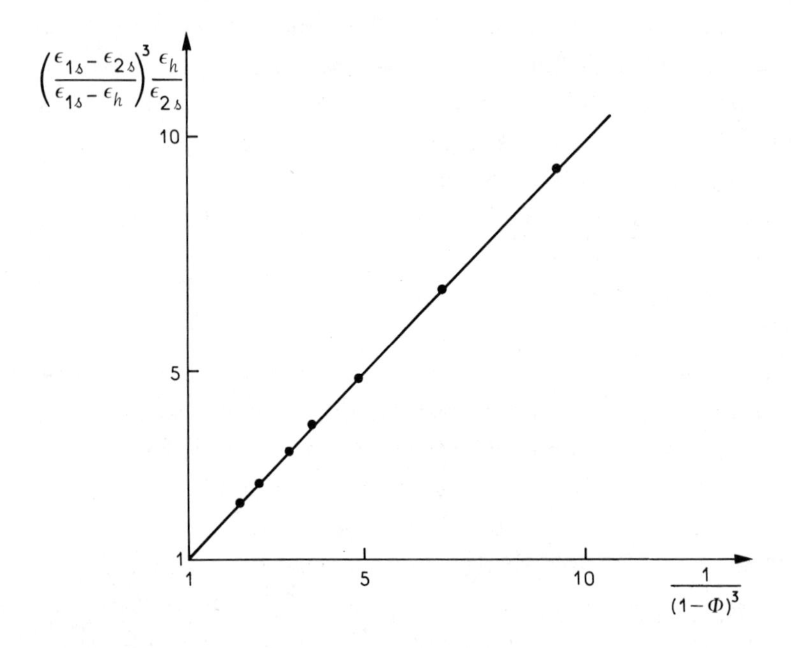


Figure 30. Dependence of $\varepsilon_h$ upon $\Phi$ for emulsions of water in vaseline oil. T = 295 K. The solid straight line represents the theoretical curve corresponding to Eq. (378). (From Ref. 77, Gauthier-Villars, Paris.)

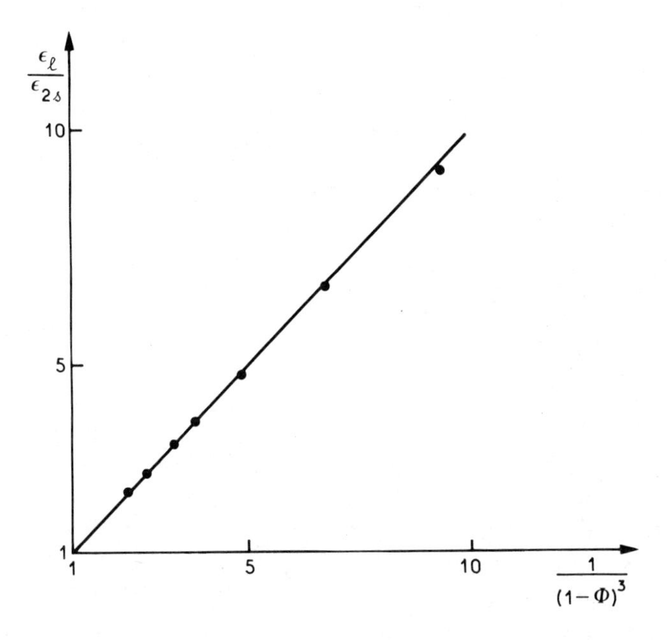


**Figure 31.** Dependence of $\varepsilon_\ell$ upon $\Phi$ for emulsions of water in vaseline oil. T = 295 K. The solid straight line represents the theoretical curve corresponding to Eq. (379). (From Ref. 77, Gauthier-Villars, Paris.)

permittivity can be represented by a Cole-Cole-type formula, in conformity with the theoretical predictions:

$$\varepsilon^* = \varepsilon_h + \frac{\varepsilon_\ell - \varepsilon_h}{1 + (jf/\bar{f}_c)^{1-h}} \tag{377}$$

As illustrated by Table 2 and Figs. 30 and 31, $\varepsilon_h$ and $\varepsilon_\ell$, the limiting permittivities at higher and lower frequencies, satisfy respectively Eqs. (330) and (351), namely,

$$\left(\frac{\varepsilon_{1s} - \varepsilon_{2s}}{\varepsilon_{1s} - \varepsilon_h}\right)^3 \frac{\varepsilon_h}{\varepsilon_{2s}} = \frac{1}{(1 - \Phi)^3} \tag{378}$$

$$\varepsilon_\ell = \varepsilon_{2s} \frac{1}{(1 - \Phi)^3} \tag{379}$$

Experiments showed that the values of $\varepsilon_h$ and $\varepsilon_\ell$ are almost temperature-independent, the slight variations observed being induced preponderantly by the change in $\Phi$ upon changing the temperature.

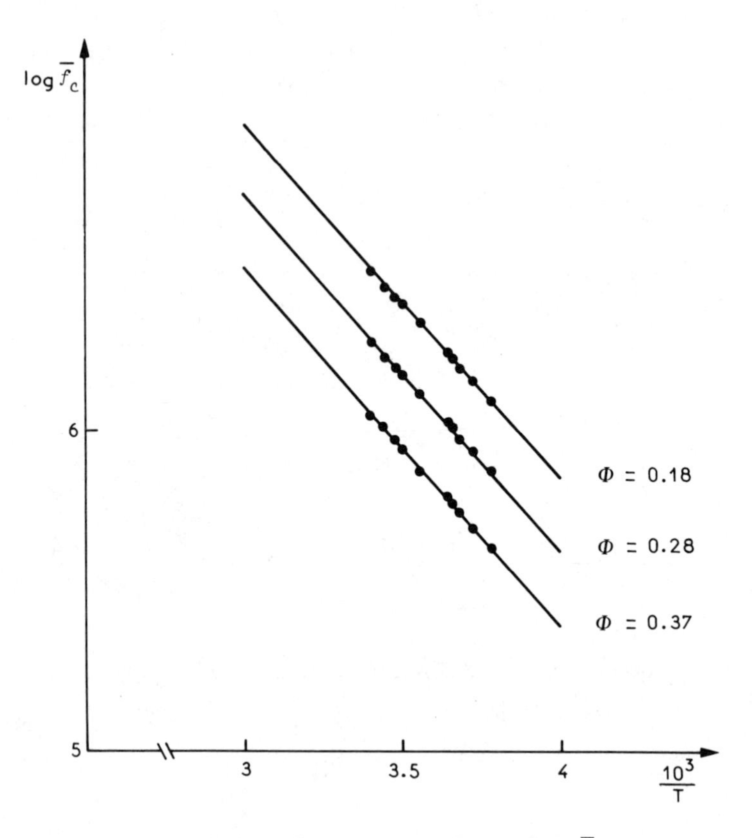


**Figure 32.** Variation of the critical frequency $\bar{f}_c$ versus T (in kelvin) for emulsions of water in vaseline oil characterized by different values of $\Phi$. (From Ref. 75, Gauthier-Villars, Paris.)

The validity of Eq. (378) can be checked also by means of the data reported by Naiki, Fujita, and Matsumura who investigated, at 2 MHz, the dielectric properties of emulsions of water-in-kerosene, terpene, and transformer oil [432]. For low values of $\Phi$, Eq. (379) can be approximated by

$$\varepsilon_{\ell} = \varepsilon_{2s}(1 + 3\,\Phi) \tag{380}$$

This equation is also an approximation for Eq. (325). Fradkina [52] found that the low-frequency permittivities of emulsions of electrolytic solutions in petroleum correctly satisfied Eq. (380).

Figure 32 and 33 illustrate the dependence of $\bar{f}_c$ upon $\Phi$ and $\sigma_1$, the volume fraction and conductivity of the disperse aqueous phase. For a given kind of disperse phase (Fig. 32), and at a given temperature, $\bar{f}_c$ is lower as $\Phi$ is higher. For a fixed value of $\Phi$ (Fig. 33), and at a given temperature, $\bar{f}_c$ is shifted toward

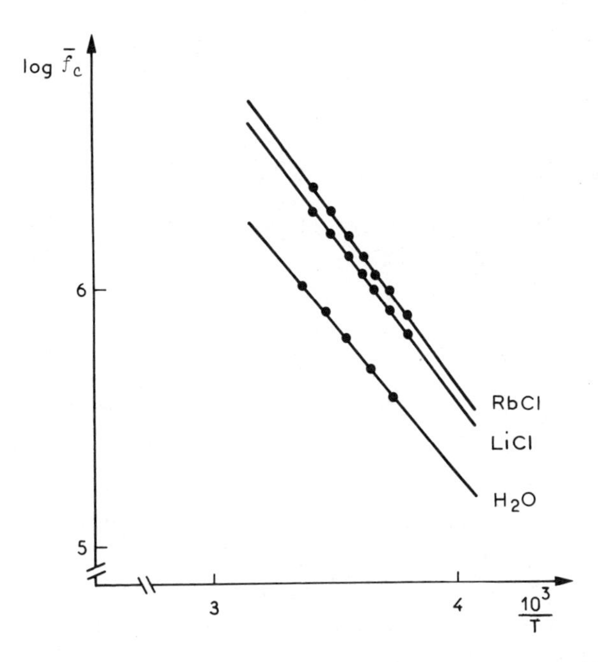


Figure 33. Variation of the critical frequency $\bar{f}_c$ versus T (in kelvin) for emulsions of water or of RbCl and LiCl dilute aqueous solutions in vaseline oil. Disperse phase volume fraction: $\Phi = 0.37$. Salt molar fraction: $C = 1.8 \times 10^{-5}$. (From Ref. 4.)

higher frequencies when an aqueous saline solution is substituted for water as the disperse phase. Figures 32 and 33 show that, whatever the nature and the volume fraction of the disperse aqueous phase, the relaxation critical frequency $\bar{f}_c$ follows an Arrhenius-type law

$$\bar{f}_c = \bar{A}e^{-\bar{u}/kT} \tag{381}$$

This experimental result is in strict conformity with the theoretical predictions derived from the numerical analysis of Hanai's formula. Experiment showed that whatever the value of $\Phi$, $\bar{u}$ the dielectric relaxation activation energy is equal to $u_1$, the conductivity activation energy of the disperse phase. The frequency factor $\bar{A}$ is a decreasing function of $\Phi$ and an increasing function of $\sigma_{1\infty}$, the pre-exponential factor appearing in the Arrhenius-type formula Eq. (326).

The investigation of the dielectric behavior of W/O-type emulsions at low frequencies has been completed by the study of emulsions of mercury-in-oil phases, such as highly viscous silicone oils or blends of vaseline oil and lanolin [4,77,104]. The emulsions were obtained by following the same experimental procedure as in the case of W/O emulsions. Owing to the high viscosity of the continuous phase,

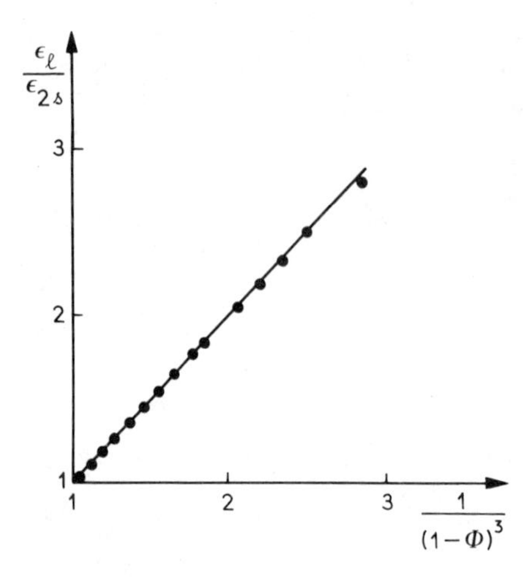


**Figure 34.** Mercury-in-oil emulsions. Variation of $\varepsilon_\ell/\varepsilon_{2s}$ versus $\Phi$. T = 295 K. (From Ref. 77, Gauthier-Villars, Paris.)

mechanically stable systems were obtained up to mercury volume fractions equal to 0.30 or so. In this situation, the theoretical predictions derived for Hanai's general formula state that the emulsion complex permittivity is given simply by

$$\varepsilon^* = \varepsilon_2^* \frac{1}{(1-\Phi)^3} \tag{382}$$

If the continuous phase is nonconducting and does not display any intrinsic relaxation, $\varepsilon_2^*$ is equal to a real value $\varepsilon_{2s}$, implying that $\varepsilon^*$ is equal, whatever the frequency, to a real value $\varepsilon_\ell$ given by

$$\varepsilon_\ell = \varepsilon_{2s} \frac{1}{(1-\Phi)^3} \tag{383}$$

The validity of Eq. (383) can be checked with a great accuracy, as shown in Fig. 34. These results are in good agreement with earlier data reported by Guillien [361] and Naiki et al. [432].

If the continuous phase exhibits an intrinsic dielectric relaxation, $\varepsilon_2^*$ being given either by the Debye or the Cole-Cole formula, Eq. (382) tells us that the emulsion will display a dielectric relaxation which is a magnified image of that of the disperse phase, the magnification factor being equal to $(1-\Phi)^{-3}$. At any frequency, the real and imaginary parts of $\varepsilon^*$ will be expressed by

$$\varepsilon' = \varepsilon_2' \frac{1}{(1-\Phi)^3} \tag{384}$$

and

$$\varepsilon'' = \varepsilon_2'' \frac{1}{(1 - \Phi)^3} \tag{385}$$

$\varepsilon_2'$ and $\varepsilon_2''$ representing the real and imaginary parts of $\varepsilon_2^*$. Moreover, the critical frequency $f_c$ will be equal to $f_{c1}$, whatever the value of $\Phi$. Such behavior was observed experimentally for mercury emulsions whose continuous oil phase exhibited an intrinsic Cole-Cole-type dielectric relaxation arising most probably from the presence of adsorbed water [104].

As concerns the steady conductivity, very few data are available since oil phases are generally nonconducting, which implies, according to Eq. (331), a zero bulk conductivity for the emulsion. If the oil phase is characterized by a conductivity which is small compared to that of the disperse phase, the emulsion effective steady conductivity should satisfy Eq. (350). Meredith and Tobias [59] investigated, at 298 K, the low-frequency conductivity of emulsions of water or aqueous solutions of KCl in propylene carbonate. Up to $\Phi$ equal to 0.2 or so, their data can be fitted either by Eq. (322) or by Eq. (350). At higher values of $\Phi$, the experimental points fall between the curves representative of Eqs. (322) and (350). Meredith and Tobias derived a modified form of Wiener's equation by assuming that their systems were bidisperse and found good agreement between this new formula and their experimental results. This treatment is rather an artificial one since the authors indicate that "in any given emulsion the size range of droplets was within less than an order of magnitude." The discrepancies observed may arise from frequency effects and/or to the fact that water-propylene carbonate systems do not form typical emulsions since the constituents are partially miscible.

*(2) High-frequency studies.* As was reported earlier, theoretical models have been derived by Clausse [4,78] concerning the total dielectric behavior of W/O-type emulsions over the entire radio-frequency range. Owing to the fact that the disperse phase exhibits, along with a conduction absorption, a dielectric relaxation of dipolar origin located around 20 GHz (see Fig. 2), it may be expected that the total dielectric behavior of W/O emulsions will be characterized by the existence of two dielectric relaxations, one located between 10 and 100 GHz and connected with the disperse-phase intrinsic dielectric relaxation, the other arising from interfacial polarization phenomena. The figures reported in Table 1 indicate that, as the disperse-phase conductivity increases, the MWS dielectric relaxation is shifted toward higher frequencies and can interfere with or even merge into the dielectric relaxation of dipolar origin. These theoretical predictions have been verified successfully by Le Petit and coauthors [80] on emulsions of aqueous solutions of potassium chloride. Microwave studies of W/O emulsions have been reported also by Noguchi and Maeda [66] at 9.4 GHz, and by Mudgett et al. [67] at

3 GHz. Although the results found by these authors conform qualitatively with the above theoretical scheme no quantitative comparisons can be made because of the lack of numerical data in their papers.

By investigating at seven discrete frequencies, between 100 kHz and 24 GHz, the dielectric behavior of model systems consisting of Plexiglass lumps containing fairly spherical water droplets, De Loor [451] found dielectric relaxations connected with the intrinsic Debye relaxation of water. His data show that the relaxation intensity of the system increases with the water content and that the system relaxation critical frequency is higher than that of bulk water, in conformity with the theoretical predictions which can be derived from Wagner's and Hanai's formulas. Dielectric relaxations located in the 10 GHz region were found also on hardened Portland cement paste by De Loor [198,199], who ascribed them to the dielectric relaxation of free water trapped in the pores of the material.

b. Out-of-Line Dielectric Behaviors. Hanai et al. [2,3,61,71] reported quite interesting results concerning the low-frequency dielectric behavior of W/O emulsions using a mixture of Nujol and carbon tetrachloride as the continuous phase, the surfactant being a blend of 60% (w/w) Arlacel, 20% (w/w) Span 20 and 20% (w/w) Span 60, all of them from Atlas Chemical Industries (ICI Americas Inc.) [470]. The proportions of surfactant present in both phases before carrying out the emulsification were 5% (w/w) in oil and 0.5% (w/w) in water. The disperse droplet sizes fell between 0.5 μm and 1.5 μm. The experiments were performed at 303K. Hanai and his co-workers having found that the dielectric properties of their emulsions were sensitive to agitation, the dielectric measurements were made by means of a Green-type double cylindrical rotational viscometer used as a capacitor cell. Thus it was possible to determine the sample complex permittivities under various conditions of shear.

Figure 35 gives a good example of the strong dependence of emulsion dielectric behavior upon the applied shearing stress, as found by Hanai and his co-workers. When the rotation frequency $N$ (in revolutions per minute, rpm) of the viscometer cup increases, the emulsion dielectric behavior undergoes spectacular changes which can be characterized mainly by the decrease of the dielectric increment, resulting from that of the low-frequency limiting permittivity, and by the increase of the critical frequency $\bar{f}_c$. The complex permittivity of the emulsion systems considered can be written

$$\varepsilon^* = \varepsilon_h + \frac{\varepsilon_{LF} - \varepsilon_h}{1 + (jf/\bar{f}_c)^{1-h}} + \frac{\sigma_{LF}}{2\pi j\varepsilon_0 f} \tag{386}$$

The high frequency limiting permittivity $\varepsilon_h$ is practically insensitive to the rate of shear and correctly satisfies Eq. (378), as shown in Table 3. On the contrary,

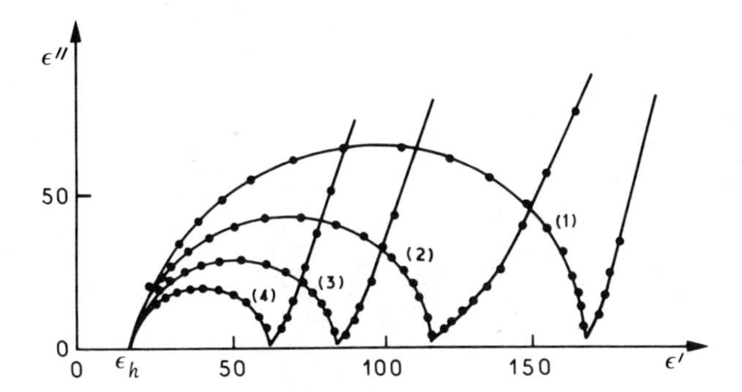


**Figure 35.** Cole-Cole plots recorded by Hanai et al. [2], under various conditions of shear, for a W/O emulsion characterized by $\Phi = 0.60$. (1) $N = 0$, $\bar{f}_c = 460$ kHz, $\varepsilon_{LF} = 166.3$; (2) $N = 100$, $\bar{f}_c = 620$ kHz, $\varepsilon_{LF} = 116.5$; (3) $N = 200$, $\bar{f}_c = 800$ kHz, $\varepsilon_{LF} = 85.7$; (4) $N = 300$, $\bar{f}_c = 1$ MHz, $\varepsilon_{LF} = 63.3$. $N$ (in revolutions per minute) represents the rotation frequency of the Green-type viscometer cup. (From Ref. 2.)

the dependence of $\varepsilon_{LF}$, the low-frequency limiting permittivity, upon $N$ is very drastic, as can be seen from Table 3. When $N$ increases, $\varepsilon_{LF}$ decreases toward the theoretical values given by Eq. (379). Hanai et al. suggested an empirical equation linking $\varepsilon_{LF}$ to $N$, namely,

$$\varepsilon_{LF} - \varepsilon_{\ell} = Be^{-N/N_0} \tag{387}$$

$B$ appeared to be an increasing function of $\Phi$, while $N_0$ decreased with $\Phi$ for values of $\Phi$ equal to or higher than 0.3. As concerns $\sigma_{LF}$ and h, no definite conclusions can be drawn as to their dependence upon $N$. It is worth mentioning that the values of $\sigma_{LF}$ fell below the theoretical values of $\sigma_{\ell}$ predicted by Eq. (350).

Hanai, Koizumi, and Gotoh explained the discrepancies observed between the actual emulsion dielectric behavior and the theoretical model on the basis of agglomeration effects in emulsions, the dependence of $\varepsilon^*$ upon $N$ being ascribed to the breaking up of droplet agglomerates under the influence of the applied shear. These authors performed a complementary study, investigating the rheological behavior of their emulsions by means of a MacMichael viscometer. While both of the constituents behaved as Newtonian liquids, the W/O emulsions made from them did not, as proved by the deviations of the flow curves from straight lines characterized by a $\pi/4$ slope. Hanai and his coworkers concluded that the results of their rheological measurements were consistent with the observations made as to the influence of the rate of shear upon the emulsion dielectric behavior.

**Table 3.** Influence of the Rate of Shear upon $\varepsilon_h$ and $\varepsilon_{LF}$ [71][a]

| | $\Phi$ | 0 | 0.1 | 0.2 | 0.3 | 0.4 | 0.5 | 0.6 | 0.7 | 0.8 |
|---|---|---|---|---|---|---|---|---|---|---|
| $(1 - \Phi)^{-3}$ | | 1 | 1.37 | 1.95 | 2.92 | 4.63 | 8 | 15.63 | 37.04 | 125 |
| $\varepsilon_h$ | | 2.5 | | | 6.3 | 8.7 | 13.3 | 18.8 | 28.2 | 39.0 |
| $\left(\frac{\varepsilon_{1s} - \varepsilon_{2s}}{\varepsilon_{1s} - \varepsilon_h}\right)^3 \frac{\varepsilon_h}{\varepsilon_{2s}}$ | | 1 | | | 2.95 | 4.52 | 8.52 | 15.81 | 40.31 | 118.47 |
| $N = 0$ | $\varepsilon_{LF}$ | 2.50 | 3.86 | 10.93 | 21.60 | 45.00 | 126.50 | 166.30 | 346.00 | 754.00 |
| | $\varepsilon_{LF}/\varepsilon_{2s}$ | 1 | 1.54 | 4.37 | 8.64 | 18.00 | 50.60 | 66.52 | 138.40 | 301.60 |
| $N = 100$ | $\varepsilon_{LF}$ | 2.50 | 3.60 | 8.17 | 19.60 | 42.60 | 78.50 | 116.50 | 218.00 | 450.00 |
| | $\varepsilon_{LF}/\varepsilon_{2s}$ | 1 | 1.44 | 3.27 | 7.84 | 17.04 | 31.40 | 46.60 | 87.20 | 180.00 |
| $N = 200$ | $\varepsilon_{LF}$ | 2.50 | 3.43 | 7.23 | 15.90 | 30.00 | 55.00 | 85.70 | 159.00 | 341.00 |
| | $\varepsilon_{LF}/\varepsilon_{2s}$ | 1 | 1.37 | 2.89 | 6.36 | 12.00 | 22.00 | 34.28 | 63.60 | 136.40 |
| $N = 300$ | $\varepsilon_{LF}$ | 2.50 | 3.34 | 6.91 | 13.80 | 24.40 | 42.60 | 63.50 | 119.00 | 258.00 |
| | $\varepsilon_{LF}/\varepsilon_{2s}$ | 1 | 1.34 | 2.76 | 5.52 | 9.76 | 17.04 | 25.40 | 47.60 | 103.20 |

[a] $\varepsilon_{1s} = 76.8$; $N$ is expressed in rpm.

*Source:* Modified from Ref. 71.

Although this general conclusion is somewhat simplistic, because the non-Newtonian rheological behavior of emulsions can arise from several factors, as Sherman clearly stated [471], the results reported by Hanai and his co-workers have great significance, since they show that dielectric studies can make a valuable contribution to the investigation of emulsion rheological properties, in particular when agglomeration and reticulation phenomena are expected to exist. Because the dielectric method, as applied to the investigation of thixotropy and dilatancy in disperse systems, is a nondestructive one, it has attracted the attention of several scientists. Parts [249] found a relationship between the degree of thixotropy and the dielectric constant of printing inks. Voet and Suriani [250-256] made a series of studies concerning the connection of dielectric behavior with rheological properties in pigment suspensions, metal powders, and carbon black dispersions in oil, and reported very interesting experimental results concerning permittivity and conductivity drifts arising from the progressive development of agglomeration and reticulation processes, permittivity and conductivity responses to a change from one shear rate to another, and electrical detection of hysteresis phenomena. Mewis and others [271] investigated also through dielectric measurements the structural hysteresis occuring in dispersions of carbon black in mineral oil, and Helsen and co-workers [272] designed a versatile apparatus to pick up the dielectric spectrum of such systems under transient shear conditions. Nasuhoglu [260] studied the effect of shear upon the dielectric and conductive properties of dispersions of iron powders in oil and of emulsions of water in oil or in a mixture of benzene and carbon tetrachloride, the surfactant being magnesium oleate, and explained his results on the basis of agglomeration effects. Recent data have been reported by Kuo et al. [472] on dielectric relaxation phenomena in sheared organobentonite dispersions. These authors found that, as the shear rate increased, the low-frequency limiting permittivity of their systems decreased while the high-frequency limiting permittivity and the critical frequency remained unchanged. An increase of the dielectric loss maximum and of the critical frequency with system aging was observed.

Dukhin and Shilov [8] thoroughly discussed the influence of particle aggregation upon the dielectric behavior and suggested the use of an iterated formula involving Bruggeman's equation in order to interpret the abnormally high values of the low-frequency permittivity exhibited by aggregated disperse systems. The Dukhin and Shilov analysis is somewhat too restrictive, as these authors consider only agglomeration effects and disregard reticulation phenomena. It is most likely that reticulation occurred in the emulsion systems studied by Hanai, Koizumi, and Gotoh, as proved by the existence of a nonzero low-frequency conductivity which can be ascribed to the existence of conducting paths stretching across the emulsions. In a later paper [72], Hanai and Koizumi reported results of a dielectric study of emulsions of water in a mixture of kerosene and carbon tetrachloride, with a minimal amount of emulsifier. They found for these systems a better agreement of the low-frequency limiting permittivity values with Eq. (379) than for the systems they had studied previously [2,3,61,71] and concluded that the use of excessive amounts of emulsifiers could explain the effect of shear upon the dielectric properties of emulsions. Although Hanai and Koizumi's contribution is of some significance as to the effect of surfactant interfacial films on rheological and dielectric behavior of emulsions, their demonstration is not entirely satisfactory since it appears from the data reported in [72] that the low-frequency limiting permittivity values are still much higher than those predicted by the theory. Chapman [65] investigated the effect of emulsifier on the dielectric properties of emulsions of water in paraffin oil, using as the surfactant either sorbitan monooleate (Span 80), or magnesium stearate, the emulsification being carried out by ultrasonic techniques. He observed that, in contrast to emulsions stabilized by means of magnesium stearate, those using sorbitan monooleate exhibited an abnormal dielectric behavior. A microphotographic study of samples of both emulsion types proved that droplets in the Span 80 stabilized emulsions tend to form interlinking clusters while there appears to be little aggregation in the emulsions using magnesium stearate. Chapman's observations confirm the influence of particle agglomeration upon the dielectric properties of emulsions and throw some light on the effect of surfactants. It appears, contrary to Hanai and Koizumi's opinion, that the preponderant factor would be the surfactant chemical nature rather than its proportion in the systems. In that respect, it is worth pointing out that an optimum stabilizing of emulsions involving water is obtained when using simultaneously, in adequate proportions, surfactants characterized by low and high HLB values [470]. This general recipe was not followed by Hanai and his co-workers, who used only blends of emulsifiers with low HLB values, ranging from 2 to 9, without any incorporation of some surfactant with a high HLB value such as, for instance, Tween 20, 40, 60, and 80, Myrj 52, Brij 35, 58, 78, and 98 (ICI Americas Inc.). Consequently, it can be inferred that flocculation phenomena occurred in Hanai and Koizumi's systems, which, consequently, exhibited abnormal dielectric properties, by contrast with the typical dielectric behavior displayed by the emulsions studied by Chapman [65] and Clausse [4,75]. Equation (387) shows that the degree of droplet agglomeration at rest can be evaluated from the preexponential factor $B$, $N_0$ being a measure of the agglomerate breakup when a shearing stress is applied. The observation that $B$ is an increasing function of $\Phi$ indicates that flocculation is more pronounced as the disperse water content increases, a reasonable conclusion.

As a final remark, it is worth mentioning that the discrepancies observed by several authors between the experimental values of $\overline{f}_c$ and its theoretical values calculated on the basis of experimental determinations of the conductivity of the aqueous phases, prior to their emulsification, can be ascribed to the effect of ionic impurities present in the surfactants used [4,65].

*2. Emulsions of the Oil-in-Water Type*

a. The Oil-Type Phase is Nonpolar and Nonconducting. *(1) Low-frequency studies.* As has been demonstrated earlier, from both Wagner's and Hanai's formulas, it can be expected that the dielectric relaxations arising from migration polarization in emulsions of nonpolar and nonconducting oil-type liquids in aqueous phases will be too small to be observable experimentally, because of the preponderant conduction contribution and because of the usual uncertainties attached to dielectric measurements. Consequently, it may be considered that in practice the complex permittivity of such systems is expressed, to a good approximation, simply by

$$\varepsilon^* = \varepsilon' + \frac{\sigma}{2\pi j \varepsilon_0 f} \tag{388}$$

with $\varepsilon'$ and $\sigma$ given by

$$\frac{2\varepsilon' - 3\varepsilon_{1s}}{2\varepsilon_{2s} - 3\varepsilon_{1s}} = (1 - \Phi)^{3/2} \tag{389}$$

$$\sigma = \sigma_2(1 - \Phi)^{3/2} \tag{390}$$

Experiments performed by Hanai and co-workers [2,3,60,69,71] and by Clausse and co-workers [79,473,474] have confirmed the validity of Eqs. (388) to (390) in the case of emulsions of nonpolar and nonconducting oil-type phases in water, as shown by Table 4 and Fig. 36. Low-frequency permittivity measurements were performed by Hanai et al., at 303 K, on emulsions of a mixture of Nujol and carbon tetrachloride in water, the emulsifier (weight concentration in the oil phase equal to 1%) being a blend of 50% (w/w) Tween 20 and 50% (w/w) Span 20. After allowance for electrode polarization phenomena, these authors found that the dielectric behavior of the O/W emulsions under investigation conformed to Eq. (388) and that $\varepsilon'$ satisfied Eq. (389) fairly well, as shown by the data reported in Table 4. A similar behavior was reported by Clausse and others [79,473,474] for Nujol-in water and benzene-in-water systems, the surfactant being a blend of Tween 20 and Span 20. Figure 36 shows the good agreement existing between the experimental results and Eq. (389). As concerns conductivity, Hanai and his co-workers used, for the sake of experimental reproducibility, aqueous solutions of KCl as the continuous phase, the molar fraction C of KCl being equal to $9 \times 10^{-4}$. The disperse phase was either a mixture of Nujol and carbon tetrachloride, the emulsi-

**Table 4.** Observed and Calculated Values of the Low-Frequency Permittivity and Conductivity of O/W-type Emulsions

| $\Phi$ | 0 | 0.1 | 0.2 | 0.3 | 0.4 | 0.5 | 0.6 | 0.7 | 0.8 | 0.85 | 1 |
|---|---|---|---|---|---|---|---|---|---|---|---|
| $\varepsilon'_{exp}$ [a] | 76.80 | 65.42 | 55.48 | 46.36 | 38.06 | 29.48 | 22.28 | 15.72 | 9.78 | 7.73 | 2.50 |
| $\varepsilon'_{calc}$ Eq. (389) | 76.80 | 66.12 | 56.02 | 46.53 | 37.70 | 29.58 | 22.23 | 15.75 | 10.28 | 7.99 | 2.50 |
| $(\sigma/\sigma_2)_{exp}$ [b] | 1 | 0.855 | 0.714 | 0.584 | 0.465 | 0.353 | 0.268 | 0.177 | 0.100 | | |
| $(\sigma/\sigma_2)_{exp}$ [c] | 1 | 0.849 | 0.694 | 0.584 | 0.460 | 0.354 | 0.262 | 0.181 | 0.108 | | |
| $(\sigma/\sigma_2)_{calc}$ Eq. (390) | 1 | 0.854 | 0.716 | 0.586 | 0.465 | 0.354 | 0.253 | 0.164 | 0.089 | | |

[a] Emulsions of a mixture of Nujol and $CCl_4$ in water.
[b] Emulsions of a mixture of Numol and $CCL_4$ in aqueous solutions of KCl.
[c] Emulsions of 1,2,3,4-tetrahydronaphtalene in aqueous solutions of KCl.
*Source:* Modified from Ref. 71.

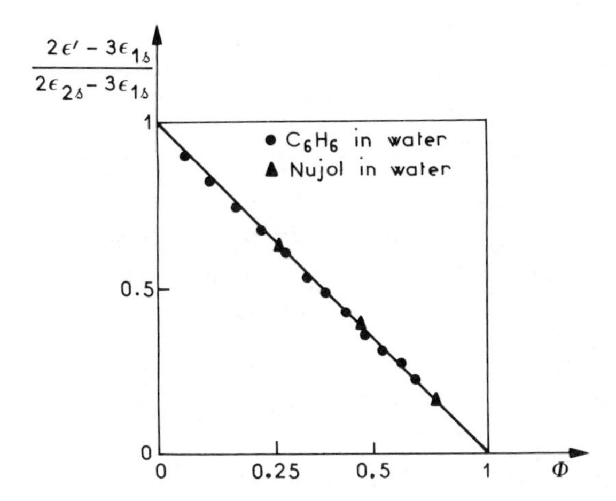


**Figure 36** Variation of ε' versus Φ for emulsion systems of nonpolar oil-type phases in water, at 1 MHz. T = 295 K. (Modified from Ref. 79.)

fier being a blend of 50% (w/w) Tween 20 and 50% (w/w) Span 20, or tetralin, the surfactant being polyoxyethylene cetyl ether. The figures reported in Table 4 show that the experimental data are correctly fitted by Eq. (390). As opposed to the W/O emulsions, the O/W emulsions investigated by Hanai et al. did not display changes in permittivity and conductivity when a shearing stress was applied. A complementary study made by means of a MacMichael viscometer showed that, up to $\Phi = 0.6$, the O/W emulsions exhibited a Newtonian behavior, the flow curve becoming steeper at upper values of $\Phi$. Possible relationships between dielectric and rheological properties were suggested by Clausse and others in the case of benzene-in-water systems [473,474].

*(2) High-frequency studies.* From both the Wagner and Hanai formulas, it may be expected that emulsions of nonpolar oils in water will exhibit a dielectric relaxation connected with the intrinsic Debye relaxation of water. Although no serious quantitative verification can be made because of the lack of systematic experimental data, it is possible to bring some support to the theory on the basis of results scattered throughout the literature. The results reported in Refs. 66 and 67 as to the microwave dielectric behavior of emulsions of the O/W type cannot be commented upon profitably since the authors restricted their studies to a single frequency and did not give numerical data. More interesting are the experiments performed at 9.4 GHz by Kraszewski, Kulinski, and Matuszewski [386] on suspensions of fine-milled mineral particles in water. Kraszewski [387] compared the results thus obtained with Eqs. (181), (182), (194), (222), and (225), and claimed that Beer's Eq. (182) is the most adequate although the agreement with Bruggeman's Eq. (225) is also

fairly good. In fact, Kraszewski's discussion is not reliable because the author should have compared experimental data with theoretical formulas in terms of complex permittivity, and not in terms of real permittivity on the one hand and of loss factor on the other. Had the correct procedure been followed, good agreement might have been found between experimental complex permittivity data and Hanai's equation. De Loor and Meijboom [200] investigated, between 2 GHz and 18 GHz, the dielectric properties of systems with high water contents, such as potato pulp and gels containing 4 or 8% Agar or 26% potato starch and 2% Agar, and of emulsion- or suspension-type systems such as milk and potato liquor. For all of these systems, De Loor and Meijboom found dielectric relaxations of the Debye type which became closer to that of pure water as the system water content was increased. The authors ascribed this dielectric behavior to the intrinsic relaxation of free water present in their systems and suggested the use of microwave measurements to determine the amount of free and bound water present in moist materials. In addition, it may be mentioned that Cook [475,476] reported that blood and biological tissues exhibit Debye-type dielectric relaxations in the gigahertz region.

b. The Oil-Type Phase Is Polar and Conducting. When both emulsion constituents have comparable properties (static permittivity and steady conductivity), it is not possible to predict the emulsion dielectric behavior by means of simplified equations, as has been done previously in the special cases of emulsions involving water and nonpolar and nonconducting oil phases. From the general numerical analysis of Eq. (329), performed by Boned and others [83], it can be predicted that, in contrast to emulsions of nonpolar oils in water whose dielectric relaxations are very small, emulsions of polar oil-type phases in water will exhibit in the kilo- to megahertz frequency range, along with conduction absorption, striking dielectric relaxation arising from migration polarization phenomena. For such systems, the high-frequency limiting permittivity $\varepsilon_h$, the low-frequency limiting permittivity $\varepsilon_\ell$, and the steady conductivity $\sigma_\ell$ will be given respectively by general formulas, Eqs. (330), (331), and (332). These theoretical predictions are supported by the data concerning nitrobenzene-in-water emulsions and reported by Hanai et al. [2,3,70,71]. These authors investigated over the kHz frequency range, at 293 K and for disperse volume fractions equal to or higher than 0.5, the dielectric behavior of nitrobenzene-in-water emulsions incorporating as the surfactant 2 per cent (w/w) Tween 20 (ICI Americas Inc.). They proved that these systems exhibit dielectric relaxations of the Cole-Cole type, as illustrated by Fig. 37 which shows the variation of $\varepsilon_R''$ versus $\varepsilon'$, the relaxation loss factor $\varepsilon_R$ being obtained by subtracting the conduction contribution from $\varepsilon''$, the total loss factor. Hanai and his co-workers found a good agreement between the ex-

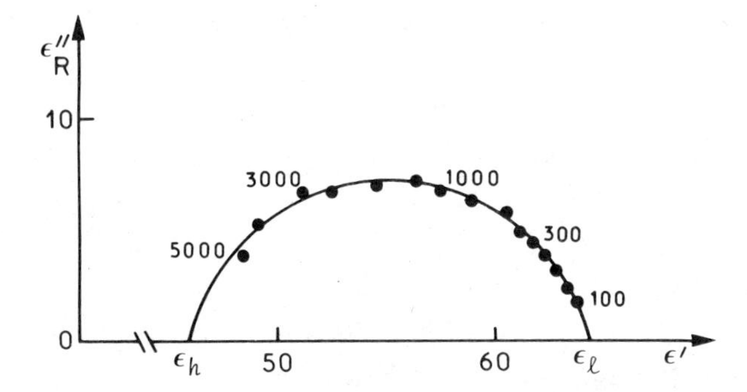


**Figure 37** Cole-Cole plot of the relaxation exhibited by a nitrobenzene-in-water emulsion. $\Phi = 0.70$, $T = 293$ K. The frequencies are indicated in kilohertz. [From Ref. 3, reprinted with permission from T. Hanai, Electrical Properties of Emulsions, in *Emulsion Science* (P. Sherman, ed.), 1968, p. 438. Copyright by Academic Press Inc. (London) Ltd.]

perimental values of $\varepsilon_h$ and Eq. (330). As concerns $\varepsilon_\ell$ and $\sigma_\ell$, discrepancies were noted that could be ascribed to parasitic electrode phenomena and/or to the fact that the emulsions were somewhat unstable because of the lack of some low HLB surfactant in the emulsifier blend used. No effect of shear upon the dielectric properties of nitrobenzene-in-water emulsions was reported.

## B. Dielectric Investigation of Disperse-Phase Supercooling and Crystallization in W/O-Type Emulsions and of Subsequent Dispersions of Solid Particles

### 1. *General Background*

It is well known that minute liquid droplets can undergo large supercooling while bulk samples do not supercool more than a few degrees Kelvin. The study of the supercooling and of related phenomena (supersaturation for instance) is of paramount importance in liquid and solid state physics as an essential part of nucleation and crystal growth investigations. They deserve particular attention as well in the applied sciences because of the many technological implications in various fields such as, for example, weather physics, catalysis efficiency, metallurgy, polymer processing, resistance to frost of building materials, lyophilization techniques, and preservation of foods and biological tissues at low temperatures. In that respect, because of the finely divided state of the disperse phase, the study of emulsions undergoing disperse phase freezing and of the subsequent frozen emulsions can yield valuable information from both the fundamental and the technological standpoints.

The investigation of supercooling and related phenomena through studies of emulsion physicochemical properties presents several complementary aspects and can be carried out by means of various techniques and methods. One field of interest concerns the determination of the temperature range within which most of the disperse droplets undergo crystallization when emulsion samples are submitted to a steady cooling. Other experiments consist of studying, as a function of time, the progressive crystallization of the disperse phase occurring in emulsion samples held at a fixed temperature below the disperse-phase melting point. This kind of investigation, aiming at the determination of nucleation rates in liquids, can be performed either at temperatures close to the most probable supercooling breakdown temperature $T^*$ (determined by steady cooling experiments, as indicated previously), or at temperatures remote from $T^*$, though situated below the disperse-phase melting point. Moreover, studies of emulsions in which the disperse phase remains supercooled are of interest for a better understanding of the structural properties of liquid media. It is the same with dispersions of solid particles obtained from the breakdown of supercooling of emulsions, since studies of the properties of such systems can provide valuable information as to the structure of solids formed under high degrees of thermodynamic irreversibility. Among the numerous techniques available to carry out these investigations, the most frequently used are direct microscopic observation or microphotography, dilatometry, x-ray diffraction, and differential thermal analysis (DTA) or differential scanning calorimetry (DSC), at atmospheric pressure or under high pressures. Nuclear magnetic resonance and dielectrometry have also been used, as indicated in recent papers. To illustrate this, some of the numerous papers available in the literature will be reported and commented upon briefly.

Vonnegut [127] was one of the first to recommend using emulsions in nucleation studies. He described a dilatometric technique and used it to investigate nucleation rates in systems of supercooled liquid metal droplets dispersed in oil, the droplets being isolated from each other by thin oxide coatings. Turnbull [128] also carried out dilatometric experiments on emulsions of mercury-in-oil-type phases, either silicone oils or mineral oils or ethanol saturated with sodium oleate, with incorporation of various film-forming agents such as sulfur or iodine. Depending upon the emulsion recipe, the droplets were from 2 to 100 μm in diameter. He claimed that the solidification rate of supercooled liquid mercury droplets is strongly dependent upon the nature of foreign substances either suspended in them or present on their surface. The supercooling and the breakdown of supercooling of liquid mercury was studied also by Monge [139], through either microscopic observations of individual droplets submitted to steady cooling, or by DTA experiments performed on lumps of porous material filled with mercury and on emulsions of mercury in vaseline oil, lanolin being used as the surfactant. Monge found that the breakdown of supercooling occurs statistically in the vicinity of defined discrete temperatures ranging from 220 to 174 K. These temperatures could be related to the structural properties of liquid mercury.

Pound and La Mer [129], also using dilatometry, found a 112 K supercooling degree for oxide-coated molten tin droplets dispersed in oil with a narrow size range centered about 5 μm. They pointed out that the solidification rate increases rapidly with temperature decrease and droplet-size increase. They suggested that the nucleation was inhomogeneous because of the presence of solid impurities. Lemercier and D. Clausse [154,155], from DSC studies of emulsions of molten tin in silicone oils, showed that the degree of supercooling becomes larger as the peak rewarming temperature above the melting point of tin is higher. Supercooling and crystallization phenomena were investigated by Rasmussen and Loper [176] in emulsions in poly(phenylether) of several low-melting point metals and alloys, using as surface coating agents sulfur, sulfuric acid, atmospheric oxygen, *p*-toluene sulfonic acid, Cu-isophthalate, Ag-isophthalate, and cumene-hydroperoxide, benzoyl peroxide, and di-tertiary butylperoxide in the presence of isophthalic acid. The emulsions thus obtained had spherical droplets with narrow diameter distributions in the range 5 to 10 μm. Owing to the varied droplet coatings, it was possible to study the influence of surface state upon the supercooling. The authors claimed that each emulsion type had a characteristic range of supercooling, depending upon the nature of the droplet surface and that certain droplet coatings permitted maximum supercooling consistent with homogeneous nucleation. In a fairly recent paper, Rasmussen and Loper [177] described a rapid method for isothermal nucleation rate measurements based on DSC and applied it to the case of emulsions of alloys of 90% tin and 10% bismuth and found results in agreement with a homogeneous nucleation mechanism. Because of the peculiar physical properties of gallium, a considerable amount of work has been devoted to the study of the supercooling of this metal, in particular by Bosio, Defrain, and others [132-136] who used droplet samples extracted from emulsions of gallium alcoholic solutions saturated with sodium oleate. These authors found that gallium can supercool down to temperatures as low as 151 K and reported the existence of metastable species at atmospheric pressure. Miyazawa and Pound [175] made dilatometric studies of emulsions of gallium droplets in silicone oils, the surfactant being Span 80 (ICI Americas Inc.). They found that supercooling ceased between 201 and 208 K and determined isothermal nucleation rates at different temperatures included between these limits.

Experiments similar to those performed on metals have been carried out to study the crystallization of polymer droplets dispersed in oil and of emulsions of organic compounds. For instance, Cormia and co-workers [131] used microscopic observation under slow cooling and warming conditions to study the freezing behavior of droplets of a linear polyethylene (Marlex-50, from Phillips Petroleum Company), extracted by centrifugation from a cooled solution in nitrobenzene and redispersed in the wetting agent Igepal CA-630 (GAF Corporation). From their observations, the authors suggested that the Marlex-50 solid spherulites which formed near 393 K were nucleated heterogeneously since it was possible to cool the molten droplet suspensions to 360 K or so before the majority of the droplets froze, provided that adequate precautions were taken. Complementary experiments on the same systems have been reported by Gornick and co-workers [477]. As concerns the supercooling of organic compounds, a considerable amount of work has been done by Dumas [160-165]. This author studied through DSC measurements the freezing behavior of emulsions using mixtures of glycerol and D-sorbitol as the continuous phase and sodium laurylsulfate as the surfactant, the disperse phase being either benzene, chlorobenzene, *p*-xylene, aniline, nitrobenzene, trichloromethane, 1,2-dichlorobenzene, *o*-xylene, cyclohexane, carbon tetrachloride, acetamide, or *m*-nitrochlorobenzene. He found that the disperse phase state of division strongly influences the supercooling phenomena, the degrees of supercooling being the larger as the disperse droplets are the smaller, and reported the existence of crystal species which are metastable under atmospheric pressure. He showed also that crystal-crystal transformations could undergo delays because of the finely divided state of the disperse solid phases obtained from the breakdown of supercooling. These investigations have been extended to the case of systems subjected to high pressures [166-168].

A thorough knowledge of the physicochemical properties of water being of paramount importance in many fields of science and technology, the study of aqueous phase supercooling has attracted the attention of many scientists for more than a century. Among the numerous studies related to atmospheric physics, a great number of experiments using various techniques have been, and still are, devoted to the investigation of supercooling and nucleation phenomena taking place in natural or artificial clouds and in individual droplets suspended on spider threads or synthetic fibers. The utilization of emulsions in which the disperse aqueous phase is divided into numerous tiny droplets allows one easily to obtain statistical information about the supercooling and the freezing behavior of water and aqueous solutions. Earlier dilatometric experiments in this direction were carried out by Fox [130], who used emulsions of water in Nujol containing 5% surfactant, either lanolin or sorbitan monooleate, trioleate, and sesquioleate. The distribution of droplet sizes presented a sharp maximum between 2 and 4 μm. This author found supercooling degrees of 14 K or so, which are much smaller than those currently reported in the literature. He attributed his results to "the formation of ice crystals on the inside surface of the droplets and catalyzed by this surface." It is more likely that the phenomena found by Fox can be ascribed to the fact that, owing to emulsion mechanical instability, the separated aqueous phase froze in bulk. This is consistent with the data obtained for water samples of a few cubic centimeters which are found statistically to undergo supercoolings of 14 K or so [137,138,142,149]. Later, Wood and Walton [169] made microscopic observations on emulsions of water in silicone or mineral oils and deposited on different substrates. The droplet-size distribution covered the range 2 to 50 μm. From steady cooling experiments these authors showed that 75% water droplets were frozen when the temperature had reached a value equal to 236.5 K. The consistency of this temperature value led them to conclude that the environment did not influence the nucleation process. By contrast, from isothermal experiments performed at 237.6 K, they pointed out the importance of considering droplet size in determining nucleation rates. Experiments of the same type were carried out by Kozlov and Ravdel [170], using polarized light in conjunction with synchronous rotation of crossed nicols to study the kinetics of isothermal crystallization of supercooled water droplets dispersed in silicone or castor oils. Supercooling degrees of 34 to 38 K were found by Butorin and Skripov [478] from DTA measurements on individual water droplets imbedded in oil. Following Babin's earlier investigations [137,138], systematic studies using DTA and DSC techniques were undertaken relating to the supercooling of water and aqueous saline solutions dispersed within W/O-type emulsions. Aqueous saline solution supersaturation phenomena were also considered. In the first stage, attention was focused on the determination of the temperatures at which the breakdown of metastability (supercooling or supersaturation) occurs when the systems are submitted to steady cooling. It was found that the breakdown of metastability follows definite rules, the crystallization temperatures defining curves geometrically analogous to the cryometric and solubilization curves in the phase diagram [140-152]. More specific studies in this field have been made as concerns the influence of the pressure [148], and of the incorporation of nucleating agents such as silver iodide [152]. The temperature at which the breakdown of supercooling occurs $T^*$ is the temperature corresponding to the apex of the recorded thermal signal, as shown in Fig. 38. This choice was fixed after a thorough theoretical and experimental study of thermograms of emulsions obtained through differential scanning calorimetry [153]. Figure 39 shows that it is possible [142] to observe two (sometimes three) different $T^*$ values, each of them corresponding to the breakdown of supercooling of a specific category of disperse droplets. Research is still in progress to determine the different factors ruling this phenomenon. In a second stage, research has been concentrated on investigations concerning isothermal (or, preferably, monothermal) crystallization rates in W/O-type emulsions maintained at subzero or subeutectic temperatures, remote from the supercooling breakdown temperature [156-159]. The rate of crystallization is obtained by determining, at regular time intervals, the propor-

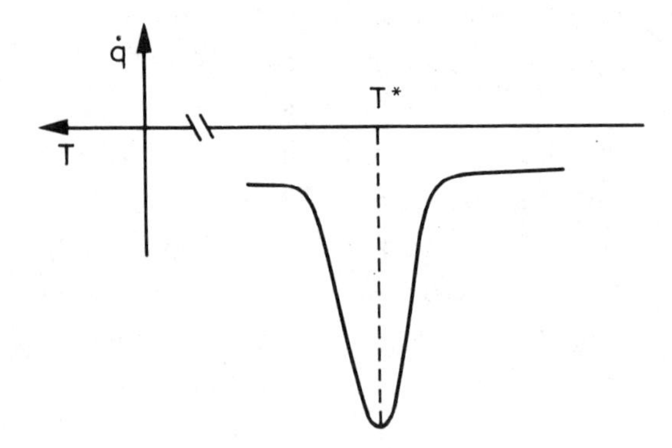


Figure 38 Thermogram of an emulsion of water in vaseline oil submitted to steady cooling. $\Phi = 0.23$, $dT/dt = -1.25\ K\ mn^{-1}$. $T^* = 233.9$ K. (From Ref. 156. Copyright by the Institute of Physics, Bristol, England.)

tion of frozen water in the emulsion. This is achieved by warming the samples from their maintenance temperature $T_c$ up to past the disperse-phase melting point and measuring q, the heat exchanged during the melting, then cooling the samples down past T* so as to obtain complete crystallization of the disperse phase, and rewarming them as before and measuring $q_0$, the heat exchanged during melting. The values of $q/q_0$ yield the values of p, the proportion of frozen water at instant t. Figure 40 shows, as examples, curves of p versus t from which the

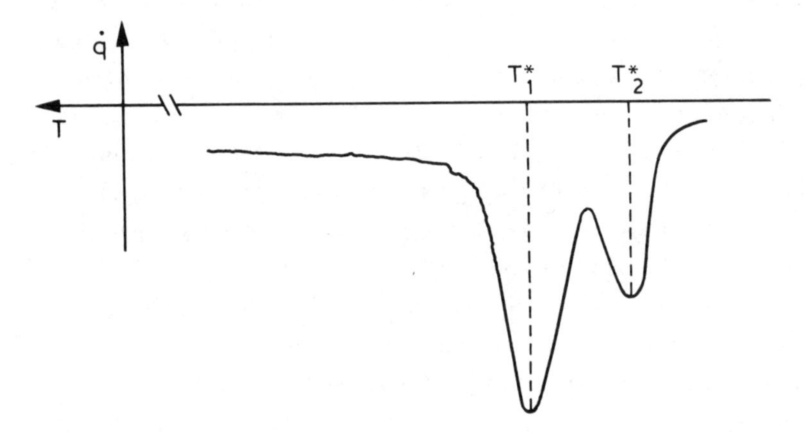


Figure 39 Thermogram of an emulsion ($\Phi = 0.5$) of an aqueous $NH_4Cl$ solution ($C = 0.024$) in vaseline oil submitted to steady cooling. Two different breakdown of supercooling temperatures are displayed $T_1^* = 232.7$ K, $T_2^* = 225.7$ K. (From Ref. 142, courtesy of D. Clausse, Pau, France.)

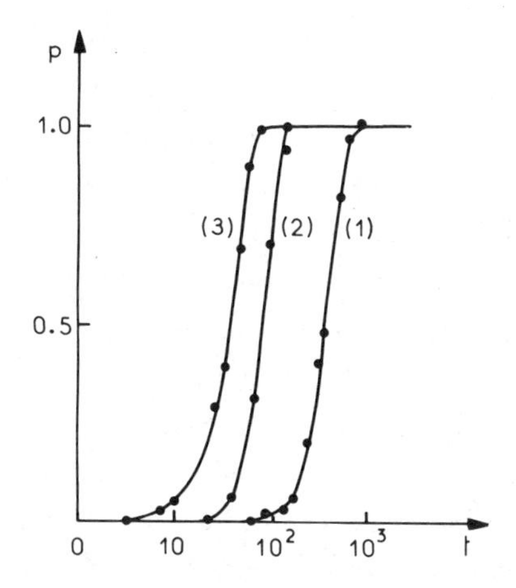


Figure 40 Crystallization rates observed in an emulsion of water in vaseline oil. $\Phi = 0.18$, $T^* = 234.2$ K. t, in hours, represents the time during which the samples are held at $T_c$. (1) $T_c = 252.2$ K; (2) $T_c = 246.7$ K; (3) $T_c = 240$ K. (From Ref. 156. Copyright by the Institute of Physics, Bristol, England.)

monothermal crystallization rates at different maintenance temperatures can be determined.

Other experiments in this field have been performed by Rasmussen and MacKenzie [171], who examined, on the basis of homogeneous nucleation studies, the effect of solute on ice solution interfacial free energy, and by Kano et al. [178], who studied the influence of pressure on the supercooling breakdown temperature, and found that the commonly reported value of T* for emulsified water under atmospheric pressure, namely 234.7 K, is lowered to 181 K when a pressure equal to $2 \times 10^8$ Pascal is applied. In addition, certain physicochemical properties of supercooled aqueous phases dispersed within emulsions have been investigated by several authors using various techniques. For instance, the densities and specific heats of water and heavy water were measured, from room temperature down to the breakdown of supercooling temperature, by Rasmussen and others [172,173], using W/O emulsions with heptane as the continuous phase, and sorbitan tristearate as the surfactant. Densities were determined by means of a dilatometer and heat capacities by drift or differential scanning calorimetry. These authors reported striking rises in the constant-pressure heat capacity below 253 K. Proton magnetic resonance chemical-shift measurements performed on supercooled water trapped in capillaries [174] suggested that the anomalous thermal behavior of supercooled water could be associated with a cooperative acceleration in hydrogen-bond strength or formation rate at low temperatures.

As has been already mentioned, dielectric measurements performed on emulsions involving disperse aqueous phases are of interest in supercooling and crystallization studies. For instance, the temperature dependence of both the conductivity and static permittivity of supercooled aqueous media can be determined from temperature investigations of the dielectric relaxations originating from migration polarization phenomena in W/O-type emuslions. Similarly, liquid-solid transformations under steady cooling, or progressive freezing rates at fixed subzero temperatures occurring in W/O-type emulsions, can be detected by means of dielectrometry, because of the striking discrepancies existing between the dielectric properties of water and those of ice. Moreover, studies of the dielectric behavior of dispersions of pure or doped ice microcrystals (frozen W/O emulsions) can bring valuable information about the structural properties of ice formed under special thermodynamic conditions. It must be pointed out that such dielectric investigations are also of importance for practical problems since they can provide, in parallel with other techniques, experimental procedures applicable to the study of the behavior of moist materials submitted to low temperatures. Straightforward applications can be foreseen, for instance, in the fields of geophysics, of building materials and frozen food technologies, and of cryobiology. In addition, conductivity and permittivity measurements can yield significant results about the freeze-thaw stability of emulsions, reviewed by Sherman [471] 15 years ago.

### 2. *Dielectric Studies of Disperse Aqueous Phases in the Supercooled State*

Dielectric investigations on aqueous phases in the supercooled state are of great interest in the background of water-structure temperature studies. One interesting problem in this field, to which clear theoretical and experimental answers are still to be given concerns the possible existence, suggested by Drost-Hansen [479] of thermal anomalies (kinks) in the physical properties of water. As pointed out by Franks [180], objective discussions of this interesting issue, based on experimental evidence, often have been obscured by dogmatic statements and bitter polemics, so that the problem is still unresolved.

Among the physical properties of water, the so-called transport coefficients such as viscosity, static permittivity, and steady conductivity have attracted special attention. As concerns static permittivity, the commonly accepted temperature determinations under normal pressure are those of Malmberg and Maryott [480], who proposed, after a thorough study, the following polynomial,

$$\varepsilon_s = 87.740 - 0.4008(T - T_0) + 9.398 \times 10^{-4}(T - T_0)^2 - 1.410 \times 10^{-6}(T - T_0)^3 \tag{391}$$

where T is the Kelvin temperature and $T_0 = 273.15$ K.

More recent experiments have been reported by Rusche and Good [481], who extended the investigations past $T_0$ to 268.2 K. From their data, which are in good agreement with Malmberg and Maryott's determinations, there is no evidence for the existence of kinks, in particular at the normal melting point of ice. Similar conclusions can be derived from the observations made by Clausse and others [4,75] as to the influence of temperature upon the MWS dielectric relaxation exhibited by W/O-type emulsions. These authors did not find, when the temperature was decreased from 298 down to 263 K, any anomaly in the emulsion dielectric parameters, in particular at the normal ice melting point.

Owing to the fact that emulsified micrometer-size water droplets can undergo supercoolings of 39 K or so, as proved by microscopic observations, dilatometry, and differential calorimetry, several authors have carried out low temperature permittivity studies on W/O-type emulsions with the view of obtaining information about the dielectric behavior of aqueous phases in the supercooled state. From experiments performed at 10 GHz on dispersions of water or ammonium chloride aqueous solutions in purified lanolin, it was reported [482,483] that the real part of the permittivity of these systems exhibited discontinuities in its first-order temperature derivative. This phenomenon, ascribed by the authors to higher order transitions occurring in liquids, should be considered in connection with the problem of kinks, commented upon by Hasted [291] as concerns dielectric phenomena. Hasted and Shahidi [469] used emulsions with low water contents (0.5 and 1%), the continuous phase being *n*-hexane and the emulsifier Tween 65 (Honeywell-Atlas Ltd.), to investigate, at 1.652 kHz, the temperature dependence of the static permittivity of water down to 237 K. These authors found that their experimental values, computed with the aid of Wieners' formula, lay on a smooth curve and that no ferroelectric transition could be observed. Hasted and Shahidi pointed out that their data fell well outside the polynomial fit expressed by Eq. (391), a feature which has been noted as well in other supercooled polar liquids [484]. In the same direction, Hodge and Angell [68] studied down to 238 K the dielectric relaxation originating from migration polarization phenomena in emulsions of water in a mixture of heptane and carbon tetrachloride, using sorbitan tristearate as the surfactant. Their experiments were aimed at determining whether the static dielectric constant and the steady conductivity of water conformed to the following general phenomenological equation:

$$\mathcal{P}_i = \mathcal{P}_i^0 \left(\frac{T}{T_s} - 1\right)^{\gamma_i} \tag{392}$$

where $\mathcal{P}_i$ represents the property and $T_s$ is equal to 228 K, $\mathcal{P}_i^0$ and $\gamma_i$ being adjustable parameters. The general equation (392) has proved suitable to describe, within the experimental scatter, the temperature dependence of several physicochemical properties of water below the normal melting point of ice, namely, isobaric heat capacity, isothermal compressibility, density, diffusion coefficient, bulk and shear viscosity, sound velocity and absorption, proton spin-lattice relaxation time, and oxygen spin-lattice relaxation time, and also of the Debye-type dielectric relaxation time, between 273.2 and 333 K [485]. Should the $\varepsilon_{1s}(T)$ curve conform to Eq. (392), a presumption would exist in favor of a ferroelectric transition at $T = T_s$ whose connection with other anomalies observed with supercooled water has been mentioned by Stillinger on the basis of computer-dynamical calculations of water structure [486,487]. After a clever treatment of the experimental data obtained on emulsions, Hodge and Angell found that, within experimental uncertainties, the temperature dependence of the static dielectric constant of water could be fitted reasonably by Eq. (392) when $P_i^0 = 72.94$ and $\gamma_i = -0.1256$. The authors pointed out that the value of the critical exponent $\gamma_i$ thus obtained is the smallest that has been found for water so far, and suggested that the anomalous temperature behavior of water static permittivity "should probably be regarded as incidental to, rather than causative of, the other anomalies." As concerns conductivity, whose temperature dependence was derived from those of both $\varepsilon_s$ and $f_c$ ($f_c$, emulsion dielectric relaxation frequency), a critical exponent equal to 1.133 was computed, which is to be compared to those found for the diffusivity (1.454), and for the shear viscosity (1.476).

The few examples reported above show that, in spite of the great inherent experimental difficulties, dielectric studies of emulsions deserve attention and should be developed in the future as an additional method of investigation of the physical properties of liquids in the supercooled state.

### 3. *Dielectric Detection of Disperse-Phase Crystallization Phenomena in W/O-Type Emulsions*

a. Introductory Considerations. As a result of the discrepancies existing between the dielectric properties of liquid water and of ice, it is assured that disperse-phase freezing in W/O-type emulsions can be investigated through emulsion dielectric behavior measurements, either in the low frequency region (up to a few megahertz), or at high frequencies (1- to 100-GHz range).

At low frequencies, W/O-type emulsions exhibit dielectric relaxations arising from interfacial polarization phenomena, the intrinsic Debye-type dielectric relaxation of disperse water giving a frequency-independent contribution which is characterized by the value of the emulsion limiting permittivity $\varepsilon_h$. If the disperse aqueous phase undergoes freezing (either abruptly or progressively), the emulsion dielectric features will display, in the general case, striking alterations arising from the changes in the contribution of the disperse phase to the emulsion dielectric properties. When the freezing process is completed, the dielectric behavior of the frozen emulsion will reflect that of the disperse ice particles, which is characterized by the existence, in the kHz frequency region, of a Debye-type dielectric relaxation along with that of a conduction absorption [488-494]. More details about ice dielectric properties can be found in Refs. 9, 180, 293, 294, and 295.

Consequently, in an Argand diagram of the complex plane, the point representative of the emulsion complex permittivity $\varepsilon^*$ at a given frequency f will be shifted from its initial location on the Cole-Cole plot of the unfrozen emulsion MWS dielectric relaxation to a final position on the Cole-Cole plot characteristic of the frozen emulsion dielectric behavior. Should the observation frequency f be high enough, 1 MHz, for instance, the final value of $\varepsilon^*$ will be equal to $\varepsilon_d$, the frozen emulsion limiting permittivity at higher frequencies, for which ice globule conductivity and intrinsic dielectric relaxation contribute nothing to the dielectric behavior. In particular, if the observation frequency f is sufficiently high, the freezing phenomenon can be detected through the study of the high-frequency limiting permittivity, which will vary from $\varepsilon_h$, its initial value when the crystallization starts, to $\varepsilon_d$, its final value when the emulsion disperse phase is totally frozen.

A similar scheme is valid at high frequencies although the dielectric phenomena taking place in the unfrozen emulsions do not have the same origin as in the preceding case. In the 10-GHz region, W/O-type emulsions exhibit dielectric relaxations connected with the intrinsic Debye-type relaxation of the disperse aqueous phase. If the disperse aqueous phase undergoes freezing, the contribution of the frozen droplets to the total complex permittivity of the system is reduced to their distortion polarization, and consequently, the emulsion dielectric relaxation features will display changes concomitant with the freezing process. After completion of crystallization, the frozen emulsion dielectric behavior will reflect that of the disperse ice globules, reduced in this case to their distortion polarization. In an Argand diagram, the point representative of the emulsion complex permittivity $\varepsilon^*$ at a fixed frequency f will move from its initial position on the Cole-Cole plot of the unfrozen emulsion dielectric relaxation, connected with the disperse-phase Debye-type intrinsic relaxation, to a final position which is determined by $\varepsilon_d$, the value of the frozen emulsion limiting permittivity at high frequency, to which there is no contribution of the intrinsic dielectric relaxation and conductivity of ice.

b. Detection of Supercooling Breakdowns. It has been reported earlier that the disperse droplets in a W/O-type emulsion submitted to a steady cooling remain supercooled until the temperature T approaches a value T*, and they then freeze rapidly over a narrow temperature range when further cooled. This breakdown of supercooling can be detected by several methods, in particular through differential scanning calorimetry measurements [153]. Owing to the great number of droplets dispersed in the emulsions, T* is determined on a statistical basis. From a thorough analysis of DSC thermograms, it has been concluded [153] that T* can be given the value of the temperature corresponding to the apex of the thermal signal, as represented in Fig. 38. For micrometer-size water droplets dispersed in well-stabilized emulsions, the commonly reported value for T* is equal to 234.7 K to within 1 K [138,142,156,173,178]. For similar emulsions of aqueous saline solution, the breakdown of supercooling occurs at temperatures which are lower as the salt concentration is higher. In certain cases, as shown in Fig. 39, it is possible to identify several discrete breakdowns of supercooling temperatures.

Figures 41a and b and 42a and b show the consistency of results obtained in this field from dielectric measurements performed either at low or high frequencies with data established by using other usual techniques. In Fig. 41a is shown the variation versus T of the real part $\varepsilon'$ of the complex permittivity of an emulsion of water in paraffin oil characterized by $\Phi = 0.18$ (at room temperature), the proportion of surfactant (lanolin), in the oil phase being equal to 4.5% (w/w). The observation frequency f is equal to 100 kHz. From the normal melting point of ice down to the temperature T*, the variation of $\varepsilon'$ with T is represented by an almost horizontal smooth curve. For temperatures lower than T*, a drastic decrease in $\varepsilon'$ is observed over a narrow temperature range and eventually $\varepsilon'$ again lies on an almost horizontal curve. It must be noted that, upon disperse phase freezing, $\Phi$ undergoes an abrupt variation arising from the change from one disperse-phase density value to another. This phenomenon adds a contribution to the magnitude of the $\varepsilon'$ fall, which, however, is determined predominantly by the change in disperse phase dielectric behavior. Figure 41b gives an example for the dielectric detection of two distinct supercooling breakdowns occurring within an emulsion and characterized by two different values of T*. Typical features for high-frequency dielectric signals induced by supercooling breakdowns occurring in W/O-type emulsions are shown in Figs. 42a and b. In Fig. 42a is shown the $\varepsilon'$ versus T curve corresponding to a steadily cooled emulsion containing 0.5% (w/w) of a $NH_4Cl$ aqueous solution characterized by C = 0.095. As in the case of low frequency studies, the sudden disperse phase freezing is detected by the abrupt change in the value of $\varepsilon'$, the real part of the emulsion complex permittivity. Figure 42b gives an example of the phenomena observed when supercooling breakdowns occur at two distinct temperatures.

It is worth pointing out that supercooling breakdowns can also be detected from the temperature variations of $\varepsilon''$, the emulsion loss factor, but, in most cases, the $\varepsilon'$ method is more straightforward and accurate than the $\varepsilon''$ method.

c. Detection of Progressive Freezing Processes. When a W/O-type emulsion is maintained at a temperature located between the solid disperse-phase melting

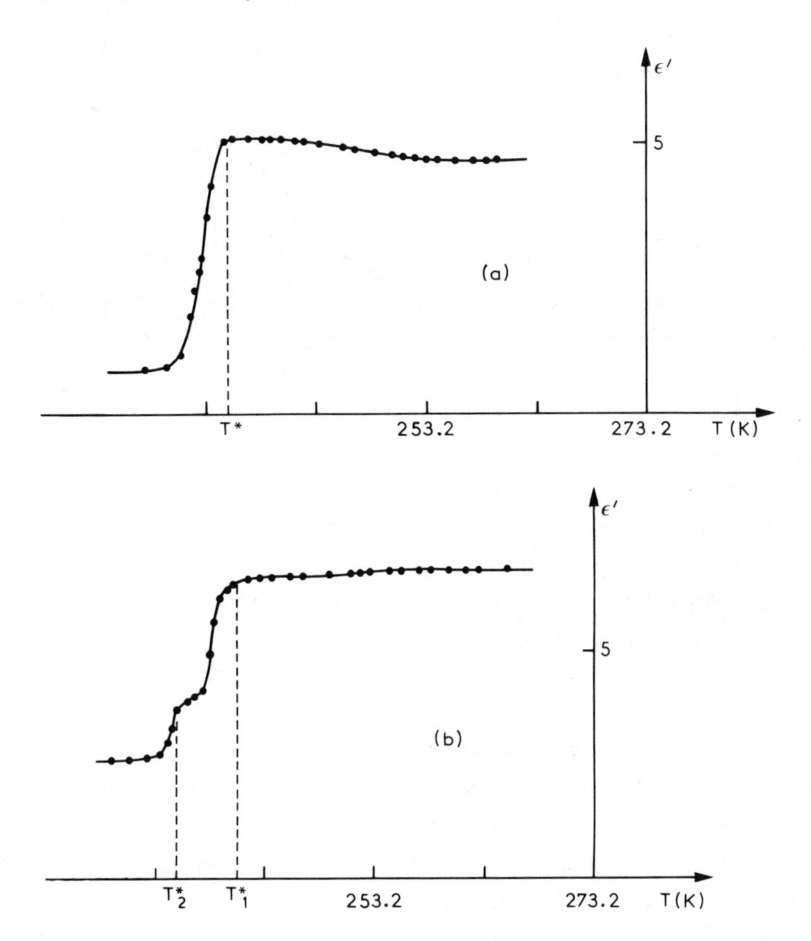


**Figure 41.** Dielectric detection at 100 kHz of the disperse-phase supercooling breakdown occurring in an emulsion of water in paraffin oil. Initial water volume fraction at room temperature: $\Phi = 0.18$. Proportion of surfactant in the oil phase: 4.5% (w/w). Cooling rate: $dT/dt = -0.5$ K $mn^{-1}$. $T^* = 234.7$ K. (From Ref. 93, courtesy of B. Lagourette, Pau, France.) (b) Dielectric detection at 100 kHz of two successive disperse-phase dielectric breakdowns occurring in an emulsion of water in paraffin oil. Initial water volume fraction at room temperature: $\Phi = 0.28$. Other specifications identical to those of Fig. 41a. $T_1^* = 240$ K, $T_2^* = 234.7$ K. (From Ref. 93, courtesy of B. Lagourette, Pau, France.)

point $T_0$ and the supercooling breakdown temperature T*, a progressive freezing process develops, the rate of which is dependent upon the temperature $T_c$ at which the sample is held, as shown in Fig. 40. In parallel with DSC studies [156, 157,159], low-frequency dielectric experiments have been carried out to investigate this phenomenon [86-92]. Because of the magnitude of the time-constants usually displayed (ranging from a few tens to several thousands of hours), it is possible

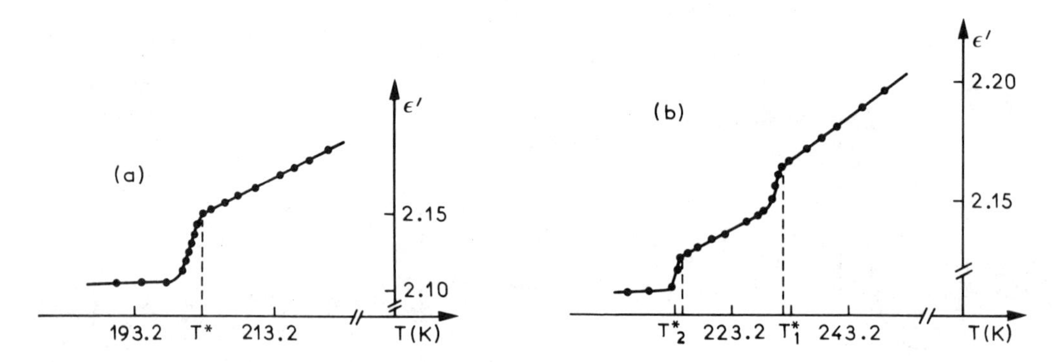


Figure 42. (a) Dielectric detection at 10 GHz of the disperse-phase supercooling breakdown occurring in an emulsion of a $NH_4Cl$ aqueous solution in purified lanolin [483]. Initial water volume fraction at room temperature: $\Phi = 0.05$; molar concentration of $NH_4Cl$: $C = 0.095$; $T^* = 202.7$ K. (b) Dielectric detection at 10 GHz of two successive disperse-phase supercooling breakdowns in an emulsion of a $NH_4Cl$ aqueous solution in purified lanolin [483]. $T_1^* = 231.4$ K. $T_2^* = 214.2$ K. (From Ref. 483, with permission from Societé de Chimie Physique, Paris.)

to perform repetitive detailed runs of experiments so as to determine completely, at regular time intervals, the features of the dielectric behavior of the system over the entire measuring frequency range. The time needed to carry out an experimental run being short enough compared to the duration of the phenomenon, the relaxation diagrams thus obtained can be considered as "snapshots."

*(1) Experimental observations.* Dielectric measurements in the frequency range 100 Hz to 1 MHz demonstrate time drifts in the parameters characteristic of the system dielectric behavior, as shown in Figs. 43 and 44. With time, the low-frequency limiting permittivity $\varepsilon_{LF}$ increases smoothly from $\varepsilon_\ell$, the low-frequency limiting permittivity of the unfrozen emulsion MWS dielectric relaxation, up to a final plateau value which is the low-frequency limiting permittivity of the totally frozen emulsion. Simultaneously, the high-frequency limiting permittivity $\varepsilon_{HF}$ decreases smoothly from $\varepsilon_h$, the high-frequency limiting permittivity of the unfrozen emulsion, down to a final plateau value which is the high-frequency limiting permittivity of the totally frozen emulsion $\varepsilon_d$. At the same time, the maximum value $\varepsilon''_m$ of the loss factor increases while the frequency $f_m$ corresponding to it decreases (Fig. 44).

These phenomena are quite reproducible and are general ones, as proved by the numerous complementary investigations made as concerns the influence of the temperature at which the emulsion is held and the surfactant amount, and also of electrolytes dissolved in the aqueous phase [89,92]. Figure 45 shows that the duration of the transformation is shorter as $T_c$ is the closer to $T^*$, the supercool-

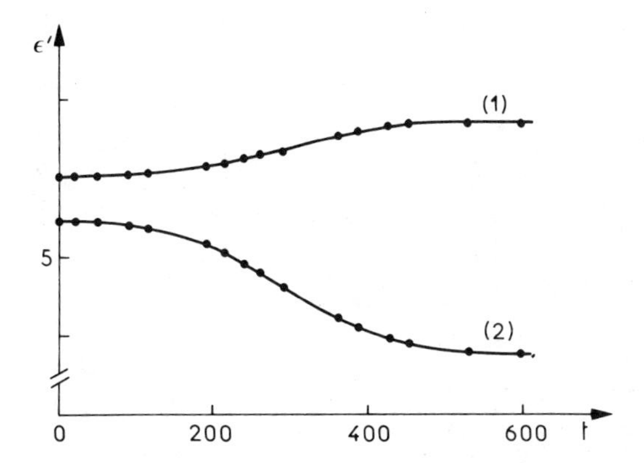


Figure 43. Time variation of the low-frequency (1) and high-frequency (2) limiting permittivities of a W/O emulsion maintained at $T_c$ = 253.2 K. T* = 234.7 K. Initial volume fraction of water at room temperature: Φ = 0.30. t is indicated in hours. (From Ref. 86, Gauthier-Villars, Paris.)

ing breakdown temperature. As a general rule, the lower the surfactant content, the shorter the duration of the transformation (Fig. 46). As concerns the influence of dissolved electrolytes, the situation is rather complex, some having no marked influence, others retarding or advancing the transformation. It is worth mentioning that compounds presenting crystallographic analogies with ice, such as lead iodide and ammonium fluoride, displayed an accelerating effect.

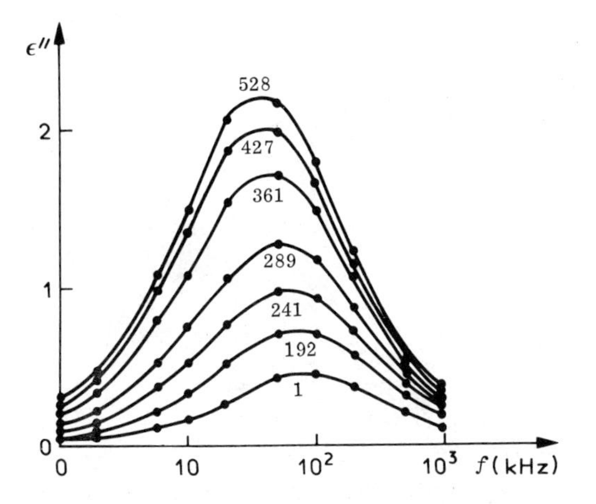


Figure 44. Time variation of the dielectric loss diagram of a W/O emulsion maintained at $T_c$ = 253.2 K. Specifications identical to those of Fig. 43. (From Ref. 86, Gauthier-Villars, Paris.)

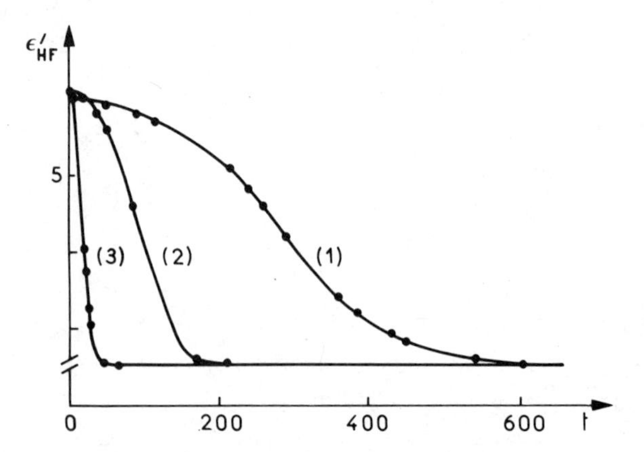


**Figure 45.** Comparative time variations of the high-frequency limiting permittivity of identical W/O emulsion samples held at different temperatures. Initial water content at room temperature: $\Phi = 0.30$. Surfactant proportion in the oil phase: 25% (w/w). $T^* = 234.7$ K. (1) $T_c = 253.2$ K; (2) $T_c = 246$ K; (3) $T_c = 242$ K. (From Ref. 86, Gauthier-Villars, Paris.)

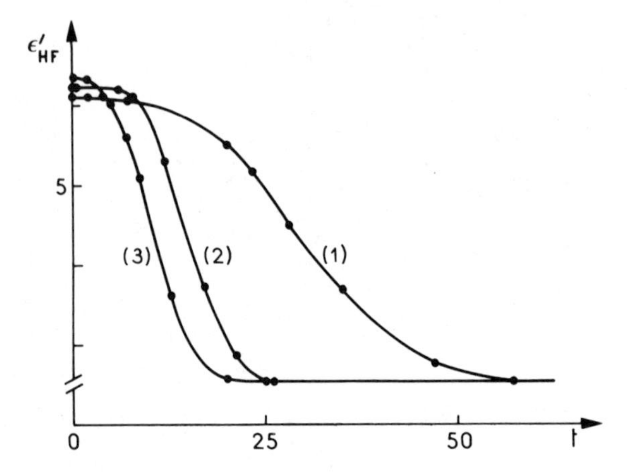


**Figure 46.** Influence of the proportion of surfactant on the time variation of the high-frequency limiting permittivity of emulsions with identical water contents ($\Phi = 0.305$), and held at $T_c = 246$ K. $T^* = 234.7$ K. Surfactant proportion (w/w) in the oil phase: (1) 22% (2) 15% (3) 12%. (From Ref. 86, Gauthier-Villars, Paris.)

Thus, whatever the experimental conditions, a W/O-type emulsion maintained at a fixed temperature lying between the solid disperse-phase melting point and the supercooling breakdown temperature exhibits a transformation in its dielectric properties. The system dielectric behavior, initially identical to the MWS dielectric relaxation of the unfrozen emulsion, transforms with time toward the dielectric behavior characteristic of a dispersion of ice globules. This transformation phenomenon, which arises from the progressive freezing of the disperse supercooled aqueous droplets, can be followed by means of the variations of the dielectric parameters of the system, namely, the high frequency limiting permittivity $\varepsilon_{HF}$, the low frequency limiting permittivity $\varepsilon_{LF}$, and the frequency corresponding to the maximum value of the loss factor $f_m$. From the theoretical analysis performed on Hanai's formula, it can be inferred that the most convenient parameter is $\varepsilon_{HF}$ since it depends only upon the permittivity of the continuous phase, the static permittivity of water, and the high-frequency limiting permittivity of ice, while $\varepsilon_{LF}$ and $f_m$ depend also upon the conductivity of the unfrozen supercooled aqueous phase and upon that of the ice being formed, neither of which can be known beforehand. Moreover, it is readily seen from Figs. 43 and 44 that $\varepsilon_{HF}$ is the parameter which undergoes the most marked variations, which fact is in favor of $\varepsilon_{HF}$ being retained as the most characteristic parameter for the transformation phenomenon. It will be shown below that this choice is quite consistent with the predictions derived from the theoretical model proposed with the view of interpreting the experimental observations reported above.

*(2) Interpretative model.* Because of the good agreement existing between theoretical predictions derived from Hanai's formula and data obtained through experimental studies of emulsion dielectric properties, it was thought possible to build up an interpretative model for the low-frequency dielectric behavior of W/O-type emulsions undergoing progressive freezing of the disperse phase [87,88]. The main hypotheses underlying the model are indicated briefly as follows.

On the basis of investigations made through DSC experiments [156,157,159], it can be assumed that the freezing process is nonsimultaneous for the whole population of droplets and is quasi-instantaneous for each individual droplet. Consequently, p the frozen disperse-phase weight proportion at time t, can be considered as representing also the proportion of frozen droplets at t, since droplet-size distributions in emulsions are usually fairly selective. Furthermore, it is assumed that, upon freezing, the droplets retain their spherical shape and the system its statistical isotropy and random geometry, in conformity with results obtained from electron microscopic observations [158], and that all the unfrozen droplets, on the one hand, and all the frozen globules, on the other hand, have identical dielectric properties.

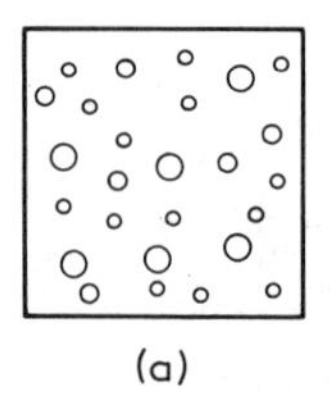

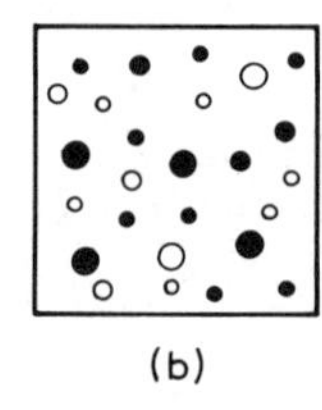

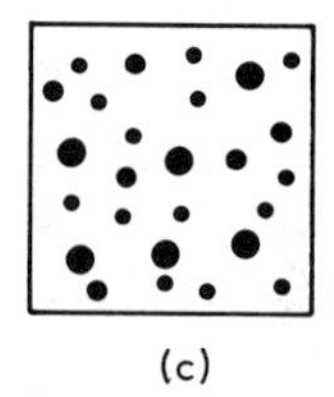


Figure 47. Model representation of disperse-phase freezing processes occurring in W/O-type emulsions held at temperatures between the solid disperse-phase melting point and the supercooling breakdown temperature. (a) Initial situation: totally unfrozen emulsion. (b) Intermediate situation: partially frozen emulsion. (c) Final situation: totally frozen emulsions.

On this basis, it is possible to consider a partially frozen emulsion as a system of the double spherical dispersion type (Fig. 47), and to depict its dielectric behavior by using an iterative-type application of Hanai's formula. Consequently, if $\varepsilon_1^{*a}$ and $\varepsilon_1^{*b}$ represent the disperse-phase complex permittivity respectively in the supercooled state and in the frozen state, the complex permittivity of a partially frozen emulsion is given by either of the two following formulas, obtained from Eqs. (239) and (240):

$$(\varepsilon^*)^{ba} = \mathfrak{T}[\varepsilon_1^{*a}, \mathfrak{T}(\varepsilon_1^{*b}, \varepsilon_2^*, \Psi_b(P_w, p)), \Phi_a(P_w, p)] \tag{393}$$

$$(\varepsilon^*)^{ab} = \mathfrak{T}[\varepsilon_1^{*b}, \mathfrak{T}(\varepsilon_1^{*a}, \varepsilon_2^*, \Psi_a(P_w, p)), \Phi_b(P_w, p)] \tag{394}$$

In the first case, the system is considered as an emulsion of supercooled aqueous droplets dispersed within a matrix consisting of a suspension of ice globules in the oil phase. $\Psi_b$ represents the volume fraction of the ice globules in the matrix, and $\Phi_a$ the volume fraction of the liquid droplets in the system. In the second case, the system is considered as a suspension of ice globules in a matrix consisting of an emulsion of aqueous droplets dispersed in the oil phase. $\Psi_a$ represents the volume fraction of the liquid droplets in the matrix and $\Phi_b$ the volume fraction of the ice globules in the system. If $\varepsilon_1^{*a}$, $\varepsilon_1^{*b}$, and $\varepsilon_2^*$ are given representations characteristic of the dielectric phenomena taking place respectively in the supercooled liquid disperse phase, frozen solid disperse phase, and oil-type continuous medium, it is possible to compute at each frequency $(\varepsilon^*)^{ba}$ and $(\varepsilon^*)^{ab}$ as functions of p, the weight proportion of the frozen aqueous phase. This computation can be repeated for different values of $P_w$, the disperse-phase total mass fraction in the system. For a given value of $P_w$, it is possible to calculate $\Psi_b$, $\Phi_a$, $\Psi_a$, and $\Phi_b$, at each value of p from the values of the density of the oil phase at $T_c$ (determined experimentally), and of the density of the supercooled liquid disperse phase and of the solid disperse phase (obtained from the literature). In the case of pure water, data have been reported by Kell [495,496] as concerns supercooled water, and by Lliboutry [497] as concerns ice I. A problem arises as to the choice of the functions $\varepsilon_1^{*a}(f)$ and $\varepsilon_1^{*b}(f)$. For $\varepsilon_1^{*a}(f)$, the problem can be solved easily since it is granted that, at low frequencies, the dielectric behavior of the supercooled disperse aqueous phase satisfies the general Eq. (243) in which the static permittivity can be ascribed values obtained or extrapolated from data available in the literature as to the static permittivity of water and aqueous solutions. As concerns the steady conductivity, arbitrary values can be taken within reasonable limits. For $\varepsilon_1^{*b}(f)$, the situation is more complicated since the actual dielectric properties of the forming ice cannot be known beforehand. It may just be stated that the ice globules exhibit a dielectric relaxation along with a conduction, the only dielectric parameter known with some certainty being the high-frequency limiting permittivity, which can be given the value commonly reported for bulk ice, namely $\varepsilon_{1d} = 3.08$. As concerns the static permittivity, the relaxation time and the frequency spread, no definite choice can be made since the structure of the ice microcrystals is not known, and these parameters must be considered as adjustable ones. The same applies to the conductivity of the ice globules. Fortunately, as explained below, exact values for these parameters are not necessary for the computation of a quantitative relationship connecting the emulsion dielectric behavior time drift with the rate of the disperse freezing process.

Computer calculations have been used [88] to obtain behavior models corresponding to different hypotheses relative to the dielectric properties of the solid disperse phase. In Figs. 48 and 49 are reported the dispersion and absorption curves obtained from Eqs. (393) and (394) for the case of a system characterized by $P_w = 0.28$. It was assumed that the frozen solid phase exhibited the Debye-type dielectric relaxation of pure ice at 246.2 K [489]. As could have been predicted on the basis of the theory of upper and lower bounds, the exact solution lies somewhere between that of Eq. (393) and that of Eq. (394). Nevertheless, the curves reported give a good qualitative description of the progressive changes observed experimentally as to the dielectric behavior of W/O-type emulsions undergoing progressive freezing of the disperse phase. The fit of numerical results with experimental data can be optimized at will by adjusting the values of the unknown parameters. It is not necessary to do this to establish a quantitative relationship between the transformation of the system dielectric behavior and the freezing process, because it appears that Eqs. (393) and (394) yield, to within the experimental uncertainties, identical values for the high-frequency limiting permittivity, as shown in Fig. 48. This result, which is a general one, is of great importance since it shows that disperse phase freezing processes occurring in W/O-

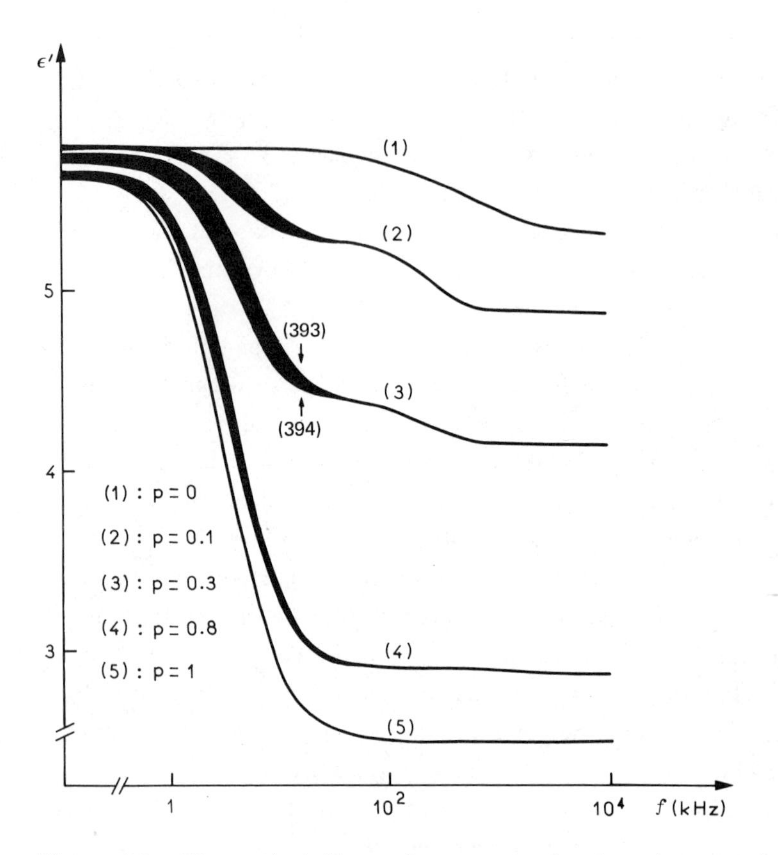


**Figure 48.** Theoretical dispersion curves showing the transformation of the dielectric behavior of a W/O-type emulsion undergoing disperse-phase progressive freezing. $P_w = 0.28$; $T_c = 246.2$ K. p represents the weight proportion of frozen disperse phase which is assumed to exhibit the Debye-type dielectric relaxation of polycrystalline ice I. (393) indicates the curves corresponding to $(\varepsilon^*)^{ba}$ and (394) those corresponding to $(\varepsilon^*)^{ab}$. (From Ref. 88, Gauthier-Villars, Paris.)

type emulsions can be safely followed by means of a single parameter, namely the high-frequency limiting permittivity of the system $\varepsilon_{HF}$, which depends only, from the dielectric standpoint, upon the static permittivity of the supercooled disperse aqueous phase and the high-frequency limiting permittivity of ice. As has been mentioned earlier, these conclusions are consistent with experimental observations which show that $\varepsilon_{HF}$ is the most reliable parameter.

The numerical calculations yield the relation linking the frozen disperse phase weight proportion p with $\varepsilon_{HF}$, as indicated below,

$$\varepsilon_{HF} = F(p) \tag{395}$$

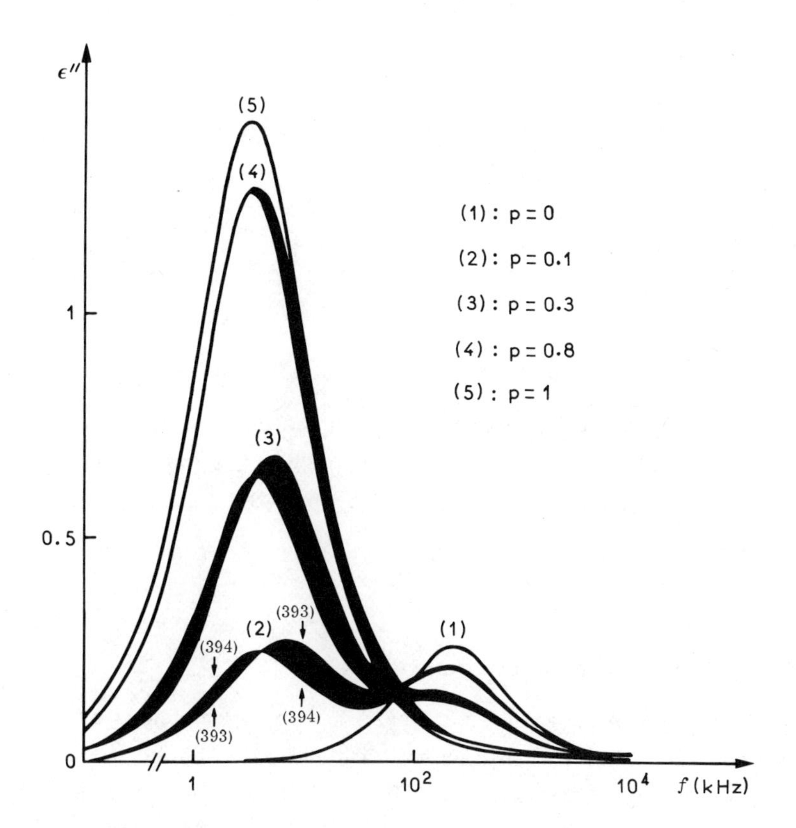


**Figure 49.** Theoretical absorption curves showing the transformation of the dielectric behavior of a W/O-type emulsion undergoing progressive disperse-phase freezing. All specifications identical to those of Fig. 48. (From Ref. 88, Gauthier-Villars, Paris.)

while the experiments give the relation between t, the sample holding duration at $T_c$, and $\varepsilon_{HF}$,

$$\varepsilon_{HF} = G(t) \tag{396}$$

The equation representative of the progressive freezing rate is obtained by combining Eqs. (395) and (396):

$$p = F^{-1}[G(t)] \tag{397}$$

Figure 50 shows an example of the dependence of p upon t, as determined using the dielectric method. It is readily seen that the agreement with the determination obtained from DSC experiments is excellent. The sigmoidal shape of the p versus t curve suggests that, at temperatures $T_c$ sufficiently remote from T*, the disperse-phase monothermal freezing process obeys the following differential equation [159]:

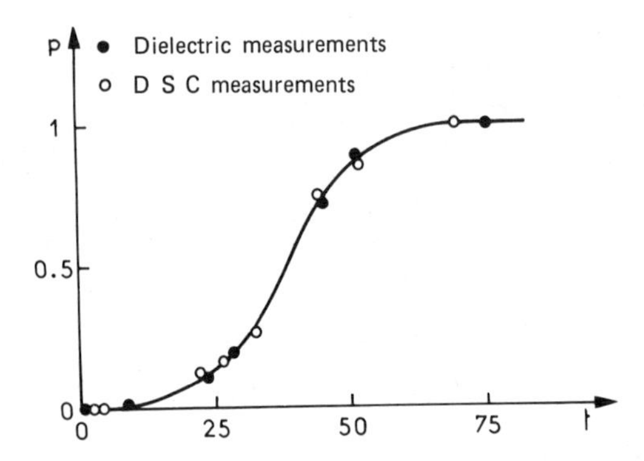


**Figure 50.** Disperse-phase progressive freezing rate observed on an emulsion of water in vaseline oil. $P_W = 0.34$. $T_c = 247.2$ K. $T^* = 234.7$ K. t is indicated in hours. (From Ref. 90. Copyright by the Institute of Physics, Bristol, England.)

$$dp = Kp(1 - p)\, dt \tag{398}$$

### 4. *Dielectric Studies of Frozen W/O-Type Emulsions*

Earlier investigations [99-103] have revealed that frozen W/O-type emulsions exhibit dielectric relaxations located below 1 MHz or so. Experiments showed that, upon aging, these systems displayed drifts in their dielectric properties, the most remarkable evolutive phenomenon being the irreversible decrease of the dielectric relaxation activation energy [99,498]. This evolution phenomenon, which was observed also in macroscopic samples of polycrystalline ice [104-109], has been given a model phenomenological representation by Boned [110]. Recently, a thorough study of the dielectric behavior of dispersions of ice microcrystals in oil, obtained from the supercooling breakdown of W/O-type emulsions, was achieved by Lagourette and co-workers [93-98], who gave special attention to the dielectric property alterations observed in samples held at a temperature close to the solid disperse-phase melting point. Experiments in the same field have been performed by other authors [499], but the data reported are rather questionable since only three samples were studied.

a. Experimental Observations. Dispersions of ice globules in oil were obtained by steadily cooling W/O-type emulsions down to past T*, the disperse phase supercooling breakdown temperature, determined by dielectric measurements, as has been described earlier. The samples were then warmed up to a temperature $T_c$ at which they were maintained during variable time periods. Dielectric measurement runs were made at regular time intervals, for different values of $T_c$, so as to study the alterations in dielectric behavior with time. Series of experiments were also performed at several discrete temperatures so as to investigate the influence of temperature upon the dielectric properties of systems for which the transformation mentioned above had ceased. The influence of the disperse phase content (weight fraction $P_W$ or volume fraction $\Phi$), and that of the surfactant nature and proportion were considered [93,94]. All these investigations were achieved through complex permittivity measurements of the fixed-temperature and variable-frequency type (FTVF) over the working frequency range 20 Hz to 3 MHz. Additional experiments of the fixed-frequency and variable-temperature type (FFVT) were also carried out [97,98]. Complementary studies were devoted to the influence of ion incorporation upon the dielectric properties of ice microcrystals suspended in dispersions obtained from the supercooling breakdown of emulsions of aqueous electrolyte solutions in oil [111-114].

*(1) Holding temperatures lower than 270 K.* As an example, Fig. 51 shows the dielectric behavior of a frozen W/O-type emulsion maintained at a temperature $T_c$ equal to 265.2 K. Two distinct dielectric relaxations are observable. The first, located in the higher frequency range, arises from the Debye-type intrinsic relaxation exhibited by the disperse ice globules, as proved by the good agreement existing between its features and the theoretical predictions derived from the Hanai formula (see Fig. 21). On the basis of the theoretical models obtained from the analysis of Hanai's formula, the second relaxation, located in the lower frequency range, could be ascribably a priori to interfacial polarization phenomena connected with the conductivity of the disperse ice globules. However, a more precise analysis of the experimental results shows that the second relaxation does not conform to models established by assuming that the disperse ice microcrystals display merely an ohmic conductivity. As explained below, a more sophisticated treatment is necessary to fit the experimental data properly. It can be seen from Fig. 51 that an evolution of the dielectric behavior is noticeable when the samples are held at $T_c$ during long periods of time. While the limiting permittivity at higher frequencies remains almost unchanged, the dielectric increments of both relaxations decrease slightly and the critical frequencies increase. The features of this transformation, which does not deeply affect the total dielectric behavior of the system, are consistent with previous observations reported by several authors [99,105] as concerns the decrease with time of the dielectric relaxation activation energy of polycrystalline ice. FFVT measurements, performed immediately after the breakdown of supercooling, yielded results quite comparable to FTVF determinations, as proved by the $\varepsilon''$ versus $\varepsilon'$ plot reported in Fig. 52. The diagram distortions and the loop noticeable in the vicinity of 250 K are perfectly normal and arise from the temperature variations of the static permittivity of ice. Detailed comments about this peculiar phenomenon can be found in Ref. 98. The surfactant proportion in the oil phase being fixed, the dependence of the system dielectric parameters upon $P_W$ is characterized, with increasing $P_W$, by an increase

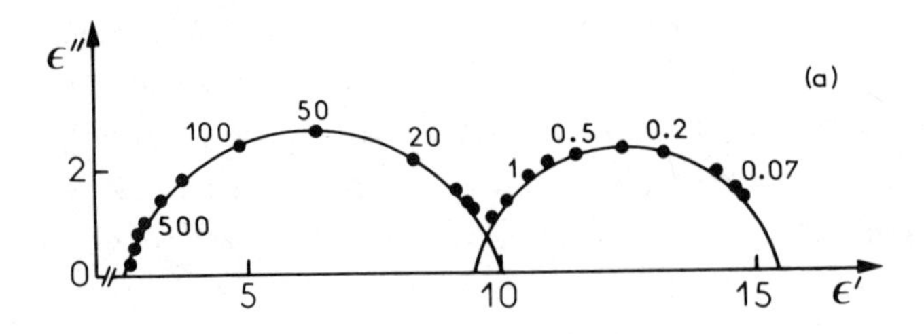


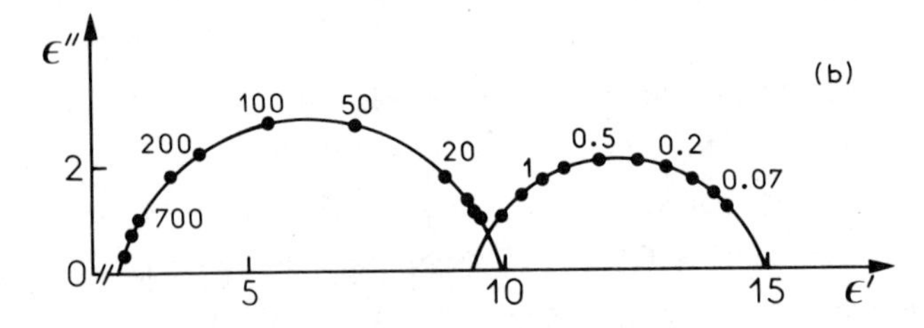


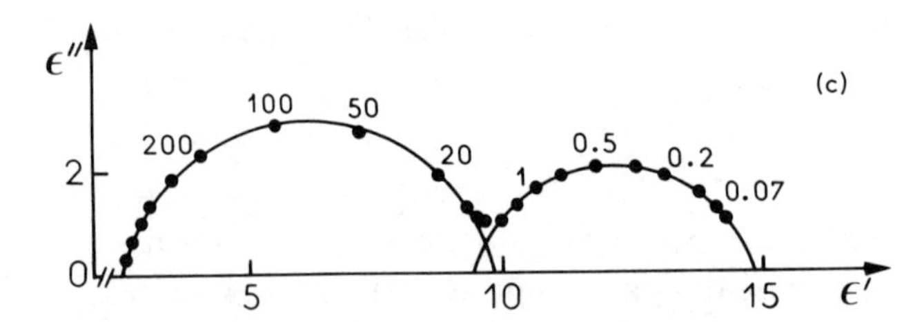


**Figure 51.** Time drift in the dielectric behavior of a frozen emulsion of water in vaseline oil held at $T_c = 265.2$. $T^* = 234.7$. $\Phi = 0.40$. Surfactant proportion in the oil phase: 33% (w/w). The frequencies are indicated in kilohertz. Duration of maintenance at $T_c$ after the supercooling breakdown: (a) 1 h, (b) 4 days, and (c) 12 days. [(a) from Ref. 98, reproduced from the *Journal of Glaciology* by permission of the International Glaciological Society.]

of the limiting permittivities and dielectric increments of both relaxations, and by a decrease of their maximum loss frequencies, in qualitative conformity with theoretical predictions. At a given disperse phase content, the dielectric parameters are strongly dependent upon the proportion of surfactant in the oil phase. While the higher frequency limiting permittivity and the dielectric increment of the high-frequency dielectric relaxation are not much affected, the dielectric incre-

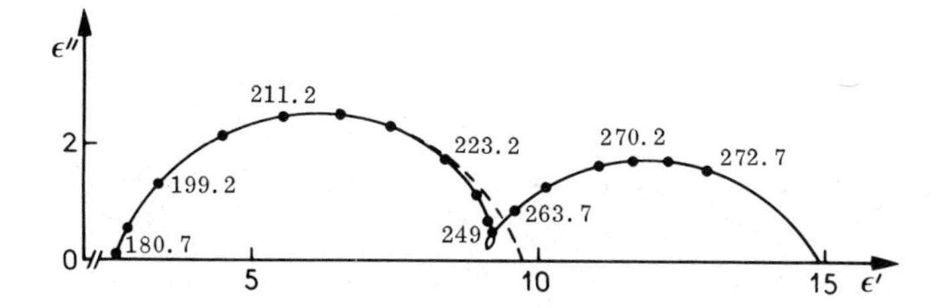


Figure 52. ε'' versus ε' plot obtained through FFVT measurements performed on a frozen W/O emulsion submitted to steady warming from 180.7 up to 272.7 K, immediately after the supercooling breakdown. $\Phi = 0.40$. $T^* = 234.7$ K. Working frequency: $f = 0.4$ kHz. The temperatures are expressed in kelvin. (From Ref. 98, reproduced from the *Journal of Glaciology* by permission of the International Glaciological Society.)

ment of the low-frequency dielectric relaxation undergoes drastic increases when the surfactant proportion in the oil phase is lowered. Correlatively, the maximum loss frequencies of both relaxations decrease, the phenomenon being particularly marked for the low-frequency dielectric relaxation. These results show that the existence of special phenomena occurring within the transition region between the disperse ice globules and the continuous medium must be taken into account in designing a model for the dielectric behavior of ice microcrystal dispersions. This conclusion is supported by the data derived from dielectric studies of these systems performed at holding temperatures closer to $T_0$, the ice melting point.

*(2) Holding temperatures included between 270 and 273 K.* If, after the disperse-phase breakdown of supercooling, the samples are warmed up to and held at a temperature $T_c$ close to $T_0$, the solid disperse-phase melting point, their dielectric properties undergo a rapid change whose effects are qualitatively similar to those described above, but of greater amplitude. As an illustration, Fig. 53 compares two Cole-Cole plots recorded on the same system after being held one hour or so at 272.2 K and after 15 days at the same temperature. It is readily seen from these diagrams that the dielectric behavior alterations are very drastic, in particular as concerns the low-frequency dielectric relaxation. Maintaining the sample between 270 and 273 K demonstrates additional phenomena that were not observed for $T_c < 270$ K. For frozen emulsions with a high proportion of surfactant in oil, it was found that the low-frequency dielectric relaxation resolved with time into two distinct dielectric relaxations, so that the total system behavior was representable eventually by three connected circular arcs, as shown in Figs. 54a and b.

Once the systems no longer displayed drifts in their dielectric properties, runs of FTVF measurements were performed at several discrete temperatures in order to

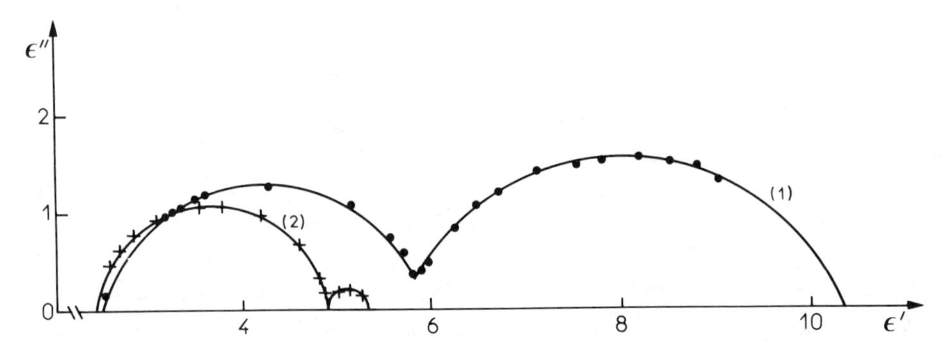


Figure 53. Cole-Cole plots showing the transformation of the dielectric behavior of a frozen W/O emulsion held at $T_c = 272.2$ K. $\Phi = 0.24$. Surfactant concentration in the oil phase: 10% (w/w). Duration of holding at $T_c$, after the supercooling breakdown: (1) 1 h and (2) 15 days. (From Ref. 94, with permission from Les Editions de Physique, Orsay, France.)

investigate the influence of temperature on their final dielectric behavior. Figure 55 shows a typical set of results obtained through such experiments. It must be mentioned that the effect of temperature was found to be symmetrical, repetitive cooling and warming runs yielding identical results.

Over an intermediate temperature range, 253 to 270 K or so, the dielectric behavior of systems with low surfactant contents is characterized by two adjacent

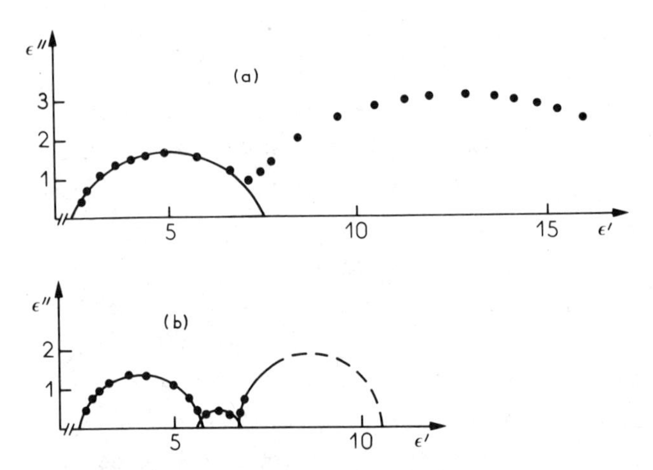


Figure 54. Cole-Cole plots showing the transformation of the dielectric behavior of a frozen W/O emulsion with a high proportion of surfactant in the oil phase, 33% (w/w). $T_c = 272.2$ K. $\Phi = 0.28$. Duration of holding at $T_c$, after the breakdown of supercooling: (a) 1 h and (b) 3 days. (From Ref. 94, with permission from Les Editions de Physique, Orsay, France.)

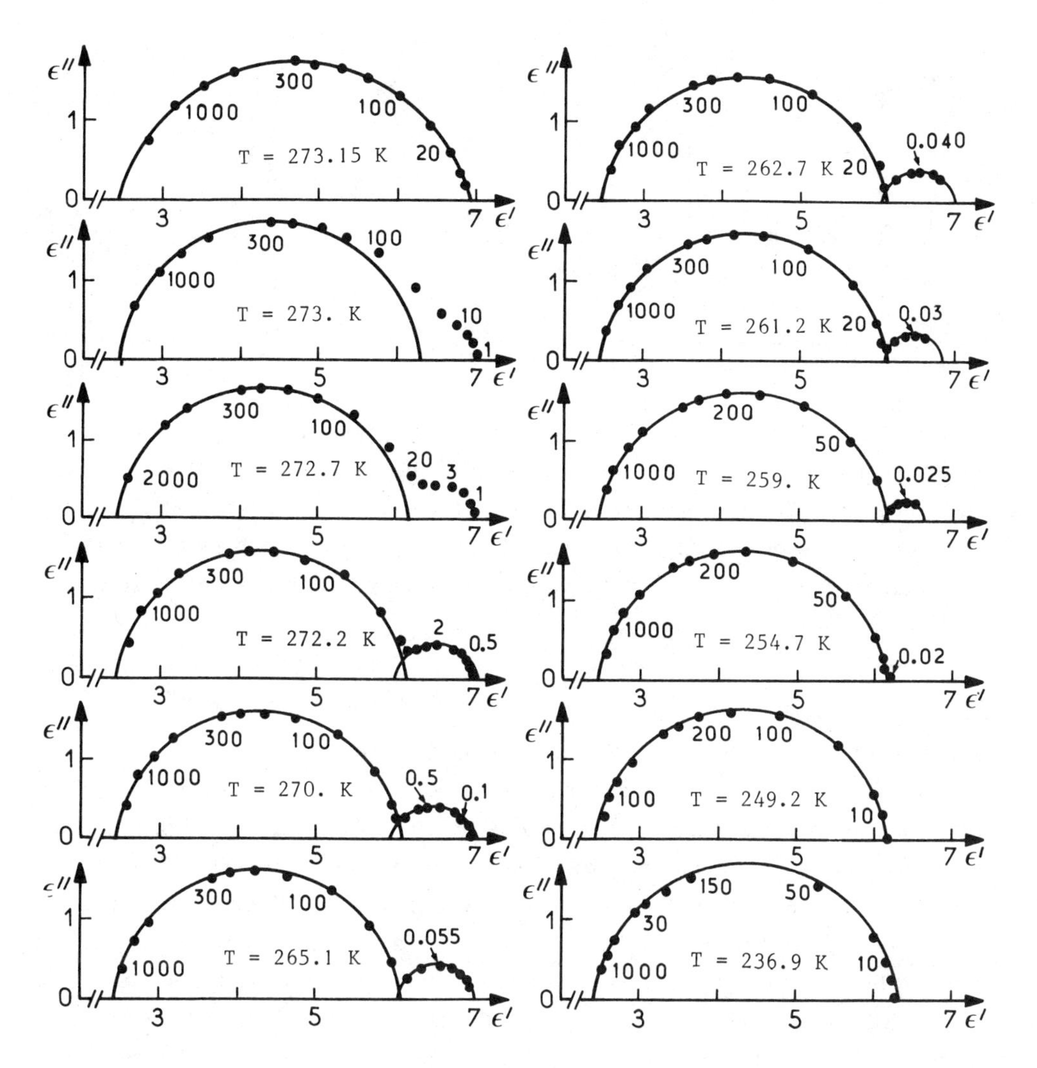


Figure 55. Temperature dependence of the dielectric behavior of a frozen W/O emulsion after completion of its transformation at $T_c$ = 272.2 K. Φ = 0.33. Surfactant proportion in the oil phase: 2.5% (w/w). The frequencies are expressed in kilohertz. (From Ref. 94, with permission from Les Editions de Physique, Orsay, France.)

dielectric relaxations. The higher frequency limiting permittivity and the dielectric increment of the high-frequency relaxation are almost temperature independent, except for the expected slight variations arising from the effect of temperature upon the static permittivity and volume fraction of disperse ice globules. By contrast, the low-frequency dielectric increment decreases as T decreases and becomes no longer detectable at temperatures equal to or lower than 253 K or so. Below 253 K, the systems exhibit a unique dielectric relaxation aris-

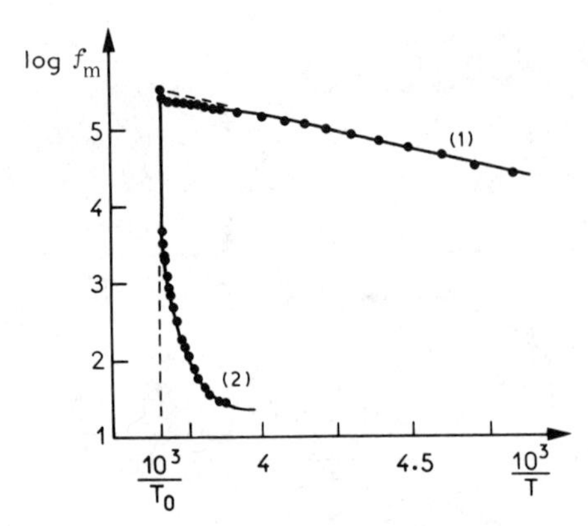


Figure 56. Variation of the maximum loss frequency $f_m$ versus T for (1) the high-frequency dielectric relaxation and (2) the low-frequency dielectric relaxation exhibited by a frozen W/O emulsion after completion of its transformation at $T_c$ = 272.2 K. (Specifications identical to those of Fig. 55). $f_m$ is expressed in hertz and T in kelvin. (From Ref. 94, with permission from Les Editions de Physique, Orsay, France.)

ing from the intrinsic Debye-type dielectric relaxation of the ice globules. Above 270 K or so, both relaxations present increasing coupling at T increases and eventually merge into one another when T is almost equal to $T_0$, but slightly lower.

In Fig. 56 the maximum loss frequency variations versus T are plotted for both of the relaxations. Below 253 K or so, the log $f_m$ versus $T^{-1}$ diagram corresponding to the high-frequency relaxation is a straight line, which proves that $f_m$ satisfies an Arrhenius-type law. Thus, over this temperature range, $f_m$ becomes identified with the critical frequency $f_c$ of the Cole-Cole-type dielectric relaxation of the system connected with the disperse ice microcrystal intrinsic dielectric relaxation, whose activation energy can be determined from the variations of $f_m$ with temperature. Between 253 K and $T_0$, because of the coupling of the two relaxation, the log $f_m$ versus $T^{-1}$ plots of the two of the relaxations exhibit marked curvature that prevent the determination of activation energies, since the phenomena can no longer be described by the Arrhenius law.

Systems with higher surfactant contents in the oil phase exhibit a behavior similar to that reported above as concerns the first two relaxations, but the situation is more complicated, owing to the existence of a third relaxation taking place in the lower frequency range. Figure 57 shows a typical set of results obtained between $T_0$ and 262 K on a system characterized by $\Phi$ = 0.29, the surfactant concentration in the oil phase being equal to 66% (w/w). The system dielectric behavior at high and low frequencies is similar to that observed in the case of systems with low surfactant contents. The additional dielectric relaxation displayed

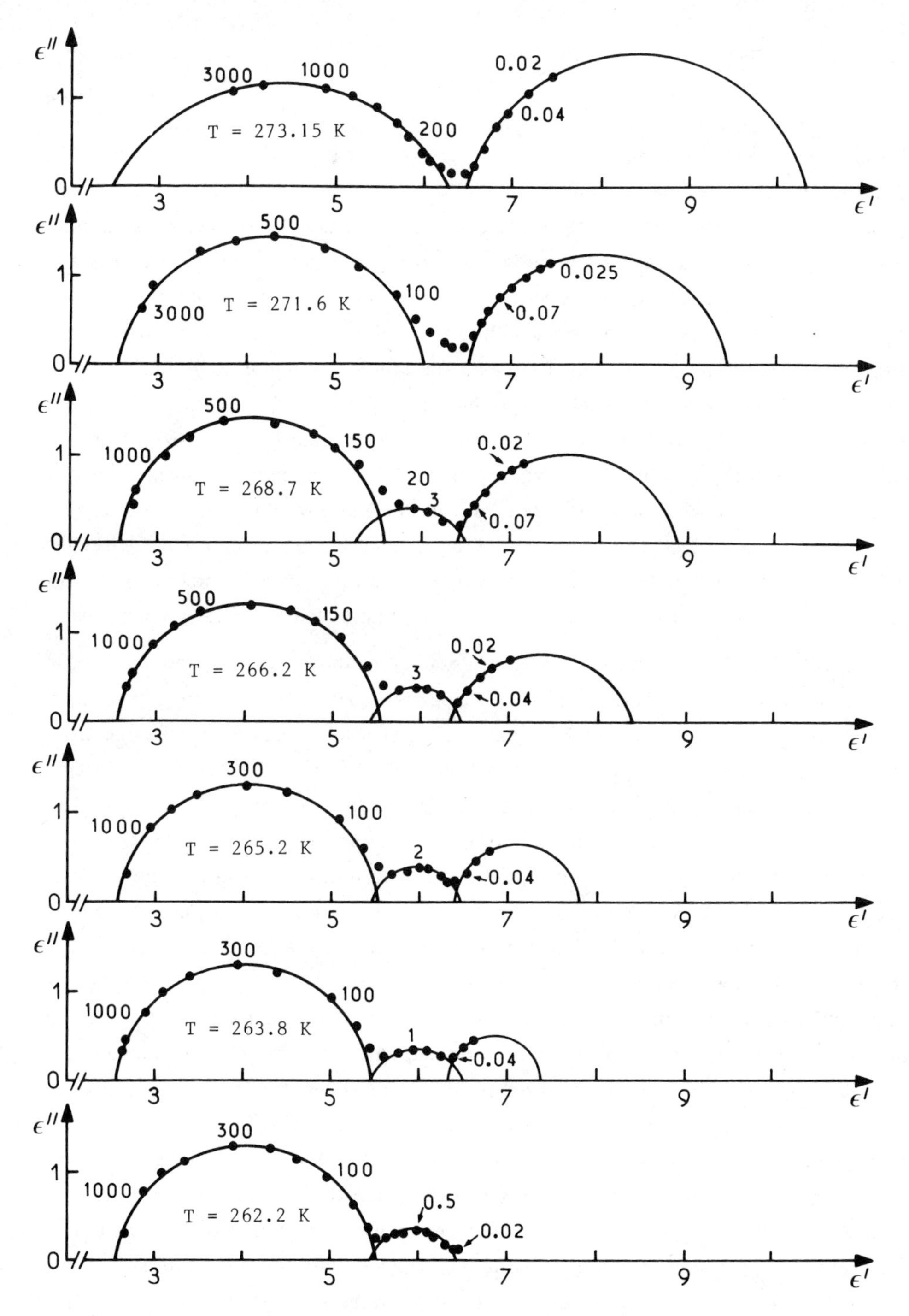


Figure 57. Temperature dependence of the dielectric behavior of a frozen W/O emulsion after completion of its tranformation at $T_c$ = 272.2 K. $\Phi$ = 0.29. Surfactant proportion in the oil phase: 66% (w/w). The frequencies are expressed in kilohertz. (From Ref. 94, with permission from Les Editions de Physique, Orsay, France.)

in the lower frequency range undergoes drastic alterations with decreasing T, the dielectric increment decreasing and eventually vanishing around 262 K. At lower temperatures, only two relaxations are detectable, the low frequency one vanishing as well around 253 K, as in the case of systems with low surfactant contents. These phenomena seem to depend predominantly rather upon the amount of surfactant than upon its nature, as proved by the similarity of results obtained through comparative measurements using either lanolin or sorbitan trioleate.

Dispersions of doped ice microcrystals obtained from the breakdown of supercooling of aqueous electrolytic solutions exhibit a dielectric behavior which is roughly similar to that of frozen water-in-oil emulsions, except for some deviations arising mainly from the incorporation of ions in the ice globules. In particular, the phenomena are shifted toward higher frequencies in accordance with the nature and concentration of the electrolyte and also toward lower temperatures, since eutectic melting temperatures are substituted for the melting point of ice [97,98].

b. Interpretative Model. From the results reported above, it appears that the dielectric behavior of dispersions of ice microcrystals is rather complex in the higher temperature range and cannot be depicted by simple models derived from the Hanai formula. Although at temperatures below 253 K or so the system dielectric relaxation arises simply from the intrinsic Debye-type dielectric relaxation of the disperse ice microcrystals, the additional low-frequency dielectric relaxation cannot be ascribed to the existence of a classical ohmic conduction in the ice microcrystals, as proved in particular by the abnormal features of the log $f_m$ versus $T^{-1}$ diagram. Moreover, experiments have demonstrated the strong influence of the environment close to the disperse ice globule, which is reflected in the dependence of the dielectric behavior upon the surfactant concentration and by the existence of a third dielectric relaxation observed in systems with high surfactant contents.

On the basis of these results and on the basis of theoretical and experimental data available in the literature as to the structure of polycrystalline ice, Lagourette and co-workers [93,95] designed a theoretical model which is described schematically in Fig. 58. Frozen W/O emulsions are considered to be dispersions of spherical particles covered with shells that represent the transition region between the disperse ice globules and the continuous matrix. The particle spherical cores are assumed to be made up of disturbed clusters imbedded within the normal ice lattice. It is worth mentioning, without entering into detailed considerations of solid state physics, that the existence of microscopic structural inhomogeneities (disturbed clusters) in solids is a subject of growing interest, since many anomalous phenomena observed in polymorphic materials may be explained on this basis with the aid of the *effective medium theory* [419]. In the case of ice, contributions by different authors and relevant discussions on "core structure" can be found in [295].

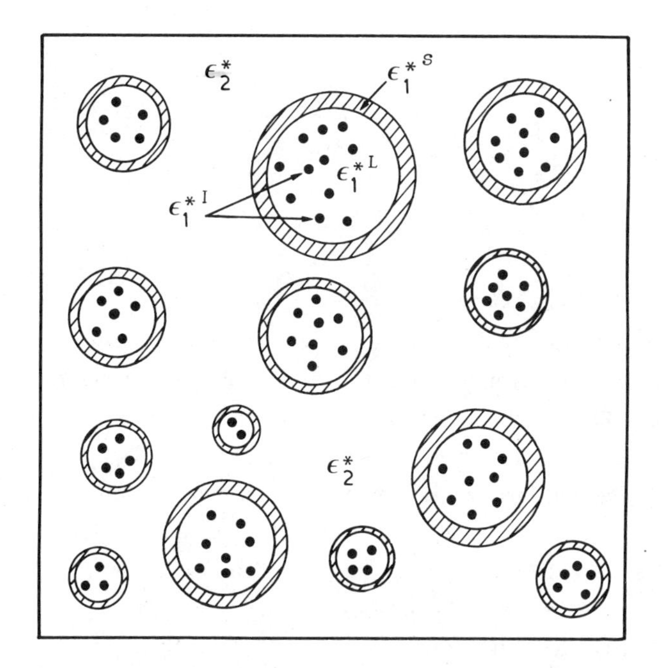


**Figure 58.** Schematic description of the model designed to account for the dielectric behavior of frozen W/O emulsions. The black dots represent structural inhomogeneities existing within the ice lattice.

Following the model suggested by Lagourette and coauthors, the complex permittivity of frozen W/O-type emulsions may be expressed by the following formula which combines the Maxwell-Wagner Eq. (233) and the Hanai Eq. (234):

$$\varepsilon^* = \mathfrak{T}[\mathfrak{W}[\mathfrak{T}(\varepsilon_1^{*I}, \varepsilon_1^{*L}, \psi), \varepsilon_1^{*S}, \underline{\phi}], \varepsilon_2^*, \Phi] \tag{399}$$

$\varepsilon_2^*$ represents the complex permittivity of the continuous matrix, $\varepsilon_1^{*S}$ that of the shells covering the particles, $\varepsilon_1^{*L}$ that of the normal ice lattice and $\varepsilon_1^{*I}$ that of the structural inhomogeneities. $\psi$ is the volume fraction of inhomogeneities imbedded in the ice lattice, $\underline{\phi}$ the volume fraction corresponding to the ice core in the disperse spherical particles, and $\Phi$ the disperse particle volume fraction, which is related to $P_w$, the disperse-phase weight fraction in the system. If the shell thickness can be neglected, $\underline{\phi}$ is equal to 1 and Eq. (399) reduces to

$$\varepsilon^* = \mathfrak{T}[\mathfrak{T}(\varepsilon_1^{*I}, \varepsilon_1^{*L}, \psi), \varepsilon_2^*, \Phi] \tag{400}$$

In a first analysis, it was assumed that the shell was infinitesimally thin, so that Eq. (400) could be used. On the basis of the experimental results and of data available in the literature, $\varepsilon_2^*$, $\varepsilon_1^{*I}$, and $\varepsilon_1^{*I}$ were given the formulations

$$\varepsilon_2^* = \varepsilon_{2s} \tag{401}$$

$$\varepsilon_1^{*I} = \varepsilon_{1d}^{I} + \frac{\sigma_1^I}{2\pi j \varepsilon_0 f} \tag{402}$$

$$\varepsilon_1^{*L} = \varepsilon_{1d}^{L} + \frac{\varepsilon_{1s}^{L} - \varepsilon_{1d}^{L}}{1 + jf/f_{c_1}} \tag{403}$$

Equation (401) implies that the continuous medium is nonpolar and nonconducting, and $\varepsilon_{2s}$ was given experimental values determined through dielectric measurements performed on the oil phases used. Equation (402) describes the dielectric behavior of the structural inhomogeneities which were considered to be essentially conductive, and Eq. (403) that of the ice lattice, assumed to exhibit merely the Debye-type dielectric relaxation of ice I. $\varepsilon_{1s}^{L}$ was given the values of polycrystalline ice I static permittivity at different temperatures [489], and the value of the high-frequency limiting permittivity of ice, namely 3.08, was ascribed to both $\varepsilon_{1d}^{L}$ and $\varepsilon_{1d}^{I}$. The variations of $\sigma_1^I$ and $f_{c_1}$ with temperature were adjusted by using data obtained from the experimental study of the dielectric behavior of ice microcrystal dispersions. To account for the observations made involving the decrease of the low frequency dielectric increment with decreasing temperature, $\psi$ was assumed to decrease linearly with T, from 0.30 at 263 K down to zero at 253 K or so. Figure 59 shows, for $\Phi = 0.33$, a typical set of results derived from the numerical solution of Eq. (400) performed on the basis of the above hypotheses. The comparison of Fig. 59 with Fig. 55 demonstrates the good agreement, both qualitative and quantitative, existing between the experimental results and the numerical simulation data. It must be noticed in particular that the increasing coupling of both relaxations as T increases and their eventual merging into one another when T is very close to $T_0$ is remarkably depicted by the model. In addition, Fig. 60 shows that the temperature behavior of the maximum loss frequency of both relaxations is described adequately, in particular as concerns the curvature observed at higher temperatures in the log $f_m$ versus $T^{-1}$ diagram corresponding to the high-frequency relaxation.

In a second stage, dispersions of particles covered with shells of finite thickness were considered. The hypotheses and the procedures used previously were retained and it was assumed, in addition, that the shell dielectric behavior could be described by the formula

$$\varepsilon_1^{*S} = \varepsilon_{1d}^{S} + \frac{\sigma_1^S}{2\pi j \varepsilon_0 f} \tag{404}$$

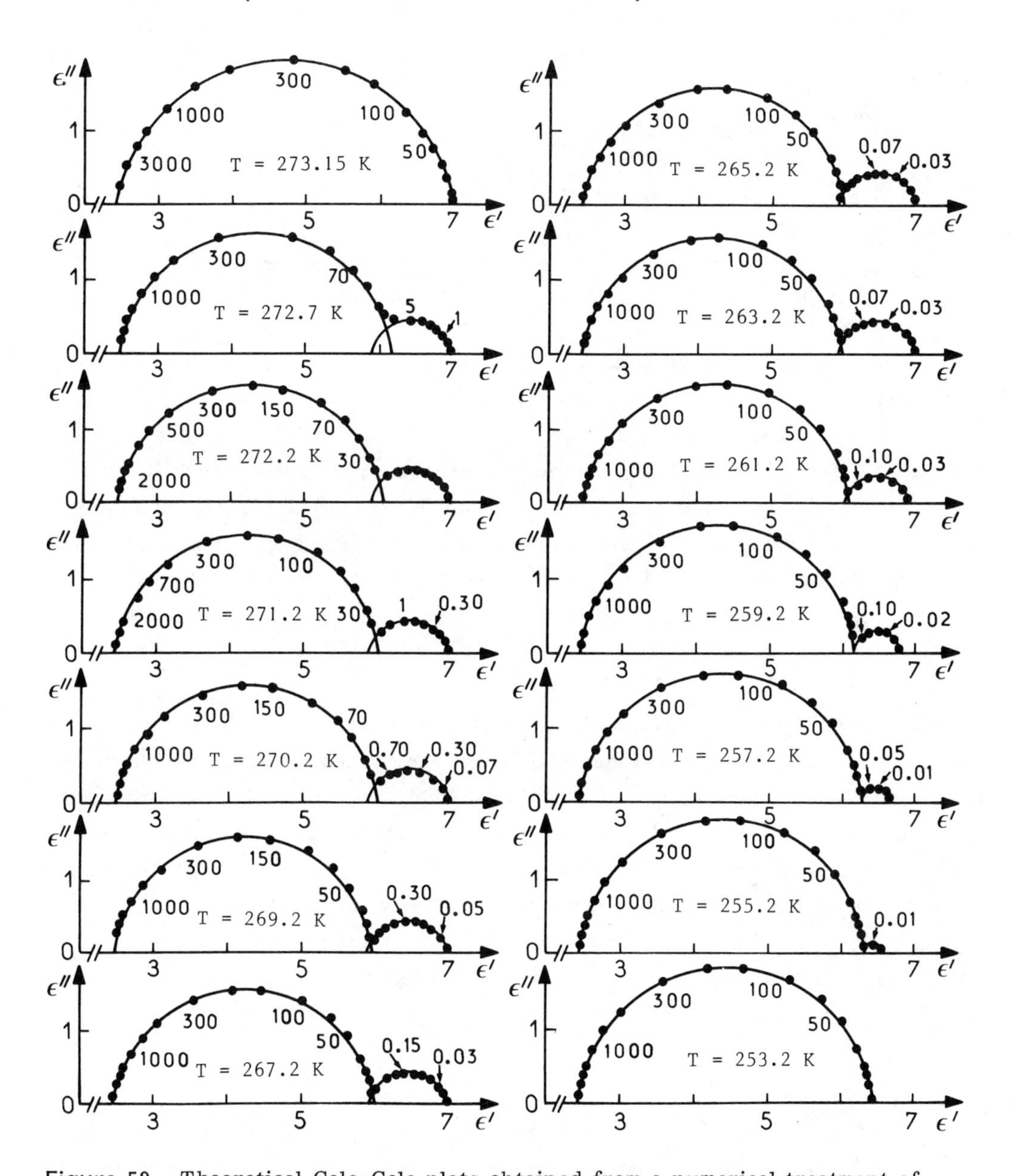


**Figure 59.** Theoretical Cole-Cole plots obtained from a numerical treatment of Eq. (400) and representing the influence of temperature upon the dielectric behavior of a frozen W/O emulsion after completion of the transformation of its dielectric properties. The shell was assumed to be infinitesimally thin. $\Phi = 0.33$. The frequencies are expressed in kilohertz. (From Ref. 95, with permission from Les Editions de Physique, Orsay, France.)

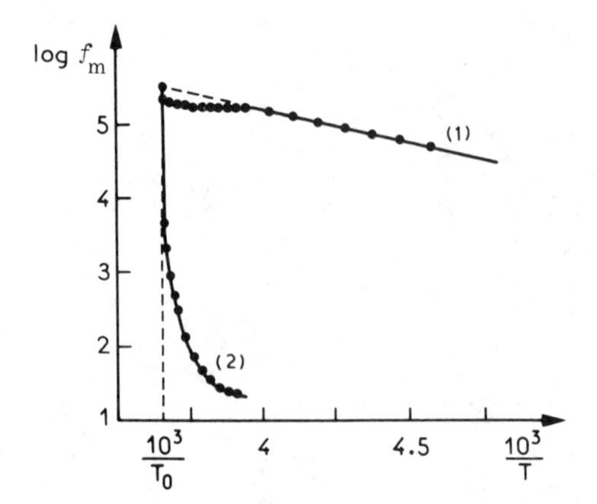


Figure 60. Theoretical variation, as derived from a numerical treatment of Eq. (400), of the maximum loss frequency $f_m$ versus T for (1) the high-frequency and (2) the low-frequency relaxations exhibited by a dispersion of ice microcrystals characterized by $\Phi = 0.33$. $f_m$ is expressed in hertz and T in kelvin. (From Ref. 95, with permission from Les Editions de Physique, Orsay, France.)

Equation (404) implies that the shells covering the disperse particles can be treated as made up of an essentially conductive homogeneous medium. This hypothesis is rather a crude one since it is most likely that the dielectric parameters vary continuously from point to point along the shell thickness direction. Nevertheless, by assuming that the relative shell thickness is the same for all the disperse particles, the numerical solution of Eq. (399) yields, either with $\varepsilon_{1d}^{S} = \varepsilon_{2s}$ or with $\varepsilon_{1d}^{S} = \varepsilon_{1d}^{L}$, results in good qualitative agreement with the experimental observations, as proved by Fig. 61, showing the existence of a third dielectric relaxation located at lower frequencies. $\phi$ being assumed to decrease with decreas-

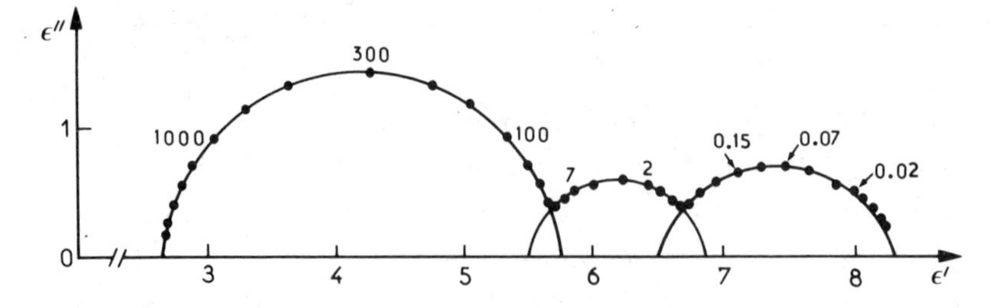


Figure 61. Theoretical Cole-Cole plots obtained from a numerical solution of Eq. (399), by using the hypotheses expressed by Eq. (401) to (404). $\Phi = 0.29$, $\underline{\phi} = 0.27$. The frequencies are expressed in kilohertz. (From Ref. 95, with permission from Les Editions de Physique, Orsay, France.)

ing T, a concomitant decrease in the third relaxation dielectric increment was found, in conformity with the experimental observations reported in Fig. 57. The values of the shell thicknesses involved were a few hundreds of angstroms, a reasonable magnitude by comparison with the mean value of the disperse particle diameter which is equal to 1 $\mu m$ or so.

From the preceding developments, it can be concluded that the model designed by Lagourette and co-workers on the basis of Hanai's formula yields a good theoretical description of the dielectric behavior of dispersions of ice microcrystals obtained from disperse phase supercooling breakdowns in W/O-type emulsions, in particular as concerns the anomalies observed at temperatures close to the solid disperse-phase melting point. Further information about these anomalies which were ascribed by the authors to so-called premelting phenomena connected with the existence of structural inhomogeneities in the ice particles can be found in the original papers published on the subject [94,95,97,98,110].

### 5. *Dielectric Investigation of Thawing of Frozen W/O Emulsions*

In addition to their studies of dispersions of ice microcrystals, Lagourette and co-workers [93,96] carried out dielectric investigations as concerns thawing phenomena in frozen W/O-type emulsions. This extension was quite a logical one, since such experiments could yield valuable information complementary to those obtained on the dielectric behavior of frozen W/O-type emulsions at temperatures close to the solidified disperse-phase melting point.

The systems studied were similar to those used in the previous investigations. The measurements were performed according to the following procedure. The dielectric relaxation arising from interfacial polarization phenomena in a freshly made emulsion was studied down to 260 K or so. The emulsion was cooled immediately so as to obtain disperse phase supercooling breakdown and the dielectric properties of the subsequent frozen system were investigated until completion of its transformation, as reported previously. The sample was then thawed and the dielectric behavior of the resulting W/O emulsion was studied. The results obtained through this experimental procedure are quite novel and surprising. They are summarized by means of Figs. 62 to 65. The diagrams reported in Fig. 62 show, over a limited temperature range below the ice melting point, the variation of the maximum loss frequency versus temperature, comparing the unfrozen emulsion MWS dielectric relaxation with the two relaxations exhibited by the frozen emulsion at the end of the transformation of its dielectric properties. Figure 63 describes the analogous situation in which the diagram labeled (3) refers now to the dielectric relaxation exhibited by the system obtained upon thawing the ice microcrystal dispersion. It is readily seen from these figures that the log $\bar{f}_c$ versus $T^{-1}$ diagram characteristic for the emulsion MWS dielectric relaxation is shifted downwards just

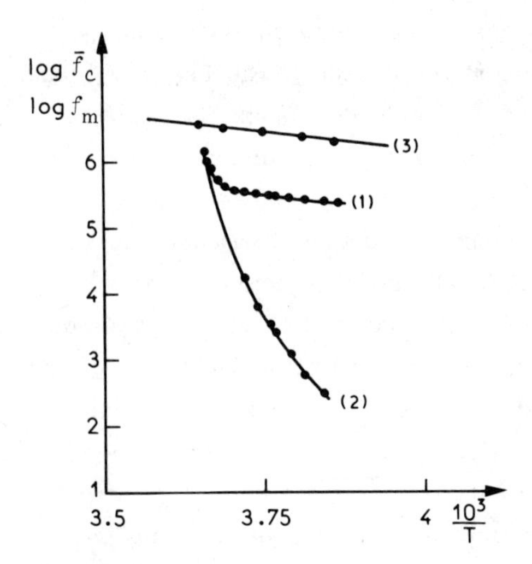


Figure 62. Variation of $\bar{f}_c$ and $f_m$ observed in the vicinity of $T_0$ on a W/O system characterized by $P_W = 0.28$. $\bar{f}_c$ and $f_m$ are indicated in hertz, and T in kelvin. (1) Frozen system high-frequency dielectric relaxation, (2) frozen system low-frequency dielectric relaxation, and (3) MWS-type dielectric relaxation exhibited by the emulsion prior to freezing. (From Ref. 96, with permission from Les Editions de Physique, Orsay, France.)

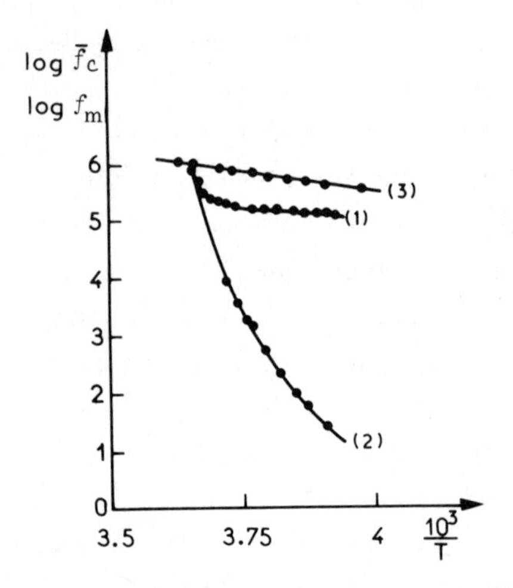


Figure 63. Specifications identical to those of Fig. 62. (1) Frozen system high-frequency dielectric relaxation, (2) frozen system low-frequency dielectric relaxation, and (3) MWS-type dielectric relaxation exhibited by the emulsion obtained from thawing. (From Ref. 96, with permission from Les Editions Physiques, Orsay, France.)

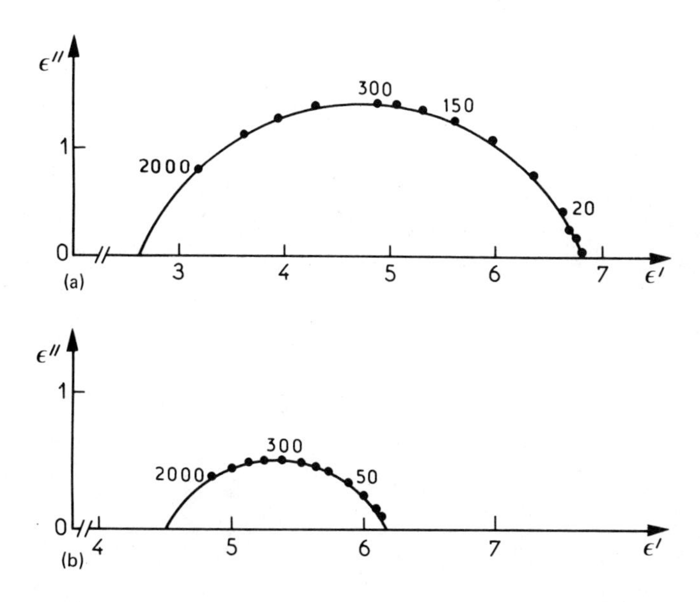


**Figure 64.** Comparative Cole-Cole diagrams of a dispersion of (a) ice microcrystals and (b) the corresponding thawed emulsion at temperatures (a) slightly lower and (b) slightly higher than $T_0$, the solidified disperse-phase melting point. $P_w = 0.30$. Surfactant concentration in the oil phase: 25% (w/w). The frequencies are expressed in kilohertz. (From Ref. 96, with permission from Les Editions Physiques, Orsay, France.)

after thawing, in such a way that $\bar{f}_c$ is equal to $f_m$ at $T_0$. The Cole-Cole diagrams reported in Fig. 64 show, for a dispersion of ice microcrystals and the corresponding thawed emulsion, the coincidence of $\bar{f}_c$ with $f_m$ at temperatures very close to each other although located on both sides of $T_0$, and, by contrast, the striking thaw-induced changes in the system limiting permittivities and dielectric increment. The diagrams of log $\bar{f}_c$ versus $T^{-1}$ reported in Fig. 65 show that the activation energy of the MWS dielectric relaxation exhibited by a thawed W/O emulsion is larger than that of the MWS dielectric relaxation displayed by the original freshly made emulsion. When a thawed emulsion was maintained during 10 days at a few kelvin above $T_0$, experiments proved that its log $\bar{f}_c$ versus $T^{-1}$ diagram shifted progressively towards the original emulsion diagram and eventually became identical to it.

These data show that W/O emulsions obtained by thawing dispersions of micrometer-size particles of ice display hysteresis effects in their dielectric behavior, namely as concerns their MWS dielectric relaxation activation energy $\bar{u}$, and the preexponential coefficient $\bar{A}$ in the Arrhenius-type equation ruling the dependence of $\bar{f}_c$ upon T. Since, according to the theoretical models, $\bar{u}$ is equal to the conductivity activation energy of the disperse phase, $\bar{A}$ being connected with the pre-

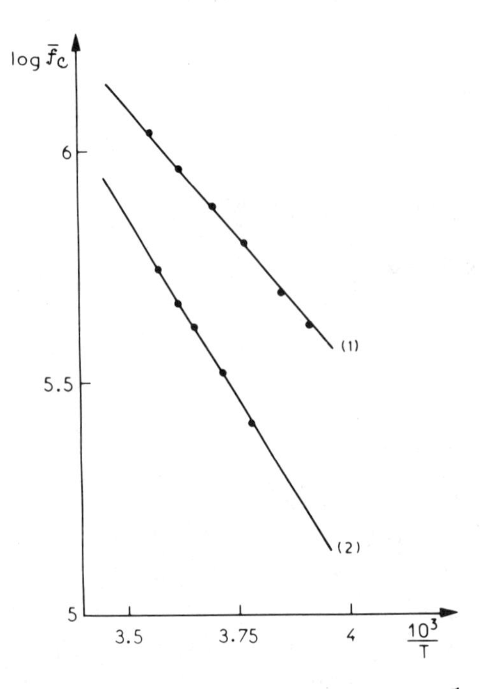


Figure 65. Plots of log $\bar{f}_c$ versus $T^{-1}$ obtained for a W/O emulsion, (1) prior to disperse-phase freezing; (2) just after thawing; and 10 days or so after holding at $T_c$ = 279 K (1). $P_w$ = 0.22. $\bar{f}_c$ is expressed in hertz and T in kelvin. Surfactant proportion in the oil phase: 25% (w/w). (1) $\bar{u}$ = 0.22 eV, and (2) $\bar{u}$ = 0.31 eV. (From Ref. 96, with permission from Les Editions de Physique, Orsay, France.)

exponential factor in the conductivity Arrhenius law, the existence of hysteresis phenomena could suggest that the molten disperse phase "remembers" its former solidified state, this memory effect diminishing with time. Suggestions as to the mechanisms underlying these phenomena and comments about their implications in the fields of solid and liquid state physics and of surface science can be found in the original papers by Lagourette and his coauthors [93,96].

## VII. Permittivity and Conductivity Studies of Microemulsion Systems

So-called microemulsions are macroscopically monophasic, fluid, transparent and isotropic media that can be obtained by mixing, within definite proportion ranges, two normally nonmiscible liquids in presence of a suitable surface-active agent or combination of surface-active agents. Most of the microemeulsions commonly studied are made up of water and hydrocarbon (either aliphatic or aromatic), and incorporate as surface active agents combinations of an ionic surfactant (either anionic or cationic), with a so-called cosurfactant that is generally an alkanol. Quite similar systems can be formed as well without addition of a cosurfactant, by using double-chained ionic surfactants such as Aerosol OT (AOT) or nonionic surfactants such as polyoxyethylene alkylphenols, alcohols, or esters of fatty acids. In that case, the microemulsion realm of existence and stability is highly temperature sensitive. It has been also reported that microemulsion-type media can be obtained from ternary systems involving water, hexane or toluene, and 2-propanol, with no surfactant added. In any case, the microemulsions can be either of the water-in-hydrocarbon type (W/O) or of the hydrocarbon-in-water type (O/W), depending upon the constituent chemical nature and proportions, and upon the thermodynamic conditions.

During the past decade or so, a great amount of studies have been devoted to microemulsions and related systems because of their numerous potential applications as versatile fluid agents in many technical processes, more especially in enhanced oil recovery. Moreover, microemulsions belonging to the class of short-range organized fluids, they deserve being studied per se by condensed matter physicochemists. Since the subject is treated elsewhere in the present volume (Chap. 4), microemulsion formation, phase behavior, stability, and structural properties will not be analyzed and discussed in detail in the following developments, but for notions essential to the interpretation of results concerning their electrical properties. In addition, comprehensive articles about microemulsions can be found in recent books and symposium proceedings [500-508].

Schulman and co-workers investigated through various techniques the conditions of formation, the phase behavior and the structural properties of monophasic fluid transparent isotropic media involving water, hydrocarbons, and soap/alcohol combinations [509-518]. They arrive at the conclusion that these media consisted of stable, almost monodisperse populations of composite microglobules suspended in a continuous phase, and, consequently, coined the term "microemulsion." In agreement with the Schulman school concepts, Fig. 66a and b shows tenta-

Figure 66. Schematic representation of swollen composite globules in microemulsions: (a) water in oil and (b) oil in water.

tive descriptions of W/O and O/W composite globules using a surfactant/cosurfactant combination as the surface-active agent. For oil-external microemulsions (W/O), the composite globules are made up of an inner spherical aqueous core surrounded by a concentric shell of mixed surfactant and alcohol and are suspended in an organic phase (hydrocarbon plus dissolved alcohol). The reverse situation prevails in the case of water-external microemulsions (O/W), the composite globules, suspended in an aqueous phase, consisting then of an inner spherical organic core surrounded by a surface active agent shell. In both situations, the shell (or film, or membrane, or interphase), delimiting the globules is represented as a spherical monolayer of regularly interspersed surfactant and cosurfactant molecules whose polar heads are oriented toward the aqueous phase.

The approach followed by Schulman and his co-workers was based essentially on surface physicochemistry considerations that emphasized the interfacial aspects of microemulsion formation, stability, and structure. In contrast, other groups of authors stressed the micellar nature of the monophasic fluid transparent and isotropic media made up of water, hydrocarbons, and surface-active agents. Their results were reported in terms of solubilization and micellization rather than in terms of microemulsification, and Schulman's microemulsions were considered, not as "extraordinary" emulsions, but as one among the different structures that can be obtained from the mutual solubilization of water and hydrocarbon in presence of amphiphilic agents [519-551]. This point of view led to describe microemulsion composite globules as water-swollen (W/O) or oil-swollen (O/W) spherical micelles, and some authors even argued that the term *microemulsion* was a misnomer that should not be used to label thermodynamically stable solutions of swollen spherical micelles [538]. In that respect, on the basis of phase diagram studies and structural data gained from various techniques, Friberg and co-workers [546-551], claimed that W/O microemulsions incorporating soap/alcohol combinations as surface-active agents are not different in nature from the inverse micellar media that exist in the corresponding water/soap/alcohol ternary systems. The structure of the medium in both situations is considered to depend basically upon the so-called structure-forming components, i.e., water, soap, and alcohol, the addition of a hydrocarbon affecting only the extension of the W/O microemulsion domain. According to the same authors, the connection of O/W microemulsion-type systems with direct ternary micellar solutions appears to be less direct [547]. Moreover, recent studies [552] tend to indicate that O/W microemulsions could be only kinetically stable, in contrast with W/O microemulsions whose thermodynamic stability can be considered as a well-established fact. It is also worth mentioning that, in a recent paper [553], Sagitani and Friberg reported, for systems incorporating nonionic surfactants, that the W/O microemulsion region is not a direct continuation of the inverse micellar area at zero content of hydrocarbon, as in systems incorporating soap/alcohol combinations.

Recent structural data obtained through sophisticated techniques [554-562], have confirmed, in the W/O case, that microemulsion systems can be depicted satisfactorily with the composite globule model to which quite significant refinements have been brought. By using ultracentrifugation and small angle neutron scattering (SANS) to investigate the properties of water/sodium dodecylsulfate/1-pentanol/cyclohexane W/O microemulsions, Dvolaitzky et al. [554,555] showed that it is possible to define for composite W/O globules three radii, $r_h$, $r_w$, and $r_c$. $r_h$, evaluated from ultracentrifugation data, is the hydrodynamic radius of the globule considered as an hydrodynamical object. It corresponds to the volume displaced with the droplet during its motion. The hydrodynamic thickness of the interphase, i.e., the difference between $r_w$, the aqueous core radius, and $r_h$, the hydrodynamic radius, can be determined from the absolute value of the sedimentation coefficient. The hydrodynamic thickness $r_h - r_w$ was found to vary from 18 to 27 Å, which is to be compared with 21 Å, the length of the extended molecule of sodium dodecylsulfate. Having thus defined the hydrodynamic thickness, the geometrical distribution of the constituents in it can be investigated through small angle neutron scattering, which is a technique sensitive to chemical composition when associated with the variable contrast method using deuterated molecules [556]. Through their SANS experiments, Dvolaitzky et al. found values of $r_w$, the water core radius, quite consistent with those obtained from their ultracentrifugation data and were able to define $r_c$, the so-called *chemical radius*, that limits the part of the interphase not penetrated by the continuous organic phase. $r_c$, whose value lies between those of $r_w$ and $r_h$, is a physically important parameter since it defines the hard-sphere volume of the composite globule. Dvolaitzky et al. reported for the "chemical" thickness values around 10 Å which are compatible with the 1-pentanol molecular length (7Å) and the sodium dodecyl sulfate one (21Å). Similar conclusions were arrived at by other authors [562-564], in particular by Cebula et al. [564], on the system water/potassium oleate/1-hexanol/toluene whose properties were investigated by means of SANS and photon correlation spectroscopy (PCS) techniques. In addition, several studies [554, 565-572] have clearly shown that the second virial coefficient of the osmotic compressibility in W/O microemulsions incorporating surfactant/alcohol combinations is negative, which indicates the existence of attractive interactions between disperse globules. This result is consistent with the observations made as to the high stability of W/O microemulsions over wide composition ranges and comes in strong support to the suggestions made by different groups of authors that micelle clustering processes could take place in W/O microemulsions [560,573,574,575,576]. As concerns O/W microemulsions, osmotic pressure and light beating spectroscopy measurements have led to positive values of the second virial coefficient linked with the electrical charge of the direct micelles, which indicates the existence of repulsive interactions in such systems [566,577].

To sum up, it appears, in the light of the data briefly reported and commented upon above, that the composite globule model originally proposed by Schulman and his co-workers can be retained as a basis for the structural description of microemulsions. In that respect, controversies between the adherents of the interfacial approach and the adherents of the micellar approach can be considered as being rather on the semantic side. In addition, recent theoretical works [578-586], have shown that microemulsion formation and stability could be accounted for without resorting to the "transient negative interfacial tension" concept developed by Prince [587-589]. For quaternary water/ionic surfactant/cosurfactant/hydrocarbon systems, microemulsions can be depicted as suspensions of composite microglobules (about 200 Å in diameter), consisting of an inner spherical core surrounded by a concentric film of mixed surfactant and cosurfactant molecules whose polar heads are oriented toward the aqueous phase. In the W/O case, the inner spherical core is made up essentially of water and the suspending medium is an organic phase (hydrocarbon component in which part of the cosurfactant and a small amount of water are dissolved [565]). The composite globules are characterized by three parameters $r_w$, $r_c$, and $r_h$. $r_w$ is the radius of the inner aqueous core, comprising the surfactant and cosurfactant polar heads. $r_c$, the chemical radius, defines the *hard-sphere* part of the globule, that is to say the part of the globule which is not penetrated by molecules of the suspending organic phase. $r_c$ is greater than $r_w$ by 10 Å or so, which is to be compared with the molecular length of common cosurfactants. $r_h$ is the hydrodynamic radius whose definition and value depend upon the investigation method followed. The hydrodynamic thickness $r_h - r_w$ differs from the surfactant molecular length by ±5 Å or so, which indicates that the outer regions of the surfactant molecule chains retain some flexibility. Thus, the peripheral part of the composite globules appear to be compressible and to present no sharp boundary with the suspending organic phase. As revealed by negative values of the second virial coefficient of the osmotic pressure, the composite globules are submitted to attractive interactions whose potential can be approximated by a hard-sphere term corrected with a small attraction term. As individual entities, the globules form fairly monodisperse populations, but they can undergo aggregation and clustering processes, owing to the existence of attractive interactions. The composite globule model, as characterized by the parameters $r_w$, $r_c$, and $r_h$, represents a time-average structure of microemulsion disperse phase. The actual structure is a dynamic one and continuous exchanges take place between the surface-active agent molecules engaged in the interphasic film and dissolved in both the aqueous and organic phases, these exchanges being particularly fast as concerns cosurfactant molecules [590-593]. It is most likely that, but for some variations, the dynamic composite globule model can be retained as well to depict other W/O type microemulsion systems formed with either nonionic surfactants [594-599] or well-balanced ionic surfactants such as Aerosol OT [600-616]. In spite of the scarcity of structural data concerning microemulsions of the O/W type, it can be reasonably assumed that the composite globule model is still applicable [617-619], but these systems appear to be fairly polydisperse and to exhibit internal repulsive interactions between their disperse oil-swollen globules.

On the basis of both experimental results and theoretical considerations, several groups of authors have stressed that the model of monodisperse populations of swollen composite globules cannot be taken as a universal structural description that would be valid all over the realm of existence of monophasic, fluid, transparent and isotropic media made up of water, hydrocarbons, and surface-active agents. On the theoretical side, an analysis carried out by Bothorel et al. [620], has clearly shown that, inside the theoretical pseudoternary phase diagram of four-component systems, composition regions are forbidden to the presence of monodisperse populations of swollen globules. Owing to the existence of upper critical volume fractions corresponding to close packings in dispersions of identical spheres, it can be taken for granted that the composite globule monodisperse model cannot be retained when the volume fraction $\Phi$ of the disperse phase exceeds either $\Phi_1^c = 0.637$, in the random close-packing case [621], or $\Phi_2^c = 0.741$, in the cubic close-packing case [429]. This allows us to define geometrically, inside the theoretical phase diagram, two excluding domains for either direct or inverse swollen spherical micelles, and, for both type of micelles, an excluding domain that results from the intersection of the former two ones. In addition, the existence of a minimum value of the mean area per surfactant molecule polar head, for instance $20 Å^2$ for sodium dodecyl sulfate in planar layers [622] implies, for inverse swollen spherical micelles, an additional excluding zone located in the water-poor region of the phase diagram. All these conclusions are summarized by the schematic phase diagram of Fig. 67. It can be remarked that the phase inversion, i.e., the transition from inverse spherical micelles to direct ones (or vice versa), can occur theoretically either through the buildup of nonspherical structures (path a) or in a diffuse way that implies the coexistence of inverse and direct micelles (path b). These theoretical considerations are consistent with conclusions derived by different authors from studies performed on microemulsion-type media by means of various techniques. In the case of systems incorporating ionic surfactants, several experimental data converge to indicate that, at low water contents, the monophasic fluid transparent isotropic media could be suspensions of premicellar entities made up of hydrated surfactant aggregates [547-551,573,574,603,606-614,623-627] or even, for compositions rich in surface-active agents, lamellar structures [592]. Upon increasing the water content above a threshold value corresponding to a water/surfactant molecular ratio of 8 to 12 or so, these premicellar entities yield

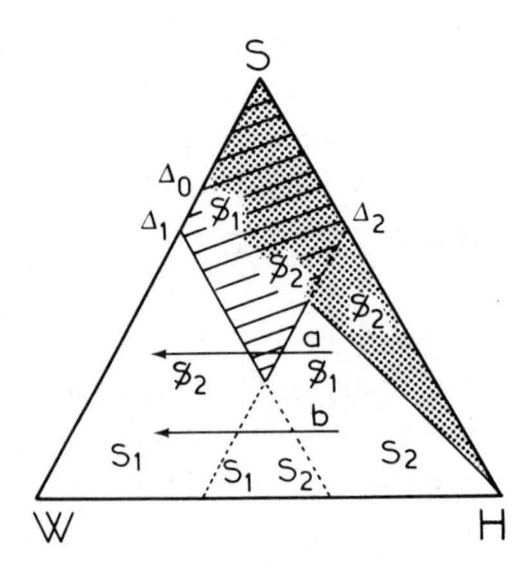


Figure 67. Schematic pseudoternary phase diagram of four component systems. $\Delta_1$: boundary of $\$_1$, the excluding domain for direct spherical micelles, associated with a close-packing volume fraction. $\Delta_2$: boundary of $\$_2$, the excluding domain for inverse spherical micelles, associated with a close-packing volume fraction. The hatched area labeled $\$_1, \$_2$ represents the excluding domain for both types of micelles. $\Delta_0$: boundary of the additional excluding domain for inverse micelles, associated with a minimum value of the surfactant molecule polar head area (dotted area). (From Ref. 620, Gauthier-Villars, Paris.)

globular objects with a central core of free water, i.e., water-swollen spherical micelles forming proper W/O microemulsions. For several systems, results gained through different methods suggest that, at intermediate water contents, the media are in a hybrid state that cannot be characterized as being typically either of the W/O type or of the O/W type [628-632]. Such situations, which could be characteristic for a diffuse phase inversion mechanism [633-636] have been tentatively described by means of models postulating the existence of interspersed aqueous and organic microdomains [637-640]. Besides, it has been shown [574,635,641-644] that slight formulation changes induce drastic alterations in some important physicochemical properties of microemulsion-type systems. For instance, using a combination of physical methods such as conductometry, viscometry, high-resolution NMR, and spin-spin relaxation time measurements, Shah et al. [641] have shown that two monophasic, fluid, transparent, and isotropic systems with identical compositions, except that one incorporates 1-pentanol as the cosurfactant and the other 1-hexanol, exhibit striking discrepancies in their behavior, which suggests that they are structurally quite different media. To account for these experimental observations, Shah et al. proposed to establish a distinction between true microemulsions, to which the composite globule model is applicable, and so-called cosolubilized systems which are assimilated to quarternary molecular solutions. Further studies [642-643] have led the same group of authors to suggest an alkyl chain-length compatibility effect, which could be quite instrumental as a guideline to explain the correlated changes observed in water solubilization capacity and electrical properties upon varying the nature of hydrocarbon, surfactant, and alcohol in microemulsion-type systems. Quite consistent results were obtained by Clausse et al. [574,635,644], who followed a somewhat different approach. These authors studied microemulsion electrical properties in connection with phase diagram features that, other things equal, are strongly dependent upon alcohol molecular structure. Such a strong influence of the alcohol nature has been reported as well in recent papers by Cazabat and Langevin [571] and Lindman et al. [632]. According to the latter group of authors, monophasic fluid transparent isotropic media incorporating short-chained alkanols are rather on the molecular solution side while those incorporating a long-chained alkanol 1-decanol or double-chained surfactant (AOT) appear to be proper microemulsions. To add a final touch, it is worth mentioning that Gulari and Chu [618] reported, on the basis of x-ray and light-scattering studies, that W/O microemulsion type systems with fairly large globule size immediately after their preparation (200 Å or so) can change into a molecular solution after some time.

In the light of the considerations developed above, it is clear that one should be cautious when attempting to characterize monophasic fluid transparent and isotropic media made up of water, hydrocarbons and surface active agents. Depending upon composition factors, these media can be considered either as suspensions of hydrated surfactant aggregates, or proper microemulsions for which the composite globule model is relevant, or microdomain dynamic structures, or even quaternary molecular solutions. In that respect, the nomenclature proposed by Winsor [519] is useful since its neutrality toward medium structure can help in avoiding confusion that would not be of only semantic origin and could lead to serious misinterpretations. Winsor's classification is depicted in Fig. 68. Winsor IV systems contain no free aqueous or organic layers and are macroscopically mono-

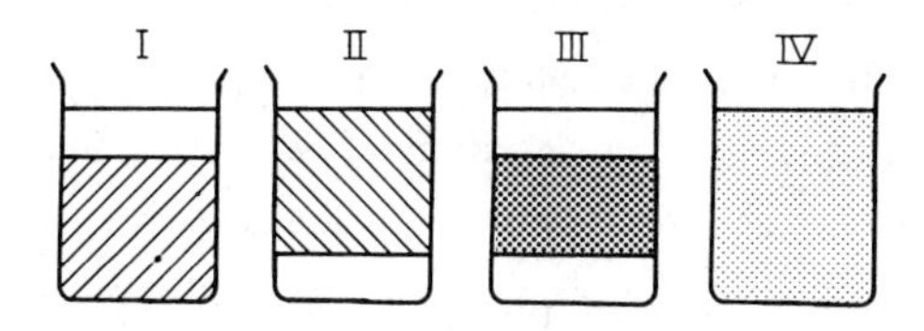


**Figure 68.** Schematic representation of Winsor's classification for fluid nonordered media made up of water, hydrocarbon, and surface-active agents.

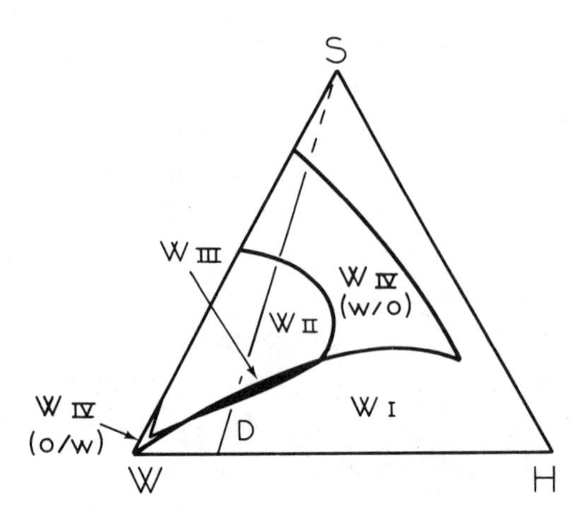


Figure 69. Pseudoternary mass phase diagram of the quaternary system NaCl brine/sodium octylbenzene sulfonate/1-pentanol/dodecane. T = 297 K; NaCl weight concentration in brine: 2%; surfactant/alcohol mass ratio: k = 1/2. W, 100% brine; H, 100% hydrocarbon; S, 100% surfactant/alcohol combination. I, II, III, and IV label the composition regions corresponding to the different Winsor types. Upon varying system composition along a line such as D, the following transition sequence is observable: WI → WIII → WII → WIV. (From Ref. 647, Gauthier-Villars, Paris.)

phasic fluid media made up of water, hydrocarbon, and surface-active agents. Microemulsion-type media are Winsor IV systems. In Winsor I systems, an organic phase containing only small proportions of water and surface-active agents is in equilibrium with a Winsor IV medium. The symmetrical situation prevails in Winsor II systems that are formed of an aqueous phase in equilibrium with a Winsor IV phase. In Winsor III systems, a type IV medium coexists with both an organic phase and an aqueous one. In a recent article, Chan and Shah [645] have recalled the main factors influencing the transition between the different types of systems. As illustrated by Fig. 69, the realms of existence of the different types of Winsor systems can be delimited within a three-component phase diagram that is either a proper ternary one for systems incorporating a unique surface-active agent or a pseudoternary one for systems involving a surfactant/cosurfactant combination characterized by k, the surfactant-cosurfactant mass ratio. Friberg [547] stressed that the three-component phase diagram is a practical tool for understanding the association phenomena of importance to microemulsion-type media. As proved by the detailed analyses carried out by Bothorel et al. [620,646-648], this representation is particularly convenient in the study of water or brine/ionic surfactant/co-surfactant/hydrocarbon systems which are not highly temperature sensitive. For

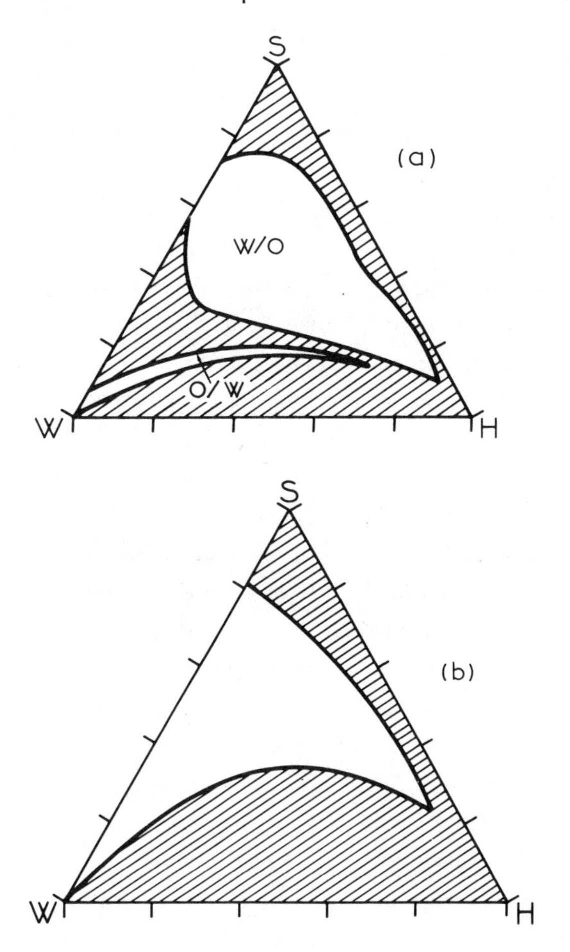


**Figure 70.** Winsor IV domains (blank areas), in pseudoternary mass phase diagrams of quaternary systems. (a) Water/potassium oleate/1-hexanol/*n*-hexadecane. Surfactant-alcohol mass ratio: k = 3/5. T = 295 K. (b) Water/sodium dodecyl sulfate/2-methyl-2-butanol/benzene. Surfactant-alcohol mass ratio: k = 1/2. T = 295 K. (From Ref. 635.)

systems incorporating nonionic surfactants, the phase behavior depends strongly upon temperature, and classical temperature-concentration diagrams, such as those used by Shinoda et al. [539-543] may be preferable.

Depending upon composition factors, such as the molecular structure of the nonaqueous components or the ionic surfactant-alcohol ratio, the realm of existence of Winsor IV-type media in a pseudoternary phase diagram may exhibit various geometrical features. Figure 70a and b depicts two typical situations that are both quite common. In the first case (Fig. 70a), the Winsor IV domain is split into two disjointed subdomains that are separated by a composition region over which viscous, turbid, and sometimes birefringent structures are encountered. The more extended subdomain corresponds to W/O-type media while the smaller one, springing from the W vertex of the phase diagram, corresponds to O/W-type media. The W/O ↔ O/W transition (phase inversion) is characterized by macroscopically detectable striking alterations of mechanical, electrical and optical properties of the medium [573,574,649-652]. According to Shah and co-workers [649-652], the long-range organized structures existing over the intermediary composition region are directly connected to the phase inversion mechanism. In the second case (Fig. 70b), the W/O and O/W areas merge into each other so as to form a unique domain that has the shape of a curvilinear triangle adjacent to a large portion of the SW side of the phase diagram. In such a situation, the W/O ↔ O/W transition occurs in a progressive diffuse way, the medium remaining fluid, transparent and isotropic as the composition varies. It is worth recalling here that these experimental observations about phase inversion phenomena in Winsor IV systems are consistent with conclusions derived from the geometrical model of microemulsions proposed by Bothorel et al. [620]. Hybrid situations can also be observed in which the Winsor IV domain consists of a unique area presenting a more or less deep indentation [635]. In such cases, depending upon the composition path followed, the phase inversion may occur either through the buildup of long-range organized structures or in a progressive diffuse way. It should be noted that, in any situation, the upper branch of the Winsor IV domain boundary is distinct from the SH side of the phase diagram, which proves that a minimal amount of water is required to obtain stable monophasic media.

The correct investigation of the electrical properties of Winsor IV media needs to be performed according to well-defined experimental procedures, so as to put clearly into evidence the influence of composition factors. Figures 71 and 72 show

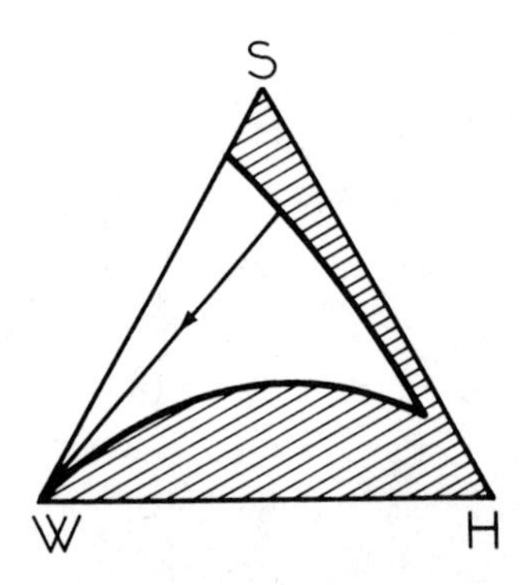


**Figure 71.** Experimental path defined, inside the Winsor IV domain, by a constant value of r, the mass ratio of hydrocarbon to surface-active agent (r path).

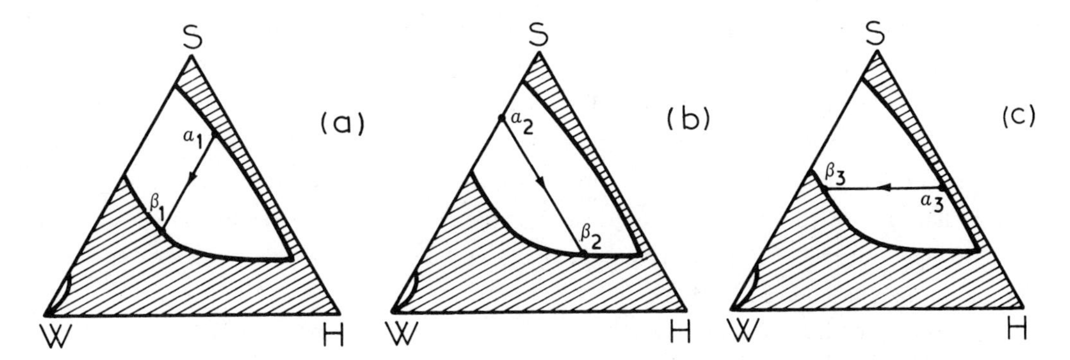


Figure 72. Experimental paths inside the Winsor IV domain. (a) $p_h$, the hydrocarbon mass fraction, is kept constant ($p_h$ path). (b) $p_w$, the water mass fraction, is kept constant ($p_w$ path). (c) $p_s$, the surface active agent mass fraction, is kept constant ($p_s$ path).

examples of typical experimental paths within three-component phase diagrams. The experimental path displayed in Fig. 71 is defined by a constant value of r, the mass ratio of hydrocarbon to surface-active agent (surfactant/cosurfactant combination, in the case of quaternary systems). The three diagrams of Fig. 72 show other possible experimental paths corresponding to constant values of either $p_h$, the hydrocarbon mass fraction (Fig. 72a), or $p_w$, the water mass fraction (Fig. 72b), or $p_s$, the surface-active agent mass fraction (Fig. 72c). By varying the values of the parameters r, $p_h$, $p_w$, and $p_s$, and combining the data thus gained, it is possible to draw, within the Winsor IV domain, isoconductivity and isopermittivity lines that form maps of the electrical behavior of Winsor IV media [635]. It has been shown [573,574] that the r-path and $p_h$-path processes yield quite identical information, but it clear from Fig. 71 that the r-path process is more convenient when studying systems with a unique Winsor IV domain since it allows complete composition variations down to the W apex of the pseudoternary phase diagram. Different experimental procedures can be followed as well. For instance, Lagües and co-workers [557,558,561] measured the electrical conductivity of microemulsion systems characterized by a constant value of the water-ionic surfactant ratio, obtained by titrating coarse emulsions to transparency with added alcohol.

Electrical methods have been applied for a long time to the investigation of the behavior and structural properties of micellar systems. Papers in this field have been published since the beginning of the century, as proved by the selection of articles corresponding to Refs. 653 to 695. Classical conductivity and permittivity measurements were aimed at determining critical micelle concentrations, as influenced by several parameters and factors such as the solute and solvent chemical structure, the temperature, the concentration, and chemical structure of various inorganic or organic additives. Conductometry and dielectrometry have been used as well in conjunction with other techniques to gain information about micelle structure, ionization, local environment, and stability. Some of the authors were not aware that the systems they studied were in fact microemulsion-type media, because of the nature of the additives (alcohols and hydrocarbons) incorporated to the surfactant aqueous solutions.

In a way similar to that followed in the case of macroemulsions, early electrical conductance measurements were performed by Schulman and co-workers [510,511] in view of determining the aqueous or organic nature of the external continuous phase of microemulsion systems. The basic crude assumption behind these experiments was that O/W-type microemulsions should exhibit a fairly high conductivity, in contrast with the W/O type, which should be non- or poorly conducting. Schulman and McRoberts [510] reported observations about the influence of the cosurfactant nature, of the surfactant and cosurfactant concentrations and of the water-hydrocarbon ratio, upon the conductive behavior of several microemulsion systems. In mixtures defined by a water-benzene volume ratio equal to 1 and sodium oleate concentrations ranging from 0 to $2 \times 10^{-3}$ moles $cm^{-3}$ of water, the titration to transparency yielded conducting systems when saturated primary alkanols with small carbon numbers were used ($C_2$ to $C_5$) and nonconducting systems with 1-hexanol, 1-hexanol-2-ethyl and 1-decanol. As concerns cyclic cosurfactants, either aliphatic or aromatic, conducting systems were obtained with cyclohexanol and phenol and nonconducting ones with cyclohexanol-*p*-methyl and *m*-cresol. When, under the same preparative conditions, the more hydrophobic Nujol was used instead of benzene, no transparent media could be formed with shorter chain alkanols ($C_2$ to $C_4$). Conducting systems were obtained with 1-pentanol, 1-hexanol and the cyclic cosurfactants used, and nonconducting ones with 2-ethyl-1-hexanol and 1-decanol. The authors noted that important changes in conductance were induced by varying the water-hydrocarbon ratio. In a following paper, Schulman and Riley [511] reported results gained from a detailed study performed essentially on the systems water/potassium oleate/*p*-methyl cyclohexanol/benzene, which were of the W/O type, and water/potassium oleate/*p*-methyl cyclohexanol/Nujol, which were of the O/W type. For certain compositions, W/O-type systems appeared to be conducting, and the authors labeled them as "anomalous." The existence of such "anomalous" systems was ascribed to the formation of elongated micelles, which was thought to be consistent with results gained from x-ray diffraction experiments. In spite of their historical interest, the conductivity experiments carried out by Schulman and co-workers are lacking in generality since they do not establish any actual connection between conductive behavior and internal structure of microemulsion systems. Moreover, the recognition that so-called anomalous W/O-type media can be fairly conductive weakens the guiding line that conductometry used alone is a reliable method to determine the nature of the continuous phase in microemulsions. In that respect, several authors [696, 697] stressed that isolated conductivity measurements yield somewhat ambiguous results as concerns the phase inversion phenomenon in microemulsions.

More interesting experiments were performed by Winsor et al. [519-530] on various systems involving water, hydrocarbons, and amphiphilic compounds, Such systems, e.g., water/undecane-3 sodium sulfate/1-hexanol or 1-octanol/*n* hexane or cyclohexane, are quite similar in composition to those from which Schulman et al. obtained microemulsions. Winsor and co-workers extensively studied solubilization phenomena as influenced by a great number of factors and focused their attention on structural changes induced by composition and temperature variations. These phase behavior studies were paralleled with an analysis of data gained by means of several methods, i.e., conductometry, viscometry, refractometry, polarized light microscopy, x-ray diffraction, and NMR. In a first paper devoted to conductivity studies [520], Winsor reported results concerning the transition from type IV W/O media to O/W ones, as induced by alcohol addition to water/ionic surfactant/hydrocarbon mixtures. In certain cases, the conductivity versus alcohol concentration plots were smooth curves, the type IV systems undergoing a continuous O/W to W/O transformation. This observation led Winsor to suggest the possibility of an equilibrium between $S_1$ micelles (O/W) and $S_2$ micelles (W/O) over certain composition ranges. In other cases, the O/W to W/O transition occurred through a gel stage and the conductivity plots exhibited anomalous features over the composition region corresponding to gel formation. Correlatively, anomalies were observed on the refractive index plots. Winsor observed that on both the conductivity and refractive index curves, the branches corresponding to type IV O/W and W/O media could always be connected by an imaginary smooth portion of curve, so as to form a continuous plot. He noted also that some W/O media close in composition to the gel region were fairly conductive, which could indicate that they did not consist exclusively of $S_2$ micelles. It is noteworthy that, other things equal, the occurrence of a gel stage depended upon the chemical structure of the alcohol added. For instance, with "hydrocarbon B" (a petroleum fraction which had been exhaustively extracted with oleum to remove aromatic compounds), the O/W to W/O transition occurred continuously, when either 1-hexanol or cyclohexanol was used, while addition of 1-octanol or 1-dodecanol induced the formation of a gel. Winsor studied also the influence of several factors such as the nature of the hydrocarbon, the addition of either water, ethylene glycol or hydrocarbon to an $S_2$-type medium, the incorporation of an inorganic salt ($Na_2SO_4$), and the temperature. All the results thus gained were mutually consistent and were interpreted

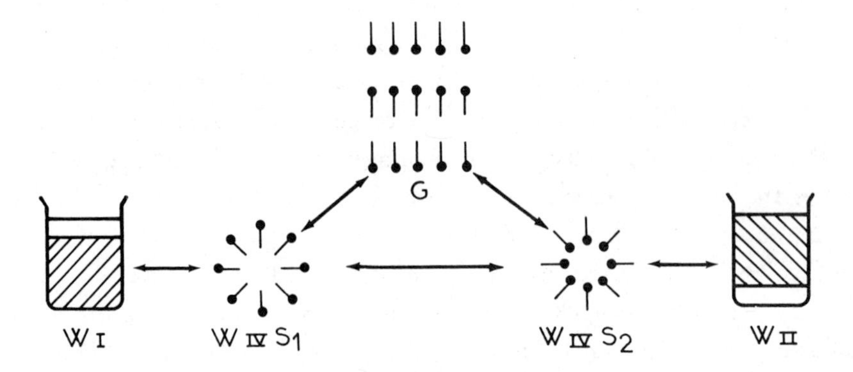


Figure 73. Schematic representation of Winsor's intermicellar equilibrium and associated phase changes.

by the authors in terms of micellar equilibrium displacement, as summarized by the following scheme (see also Fig. 73):

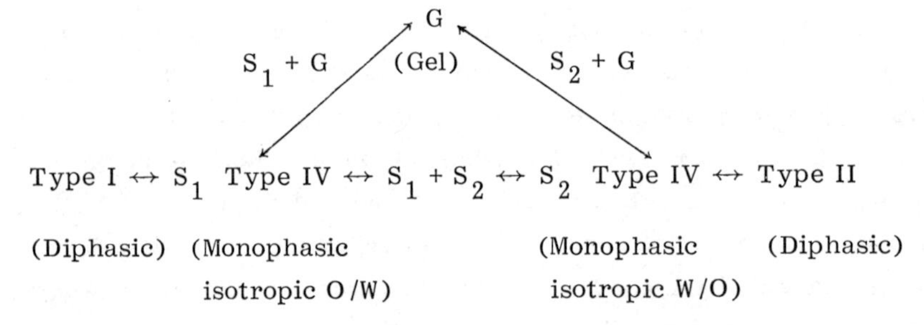


According to Winsor, the phase sequence is linked to the micellar equilibrium which is shifted to the right "by any change which would be expected to increase the ratio of the solvent attraction between the amphiphilic salt and organic liquid to the solvent attraction between the amphiphilic salt and aqueous liquid." Conversely, the equilibrium is shifted to the left upon diminution of the "solvent attraction ratio." As concerns the influence of inorganic salt incorporation, Winsor reported that the conductivity anomaly corresponding to the formation of a gel was progressively rubbed out as $Na_2SO_4$ concentration was increased in mixtures of hydrocarbon B, 1-octanol, and undecane-3 sodium sulfate aqueous solution. In the same systems, the onset of the conductivity anomaly associated with gel formation occurred at higher alcohol concentrations, as the temperature was increased. Winsor observed that some W/O media ($S_2$ systems) exhibited a marked temperature dependence of their conductive properties. Over a narrow temperature range, 2° or so, the conductivity increased with temperature by a ratio of 30 or so. Repeated temperature-cycling experiments proved that this effect was quite reversible, which led Winsor to suggest that it could be of use for thermistor design. In a subsequent paper [522], Winsor supported his views on intermicellar equilibrium with the aid of published x-ray measurements performed on solutions of amphiphiles. He agreed that, in some cases, $S_1$- and $S_2$-type media could be identified as Schulman's O/W and W/O microemulsions, since the Bragg spacings could be reasonably interpreted in terms of intermicellar distance in populations of spherical globules. On the other hand, he stressed that, in systems of intermediate conductivity, the weakness of the x-ray diffraction patterns was in favor of the existence of hybrid media. The results obtained with undecane-3 sodium sulfate as the surfactant were further generalized by Bromilow and Winsor [523], who studied conductivity variations associated with phase changes in various systems incorporating *n*-hexane or cyclohexane as the hydrocarbon, 1-octanol or cyclohexanol as the alcohol, and various surface active compounds used alone or in combinations, (Aerosol OT, cyclohexylammonium chloride, monoethanolamine laurate or oleate). The authors concluded that the intermicellar equilibrium scheme of Fig. 73 is "operative in all solutions of amphiphilic substances from the completely aqueous to those based completely on hydrocarbon." They pointed out that conductivity anomalies over the composition range corresponding to the existence of a gel indicate, along with other special properties, that the micellar organization is quite different from that of the isotropic $S_1$ and $S_2$ phases. In later papers [525,527,529], the attention was focused on the gel state in different systems and conductometry was used in conjunction with other techniques to investigate phase transitions and structural changes. Winsor [525] demonstrated an interesting electrical effect in aqueous solutions of amphiphilic salts. Upon application of a static electrical field ($3 \times 10^3$ V m$^{-1}$), the optical appearance of aqueous solutions of either Aerosol OT (20% w/w), tetradecane-7 sodium sulfate (20% w/w), or pentadecane-8 sodium sulfate (3.1% w/w), changed almost instantaneously from clear isotropic to turbid. A quaternary system made of water, undecane-3 sodium sulfate, 1-octanol, and hydrocarbon B exhibited the same effect. On removal of the applied electrical field, the reverse turbid to clear transformation was observed. These experiments came in support to former observations [524], concerning the liquid crystal character of certain systems incorporating amphiphilic compounds.

Though being at times rather confusing, the data reported by Winsor et al. are quite instrumental and valuable ones. They bear more generality than those obtained by Schulman and his co-workers in that they correlate conductive properties with structural changes and behavior in systems incorporating amphiphilic compounds. The main results to be kept in mind concern the existence of an anomalous conductive behavior upon formation of a gel, in contrast with the smooth conductivity variations that characterize continuous O/W $\leftrightarrow$ W/O transformations, and the striking influence of composition factors, such as the alcohol chemical structure. Other groups of authors have used as well conductometry and, more recently, dielectrometry to detect structural changes. Hyde and co-workers [698] followed phase equilibria in Teepol/water systems incorporating various amphiphiles through electrical conductivity, viscosity, refractive index, and surface tension measurements. The authors showed that these properties exhibited correlated anomalies as composition changes induced gel formation. The results obtained were quite similar to those reported by Winsor. In particular, it was shown that, other things equal, the chain length of the added straight alkanol was a key factor for gel formation. The "solubilization" of water in hydrophobic organic compounds by cationic surfactants was investigated by Palit and co-workers [531], who used conductometry and polarized light refractometry and microscopy to detect the formation of long-range organized media. The hydrocarbons used were chloroform, carbon tetrachloride, benzene, toluene, *p*- and *o*-xylene, tetrachloroethane, and chlorobenzene, the cationic surfactants being quaternary ammonium soaps such as cetyltrimethylammonium bromide. The results obtained were quite consistent with those of Winsor et al. Conductivity anomalies were detected when gel-type media were formed and very drastic temperature-dependence effects were put into evidence for conductivity near gel-transition points. Palit et al. showed that in a three-component phase diagram, Winsor's $S_1$ and $S_2$ regions could be connected in some cases by a narrow composition channel along which the $S_1 \leftrightarrow S_2$ transition occurred smoothly, the medium remaining clear and isotropic. Conductometry was used in conjunction with other techniques by Shah and co-workers [649-652] to detect phase changes in the system water/potassium oleate/1-hexanol/*n*-hexadecane whose monophasic domain is displayed in the pseudoternary phase diagram of Fig. 70a. The variations of conductivity, viscosity, turbidity, birefringence, and bandwidth at half-height and chemical shift of water, methylene and methyl protons were followed along experimental paths of the r type (cf. Fig. 71). Over the W/O subdomain, addition of water induced smooth nonmonotonous variations of the conductivity, followed by a steep increase as the boundary of the subdomain was approached. Over the composition region separating the two monophasic subdomains and corresponding to viscous turbid and birefringent media, the conductivity displayed anomalous variations. Finally, a smooth conductivity increase was found over the O/W subdomain. On the basis of these observations and of the parallel data gained through the other methods used, Shah et al. concluded that the phase changes induced by water addition could be depicted by the following sequence, which describes a phase inversion mechanism: isotropic suspensions of W/O globules $\leftrightarrow$ birefringent systems of W/O cylinders $\leftrightarrow$ birefringent lamellar structures of interspersed water, surface-active agents, and hydrocarbon $\leftrightarrow$ isotropic suspensions of O/W globules. The same descriptive scheme can be retained to account for dielectric data gained on the same quaternary system [699] or on similar ones [700-701]. Clausse et al. [699] carried out low-frequency complex permittivity determinations over the W/O subdomain of the water/potassium oleate/1-hexanol/*n*-hexadecane system. Dielectric relaxations of the Cole-Cole type were put into evidence in the 10-MHz region. The authors pointed out that these relaxations could not be depicted by means of the theoretical models that had proved suitable for macroemulsion dielectric behavior [81-83]. In particular, upon increasing water content along a r-type composition path, the low frequency limiting permittivity was found to exhibit, near the transparency-to-turbidity transition, a divergent behavior. This phenomenon, illustrated by the $\varepsilon_\ell$ versus $p_W$ plot reported in Fig. 74, could not be accounted for by means of Hanai's formula (351). Correlatively, a divergent behavior was found for conductivity, in agreement with the observations made by Shah et al. [649-651]. Above the water mass fraction critical value marking the transparency-to-turbidity transition, i.e., over the composition region corresponding to the existence of viscous turbid media, drifts in time were observed for the complex permittivity, which indicated instability. Quite similar results were reported later by Senatra and Guibilaro [700,701] on the system water/potassium oleate/1-hexanol/*n*-dodecane.

In their preliminary report on the dielectric properties of water-in-hexadecane microemulsions Clausse et al. [699] stressed that the peculiar conductive and dielectric behavior put into evidence ought to be investigated in detail, in connection with quantitative determinations of three-component phase diagrams. On

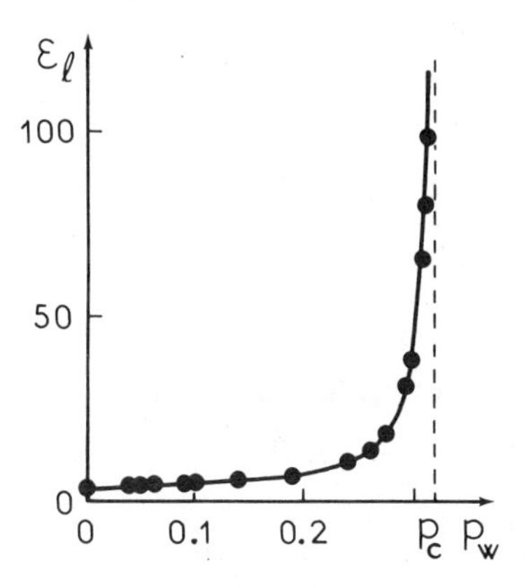


**Figure 74.** Variations, along a r-type composition path, of $\varepsilon_\ell$, the low-frequency limiting permittivity, versus $p_W$, the water mass fraction in water/potassium oleate/1-hexanol/*n*-hexadecane W/O microemulsions. T = 295 K. Surfactant-alcohol mass ratio: k = 3/5. Hydrocarbon/surface-active combination mass ratio: r = 3/2. (From Ref. 699.)

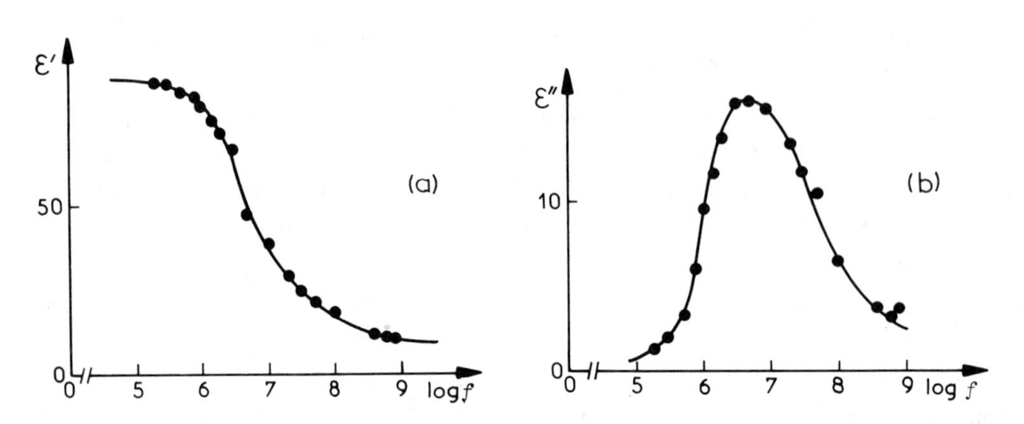


**Figure 75.** Dielectric dispersion and absorption plots of a water/potassium oleate/1-hexanol/*n*-hexadecane microemulsion, near the transparency-to-turbidity transition. T = 295 K; $p_c = 0.32$; $p_w = 0.308$. f is given in hertz. Other specifications as in Fig. 74. (From Ref. 699.)

the other hand, Bansal et al. [702] pointed out that along with other techniques, dielectrometry proved to be sensitive to structural changes occurring in surfactant solutions and microemulsions. It has been already mentioned that Cole-Cole-type dielectric relaxations were put into evidence in water-in-hexadecane microemulsions [699]. The features of these relaxations are illustrated by the dielectric dispersion and absorption plots reported in Fig. 75a and b. Clausse et al. reported that such curves could be fitted by the Cole-Cole formula, as modified by an ohmic term, Eq. (173). This was confirmed later by Peyrelasse et al. [703], who used curve-fitting procedures derived from the method proposed by Marquardt et al. [287-288]. The same group of authors made an attempt to study the influence of temperature upon the dielectric relaxation. For water contents close to the transparency-to-turbidity transition critical value, both the dielectric increment $\varepsilon_\ell - \varepsilon_h$ and critical frequency $f_c$ displayed nonmonotonous variations, in contrast with the phenomena observed for macroemulsions. These observations, joined to the fact that the variations of $\varepsilon_\ell - \varepsilon_h$ did not conform to the Hanai model, led Peyrelasse et al. to conclude that the dielectric behavior of W/O microemulsions was related to complicated mechanisms involving the surface admittance of the mixed surfactant and cosurfactant shell, the effect of temperature upon component partitioning in the systems, and a possible electrophoretic contribution to the total conductivity. Structural aspects of microemulsions were investigated by Bansal et al. [642] through titration studies and conductivity, permittivity, and electron spin resonance (ESR) measurements. The microemulsions studied were made of water, sodium stearate, and hexadecane, the cosurfactant being either 1-pentanol,

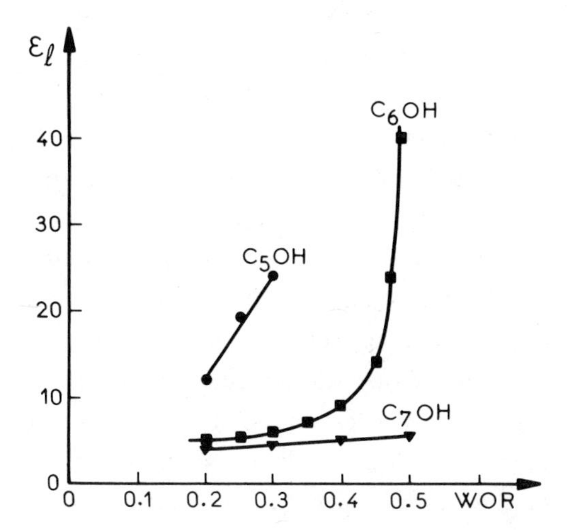


**Figure 76.** Variations of $\varepsilon_\ell$, the low-frequency limiting permittivity of microemulsions, with the water-oil volume ratio (WOR) along a r-type composition path. Influence of the alcohol chain length. T = 300 K. Measuring frequency: f = 2 MHz. System composition: sodium stearate (1g), alcohol (4 $cm^3$), *n*-hexadecane (10 $cm^3$). (From Ref. 642.)

1-hexanol, or 1-heptanol. The experiments were performed by varying the water-hydrocarbon ratio, along r-type composition paths. In the case of 1-heptanol microemulsions, no dielectric relaxation was observable in the frequency range 0.5 tc 150 MHz, even at higher water contents. For water-hydrocarbon ratios greater than 0.4 or so, 1-hexanol microemulsions exhibited dielectric relaxations located around 10 MHz. In 1-pentanol microemulsions, dielectric relaxations, located also around 10 MHz, were observed at moderate water-hydrocarbon ratios. It was found that the relaxation critical frequency decreased slightly as the water-hydrocarbon ratio increased. As concerns the low-frequency limiting permittivity $\varepsilon_\ell$, its variations with the water-hydrocarbon ratio depended strongly upon the nature of the alcohol used as the cosurfactant. This is illustrated by the plots reported in Fig. 76. For 1-heptanol, $\varepsilon_\ell$ increased very slightly with water content. In the case of 1-hexanol, $\varepsilon_\ell$ first increased slowly and then, beyond the water-hydrocarbon ratio of 0.4 or so, followed a sharply ascending branch as the transparency-to-turbidity transition point was approached. It is readily seen that this phenomenon is quite similar to that put into evidence by Clausse et al. [699] on water-hexadecane systems (see Fig. 74). In 1-pentanol microemulsions, $\varepsilon_\ell$ increased steeply and monotonically up to the water-hydrocarbon ratio of 0.3, beyond which no reliable dielectric measurements could be taken, owing to the system high conductivity. Correlatively, it was shown that the nature of the alcohol had a strong influence upon the conductive behavior. The 1-heptanol systems could be considered as being almost nonconducting for all water-hydrocarbon ratios. With 1-pentanol, the conductivity was fairly high and increased smoothly and steeply as the water content increased up to the critical value corresponding to the transparency-to-turbidity transition. Upon further increase of the water content beyond the transition point, the conductivity varied slowly and reached a plateau value. In contrast, the conductivity curve of 1-hexanol systems displayed a sharp increase as the transition point was approached and kinks over the turbid region. On the basis of these results and of data gained by ESR experiments, Bansal et al. proposed for the structure of microemulsion globules a model in which the straight alkanol chain length strikingly influences the ordering and the ionization of surfactant molecules at the water/hydrocarbon interface. Higher chain length alcohols (in the case considered, 1-heptanol and, to a lesser extent, 1-hexanol) would preferably partition into the organic suspending medium, which would lead to a relatively alcohol-poor interphasic film. For shorter chain length alcohols (1-pentanol), more cosurfactant molecules per surfactant molecule would be engaged in the interphasic film which would be more "fluid" and in which the disorder among the alkyl tails of the surfactant molecules and the ionization degree of the carboxyl groups would be greater, as compared to the 1-hexanol and 1-heptanol cases. Similar views were expressed by Cazabat and Langevin [571], who stressed that an important aspect of the role of alcohol is of a geometrical nature. In a subsequent paper [643], Bansal et al. showed that the influence of the alcohol chemical structure is correlated to those of both the surfactant and hydrocarbon. The authors suggested the existence of an alkyl chain-length compatibility effect that could be expressed by the formula

$$\ell_o + \ell_a \gtrless \ell_s \tag{405}$$

where $\ell_o$, $\ell_a$, and $\ell_s$ are the chain lengths of, respectively, the hydrocarbon, alcohol, and surfactant molecules. Their demonstration was based on titration studies and both permittivity and conductivity measurements in systems involving as surfactants either sodium myristate ($C_{14}$) or stearate ($C_{18}$), and various normal paraffinic alcohols ($C_4$ to $C_7$) and hydrocarbons ($C_7$ to $C_{16}$). For a given initial surfactant, alcohol, and hydrocarbon composition, it was observed that the amount of water solubilized reached a maximum when

$$\ell_o + \ell_a = \ell_s \tag{406}$$

Accordingly, it was shown that the conductive and dielectric behavior was very sensitive to the chemical structure of the compounds used, as illustrated by the conductivity plots reported in Fig. 77, from which it can be readily seen that the

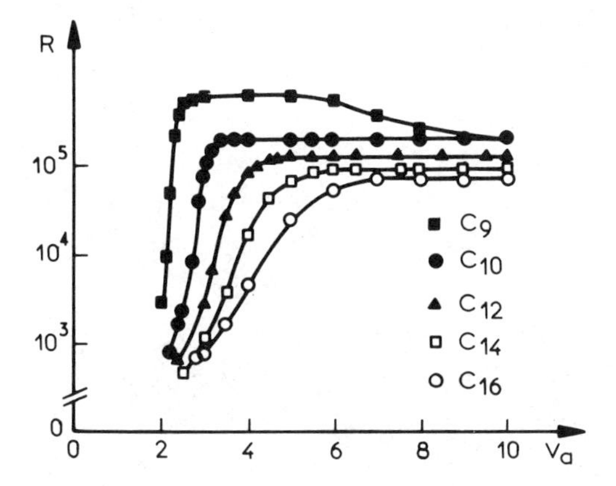


**Figure 77.** Variations of R, the microemulsion sample electrical resistance, with $v_a$, the volume of added alcohol. Influence of the hydrocarbon chain length. T = 298 K. Measuring frequency: f = 1 kHz. System composition: water (3 $cm^3$), sodium stearate (1 g), hydrocarbon (10 $cm^3$), 1-pentanol. R in ohms, $v_a$ in cubic centimeters. (From Ref. 643.)

paraffinic chain length greatly influences conductivity trend and value. Similar observations were made as to the dielectric relaxation whose characteristics, such as the dielectric increment, appeared to be strongly dependent, other things equal, upon the hydrocarbon chain length. For instance, the substitution of *n*-undecane for *n*-dodecane was found to lower $\varepsilon_\ell - \varepsilon_h$ by 50% or so, which is a striking effect as compared to the 0.01 difference between the dielectric constants of *n*-dodecane ($\varepsilon_s$ = 2.015) and *n*-undecane ($\varepsilon_2$ = 2.005). Bansal et al. explained their results in terms of hydrocarbon chain-length influence upon the partitioning of surfactant and alcohol that, in turn, modifies the ionization degree of surfactant molecules and the thickness of the electrical double layer at the periphery of the water core of the dispersed inverted globules, to which the O'Konski model was applied [14]. As mentioned by the authors themselves, no evident correlation was found between the water solubilization capacity and the electrical properties. It will be shown further that more systematic studies [635,644,704] allowed the correlation of electrical behavior and phase diagram features of microemulsion systems. In a recent article [705], Chou and Shah reported data gained from a dielectric relaxation study of W/O microemulsions made up of 1.5% NaCl-$H_2O$ or $D_2O$, isobutanol, and dodecane, the surfactant being the TRS 10-410 petroleum sulfonate from Witco Chemical Co. They found that their systems exhibited over the 0.5 to 100-MHz frequency range two consecutive dielectric relaxation processes. This behavior was particularly pronounced for $H_2O$ microemulsions and less evident for $D_2O$ ones. Though $H_2O$ and $D_2O$ have nearly the same dielectric constants (78.54 and 78.25, respectively, at

298 K), the dielectric properties of $H_2O$ or $D_2O$ microemulsions were found to be strikingly different, the dielectric increments and relaxation times for $H_2O$ systems being much higher than those of the $D_2O$ systems at identical compositions. The authors ascribed the existence of the dielectric relaxations to migration polarization phenomena and made an attempt to fit the experimental data with appropriate models. The simple Maxwell-Wagner-Sillars model was promptly recognized as being irrelevant, and the treatment proposed by Schwarz [458] which is based on the concept of bound ions in the double layer was retained. It was thus possible to establish correlations of the higher relaxation time with the inverse micelle mean radius, as determined from combined ultracentrifugation, quasi-elastic light scattering, and membrane diffusion experiments, and of the lower relaxation time with the square of the radius. In contrast, the dielectric increment experimental values could not be explained by the model used nor by any other one within the framework of migration polarization theory. This was ascribed tentatively to the formation of micelle clusters. As concerns the discrepancies between the behavior of $H_2O$ microemulsions and of $D_2O$ ones, the authors pointed out the importance of hydration effects on interfacial and double-layer polarizations. Results thereabout were reported in detail by Chou and Shah in two subsequent papers [706, 707], devoted to the effects of hydrocarbon, valency and concentration of counterions on the solubilization capacity of Winsor IV W/O microemulsions and Winsor III middle-phase microemulsions, and on the coacervation of aqueous petroleum sulfonate solutions. In particular, it was found through conductometry and dielectrometry experiments that the surface charge density in $H_2O$ W/O microemulsions was three times higher than that in $D_2O$ W/O microemulsions at same compositions.

It has been already mentioned that the influence of alcohol chemical structure upon microemulsion physical properties was put initially into evidence by Shah and co-workers [641], by substituting 1-pentanol for 1-hexanol in systems incorporating *n*-hexadecane as the hydrocarbon and potassium oleate as the surfactant. In Fig. 78, the conductivity plots recorded on both systems along a given r-type composition path show clearly the striking discrepancies observed in conductivity value and variation trend. The highly conducting 1-pentanol systems were labeled by Shah et al. as "cosolubilized" systems, that is, in fact, quaternary molecular solutions, while the poorly conducting 1-hexanol systems were considered as proper microemulsions. The conductive and dielectric behavior of the same two systems was investigated as well by Clausse and co-workers [573,574], who analyzed in detail conductivity and permittivity plots. Figure 79 shows the striking discrepancies exhibited by the conductivity plots recorded for both systems along a given $p_s$-type composition path. In agreement with the data gained by Shah et al., the 1-hexanol systems were found to be much less conducting than the 1-pentanol ones. From a careful analysis of the conductivity plots, it was proved that the conductive

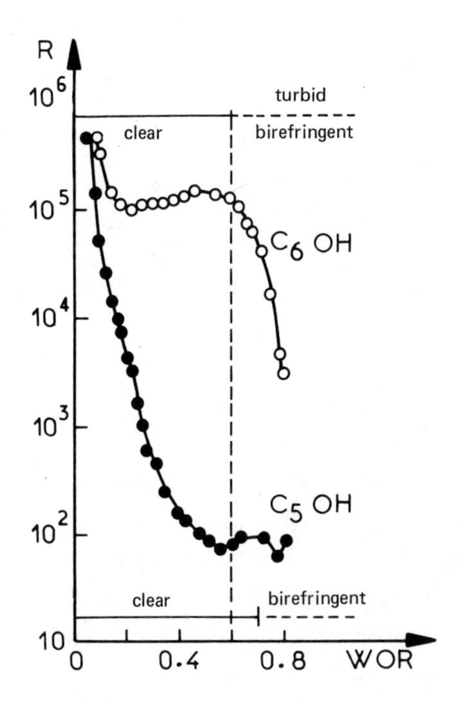


Figure 78. Optical appearance and electrical resistance of samples of water/potassium oleate/1-pentanol or 1-hexanol/$n$-hexadecane Winsor IV media. Influence of WOR, the water-hydrocarbon volume ratio, along an r-type composition path. R is expressed in ohms. (From Ref. 641.)

behavior of the 1-pentanol systems was of the percolative type, as illustrated by Fig. 80 which shows that the conductivity variations can be fitted by using equations derived from the percolation and effective medium theories [342,357-359]. This phenomenon, which will be analyzed and commented upon later on, is similar to that put into evidence previously by the same group of authors on the system water/potassium oleate/1-butanol/toluene [633], and by Lagües et al. [557], on the system water/sodium dodecyl sulfate/1-pentanol/cyclohexane.

In contrast, the variations of the low conductivity of the 1-hexanol systems could not be accounted for through the percolation and effective medium theories. It appears so that, as far as electrical conductive properties are concerned, the distinction between cosolubilized systems and microemulsions proposed by Shah et al. [641] can be expressed in terms of percolating and nonpercolating systems. Starting from these data, an extensive investigation has been performed recently [635], aimed at establishing reliable correlations between the phase behavior and the electrical properties of quaternary Winsor IV media, as influenced by the chemical structure of the nonaqueous components (surfactant, alcohol, and

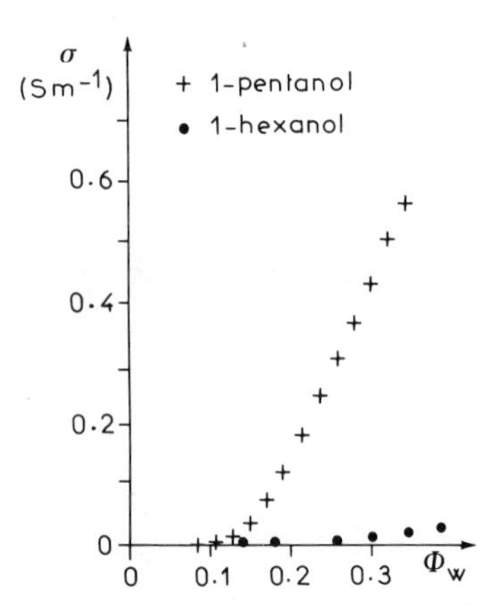


**Figure 79.** Comparative conductive behavior of water-in-hexadecane systems using potassium oleate as the surfactant and either 1-pentanol or 1-hexanol as the co-surfactant. Variations of the electrical conductivity $\sigma$, with the water volume fraction $\Phi_W$ along a $p_S$-type composition path. T = 298 K. Surfactant-alcohol mass ratio: k = 3/5. Surfactant/alcohol combination mass fraction: $p_S = 0.4$. (From Ref. 574.)

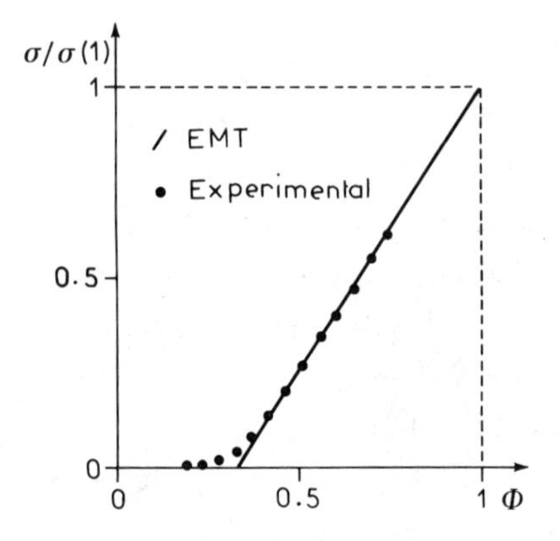


**Figure 80.** Water/potassium oleate/1-pentanol/*n*-hexadecane system (specifications identical to those of Fig. 79). Plots of reduced conductivity versus disperse volume fraction, as obtained from the application of the percolation and effective medium theories. The solid line represents the effective medium theory formula. (From Ref. 574.)

hydrocarbon). Thirty or so quaternary systems involving various anionic surfactants, straight or branched alkanols, and aliphatic or aromatic hydrocarbons were studied, which led to quite significant and general results. The hydrocarbon and the surfactant being chosen and the surfactant-alcohol mass ratio being fixed, the geometrical features of the Winsor IV domain in the pseudoternary phase diagram depended critically upon the chemical structure of the alcohol used as the cosurfactant. With normal alkanols, it was observed that the Winsor IV domain was split into two disjointed areas when the alcohol carbon number $n_a$ was greater than a critical value $n_c$. On the contrary, for $n_a$ equal or smaller than $n_c$, the Winsor IV domain consisted of a unique area with no macroscopically detectable internal boundary separating the W/O and O/W region. In any case, the extension of the total Winsor IV domain was found to be the greatest at $n_c$. The value of $n_c$ appeared to depend upon the chemical structure of both the hydrocarbon and surfactant. For instance the splitting of the Winsor IV domain occurred above $n_c = 4$ (i.e., for 1-pentanol), for systems involving benzene and sodium dodecylsulfate, and above $n_c = 5$ (i.e., for 1-hexanol) for systems involving dodecane instead of benzene. Referring to the molecular length compatibility effect suggested by Bansal et al. [643], Clausse et al. [644] pointed out that the value of 5 found for $n_c$ in the case of dodecane and sodiumdodecyl sulfate was consistent with formula (406), since 1-pentanol is the alkanol whose molecular length ($\ell_a \simeq 7$ Å) added to the dodecane one ($\ell_0 \simeq 14$ Å) fits best that of sodium dodecylsulfate ($\ell_s \simeq 21$ Å). The fact that in the pseudoternary phase diagram, the Winsor IV domain took its greatest extension when 1-pentanol was used as the cosurfactant is also consistent with the observation made by Bansal et al. that the water solubilization capacity is maximum when molecular length compatibility is obtained. However tempting and instrumental as a guiding line for microemulsion system optimization be the molecular length compatibility rule proposed by Bansal et al., it should be used cautiously since it cannot be applied straightforwardly in every situation, in particular when cyclic (or branched) aliphatic or aromatic compounds are involved. In that respect, the study made by Heil et al. [635,704] has shown that the transition from one area to two areas can be observed as well with isomeric alcohols. The demonstration was made on systems made up of water, benzene, and either potassium oleate or sodium dodecyl sulfate, the cosurfactnat being in turn one among the eight amylic alcohol isomers. For both of the surfactants used, it was found that Winsor IV domains forming a unique area were associated with 2-butanol-2-methyl, that is, the most water-soluble alcohol of the series, while Winsor IV domains split into two disjointed areas were associated with 1-pentanol, that is, the least water-soluble alcohol. The transition from one area to two areas occurred progressively, other things equal, upon substituting, as a general rule, a less water-soluble isomer for a more water-soluble one. At intermediary water affinities, hybrid situations were encountered, characterized by Winsor IV domains forming a unique area with a more or less deep indentation. The extension of the whole Winsor IV domain was found to increase with the solubility of alcohol in water, except for the case of 3-pentanol which displayed a slightly anomalous behavior. Anyhow, with both of the surfactants, the least water-soluble 1-pentanol yielded for the Winsor IV domain the smallest extension, and the most water-soluble 2-butanol-2-methyl the greatest extension, which was found to be slightly smaller than in the case of systems incorporating the more water-soluble 2-butanol. These diagram studies, that allowed to put very clearly into evidence the striking influence of alcohol chemical structure, were paralleled by thorough investigations into Winsor IV media electrical behavior [635,704]. For all systems that displayed a unique Winsor IV area (see Fig. 70b), the Winsor IV media conductivity was found to be fairly high and to undergo drastic variations upon composition varying along either r-type paths (Fig. 78) or $p_s$-type paths (Fig. 79). Upon $p_w$, the water mass fraction, increasing up to 0.4 or so, the conductivity variations along r-paths or $p_s$-paths appeared in all cases to be of the percolative type. In contrast, for all systems that displayed a Winsor IV domain split into two distinct areas (see Fig. 70a), the conductivity was found to keep low values and to undergo variations that could not be depicted by means of percolation and effective medium formulas. In particular, as the water content was increased up to 0.40 or so along r-type paths (or $p_h$-type ones), the conductivity displayed nonmonotonous smooth variations. This is illustrated by the 1-hexanol resistance curve of Fig. 78 that shows, over the clear region, a minimum at WOR = 0.2 or so and a maximum at WOR = 0.5 or so. This study, which generalized the comparative data gained by Shah et al. [641] and Clausse et al. [573,574] on the water/potassium oleate/1-pentanol or 1-hexanol/*n*-hexadecane systems, proved that the electrical behavior of quaternary Winsor IV media made up of water, an ionic surfactant, an alkanol, and either an aliphatic or aromatic hydrocarbon could be correlated directly with the Winsor IV domain geometrical features, these depending, other things equal, upon the alcohol chemical structure [704]. These very general results are quite useful ones since they introduce a clear classification among microemulsion-type systems and so help in reconciling and understanding apparently contradictory data scattered throughout the literature [644]. In that respect, it appears that Schulman's and Winsor's anomalous systems, and Shah's cosolubilized systems can be all identified as percolating Winsor IV media, while proper microemulsions can be characterized as nonpercolating systems. From the fundamental standpoint of the structure of Winsor IV media, these results are also of significance in that they prove that phase diagram features and electrical properties reflect structural discrepancies related to composition factors, as explained in the following developments.

Clausse and co-workers [573,574,635,704,708], studied in detail the variations of both the low frequency conductivity $\sigma$ (at 1 kHz or so) and permittivity $\varepsilon_\ell$ (at 1 MHz or so) over the Winsor IV W/O-type area of the following quaternary systems:

Water/potassium oleate/1-hexanol/*n*-hexadecane (k = 3/5)
Water/sodium dodecyl sulfate/1-hexanol/*n*-dodecane (k = 1/2)
Water/sodium dodecyl sulfate/1-heptanol/*n*-dodecane (k = 1/2)
Water/potassium oleate/1-pentanol/benzene (k = 1/2 or 3/5)
Water/sodium dodecyl sulfate/1-pentanol/benzene (k = 1/2)
Water/sodium dodecyl sulfate/1-hexanol/benzene (k = 1/2)
Water/sodium dodecyl sulfate/1-heptanol/benzene (k = 1/2)

All of these systems are of the first type, that is their Winsor IV domains are split into two disjointed areas, one corresponding to W/O-type media and the other to O/W ones, as illustrated by Fig. 70a. Depending upon the chemical structure of the nonaqueous components used, both the W/O and O/W areas are more or less extended, in accordance with the observations reported above as concerns the influence of the alcohol chemical structure. With a view to investigate thoroughly the way in which the dielectric and conductive behavior of Winsor IV W/O-type media depends upon system composition, systematic determinations of both $\varepsilon_\ell$ and $\sigma$ were carried out all over the W/O Winsor IV area, according to the experimental procedures described in Fig. 72a to c. Details about the experimental procedures can be found in the original papers [573,574]. By combining all the data gained in turn along $p_h$, $p_w$, and $p_s$ paths, it was possible to cover the Winsor IV W/O areas with tight networks of permittivity and conductivity values. Figures 81 to 83 display typical permittivity and conductivity plots obtained on the water/potassium oleate/1-hexanol/*n*-hexadecane system which was the first to be investigated [573,574]. In $p_h$-type experiments, the points representing the sample compositions lie on a portion of straight line parallel to the zero-hexadecane side of the pseudoternary phase diagram and bounded by a couple of points $\alpha_1$ and $\beta_1$ belonging to the boundary $\Gamma$ of the Winsor IV W/O area (Fig. 81a). Upon $p_w$, the water mass fraction, increasing along $\alpha_1\beta_1$ from $(p_w)_{\alpha 1}$ to an intermediary value $(p_w)_i$, $\varepsilon_\ell$ increases almost linearly. For values of $p_w$ ranging from $(p_w)_i$ up to $(p_w)_{\beta 1}$, the $\varepsilon_\ell$ versus $p_w$ plot is a sharply ascending branch resulting from the steep increase exhibited by $\varepsilon_\ell$ as the boundary $\Gamma$ is approached, (Fig. 81b). Concerning $\sigma$, its variations with $p_w$ are represented by a nonmonotonous curve displaying first a maximum for $p_w = (p_w)_c$ and then a minimum for $p_w = (p_w)_j$ and beyond which $\sigma$ increases sharply with $p_w$ (Fig. 81c).

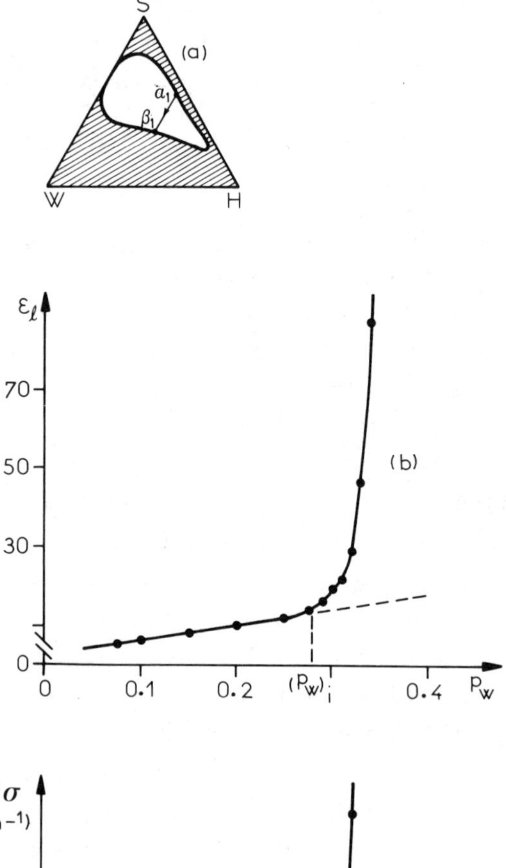


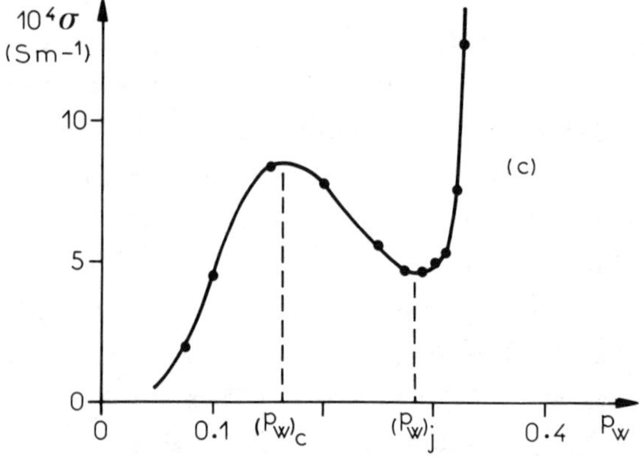


**Figure 81.** Water/potassium oleate/1-hexanol/*n*-hexadecane system. Surfactant-alcohol mass ratio: k = 3/5. T = 298 K. (a) Experimental composition path followed within the Winsor IV W/O area. $p_h = 0.4$. (b) Variations of $\varepsilon_\ell$ with the water mass fraction $p_w$. (c) Variations of $\sigma$ with $p_w$. (From Ref. 573, reprinted with permission from C. Boned et al., *J. Phys. Chem.* *84*:1520-1525. Copyright 1980 American Chemical Society.)

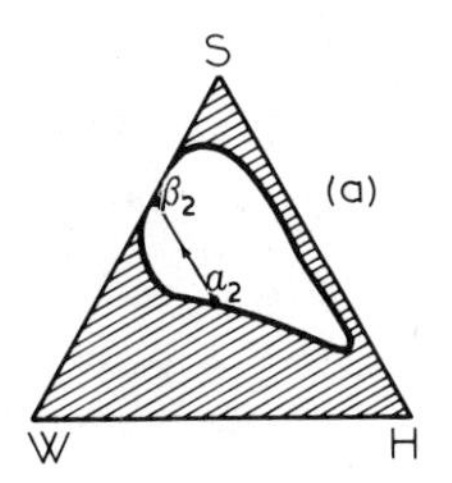


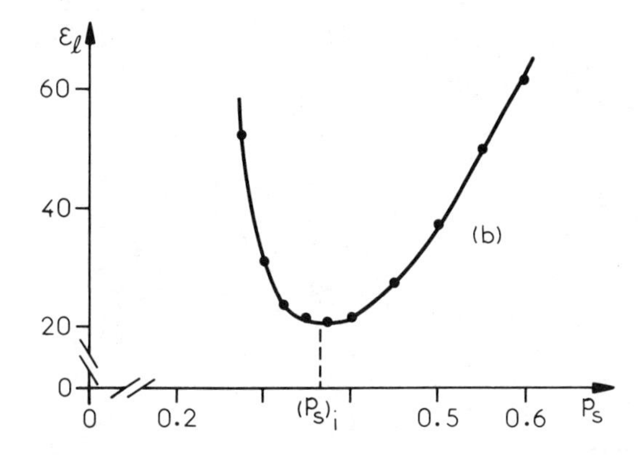


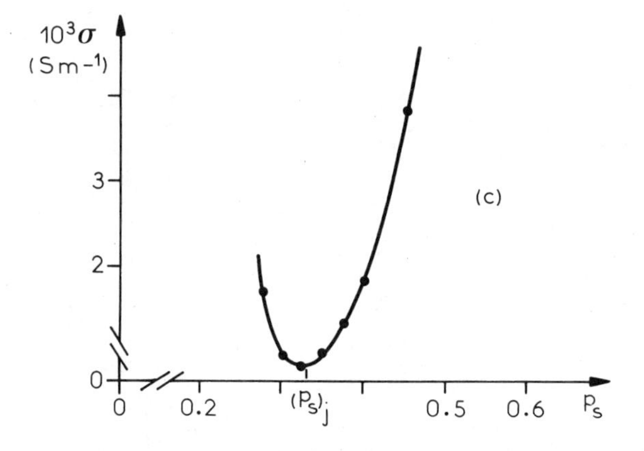


**Figure 82.** Water/potassium oleate/1-hexanol/*n*-hexadecane system (specifications as in Fig. 81). (a) Experimental composition path followed within the Winsor IV W/O area. $p_W = 0.35$. (b) Variations of $\varepsilon_\ell$ with the mass fraction $p_S$ of combined surfactant and alcohol. (c) Variations of $\sigma$ with $p_S$. (From Ref. 573, reprinted with permission from C. Boned et al., *J. Phys. Chem. 84*:1520-1525. Copyright 1980 American Chemical Society.)

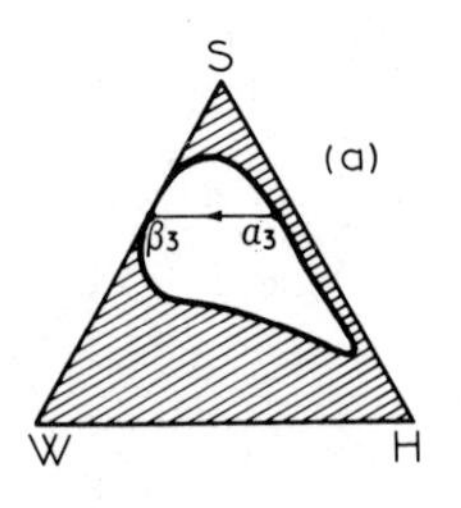


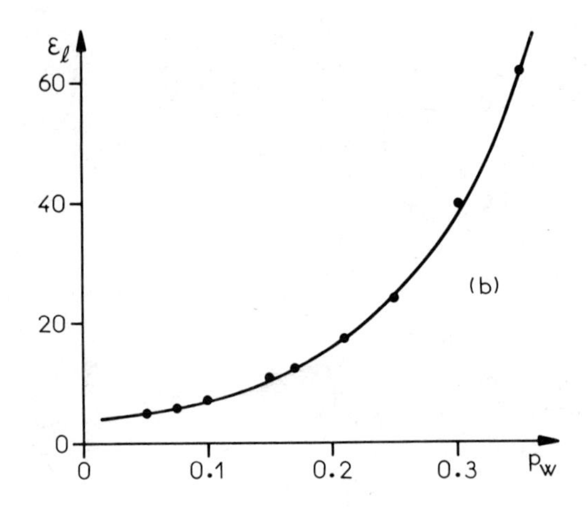


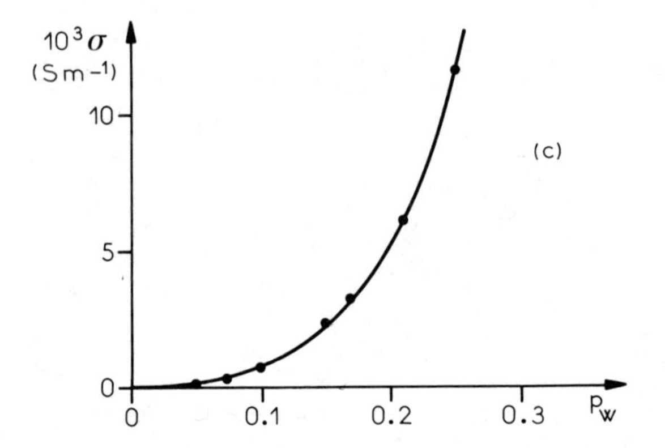


**Figure 83.** Water/potassium oleate/1-hexanol/*n*-hexadecane system (specifications as in Fig. 81). (a) Experimental composition path followed within the Winsor IV W/O area. $p_S = 0.6$. (b) Variations of $\varepsilon_\ell$ with the water mass fraction $p_W$. (c) Variations of $\sigma$ with $p_W$. (From Ref. 573, reprinted with permission from C. Boned et al., *J. Phys. Chem.* *84*:1520-1525. Copyright 1980 American Chemical Society.)

Whatever the value of $p_h$, the same behavior is evident all over the Winsor IV O/W area, the values of $(p_w)_i$ and $(p_w)_j$, which are almost equal, and of $(p_w)_c$ depending of course upon that of $p_h$. Similar phenomena can be observed when crossing the Winsor IV W/O area along a r-type path, as illustrated by the permittivity plot of Fig. 74 and the 1-hexanol conductivity curve of Fig. 78. Clausse et al. [573,574] checked that the values of $(p_w)_i$, $(p_w)_j$, and $(p_w)_c$ determined through both these procedures are consistent. With $p_w$ fixed, the points representing the sample compositions define a portion of straight line parallel to the zero-water side of the pseudoternary phase diagram and bounded by a couple of points, $\alpha_2$ and $\beta_2$, belonging to the boundary $\Gamma$ of the Winsor IV W/O area (Fig. 82a). As illustrated by Fig. 82b and c, whatever the value of $p_w$, as $p_s$ increases from $(p_s)_{\alpha 2}$ to $(p_s)_{\beta 2}$, both $\varepsilon_\ell$ and $\sigma$ begin to decrease, reach a minimum and then increase. $(p_s)_i$ and $(p_s)_j$, which are the values of $p_s$ corresponding respectively to the minimum of $\varepsilon_\ell$ and $\sigma$, depend upon $p_w$, while remaining very close to each other, the discrepancy being equal or smaller than 0.03. If $p_s$ is taken as the variable parameter, the points representing the sample compositions are located on a portion of straight line parallel to the zero-surfactant/alcohol combination side of the pseudoternary phase diagram and bounded by a couple of points $\alpha_3$ and $\beta_3$ belonging to the boundary $\Gamma$ of the Winsor IV W/O area (Figure 83a). In contrast with the other two experimental processes, no special phenomena are encountered, both $\varepsilon_\ell$ and $\sigma$ increasing smoothly with $p_w$, as shown in Fig. 83b and c. The absence of extrema or kinks on the $\varepsilon_\ell$ and $\sigma$ versus $p_w$ plots is consistent with the results gained with $p_w$-type experiments since, in that case, the $\varepsilon_\ell$ and $\sigma$ versus $p_s$ curves do not intersect and are stacked regularly, the top curve corresponding to the highest value of $p_w$ [574]. In addition, it must be noted that, in contrast with the case of systems displaying a unique Winsor IV area (see Fig. 85), the increase of $\sigma$ with $p_w$ is a fairly moderate one and is not linear. Plotting in the pseudoternary phase diagram the compositions corresponding to $(p_W)_c$, $(p_W)_i$, $(p_W)_j$, $(p_S)_i$ and $(p_S)_j$ reveals that the representation points are not distributed at random but define two lines, $\Gamma_1$ and $\Gamma_2$, which partition the Winsor IV W/O area into three adjacent regions labeled PM, ME, and MC in Fig. 84. $\Gamma_1$ is the line defined from the values of $(p_W)_c$ that form an isolated class. Starting from the water-poor region, $\Gamma_1$ runs linearly while stepping away gradually from the upper branch of the Winsor IV W/O area boundary $\Gamma$, bends and becomes roughly parallel to the zero-hexadecane side of the pseudoternary diagram. As shown in Fig. 84, the prolongation of the linear part of $\Gamma_1$ is directed towards the H vertex of the phase diagram, which indicates a constant water-to-surface active combination ratio and, consequently, a constant water-surfactant ratio. Since the values of $(p_w)_i$, $(p_w)_j$, $(p_s)_i$, and $(p_s)_j$ are correlated [573], they together define a second line, labeled $\Gamma_2$ in Fig. 84. Quite similar results were obtained on the sys-

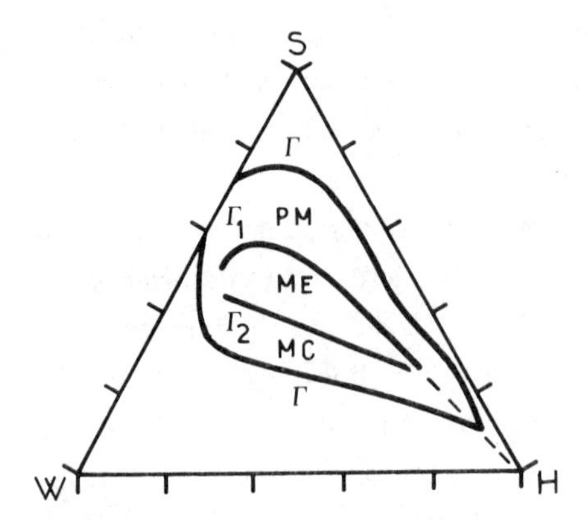


Figure 84. Partition of the Winsor IV W/O area of the quaternary system water/potassium oleate/1-hexanol/*n*-hexadecane. Surfactant-alcohol mass ratio: k = 3/5. T = 298 K. $\Gamma_1$ is the line defined from the values of $(p_w)_c$. $\Gamma_2$ is the line defined from the values of $(p_w)_i$, $(p_w)_j$, $(p_s)_i$, and $(p_s)_j$. (From Ref. 573, reprinted with permission from C. Boned et al., *J. Phys. Chem. 84*:1520-1525. Copyright 1980 American Chemical Society.)

tems water/potassium oleate/1-pentanol/benzene, water/sodium dodecyl sulfate/1-pentanol/benzene and water/sodium dodecyl sulfate/1-hexanol/*n*-dodecane, whose Winsor IV W/O areas can be consequently partitioned as well into three adjacent subareas. A similar phenomenon was also put into evidence in the case of the ternary system water/AOT/*n*-dodecane [635,708]. For three other systems, namely, water/sodium dodecyl sulfate/1-heptanol/dodecane and water/sodium dodecyl sulfate/1-hexanol or 1-heptanol/benzene, which incorporate longer alkanols than the preceding corresponding ones, the line $\Gamma_2$ cannot be put into evidence because the lower branch of the Winsor IV W/O area boundary Γ is reached at or before the composition points corresponding to $(p_w)_i$, $(p_w)_j$, $(p_s)_i$, and $(p_s)_j$ [635,708]. These results can be interpreted in terms of structural changes induced by composition variations within the Winsor IV W/O area [573,574]. In the upper region labeled PM in Fig. 84, the surfactant is in excess compared to the content of water to be solubilized, which results in the preferred association of water with surfactant. Consequently, the media can be considered as consisting of bound water premicellar entities built up from hydrated surfactant aggregates that swell progressively as the water content increases. Line $\Gamma_1$ defines compositions at which all the available surfactant molecules are engaged in shells of W/O micelles possessing a core of "free" water. A good assessment for this interpretation is the fact that, along the linear portion of $\Gamma_1$, the water to surfactant molecular ratio is found to be constant and equal to 12 ± 2 [635]. This is consistent with both theoretical predictions derived from the geometrical model of Bothorel et al. [620] (see line $\Delta_0$ in the schematic pseudoternary phase diagram of Fig. 67), and with data gained by other groups of authors through different techniques. For instance, on the basis of density, light-scattering, and electron microscopy experiments performed on systems made up of potassium oleate, 1-pentanol and, as the hydrocarbon, either *n*-decane, benzene, or phenyldodecane, Sjöblom and Friberg reported that the onset of formation of W/O swollen micelles corresponds to a water-surfactant molecular ratio of 8 or so [550]. A light scattering study carried out on the systems water/potassium oleate/1-hexanol/*n*-dodecane and water/sodium dodecyl sulfate/1-butanol/toluene led Fourche and Bellocq [626-627] to retain the value of 8 to 10 for the minimum water-surfactant molecular ratio at which the Winsor IV media can be considered as consisting of inverse micelles. Incidentally, it is worth pointing out that these authors determined from their light-scattering data $\Gamma_1$ lines quite similar to the ones evidenced from conductivity data by Clausse et al. [573,574,635,708]. In particular both the light-scattering lines $\Gamma_1$ present in the region rich in surface-active agent combination a curvature similar to the one displayed by the conductivity lines $\Gamma_1$ (see Fig. 84). This curvature can be interpreted as reflecting a lamellar organization of the medium, owing to the high surface-active-agent content [592]. For ternary systems of the type water/AOT/hydrocarbon, it has been reported as well that the onset of formation of inverse micelles is characterized by water-to-surfactant molecular ratio values of 10 or so. Eicke and co-workers found a value of 9 to 10 in the system water/AOT/isooctane studied by means of several techniques, ultracentrifugation and light scattering [603], nanosecond spectroscopy [607], photon correlation spectroscopy [610], and fluorescence and polarization decay [611]. They showed that "the phenomena characteristic for microemulsions occur when water becomes the major component in the colloidal aggregates" and stressed that there is no discontinuity in the medium property as the water content increase, which is consistent with the observations made by Clausse et al. [573,574,635], that, when crossing $\Gamma_1$ lines, the static permittivity displays no anomaly that would indicate conformational changes of the dispersed entities. So the transformation of premicellar entities, however they be termed, into inverse micelles is not a phase change, but a rather diffuse process that subtly influences structural parameters. For instance, at low water-surfactant molecular ratios, the average area per surfactant molecule increases steeply and then reaches an almost constant value for water to surfactant molecular ratios equal to or greater than 15 or so [613]. Other values around 10 for the water-surfactant molecular ratio have been reported by Rouvière et al. [623] (8, from RMN, viscometry, $Na^+$ autodiffusion, and Kerr effect experiments on the system water/AOT/*n*-hexane), by Wong et al. [624] (12, from fluorescence probe experiments on the system water/AOE/*n*-

heptane), and by Bakale et al. [625] (8, from picosecond pulse-conductivity experiments in the system water/AOT/isooctane). In that respect, it is worth mentioning as well the positron annihilation experiments carried out on several systems by Boussaha et al. [709], although these authors did not see clearly the exact implications of their data. The value of 10 to 11 obtained by Heil [635] on the system water/AOT/*n*-dodecane through conductivity measurements is quite consistent with all the results reported above. In the light of all these data, it appears that line $\Gamma_1$, as determined through conductometry experiments, marks the onset of the formation of water-swollen inverse spherical micelles and, consequently, can be considered as a kind of CMC curve stretched across the Winsor IV W/O area of systems incorporating an ionic surfactant. Below $\Gamma_1$, i.e., in the region labeled ME, the system compositions are fairly balanced. This ensures the stability of monodisperse populations of inverse spherical micelles that grow in both size and number upon addition of water, which is consistent with the moderate regular increase of $\varepsilon_\ell$ with $p_w$ along $p_h$-type paths, its decrease with $p_h$ along $p_w$-type paths, and the decrease of $\sigma$ along both $p_h$-type and $p_w$-type paths [573]. Beyond $\Gamma_2$, i.e., over the MC subarea, the sharp increase of both $\varepsilon_\ell$ and $\sigma$ along either $p_h$-type or $p_w$-type paths can be interpreted in terms of structural changes occurring in the systems as the water/surfactant ratio becomes too high to further ensure the stability of populations of inverse spherical micelles. The steep increase of $\varepsilon_\ell$ beyond $\Gamma_2$ reveals that the system undergoes local anisotropy, which can be accounted for by assuming the formation of nonspherical micelles or, most likely, of micellar clusters resulting from the aggregation of spherical micelles that tend to "flocculate" so as to offer the optimum surface-volume ratio. The latter assumption is more consistent with the increase of $\sigma$ beyond $\Gamma_2$ that can be ascribed to the formation of conducting paths resulting from the progressive interlinking of micellar clusters. This clustering stage can be considered as predictive of final system stability breakdown and reorganization into long-range ordered media which are encountered as the lower branch of the boundary $\Gamma$ of the Winsor IV W/O subarea is reached and crossed. Results gained by other authors through different techniques come in support to this interpretation. Bellocq et al. [710] have suggested, from dynamic light-scattering experiments, the existence of aggregation phenomena in Winsor IV media whose composition is close to that of gel-type media existing of the other side of the boundary $\Gamma$ of the Winsor IV area. An electron microscopy study performed by the same group of authors [592] has revealed the possible existence of micellar clusters in water/sodium dodecylsulfate/1-butanol/toluene Winsor IV media whose composition is close to $\Gamma$. Candau et al. [575,576] arrived at quite similar conclusions on the basis of data gained through electron microscopy and light-scattering experiments on Winsor IV media stabilized by polymeric surfactants. Moreover, as already mentioned, the negative values of the second virial coefficient of the osmotic compressibility in Winsor IV W/O media found by several authors [554,565-572], lend support to the existence of aggregation phenomena resulting from attractive interactions between inverse micelles. It is worth pointing out here that the existence of clustering processes is put forth as well to account for the conductive behavior exhibited by Winsor IV media of systems displaying a unique Winsor IV area [633-635]. The distinction between the two cases stands in the value of the second virial coefficient that depends upon system composition factors. This is clearly illustrated by the data reported recently by Cazabat and Langevin [571], who showed that the second virial coefficient is the more negative as the alcohol used as the cosurfactant is the shorter, other things equal. This finding is consistent with the general observations reported as to the influence of alkanol chemical structure upon phase diagram features and electrical properties of Winsor IV media [635, 644,704]. For the systems considered here that incorporate higher alkanols and, consequently, display a Winsor IV domain split into two disjointed areas, the clustering process is a moderate one which leads to limited $\varepsilon_\ell$ and $\sigma$ increases, in contrast with the drastic variations observed in systems that incorporate shorter alkanols and display a unique Winsor IV area (see Figs. 83 and 85 for a comparison). Besides, the clustering process appears to be sometimes inhibited since the line $\Gamma_2$, marking the onset of the increase in $\varepsilon_\ell$ and $\sigma$, cannot be determined. This is the case for the systems water/sodium dodecylsulfate/1-heptanol/*n*-dodecane and water/sodium dodecyl sulfate/1-hexanol or 1-heptanol/benzene for which the MC subarea does not exist [708]. To sum up and conclude, it can be said that conductivity and permittivity studies yield quite significant information concerning the "solubilization" of water in systems whose Winsor IV domain consists of two disjointed areas. In the general case, the W/O area can be divided by two lines $\Gamma_1$ and $\Gamma_2$ into three adjacent composition regions, labeled PM, ME, and MC (see Fig. 84), that represent the realms of existence of, respectively, premicellar entities, inverse spherical micelles, and micellar clusters. In some cases, the micellar cluster region does not exist. These general results are consistent with theoretical considerations and other experimental data gained from various techniques. They suggest in particular that the model of a monodisperse population of composite globules cannot be retained to depict the structure of the medium all over the Winsor IV W/O area and that only the phases belonging to the composition region labeled ME can be identified strictly as Schulman's microemulsions.

A quite different electrical behavior is displayed by systems of the second type, i.e., systems whose Winsor IV domain consists of a unique area with no apparent boundary separating the W/O and O/W regions. Belonging to this category are the following systems which have been studied by Clausse et al. [633-635]:

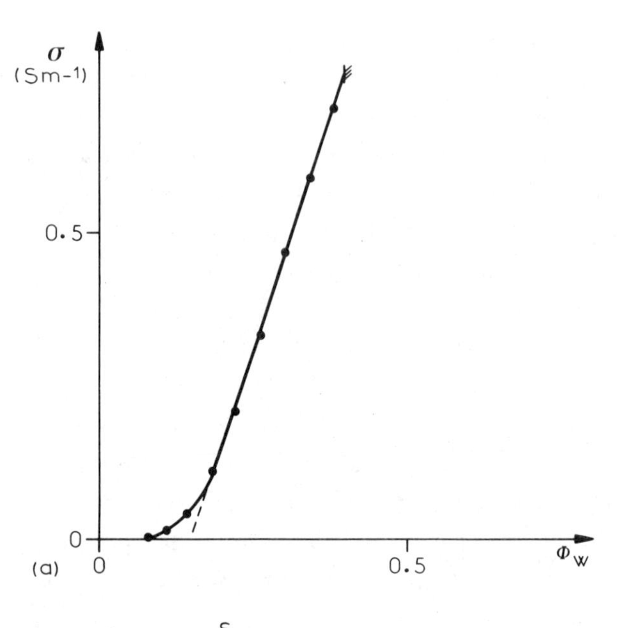


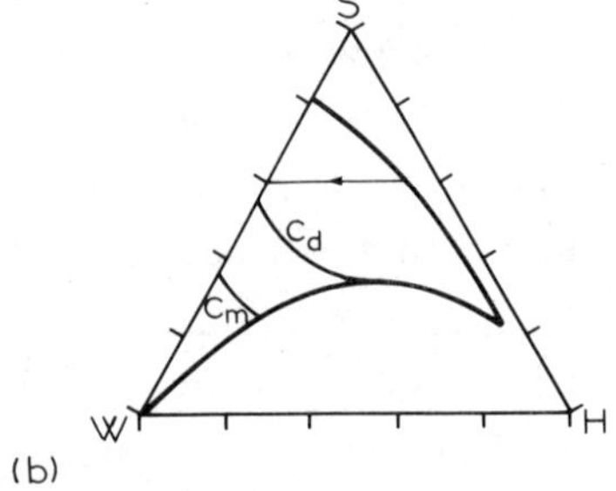


**Figure 85.** Water/sodium dodecyl sulfate/2-butanol-2-methyl/benzene system. Surfactant-alcohol mass ratio: k = 1/2. T = 298 K. (a) Variations of $\sigma$ with $\Phi_W$, the water volume fraction, $p_S$ being kept at 0.6. (b) Experimental composition path followed within the Winsor IV area of the system pseudoternary phase diagram. (For $C_d$ and $C_m$ see text.) (From Ref. 634, reprinted by permission from *Nature* *293*:636-638. Copyright 1981 Macmillan Journals Limited.)

Water/sodium dodecyl sulfate/1-pentanol/*n*-dodecane (k = 1/2)
Water/sodium dodecyl sulfate/1-propanol or 1-butanol/benzene (k = 1/2)
Water/sodium dodecyl sulfate/2-butanol-2-methyl/benzene (k = 1/2)
Water/potassium oleate/1-propanol/benzene (k = 3/5)
Water/potassium oleate/2-butanol-2-methyl/benzene (k = 3/5)
Water/potassium oleate/1-butanol/toluene (k = 1/2)

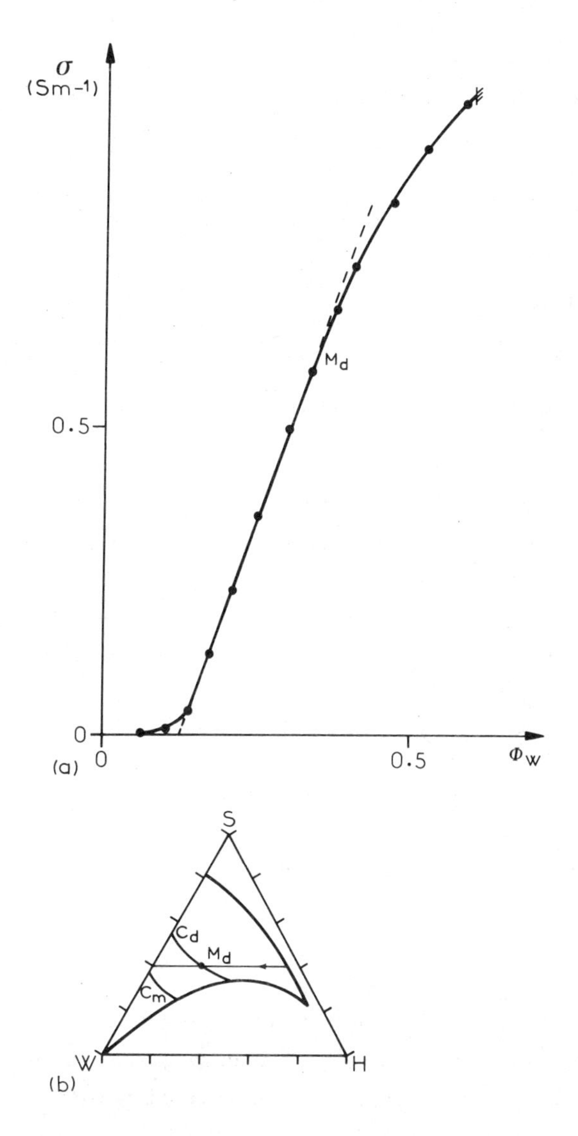


**Figure 86.** Water/sodium dodecyl sulfate/2-butanol-2-methyl/benzene system (specifications as in Fig. 85). (a) Variations of $\sigma$ with $\Phi_W$, at $p_S = 0.40$. (b) Experimental path within the Winsor IV area. (From Ref. 634, reprinted by permission from *Nature 293*:636-638. Copyright 1981 Macmillan Journals Limited.)

The experimental procedures followed were quite similar to those used in the case of systems displaying a Winsor IV domain split into two disjointed areas. Figures 85 to 87 give typical conductivity plots recorded along either $p_S$-type or r-type composition paths within the Winsor IV domain of the system water/sodium dodecyl sulfate/2-butanol-2-methyl/benzene [634]. Identical results were obtained for the

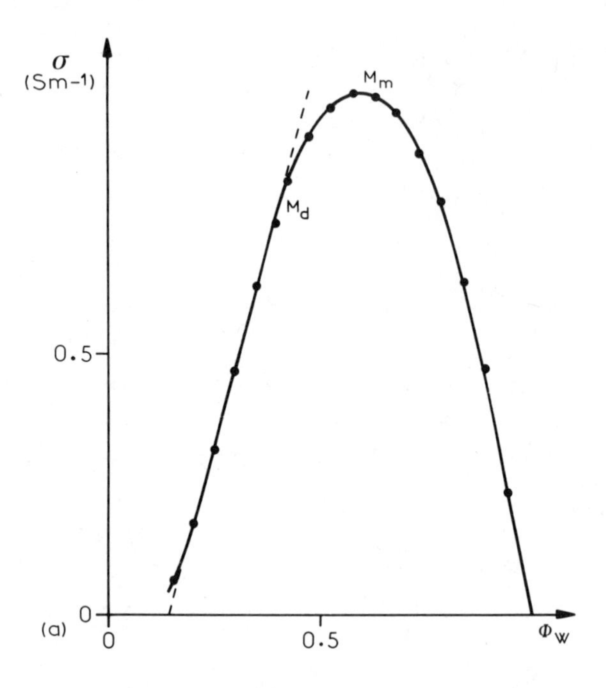


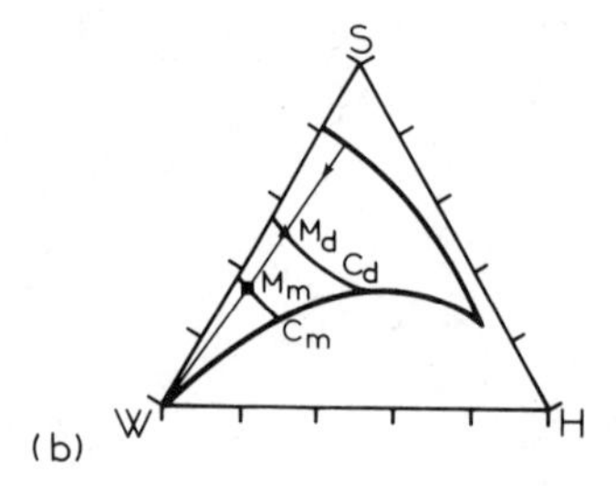


**Figure 87.** Water/sodium dodecyl sulfate/2-butanol-2-methyl/benzene system (specifications as in Fig. 85). (a) Variations of $\sigma$ with $\Phi_W$, r being kept at 1/9. (b) Experimental path within the Winsor IV area. (From Ref. 634, reprinted by permission from *Nature 293*:636-638. Copyright 1981 Macmillan Journals Limited.)

other five systems [633,635]. Figure 85 shows the variations of the electrical conductivity with $\Phi_W$, the water volume fraction, increasing along a $p_S$-type path characterized by $p_S = 0.6$. Similar plots were obtained for other values of $p_S$ in the range 0.55 to 0.8 or so. This conductive behavior, which differs strikingly from that put into evidence in the case of systems displaying two disjointed Winsor IV areas (see Fig. 83b for a comparison), has been accounted for by Lagourette et al. [633], using formulas derived from the percolation and effective medium theories. On the basis of previous formal studies (see the review papers by

Frisch and Hammersley [340] and Shante and Kirkpatrick [341]), Kirkpatrick [342] developed, through Monte-Carlo numerical simulations applied to resistor network models, a workable approach toward percolative conduction phenomena in inhomogeneous media. The existence of percolation thresholds have been reported by several authors interested in the electrical properties of composite or disordered materials. The percolation phenomenon has been described concisely by Webman et al. in the case of electrical conduction [343]. A conductor-insulator composite material of random geometry and constituent distribution being given, when the conductor volume fraction $\Phi$ is sufficiently large, continuous conducting paths (infinite clusters, in percolation terms) stretch across the material which then behaves as a conductor. If $\Phi$ decreases past some critical value $\Phi^P$, called the *percolation threshold,* extended conducting paths cease to exist and conduction is "cut off." The value of the percolation threshold $\Phi^P$ depends upon the geometry and physical details of the system. As reported by Kirkpatrick [342] different percolation threshold theoretical values have been proposed, in accordance with the system geometry. Of particular relevance to microemulsion studies is the value $\Phi^P = 0.29$ [342], derived from a numerical study of a continuum percolation model in which the allowed volumes surrounding percolation sites consist of identical spheres permitted to overlap and with centers distributed at random. This value is roughly twice as large as that obtained by associating to each percolation site an allowed volume consisting of a hard sphere with a radius equal to half the nearest-neighbor separation, the sites being assumed to lie on a regular lattice. A value of 0.25 has been suggested by Kirkpatrick for systems in which the conducting and nonconducting regions have, on the average, similar shapes. Close values, ranging from 0.27 to 0.31, have been proposed as well by Janzen for populations of spherical particles of low polydispersity [711]. If a rigorous percolative conduction regime is assumed, the effective conductivity $\sigma$ of a conductor-insulator random composite material, null as long as $\Phi < \Phi^P$, suddenly takes nonzero values when $\Phi$ becomes slightly greater than $\Phi^P$ and then increases with $\Phi$. It has been established [342,343] that in the vicinity of the percolation threshold, the dependence of $\sigma$ upon $\Phi$ can be depicted by a power law (or scaling law) as formulated:

$$\sigma \propto (\Phi - \Phi^P)^\gamma \tag{407}$$

where $\gamma$ is a critical exponent whose value depends only upon the dimensionality of the system under consideration. For three-dimensional systems, $\gamma$ was estimated to $1.55 \pm 0.20$ [342,343]. For higher values of $\Phi$, the variations of $\sigma$ upon increasing $\Phi$ can be fitted by using Bottcher-Landauer's formula (221) or (222) applied to conductivities, which yields, in the case where $\sigma_1 >> \sigma_2$, the following equation:

$$\sigma = \frac{3}{2}\sigma_1\left(\Phi - \frac{1}{3}\right) \tag{408}$$

It is noteworthy that the cutoff value in equation (408), namely, $\Phi_0^P = 1/3 \simeq 0.33$, is greater than any of the percolation threshold values reported in the case of spherical dispersions, and that the applicability domains of Eqs. (407) and (408) overlap each other. In the case of dispersion of identical spheres, the validity range of formula (408) is bounded above by values of $\Phi$ corresponding to close packing, either $\Phi_1^c = 0.637$ in the case of a random close packing [621], or $\Phi_2^c = 0.741$ in the case of a cubic close packing [429]. For $\Phi$ greater than $\Phi_1^c$ or $\Phi_2^c$, depending on the geometrical situation, the system cannot any longer exist as a population of identical spheres. In the case of fluid systems, an increase in $\Phi$ beyond the close-packing value induces geometrical and physical transformations leading to phase inversion. Figure 88 summarizes all the preceding considerations on percolative conduction in binary insulator-conductor composites by showing the features of the conductivity plot and the relative positions of the conductor volume fraction remarkable values. The comparison of the experimental plot reported in Fig. 85 with the theoretical curve of Fig. 88 proves clearly that, for

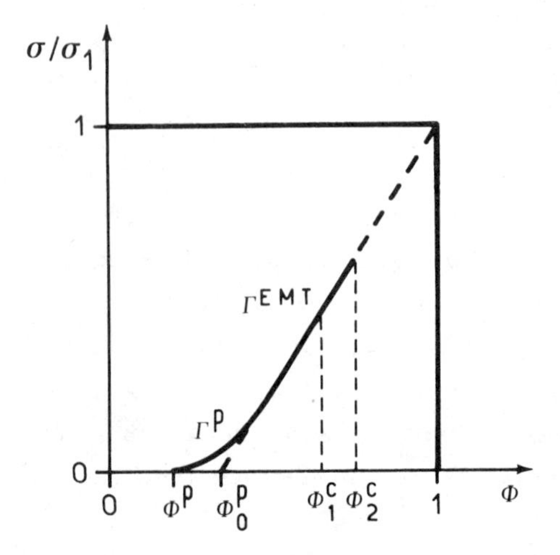


Figure 88. Model for the variations of the conductivity of a binary conductor-insulator composite sample, according to the percolation and effective medium theories. $\Phi^P$ is the percolation threshold, Eq. (407). $\Gamma^P$ is the conductivity plot in the upper vicinity of $\Phi^P$, Eq. (407). $\Phi_0^P$ is the percolation threshold for the EMT equation, Eq. (408). $\Gamma^{EMT}$ is the conductivity plot beyond $\Phi^P$, Eq. (408). $\Phi_1^c$ is the disperse-phase critical volume fraction corresponding to the random close packing of identical spheres; $\Phi_1^c = 0.637$. $\Phi_2^c$ is the disperse-phase critical volume fraction corresponding to the cubic close packing of identical spheres; $\Phi_2^c = 0.741$.

$p_s$ greater than 0.55 or so, the variations of $\sigma$ with $\Phi_w$ represents a percolative behavior. This phenomena was put previously into evidence on the system water/potassium oleate/1-butanol/toluene by Lagourette et al. [633], who made a detailed analysis concerning the determination of the percolation threshold $\Phi^P$ and critical exponent $\gamma$. These authors reported $\gamma$ values consistent with the theoretical value 1.55 proposed by Kirkpatrick [342] and, for the percolation threshold in terms of water volume fraction, values around 0.18 which were considered as being consistent with the theoretical value of 0.29 suggested by Kirkpatrick. Percolation phenomena were put into evidence as well by Lagües et al. [557,561] on the system water/sodium dodecyl sulfate/1-pentanol/cyclohexane. Their conductivity study was performed on systems in which the water-surfactant ratio was fixed, this in view of ensuring that the micellar radius remained constant upon increasing water content. The microemulsion conductivity was considered as being equal to the conductivity maximum recorded at the clearing point of water/surfactant/hydrocarbon mixtures titrated with added alcohol. The authors observed that, over narrow ranges of the water content, the conductivity increased steeply, by two orders of magnitude or so, which indicated clearly a percolative behavior. They found for the critical exponent $\gamma$ values close to 1.55, in agreement also with the theoretical value proposed by Kirkpatrick. As concerns the percolation threshold, they determined values ranging satisfactorily from 0.10 to 0.26, in terms of disperse phase volume fraction (from 0.059 to 0.16, in terms of water volume fraction). The authors complemented these results by conductivity data gained in the low water concentration range, that is, below the percolation threshold. At very low water contents ($\Phi_w < 0.05$ or so), the conductivity was observed to be approximately proportional to $\Phi_w$, as expressed by the formula

$$\sigma = k\Phi_w = Nq^2\bar{\mu} \qquad (409)$$

where k is the proportionality coefficient, N is the number of micelles per unit volume, q is the micelle mean charge, and $\bar{\mu}$ is the average mobility.

This phenomenon was ascribed to electrophoretic phenomena. Using previous determinations of micelle mobility gained from ultracentrifugation experiments [554], the authors were able to evaluate the value of the mean micellar electrical charge that was found, at vanishing values of $\Phi_w$, to be roughly proportional to the number of surfactant molecules in each micelle, in the approximate ratio of one electronic charge for 500 molecules of surfactant. In the lower vicinity of the percolation threshold, Lagües et al. observed that the conductivity exhibited a divergent behavior, as represented by the scaling law

$$\sigma \propto (\Phi_w^P - \Phi_w)^{-s} \qquad (410)$$

with s equal to 1.1 to 1.3. Such a behavior can be predicted from the phase transition theory [712] with value of s in the range 0.5 to 0.78 [713,714]. To account for the discrepancies between experimental and estimated theoretical values of s, Lagües [558] carried out an analysis showing that theoretical values of s around 1.2 can be derived from the "stirred percolation" model that is more suitable for liquid systems than the "frozen percolation" model. The main implication of the percolation theory is that, at the percolation threshold $\Phi^P$, an "infinite cluster" of conducting particles appears in the conductor-insulator composite system which is then crossed through by a conducting path allowing electrical conduction. Upon increasing the conductor concentration, other clusters form and interconnect, which progressively increases the system conductivity until the relevant close-packing volume fraction is reached. Transposed to the case of Winsor IV media, this means that beyond a critical water content, disperse inverse micelles undergo clustering processes leading to the formation of conducting paths that are, of course, of a dynamic nature. Progressive addition of water enhances the clustering process and, by way of consequence, the system conductivity. This interpretative scheme is consistent with the existence of micellar attractive interactions as revealed by negative values of the second virial coefficient of the osmotic pressure. As already mentioned, a key factor for the occurrence of percolative conduction phenomena is, other things equal, the chemical structure of the alcohol, which influences the geometrical features of the Winsor IV domain [644] and the value of the second virial coefficient [571]. For systems incorporating shorter alkanols, the Winsor IV domain forms a unique area, the values of the second virial coefficient are highly negative and, owing to strong clustering processes, the conductive behavior is of the percolative type. In that respect, it is noteworthy that the occurrence of aggregation phenomena has been put forth by Ober and Taupin [560] to explain the presence in water/sodium dodecyl sulfate/1-pentanol/cyclohexane microemulsions of anisotropic objects detected through SANS experiments. It is worth mentioning also that Cazabat et al. [570] derived from light-scattering studies that long-range (500 Å or so) attractive interactions take place in microemulsion near a percolation threshold. As pointed out by Lagües et al. [557] and Lagourette et al. [633], it is most likely that the percolation conduction mechanism is essentially interfacial.

Figure 86a shows that, for values of $p_s$ lower than 0.55 or so, the conductivity versus water volume fraction plots display a final nonlinear part that follows the linear section depictable by means of formula (408). Referring to Fig. 88, the occurrence of anomalous conductivity variations following the percolative regime can be interpreted as reflecting structural transformations undergone by the system past a critical water content corresponding to micelle close packing. Plotting in the phase diagram (Fig. 86b), the compositions computed from the critical volume fractions (defined by means of points such as $M_d$ in Fig. 86a), allows us to determine the curve labeled $C_d$ in Fig. 86b. Figure 87a shows the typical variations of $\sigma$ versus $\Phi_w$ recorded along a r-type composition path. The conductivity plot can be segmented into three parts [634,635]. The initial toe and linear section (up to $M_d$) are characteristic for a percolation regime, as in the case of $p_s$-type conductivity plots. As clearly demonstrated by Heil [635], the descending branch that starts at $M_m$ depicts the conductive behavior of O/W microemulsions whose conductivity decreases because of the progressive dilution of the external aqueous phase with added water. In between, the nonlinear portion of curve delimited by $M_d$ and $M_m$ represents an anomalous conductive behavior that may be ascribed to structural transformations taking place in the systems over a narrow $\Phi_w$ range [634,635]. As shown in Fig. 87b, the points representing the compositions computed from the $\Phi_w$ values marking the end of the percolation regime fall on the curve $C_d$ determined from $p_s$-type experiments, which is quite satisfactory and consistent. On the other hand, the points representing the compositions computed from the values of $\Phi_w$ at which the conductivity reaches its maximum define the curve labeled $C_m$ in Fig. 87b. Similar results were obtained for the other five systems investigated and also for systems displaying a unique Winsor IV domain with an indentation [635]. It appears that the Winsor IV area of systems of the second type can be partitioned into three adjacent composition regions by means of the two curves $C_d$ and $C_m$ which can be considered as internal boundaries [634]. The water-rich triangular region springing from the W vertex of the phase diagram and delimited by $C_m$ is the realm of existence of O/W microemulsion-type media. The upper region, delimited by $C_d$ and the upper branch of the boundary $\Gamma$ of the Winsor IV domain are W/O microemulsion-type media in which the disperse inverse micelles undergo aggregation, as revealed by their percolative-type electrical conductivity. The narrow composition strip bounded by $C_d$ and $C_m$ represents most likely the phase inversion region over which the W/O to O/W transformation takes place in a diffuse progressive way, with no discontinuities in the optical, electrical, and mechanical properties of the medium, in contrast with the case of system of the first type whose W/O and O/W Winsor IV areas are separated by a composition region within which turbid, sometimes birefringent, and highly viscous structures are encountered. It has been suggested [633-635] to depict the hybrid systems belonging to the $C_d$ to $C_m$ region as equilibrium dynamic bicontinuous structures, a concept developed by Scriven [637,638], and modeled by Talmon and Prager [639,640], who treated the microemulsion structure as a "random geometry of interspersed hydrocarbon and water domains generated by a Voroni tesselation" [715], the surface-active agent being assumed to be entirely adsorbed at the water/hydrocarbon interface. This implies that Winsor IV media in the bicontinuous state should appear as being typically neither of W/O nor of O/W type, but rather as aqueous-organic fluid compounds "related to ordinary liquids as porous media are to homogeneous solids" [637,638]. In the present state of knowledge, there is no positive theoretical or experimental evidence that the bicontinuous structure model may represent Winsor IV media over certain composition ranges. However, several theoretical considerations and experimental data stand in favor of the possible existence of so-called bicontinuous structures. On the theoretical side, it should be noted that both the intermicellar equilibrium scheme suggested by Winsor [520] and the geometrical model designed by Bothorel et al. [620] predict that over limited composition ranges, inverse and direct swollen micelles may coexist in Winsor IV media. In addition, the "infinite cluster" concept of the percolation theory, which is of relevance for Winsor IV systems, has strong connotations with the bicontinuous structure concept. In that respect, it is noteworthy that Landauer [359] derived the EMT formula (221) for binary composites consisting of randomly interspersed conductor and insulator microdomains. On the experimental side, the conclusions derived from the study of the conductive behavior of Winsor IV media of systems with a unique Winsor IV area are supported by results gained through various techniques. Tondre and Zana [628] investigated the rate of dissolution of $n$-alkanes and water in water/sodium dodecyl sulfate/1-pentanol/$n$-dodecane ($k = 1/2$) in rapid mixing experiments. At intermediate compositions, fast dissolution rates were found for both water and alkane, which led the authors to retain as a possible explanation for these observations that the microemulsion system could be in a bicontinuous state. Larche et al. [629] made a comparative self-diffusion and conductivity study within the unique Winsor IV area of the system 0.3% NaCl brine/sodium $p$-octylbenzene sulfonate/1-pentanol/$n$-decane. They showed that, at intermediate water contents, the diffusion coefficients of water, $n$-decane, sodium, and chlorine were all greater than that of the surfactant, which was considered as being incompatible with structures where one of the phases is totally enclosed in the other. In addition, they found a conductive behavior similar to the one demonstrated by Clausse et al. [634,635] along $p_s$-type paths, which was ascribed to the "opening" of the inverse micelles. On the basis of these results, the authors concluded that, at intermediate water contents, the microemulsion organization could be of the bicontinuous type, comparable to the structure of porous media or or phase-separated glasses [716-718]. In a series of three articles [630-632], Lindman et al. reported self-diffusion data obtained through the open-ended capillary and NMR spin-echo pulsed-field gradient methods. The systems investigated incorporated various ionic or nonionic surfactants, normal alkanols (from $C_4$ to $C_{10}$), and either aliphatic or aromatic hydrocarbons. The authors were able to establish a distinction between two types of systems. Systems incorporating AOT or combinations of an ionic surfactant with a long-chain alcohol showed a "pronounced separation into hydrophobic and hydrophilic regions," which is in agreement with the results gained by Clausse et al. [644,704,708] from phase diagram and conductivity studies performed on systems with separated W/O and O/W areas. In contrast, systems incorporating short-chain alcohols displayed composition zones over which the nature of the Winsor IV media could not be characterized clearly, since they appeared to "have flexible and highly dynamic disorganized internal interfaces." This result is quite consistent with the bicontinuous structure concept and conclusions arrived at on the basis of phase diagram and conductivity studies of systems with a unique Winsor IV area [634,635,644,704]. Microemulsions formed with nonionic surfactants of the polyethyleneoxide type exhibited properties similar to microemulsions incorporating short-chain alcohols. A recent study by the same group of authors [719], proved that the same conclusions apply in the case of the ternary systems water/sodium *n*-octanoate/1-decanol, water/sodium *n*-octanoate/*n*-octanoic acid and water/sodium cholate/1-decanol. The possible existence of bicontinuous structures associated with the W/O $\leftrightarrow$ O/W phase inversion in Winsor IV media has been recently reported by Dvolaitzky et al. [636]. These authors carried out spin-label experiments that clearly demonstrated three distinct composition zones in water/sodium dodecyl sulfate/1-pentanol/cyclohexane microemulsions. The spin-label spectra were interpreted as reflecting a progressive modification of the curvature of the interfacial film. Over the intermediate composition zone ($\Phi_W$ in the range 0.22 to 0.355), the state of the interfacial film appeared to be stationary. Finally, it should be mentioned that an analysis carried out by Heil [635], identified the curve $C_d$ as defining a random close-packing locus ($\Phi_d = \Phi_1^c = 0.637$) compatible with the representation of inverse micelles as composite globules made up of an inner spherical hard core surrounded by a compressible peripheral region, which is consistent with structural data reported by several authors [554,555,561-564]. The micelle structural parameters computed along $C_d$ by Heil are realistic ones and agree with values determined through neutron and light-scattering experiments. It appears that results gained from phase diagram and conductivity studies converge with chemical relaxation, self-diffusion and spin-label data to indicate that, for systems displaying a unique Winsor IV area, the Winsor IV media of intermediate compositions cannot be characterized clearly as being of the W/O or O/W type. In addition, the phase inversion process is a progressive and diffuse one that is not detectable, contrary to the case of systems with disjointed W/O and O/W areas, from macroscopic anomalies affecting the medium optical, electrical, and mechanical properties. This behavior can be accounted for by considering that the medium structure is a highly dynamic one that is depictable tentatively as a cosolubilized state [641], i.e., a quaternary molecular solution, or as a dynamic equilibrium bicontinuous state [637-640], i.e., a random pattern of aqueous and organic microdomains separated by flexible interfaces. Both these tentative interpretations are compatible with chemical relaxation and self-diffusion experimental data, but the existence of percolative conduction phenomena and the results of the recent spin-label experiments performed by Dvolaitzky et al. [636] could stand more in favor of the bicontinuous model. In that respect, it is noteworthy that several authors [720,721] have put into evidence percolative-type conduction phenomena associated with transitions in polyphasic systems. Transitions between the different types of polyphasic systems of the Winsor's classification (see Fig. 68), can be obtained by varying the salinity in the brine used as the aqueous component. Upon increasing the salinity, the following transition sequence is observed: WI → WIII → WII. This transition sequence can be followed by measuring the conductivity of the surfactant-rich phase, i.e., the Winsor IV-type medium which is in equilibrium by turns with either an organic phase (WI, lower phase), or both an organic and an aqueous phase (WIII, middle phase), or eventually, with an aqueous phase (WII, upper phase). Healy and co-workers [720] made such experiments on systems made up of NaCl brine, a mixture of 90% Isopar M and 10% Heavy Aromatic Naphtha (trade names of Exxon Co., U.S.A.) as the hydrocarbon component, Sulfonate FA-400 (Exxon Chemical Co.) as the surfactant, and 3-pentanol as the cosurfactant. They reported a resistivity versus salinity plot characteristic for a percolation phenomenon, but they did not analyze it. Bennett et al. [721] reported results of extensive viscosity and conductivity measurements carried out along with studies of phase number, volume and opacity, on polyphasic systems involving NaCl brine, either sodium 4-(1-heptylnonyl)benzene sulfonate or TRS10-80 (Witco Chemical Co., U.S.A.) as the surfactant, either isobutanol or 3-pentanol as the cosurfactant, and either *n*-octane, *n*-decane, or *n*-tetradecane as the hydrocarbon. The authors found that the WII → WIII → WI transition sequence was in any case paralleled by smooth variations of the surfactant-rich phase conductivity versus the volume fraction of brine, in contrast with the viscosity, which exhibited local maxima and minima, and the opacity, which presented maxima at the WII → WIII and WIII → WI transitions. The conductivity plots were recognized as depictable by means of the Percolation and Effective Medium formulas (407) and (408). The results were analyzed in the light of the statistical treatment proposed by Talmon and Prager [639,640] which predicts sequences of ternary phase diagrams qualitatively similar to those depicting transitions between polyphasic systems of the Winsor's classification. Bennett et al. concluded that the agreement of the experimental conductive behavior with the Voronoi random interspersion model was significant and argued strongly for the existence of bicontinuous structures in microemulsion-type systems. They conjectured that the polyphasic transition sequence WII → WIII → WI can be considered as reflecting microstructure changes in the system surfactant-rich phase that progresses, with decreasing salinity, from a solution of water-swollen micelles to a bicontinuous microstructure rich in hydrocarbon to a bicontinuous structure rich in water and, eventually, to a solution of hydrocarbon-swollen micelles. The coincidence of the postulated bicontinuous region with the realm of existence (in terms of salinity) of the Winsor III systems tends to indicate that, as suggested by Scriven [637,638], the surfactant-rich middle phase could be in a bicontinuous state, which would not be inconsistent with the fact that the middle phase is in equilibrium with both an aqueous phase and an organic phase.

All the preceding developments prove clearly that, as far as quaternary water/ionic surfactant/cosurfactant/hydrocarbon systems are concerned, conductivity and permittivity studies yield (concerning more especially the influence of composition factors upon the electrical behavior as correlated to the configuration of the Winsor IV domain) quite significant results that are consistent with structural data obtained by means of other physicochemical methods. As concerns the electrical properties of other types of microemulsion systems, only a limited number of data are available in the literature. However, they are generally interesting ones that deserve being analyzed and commented upon.

By using sodium sulfosuccinic acid esters (Aerosols), it is possible to form, without addition of a cosurfactant, Winsor IV media that display physicochemical properties similar to those of quaternary Winsor IV media made up with classical ionic surfactant/alcohol combinations [519-530,532,538,600-616], more particularly when the alcohol is a higher one [632]. The electrical properties of Winsor IV phases formed with Aerosols have been investigated by several groups of authors. It has been already mentioned that earlier experiments in this field were carried out by Bromilow and Winsor [523]. Winsor [525] used conductometry to detect phase changes, as related to "micellar equilibrium displacement," in systems incorporating as the surfactant Aerosol OT [AOT, sodium di-(2-ethylhexyl)sulfosuccinate]. Matthews and Hirschhorn [532] investigated the solubilization and micellization of AOT in *n*-dodecane, as influenced by the addition of water. These authors found striking discrepancies in the conductive behavior at lower and higher water-surfactant ratios. The results obtained were not analyzed but were considered as being consistent with the data gained from the other methods used, (volumetry, viscosity, and ultracentrifugation). It was suggested that addition of water induced structural changes from solutions of platelike surfactant aggregates, (at zero-water content), to solutions of "spherical or near-spherical" inverse micelles made up of a water core surrounded by a monolayer of surfactant. Micellar solutions characterized with low water-surfactant mass ratios appeared to be fairly monodisperse, while those with water-surfactant ratios of 1 or so appeared to be polydisperse "with respect to particle size and/or shape." Hanai and Koizumi [722] studied the dielectric behavior of water/AOT/kerosene mixtures, over the frequency range 20 Hz to 5 MHz and the temperature range 278 to 333 K. They found, located at frequencies well above 100 kHz, dielectric relaxations whose features were quite different from those of both W/O emulsions and binary aqueous-organic molecular solutions, which was considered as "indicating a complex structure in the solubilized systems, due to AOT-water complexes or to micelle formation." More systematic conductivity and permittivity measurements were carried out by Eicke and co-workers, generally in conjunction with experiments performed by means of other various techniques [600-614,723,724]. Eicke and Shepherd [600] studied the low-frequency conductivity and, over the frequency range 200 kHz to 10 MHz, the complex permittivity of water-in-benzene Winsor IV media incorporating as the surfactant Aerosol Y [AY, sodium di-(2-pentyl)sulfosuccinate]. This compound was selected because its conductive behavior had been investigated previously within the framework of a study [723] devoted to interactions in apolar solvents of proton donors (metal- and metalhydroxy-3,5-diisopropylsalicylates) with colloidal electrolytes [AOT, AY, and sodium di-(2-pentyl)sulfosuccinate]. This In this previous study, striking interaction effects were found, concerning the considerable lowering of the colloidal electrolyte CMC and the large nonadditive conductivity enhancement induced in the systems by the addition of the different metal-chelates. These results led Eicke [724] to propose electrostatic models for micelle stabilization in apolar solvents by solubilized ions and polar liquids. In the case of water/AY/benzene systems, Eicke and Shepherd demonstrated dielectric relaxations of the Davidson-Cole type [see Eq. (171)], whose features were strongly dependent upon the composition. Upon addition of water, the dielectric increment $\varepsilon_\ell - \varepsilon_h$ increased, the critical frequency $f_c$ began to increase and remained roughly constant at higher water contents, while the spread parameter $\beta$ took values around 0.35. For a system containing 5% or so water, both $\varepsilon_\ell - \varepsilon_h$ and $\beta$ decreased, while $f_c$ increased, upon increasing the temperature in the range 273 to 303 K. Analyzing these results, Eicke and Shepherd found that the dielectric increment varied nonlinearly with the amount of solubilized water and with the "hydrated micelle" concentration, in contrast with the case of water-free AY-benzene solutions whose dielectric increment exhibited a linear increase with micelle concentration. In systems where an aqueous saline solution (NaCl, KCl, $MgCl_2$, or $CrCl_3$), was substituted for pure water, nonlinear variations of $\varepsilon_\ell - \varepsilon_h$ were put into evidence as well, but with a shift towards higher aqeuous component concentrations. These phenomena were ascribed by the authors to association processes leading to the formation of hydrated micelles whose dipole moment increases with the amount of aggregated water molecules. At higher aqueous concentrations, a quasi-sudden steep increase in conductivity was found which was interpreted as reflecting a change of the conformation of the suspended micellar entities. In subsequent articles [601-603], Eicke and co-workers stated more precisely their views on inverted micelle formation in water/aerosol/hydrocarbon-type systems. From correlated vapor-pressure osmometry, dielectrometry, and conductivity experiments performed on systems involving either AY, AOT, or sodium di-(2-ethylhexyl)phosphate as the surfactant and various organic solvents (benzene, cyclohexane,1,4-dioxane, and 2,2-dimethyl butane), Eicke and Christen [601] inferred that so-called nuclei existed in the systems at concentrations lower than the CMC, and the authors developed a model for the micellization process. On this basis, Eicke et al. [602] assigned to surfactant trimeric units (nuclei) and to micelles the rotational relaxations demonstrated in AOT solutions in benzene, cyclohexane, and 1,4-dioxane, by means of a sensitive differential dielectric method. They suggested that micelle formation is ruled by a three-step mechanism, the system progressing from a sub-CMC solution of coexisting surfactant monomers, dimeric, and trimeric units to a solution of premicellar aggregates resulting from the binding of further surfactant molecules and the aggregation of nuclei, and eventually, to a stable micellar solution. A similar scheme was retained by Eicke and Rehak [603] to describe the formation of water/AOT/isooctane W/O microemulsions. The authors considered the results they obtained on these systems through ultracentrifugation and light scattering experiments as reflecting "repeated aggregational processes of micellar aggregates containing water," which "makes it possible for the microemulsion to take up increasing amounts of water." It is noteworthy that Eicke and Rehak found in the variation trend of the micellar average apparent molecular weight a slope break at the water-surfactant ratio of 9 that could be considered as the water concentration threshold marking the onset of the formation of water-swollen micelles. This result is to be compared with similar ones gained from other techniques by several authors [623-627], and with the value of 10 to 11 determined by Heil et al. [635,708] through a study of the conductive behavior of water/AOT/*n*-dodecane Winsor IV media. In this connection, it is worth mentioning that Eicke and Denss [609] derived a relatively simple model depicting dissociation and charge transfer processes in W/O microemulsion-type systems, at low water contents. The model developed assumes the existence of a dissociation process of surfactant molecules inside the micellar entities that, consequently, can exchange mutually anionic charges. The system may be considered as made up of two coexisting pseudophases, one being the micellar phase (i.e., all the neutral micellar entities), the other being the solvent pseudophase consisting of a very dilute solution of charged micellar entities associated in pairs. The computation of this model yields for the conductivity of the system an expression that predicts, with increasing water content, an initial increase followed by a decrease that arises from hindering effects of the medium viscosity. Eicke and Denss reported a fair agreement between their model and conductivity data gained on water/AOT/benzene microemulsions for water-AOT molecular ratios lower than 15 or so. More generally, it is noteworthy that the features of the theoretical conductivity curves that can be derived from the model, for water-surfactant molecular ratios up to 15 or so, are consistent with those of the experimental conductivity plots determined, along r-type or $p_h$-type composition paths, by Shah et al. [649-651] and Clausse et al. [573,574,635,708], in the case of systems whose Winsor IV domain is split into two disjointed areas (see Figs. 78 and 81). It appears so that the model proposed by Eicke and Denss is consistent with the existence of the lines $\Gamma_1$ that have been considered by Clausse et al. as reflecting the transition from premicellar systems to proper microemulsions in which all the available surfactant molecules are engaged in the interfacial shells of the water-swollen micelles. In fact, it is probable that the model may be extended as well to systems displaying a unique Winsor IV area, since it is most likely that $\Gamma_1$ lines exist also in that case but cannot be determined easily, owing to the occurrence of the percolative-type conduction phenomena whose effects shadow completely the tiny conductivity variations associated with the progressive hydration of the premicellar entities. All the preceding developments show clearly that, as far as electrical properties are concerned, W/O Winsor IV media formed with double-chained surfactants such as AOT or AY present strong similarities with W/O Winsor IV media incorporating combinations of a ionic surfactant with a higher alkanol, which is confirmed by conclusions arrived at by other authors through self-diffusion studies [632]. Conductivity and permittivity data indicate that, in both cases, the Winsor IV media containing small amounts of water are in fact not microemulsions but dispersions of hydrated surfactant premicellar aggregates. Upon water content increasing beyond a threshold value corresponding to a water-surfactant molecular ratio of 10 or so, the dispersed micellar entities can be considered as inverse micelles with a central core of "free" water. The occurrence of such a transition, especially in systems incorporating AOT as the surfactant, is confirmed by data obtained through other methods [603,610,613,614,623-627]. At high water-surfactant ratios, conductivity and permittivity data suggest the existence, in both types of systems, of anisotropic objects that could be micelle clusters predictive for the final stability breakdown of the system [573,574,635,704, 708].

Only a few studies have been devoted to the electrical properties of microemulsion-type systems formed with nonionic surfactants. It has been already mentioned that water-hydrocarbon systems incorporating nonionic surfactants are highly temperature sensitive, as proved by the phase diagram determinations performed by Shinoda and co-workers [539-543]. Peyrelasse et al. [725,726] investigated the low-frequency conductive and dielectric behavior of W/O-type Winsor IV media made up of water, *n*-undecane, and, as the surfactant, a blend of two octylphenylether polyoxyethylenes of different chain lengths, Octarox 1 (10%) and Octarox 5 (90%), both from Montanoir, France. They found that, all over their realm of existence, the W/O Winsor IV media exhibited, along with a conduction absorption, a dielectric relaxation of the Cole-Cole type [see Eq. (173)] located in the 1-MHz region. The occurrence of this relaxation phenomenon was ascribed to interfacial polarization processes, which would be consistent with the existence of water-swollen micelles. To explain the rather complicated variations of the dielectric relaxation parameters with both the water content and temperature, the authors suggested tentatively that repetitive aggregation processes took place within the systems which were considered as being the most "emulsion-like" upon reaching the water-solubilization end curve. This suggestion was supported by the results obtained concerning the conductive behavior. It was observed that, at any fixed water content, the low-frequency conductivity decreased regularly with increasing the temperature and eventually reached a minimum at the temperature corresponding to the solubilization end point. An interesting application of this unexpected conductive behavior consisted in redetermining the solubilization end curve from systematic conductivity versus temperature measurements, which yielded temperature-water content data in excellent agreement with those obtained from visual observations. A similar conductive and dielectric behavior was put into evidence as well in the case of systems incorporating *n*-dodecane instead of *n*-undecane. Mackay and co-workers [727-728] studied the electrical properties of O/W-type microemulsion systems involving either NaCl or NaF brine as the aqueous component, either *n*-hexadecane or Nujol as the hydrocarbon component and various nonionic surfactants of the ICI Tween series (Tween 40, 60, and 81) [470]. In their first paper [727], specific ion electrode, conductivity, and polarographic data were reported. The authors found that the behavior of the specific electrodes for $Na^+$, $Cl^-$, and $F^-$ was Nernstian in the O/W microemulsions, as in water. The values of the disperse-phase volume fraction determined from specific ion electrode experiments were found to be greater than the values computed from the system compositions and appeared to be also greater than the actual disperse phase volume fractions. The $Cl^-$ and $F^-$ electrodes reflected the continuous phase composition better than the $Na^+$ electrode with which it was observed a roughly constant difference between measured and computed disperse phase volume fractions. Conductivity data were expressed in terms of equivalent conductance. The authors showed that the equivalent conductance variations with the disperse-phase volume fraction (either computed or measured) did not follow Bruggeman's limiting formula (347), which is applicable in the case of O/W macroemulsions [71]. The authors showed that a better fit was obtained with an exponent of 5/2 instead of 3/2 in formula (347), and with the introduction in Eq. (347) of a premultiplying factor to take into account the difference between the computed and the actual values of the disperse phase volume fraction. Accordingly, polarographic measurements yielded diffusion coefficient values that could be fitted with a formula similar to that used for the conductivity data. More details about the conductive behavior were given in a second paper [728]. It was confirmed that the variations of the conductivity of the O/W microemulsion-type media investigated could be fitted correctly with the following formula:

$$\sigma = \sigma_2(1 - a\Phi)^n \tag{411}$$

The disperse phase volume fraction correcting factor was found to be approximately constant and equal to 1.2, and was interpreted as reflecting the binding of water molecules to the disperse hydrocarbon micelles, at a constant water-surfactant ratio. At low hydrocarbon contents, a limiting value of 2.5 was obtained for the exponent n. Upon water content increasing, n decreased down to 1.5 which is the limiting value relevant to O/W-type microemulsions, as expressed by formula (347). In a recent paper [729], the same group of authors has reported low- and high-frequency conductivity data suggesting that in O/W microemulsions formed with nonionic surfactants a great proportion of water (40 to 100%) could be associated with the disperse hydrocarbon micelles. For a mean micelle size of 74 Å or so, this would be equivalent to a shell of two water monolayers, most likely as water of hydration of the surfactant oxyethylene tails. Bostock et al. [730,731] carried out permittivity, conductivity, and pulsed NMR self-diffusion measurements on micellar solutions and microemulsions incorporating the nonionic surfactant tetraethylene glycol dodecylether, the hydrocarbon being a normal alkane (*n*-decane, or *n*-hexadecane). For binary water-surfactant mixtures, no dielectric relaxation was observed over the frequency range 300 kHz to 12 MHz. The frequency independent permittivity values recorded were found to conform satisfactorily with Hanai's limiting formula (346), valid for O/W-type macroemulsions, but deviations appeared at low surfactant contents and high temperatures, [730]. From these results, the authors inferred that all the isotropic regions of the binary phase diagram corresponded to water-external systems with spherical surfactant micelles, except for systems with low surfactant concentrations at high temperatures that could contain ellipsoidal micelles. For ternary water/nonionic surfactant/*n*-heptane systems, permittivity data suggested the existence of separated isotropic regions corresponding to either W/O or O/W-type media. Again, a fair agreement between the experimental data and formulas derived from Hanai's general equation (234) was reported. However, the authors failed to observe, on the supposed W/O media, dielectric relaxations similar to the ones put into evidence by Peyrelasse et al. on close systems [725,726]. A subsequent study of transport properties (conductivity and self-diffusion) in water/nonionic surfactant/*n*-alkane systems yielded somewhat peculiar results that are rather difficult to interpret, in the present state of knowledge [731]. It is readily seen from the preceding developments that it is not possible to arrive at general conclusions on the basis of the few isolated data presently available in the literature as concerns the electrical properties of microemulsion-type systems incorporating nonionic surfactants. The most promising results might be those recently reported by Foster et al. [729], which suggest that hydration phenomena are of importance in nonionic microemulsion systems. It can be hoped that further conductometry and dielectrometry experiments will yield more consistent data, as in the case of microemulsion systems incorporating ionic surfactants.

Finally, it is worth mentioning results obtained on ternary water/alcohol/hydrocarbon systems, with no surfactant added. In the first article of a series of four [732-735], Smith et al. [732] reported conductivity and ultracentrifugation results gained on the ternary system water/2-propanol/*n*-hexane. They claimed that certain compositions were W/O-type microemulsions, though no surfactant was present. The authors suggested that the Winsor IV-type region of the ternary phase diagram could be divided into three adjacent subregions corresponding successively, upon increasing alcohol content at fixed water-hydrocarbon ratios, to W/O-type microemulsions, suspensions of mixed water and alcohol aggregates in hydrocarbon and ternary water-alcohol-hydrocarbon molecular solutions. The boundaries between the three subregions were determined by following the conductivity versus composition variations that exhibited systematic kinks at certain compositions. The existence of boundaries marking structural changes was confirmed later by NMR data [733]. The incorporation to the system of various additives, either a cationic surfactant [732], an alkaline salt [733], or an alkaline hydroxide [734], and the substitution of toluene for *n*-hexane [735] did not change fundamentally the phenomena observed and induced only shifts in the positions of the different boundaries. The authors checked that besides 2-propanol only 1-propanol had comparable effects, which was ascribed to a favorable partitioning of these alcohols [732]. Although they concern very particular systems, the results reported by Smith et al. are quite interesting ones with respect to the electrical properties of Winsor IV media. They show in particular that conductivity variations with system composition reflect structural changes, as in the case of Winsor IV media incorporating ionic surfactants.

## VIII. Concluding Remarks

By following the composite material approach, the dielectric properties of emulsions and closely related systems can be given reliable theoretical descriptions on the basis of the general laws of electromagnetism and of general methods established to study the physical properties of heterogeneous materials. The agreement between experimental data and theoretical predictions derived from Bruggeman's and Hanai's formulas is fairly good when the emulsion systems are formulated in such a way that they consist of well-stabilized dispersions of numerous individual globules imbedded in a continuous matrix. Unstable systems exhibit abnormal dielectric features which are sensitive to the shearing stress applied to them. This cross-linking of dielectric with rheological properties allows one to envisage the use of nondestructive experimental procedures, based on complex permittivity measurements, in view of investigating agglomeration and reticulation processes developing in emulsions. Dielectrometry combined with emulsion technique can yield valuable information as concerns physical properties of liquid aqueous media in the supercooled state, supercooling breakdown phenomena and progressive freezing rates, and structural properties of polycrystalline solids formed with high degrees of thermodynamic irreversibility. From the practical standpoint, emulsion complex permittivity studies provide behavior models and investigation methods applicable to either abiotic or biotic composite materials which are of growing importance in numerous advanced branches of science, technology, and industry. In that respect, further research in the field is still needed, to both refine theoretical models [736] and gain new experimental information on the electrical behavior of emulsions, as correlated to their formulation and thermodynamic conditions.

As revealed by significant new data gained in the past 5 years, so-called microemulsions (more correctly, Winsor IV media), which are submicrometer heterogeneous fluids of higher stability than emulsions, display electrical properties quite different from those of emulsions. However, the composite materials approach appears to be still useful in this case, as demonstrated by the good description of the conductive behavior of certain types of Winsor IV media by means of formulas derived from the percolation and effective medium theories. Conductivity and permittivity data converge with results obtained through other techniques to put into evidence the great structural diversity in Winsor IV media. Depending upon composition factors (but not upon preparative procedures), Winsor IV media appear to behave as either multicomponent molecular solutions, dispersion of hydrated surface-active agent premicellar aggregates, suspensions of inverse or direct swollen micelles, interspersions of aqueous and organic microdomains separated by flexible interfaces and that could be modelized as bicontinuous structures. All these different structures are highly dynamic ones, in contrast with the case of emulsions. Thus, conductometry and dielectrometry have proved to be valuable investigation methods that allowed in particular the demonstration of the existence of structural transitions and to stress the importance of the cosurfactant chemical structure, which has been corroborated through other methods. Many problems are still to be solved as concerns the formation, structure, and behavior of Winsor

IV media, whose potential applications in many industrial fields are numerous and very promising, especially in enhanced oil recovery technology. Within the framework of both fundamental research programs devoted to Winsor IV media physicochemical properties and technical research programs aimed at promoting industrial applications, it is most likely that conductometry and dielectrometry, as well as other electrical methods [737-738], used in conjunction with other techniques, will continue to yield quite useful information [739].

The results reported in the present contribution and summarized above being rather significant and promising, it is to be hoped that they will attract the attention of many scientists and stimulate research in the field of emulsion and microemulsion electrical properties, where there is still a lot to be done.

## Symbols

Several symbols have been deliberately omitted in the following list because they appear in the text casually and with an unambiguous meaning. With a few exceptions, the Systeme International (S.I.) rationalized system of units is used throughout the text.

### General Symbols

*Electromagnetic Symbols*

Vectors

| | |
|---|---|
| $\mathbf{A}$ | magnetic vector potential |
| $\mathbf{B}$ | magnetic induction vector |
| $\mathbf{B}_{n_1}$, $\mathbf{B}_{n_2}$ | normal components of $\mathbf{B}$ on each side of an interface |
| $\mathbf{D}$ | electric induction (or displacement) vector |
| $\mathbf{D}_{n_1}$, $\mathbf{D}_{n_2}$ | normal components of $\mathbf{D}$ on each side of an interface |
| $\mathbf{E}$ | electric field |
| $\mathbf{E}_{n_1}$, $\mathbf{E}_{n_2}$ | normal components of $\mathbf{E}$ on each side of an interface |
| $\mathbf{E}_{T_1}$, $\mathbf{E}_{T_2}$ | tangential components of $\mathbf{E}$ on each side of an interface |
| $\mathbf{E}_0$ | initial electric field created by fixed true electrical charges |
| $\mathbf{E}_i$ | effective internal electric field vector acting upon polarizable molecules |
| $\mathbf{E}_r$ | effective directing electric field vector acting upon polar molecules |
| $\mathbf{H}$ | magnetic field vector |
| $\mathbf{H}_{n_1}$, $\mathbf{H}_{n_2}$ | normal components of $\mathbf{H}$ on each side of an interface |

| | |
|---|---|
| $H_{T1}$, $H_{T2}$ | tangential components of H on each side of an interface |
| J | current density vector |
| $J_{n_1}$, $J_{n_2}$ | normal components of J on each side of an interface |
| P | polarization vector |
| $P_{n_1}$, $P_{n_2}$ | normal components of P on each side of an interface |
| $P_d$ or $P_\infty$ | distortion polarization vector |
| $P_o$ | orientation polarization vector |
| $P_c$ | migration polarization vector |
| $P_s$ | static polarization vector |
| $P_{os}$ | static orientation polarization vector |
| $P_{cs}$ | static migration polarization vector |
| $\mu$ | permanent electric dipole moment of polar molecules |
| p | electric dipole moment of a particle |
| $p_d$ | electric dipole moment of a particle, arising from distortion polarization |
| $p_o$ | electric dipole moment of a particle, arising from orientation polarization |
| $p_c$ | electric dipole moment of a particle, arising from migration polarization |
| $p_s$ | static value of p |

Tensors and Scalars

| | |
|---|---|
| $\phi$ | electric potential |
| $\phi_0$ | initial electric potential created by fixed true electrical charges |
| $\rho$ | true electrical charge volume density |
| $\rho'$ | bound electrical charge volume density |
| $\rho_T$ | $\rho + \rho'$ |
| $\underline{\sigma}$ | true electrical charge surface density |
| $\underline{\sigma}'$ | bound electrical charge surface density |
| $\underline{\sigma}_T$ | $\underline{\sigma} + \underline{\sigma}'$ |
| $\alpha_d$ | distortion polarizability of a molecule |
| $\alpha_0$ | orientation polarizability of a molecule |
| $\varepsilon$ | permittivity (general symbol) |
| $\|\|\hat{\varepsilon}^*\|\|$ | complex permittivity tensor |
| $\hat{\varepsilon}^*$ | complex permittivity in linear, homogeneous, and isotropic (LHI) media |
| $\varepsilon_0$ | permittivity of free space |
| $\varepsilon^*$ | relative complex permittivity in LHI media ($\varepsilon^* = \hat{\varepsilon}^*/\varepsilon_0$) |
| $\varepsilon'$ | real part of $\varepsilon^*$ |
| $\varepsilon''$ | imaginary part of $\varepsilon^*$ (loss factor) |

| | |
|---|---|
| $\varepsilon''_R$ | loss factor arising from dielectric relaxation phenomena |
| $\varepsilon''_C$ | loss factor arising from conduction phenomena |
| $\varepsilon''_T$ | $\varepsilon''_R + \varepsilon''_C$ |
| $\varepsilon_{HF}$ | limiting value of $\varepsilon^*$ at the high-frequency end of a relaxation domain |
| $\varepsilon_{LF}$ | limiting value of $\varepsilon^*$ at the low-frequency end of a relaxation domain |
| $\varepsilon_d$ | distortion polarization phenomena contribution to $\varepsilon^*$ |
| $\varepsilon_s$ | static permittivity |
| $\varepsilon_s - \varepsilon_d$ | dielectric increment, representing the contribution of orientation polarization phenomena to $\varepsilon^*$ |
| $\varepsilon''_m$ | maximum loss factor [$\varepsilon''_m = (\varepsilon_s - \varepsilon_d)/2$, in the case of single Debye-type dielectric relaxations] |
| $\varepsilon'_m$ | value of $\varepsilon'$ corresponding to $\varepsilon''_m$ [$\varepsilon'_m = (\varepsilon_s + \varepsilon_d)/2$, in the case of single Debye-type or Cole-Cole-type dielectric relaxations] |
| $\varepsilon_1$ | local value of $\varepsilon_s$ in the case of inhomogeneous media |
| $\bar{\varepsilon}_1$ | averaged value of $\varepsilon_1$ in the ensemble sense |
| $\tilde{\varepsilon}_1$ | averaged value of $\varepsilon_1$ in the space sense |
| $\varepsilon_s^+$ | upper bound for the effective static permittivity of an inhomogeneous medium |
| $\varepsilon_s^-$ | lower bound for the effective static permittivity of an inhomogeneous medium |
| $\varepsilon_h$ | effective limiting permittivity at the high-frequency end of a migration polarization relaxation domain |
| $\varepsilon_\ell$ | effective limiting permittivity at the low-frequency end of a migration polarization relaxation domain |
| $\varepsilon_\ell - \varepsilon_h$ | dielectric increment of a relaxation arising from migration polarization phenomena |
| n | refractive index |
| $\mu$ | magnetic permeability (general symbol) |
| $\|\|\hat{\mu}^*\|\|$ | complex magnetic permeability tensor |
| $\hat{\mu}^*$ | complex magnetic permeability in LHI media |
| $\mu_0$ | magnetic permeability of free space |
| $\mu^*$ | relative complex magnetic permeability in LHI media ($\mu^* = \hat{\mu}^*/\mu_0$ |
| $\mu'$ | real part of $\mu^*$ |
| $\mu''$ | imaginary part of $\mu^*$ (magnetic loss factor) |
| $\sigma$ | conductivity general symbol |
| $\|\|\sigma^*\|\|$ | complex conductivity tensor |

| | |
|---|---|
| $\sigma^*$ | complex conductivity in LHI media |
| $\sigma'$ | real part of $\sigma^*$ |
| $\sigma''$ | imaginary part of $\sigma^*$ (In the present situation, it is assumed that $\sigma^*$ is real and frequency-independent, and $\sigma$ is used to represent the steady conductivity of homogeneous media.) |
| $\sigma_\ell$ | effective steady conductivity in heterogeneous media |
| $\sigma_\infty$ | preexponential factor in conductivity Arrhenius law |
| $T$ | time constant characteristic for a LHI medium ($T = \varepsilon_0\varepsilon_s/\sigma$) |
| c | phase velocity of electromagnetic signals traveling in free space ($\varepsilon_0\mu_0 c^2 = 1$) |
| v | phase velocity of electromagnetic signals traveling in material media |
| $\omega$ | pulsation or angular frequency of sinusoidal monochromatic electromagnetic signals |
| $\omega_c$ | value of $\omega$ corresponding to maximum dielectric loss in the case of Debye- or Cole-Cole-type relaxations |
| f | frequency ($\omega = 2\pi f$) |
| $f_m$ | value of f corresponding to $\varepsilon''_m$ |
| $f_c$ | critical frequency ($\omega_c = 2\pi f_c$). $f_c = f_m$ in the case of Debye- or Cole-Cole-type dielectric relaxations |
| $\bar{f}_c$ | critical frequency of dielectric relaxations arising from migration polarization phenomena |
| $\tau$ | macroscopic relaxation time ($\omega_c\tau = 1$, or $2\pi\tau f_c = 1$) |
| $\bar{\tau}$ | macroscopic relaxation time corresponding to dielectric relaxation arising from migration polarization phenomena ($2\pi\bar{\tau}\bar{f}_c = 1$) |
| h | frequency-spread parameter in the Cole-Cole formula |
| $\beta$ | frequency-spread parameter in the Davidson-Cole formula |
| u | activation energy (general symbol) |
| $\bar{u}$ | activation energy of dielectric relaxations arising from migration polarization phenomena |
| A | preexponential factor in critical frequency Arrhenius law |
| $\bar{A}$ | preexponential factor in Arrhenius law for $\bar{f}_c$ |
| W | electrostatic energy |
| t | general symbol for time |
| C | molar fraction |
| D | diffusion coefficient |
| M | molar mass |
| N | general symbol for numbers of molecules or particles |
| $\mathcal{N}$ | Avogadro constant |

| | |
|---|---|
| V | volume |
| $\beta^*$ | Trukhan's correcting factor in the Wagner formula |
| $\kappa$ | Debye screening distance |
| $\gamma$ | $\gamma^2 = \kappa^2 + \frac{j\omega}{D}$ |
| $\gamma_i$ | critical exponent in Eq. (392) |
| $\bar{\omega}$ | specific mass |
| $\mathcal{J}$ | flux density vector |
| $\mathcal{J}_{n_1}$, $\mathcal{J}_{n_2}$ | normal components of $\mathcal{J}$ on each side of an interface |
| $\mathcal{X}$ | thermodynamic force |
| $\mathcal{X}_{n_1}$, $\mathcal{X}_{n_2}$ | normal components of $\mathcal{X}$ on each side of an interface |
| $\mathcal{X}_{T_1}$, $\mathcal{X}_{T_2}$ | tangential components of $\mathcal{X}$ on each side of an interface |
| $\Theta$ | phenomenological coefficient in LHI media |
| $\underset{\sim}{\Theta}_{ij}$ | ijth component of the phenomenological coefficient tensor |
| $\Theta_{ij}$ | homogenized value of $\Theta_{ij}$ |
| $\mathcal{P}_i$ | general symbol for physical properties, used in Eq. (392) |
| $\mathcal{P}_i^0$ | adjustable parameter in Eq. (392) |

*Nonelectromagnetic Symbols*

Temperatures

| | |
|---|---|
| T | temperature (general symbol) |
| T* | supercooling breakdown temperature |
| $T_c$ | sample holding temperature |
| $T_0$ | normal melting temperature of a solid (in the case of pure ice, $T_0$ = 273.15 K) |
| $T_s$ | critical temperature in Eq. (392) |

Miscellaneous

| | |
|---|---|
| a | general symbol for the radius of spherical particles |
| j | imaginary number defined by $j^2 = -1$ |
| k | Boltzmann constant |
| $\ell$ | general symbol for the linear dimension of particles |
| p | mass proportion of disperse phase in the frozen state at time t |
| q | heat exchanged during disperse-phase melting process in partially frozen W/O-type emulsions, and recorded through DSC measurements |
| $q_0$ | heat exchanged during disperse-phase melting process in totally frozen W/O-type emulsions, and recorded through DSC measurements |
| $\dot{q}$ | time derivative of q |

Special Symbols Used in the Dielectric Studies of Heterogeneous Systems and Emulsions

Where binary heterogeneous systems are concerned, any of the general electromagnetic symbols reported above refers to component 1 when labeled with subscript 1 and to component 2 when labeled with subscript 2. In the case of disperse systems, component 1 is the disperse phase, and component 2 is the continuous one.

*Special Permittivity and Conductivity Symbols*

| | |
|---|---|
| $\varepsilon_{1s}^{a}$, $\varepsilon_{1s}^{b}$ | static permittivities characteristic for disperse phases 1a and 1b, respectively, in the case of systems of the double spherical dispersion type |
| $(\varepsilon_s)^{ba}$, $(\varepsilon_s)^{ab}$ | bounds for the static permittivity of systems of the double spherical dispersion type |
| $\varepsilon_{1s}^{c}$ | static permittivity characteristic for the core of a spherical particle covered with a shell |
| $\varepsilon_{1s}^{s}$ | static permittivity characteristic for the shell of a spherical particle covered with a shell |
| $\varepsilon_{1d}^{I}$ | real part of $\varepsilon_1^{*I}$, the complex permittivity characteristic for structural inhomogeneities imbedded in ice I lattice |
| $\varepsilon_{1d}^{L}$ | distortion polarization permittivity characteristic for ice I lattice |
| $\varepsilon_{1s}^{L}$ | static permittivity characteristic for ice I lattice |
| $\varepsilon_{1d}^{S}$ | real part of $\varepsilon_1^{*S}$, in the case of dispersions of ice microcrystals |
| $\varepsilon_d$ | high-frequency limiting permittivity of a spherical dispersion system whose disperse phase exhibits an intrinsic dielectric relaxation |
| $\varepsilon_h$ | low-frequency limiting permittivity of a spherical dispersion system whose disperse phase exhibits an intrinsic dielectric relaxation, or high-frequency limiting permittivity of the dielectric relaxation arising from migration polarization phenomena in heterogeneous systems |
| $\varepsilon_\ell$ | low-frequency limiting permittivity of the dielectric relaxation arising from migration polarization phenomena in heterogeneous systems |
| $\varepsilon_i$ | value of the real permittivity at the joining of the two relaxations displayed by dispersions of spherical particles covered with shells |
| $\varepsilon^*$ | general symbol for the complex permittivity of heterogeneous systems and more particularly of emulsions |

$\varepsilon_1^{*a}$, $\varepsilon_1^{*b}$ complex permittivities characteristic for disperse phases 1a and 1b, respectively, in the case of systems of the double spherical dispersion type

$(\varepsilon^*)^{ba}$, $(\varepsilon^*)^{ab}$ bounds for the complex permittivity of systems of the double spherical dispersion type

$\varepsilon_1^{*c}$ complex permittivity characteristic of the core of a spherical particle covered with a shell

$\varepsilon_1^{*s}$ complex permittivity characteristic for the shell of a spherical particle covered with a shell

$\varepsilon_1^{*I}$ complex permittivity characteristic for structural inhomogeneities imbedded in ice I lattice

$\varepsilon_1^{*L}$ complex permittivity characteristic for ice I lattice

$\sigma_1^c$ steady conductivity characteristic for the core of a spherical particle covered with a shell

$\sigma_1^s$ steady conductivity characteristic for the shell of a spherical particle covered with a shell

$\sigma_1^I$ steady conductivity characteristic for structural inhomogeneities imbedded in ice I lattice

$\sigma_\ell$ effective steady conductivity of heterogeneous systems

$\sigma_{LF}$ low-frequency conductivity in emulsion systems sensitive to mechanical agitation

*Frequencies*

$f_c$ critical frequency of the relaxation displayed by a spherical dispersion system whose disperse phase exhibits an intrinsic dielectric relaxation

$f_c(0)$, $f_c(1)$ values of $f_c$ for $\Phi = 0$ and $\Phi = 1$

$\bar{f}_c$ critical frequency of the relaxation arising from migration polarization phenomena in spherical dispersion systems

$\bar{f}_c(0)$, $\bar{f}_c(1)$ values of $\bar{f}_c$ for $\Phi = 0$ and $\Phi = 1$

$f_m$ frequency corresponding to maximum dielectric loss

*Miscellaneous*

$a$ external radius of a spherical particle

$a_c$ core radius of a shell-covered spherical particle

$d$ shell thickness

$k$ surfactant/cosurfactant mass ratio in microemulsion-type systems

$\ell_a$, $\ell_o$, $\ell_s$ molecular lengths of alcohol, oil, and surfactant, respectively

| | |
|---|---|
| $p_h$, $p_s$, $p_w$ | in microemulsion-type systems, mass fractions of hydrocarbon, surface active agent, and water, respectively |
| $p_c$ | critical value of $p_w$ corresponding to a clear-to-turbid transition in microemulsion-type systems |
| $r$ | hydrocarbon/surface-active agent mass ratio in microemulsion-type systems |
| $r_c$ | inverse swollen-micelle "chemical" radius (hard-sphere radius) |
| $r_h$ | inverse swollen-micelle hydrodynamic radius |
| $r_w$ | inverse swollen-micelle aqueous core radius |
| $P_w$ | aqueous-phase mass fraction in emulsions |
| $\Phi$ | component 1 volume fraction in heterogeneous systems (in spherical dispersion-type systems, disperse-phase volume fraction) |
| $1 - \Phi$ | component 2 volume fraction in heterogeneous systems (in spherical dispersion-type systems, continuous-phase volume fraction) |
| $\Phi_w$ | disperse water volume fraction in microemulsion-type systems |
| $\Phi_1^c$ | critical volume fraction corresponding to the random close packing of identical spheres in disperse systems ($\Phi_1 = 0.637$) |
| $\Phi_2^c$ | critical volume fraction corresponding to the cubic close packing of identical spheres in disperse systems ($\Phi_2^c = 0.741$) |
| $\Phi^P$ | disperse-phase volume fraction corresponding to a percolation threshold |
| $\Phi_0^P$ | cutoff value of $\Phi$ in Eq. (408): $\Phi_0^P = 1/3$ |
| $\Phi_a$ | disperse component 1a volume fraction in systems of the double spherical dispersion type |
| $\Phi_b$ | disperse component 1b volume fraction in systems of the double spherical dispersion type |
| $\psi_a$ | disperse component 1a volume fraction in the matrix within which component 1b is considered to be dispersed, in the case of systems of the double spherical dispersion type |
| $\psi_b$ | disperse component 1b volume fraction in the matrix within which component 1a is considered to be dispersed, in the case of systems of the double spherical dispersion type |
| $\underline{\phi}$ | core volume fraction in a shell-covered spherical particle [$\underline{\phi} = (a_c/a)^3$] |
| $\psi$ | volume fraction of structural inhomogeneities imbedded in ice I lattice |
| $B$ | preexponential factor in Eq. (387) |
| $N$ | Green-type viscometer cup rotation frequency |
| $N_0$ | decay factor in Eq. (387) |

*Symbols for Formulas*

| | |
|---|---|
| $\mathfrak{W}$ | Wiener's or Wagner's formula |
| $\mathfrak{B}$ | Bruggeman's formula |
| $\mathfrak{C}$ | Hanai's formula |

Abbreviations

| | |
|---|---|
| CMC | critical micelle concentration |
| DSC | differential scanning calorimetry |
| DTA | differential thermal analysis |
| EMT | effective medium theory |
| ESR | electron spin resonance |
| FFVT | fixed frequency-variable temperature measurements |
| FTVF | fixed temperature-variable frequency measurements |
| HLB | hydrophile-lipophile balance |
| LHI | linear homogeneous and isotropic media |
| MWS | Maxwell-Wagner-Sillars effect |
| NMR | nuclear magnetic resonance |
| O/W | oil-in-water type systems |
| PCS | photon correlation spectroscopy |
| PIT | phase-inversion temperature |
| rpm | revolutions per minute |
| SANS | small-angle neutron scattering |
| SCS | self-consistent scheme |
| W/O | water-in-oil type systems |
| WOR | water-oil volume ratio |
| WI, WII<br>WIII, WIV | symbols in Winsor's classification for fluid nonordered multicomponent media |

## REFERENCES


1. P. Becher, *Emulsions: Theory and Practice*, 2d ed., Reinhold, New York, 1965, Chap. 3, pp. 85-94.
2. T. Hanai, N. Koizumi, and R. Gotoh, in *Rheology of Emulsions* (P. Sherman, ed.), Pergamon, Oxford, 1963, pp. 91-113.
3. T. Hanai, in *Emulsion Science* (P. Sherman, ed.), Academic, London, 1968, Chap. 5, pp. 353-478.
4. M. Clausse, *Contribution à l'Etude des Propriétés Diélectriques des Emulsions*, Thèse de Doctorat ès Sciences Physiques, Université de Pau, France, 1971, (Ref. B.S. CNRS 32-160-17334).
5. L. K. H. van Beek, in *Progress in Dielectrics* (J. B. Birks, ed.) Vol. 7, Heywood, London, 1967, pp. 69-114.
6. R. J. Meakins, in *Progress in Dielectrics* (J. B. Birks, ed.), Vol. 3, Heywood, London, 1961, pp. 151-202.

7. S. S. Dukhin, in *Surface and Colloid Science* (E. Matijevic, ed.), Vol. 3, Wiley-Interscience, New York, 1971, pp. 83-165.
8. S. S. Dukhin and V. N. Shilov, *Dielectric Phenomena and the Double Layer in Disperse Systems and Polyelectrolytes*, Wiley, New York, 1974.
9. J. B. Hasted, *Aqueous Dielectrics*, Chapman and Hall, London, 1973.
10. N. E. Hill, W. E. Vaughan, A. H. Price, and M. Davies, *Dielectric Properties and Molecular Behavior*, Van Nostrand Reinhold, London, 1969.
11. J. B. Miles, Jr. and H. P. Robertson, *Phys. Rev. 40*:583 (1932).
12. E. Heymann, *Kolloid Z. 66*:229 (1934).
13. R. S. Alwitt, *J. Phys. Chem. 73*:1052 (1969).
14. C. T. O'Konski, *J. Phys. Chem. 64*:605 (1960).
15. H. P. Schwan, G. Schwarz, J. Maczuk, and H. Pauly, *J. Physiol. Chem. 66*:2626 (1962).
16. G. P. South and E. H. Grant, *Biopolymers 13*:1777 (1974).
17. F. van der Touw and M. Mandel, *Biophys. Chem. 2*:218 (1974).
18. F. van der Touw and M. Mandel, *Biophys. Chem. 2*:230 (1974).
19. O. F. Schanne and E. R. P.-Ceretti, *Impedance Measurements in Biological Cells*, Wiley-Interscience, New York, 1978.
20. H. Baessler, R. B. Beard, and M. M. Labes, *J. Chem. Phys. 52*:2292 (1970).
21. J. P. Poley, *Appl. Sci. Res. 4*:337 (1955).
22. H. Looyenga, *Molec. Phys. 9*:501 (1965).
23. F. Franks and D. J. G. Ives, *Quarterly Rev. 20*:1 (1966).
24. P. Sixou, P. Daumezon, and P. Dansas, *J. Chem. Phys. 64*:824 (1967).
25. D. C. Dube and R. Parshad, *J. Phys. Chem. 73*:3236 (1969).
26. R. Heitz and P. Daumezon, *J. Chem. Phys. 68*:1 (1971).
27. Y. N. Rao and T. S. Ramu, *IEEE Trans. EI-7*:195 (1972).
28. T. S. Ramu and Y. N. Rao, *IEEE Trans. EI-8*:55 (1973).
29. M. Planck, *Electromagnétisme*, Felix Alcan, Paris, 1939.
30. J. A. Stratton, *Electromagnetic Theory*, McGraw-Hill, New York, 1941.
31. J. R. Reitz and F. J. Milford, *Foundations of Electromagnetic Theory*, Addison-Wesley, Reading, Mass., 1962.
32. W. K. H. Panofsky and M. Phillips, *Classical Electricity and Magnetism*, 2d ed., Addison-Wesley, Reading, Mass, 1964.
33. G. Bruhat, *Electricité*, 8ème éd., Masson et Cie, Paris, 1963.
34. E. Durand, *Electrostatique et Magnétostatique*, Masson et Cie, Paris, 1953.
35. L. D. Landau and E. M. Lifshitz, *Electrodynamics of Continuous Media*, Pergamon, London, 1960.
36. C. J. F. Bottcher, *Theory of Electric Polarisation*, Elsevier, Amsterdam, 1952.
37. C. P. Smyth, *Dielectric Behavior and Structure*, McGraw-Hill, New York, 1955.
38. J. Barriol, *Les Moments Dipolaires*, Gauthier-Villars, Paris, 1957.
39. H. Fröhlich, *Theory of Dielectrics*, 2d ed., Clarendon, Oxford, 1963.
40. V. V. Daniel, *Dielectric Relaxation*, Academic, London, 1967.
41. R. H. Cole, in *Progress in Dielectrics* (J. B. Birks ed.), Vol. 3, Heywood, London, 1961, pp. 47-99.
42. J. J. O'Dwyer and E. Harting, in *Progress in Dielectrics* (J. B. Birks, ed.), Vol. 7, Heywood, London, 1967, pp. 1-44.
43. A. R. von Hippel, *Dielectrics and Waves*, Wiley, New York, 1954.
44. A. R. von Hippel, *Dielectric Materials and Applications*, Wiley, New York, 1954.
45. A. R. von Hippel, *Les Diélectriques et leurs Applications*, Dunod, Paris, 1961.

46. R. Freymann and M. Soutif, *La Spectroscopie Hertzienne Appliquée à la Chimie*, Dunod, Paris, 1960.
47. A. Lebrun, *Méthodes de mesure de la permittivité complexe des diélectriques du continu à l'infrarouge*, Contrat DGRST 62FR107, Rapport IREL, Faculté des Sciences, Université de Lille, France, 1963.
48. A. F. Dunn, in *Progress in Dielectrics* (J. B. Birks, ed.), Vol. 7, Heywood, London, 1967, pp. 45-67.
49. J. Chamberlain and G. W. Chantry, *High Frequency Dielectric Measurement*, IPC Science and Technology, Guilford, England, 1972.
50. Y. I. Frenkel, *Kolloid. Zhur. 10*:148 (1948).
51. Y. I. Frenkel and E. M. Fradkina, *Kolloid. Zhur. 10*:241 (1949).
52. E. M. Fradkina, *Zhur. Eksptl. Teoret. Fiz. 20*:1011 (1950).
53. E. M. Fradkina and S. F. Khumunin, *Kolloid. Zhur. 18*:604 (1956).
54. S. F. Khumunin, *Kolloid. Zhur. 21*:731 (1959).
55. M. Kubo and S. Nakamura, *Bull. Chem. Soc. Jap. 26*:318 (1953).
56. B. G. Ben'kovskii, *Kolloid. Zhur. 14*:10 (1952).
57. S. S. Bhatnagar, *J. Chem. Soc. 117*:542 (1920).
58. S. G. Lifshits and V. P. Teodorovich, *Energet. Byull. 8*:16 (1947).
59. R. E. Meredith and C. W. Tobias, *J. Electrochem. Soc. 108*:286 (1961).
60. T. Hanai, N. Koizumi, and R. Gotoh, *Kolloid-Z. 167*:41 (1959).
61. T. Hanai, *Kolloid-Z. 177*:57 (1961).
62. T. Hanai, *Kolloid-Z. 171*:23 (1960).
63. C. A. R. Pearce, *Brit. J. Appl. Phys. 6*:113 (1955).
64. J. S. Dryden and R. J. Meakins, *Proc. Phys. Soc. London 70 B*:427 (1957).
65. I. D. Chapman, *J. Phys. Chem. 75*:537 (1971).
66. S. Noguchi and Y. Maeda, *Agr. Biol. Chem. 37*:1531 (1973).
67. R. E. Mudgett, D. I. C. Wang, and S. A. Goldblith, *J. Food Sci. 39*:632 (1974).
68. I. M. Hodge and C. A. Angell, *J. Chem. Phys. 68*:1363 (1978).
69. T. Hanai, N. Koizumi, T. Sugano, and R. Gotoh, *Kolloid-Z. 171*:20 (1960).
70. T. Hanai, N. Koizumi, and R. Gotoh, *Kolloid-Z 184*:143 (1962).
71. T. Hanai, N. Koizumi, and R. Gotoh, *Bull. Inst. Chem. Res. Kyoto Univ. 40*:240 (1962).
72. T. Hanai and N. Koizumi, *Bull. Inst. Chem. Res. Kyoto Univ. 53*:153 (1975).
73. T. Hanai, *Kolloid-Z 175*:61 (1961).
74. T. Hanai, *Bull. Inst. Chem. Res. Kyoto Univ. 39*:341 (1962).
75. C. Lafargue, M. Clausse, and J. Lachaise, *C. R. Acad. Sci., Ser. B. 274*:540 (1972).
76. M. Clausse, *C. R. Acad. Sci., Sér. B 274*:887 (1972).
77. M. Clausse, *C. R. Acad. Sci., Sér. B 277*:261 (1973).
78. M. Clausse, *C. R. Acad. Sci., Sér. B 275*:427 (1972).
79. M. Clausse, *Colloid Polymer Sci. 255*:40 (1977).
80. J. P. Le Petit, G. Delbos, A. M. Bottreau, Y. Dutuit, C. Marzat, and R. Cabanas, *J. Microwave Power 12*:335 (1977).
81. M. Clausse, *C. R. Acad. Sci., Sér. B 274*:649 (1972).
82. M. Clausse and R. Royer, in *Colloid and Interface Science* (M. Kerker, ed.), Vol. 2, Academic, New York, 1976, pp. 217-231.
83. C. Boned, J. Peyrelasse, M. Clausse, B. Lagourette, J. Alliez, and L. Babin, *Colloid Polymer Sci. 257*:1073 (1979).
84. T. Hanai and N. Koizumi, *Bull. Inst. Chem. Res. Kyoto Univ. 54*:248 (1976).
85. T. Hanai, A. Ishikawa, and N. Koizumi, *Bull. Inst. Chem. Res. Kyoto Univ. 55*:376 (1977).

86. J. Lachaise, *C. R. Acad. Sci., Sér. B 274*:1095 (1972).
87. M. Clausse and J. Lachaise, *C. R. Acad. Sci., Sér. B 275*:797 (1972).
88. J. Lachaise and M. Clausse, *C. R. Acad. Sci., Sér. B 276*:287 (1973).
89. J. Lachaise, *Contribution à l'Etude par Spectroscopie Hertzienne de la Congélation Monotherme de l'Eau Dispersée au Sein d'Emulsions*, Thèse de Doctorat ès Sciences Pysiques, Université de Pau, France, 1973.
90. J. Lachaise and M. Clausse, *J. Phys. D: Appl. Phys. 8*:1227 (1975).
91. C. Lafargue, J. Lachaise, and M. Clausse, *Proc. 8th I.C.P.S.*, Vol. 1, Giens, France, 1974, pp. 492-504.
92. M. Mathurin, *Contribution à l'Etude des Propriétés Diélectriques des Emulsions de Solutions Salines Aqueuses Sufondues*, Thèse de Doctorat de Spécialité, Université de Pau, France, 1972.
93. B. Lagourette, *Etude des Propriétés Diélectriques de Microcristaux de Glace au Voisinage de la Température de Fusion*. Thèse de Doctorat ès Sciences Physiques, Université de Pau, France, 1977.
94. B. Lagourette, *J. Physique 37*:945 (1976).
95. B. Lagourette, C. Boned, and R. Royer, *J. Physique 37*:955 (1976).
96. B. Lagourette, C. Boned, and L. Babin, *J. Physique 38*:825 (1977).
97. C. Boned, B. Lagourette, and M. Clausse, *J. Glaciology 21*:696 (1973).
98. C. Boned, B. Lagourette, and M. Clausse, *J. Glaciology 22*:145 (1979).
99. C. Lafargue and L. Babin, *Proc. 12 Colloque Ampère* (R. Servant and A. Charru, eds.), Amsterdam: North-Holland, 1964, pp. 374-379.
100. I. D. Chapman, *J. Phys. Chem. 72*:33 (1968).
101. G. Evrard, B. Lagourette, and J. P. Montfort, *C. R. Acad. Sci., Sér. B 279*:461 (1974).
102. G. Evrard, B. Lagourette, and J. P. Montfort, *C. R. Acad. Sci., Sér. B 279*:491 (1974).
103. J. P. Montfort, B. Lagourette, G. Evrard, and A. Le Traon, *C. R. Acad. Sci., Sér. B 280*:397 (1975).
104. C. Boned, *Contribution à l'Etude des Propriétés Diélectriques d'Echantillons Obtenus par Solidification de Solutions Aqueuses de Sels d'Ammonium*, Thèse de Doctorat ès Sciences Physiques, Université de Pau, France, 1973.
105. C. Boned and A. Barbier, in *Physics and Chemistry of Ice* (E. Whalley, S. J. Jones, and L. W. Gold, eds.), Royal Society of Canada, Ottawa, 1973, pp. 209-211.
106. C. Boned, *C. R. Acad. Sci., Sér. B 275*:801 (1972).
107. C. Lafargue and C. Boned, *C. R. Acad. Sci., Sér. B 276*:315 (1973).
108. C. Boned, *C. R. Acad. Sci., Sér. B 276*:539 (1973).
109. C. Boned, *C. R. Acad. Sci., Sér. B 282*:125 (1976).
110. C. Boned, *J. Physique 37*:165 (1976).
111. C. Boned, H. Saint-Guirons, and R. Cazaban-Marque, *J. Chim. Phys. 73*:367 (1976).
112. H. Saint-Guirons, *Etude de l'Energie d'Activation de Relaxation Dipolaire de Microcristaux de Glace Dopée de Chlorure d'Ammonium et de Chlorures Alcalins*, Thèse de Doctorat de Spécialité, Université de Pau, France, 1976.
113. H. Saint-Guirons, *C. R. Acad. Sci., Sér. B 282*:189 (1976).
114. H. Saint-Guirons, *J. Phys. C.: Solid State Phys. 11*:L-343 (1978).
115. J. C. Maxwell, *A Treatise on Electricity and Magnetism*, Clarendon, Oxford, 1892.
116. K. W. Wagner, *Arch. Elektrotech. 2*:371 (1914).
117. K. W. Wagner, *Arch. Elektrotech. 3*:100 (1914).
118. R. W. Sillars, *Proc. Inst. Elec. Engrs. London 80*:378 (1937).

119. E. M. Trukhan, *Soviet Physics-Solid State 4*:2560 (1963).
120. C. T. O'Konski, *J. Chem. Phys. 23*:1559 (1955).
121. V. N. Shilov and S. S. Dukhin, *Kolloid. Zh. 31*:706 (1969).
122. V. N. Shilov and S. S. Dukhin, *Kolloid. Zh. 32*:117 (1970).
123. V. N. Shilov and S. S. Dukhin, *Kolloid. Zh. 32*:293 (1970).
124. P. Debye and H. Falkenhagen, *Phys. Z. 29*:121 (1928).
125. H. Falkenhagen, *Electrolytes*, Oxford University Press, 1934.
126. L. Babin, D. Clausse, I. Sifrini, F. Broto, and M. Clausse, *J. Physique 39*:L-359 (1978).
127. B. Vonnegut, *J. Colloid Sci. 3*:563 (1948).
128. D. Turnbull, *J. Chem. Phys. 20*:411 (1952).
129. G. Pound and V. K. La Mer, *J. Amer. Chem. Soc. 74*:2323 (1952).
130. P. G. Fox, *Nature 184*:546 (1959).
131. R. L. Cormia, F. P. Price, and D. Turnbull, *J. Chem. Phys. 37*:1333 (1962).
132. L. Bosio and A. Defrain, *C. R. Acad. Sci. 254*:1020 (1962).
133. L. Bosio and A. Defrain, *C. R. Acad. Sci. 258*:4929 (1964).
134. L. Bosio, *Métaux, Corrosion, Industries 483*:421 (1965).
135. L. Bosio, *Métaux, Corrosion, Industries 484*:451 (1965).
136. L. Bosio, A. Defrain, and M. Dupont, *J. Chim. Phys. 68*:542 (1971).
137. L. Babin, *C. R. Acad. Sci. 256*:3009 (1963).
138. L. Babin, *Contribution à l'Etude de la Surfusion de Solutions Salines Aqueuses*, Thèse de Doctorat ès Sciences Physiques, Université de Bordeaux, France, 1966.
139. P. Monge, *Contribution à l'Etude de la Surfusion du Mercure*, Thèse de Doctorat ès Sciences Physiques, Université de Bordeaux, France, 1968.
140. P. Xans, *Contribution à l'Etude de la Surfusion et de la Sursaturation de Solutions Salines Aqueuses*, Thèse de Doctorat ès Sciences Physiques, Université de Bordeaux, Faculté des Sciences de Pau, France, 1970.
141. M. Lère-Porte, *Contribution à l'Etude de la Surfusion de L'Eau Lourde*, Thèse de Doctorat ès Sciences Physiques, Université de Bordeaux, Faculté des Sciences de Pau, France, 1970.
142. D. Clausse, *Contribution à l'Etude de la Surfusion et de la Sursaturation de Solutions Aqueuses de Chlorure d'Ammonium*, Thèse de Doctorat ès Sciences Physiques, Université de Pau, France, 1972.
143. C. Lafargue, L. Babin, P. Xans, D. Clausse, and M. Lère-Porte, *C. R. Acad. Sci., Sér. B 277*:245 (1973).
144. P. Xans, *C. R. Acad. Sci., Sér. B 277*:321 (1973).
145. P. Xans, *C. R. Acad. Sci., Sér. B 277*:533 (1973).
146. D. Clausse, *C. R. Acad. Sci., Sér. C 278*:985 (1974).
147. D. Clausse, J. P. Dumas, and F. Broto, *C. R. Acad. Sci., Sér. B 279*:415 (1974).
148. P. Xans and G. Barnaud, *C. R. Acad. Sci., Sér. B. 280*:25 (1975).
149. D. Clausse, *J. Chim. Phys. 72*:229 (1975).
150. D. Clausse, *J. Chim. Phys. 73*:333 (1976).
151. D. Clausse, P. Xans, and G. Barnaud, *J. Chim. Phys. 73*:829 (1976).
152. M. Lère-Porte and D. Clausse, *C. R. Acad. Sci., Sér. B 281*:77 (1975).
153. J. P. Dumas, D. Clausse, and F. Broto, *Thermochim. Acta 13*:261 (1975).
154. M. Lemercier, *Contribution à l'Etude de la Surfusion de l'Etain*, Thèse de Doctorat de Spécialité, Université de Pau, France, 1975.
155. M. Lemercier and D. Clausse, *Mem. Sci. Rev. Metallurg. 72*:753 (1975).
156. F. Broto and D. Clausse, *J. Phys. C: Solid State Phys. 9*:4251 (1976).
157. F. Broto, D. Clausse, L. Babin and M. Clausse, *Colloid Polymer Sci. 257*:302 (1969).

158. F. Broto, D. Clausse, L. Babin, and M. Mercier, *J. Chim. Phys. 75*:908 (1978).
159. F. Broto, *Etude de la Cristallisation Monotherme de l'Eau et de l'Eau Lourde Surfondues, à l'Etat Dispersé*, Thèse de Doctorat ès Sciences Physiques, Université de Pau, France, 1979.
160. J. P. Dumas, *Etude de la Rupture de Métastabilitié et du Polymorphisme de Corps Organiques*, Thèse de Doctorat ès Sciences Physiques, Université de Pau, France, 1976.
161. J. P. Dumas, *J. Phys. C: Solid State Phys. 9*: L-143 (1976).
162. J. P. Dumas, *C. R. Acad. Sci., Sér. C 284*:257 (1977).
163. J. P. Dumas, *C. R. Acad. Sci., Sér. C 284*:549 (1977).
164. J. P. Dumas, *C. R. Acad. Sci., Sér. C 284*:817 (1977).
165. J. P. Dumas, *C. R. Acad. Sci., Sér. C 284*:857 (1977).
166. H. Saint-Guirons and P. Xans, *C. R. Acad. Sci., Sér. C 285*:455 (1977).
167. P. Xans and J. P. Dumas, *J. Chim. Phys. 74*:751 (1977).
168. P. Xans, H. Saint-Guirons, and J. P. Dumas, *J. Phys. C: Solid State Phys. 10*:L-267 (1977).
169. G. R. Wood and A. G. Walton, *J. Appl. Phys. 7*:41 (1970).
170. G. A. Kozlov and A. A. Ravdel, *Kolloid. Zh. 33*:847 (1971).
171. D. H. Rasmussen and A. P. MacKenzie in *Water Structure at the Water-Polymer Interface* (H. H. G. Jellinek, ed.), Plenum, New York, 1972, pp. 126-145.
172. D. H. Rasmussen, and A. P. MacKenzie, *J. Chem. Phys. 59*:5003 (1973).
173. D. H. Rasmussen, A. P. Mackenzie, C. A. Angell, and J. C. Tucker, *Science 181*:342 (1973).
174. C. A. Angell, J. Shuppert, and J. C. Tucker, *J. Phys. Chem. 77*:3092 (1973).
175. Y. Miyazawa and G. M. Pound, *J. Cryst. Growth 23*:45 (1974).
176. D. H. Rasmussen, and C. R. Loper, *Acta Metallurgica 23*:1215 (1975).
177. D. H. Rasmussen and C. R. Loper, *Acta Metallurgica 24*:117 (1976).
178. H. Kanno, R. J. Speedy, and C. A. Angell, *Science 189*:880 (1975).
179. J. A. Koutsky, A. G. Walton, and E. Baer, *J. Appl. Phys. 38*:1832 (1967).
180. F. Franks, *Water: A Comprehensive Treatise*, 6 vols., Plenum, New York and London, 1977.
181. R. L. Smith-Rose, *Proc. Roy. Soc. Lond 140*:359 (1933).
182. R. L. Smith-Rose, *J. Inst. Elect. Engr. 75*:221 (1934).
183. R. L. Smith-Rose, *Proc. Phys. Soc. 47*:923 (1935).
184. A. W. Straiton and C. W. Tolbert, *J. Franklin Inst. 246*:13 (1948).
185. A. Cownie and L. S. Palmer, *Proc. Phys. Soc. 65*:295 (1952).
186. M. Salaruddin and B. V. Khasbardar, *Indian J. Technol. 5*:296 (1967).
187. N. Suresh, J. C. Callaghan, and A. E. Creelman, *J. Microwave Power 2*:129 (1967).
188. W. A. Cummings, *J. Appl. Phys. 23*:768 (1952).
189. Y. Ozawa and D. Kuroiwa, in *Microwave Propagation in Snowy Districts* (Y. Asami, ed.), Research Inst. Hokkaido University, Sapporo, Japan, 1958, p. 51.
190. S. Evans, *J. Glaciology 5*:773 (1965).
191. K. Fujino, *Low Temp. Sci. 25*:127 (1967).
192. J. W. Glen and J. G. Paren, *J. Glaciology 15*:15 (1975).
193. W. J. Fitzgerald and J. G. Paren, *J. Glaciology 15*:39 (1975).
194. J. G. Paren and J. W. Glen, *J. Glaciology 21*:173 (1978).
195. M. Steru, *Mesures et Contrôle Industriel 258*:865 (1958); *259*:957 (1958); *260*:33 (1959).; *261*:105 (1959).

196. M. Steru, *Mesures et Contrôle Industriel June*:91 (1964).
197. M. Steru, *Ind. Chim. Belge 32*:147 (1967).
198. G. P. De Loor, in *Proc. 11 Colloque Ampère* (J. Smidt, ed.), North-Holland, Amsterdam, 1963, pp. 288-292.
199. G. P. De Loor, *Appl. Sci. Res. 9*:297 (1962).
200. G. P. De Loor and F. W. Meijboom, *J. Food Technol. 1*:313 (1966).
201. G. P. De Loor, *J. Microwave Power 3*:67 (1968).
202. J. Paquet, *Onde électrique 44*:940 (1964).
203. J. B. Hasted and M. A. Shah, *Brit. J. Appl. Phys. 15*:825 (1964).
204. H. B. Taylor, *Ind. Electronics 3*:66 (1965).
205. L. E. Way, *Aust. J. Instrum. Contr. 23*:19 (1967).
206. E. D. Ponomarenko, *Trudy Leningrad. Politekh. Inst. 276*:39 (1967).
207. R. Bartnikas, *I.E.E.E. Trans., Elec. Insul. 2*:33 (1967).
208. M. Beyer and H. Langer, *Elektrotech. Z. 88*:569 (1967).
209. V. S. Roife, *Pribory Sistemy Upravl. 5*:57 (1967).
210. R. T. Lin, *Forest Prod. J. 17*:61 (1967).
211. M. A. Berliner and V. A. Ivanov, *Pribory Sistemy Upravl. 3*:14 (1967).
212. Y. A. Chernyshov and V. G. Gruzintsev, *Koks. Khim. 4*:10 (1967).
213. W. A. G. Voss, *J. Microwave Power 4*:165 (1969).
214. W. A. G. Voss, *J. Microwave Power 4*:323 (1969).
215. V. K. Benzar, *Proc. Conf. on Radiophysical Methods Applied in Building Industry*, Minsk, USSR, 1970, pp. 27-36.
216. J. E. Algie and R. A. Gamble, *Kolloid. Zh. 251*:554 (1973).
217. A. Kraszewski, *J. Microwave Power 8*:323 (1973).
218. A. Kraszewski, *J. Microwave Power 9*:295 (1974).
219. N. B. Troitskii, *Soviet Physics J. (USA) 16*:1335 (1975).
220. I. A. Romanovskii, *Ind. Lab. (USA) 41*:1845 (1975).
221. E. F. Burton and L. G. Turnbull, *Proc. Roy. Soc. London 158*:182 (1937).
222. M. T. Rolland and R. Bernard, *C. R. Acad. Sci. 232*:1098 (1951).
223. R. Freymann, *J. Chim. Phys. 50*:C-27 (1953).
224. P. Abadie, R. Charbonnière, A. Gidel, P. Girard, and A. Guilbot, *J. Chim. Phys. 50*:C-47 (1953).
225. A. Guilbot, R. Charbonnière, P. Abadie, and P. Girard, *Die Stärke 11-12*:327 (1960).
226. B. J. Goldsmith and J. Muir, *Trans. Faraday Soc. 56*:1656 (1960).
227. R. M. Barrer and E. A. Saxon-Napier, *Trans. Faraday. Soc. 58*:145 (1962).
228. D. Rosen, *Trans. Faraday Soc. 59*:2178 (1963).
229. S. Takashima and H. P. Schwan, *J. Phys. Chem. 69*:4176 (1965).
230. R. A. Weiler and J. Chaussidon, *Clays and Clay Minerals 16*:147 (1968).
231. A. Chapoton, A. Lebrun, and G. Ravalitera, *C. R. Acad. Sci., Sér. C 271*:525 (1970).
232. K. Umeya and T. Kanno, *Bull. Chem. Soc. Jap. 46*:1660 (1973).
233. V. P. Tomaselli and M. H. Shamos, *Biopolymers 12*:353 (1973).
234. R. Martens, H. Nagerl, and F. Freund, *Ind. Chim. Belge 38*:514 (1973).
235. R. Martens, H. Nagerl, and F. Freund, *Ind. Chim. Belge 38*:519 (1973).
236. T. L. Chelidze, *Colloid J. USSR (USA) 35*:573 (1973).
237. C. Guyot, *Onde Électrique 33*:421 (1953).
238. T. D. Callinan, R. M. Roe, and J. B. Romans, *Anal. Chem. 28*:1911 (1956).
239. T. Fukushima and T. Ichimura, *Bull. Jap. Petrol. Inst. 1*:3 (1959).
240. W. C. Wolfe, *Anal. Chem. 35*:1884 (1963).
241. T. Fukushima and T. Itoh, *Bull. Jap. Petrol. Inst. 9*:53 (1967).
242. A. D. Dzhabrailov, *Neft Khozyaistvo 45*:46 (1967).

243. I. S. Liderman, *Trudy GIPKh 57*:59-68 (1967).
244. V. G. Pustynnikov and A. D. Dzhabrailov, *Zavodsk. Lab. 33*:582 (1967).
245. I. S. Liderman, *Neftezavod. Neftekhim. Oborudovanie 2*:7 (1967).
246. R. A. Lipshtein and E. N. Shtern, *Khim. Tekhnol. Topl. Masel'. 12*:51 (1967).
247. C. A. R. Pearce, *Brit. J. Appl. Phys.* 5:136 (1954).
248. C. R. Kaye and H. Seager, *J. Pharm. Pharmacol. 17*:S-92 (1965).
249. A. Parts, *Nature 155*:236 (1946).
250. A. Voet, *J. Phys. Colloid. Chem. 51*:1037 (1947).
251. A. Voet and L. R. Suriani, *J. Colloid Sci. 6*:155 (1951).
252. A. Voet and L. R. Suriani, *J. Colloid Sci. 7*:1 (1952).
253. A. Voet, *Am. Ink Maker 31*:34 (1953).
254. A. Voet, *Am. Ink Maker 35*:34 (1957).
255. A. Voet, *J. Phys. Chem. 61*:301 (1957).
256. A. Voet, *J. Phys. Chem. 66*:2259 (1962).
257. M. J. Foster and D. J. Mead, *J. Appl. Phys. 22*:705 (1951).
258. A. Bondi and C. J. Penther, *J. Phys. Chem.* 57:72 (1953).
259. A de Waele and G. L. Lewis, in *Proc. 2d Int. Congress Surface Activity*, Vol. 3, pp. 21-27, 1957.
260. R. Nasuhoglu, *Comm. Fac. Sc., Univ. Ankara 4*:108 (1952).
261. A. Bondi and H. Diamond, *J. Colloid. Sci 12*:510 (1957).
262. S. S. Vysotsky, A. D. Zayochkhovsky, and V. A. Kargin, *Kolloid. Zh. 13*:333 (1951).
263. Y. F. Deinega, A. V. Dumansky, and O. D. Kurilenko, *Kolloid. Zh. 15*: 351 (1953).
264. Y. F. Deinega, A. V. Dumansky, G. V. Vinogradov, and V. P. Pavlov, *Kolloid. Zh. 22*:16 (1960).
265. Y. F. Deinega, A. V. Dumansky, and G. V. Vinogradov, *Kolloid Zh. 23*; 25 (1961).
266. Y. F. Deinega and G. V. Vinogradov, *Kolloid. Z. 198*:77 (1964).
267. G. S. Bright, *NLGJ Spokesman 30*:84 (1966).
268. Y. F. Deinega, A. V. Dumansky, and O. D. Kurilenko, *Kolloid. Zh. 15*: 234 (1953).
269. G. G. Petrzhik, A. A. Trapeznikov, and T. I. Korotina, *Colloid J. USSR (USA) 34*:324 (1972).
270. J. M. P. Papenhuyzen, in *Proc. 5th Int. Congress on Rheology* (S. Onogi, ed.), Vol. 2, Tokyo University Press, 1970, pp. 353-374.
271. J. Mewis, A. J. B. Spaull, and J. Helsen, *Nature 203*:618 (1975).
272. J. A. Helsen, R. Govaerts, G. Schoukens, J. DeGraeuwe, and J. Mewis, *J. Phys. E: Sci. Instrum. 11*:139 (1978).
273. Y. A. Atanov and D. I. Kusnetzov, *Inzh. Fiz. Zhur. 29*:620 (1972).
274. D. Goldschmidt and P. Le Goff, *Chem. Eng. Sci., 18*:805 (1963).
275. C. Prost and P. Le Goff, *Génie Chimique 91*:6 (1964).
276. P. Le Goff and C. Prost, *Génie Chimique 95*:1 (1966).
277. C. Prost, *Chem. Eng. Sci. 22*:1283 (1967).
278. J. Bordet, F. Coeuret, P. Le Goff, and F. Vergnes, *Powder Technology 6*:253 (1972).
279. P. Debye, *Phys. Z. 13*:97 (1912).
280. A. F. Chudnovsky, *Thermophysical Characteristics of Disperse Materials*, Fizmatgiz, Moscow, 1962.
281. V. I. Odelevsky, *Zh. Tekhn. Fiz. 21*:667 (1951); *21*:678 (1951).
282. G. N. Dul'nev, Y. P. Zarichnyak, and V. V. Novikov, *Inz. Fiz. Zh. 31*: 150 (1976).

283. R. W. P. King, in *Handbuch der Physik* (S. Flügge, ed.), Vol. 16, Springer-Verlag, Berlin, 1958, pp. 165-284.
284. K. S. Cole and R. H. Cole, *J. Chem. Phys.* *9*:341 (1941).
285. D. W. Davidson and R. H. Cole, *J. Chem. Phys.* *19*:1484 (1951).
286. D. J. Denney, *J. Chem. Phys.* *27*:259 (1957).
287. D. W. Marquardt, R. G. Bennett, and E. J. Burrell, *J. Molec. Spectrosc.* *7*:269 (1961).
288. D. W. Marquardt, *J. Soc. Ind. Appl. Math.* *11*:431 (1963).
289. R. J. Sheppard, B. P. Jordan, and E. H. Grant, *J. Phys. D: Appl. Phys.* *3*:1759 (1970).
290. E. H. Grant, T. J. Buchanan, and H. F. Cook, *J. Chem. Phys.* *26*:156 (1957).
291. J. B. Hasted, in *Water: A Comprehensive Treatise* (F. Franks, ed.), Vol. 1, Plenum, New York and London, 1972, Chap. 7. pp. 255-305.
292. R. J. Sheppard and E. H. Grant, *Adv. Molec. Relaxation Processes* *6*: 61 (1974).
293. N. Riehl, B. Bullemer, and H. Engelhardt, *Physics of Ice*, Plenum, New York, 1969.
294. E. Whalley, S. J. Jones, and L. W. Gold, *Physics and Chemistry of Ice*, Royal Soc. of Canada, Ottawa, 1973.
295. J. W. Glen, R. J. Adie, D. M. Johnson, D. R. Homer, and A. D. Macqueen, *J. Glaciology* *21*: n°85 (1978).
296. J. A. Krumhansl, in *Amorphous Magnetism* (H. O. Hooper and A. M. De Graaf, eds.), Plenum, New York and London, 1973, pp. 15-25.
297. H. Fricke, *Phys. Rev.* *24*:575 (1924).
298. H. Fricke, *J. Gen. Physiol.* *6*:741 (1924).
299. H. Fricke, and S. Morse, *Phys. Rev.* *25*:361 (1925).
300. H. Fricke, *Phys. Rev.* *26*:678 (1925).
301. H. Fricke, *J. Appl. Phys.* *24*:644 (1953).
302. H. Fricke, *J. Phys. Chem.* *57*:934 (1953).
303. H. Fricke, *J. Phys. Chem.* *59*:168 (1955).
304. H. Fricke, *Phys. Rev.* *26*:682 (1925).
305. H. Fricke and S. Morse, *J. Gen. Physiol.* *9*:153 (1926).
306. H. Fricke and H. J. Curtis, *Nature* *133*:651 (1934).
307. H. Fricke and H. J. Curtis, *Nature* *134*:102 (1934).
308. H. Fricke and H. J. Curtis, *Nature* *135*:436 (1935).
309. H. Fricke, *Nature* *172*:731 (1953).
310. W. Woodside and J. H. Messmer, *J. Appl. Phys.* *32*:1688 (1961).
311. H. W. Russel, *J. Amer. Ceram. Soc.* *18*:1 (1935).
312. R. H. Wang and J. G. Knudsen, *Ind. Eng. Chem.* *50*:1667 (1958).
313. L. E. Nielsen, *Polymer Rheology*, Dekker, New York and Basel, 1977, Chap. 9 and 10, pp. 133-177.
314. E. H. Kerner, *Proc. Phys. Soc.* *69*:808 (1956).
315. Z. Hashin and S. Shtrikman, *J. Mech. Phys. Solids* *11*:127 (1963).
316. R. Hill, *J. Mech. Phys. Solids* *13*:213 (1965).
317. J. C. Halpin, *J. Composite Materials* *3*:732 (1969).
318. S. W. Tsai, *Formulas for the elastic properties of fiber-reinforced composites*, Rep. AD 834851, U.S. Dept. of Commerce, 1968.
319. M. N. Miller, *J. Math. Phys.* *10*:2005 (1969).
320. J. Korringa, *J. Math. Phys.* *14*:509 (1973).
321. T. B. Lewis and L. E. Nielsen, *J. Appl. Polymer Sci.* *14*:1449 (1970).
322. L. E. Nielsen, *Appl. Polymer Symp.* *12*:249 (1969).
323. L. E. Nielsen, *J. Appl. Phys.* *41*:4626 (1970).

324. L. E. Nielsen, *Ind. Eng. Chem. Fund. 13*:17 (1974).
325. L. E. Nielsen, *J. Phys. D: Appl. Phys. 7*:1549 (1974).
326. V. E. Legg and F. J. Given, *Bell Syst. Techn. J. 19*:385 (1940).
327. R. C. Mildner and P. C. Woodland, in *Proc. 13th Symp. Technical Progress in Communication Wires and Cables*, Atlantic City, N.J., 1964.
328. G. Bahder and F. G. Garcia, *I.E.E.E. Trans. PAS-3*:917 (1971).
329. R. M. Scarisbrick, *J. Phys. D: Appl. Phys. 6*:2098 (1973).
330. F. Bueche, *J. Appl. Phys. 44*:532 (1973).
331. W. E. Kock, *Bell System Techn. J. 27*:58 (1948).
332. J. M. Kelly, J. O. Stenoien, and D. W. Isbell, *J. Appl. Phys. 24*:258 (1953).
333. I. L. Jashnani and R. Lemlich, *Ind. Eng. Chem. Fund. 14*:131 (1975).
334. R. Lemlich, *J. Colloid Interface Sci. 54*:107 (1978).
335. E. F. Goodridge, *Chem. Proc. Eng. 49*:93 (1968).
336. P. Le Goff, F. Vergnes, F. Coeuret, and J. Bordet, *Ind. Eng. Chem. 61*:8 (1969).
337. D. C. Eardley, D. Handley, and S. P. S. Andrew, *Electrochim. Acta 18*:839 (1973).
338. Z. Hashin and S. Shtrikman, *Phys. Rev. 130*:129 (1963).
339. M. H. Cohen and J. Jortner, *Phys. Rev. Lett. 30*:696 (1973).
340. H. L. Frisch and J. M. Hammersley, *J. Soc. Ind. Appl. Math. 11*:894 (1963).
341. V. K. S. Shante and S. Kirkpatrick, *Adv. Phys. 20*:325 (1971).
342. S. Kirkpatrick, *Rev. Modern Phys. 45*:574 (1973).
343. I. Webman, J. Jortner and M. H. Cohen, *Phys. Rev. B 11*:2885 (1975).
344. D. Stroud, *Phys. Rev. B 12*:3368 (1975).
345. I. Webman, J. Jortner, and M. H. Cohen, *Phys. Rev. B 15*:5712 (1977).
346. B. E. Springett, *Phys. Rev. Lett. 31*:1463 (1973).
347. Lord Rayleigh, *Phil. Mag. 34*:481 (1892).
348. J. C. Garnett, *Phil. Trans. Roy. Soc. London 203*:385 (1904).
349. J. C. Garnett, *Phil. Trans. Roy. Soc. London 205*:237 (1906).
350. O. Wiener, *Abh. Sächs. Akad. Wiss. 32*:509 (1912).
351. K. Lichtenecker, *Phys. Zeits. 10*:1005 (1909).
352. K. Lichtenecker, *Phys. Zeits. 19*:374 (1918).
353. K. Lichtenecker, *Phys. Zeits. 25*:169,193,225,666 (1924).
354. K. Lichtenecker, *Phys. Zeits. 27*:115 (1926).
355. K. Lichtenecker, *Phys. Zeits. 28*:417 (1927).
356. K. Lichtenecker and K. Rother, *Phys. Zeits. 32*:255 (1931).
357. D. A. G. Bruggeman, *Ann. Phys. Lpz. 24*:636 (1935).
358. C. J. F. Bottcher, *Rec. Trav. Chim. Pays-Bas 64*:47 (1945).
359. R. Landauer, *J. Appl. Phys. 23*:779 (1952).
360. J. Brown, in *Progress in Dielectrics* (J. B. Birks, ed.), Vol. 2, Heywood, London, 1960, pp. 193-225.
361. R. Guillien, *Ann. Physique 16*:205 (1941).
362. J. B. Birks, *Proc. Phys. Soc. 60*:282 (1948).
363. J. C. Van Vessem and J. M. Bijvoet, *Rec. Trav. Chim. Pays-Bas 67*:191 (1948).
364. A. Helaine, S. Le Montagner, and J. Le Bot, *C. R. Acad. Sci. 232*:403 (1951).
365. W. Niesel, *Ann. Physik 6*:336 (1952).
366. W. Niesel, *Ann. Physik 6*:410 (1953).
367. J. A. Reynolds, *Proc. Phys. Soc. 57*:267 (1954).
368. P. Abadie, I. Epelboin, and B. Pistoulet, *C. R. Acad. Sci. 231*:762 (1950).
369. J. C. Bluet, I. Epelboin, and D. Quivy, *C. R. Acad. Sci. 246*:246 (1958).

370. B. R. Eichbaum, *J. Electrochemical Soc. 106*:804 (1959).
371. K. S. Ramakrishna Rao, *Indian J. Pure Appl. Phys. 4*:447 (1965).
372. A. Morabin, A. Tete and R. Santini, *Rev. Gen. Elec. 76*:1504 (1967).
373. J. Paletto, R. Goutte, and L. Eyraud, *Chim. Mod. 11*:201 (1966).
374. J. Paletto, *Etude de la Permittivité de Matériaux Composites*. Thèse de Doctorat ès Sciences Physiques, Université Claude Bernard, Lyon, France, 1972.
375. J. Paletto, R. Goutte, and L. Eyraud, *J. Solid State Chem. 6*:58 (1973).
376. J. Paletto, G. Grange, R. Goutte, and L. Eyraud, *J. Phys. D: Appl. Phys. 7*:78 (1974).
377. L. Eyraud, *Diélectriques Solides Anisotropes et Ferroélectricité*, Gauthier-Villars, Paris, 1967.
378. B. P. Pradhan and R. C. Gupta, *Dielectrics 1*:195 (1964).
379. D. C. Dube, *J. Phys. D.: Appl. Phys. 3*:1648 (1970).
380. D. C. Dube and R. Parshad, *J. Phys. D: Appl. Phys. 3*:677 (1970).
381. D. R. Goyal, A. Kumar, and K. K. Srivastava, *Indian J. Pure Appl. Phys. 16*:10 (1978).
382. D. Polder and J. H. Van Santen, *Physica 12*:257 (1946).
383. A. P. Altshuller, *J. Phys. Chem. 58*:544 (1954).
384. H. Looyenga, *Physica 31*:401 (1965).
385. A. Beer, *Einleitung in die Höhere Optik*, S. 35 Vieweg, 1853.
386. A. Krasewski, S. Kulinski, and M. Matuszewski, *J. Appl. Phys. 47*:1275 (1976).
387. A. Kraszewski, *J. Microwave Power 12*:215 (1977).
388. J. P. Poley, *Physica 19*:298 (1953).
389. D. J. Bergman, *Phys. Rev. B 14*:4304 (1976).
390. D. J. Bergman, *Physics Rep. 43*:377 (1978).
391. Z. Hashin and S. Shtrikman, *J. Appl. Phys. 33*:3125 (1962).
392. W. F. Brown, *J. Chem. Phys. 23*:1514 (1955).
393. W. F. Brown, *Trans. Soc. Rheology 9*:357 (1965).
394. J. A. Reynolds and J. M. Hough, *Proc. Phys. Soc. 70*:769 (1957).
395. G. P. De Loor, *Appl. Sci. Res. 3*:479 (1954).
396. G. P. De Loor, *Arch. Sci. Genève 9*:37 (1956).
397. D. F. Rushman and M. A. Strivens, *Proc. Phys. Soc. 59*:1011 (1947).
398. C. A. R. Pearce, *Brit. J. Appl. Phys. 6*:358 (1955).
399. W. F. Brown, in *Handbuch der Physik* (S. Flügge, ed.), Vol. 17, Springer-Verlag, Berlin, 1956, pp. 104-106.
400. L. Hartshorn and J. A. Saxton, in *Handbuch der Physik* (S. Flügge, ed.), Vol. 16, Springer-Verlag, Berlin, 1958, pp. 706-710.
401. D. Quivy, *Bull. Soc. Fr. Elect. 8(88)*:241 (1958).
402. J. Loeb, *Onde Electrique 42*:613 (1962).
403. P. Sixou, *Etude des Propriétés Diélectriques des Matéraux Hétérogènes, (Couches), Influence de la Polarisation Interfaciale*, Thèse de Doctorat ès Sciences Physiques, Université de Paris-Orsay, France, 1965.
404. R. E. Meredith and C. W. Tobias, *J. Appl. Phys. 31*:1270 (1960).
405. R. E. Meredith and C. W. Tobias, in *Advances in Electrochemistry and Electrochemical Engineering* (P. Delahay and C. W. Tobias eds.), Vol. 2, Wiley-Interscience, New York and London, 1962, Chap. 2, pp. 15-47.
406. W. R. Tinga, W. A. G. Voss, and D. F. Blossey, *J. Appl. Phys. 44*:3897 (1973).
407. L. E. Nielsen, *Predicting the Properties of Mixtures*, Dekker, New York and Basel, 1978.
408. R. Landauer, in *Proc. Conf. Electrical, Transport and Optical Properties of Inhomogeneous Media* (ETOPIM), Ohio State University, Columbus, Ohio, 1977.

409. Z. Hashin and S. Shtrikman, *J. Franklin Inst. 271*:423 (1961).
410. M. Beran and J. Molyneux, *Nuovo Cimento 30*:1406 (1963).
411. M. Beran, *Nuovo Cimento 38*:771 (1965).
412. Z. Hashin, *J. Composite Materials 2*:284 (1968).
413. J. M. Peterson and J. J. Hermans, *J. Composite Materials 3*:338 (1969).
414. M. N. Miller, *J. Math. Phys. 10*:1988 (1969).
415. S. Prager, *J. Chem. Phys. 50*:4305 (1969).
416. W. E. A. Davies, *J. Phys. D: Appl. Phys. 4*:318 (1970).
417. W. E. A. Davies, *J. Phys. D: Appl. Phys. 7*:120 (1974).
418. W. E. A. Davies, *J. Phys. D: Appl. Phys. 7*:1016 (1974).
419. D. J. Bergman, in *Proc. Conf. Electrical, Transport and Optical Properties of Inhomogeneous Media* (ETOPIM), Ohio State University, Columbus, Ohio, 1977.
420. M. Hori, *J. Math. Phys. 18*:487 (1977).
421. E. Sanchez-Palencia, *C. R. Acad. Sci., Sér. A 271*:1129 (1970).
422. E. Sanchez-Palencia, *C. R. Acad. Sci., Sér. A 272*:1410 (1971).
423. E. Sanchez-Palencia, *Int. J. Eng. Sci. 12*:331 (1974).
424. J. L. Lions, in *Boundary Value Problems for Linear Evolution Partial Differential Equations* (H. G. Garnir, ed.), NATO Advanced Study Institutes Series, Vol. 29, Reidel Pub. Co., Dordrecht, Netherlands, and Boston, 1977, pp. 175-238.
425. A. Bensoussan, J. L. Lions, and G. Papanicolaou, *Asymptotic Analysis for Periodic Structures*, Elsevier, New York, 1978.
426. K. S. Cole, *J. Gen. Physiol. 12*:29 (1928).
427. I. Runge, *Z. Tech. Physik 6*:61 (1925).
428. M. M. Z. Kharadly and W. Jackson, *Proc. Inst. Elect. Eng. 100*:199 (1953).
429. K. Günther and D. Heinrich, *Z. Physik 185*:345 (1965).
430. L. Lewin, *J. Inst. Elect. Eng. 94*:65 (1947).
431. R. W. Corkum, *Proc. Inst. Radio Eng. 40*:574 (1952).
432. I. Naiki, K. Fujita, and S. Matsumura, *Mem. Fac. Ind. Arts Kyoto Tech. Univ., Jap. 8*:1 (1959).
433. R. E. De La Rue and C. W. Tobias, *J. Electrochem. Soc. 106*:827 (1959).
434. J. Volger, in *Progress in Semi-Conductors* (A. F. Gibson, ed.), Vol. 4, Wiley, New York, 1960, pp. 206-236.
435. J. D. Cross and J. Hart, *Brit. J. Appl. Phys. 17*:311 (1966).
436. J. M. Wimmer and N. M. Tallan, *J. Appl. Phys. 37*:3728 (1966).
437. P. Dansas and P. Sixou, *Rev. Gen. Elec. 76*:726 (1967).
438. S. Mounier and P. Sixou, in *Physics of Ice* (N. Riehl, B. Bullemer, and H. Engelhardt, eds.), Plenum, New York, 1969, pp. 562-570.
439. S. Dasgupta and W. P. Conner, *J. Appl. Phys. 46*:204 (1975).
440. A. Lebrun, *Rev. Gen. Elect. 74*:948 (1965).
441. M. Mandel, *Physica 27*:827 (1961).
442. R. Goffaux, *Bull. Scient. AIM 81*:9 (1969).
443. R. Goffaux and R. Coelho, *Rev. Gen. Elect. 78*:619 (1969).
444. R. Goffaux, *Rev. Gen. Elect. 81*:9 (1972).
445. M. Clausse, *Colloid Polymer Sci. 253*:1020 (1975).
446. B. V. Hamon, *Aust. J. Phys. 6*:304 (1953).
447. L. K. H. van Beek, in *Proc. 11 Colloque Ampère* (J. Smidt, ed.), North-Holland, Amsterdam, 1963, pp. 229-234.
448. L. K. H. van Beek, J. Booy, and H. Looyenga, *Appli. Sci. Res. 12*:57 (1965).
449. T. Hanai, D. A. Haydon, and J. Taylor, *Proc. Roy. Soc. London 281*: 377 (1964).

450. G. P. De Loor, *Arch. Sci. Genève 9*:41 (1956).
451. G. P. De Loor, *Appl. Sci. Res. 11*:310 (1964).
452. A. I. Derevyanko and O. D. Kurilenko, *Colloid J. USSR (USA) 33*:168 (1971).
453. A. I. Derevyanko, V. S. Sperkach, and O. D. Kurilenko, *Colloid J. USSR (USA) 37*:232 (1975).
454. S. G. Zvereva and T. D. Shermergor, *Sov. Phys. Techn. Phys. 22*:1048 (1977).
455. H. Pauly and H. P. Schwan, *Z. Naturforsch. 14*:125 (1959).
456. T. Hanai, N. Koizumi, and A. Irimajiri, *Biophys. Struct. Mechanism 1*:285 (1975).
457. W. R. Redwood, S. Takashima, H. P. Schwan, and T. E. Thompson, *Biochim. Biophys. Acta (Amst.) 255*:557 (1972).
458. G. Schwarz, *J. Phys. Chem. 66*:2636 (1962).
459. J. M. Schuur, *J. Phys. Chem. 68*:2407 (1964).
460. R. de Backer and A. Watillon, *J. Colloid Interface Sci. 54*:69 (1976).
461. R. de Backer and A. Watillon, *J. Colloid Interface Sci. 43*:277 (1973).
462. J. Briant, S. Le Montagner and G. Le Floch, in *Proc. 4 Int. Cong. Chem. Phys. Appl. Surface Active Subst.*, Vol. 2, Gordon and Breach, New York, 1964, pp. 161-169.
463. D. C. Henry, *Proc. Roy. Soc. London 133*:106 (1931).
464. D. C. Henry, *Trans. Faraday Soc. 44*:1021 (1948).
465. F. Booth, *Trans. Faraday Soc. 44*:955 (1948).
466. F. Booth, *Proc. Roy. Soc. London 203*:514 (1950).
467. J. T. G. Overbeek, *Philips Res. Rep. 1*:315 (1946).
468. J. T. G. Overbeek, *Advan. Colloid Sci. 3*:97 (1950).
469. J. B. Hasted and M. Shahidi, *Nature 262*:777 (1976).
470. Atlas Chemical Industries Inc. (ICI Americas Inc.), *The Atlas HLB System*, 2 ed., Wilmington, Del., 1963.
471. P. Sherman, in *Emulsion Science* (P. Sherman, ed.), Academic, London, 1968, Chap. 4, pp. 217-351.
472. R. J. Kuo, R. J. Ruch, and R. R. Myers, in *Emulsions, Latices and Dispersions* (P. Becher and M. N. Yudenfreund, eds.), Dekker, New York and Basel, 1978.
473. M. Clausse and P. Sherman, *C. R. Acad. Sci., Ser. C 279*:919 (1974).
474. M. Clausse, P. Sherman, and R. J. Sheppard, *J. Colloid Interface Sci. 56*:123 (1976).
475. H. F. Cook, *Brit. J. Appl. Phys. 2*:295 (1951).
476. H. F. Cook, *Brit. J. Appl. Phys. 3*:249 (1952).
477. F. Gornick, G. S. Ross, and L. J. Frolen, *Polymer Preprints 7*:82 (1966).
478. G. T. Butorin and V. P. Skripov, *Soviet Physics—Crystallography 17*: 322 (1972).
479. W. Drost-Hansen, *Ind. Eng. Chem. 61*:10 (1969).
480. C. G. Malmberg, and A. A. Maryott, *J. Res. Natl. Bur. Std. 56*:1 (1956).
481. E. W. Rusche and W. B. Good, *J. Chem. Phys. 45*:4667 (1966).
482. J. P. Le Petit, in *Physics and Chemistry of Ice* (E. Whalley, S. J. Jones, and L. W. Gold, eds.), Roy. Soc. of Canada, Ottawa, 1973, pp. 204-207.
483. J. P. Le Petit and J. Peyrelasse, *J. Chim. Phys. 71*:383 (1974).
484. C. Dodd and G. N. Roberts, *Proc. Phys. Soc. 62*:814 (1950).
485. R. J. Speedy and C. A. Angell, *J. Chem. Phys. 65*:851 (1976).
486. F. H. Stillinger and A. Rahman, *J. Chem. Phys. 57*:1281 (1972).
487. F. H. Stillinger, *Phil. Trans. Roy. Soc. London 278 B*:97 (1977).

488. C. P. Smyth and J. Hitchock, *J. Amer. Chem. Soc. 54*:4631 (1932).
489. R. P. Auty and R. H. Cole, *J. Chem. Phys. 20*:1309 (1952).
490. F. Humbel, F. Jona, and P. Scherrer, *Helvetica Physica Acta 26*:17 (1953).
491. H. Gränicher, C. Jaccard, P. Scherrer, and A. Steinemann, *Helvetica Physica Acta 30*:553 (1957).
492. C. Jaccard, *Helvetica Physica Acta 32*:89 (1959).
493. M. Eigen, L. de Maeyer, and H. C. Spatz, *Ber. Bunsenges. Phys. Chem. 68*:19 (1964).
494. B. Bullemer, I. Eisele, H. Engelhardt, N. Riehl, and P. Seige, *Solid State Commun. 6*:663 (1968).
495. G. S. Kell, *J. Chem. Eng. Data 12*:67 (1967).
496. G. S. Kell, in *Handbook of Chemistry and Physics* (R. C. Weast, ed.), F-5, 56th ed., CRC Press, Cleveland, 1975-1976.
497. L. Lliboutry, *Traité de Glaciologie*, Masson et Cie, Paris, 1964.
498. G. Evrard, in *Physics and Chemistry of Ice* (E. Whalley, S. J. Jones, and L. W. Gold, eds.), Roy. Soc. of Canada, Ottawa, 1973, pp. 199-203.
499. C. Lafargue, S. Bourgeois, and G. Evrard, *J. Glaciology 20*:359 (1978).
500. L. M. Prince, in *Emulsions and Emulsion Technology* (K. J. Lissant, ed.), *Surfactant Science Series*, Vol. 6, Dekker, New York, 1974, Part 1, Chap. 3, pp. 125-177.
501. M. Bavière, *Rev. Institut Français du Petrole, 29*(1):41 (1974).
502. L. M. Prince, *Microemulsions. Theory and Practice*, Academic, New York, 1977.
503. D. O. Shah and R. S. Schechter, *Improved Oil Recovery by Surfactant and Polymer Flooding*, Academic, New York, 1977.
504. K. L. Mittal., ed., *Micellization, Solubilization and Microemulsions*, Vols. 1 and 2, Plenum, New York, 1977; *Solution Chemistry of Surfactants*, Vols. 1 and 2, Plenum, New York, 1979.
505. M. Rosoff, in *Progess in Surface and Membrane Science* (J. F. Danielli, M. D. Rosenberg, and D. A. Cadenhead, eds.), Vol. 12, Academic, New York, 1978, pp. 405-477.
506. J. T. G. Overbeek, *Faraday Disc. Chem. Soc. 65*:7 (1978).
507. D. O. Shah, ed., *Surface Phenomena in Enhanced Oil Recovery*, Plenum, New York, 1981.
508. K. Shinoda, *Solvent Properties of Surfactant Solutions, Surfactant Science Series*, Vol. 2, Dekker, New York, 1967.
509. T. P. Hoar and J. H. Schulman, *Nature (London) 152*:102 (1943).
510. J. H. Schulman and T. S. McRoberts, *Trans. Faraday Soc. 42B*:165 (1946).
511. J. H. Schulman and D. P. Riley, *J. Colloid Sci. 3*:383 (1948).
512. J. H. Schulman and J. A. Friend, *J. Colloid Sci. 4*:497 (1949).
513. J. E. Bowcott and J. H. Schulman, *Z. Elektrochem. 59*:283 (1955).
514. J. H. Schulman, W. Stoeckenius, and L. M. Prince, *J. Phys. Chem. 63*:1677 (1959).
515. W. Stoeckenius, J. H. Schulman, and L. M. Prince, *Kolloid. Z. 169*:170 (1960).
516. J. H. Schulman and J. B. Montagne, *Ann. N.Y. Acad. Sci. 92*:336 (1961).
517. C. E. Cooke and J. H. Schulman in *Surface Chemistry*, (P. Ekwall, K. Groth, and V. Runnström-Reio, eds.), Munksgaard, Copenhagen, 1965, pp. 231-251.
518. I. A. Zlochower and J. H. Schulman, *J. Colloid Interface Sci. 24*:115 (1967).

519. P. A. Winsor, *Trans. Faraday Soc.* *44*:376 (1948); *44*:451 (1948).
520. P. A. Winsor, *Trans. Faraday Soc.* *46*:762 (1950).
521. E. C. Lumb, *Trans. Faraday Soc.* *47*:1049 (1951).
522. P. A. Winsor, *J. Phys. Chem.* *56*:391 (1952).
523. J. Bromilow and P. A. Winsor, *J. Phys. Chem.* 57:889 (1953).
524. E. D. Lumb and P. A. Winsor, *Ind. Eng. Chem.* *45*:1086 (1953).
525. P. A. Winsor, *Nature* *173*:81 (1954).
526. P. A. Winsor, *Solvent Properties of Amphiphilic Compounds*, Butterworths, London, 1954.
527. P. A. Winsor, *J. Colloid Sci.* *10*:88 (1955).
528. P. A. Winsor, *Chem. Ind. London* *23*:632 (1960).
529. C. A. Gilchrist, J. Rogers, G. Steel, E. G. Vaal, and P. A. Winsor, *J. Colloid Interface Sci.* *25*:300 (1967).
530. P. A. Winsor, *Chem. Rev.* *68*:1 (1968).
531. S. R. Palit, V. A. Moghe, and B. Biswas, *Trans. Faraday Soc.* *55*:463 (1959).
532. M. B. Mathews and E. Hirschhorn, *J. Colloid Sci,* *8*:86 (1953).
533. C. R. Singleterry, *J. Amer. Oil Chemists Soc.* *32*:446 (1955).
534. L. Mandell, K. Fontell, and P. Ekwall, *Adv. Chem. Ser.* *63*:89 (1967).
535. P. Ekwall, L. Mandell and K. Fontell, *Mol. Crystals* *8*:157 (1969).
536. P. Ekwall, *J. Colloid Interface Sci.* *29*:16 (1969).
537. P. Ekwall, L. Mandell, and K. Fontell, *J. Colloid Interface Sci.* *31*:508 (1969).
538. P. Ekwall, L. Mandell, and K. Fontell, *J. Colloid Interface Sci.* *33*:215 (1970).
539. K. Shinoda, *J. Colloid Interface Sci.* *14*:4 (1967).
540. H. Saito and K. Shinoda, *J. Colloid Interface Sci.* *24*:10 (1967).
541. K. Shinoda and T. Ogawa, *J. Colloid Interface Sci.* *24*:56 (1967).
542. K. Shinoda and H. Saito, *J. Colloid Interface Sci.* *26*:70 (1968).
543. K. Shinoda and H. Kunieda, *J. Colloid Interface Sci.* *42*:381 (1973).
544. S. I. Ahmad, K. Shinoda, and S. Friberg, *J. Colloid Interface Sci.* *47*:32 (1974).
545. K. Shinoda and S. Friberg, *Adv. Colloid Interface Sci.* *4*:281 (1975).
546. G. Gillberg, H. Lehtinen, and S. Friberg, *J. Colloid Interface Sci.* *33*:40 (1970).
547. S. Friberg, in *Microemulsions. Theory and Practice* (L. M. Prince, ed.), Academic, New York, 1977, pp. 133-146.
548. S. Friberg and I. Buraczewska, in *Micellization, Solubilization and Microemulsions* (K. L. Mittal, ed.), Vol. 2, Plenum, New York, 1977, pp. 791-799.
549. S. Friberg and I. Buraczewska, *Prog. Colloid Polymer Sci.* *63*:1 (1978).
550. E. Sjöblom and S. Friberg, *J. Colloid Interface Sci.* *67*:16 (1978).
551. I. Danielsson, M. R. Hakala, and M. Jorpes-Friman, in *Solution Chemistry of Surfactants* (K. L. Mittal, ed.), Vol. 2, Plenum, New York, 1979, pp. 659-671.
552. M. Podzimek and S. Friberg, *J. Dispersion Sci. Technol.* *1*:341 (1980).
553. H. Sagitani and S. Friberg, *J. Dispersion Sci. Technol.* *1*:151 (1980).
554. M. Dvolaitzky, M. Guyot, M. Lagües, J. P. Le Pesant, R. Ober, C. Sauterey, and C. Taupin, *J. Chem. Phys.* *69*:3279 (1978).
555. M. Dvolaitzky, M. Lagües, J. P. Le Pesant, R. Ober, C. Sauterey, and C. Taupin, in *Proc. Surface Active Agents Symp. Soc. Chem. Ind.*, London, 1979, pp. 165-171.
556. C. Taupin, J. P. Cotton, and R. Ober, *J. Appl. Cryst.* *11*:613 (1978).
557. M. Lagües, R. Ober, and C. Taupin, *J. Phys. Lett. Fr.* *39*:L-487 (1978).

558. M. Lagües, *J. Phys. Lett. Fr. 40*:L-331 (1979); *C. R. Acad. Sci., Sér. B 288*:339 (1979).
559. M. Dvolaitzky, M. Lagües, J. P. Le Pesant, R. Ober, C. Sauterey, and C. Taupin, *J. Phys. Chem. 84*:1532 (1980).
560. R. Ober and C. Taupin, *J. Phys. Chem. 84*:2418 (1980).
561. M. Lagües and C. Sauterey, *J. Phys. Chem. 84*:3503 (1980).
562. W. G. M. Agterof, J. A. J. Van Zomeren, and A. Vrij, *Chem. Phys. Lett. 43*:363 (1976).
563. A. A. Calje, W. G. M. Agterof and A. Vrij in *Micellization, Solubilization and Microemulsions* (K. L. Mittal, ed.), Vol. 2, Plenum, New York, 1977, pp. 779-790.
564. D. J. Cebula, L. Harding, R. H. Ottewill, and P. N. Pusey, *Colloid Polymer Sci., 258*:973 (1980).
565. A. Graciaa, J. Lachaise, A. Martinez, M. Bourrel, and C. Chambu, *C. R. Acad. Sci. Sér. B 282*:547 (1976).
566. A. Graciaa, J. Lachaise, A. Martinez, and A. Rousset, *C. R. Acad. Sci. Sér. B 286*:157 (1978).
567. A. Graciaa, J. Lachaise, P. Chabrat, L. Letamendia, J. Rouch, C. Vaucamps, M. Bourrel, and C. Chambu, *J. Phys. Lett. Fr. 38*:L-253 (1977).
568. A. M. Cazabat, M. Lagües, D. Langevin, R. Ober, and C. Taupin, *C. R. Acad. Sci. Sér. B. 287*:25 (1978).
569. A. M. Cazabat, D. Langevin, and A. Pouchelon, *J. Colloid Interface Sci. 73*:1 (1980).
570. A. M. Cazabat, D. Chatenay, D. Langevin, and A. Pouchelon, *J. Phys. Lett. Fr. 41*:L-441 (1980).
571. A. M. Cazabat and D. Langevin, *J. Chem. Phys. 74*:3148 (1981).
572. A. M. Cazabat, D. Langevin, J. Menier, and A. Pouchelon, in *Surface Phenomena in Enhanced Oil Recovery* (D. O. Shah, ed.), Plenum, New York, 1981, pp. 161-179.
573. C. Boned, M. Clausse, B. Lagourette, J. Peyrelasse, V. E. R. McClean, and R. J. Sheppard, *J. Phys. Chem. 84*:1520 (1980).
574. M. Clausse, C. Boned, J. Peyrelasse, B. Lagourette, V. E. R. McClean, and R. J. Sheppard, in *Surface Phenomena in Enhanced Oil Recovery* (D. O. Shah, ed.), Plenum, New York, 1981, pp. 199-228.
575. F. Candau, J. Boutillier, F. Tripier, and J. C. Wittmann, *Polymer 20*:1221 (1979).
576. S. Candau, J. Boutillier, and F. Candau, *Polymer 20*:1237 (1979).
577. A. Graciaa, J. Lachaise, P. Chabrat, L. Letamendia, J. Rouch, and C. Vaucamps, *J. Phys. Lett. Fr. 39*:L-235 (1978).
578. E. Ruckenstein and J. C. Chi, *J. Chem. Soc. Faraday Trans. II 71*:1690 (1975).
579. E. Ruckenstein, in *Micellization, Solubilization and Microemulsions* (K. L. Mittal, ed.), Vol. 2, Plenum, New York, 1977, pp. 755-778.
580. E. Ruckenstein, *Chem. Phys. Lett. 57*:517 (1978).
581. E. Ruckenstein, *J. Colloid Interface Sci. 66*:369 (1978).
582. E. Ruckenstein and R. Krishnan, *J. Colloid Interface Sci. 71*:321 (1979).
583. E. Ruckenstein and R. Krishnan, *J. Colloid Interface Sci. 75*:476 (1980).
584. E. Ruckenstein and R. Krishnan, *J. Colloid Interface Sci. 76*:188 (1980).
585. E. Ruckenstein and R. Krishnan, *J. Colloid Interface Sci. 76*:201 (1980).
586. E. Ruckenstein, *J. Colloid Interface Sci. 78*:279 (1980).
587. L. M. Prince, *J. Colloid Interface Sci. 23*:165 (1967).
588. L. M. Prince, *J. Colloid Interface Sci. 29*:216 (1969).
589. L. M. Prince, *J. Soc. Cosmetic Chemists 21*:193 (1970).

590. J. Biais, B. Clin, P. Lalanne and B. Lemanceau, *J. Chim. Phys.—Phys. Chim. Biol.* *74*:1197 (1977).
591. P. Lalanne, J. Biais, B. Clin, A. M. Bellocq, and B. Lemanceau, *J. Chim. Phys.—Phys. Chim. Biol.* *75*:236 (1978).
592. A. M. Bellocq, J. Biais, B. Clin, P. Lalanne, and B. Lemanceau, *J. Colloid Interface Sci.* *70*:524 (1979).
593. J. Lang, A. Djavanbakht, and R. Zana, *J. Phys. Chem.* *84*:1541 (1980).
594. S. Friberg and S. Lapczinska, *Progr. Colloid Polymer Sci.* *56*:16 (1975).
595. S. Friberg, I. Buraczewska, J. C. Ravey, in *Micellization, Solubilization and Microemulsions* (K. L. Mittal, ed.), Vol. 2, Plenum, New York, 1977, pp. 901-911.
596. S. Friberg and C. Solans, *J. Colloid Interface Sci.* *66*:367 (1978).
597. H. Christenson and S. Friberg, *J. Colloid Interface Sci.* *75*:276 (1980).
598. K. Shinoda, H. Kunieda, N. Obi, and S. Friberg, *J. Colloid Interface Sci.* *80*:304 (1981).
599. J. C. Ravey and M. Buzier, *Compte-Rendu DGRST*, n°76.7.1931, DGRST, Paris, 1979.
600. H. F. Eicke and J. C. W. Shepherd, *Helvetica Chimica Acta* *57*:1951 (1974).
601. H. F. Eicke and H. Christen, *J. Colloid Interface Sci.* *48*:281 (1974).
602. H. F. Eicke, R. F. W. Hopmann, and H. Christen, *Ber. Bunsenges, Phys. Chem.* *79*:667 (1975).
603. H. F. Eicke and J. Rehak, *Helvetica Chimica Acta* *59*:2883 (1976).
604. H. F. Eicke, J. C. W. Shepherd, and A. Steinemann, *J. Colloid Interface Sci.* *56*:168 (1976).
605. H. F. Eicke, *J. Colloid Interface Sci.* *59*:308 (1977).
606. H. F. Eicke in *Micellization, Solubilization and Microemulsions* (K. L. Mittal, ed.), Vol. 1, Plenum, New York, 1977, pp. 429-443.
607. H. F. Eicke and P. E. Zinsli, *J. Colloid Interface Sci.* *65*:131 (1978).
608. H. F. Eicke, *J. Colloid Interface Sci.* *68*:440 (1979).
609. H. F. Eicke and A. Denss, in *Solution Chemistry of Surfactants* (K. L. Mittal, ed.), Vol. 2, Plenum, New York, 1979, pp. 699-706.
610. H. Zulauf and H. F. Eicke, *J. Phys. Chem.* *83*:480 (1979).
611. P. E. Zinsli, *J. Phys. Chem.* *83*:3223 (1979).
612. H. F. Eicke and R. Kubik, *Ber. Bunsenges, Phys. Chem.* *84*:36 (1980).
613. H. F. Eicke, *Pure Appl. Chem.* *52*:1349 (1980).
614. H. F. Eicke, in *Surface Phenomena in Enhanced Oil Recovery* (D. O. Shah, ed.), Plenum, New York, 1981, pp. 137-149.
615. C. Cabos and P. Delord, *J. Appl. Cryst.* *12*:502 (1979).
616. C. Cabos and P. Delord, *J. Phys. Lett. Fr.* *41*:L-455 (1980).
617. C. Cabos and P. Delord, *J. Phys. Paris* *39*:432 (1978).
618. E. Gulari and B. Chu, in *Surface Phenomena in Enhanced Oil Recovery* (D. O. Shah, ed.), Plenum, New York, 1981, pp. 181-197.
619. Y. Tricot, J. Kiwi, W. Niederberger, and M. Grätzel, *J. Phys. Chem.* *85*:862 (1981).
620. P. Bothorel, J. Biais, B. Clin, P. Lalanne, and P. Maelstaf, *C. R. Acad. Sci, Sér. C* *298*:409 (1979); *J. Colloid Interface Sci.* *80*:136 (1981).
621. J. L. Finney, *Nature* *266*:309 (1977).
622. B. A. Pethica and A. Fen, *Disc. Faraday Soc.* *258*:18 (1954).
623. J. Rouvière, J. M. Couret, M. Lindheimer, J. L. Dejardin, and R. Marrony, *J. Chim. Phys.—Phys. Chim. Biol.* *76*:289 (1979).
624. M. Wong, J. K. Thomas and M. Grätzel, *J. Amer. Chem. Soc.* *98*:2391 (1976).
625. G. Bakale, G. Beck and J. K. Thomas, *J. Phys. Chem.* *85*:1062 (1981).

626. G. Fourche and A. M. Bellocq, *C. R. Acad. Sci., Sér. B 289*: 261 (1979).
627. A. M. Bellocq and G. Fourche, *J. Colloid Interface Sci. 78*:275 (1980).
628. C. Tondre and R. Zana, *J. Dispersion Sci. Technol. 1*:179 (1980).
629. F. Larche, J. Rouvière, P. Delord, B. Brun, and J. L. Dussossoy, *J. Phys. Lett. Fr. 41*:L-437 (1980).
630. B. Lindman, N. Kamenka, T. M. Kathopoulis, B. Brun, and P. G. Nilsson, *J. Phys. Chem. 84*:2485 (1980).
631. P. Stilbs, M. E. Moseley, and B. Lindman, *J. Magnetic Resonance 40*:401 (1980).
632. B. Lindman, P. Stilbs, and M. E. Moseley, *J. Colloid Interface Sci. 83*: 569 (1981).
633. B. Lagourette, J. Peyrelasse, C. Boned, and M. Clausse, *Nature 281*:60 (1979).
634. M. Clausse, J. Peyrelasse, J. Heil, C. Boned, and B. Lagourette, *Nature 293*:636 (1981).
635. J. Heil, *Contribution à l'Etude de Phases Quaternaires Winsor IV. Diagrammes de Phases et Propriétés Electriques*, Thèse de Doctorat de Spécialité, Université de Pau, France, 1981.
636. M. Dvolaitzky, R. Ober, and C. Taupin, *C. R. Acad. Sci., Sér. II 293*:27 (1981).
637. L. E. Scriven, *Nature 263*:123 (1976).
638. L. E. Scriven in *Micellization, Solubilization and Microemulsions* (K. L. Mittal, ed.), Vol. 2, Plenum, New York, 1977, pp. 877-893.
639. Y. Talmon and S. Prager, *Nature 267*:333 (1977).
640. Y. Talmon and S. Prager, *J. Chem. Phys. 69*:2984 (1978).
641. D. O. Shah, V. K. Bansal, K. Chan and W. C. Hsieh, in *Improved Oil Recovery by Surfactant and Polymer Flooding*, (D. O. Shah and R. S. Schechter, eds.), Academic, New York, 1977, pp. 293-337.
642. V. K. Bansal, K. Chinnaswamy, C. Ramachandran, and D. O. Shah, *J. Colloid Interface Sci. 72*:524 (1979).
643. V. K. Bansal, D. O. Shah, and J. P. O'Connell, *J. Colloid Interface Sci. 75*:462 (1980).
644. M. Clausse, J. Heil, J. Peyrelasse, and C. Boned. *J. Colloid Interface Sci. 87*:584 (1982).
645. K. S. Chan and D. O. Shah, in *Surface Phenomena in Enhanced Oil Recovery* (D. O. Shah, ed.), Plenum, New York, 1981, pp. 53-72.
646. A. M. Bellocq, J. Biais, B. Clin, A. Gelot, P. Lalanne, and B. Lemanceau, *J. Colloid Interface Sci. 74*:311 (1980).
647. A. M. Bellocq, J. Biais, D. Bourbon, B. Clin, P. Lalanne, and B. Lemanceau, *C. R. Acad. Sci, Sér. C 288*:169 (1979).
648. A. M. Bellocq, D. Bourbon, and B. Lemanceau, *J. Colloid Interface Sci. 79*:419 (1981).
649. D. O. Shah and R. M. Hamlin, *Science 171*:483 (1971).
650. D. O. Shah, A. Tamjeedi, J. W. Falco, and R. D. Walker, *A.I.Ch.E. J. 18*:1116 (1972).
651. D. O. Shah, *Ann. N.Y. Acad. Sci. 204*:125 (1973).
652. J. W. Falco, R. D. Walker, and D. O. Shah, *A.I.ChE.J. 20*:510 (1974).
653. J. W. McBain, *Trans. Faraday Soc. 9*:99 (1913).
654. J. W. McBain, M. E. Laing, and E. F. Titley, *J. Chem. Soc. 115*:1279 (1919).
655. J. W. McBain and O. E. A. Bolduan, *J. Phys. Colloid Chem. 47*:94 (1943).
656. J. W. McBain and A. A. Green, *J. Amer. Chem. Soc. 68*:1731 (1946).

657. A. Lottermoser and F. Püschel, *Kolloid Z. 63*:175 (1933).
658. J. L. Moilliet, B. Collie, C. Robinson, and G. S. Hartley, *Trans. Faraday Soc. 31*:120 (1935).
659. G. S. Hartley, *Trans. Faraday Soc. 31*:31 (1935).
660. G. S. Hartley, *J. Amer. Chem. Soc. 58*:2347 (1936).
661. G. S. Hartley, B. Collie, and G. Samis, *Trans. Faraday Soc. 32*:799 (1936).
662. P. Ekwall, *Z. Physik. Chem. 161A*:195 (1932).
663. O. R. Howell and H. G. B. Robinson, *Proc. Roy. Soc. London, 155A*: 386 (1936).
664. G. Schmid and E. C. Larsen, *Z. Elektrochem. 44*:651 (1938).
665. K. A. Wright, A. D. Abbott, V. Sivertz and H. V. Tartar, *J. Amer. Chem. Soc. 61*:549 (1939).
666. A. F. H. Ward, *J. Chem. Soc. London 1*:533 (1939).
667. A. F. H. Ward, *Proc. Roy. Soc. London, 176A*:412 (1940).
668. C. W. Hoerr and A. W. Ralston, *J. Amer. Chem. Sco. 65*:976 (1943).
669. A. W. Ralston and C. W. Hoerr, *J. Amer. Chem. Soc. 68*:2460 (1946).
670. A. W. Ralston and C. W. Hoerr, *J. Amer. Chem. Soc. 69*:883 (1947).
671. A. W. Ralston and D. N. Eggenberger, *J. Amer. Chem. Soc. 52*:1494 (1948).
672. A. W. Ralston and D. N. Eggenberger, *J. Amer. Chem. Soc. 70*:980 (1948).
673. A. W. Ralston, D. N. Eggenberger, and F. K. Broome, *J. Amer. Chem. Soc. 71*:2145 (1949).
674. W. C. Preston, *J. Phys. Colloid Chem. 52*:84 (1948).
675. M. Shirai and B. Tamamushi, *Bull. Chem. Soc. Jap. 28*:545 (1955).
676. M. Shirai and B. Tamamushi, *Bull. Chem. Soc. Jap. 30*:542 (1957).
677. M. F. Emerson and A. Holtzer, *J. Phys. Chem. 71*:3320 (1967).
678. R. B. Beard and T. F. McMaster, *J. Colloid Interface Sci. 48*:92 (1974).
679. E. A. S. Cavell, *J. Chem. Soc. Faraday II 70*:78 (1974).
680. E. A. S. Cavell, *J. Colloid Interface Sci. 62*:495 (1977).
681. H. N. Singh, S. Singh, and D. S. Mahalwar, *J. Colloid Interface Sci. 59*:386 (1977).
682. S. Miyagishi, *Bull. Chem. Soc. Jap. 48*:2349 (1975).
683. G. I. Mukhayer, S. S. Davis, and E. Tomlinson, *J. Pharm. Sci. 64*:147 (1947).
684. E. Tomlinson, D. E. Guveli, S. S. Davis, and J. B. Kayes, in *Solution Chemistry of Surfactants* (K. L. Mittal, ed.), Vol. 1, Plenum, New York, 1979, pp. 355-366.
685. M. Abe and K. Ogino, *J. Colloid Interface Sci. 80*:58 (1981).
686. A. K. Jain and R. P. B. Singh, *J. Colloid Interface Sci. 81*:536 (1981).
687. W. Baumüller, H. Hoffmann, W. Ulbricht, C. Tondre, and R. Zana, *J. Colloid Interface Sci. 64*:418 (1978).
688. R. Zana, *J. Colloid Interface Sci. 78*:330 (1980).
689. R. Zana, S. Yiv, C. Strazielle, and P. Lianos, *J. Colloid Interface Sci. 80*:208 (1981).
690. S. Yiv, R. Zana, W. Ulbricht, and H. Hoffmann, *J. Colloid Interface Sci. 80*:224 (1981).
691. V. K. Bansal and D. O. Shah, *J. Colloid Interface Sci. 65*:451 (1978).
692. D. O. Shah, K. S. Chan, and R. M. Giordano, in *Solution Chemistry of Surfactants* (K. L. Mittal, ed.), Vol. 1, Plenum, New York, 1979, pp. 391-406.
693. S. Vijayan, C. Ramachandran, and D. O. Shah, *J. Amer. Oil Chemists Soc. 58*:566 (1981).

694. S. Vijayan, C. Ramachandran, and D. O. Shah, *J. Amer. Oil Chemists Soc.* *58*:746 (1981).
695. E. I. Franses, H. T. Davis, W. G. Miller, and L. E. Scriven, *J. Phys. Chem.* *84*:2418 (1980).
696. K. D. Dreher and R. D. Sydansk, *J. Pet. Technol.* *23*:1437 (1971).
697. S. C. Jones and K. D. Dreher, *J. Soc. Pet. Eng.*, *16*:161 (1976).
698. A. J. Hyde, D. M. Langbridge, and A. C. S. Lawrence, *Disc. Faraday Soc.* *18*:239 (1954).
699. M. Clausse, R. J. Sheppard, C. Boned, and C. G. Essex, in *Colloid and Interface Science* (M. Kerker, ed.), Vol. 2, Academic, New York, 1976 pp. 233-243.
700. D. Senatra and G. Giubilaro, *J. Colloid Interface Sci.* *67*:448 (1978).
701. D. Senatra and G. Giubilaro, *J. Colloid Interface Sci.* *67*:457 (1978).
702. V. K. Bansal, K. Chinnaswamy, and D. O. Shah, *Ber. Bunsenges. Phys. Chem.* *82*:979 (1978).
703. J. Peyrelasse, V. E. R. Mc Clean, C. Boned, R. J. Sheppard, and M. Clausse, *J. Phys. D: Appl. Phys.* *11*:L-117 (1978).
704. J. Heil, M. Clausse, J. Peyrelasse, and C. Boned, *Colloid Polymer Sci.* *260*:93 (1982).
705. S. I. Chou and D. O. Shah, *J. Phys. Chem.* *85*:1480 (1981).
706. S. I. Chou and D. O. Shah, *J. Colloid Interface Sci.* *80*:49 (1981).
707. S. I. Chou and D. O. Shah, *J. Colloid Interface Sci.* *80*:311 (1981).
708. C. Boned, J. Peyrelasse, J. Heil, A. Zradba, and M. Clausse, *J. Colloid Interface Sci.* *88*:602 (1982).
709. A. Boussaha, B. Djermouni, L. A. Fucugauchi, and H. J. Ache, *J. Amer. Chem. Soc.* *102*:4654 (1980).
710. A. M. Bellocq, G. Fourche, P. Chabrat, L. Letamendia, J. Rouch, and C. Vaucamps, *Optica Acta* *27*:1629 (1980).
711. J. Janzen, *J. Appl. Phys* *46*:966 (1975).
712. J. W. Essam, C. M. Place, and E. M. Sondheimer, *J. Phys. C: Solid State Phys.* *7*:L-258 (1974).
713. J. P. Straley, *Phys. Rev. B* *15*:5733 (1977).
714. M. J. Stephen, *Phys. Rev. B* *17*:4444 (1978).
715. G. Voronoi, *J. Reine Angew. Math.* *134*:198 (1908).
716. M. E. Nordberg, *J. Amer. Ceram. Soc.* *27*:299 (1944).
717. R. J. Charles, *J. Amer. Ceram. Soc.* *47*:599 (1964).
718. J. W. Cahn and R. J. Charles, *Phys. Chem. Glasses* *6*:181 (1965).
719. H. Fabre, N. Kamenka, and B. Lindman, *J. Phys. Chem.* *85*:3493 (1981).
720. R. N. Healy, R. L. Reed, and D. G. Stenmark, *J. Soc. Pet. Eng.* *16*:147 (1976).
721. K. E. Bennett, J. C. Hatfield, H. T. Davis, C. W. Macosko, and L. E. Scriven, in *Microemulsions* (I. D. Robb, ed.), Plenum, New York, 1982, pp. 65-84.
722. T. Hanai and N. Koizumi, *Bull. Inst. Chem. Res. Kyoto Univ.* *45*:342 (1967).
723. H. F. Eicke and V. Arnold, *J. Colloid Interface Sci.* *46*:101 (1974).
724. H. F. Eicke, *J. Colloid Interface Sci.* *52*:65 (1975).
725. J. Peyrelasse, C. Boned, P. Xans, and M. Clausse, *C. R. Acad. Sci., Sér. B* *284*:235 (1977).
726. J. Peyrelasse, C. Boned, P. Xans, and M. Clausse, in *Emulsions, Latices and Dispersions*, (P. Becher and M. Yudenfreund, eds.), Dekker, New York, 1978, pp. 221-236.

727. R. A. Mackay, C. Hermansky, and R. Agarwal, in *Colloid and Interface Science*, Vol. 2 (M. Kerker, ed.), Academic, New York, 1976, pp. 289-303.
728. R. A. Mackay and R. Agarwal, *J. Colloid Interface Sci. 65*:225 (1978).
729. K. R. Foster, P. C. Jenin, B. R. Epstein, and R. A. Mackay, *J. Colloid Interface Sci. 88*:233 (1982).
730. T. A. Bostock, M. H. Boyle, M. P. McDonald, and R. M. Wood, *J. Colloid Interface Sci. 73*:368 (1980).
731. T. A. Bostock, M. P. McDonald, G. J. T. Tiddy, and L. Waring, *Surface Active Agents Symp., Soc. Chem. Ind.*, London, 1979, pp. 181-190.
732. G. D. Smith, C. E. Donelan, and R. E. Barden, *J. Colloid Interface Sci. 60*:488 (1977).
733. B. A. Keiser, D. Vatie, R. E. Barden, and S. L. Holt. *J. Phys. Chem. 83*:1276 (1979).
734. N. F. Borys, S. L. Holt, and R. E. Barden, *J. Colloid Interface Sci. 71*:526 (1979).
735. G. Lund and S. L. Holt, *J. Amer. Oil. Chem. Soc. 57*:264 (1980).
736. C. Grosse and J. L. Greffe, *J. Chim. Phys.—Phys. Chim. Biol. 76*:305 (1979).
737. D. Senatra, C. M. Gambi, and A. P. Neri, *Lett. Nuovo Cimento 28*:603 (1980).
738. D. Senatra, C. M. Gambi, and A. P. Neri, *J. Colloid Interface Sci. 79*:443 (1981).
739. P. G. De Gennes and C. Taupin, *J. Phys. Chem. 86*:2294 (1982).

# Index